Rotifer Symposium V

Developments in Hydrobiology 52

Series editor
H. J. Dumont

Rotifer Symposium V

Proceedings of the Fifth Rotifer Symposium,held in Gargnano, Italy,
September 11–18, 1988

Edited by
C. Ricci, T. W. Snell and C. E. King

Reprinted from Hydrobiologia, vols. 186/187 (1989)

Kluwer Academic Publishers

Dordrecht / Boston / London

Library of Congress Cataloging-in-Publication Data

International Rotifer Symposium (5th : 1988 : Gargnano, Italy)
    Rotifer Symposium V / edited by C. Ricci, T.W. Snell, and C.E.
King.
        p.   cm. -- (Developments in hydrobiology ; 52)
    "Reprinted from Hydrobiologia."

    1. Rotifera--Congresses.   I. Ricci, C. (Claudia) II. Snell,
Terry W. III. King, C. E. (Charles E.) IV. Title. V. Series.
QL391.R8I57 1988
595.1'81--dc20   89-17596

ISBN-13: 978-94-010-6694-5   e-ISBN-13: 978-94-009-0465-1
DOI: 10.1007/978-94-009-0465-1

Printed on acid-free paper

Kluwer Academic Publishers incorporates the publishing programmes of Dr W. Junk Publishers, MTP Press, Martinus Nijhoff Publishers, and D. Reidel Publishing Company.

Distributors

for the United States and Canada: Kluwer Academic Publishers, 101 Philip Drive, Norwell, MA 02061, U.S.A.
for all other countries: Kluwer Academic Publishers Group, P.O. Box 322, 3200 AH Dordrecht, The Netherlands

Contents

PART TWO: TAXONOMY AND MORPHOLOGY

PART THREE: PHYSIOLOGY

PART FOUR: GENETICS

VIII

PART FIVE: AQUACULTURE

Preface

The *Fifth International Rotifer Symposium* was organized by Dr. Claudia Ricci and held in the northern Italian town of Gargnano (Brescia) from September 12–17, 1988. Through the generosity of the Rector of Milano University, a beautiful villa on the shores of Lake Garda was made available to the 83 people from 20 countries who attended the symposium. Ten of these rotifer workers had attended the four previous meetings.

Such symposia serve three major functions, the results of which will be apparent in the papers contained in this volume. First, because of the heterogeneity of interests and absence of concurrent sessions, the attendees are exposed to an unusually large variety of research problems, approaches and modes of interpretation. Bridges are thus built between one's own investigations and developments in the field as a whole. Second, the extensive informal interactions that occur outside of the meeting room during coffee breaks, dinners and excursions provide remarkable opportunities for research planning and sharing of results of work in progress. Third, the acquaintances established at these meetings have facilitated interactions during the three-year intervals between symposia. The result has been that visits between laboratories, acquisition of research materials from distant sources and coordination of related investigations have all been greatly enhanced.

A description of the week's events may serve to convey the ambience of the meeting. The tenor of the symposium was set on Monday morning with a delightful reflection by Tommy Edmondson on a lifetime of rotifer research. Twelve papers followed focusing on competition, parasitology, biogeography and systematics. Each session was interrupted by at least one break where coffee and tea were served on the veranda overlooking the lake. A full day of eleven papers was presented on Tuesday covering physiological ecology, life history traits, interactions with cladocera and muscle ultrastructure. That night, a poster session with ten papers was presented stimulating enthusiastic discussion. Wednesday morning consisted of four papers on species concepts, molecular evolution and genetics. Wednesday afternoon, attendees were treated to a two-hour boat ride to the northernmost shore of Lake Garda and the medieval town of Riva. After a few hours of exploration, everyone climbed back aboard for a splendid return cruise which included an elegant dinner. Thursday included seven papers on taxonomy, morphological variation and cladistics along with an afternoon workshop on systematics. The conference ended on Friday with fifteen papers focusing on aquaculture and bioenergetics. Many participants attended a post-conference excursion to the beautiful old city of Verona.

On behalf of all participants, we wish to thank Dr. Giulio Melone, Dr. Mauro Mariani and his colleagues at the Civico Acquario e Stazione Idrobiologica of Milan for their invaluable organizational help both before and during the meeting. We are also grateful to Prof. Paolo Mantegazza, Rector of the State University of Milan, and to Prof. Giulio Lanzavecchia, past head of the Department of Biology, for their support of the meeting. In addition, grants-in-aid were obtained from the Italian Ministry of Public Education (MPI), the National Council of Research (CNR), and from ENEA. Facilities were provided by Comunita del Garda and by Direzione Navigazione Laghi.

CLAUDIA RICCI
TERRY W. SNELL
CHARLES E. KING

List of participants

BARBATO, Fabio
Via Monti di Creta 47
00167 Roma, Italy

BELLELLI, Enrico
Dip. Biol. Anim., Univ. La Sapienza,
00185 Roma, Italy

BERNER-FANKHAUSER, Heidi
Othmarsingerst. 18
5600 Lenzburg, Switzerland

BIANCHI, Maria Irene
Via Appiani 24
20121 Milano, Italy

BOGAERT, Geert
Kerrebroek 83
9850 Nevele, Belgium

BØRSHEIM, Knut Yngve
Dep. Microbiology Plant Physiol.
Allegt 70
5007 Bergen, Norway

BRAIONI, Gianna
Dip. Biol. Animale [Padova]
Padua, Italy

BRETT, Michael
Uppsala University, Limnology Institute,
Box 557, [15233]
Uppsala, Sweden

CARMONA NAVARRO, Maria Jose
Depto. Ecol., Univ. Valencia
46100 Burjasot, Valencia, Spain

CHENGALATH, Rama
National Mus. Nat. Sciences,
P.O. Box 3443,
Station 'D'
Canada-Ontario-Ottawa, KIP 6P4

CLEMENT, Pierre
Equipe Neuro-Ethologie Univ. Lyon 1
43 Boulevard du 11 nov.
69622 Villeurbanne Cedex, France

COUTTEAU, Pierre
Artemia Reference Center
Rozier 44
9000 Gent, Belgium

CRUZ-PIZARRO, Luis
Catedra de Ecologia
Facultad de Ciencias
18071 Granada, Spain

DE MANUEL, Jordi
Dpt. Ecologìa, Facult. Biologìa
Av. Diagonal 645
08028 Barcelona, Spain

DE RIDDER, Margaretha
V. van Malderlaan 37/3
1710 Dilbeek, Belgium

DUMONT, Henri J.
Inst. Ecology, State Univ. Ghent
K.L. Ledeganckstraat 35
9000 Gent, Belgium

DUNCAN, Annie
Dept. Zoology
Royal Holloway College
Englefield Green, Surrey TW20 9TY, U.K.

EDMONDSON, W. T.
Dept. Zoology NJ-15
Univ. Washington
Seattle, Washington 98195, U.S.A.

EJSMONT-KARABIN, Jolanta
Polish Acad. Sci., Inst. Ecology
ul. Lesna 13
11-730 Mikolajki, Poland

ESPARCIA COLLADO, Angeles
Depto Ecol., Univ. Valencia
46100 Burjasot, Valencia, Spain

FARABEGOLI, Alessandra
Istituto di Zoologia
Via Borsari 46
44100 Ferrara, Italy

FERRARI, Ireneo
Istituto di Zoologia
Via Borsari 46
44100 Ferrara, Italy

FRANCEZ, André Jean
INRA Stat. Agronomie
12 Avenue du Brezet
63039 Clermont-Ferrand, France

GALKOVSKAYA, Galina
Inst. Zool., Acad. Sci. BSSR
Academytcheskaya 27
Minsk 220072, U.S.S.R.

GILBERT, John J.
Dept. Biol. Sci.
Dartmouth College
Hanover, New Hampshire 03755, U.S.A.

GUEVARA ESPINOSA, Maria Dolores
Sagasta 100, 5° D.
35008 Las Palmas G.C., Spain

GUISANDE GONZALEZ, Cástor
Department of Ecology, Faculty of Biology
41080 Sevilla, Spain

GULATI, Ramesh
Limnol. Inst. Koninklijke
Rijkestraatweg 6
3631 AC Nieuwersluis, The Netherlands

HAGIWARA, Atsushi
Graduate School of Marine Science and
Engineering, University of Nagasaki
Bunkyo 1-14
Nagasaki City 852, Japan

HERNANDEZ CRUZ, Carmen Maria
Centro Tecnologia Pesquera
Apdo 56
35200 Telde Las Palmas G.C., Spain

HERZIG, Alois
Biologische Stat. Neusiedlersee
7142 Illmitz, Austria

HIRAYAMA, Kazutsugu
Fac. Fisheries, Nagasaki Univ.
1-14 Bunkyo Machi
Nagasaki 852, Japan

HOLLOWDAY, Eric D.
45 Manor Road
Aylesbury, Bucks HP20 1JB, U.K.

JAMES, Charles M.
Kuwait Inst. Scient. Res.
P.O. Box 1638
22017 Salmiya, Kuwait

KING, Charles E.
Dept. Zoology, Oregon State Univ.
Corvallis, Oregon 97331, U.S.A.

KORSTAD, John
Dept. Biology, Oral Roberts Univ.
Tulsa, Oklahoma 74171, U.S.A.

KOSTE, Walter
Röntgenstrasse 1/Ofen.
2903 Bad Zwischenahn, F.R.G.

KÜHN, Jürgen
Zoologische Institut
Berliner Strasse 28
3400 Göttingen, F.R.G.

KUNICKI-GOLDFINGER, Wladyslaw J. H.
Warsaw University, Inst. Microbiology
Nowy Swiat 67
00-046 Warsaw, Poland

KUTIKOVA, Ludmilla A.
Zool. Inst. Acad. Sciences
Leningrad B 34, 199034, U.S.S.R.

LAXHUBER, Ruth
Zool. Inst. LMV-München
Seidlstrasse 25
8000 München 2, F.R.G.

LUBZENS, Esther
Israel Oceanogr. Limnol. Res.
Tel-Shikmoma, P.O. Box 8030
Haifa, Israel

MARIANI, Mauro
Acquario Civico e Stazione Idrobiologica
V.le Gadio 2
20121 Milano, Italy

MASSUTI, Sofia
G.E.S.A. Cta II "Es Murterar"
Alcudìa- Mallorca, Spain

MATVEEVA, Lilian K.
Severtsov Inst. Evol. Morphol.
Leninskii pr. 33
Moscow 117071 U.S.S.R.

MAUTI, Rita
Acquario Civico e Stazione Idrobiologica
V.le Gadio 2
20121 Milano, Italy

MAY, Linda
Inst. Terrestrial Ecol.
Bush Estate, Penicuik
Midlothian EH26 0QB, U.K.

MELONE, Giulio
Dip. Biol. U. Milano
Via Celoria 26
20133 Milano, Italy

MIKSCHI, Ernst
Goethegasse 4
2380 Perchtoldsdorf, Austria

MINKOFF, Gidon
Tinamenor S.A. Pesues
Cantabria, Spain

MIRACLE, Maria Rosa
Dept. Ecol., Fac. Cien. Biol., Univ. Valencia
46100 Burjasot, Valencia, Spain

MORALES BAQUERO, Rafael
Catedra Ecologia, Fac. Cienc., Univ. Granada
18071 Granada, Spain

NAGATA, Warren D.
Malaspina College, 900 Fifth Street
Nanaimo, B.C., V9R 5S5, Canada

NOGRADY, Thomas
Dept. Biol. Queen's Univ.
Kingston K7L 3N6, Canada

OLSEN, Yngvar
SINTEF Div. Applied Chemistry
7034 Trondheim, Norway

PAGANI, Manuela
Dip. Biologia Università di Milano
Via Celoria 26
20133 Milano, Italy

PAULI, Hans-Rainer
Limnologisches Inst.
7750 Konstanz, F.R.G.

PEJLER, Birger
Inst. Limnology, Box 557
751 22 Uppsala 1, Sweden

PEREGO, Cristina
Acquario Civico e Stazione Idrobiologica
V.le Gadio 2
20121 Milano, Italy

PONTIN, Rosalind M.
26 Hermitage Woods Crescent
St. John's, Woking, Surrey GU21 1UE, U.K.

RADWAN, Stanislaw
Acad. Agric. Dept. Zool. Hydrobiol.
Akademicka 13
20-934 Lublin, Poland

RICCI, Claudia
Dip. Biol. An., Un. Torino
Via Accademia Albertina 17
10123 Torino, Italy

ROCHE, Kennedy
Inst. Zool., State Univ. Ghent
K.L. Ledeganckstraat 35
9000 Gent, Belgium

ROTHHAUPT, Karl. O.
Max Planck Inst. fur Limnologie
Postfach 165
2320 Plön, F.R.G.

RUTTNER-KOLISKO, Agnes
Biologische Station
3293 Lunz am See, Austria

SAUNDERS-DAVIES, Anthony
33 Park Road
Esher, Surrey KT10 8NP, U.K.

SCHABER, Peter
Abt. Limnol. Inst. Zool.
Technikerstrasse 13
6020 Innsbruck, Austria

SCHMID-ARAYA, Jenny M.
Dep. Biology RHBNC
Egham, Surrey TW20 9TY, U.K.

SCHRIMPF, Andreas
Wasserwirtschaftsamt Rosenheim
Königstrasse 19
8200 Rosenheim, F.R.G.

SERRA, Manuel
Dept. Ecol., Fac. Cien. Biol., Univ. Valencia
46100 Burjasot, Valencia, Spain

SERRANO MARTIN, Laura
Department of Ecology, Faculty of Biology
41080 Sevilla, Spain

SHIEL, Russell J.
Murray Darling Freshwater Res.
P.O. Box 921, Albury N.S.W.
2640 Australia

SKJOLDAL, Lillian
Insitute of Marine Research
Division of Aquaculture
Austevol, 5392 Storebø, Norway

SNELL, Terry W.
Div. Sci. Math., Univ. Tampa
Tampa, Florida 33606, U.S.A.

STARKWEATHER, Peter L.
Dept. Biol. Sci., Univ. Nevada
Las Vegas, Nevada 89154, U.S.A.

SUDZUKI, Minoru
Nihon Daigaku
Higashi-Arai 557, Omiya-shi
Saitama-Ken, 330 Japan

VADSTEIN, Olav
SINTEF Div. Applied Chemistry
7034 Trondheim, Norway

VELLANO, Camillo
Dip. Biol. Animale
Via Accademia Albertina 17
10123 Torino, Italy

WALLACE, Robert Lee
Dept. Biology, Ripon College
Ripon, Wisconsin 54971, U.S.A.

WALSH, Elizabeth J.
Dept. Zoology OSU
Corvallis OR 97331, U.S.A.

WALZ, Norbert
Zool. Inst., Univ. München
Seidlstrasse 25
8000 München 2, F.R.G.

WILMS, Anja
Limnol. Inst. Koninklijke
Rijkestraatweg 6
3631 AC Nieuwersluis, The Netherlands

YUFERA, Manuel
Inst. Cien. Mar. Andalucia CSIC
Poligono S. Pedro s/n Apdo oficial
Puerto Real, Cadiz, Spain

ZOUFAL, Wolfgang
Schweglerstrasse 4214
1150 Wien, Austria

Plate 1 (see facing page):

Upper panel: Giulio Melone, Eric Hollowday, Henri Dumont, Claudia Ricci, Manuela Pagani; Mila Kutikova, Minoru Sudzuki, Mila Kutikova, Rama Chengalath.

Upper middle panel: Terry Snell, Davi Ehlert, Charles King, Lilian Matveeva.

Lower middle panel: Peter Starkweather, Manuela Pagani, Claudia Ricci, Giulio Melone; Tom Nogrady, Alois Herzig, Walter Koste.

Lower panel: Walter Koste, Tommy Edmondson, Yvette Edmondson; Heidi Berner-Fankhauser, Birger Pejler; John Gilbert.

Plate 2 (see p. xvi):

Upper panel: coffee break in the garden of Palazzo Feltrinelli; Excursion to Verona.

Upper middle panel: Hans-Rainer Pauli, Atsushi Hagiwara, Linda May, Manuel Serra, Claudia Ricci, Nan Duncan, Manuel Yufera, Luis Cruz-Pizarro, Maria Rosa Miracle, Tom Nogrady; Charles King, Linda May, Bob Wallace, Eric Hollowday, Tony Saunders-Davies, Rosalind Pontin.

Lower middle panel: Henri Dumont, Agnes Ruttner-Kolisko, Peter Schaber, Wolfgang Zoufal, Pierre Clement, Rama Chengalath; Claudia Ricci, Gala Galkovskaya, Tommy Edmondson, Birger Pejler, Giulio Melone, Maria Rosa Miracle, Gideon Minkoff, Henri Dumont, Ruth Laxhuber.

Lower panel: Back row: Esther Lubzens, Warren Nagata, Russ Shiel, Bob Wallace. Centre row: Kennedy Roche, Margaretha De Ridder, Ruth Laxhuber, Walter Koste. Front row: Mauro Mariani, Claudia Ricci, Giulio Melone, Maria Rosa Miracle; Pierre Clément, Terry Snell, Manuel Yufera.

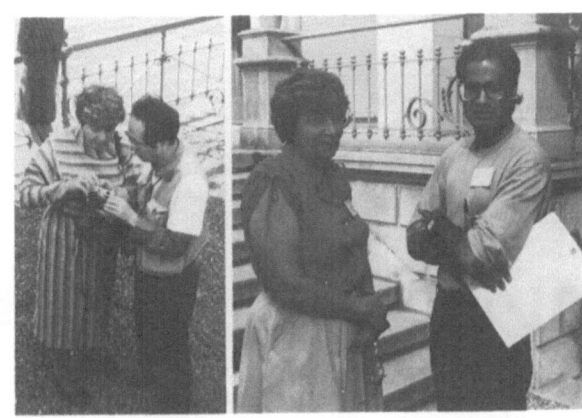

Plate 1

Plate 2

Pater Dr.h.c. Josef Donner (28.2.1909–8.1.1989) in memoriam

Suddenly and unexpectedly the "father of the bdelloids" (in Claudia Ricci's words) died in January 1989 at the age of eighty years. His life was divided between religious vocation and scientific research. He was able to fulfill both tasks to a high degree without conscientious scruple.

Josef Donner spent his youth and school years in Northern Bohemia (at that time part of the Austrian monarchy) and after ordination as Redemptorist priest he taught biology at the school of his convent. During World War II he was transferred to a small parish in Southern Bohemia. During that time he became fascinated with the microscopical life in ponds and streams. He bought a microscope of his own, which he considered his most valuable worldly possession. The first publications dating from those years deal mainly with monogonont rotifers.

After the end of the war Pater Donner – being German and a priest – was evicted from his parish, now situated in Communist Czechoslovakia. Carrying a small rucksack and his microscope he arrived on foot at the Biological Station, Lunz to meet his friend and teacher, Prof. V. Brehm, who – also a refugee – had found a simple shelter here. That was the occasion when I met Pater Donner first; the story of his escape is still vividly in my mind.

A short stay at the monastery Admont (situated in a boggy region) initiated Josef Donner's interest in the fauna of humic soils, where bdelloid rotifers are particularly abundant, a group which was practically unknown so far; he specialized and published his first papers on bdelloids.

After the interlude in Admont Pater Donner settled for several years at an extension of his own order in Vienna, where he was again teaching and supervising schoolchildren. Even here, in the surroundings of a big city, he continued his investigations on bdelloids. The result was a book "Die Ordnung Bdelloidea" (Best. Bücher zur Bodenfauna Europas, 1965), which made him renowned worldwide.

Already several years earlier a very useful little booklet "Rädertiere" had appeared in the Kosmos' Series (Einführung in die Kleinlebewelt, 1956), which was translated into English by H.G.S. Wright in 1966.

His final homestead found Pater Donner in the Redemptorist-Convent in Katzelsdorf (Lower Austria), where he taught biology in the private high school of this convent. But, after a few years, in recognition of his scientific fame, he was released from teaching by his superiors. In conversations with scientific colleagues he always very thankfully acknowledged this concession. Now completely dedicated to research, this fact became visible very soon in Josef Donner's output of papers; altogether they amount to approximately fifty.* Although he regretted losing contact with young people, he now concentrated intensively not only on taxonomy of rotifers, but included also ecological and coenological aspects. It was a lucky coincidence that he found at the same time a connection with the newly founded working group investigating the Danube (I.A.D.). Extensive papers on the litoral zone and the river meadows of the Danube, and also on the reed belt of the Neusiedlersee were produced in cooperation with this working group.

These broader aspects of Pater Donner's work were the main, but not the only reason, that he was elected in 1975 *Doctor honoris causa* by the University of Salzburg. Rightly, Josef Donner considered the election the crowning of his scientific life.

While compiling Dr. Donner's life history and his own personal notes, it became very obvious to me, how intensely physical fate, deep religious vocation and scientific aspiration have been woven together in the modest, kind and pleasant personality of our dear colleague *Pater Dr. phil. h.c. Josef Donner.*

A. RUTTNER-KOLISKO

* A complete list of publications appears in "Wasser und Abwasser', Bd. 32.

Hydrobiologia **186/187**: 1–9, 1989.
C. Ricci, T. W. Snell and C. E. King (eds), Rotifer Symposium V.
© 1989 *Kluwer Academic Publishers.*

Rotifer study as a way of life

W.T. Edmondson
Department of Zoology, University of Washington Seattle, WA 98195, USA

Introduction

When Claudia Ricci asked me to present a review of my research program on rotifers at the Fifth International Rotifer Symposium I was puzzled about what to do because I did not think I had a program. All my life various opportunities have come up and I have gone along step-by-step, doing whatever seemed most interesting at the time, within the normal constraints of existence. In fact, when I look back now I can see a sort of pattern, but it was not the result of planning. Apparently there was some selection on my part from a random assortment of opportunities.

A review of one's work necessarily involves some autobiography. To understand what I tell you about my research requires some information about my early activities. Investigation is an integral part of a scientist's life, and there is no way to discuss it intelligibly without the biographical background. Chance happenings in one's personal life can change the direction of his professional experience.

Learning

I was born in Milwaukee, Wisconsin, on the shore of Lake Michigan, on April 24, 1916 and lived there for two years. As soon as I could walk my mother would take me on strolls along the lakeshore every afternoon. We then moved to southern Indiana where there were no lakes, only ponds, creeks and rivers. They were very different, but I was fascinated by them and liked to look into them to see what was going on. Fish bored me, but crayfish were exciting, fairy shrimp beautiful, and *Chaoborus* was mysterious. There seemed to be other things, barely visible, gliding around and generating wonder.

At home I had natural history books and read everything I could find about freshwater animals in the public library. Several books mentioned rotifers, sometimes with illustrations, but gave no real information. The most that any book did was to emphasize how beautiful and interesting rotifers were, making them seem most desirable of all. This all gave me a yearning for real information about them.

A turning point came when, at about age 12, I was given a toy microscope, built something like Leeuwenhoeck's, which permitted me to see what those minute gliding points looked like. Eventually I was able to recognize that the first rotifer I had seen was *Scaridium longicaudum*. (The first sight of a rotifer must have been an important moment in the life of every rotiferologist.) An even more decisive point came a year later when I learned about Ward and Whipple's *Freshwater Biology*. My grandmother gave me money for it as a birthday gift and sewed a cloth carrying case because it was so heavy. Of course that wonderful book opened the world of rotifers for me, and a lot more. You can imagine my pleasure when, some 20 years later, in 1949, I was asked to edit a second edition. That finally appeared in 1959, one year past the 50th anniversary of the first edition.

High School

A major step forward came when we moved to New Haven, Connecticut, in 1930, just in time for me to start high school. Hillhouse High School was an excellent institution with a staff of highly professional teachers. Many of them were taking courses at Yale.

One day I went to the biology teaching rooms, entered the first one where there was a teacher, and asked whether I could use a microscope after school to look at pond water. By accident I had found the best possible person, Miss Ruth Ross, for she let me use the microscopes and helped me in many other ways.

For months I spent at least an hour after school almost every day looking at samples from ponds around New Haven. My interest was not limited to rotifers. The filamentous algae were very attractive (*Spirogyra*, *Mougeotia*, *Bulbochaete*), the protozoa and crustaceans fascinating. The only requirement was that it be small and beautiful. So with the help of Ward and Whipple I found my way around in the pond world. I got more and more interested in the things that attached to plants, and the sessile rotifers were best. All through this I think I must have been as motivated by esthetics as by any scientific consideration; I certainly was not generating any hypotheses.

Miss Ross was taking a course in histology from a very kindly, grandfatherly professor at Yale, Wesley Roswell Coe, and he agreed me to let me attend his lectures, which I enjoyed. However, my interests were obviously more toward the invertebrates, so Dr. Coe took me around to meet Professor Lorande Loss Woodruff, a protozoologist. He was interested in the history of science and had collected an amazing number of classical early works which he let me browse in. He had the wonderful Ehrenberg book of 1838, Hudson & Gosse, and many others. Once he realized what I was interested in, he took me a few doors down the hall to introduce me to the youngest instructor, G. Evelyn Hutchinson. As soon as I mentioned rotifers a dramatic thing happened. Hutchinson turned, muttering about having just been collecting, poured some water from a thermos bottle into a finger bowl and put it under a dissecting microscope. It was the densest group of *Conochilus hippocrepis* that I have ever seen. I had read about this animal in Hudson and Gosse and had been eager to see one (one! – hundreds of colonies in view at once).

Hutchinson invited me to use his microscopes. I was a freshman in high school at the time and happily took him up on the invitation. At some time he suggested that I write to F. J. Myers and I entered into a long correspondence and exchange of slides with Myers. Hutchinson taught me how to make slides with the glycerine and melted paraffin technique. I would send Myers a slide of a difficult specimen labelled with the name of what I thought it was, and he would send it back with a corrected label or with congratulations for having got it right, often accompanied by some slides of his own. He had trouble with my handwriting so he gave me a typewriter to make our correspondence easier. During this period I was concentrating on learning the rotifers and how to study them.

I read Hudson and Gosse intently, going so far as to memorize this part of a wonderful passage in the Introduction:

'... But if, retaining sense and sight, we could shrink into living atoms and plunge under the water, of what a world of wonders should we then form part! We should find this fairy kingdom peopled with the strangest creatures: – creatures that swim with their hair, that have ruby eyes blazing deep in their necks, with telescopic limbs that now are withdrawn wholly within their bodies and now stretched out to many times their own length. Here are some riding at anchor, moored by delicate threads spun out from their toes; and there are others flashing by in glass armour, bristling with sharp spikes or ornamented with bosses and flowing curves; while, fastened to a green stem, is an animal convolvulus that by some invisible power draws a never-ceasing stream of victims into its gaping cup, and tears them to death with hooked jaws deep down within its body.

Close by it, on the same stem, is something that looks like a filmy heart's-ease. A curious wheelwork runs round its four outspread petals; and a chain of minute things, living and dead, is winding in and out of their curves into a gulf at the back of the flower. What happens to them there we cannot see; for round the stem is raised a tube of golden-brown balls, all regularly piled on each other. Some creature dashes by, and like a flash the flower vanishes within its tube.

We sink still lower, and now see on the bottom slow-gliding lumps of jelly that thrust a shapeless arm out where they will, and, grasping their prey with these chance limbs, wrap themselves round their food to get a meal; for they creep without feet, seize without hands, eat without mouths, and digest without stomachs.

Time and space, however, would fail me to tell of all the marvels of the world beneath the waters. They would sound like the wild fancies of a child's fairy tale, and yet they are all literally true; and, moreover, nearly all of them are true of that rotiferous world which it is my purpose to describe.'

In high school I took two years of German and practiced by translating many pages of Remane's section on rotifers in *Bronn's Klassen und Ordnungen des Tier-Reichs*. I was particularly fascinated by the retrocerebral organ and struck by a seeming anatomical correlation: rotifers with the best developed retrocerebral-subcerebral system tended to have the simplest protonephridial system (compare Remane's figs. 119 and 266). I never pursued the point.

In 1932, Miss Ross took me down to meet Mr. Myers at the American Museum of Natural History in New York where he was an honorary Research Associate. Later Myers invited me several times to visit him in his home in Ventnor, New Jersey, which is near the Pine Barrens region, full of interesting ponds, including *Sphagnum* bogs with a remarkably diverse rotifer fauna.

Myers was independently wealthy, and was able to spend all his time on his hobby, rotifers. He had been an art student in college, which accounts for the excellent illustrations in his papers. His interest in rotifers started when he noticed strange spots on some tropical fish he had. He bought a microscope so that he could see

what they were, and noticed fascinating objects swimming past; rotifers, of course. After a period of exploring local ponds, he wrote to H. S. Jennings at the University of Pennsylvania for help in identification. Jennings had trouble because he was familiar with the fauna of alkaline waters only, and Myers had been collecting in *Sphagnum* bogs. Jennings recommended H.K. Harring as a better source of information. Harring was a specialist in precise screw instruments at the U. S. Bureau of Standards in Washington. Somehow he became interested in rotifers and built a microscope. According to Myers, Harring completed his monumental study of the literature, 'Synopsis of the Rotatoria' before seeing a live rotifer. Chancey Juday at Wisconsin learned about the work of Harring, and invited him to work on the rotifer fauna of Wisconsin, leading to the set of five papers in collaboration with Myers.

In 1932 Hutchinson was Biologist of the First Yale North India Expedition. When he came back to New Haven in January, 1933, he had a fabulous collection of material, from protozoa to fish, from many ponds, lakes and streams in a range of altitude from 528 m to 5334 m in the Himalayas and adjacent regions. He sent out most of the collections to specialists for identification but kept the rotifers and asked me to identify them. He cleared off one end of a bench in his laboratory and set me up there with a couple of microscopes for my own use to do the work on his collections. For the next six years this corner of his laboratory was the center of my existence.

Hutchinson's substitute while he was away was Dr. Richard M. Bond who had a lot of material from Hispaniola, and he, too, asked me to identify his rotifers. So my first published paper, on the rotifers of Hispaniola, came out in the spring of my last year in high school, and the second, written jointly with Hutchinson, came out later in the year, just after I entered college.

The lab was a busy place. Hutchinson's reputation in limnology was spreading, and it was becoming widely understood that there was a center in New Haven quite different intellectually from those in Wisconsin and Michigan. This difference is made clear in Beckel's (1987) account of limnology in Wisconsin (see pp. 24–25) and by Cook's (1977) review of a controversy about a paper by Lindeman (1942). As I was sitting at one end of the room looking at rotifers, Hutchinson would be running chemical analyses at the other end, calculating the data, and expressing surprise or pleasure at some new finding. It was interesting to read the formal publication of the outcome as, for instance, in the 1941 Ecological Monographs or the 1938 Proceedings of the National Academy of Sciences. Many visitors, foreign and domestic, went out of their way to visit Hutchinson's laboratory.

College

All during my undergraduate years I spent as much time in Hutchinson's laboratory as possible, working on local material and collections from other places, publishing a few new species. I continued to collect in ponds because of my interest in sessile rotifers. Inevitably I was subconsciously absorbing information about characteristics of the best ponds for various kinds of rotifers and what plants were best for the sessile ones. I helped Hutchinson's first two graduate students, Gordon A. Riley and Edward S. Deevey, Jr., with their field work, thus gaining experience with lake plankton and broadening my interests to population and community ecology.

I entered into correspondence with other rotifer investigators including E. H. Ahlstrom, P. M. de Beauchamp, B. Carlin, J. Hauer, M. Neal, A. Ruttner-Kolisko, J. Wizniewski and K. Wulfert. We exchanged drawings and slides of new and unusual species. Much later, in 1959, my wife and I visited de Beauchamp in Paris. He showed us his collection of brightly colored wooden working models of different types of trophi. Each of us could read the other's writing in his own language but not understand the other's speech in any language, so that the conversation was conducted mostly on paper with an occasional oral assist from my wife.

I took courses in a wide spread of topics including Hutchinson's invertebrates and Coe's his-

tology. In addition to auditing Histology I had taken Hutchinson's course in Invertebrate Zoology (with examinations!) while I was in high school. This time I enrolled formally so that they would appear on my college record. I liked histology, probably because of the interesting functional patterns of the cells in the tissues and the colorful stains. I still treasure my slide of *Amphiuma* spleen, vastly overstained with Mallory triple stain. Each student chose a topic of special interest to him as part of the laboratory work. I made serial sections in paraffin of *Lindia tecusa*. This was easier than it may sound because the specimens, supplied by Myers, were a full millimeter long. For orientation, I laid the animal along one of my hairs while the paraffin was still soft and transparent. I was then told that histologists went to great lengths to keep such objects out of the blocks because they caused the paraffin ribbon to tear and my project would fail. Fortunately my knife was sharp, and I have a nice clear section of hair next to each slice of rotifer.

Graduate school

When I was about to graduate, Hutchinson suggested that it would be good to get experience somewhere else, so I went for one year to the University of Wisconsin, primarily to take Juday's limnology course (Birge was no longer teaching). Another graduate student in Juday's course, Yvette Hardman, helped me collect rotifers and I helped her with her work on surface tension of lakes (Hardman, 1941). We were married in 1941. By then I had decided to write a taxonomic monograph on sessile rotifers (I still have 5 notebooks of information) and to do a Ph. D. on some aspect of their ecology. Through the kindness of Professor Juday I had an assistantship to work at the Trout Lake Laboratory in northern Wisconsin for a full summer. When I asked what my assignment would be, he said 'Study the sessile rotifers'. I collected lots of taxonomic and ecological data having to do with environmental factors (1944 paper).

I went back to Yale in 1939, and as a graduate

student had my own space so was no longer able to eavesdrop on Hutchinson's fascinating conversations with visitors.

My Ph.D. dissertation might have been limited to the kind of distributional problem covered in my 1944 paper but for a chance observation. One day I took a collection of *Utricularia* covered by *Floscularia conifera* with a striking appearance. All of the animals except the small ones had a dark band across the tube, much as in the figure on Plate V in Hudson and Gosse. I suddenly realized that it had rained heavily the day before, and the dark bands consisted of pellets made while the water was muddy. It occurred to me that this way of marking tubes could tell how fast the animals were growing and settling, because the clear part above the dark band would show how much the animals had grown between the time and the mud settled and the time they were collected. The small, young ones with no dark band must have attached after the water had cleared. I was planning to go out with a stick and stir the mud up again, when I thought of a better way: pouring in a cloud of carmine. As recorded in my 1945 paper, this permitted getting several kinds of population dynamics data. Although Hutchinson was at first skeptical, it worked, and he later included the experiment in an entry on populations in the *Encyclopedia Britannica*. During this work I realized that the eggs held inside the tube were the next generation of animals. This led to the idea that a count of eggs could be used for calculating the birth rate, an idea I was not able to implement immediately, but did for planktonic rotifers more than fifteen years later. (My official retirement in 1986 was marked by the presentation by my colleagues of an elegant bronze *Floscularia*, standing 21 cm high and complete with a band on the tube).

World War II

During the war I worked as a civilian for the U.S. Navy, mostly at Woods Hole Oceanographic Institution, on a variety of marine problems, so rotifers were not a large part of my thinking. After the war, I stayed at Woods Hole for a year working on fertilization of salt ponds with nutrients.

Harvard University

In 1946 I joined the Department of Biology at Harvard to help teach a new General Education course. While I was in Cambridge, André Burger came from Switzerland for a year to work on bdelloids and discovered the interesting anatomical structure of the stomach of *Habrotrocha*. The work became widely known when Libbie Hyman included it in the third volume of her treatise on invertebrates (1951). He also found, at my suggestion, that the growth of some bdelloids in culture was increased by adding little chips of glass because they improved conditions for egg-laying.

University of Washington

When I was invited to move to the University of Washington in 1949 I expected to resume work on rotifers, and did so, but the emphasis of the work changed from studying them for themselves to studying them as parts of communities in the context of general ecological problems.

I re-established communication with my prewar rotifer correspondents and soon added over a dozen more including B. Berzins, J. Donner, J. Evans, A. L. Galliford, E. D. Hollowday, T. Nogradi, B. Pejler, R. Pourriot, C. Rudlin, V. Sladecek, M. Sudzuki and H. G. S. Wright. Some of the British workers formed a Rotifer Society in 1949. The number of new people interested in rotifers was growing and included a large proportion with academic positions where they were able to apply advanced techniques to problems of systematics, physiology, morphology and ecology. This development eventually made possible the remarkable series of International Rotifer Symposia, five to date.

The first summer in Seattle my wife and I drove all over western Washington collecting in lakes to see what we had to work with. I had wanted to continue with carmine marking experiments, but did not find suitable populations. There were sessiles, but there was no *Utricularia* near Seattle. There was considerable interest at that time in Gause's work on competition. I had read somewhere that *Asplanchna priodonta* and *A. brightwelli* did not occur together, but I noticed that the two did in fact occur together in a collection from Hall Lake. When I went back to the lake to begin a study of the situation there was no *Asplanchna* to be found at all, but I discovered that the lake was meromictic, a condition that interested me greatly, so I started what turned out to be a three-year study. We also found the saline lakes in the Lower Grand Coulee to be interesting. Soap Lake and Lake Lenore had an usual *Hexarthra*, which Koste much later named *H. polydonta soaplakeiensis* [sic], based on my material. Through unusual circumstances one of my graduate students was able to spend two years at the Arctic Research Laboratory at Point Barrow, Alaska, incidentally providing the material on *Kellicottia* for a paper included in this volume.

In 1954 the Coulee lakes suddenly gained new interest when we found that they were being diluted by irrigation water and the plankton was changing. Likewise, the next year Lake Washington showed signs of eutrophication and the plankton changed. We started studies of the lakes as ecosystems, with emphasis on the plankton communities. Since then, much of my time and attention has been given to studies of situations that are accidental whole-lake experiments. While the entire community is the object of study, I look for interesting rotifer conditions, such as that reported for *Conchilus* at the fourth rotifer symposium.

In 1959–60, I had a Senior Postdoctoral Fellowship from the National Science Foundation for study in Europe. I wanted to develop the idea about using the counts of the eggs of planktonic rotifers to calculate birth and death rates. We went to the Instituto Italiano di Idrobiologia on Lago Maggiore in Italy to get data on the rate of egg development as a function of temperature. Armed with that information, we went on to Windermere where I found that John Lund had a long set of counts of phytoplankton that rotifers

could eat, accompanied by samples in which I could count rotifer eggs. By applying the information gained at Pallanza, I could calculate birth rates and relate them to food supply. This resulted in a long paper in the 1965 Ecological Monographs. This method of calculation can be applied to any species of animal that carries its eggs. Many people have used it for Cladocera and Copepoda, and the methods for calculating the rates have been improved.

After returning to Seattle I had several visitors who came to do research on rotifers. The first was Agostino Parise whom I had met at Pallanza. He arrived soon after we returned and worked for a year, mostly on population dynamics of *Euchlanis*. (When he returned to Italy in 1961 he left some cultures which were soon found by Charles King. Charles came to work as a graduate student with a marine invertebrate zoologist, but fortunately he discovered rotifers and did a Ph.D. study of population dynamics of *Euchlanis*.) John Gilbert finished his Ph.D. with Hutchinson and came here in 1963 with a Postdoctoral fellowship from the Public Health Service [sic]. Starting from observations by de Beauchamp, Gilbert carried out his magnificent set of experiments on the induction of spines on *Brachionus calyciflorus* by *Asplanchna brightwell*. Robert Wallace finished a Ph.D. on sessile rotifers with Gilbert in 1974 and came to Seattle for two years. He took a unique approach to substrate selection that successfully combined his knowledge of larval behavior and plant physiology. All along, as these things were happening, I continued to work on the plankton ecology of lakes.

The studies of the diluted saline lakes, the eutrophied Lake Washington and others have been fascinating. I have paid a great deal of attention of *Daphnia* which is a wonderful animal, but I have never lost sight of the rotifers. Every time my assistants make a sampling trip on any lake they bring back a live plankton haul so that I can enjoy looking at the rotifers in it. I keep my eyes open for anything unusual, and sometimes collect extra material for future attention. I have a number of small unfinished projects similar to the one with *Kellicottia*, which I hope to be finishing up in due time.

References

Beckel, A. L. 1987. Breaking new waters. A century of limnology at the University of Wisconsin. Trans. Wis. Acad. Sci. Arts & Lett. Special Issue.

Burger, A. 1948. Studies on the moss dwelling bdelloids (Rotifera) of eastern Massachusetts. Trans. am. Micro. Soc. 67: 111–142.

Cook, R. E. 1977. Raymond Lindeman and the trophic-dynamic concept in ecology. Science. 198: 22–26.

Gilbert, J. J. *Asplanchna* and postero-lateral spine production in *Brachionus calyciflorus*. Arch. Hydrobiol. 64: 1–62.

Hardman, Y. 1941. The surface tension of Wisconsin lake waters. Trans. Wis. Acad. Sci. Arts Lett. 33: 395–404.

Hudson, C. T. & P. H. Gosse. 1886. The Rotifera; or wheel animalcules. Longmans, Green.

Hutchinson, G. E. 1941. Limnological studies in Connecticut IV. Mechanism of intermediary metabolism in stratified lakes. Ecol. Monogr. 11: 21–60.

Hutchinson, G. E. 1938. Chemical stratification and lake morphology. Proc. natn. Acad. Sci. U.S.A. 24: 63–69.

King, C. E. 1967. Food, age and the dynamics of a laboratory population of rotifers. Ecology. 48: 111–128.

Lindeman, R. L. 1941. The trophic-dynamic aspect of ecology. Ecology. 23: 399–418.

Bibliography of W. T. Edmondson
Abstracts and book reviews are omitted. For a complete list see Limnology and Oceanography Vol. 33, Part 1.

1. Edmondson, W. T., 1934. Investigations of some Hispaniolan lakes. (Dr. R.M. Bond's Expedition) 1. The Rotatoria. Arch. Hydrobiol. 26: 465–471.
2. Edmondson, W. T. & G. E. Hutchinson, 1934. Yale North India Expedition. Article IX. The Rotatoria. Mem. Conn. Acad. Arts. Sci. 10: 153–186.
3. Edmondson, W. T., 1935. Some Rotatoria from Arizona. Trans. Amer. Micr. Soc. 54: 301–306.
4. Edmondson, W. T., 1936. Fixation of sessile Rotatoria. Science 84: 444.
5. Edmondson, W. T., 1936. New Rotatoria from New England and New Brunswick. Trans. Am. Micro. Soc. 55: 214–222.
6. Edmondson, W. T. & John L. Fuller, 1937. Food conditions in some New Hampshire lakes. New Hampshire Fish and Game Department Survey Report No. 2. Biological Survey of the Androscoggin, Saco and Coastal Watersheds. 95–99.
7. Edmondson, W. T., 1938. Notes on the plankton of some lakes in the Merrimack watershed. New Hampshire Fish and Game Department Survey Report No. 3. Biological Survey of the Merrimack Watershed. 107–210.
8. Edmondson, W. T., 1938. Three new species of Rotatoria. Trans. Am. Micr. Soc. 57: 153–157.
9. Edmondson, W. T., 1939. New species of Rotatoria with notes on heterogonic growth. Trans. Am. Micro. Soc. 58: 459–472.
10. Edmondson, W. T., 1940. The sessile Rotatoria of Wisconsin. Trans. Am. Micr. Soc. 59: 433–459.
11. Edmondson, W. T., 1944. Ecological studies of sessile Rotatoria. Part I. factors affecting distribution. Ecol. Monogr. 14: 31–66.
12. Edmondson, W. T., 1945. Ecological studies of sessile Rotatoria. Part II. Dynamics of populations and social structures. Ecol. Monogr. 15: 141–172.
13. Edmondson, W. T., 1946. Factors in dynamics of rotifer populations. Ecol. Monogr. 16: 357–372.
14. Clarke, G. L., W. T. Edmondson & W. E. Ricker, 1946. Mathematical formulation of biological productivity. (Part of a symposium on dynamics of production in aquatic populations). Ecol. Monogr. 16: 336–337.
15. Edmondson, W. T. & Yvette H. Edmondson, 1947. Measurements of production in fertilized salt water. Sears J. Mar. Res. 6: 228–246.
16. Bigelow, H. B. & W. T. Edmondson, 1947. Wind waves at sea, breakers and surf. U.S. Hydrographic Office Pub. 602 xi + 177.
17. Edmondson, W. T., 1948. Rotatoria from Penikese Island, Mass., with description of *Ptygura agassizi*, n. sp. Biological Bulletin 94: 169–173.
18. Edmondson, W. T., 1948. Two new species Rotatoria from sand beaches. Trans. Am. Micro. Soc. 67: 149–152.
19. Edmondson, W. T., 1948. Ecological applications of Lansing's physiological work on longevity in Rotatoria. Science 108: 123–126.
20. Edmondson, W. T., 1949. A formula key to the Rotatorian genus *Ptygura*. Trans. Am. Micro. Soc. 68: 127–135.
21. Edmondson, W. T., 1950. Centrifugation as an aid in examining and fixing rotifers. Science 112: 49.
22. Comita, G. W. & W. T. Edmondson, 1953. Some aspects of the limnology of an Arctic Lake. Stanford Univ. Pub. Univ. Ser. Biol. Sci. 11: 7–13.
23. Edmondson, W. T., 1955. Seasonal life history of *Daphnia* in an Arctic Lake. Ecol. 36: 439–455.
24. Nelson, P. R. & W. T. Edmondson, 1955. Limnological effects of fertilizing Bare Lake, Alaska. U.S. Fish & Wildlife Service. Fishery Bull. 102: 413–436.
25. Edmondson, W. T., 1955. Factors affecting productivity in fertilized salt water. Deep Sea Res. 3: 451–464. (suppl. for H. B. Bigelow)
26. Edmondson, W. T., 1956. The relationship of photosynthesis by phytoplankton to light in lakes. Ecol. 37: 161–174.
27. Edmondson, W. T., 1956. Measurement of conductivity of lake water *in situ*. Ecol. 37: 201–204.
28. Edmondson, W. T., G. C. Anderson & Donald R. Peterson, 1956. Artificial eutrophication of Lake Washington. Limnol. Oceanogr. 1: 47–53
29. Edmondson, W. T., 1956. Biological aspects of the problem. (In A new critical phase of the Lake

8

Washington problem, by R. O. Sylvester, W. T. Edmondson and R. H. Bogan). Trend in Engineering 8: 11–13.

30. Edmondson, W. T., 1957. Trophic relations of the zooplankton. Trans. Am. Micr. Soc. 76: 225–245.

31. Edmondson, W. T., 1959. Ward and Whipple's Freshwater Biology. Editor of second edition; author of Preface, Introduction and Chapters on Rotifera and Methods.

32. Edmondson, W. T., 1960. Reproductive rate of rotifers in natural populations. Me. ist. Ital. Idrobiol. 12: 21–77.

33. Edmondson, W. T., 1961a. Changes in Lake Washington following an increase in the nutrient income. Verh. Internat. Verein. Limnol. 14: 167–175. (Reprinted pp. 364–372 in Ford, R. F. and W. T. Hazen, eds. 1972. Readings in aquatic ecology, Saunders).

34. Edmondson, W. T., 1961b. Secondary production and decomposition. Verh. int. Ver. Limnol. 14: 316–339.

35. Edmondson, W. T., 1962. Food supply and reproduction of zooplankton in relation to phytoplankton population. Rapp. et Proc.-Verb. Cons. Internat. Explor. de la Mer. 153: 137–141.

36. Edmondson, W. T., G. W. Comita & G. C. Anderson, 1962. Reproductive rate of copepods in nature and its relation to phytoplankton population. Ecology 43: 625–634.

37. Edmondson, W. T., 1963. Pacific Coast and Great Basin, pp. 371–392 In D. G. Frey (ed.) Limnology in North America. University of Wisconsin Press, Madison, Wisconsin.

38. Edmondson, W. T., 1964. The rate of egg production by rotifers and copepods in natural populations as controlled by food and temperature. Proc. int. Ass. Theor. and Appl. Limnol. 15: 673–675.

39. Edmondson, W. T., 1965. Reproductive rate of planktonic rotifers as related to food and temperature in nature. Ecol. Monogr. 35: 61–111.

40. Edmondson, W. T. & G. C. Anderson, 1965. Some features of saline lakes in Central Washington. Limnol. Oceanogr. 10: (suppl.): R87–R96.

41. Edmondson, W. T., 1966. Changes in the oxygen deficit of Lake Washington. Verh. int. Ver. Limnol. 16: 153–158.

42. Oglesby, R. T. & W. T. Edmondson, 1966. Control of eutrophication. Jour., Water Pollution Control Federation. 1966: 1452–1460.

43. Edmondson, W. T., 1967. Why study blue-green algae? pp. 1–6 (introductory remarks) in Symposium: Environmental Requirements of Blue-Green Algae. Pub. Pacific Northwest Water Laboratory, Corvallis, Oregon.

44. Edmondson, W. T., 1968. Water quality management and lake eutrophication: The Lake Washington Case. Water Resources Management and Public Policy. pp. 139–178. T. H. Campbell and R. O. Sylvester (eds.) University of Washington Press, Seattle.
   Note: This paper has been translated into Hungarian by Dr Olga Sebestyen. Visminösegi ugyvetes es a tavak eutrofikacioja: A Lake Washington ügy.

45. Edmondson, W. T., 1968. A graphical model for evaluating the use of the egg ratio for measuring birth and death rates. Oecologia 1: 1–37.

46. Edmondson, W. T., 1969. Eutrophication in North America. Eutrophication: Causes, Consequences, Correctives. pp. 124–149. National Academy of Sciences Publication No. 1700.

47. Edmondson, W. T., 1969. Cultural eutrophication with special reference to Lake Washington. Mitt. int. Ver. Limnol. 17: 19–32.

48. Edmondson, W. T., 1969. The present condition of the saline lakes in the Lower Grand Coulee, Washington. Verh. int. Ver. Limnol. 17: 447–448.

49. Griffiths, M., P. S. Perrott & W. T Edmondson, 1969. Oscillaxanthin in the sediment of Lake Washington. Limnol. Oceanogr. 14: 317–326.

50. Edmondson, W. T., 1970. Phosphorus, nitrogen and algae in Lake Washington after diversion of sewage. Science 196: 690–691. (Reprinted pp. 372–374 in Ford, R. F., W. E. Hazen, eds. 1972. Readings in aquatic ecology, Saunders.).

51. Edmondson, W. T. & David E. Allison, 1970. Recording densitometry of X-radiographs for the study of cryptic laminations in the sediment of Lake Washington. Limnol. Oceanogr. 15: 138–144.

52. Shapiro, J., W. T. Edmondson & D. E. Allison, 1971. Changes in the chemical composition of sediments of Lake Washington, 1958–1970. Limnol. Oceanogr. 16: 437–452.

53. Edmondson, W. T. & G. G. Winberg (editors), 1971. A manual on methods for the assessment of secondary productivity in fresh waters. IBP handbook No. 17. Blackwell, Oxford.

54. Edmondson, W. T., 1972. Nutrients and phytoplankton in Lake Washington. pp. 172–193 in Nutrients and Eutrophication. American Society of Limnology and Oceanography, Special Symposia No. 1. G. Likens (ed.).

55. Edmondson, W. T., 1972. The present condition of Lake WAshington. Verh. int. Ver. Limnol. 18: 284–291.

56. Edmondson, W. T., 1972. Instantaneous birth rates of zooplankton. Limnol. Oceanogr. 17: 792–795.

57. Beeton, A. M. & W. T. Edmondson, 1972. The eutrophication problem. J. Fisheries. Res. Bd. Canada. 19: 673–682.

58. Edmondson, W. T., 1973. Lake Washington. pp. 281–298. In Environmental quality and water development. Ed. C. R. Goldman, James McEvoy III and Peter J. Richerson. Freeman.

59. Edmondson, W. T., 1974. Secondary production. Mitt. int. Ver. Limnol. 20: 229–272. (See 1974 in Part II).

60. Edmondson, W. T., 1974. The sedimentary record of the eutrophication of Lake Washington. Proc. Nat. Acad. Sci. USA 71: 5093–5095.

61. Edmondson, W. T., 1975. Microstratification of Lake Washington sediments. Ver. int. Ver. Limnol. 19: 770–775.

62. Griffiths, M. & W. T. Edmondson. 1975. Burial of

oscillaxanthin in the sediment of Lake Washington. Limnol. Oceanogr. 20: 945–952.

63. Edmondson, W. T., 1977. Recovery of Lake Washington from eutrophication. pp. 102–109 in Recovery and restoration of damaged ecosystems. ed. J. Cairns, Jr., K. L. Dickson & E. E. Herricks, University Press of Virginia.

64. Edmondson, W. T., 1977. Lake Washington in North American Project – A study of United States Water Bodies. A Report of the Organization for Economic Cooperation and Development. (EPA-600/3-77-086) Published by Environmental Research Laboratory, Environmental Protection Agency, Corvallis, Oregon.

65. Edmondson, W. T., 1977. Trophic equilibrium of Lake Washington. Final Report on EPA Project R 8020 82-03-1. (EPA-600/3-77-087). Published by Environmental Research Laboratory, Environmental Protection Agency, Corvallis, Oregon.

66. Edmondson, W. T., 1977. Population dynamics and secondary production. In First International Symposium on Rotifers. Ed. C. E. King. Arch. Hydrobiol. Beih. Ergebn. Limnol. 8: 56–64.

67. Edmondson, W. T., 1979. Lake Washington and the predictability of limnological events. (Symposium on Lake Metabolism and Lake Management celebrating the 500th Anniversary of the University of Uppsala). Arch. Hydrobiol. Beih. Ergebn. Limnol. 13: 234–241.

68. Edmondson, W. T., 1979. Problems of zooplankton population dynamics. In R. de Bernardi (Ed.) Proc. Symp. Biological and mathematical aspects in population dynamics. Mem. Ist. Ital. Idrobiol. Suppl. 37: 1–11. (Actually appeared in 1980).

69. Edmondson, W. T., 1980. Secchi disc and chlorophyll. Limnol. Oceanogr. 25: 378–379.

70. Edmondson, W. T. & J. T. Lehman. 1981. The effect of changes in the nutrient income on the condition of Lake Washington. Limnol. Oceanogr. 26: 1–29.

71. Edmondson, W. T. & A. H. Litt. 1982. *Daphnia* in Lake Washington. Limnol. Oceanogr. 27: 272–293.

72. Lehman, J. T. & W. T. Edmondson, 1983. The seasonality of phosphorus deposition in Lake Washington. Limnol. Oceanogr. 18: 796–800.

73. Edmondson, W. T. & A. H. Litt. 1984. Mt. St. Helens ash in lakes in the Lower Grand Coulee, Washington State. Verh. int. Ver. Limnol. 22: 510–512.

74. Edmondson, W. T. 1984. Volcanic ash in lakes. Northwest Env. Jour. 1: 139–150.

75. Edmondson, W. T. 1985. Reciprocal relations between *Daphnia* and *Diaptomus* in Lake Washington. Arch. Hydrobiol. Beih. Ergeb. Limnol. 21: 475–481.

76. Infante, A. & W. T. Edmondson. 1985. Edible phytoplankton and herbivorous zooplankton in Lake Washington. Arch. Hydrobiol. Ergebn. Limnol. 21: 161–171.

77. Edmondson, W. T. 1985. Recovery of Lake Washington from eutrophication. Proc. Internat. Congress Lake Pollution and Recovery, Rome, April, 1985. pp. 308–314 (228–234 in Preprints).

78. Wallace, R. L. & W. T. Edmondson. 1986. Mechanism and adaptive significance of substrate selection in the sessile rotifer *Collotheca gracilipes*. Ecology 67: 314–323.

79. Edmondson, W. T. & A. H. Litt. 1987. *Conochilus* in Lake Washington. Proc. Fourth Internat. Rotifer Sym. Hydrobiol. 147: 157–162.

80. Edmondson, W. T. & Sally E. B. Abella. 1988. Unplanned biomanipulation in Lake Washington. Limnologica 19: 73–79.

81. Edmondson, W. T. 1987. *Daphnia* in experimental ecology: notes on historical perspectives. Me. Ist. Ital. Idrobiol. 45: 11–30. (Actually published in 1988).

82. Edmondson, W. T. 1988. On the modest success of *Daphnia* in Lake Washington in 1965. p. 223–243 *in* Algae and the aquatic environment. Contributions in honour of J. W. G. Lund, C.B.E., F.R.S., ed. F. E. Round. Biopress.

83. Edmondson, W. T. (in press). Lessons from Washington Lakes. International Mountain Watershed Symposium, Lake Tahoe.

84. Edmondson, W. T. & Y. H. Edmondson. (in press). Pallanza as a haven for visiting limnologists. Me. ist. Ital. Idrobiol.

85. Edmondson, W. T. (in press). Perspectives in plankton research. Me. ist. Ital. Idrobiol.

86. Edmondson, W. T. 1989. Rotifer study as a way of life. Proc. Fifth Internat. Rotifer Symp. Hydrobiologia 186/187: 1–9.

87. Edmondson, W. T. & Arni H. Litt. 1989. Morphological variation in *Kellicottia longispina*. Fifth Internat. Rotifer Symp. Hydrobiologia 186/187: 109–117.

Hydrobiologia **186/187**: 11–28, 1989.
C. Ricci, T. W. Snell and C. E. King (eds), Rotifer Symposium V.
© 1989 *Kluwer Academic Publishers.*

Food limitation and body size in the life cycles of planktonic rotifers and cladocerans

Annie Duncan
Department of Biology, Royal Holloway & Bedford New College, University of London, Egham, Surrey TW20 OEX, U.K.

Key words: food limitation, body size, duration of development, fecundity, life cycle studies, rotifers, cladocerans

Abstract

This review considers what is known about the effects of food limitation upon the life cycle characteristics of rotifers and planktonic cladocerans. The characteristics considered in rotifers are the size of eggs, juveniles and adults and the durations of the juvenile phase and period of egg production. In cladocerans, the life history features dealt with are their length-weight relationships, the body size, instar stage, age and fecundity of the primiparous female and their fecundity-adult size relationship. The influence of limiting food conditions is demonstrated for these characteristics by comparison with the situation in non-limiting circumstances; the comparison is confined to experiments where food concentrations are quantified. A direct comparison is made between rotifers and cladocerans in conditions of defined food resource availability in terms of their length-weight relationships, the daily allocation of adults or near-adults to growth and reproduction and their threshold food concentrations. These comparisons are discussed in relation to the following topics: the high cost of cumulated respiration resulting from prolongation of the juvenile phase of body growth; the fundamentally different nature of growth in the two taxonomic groups; the body size of species and the size that must be attained for reproduction; the ecological implications of the very different threshold food concentrations.

Introduction

Herbivorous zooplankton from large water bodies often have to face food limitation for as long as a generation. Such limiting food conditions can affect several life cycle characteristics which are important ecologically: the duration of development, the body size at which certain stages of development are attained and reproduction begins and the fecundity of adults (King, 1967; Hrbackova, 1971; Weglenska, 1971; Hrbackova & Hrbacek, 1979). Developmental effects of food limitation are much less acknowl-

edged than the more thoroughly investigated effects of temperature on the development of rotifers, cladocerans and copepods (Bottrell *et al.*, 1976; Herzig, 1983a, b). For example, King's (1967, Fig. 2) study on *Euchlanis dilatata* shows both a doubling of the juvenile phase and a reduction in the body length of the primipara female in animals grown in ten times less algal food (16.4 and 1.64 μgDW ml^{-1}). Pilarska's (1977) eco-physiological investigation on *Brachionus rubens* presents evidence of similar size and time effects, provided that the range of test food concentrations include some limiting ones. This was not so in Leimeroth's (1980) energy budget for juvenile *B. calyciflorus* but is so in the unpublished study by Skrdla & Starkweather on the same species (Starkweather, 1987). Similarly in crustaceans, Fig. 6 in Weglenska's (1971) paper shows that both body length and the age at which four species of Cladocera from Mikolajskie Lake attained maturity were changed when reared on natural food manipulated to provide a wide range in food concentration from 0.5 to 0.02 mgC L^{-1} (converted approximately from 10 to 0.41 mgWW L^{-1}).

There is confusion in the literature on ecological studies on body sizes in cladoceran populations because changes in the size distribution detected in the field are usually attributed to the presence of size-selective predation but may also result from food limitation. Similar uncertainty may occur with rotifer populations, as, for example, in Parakrama Samudra whose small-size rotifer community was thought to be due to fish predation and not to the effects of food limitation (Duncan, 1984). In planktonic communities, these two pressures often occur simultaneously so that there is a need to find techniques to distinguish between the two effects, considering the difficulty of quantifying food availability in the field. For field populations of *Daphnia* species, Duncan (1985) proposed the use of length-carbon weight regressions whose elevations decrease significantly as the food levels become limiting. The duration of the period of food limitation can therefore be defined by regular monitoring of the length-weight regressions of field populations (Santos, 1989). In experimental and field studies on rotifers, the species' responses to limiting food conditions is usually measured in terms of changes in density (as the instantaneous rate, r or as the net reproductive rate per generation (R_0), which takes account of numbers of individuals but not their body size or nutritive state. Changes in rotifer body size, shape and spines have been reported and have generated considerable interest in relation to avoidance of predation (Gilbert & Stemberger, 1984) or cladoceran interference (Gilbert, 1985), in relation to the influence of abiotic variables (Serra & Miracle, 1987) or in

relation to geographical or seasonal morphological variation within species and to whether this has a genetic origin (Pejler, 1980; King, 1977; Snell, 1977). In aquaculture, variation in the body size of *Brachionus plicatilis* is of direct interest for finding suitable strains to feed growing fish larvae (Snell & Carrillo, 1984; Lubzens, 1987) and there is evidence that body size and biochemical composition of this species vary with food availability when reared in mass culture. There is little information on the ecological significance of changes in body size, stage or duration of development within the life cycle that might be induced environmentally.

The aim of this paper is to review the evidence for the effects of food limitation upon the life cycle characteristics of planktonic rotifers and cladocerans. The cladoceran review is based upon a comparable series of unpublished studies of temperate and tropical species of different body size and in which the experiments were designed to clarify the food effects. There appears to be only one comparable published study for rotifers, that of Pilarska (1977) for *B. rubens*, although another one, Skrdla & Starkweather on *B. calyciflorus* (Starkweather, 1987), will be available in future. Later in the paper, some comparisons will be made on the threshold food concentrations of rotifers and cladocerans as well as on how the two groups allocate scarce food resources to growth and reproduction.

In this paper, the condition of food limitation is not taken to be equivalent to starvation, which exists below the threshold food concentration for growth and reproduction and which may lead to some form of metabolic depression or diapause. There may be a series of limiting concentrations from the optimal down to the threshold value which, in long-term life cycle studies, can taken to be the minimal food level at which the life cycle can be completed (i.e. reproduction can occur). This differs from Lampert's (1977) definition that it is the concentration at which an animal just balances its metabolic losses so that it cannot and does not grow but does not lose weight. This definition, which implies different threshold values for different stages in the life cycle, is the one adopted by Schiemer (1986) in his comparison of food threshold for different aquatic invertebrates but is not so easy to use ecologically. The definition being used in this paper is rather similar to that employed by Stemberger & Gilbert (1985, 1987) which is the food concentration at which the instantaneous 'r' is zero or slightly above zero.

Fenchel (1980) used another measure of food thresholds for ciliates, which was the minimal food particle concentration needed to cover the metabolic costs when clearing water maximally at low food concentrations.

Body size of rotifers and cladocerans

Although rotifers and cladocerans differ greatly in their definitive body sizes (Fig. 1), there is a size of overlap between the two groups. The length-carbon weight regressions for the three large temperate species of *Daphnia* and for the smaller tropical *Daphnia gessneri* and *Moina reticulata* were measured by Santos (1989) and Hardy (1989) under a range of known food conditions and cover the full size range of the species. These regressions overlap with the lengths and dry weights of fourteen species of adult rotifers measured by Stemberger & Gilbert (1987) and, for this figure, converted to carbon (0.5 DW = carbon weight). These rotifers included three species of *Asplanchna*, two of *Brachionus*, one of *Keratella*, two of *Polyarthra* and three of *Synchaeta*. Also included in this figure are the length-dry weights (converted to carbon) of adult *Keratella quadrata* (Doohan & Rainbow, 1971) and juvenile and adult *Brachionus rubens* (Pilarska, 1977) as well as the carbon weights for *Brachionus angularis* and *Keratella cochlearis* (Walz, 1983). The rotifer lengths and weights are largely for well-fed adults, apart from those for *B. rubens* from Pilarska (1977) which consist of

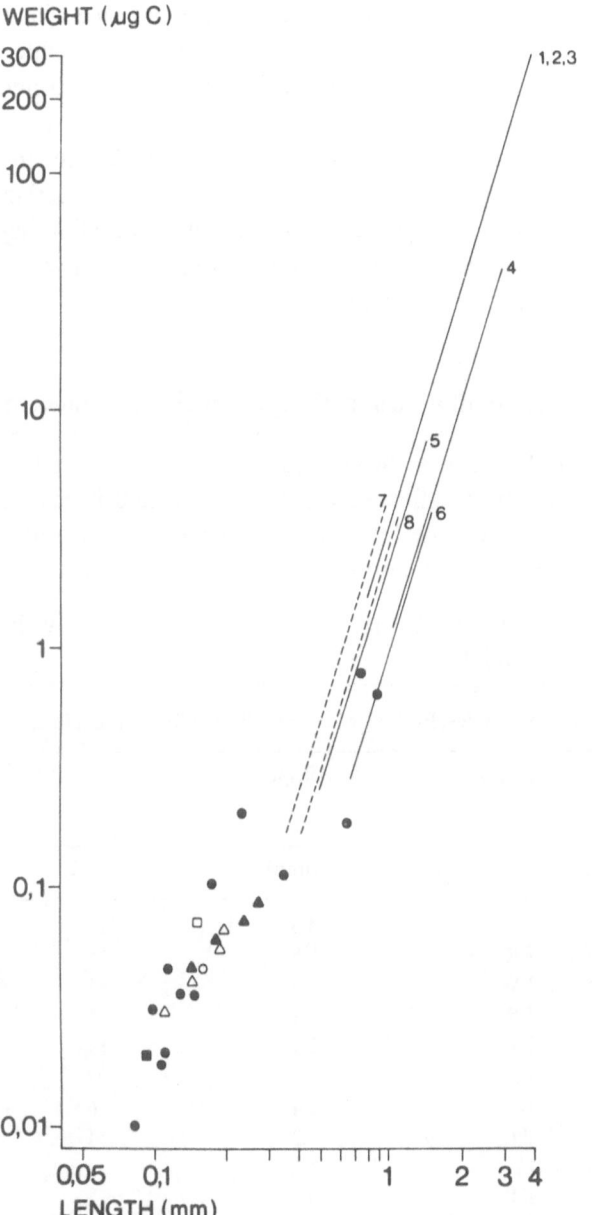

Fig. 1. The length-weight relationship in rotifers and temperate and tropical planktonic cladocerans. Regressions 1, 2 and 3 are pooled regressions for *Daphnia magna, pulicaria* and *hyalina* at food concentrations of greater than 1 mgC per

litre (1), at 1 mgC per litre (2) and at 0.1 mgC per litre (3). Regression 4 is for *Daphnia magna* reared in 0.01 mgC per litre. Regressions 5 and 6 are for the tropical *Daphnia gessneri* reared in food concentrations of 0.5 (5) and 0.05 (6) mgC per litre. Regressions 7 and 8 are for the tropical *Moina reticulata* reared in food concentrations of 0.5 (7) and limiting food levels (8). Black dots: adult rotifers from Stemberger & Gilbert (1987). Empty circle: adult *Brachionus angularis*; black square: adult *Keratella cochlearis*; from Walz (1983). Empty square: adult *Keratella quadrata* from Doohan & Rainbow (1971). Triangles: three ages of juveniles and adults of *Brachionus rubens* from Pilarska (1977); the empty triangles are for 'deficient' diets and the black triangles for 'optimal' diets.

the sizes for three ages of juveniles and for adults reared at her 'deficient' and 'optimal' food levels. Although it is hazardous to extrapolate beyond the measured ranges of the regressions, the higher elevations of the rotifer points in Fig. 1 imply heavier animals per unit length. This implication is true for the two species measured directly in carbon weight (Walz, 1983) as well as the other species determined as dry weight, with the exception of the three points for *Asplanchna* spp., and may be associated with the lorica weight. Guisande (this symposium) found that rotifers and *Daphnia magna* fell on the same length-protein weight regressions.

The range of adult sizes belonging to the rotifers in Fig. 1 can be compared with the sizes of well fed adults instars (3rd or 4th) of the species of planktonic cladocerans subjected to food limitation studies. These are given in Table 1, together with their site of origin and names of the authors who studied them. The span in the size range amongst the two groups of species is rather similar. For the rotifers, there is a 10 times span in length and a 74 times span for weight or, if the campanulate form of *A. silvestrii* is included, the span in length and weight comes to 27 times and 350 times, respectively. The comparable range for

the lengths and weights of the cladoceran species is 9 times for length and 800 times in weight. The definitive body size of species is taken to be an inherited trait, presumably constrained by the physiological possibilities of the taxonomic group, and is the result of an evolutionary 'decision' to be large or small relative to the rest. The genetic basis for the definitive size is unknown, judging from King (1977): 'to my knowledge not one single morphological feature has ever been associated with a specific single allele in (rotifers)'. Since body size seems to be a phenotypic variable responding to physiological conditions in the environment, it is interesting to explore whether being large or small is a 'trade-off' in food limiting conditions which are common in the open water of large water bodies.

Food limitation and life cycle studies in Cladocera

In Cladocera, the species' life cycle characteristics which are affected when an individual is reared from neonate to some known adult instar under food limiting conditions appear to be that:

(1) the body size is reduced, particularly in weight;

Table 1. The species of planktonic Cladocera subjected to food limitation studies. Body sizes refer to third or fourth adult instars.

Species	Author [1]	Origin [2]	Body sizes	
			Length (mm)	Weight (μgC)
Daphnia magna Straus	R	LR	4.5	322
Daphnia pulicaria Forbes	R	LR	2.8	85
Daphnia pulicaria Forbes	D	LW	2.5	36
Daphnia hyalina Leydig	R	LR	2.1	31
Daphnia thorata Forbes	D	LW	1.7	19
Daphnia gessneri Herbst	H	LJ	1.5	7
Daphnia lumholzi Sars	J	KR	1.4	6
Diaphanosoma excisum Sars	J	KR	1.2	11
Diaphanosoma brachyurum Lieven	H	LJ	1.2	11
Moina micrura Kurz	J	KR	0.8	2
Moina reticulata Daday	H	LJ	0.8	2
Ceriodaphnia cornuta Sars	J	KR	0.5	0.4

[1] Authors: Duncan (in prep.), Hardy (1989), Jayatunga (1986), Rocha (1983)
[2] Sites of origin: KR – Kalawewa Reservoir, Sri Lanka; LJ – Lake Jacaretinga, Brazil; LR – London Reservoirs, U.K.; LW – Lake Washington, U.S.A.

(2) the number of instars to a certain point in the life cycle, such as the onset of reproduction, increases;

(3) the duration of development is prolonged;

(4) fecundity is reduced.

Thus, if the primiparous female is taken as a comparable point in the life cycle which is at the end of juvenile growth and at the beginning of reproductive growth, then she is smaller in size, appears at a later instar, is older and has reduced fecundity under limiting food conditions. These effects appear simultaneously when individuals are reared under food limitation and differ in degree according to the severity of the limitation down to the threshold food level for growth and reproduction.

These effects will be illustrated for the species listed in Table 1 which come from various temperate (the London reservoirs and Lake Washington, USA) and tropical (Kalawewa and Lake Jacaretinga) sites and for all of which there is a background of ecological information. Apart from the Amazonian species which were studied in the UK, the species were studied in their country of origin and close to the water body from which they came as one aim of all the studies was to interpret field biological interactions by application of experimental results. Table 2 provides information on the temperature conditions for each of the food concentrations tested by each author, using algal food organisms which were either the single-celled *Scenedesmus acutus* (Chlorophyceae) or the flagellate *Cryptomonas* sp., both of which were adequate food for all the species, except for *C. cornuta*.

All the cladoceran results come from long-term life cycle studies where the development of a neonate is followed daily throughout its life history to a reproducing adult whilst being maintained in constant conditions of food and temperature. One of the greatest problems in such long-term experiments is to maintain constancy of food conditions, especially in batch culture which is the commonly adopted culture technique. It is commonly found that published results on how food quantity affects the life cycles of cladocerans are not comparable for methodological reasons. Incomparability of results can arise for various reasons: too small experimental volumes; using

Table 2. The concentrations of food used in cladoceran studies on food limitation.

Food levels (mgC L^{-1})	Food species		Authors
	Scenedesmus acutus	*Cryptomonas* sp.	
5.0	yes	–	R
4.5	–	yes	D
1.5	–	yes	D
1.0	yes	yes	R, D, H, J
0.5	yes	yes	D. H, J
0.25	yes	yes	D, H, J
0.125	–	yes	D
0.10	yes	–	R
0.06	–	yes	D
0.05	yes	–	H, J
0.03	yes	yes	D, H, J
0.01	yes	–	R, J
0.005	yes	–	R, D

	Key		
R O. Rocha (1983)	*D. magna, D. pulicaria, D. hyalina*	5, 10, 15 & 20 °C
D A. Duncan	*D. pulicaria, D. thorata*	9, 14 & 19 °C
H E. R. Hardy (1989)	*D. gessneri, D. brachyurum, M. reticulata*	22, 27 & 32 °C
J Y. N. A. Jayatunga (1986)	*D. lumholzi, D. excisum, M. micrura, C. cornuta*	22, 27 & 32 °C

unstirred or rotated vessels; putting too many animals in one container; too infrequent a changing of the food medium; too high a concentration of food to show food limiting effects. The most serious causes of incomparability of results is that either the experimental food levels have not been quantified or they have been quantified in terms which cannot be converted into common units. For example, many authors gives food levels as cells/ml, without indicating the food cell dimensions or volume or dry weight or carbon weight. The experimental details for this comparable series of life cycle studies are given in Rocha (1983), Duncan (1985a, b), Jayatunga (1986) & Hardy (1989).

Body size

Figure 1 shows how the length-carbon weight regressions for a cladoceran species has a significantly reduced elevation when it has been reared throughout its life cycle in a limiting compared with a non-limiting food concentration. These regressions involve, of course, all the life cycle stages. This size effect is illustrated in Fig. 1 for the large temperate *Daphnia magna* (regressions 1 and 4) and the two tropical species, the larger *Daphnia gessneri* (regressions 5 and 6) and the smaller *Moina reticulata* (regressions 7 and 8). In all cases, the slopes for the paired regressions of each species are parallel (not significantly different) but the elevations are significantly different. Details of the food concentrations in which the animals were reared are given in the legend for Fig. 1. The significant lowering of the elevations of regressions 4, 6 and 8 occurred in food concentrations of 0.01, 0.05 and 0.03 mgC L^{-1}, respectively. In all species, non-limiting levels were at concentrations greater than 0.1 mgC L^{-1} and, under such conditions, the regression 1, 2 and 3 for the three temperate species of *Daphnia* (*magna*, *pulicaria* and *hyalina*) were not different and could be pooled.

Table 3 shows the extent to which the body size of the primipara female can be reduced when she is reared in the near-threshold food concentrations of 0.01-0.05 mgC L^{-1} from her size in non-limiting food levels. The size change is presented for six species of Cladocera of different size (col-

Table 3. The ranges in length, weight, instar and fecundity of the primiparous female in temperate and tropical species of cladocerans reared under various limiting and non limiting food conditions and at various temperatures. The values in brackets refer to 20–22 °C only.

Species	Range in length [1] mm	Non-limiting to limiting food concentrations [2]				
		Primipara length mm	Primipara weight μgC	Primipara instar	Primipara age days	Primipara fecundity eggs brood^{-1}
Daphnia magna	0.90–4.50	3.70–2.20	190–12 (111–17)	(V–IX)	(7–16)	21–1
Daphnia pulicaria	0.80–2.90	2.45–1.60	53–4 (31–6)	(V–XII)	(8–26)	8–1
Daphnia hyalina	060–2.20	1.90–1.40	8–4 (13–3)	(VI–IX)	(8–16)	6–1
Daphnia lumholzi	0.50–1.50	1.30–1.10	5–1 (4–1)	(V–VI)	(6–10)	4–1
Diaphanosoma excisum	0.47–1.30	1.10–0.85	7–2 (5–1)	(XII–XIII)	(5–8)	4–1
Moina reticulata	0.45–0.81	0.70–0.50	1.5–0.5	(III–V)	(2 + –4 +)	4 + –3

[1] Neonate to 3rd or 4th adult instar.

[2] See table 2. Limiting food concentrations were 0.06 mgC L^{-1} or less.

umns 1 and 2) in terms of length (column 3) and carbon weight (column 4). In column 4, the values in brackets refer to experiments conducted at the comparable temperatures of 20-22 °C. The consistent result is that the primipara female is smaller in length and weight when reared in low food compared with optimal levels. The degree of reduction differs with size of species, which probably reflects the difficulty of detecting small change in small organisms. Weight is the better measure and is reduced by three to five times in smaller species compared with a thirteen to fifteen times reduction in larger ones.

Instar of primipara

Column 5 in Table 3 gives the instar stage at which the onset of reproduction occurs in the various cladoceran species under the experimental food conditions and at 20-22 °C. In all the species, the primiparous female produces her first brood of eggs at an earlier instar when reared under optimal food conditions but is forced to pass through several later instars (from one to seven more) before she can achieve this as food availability approaches more severely limiting levels ($0.01-0.05$ mgC L^{-1}).

Age of primipara

The age of the primipara female under the different experimental conditions is given in column 6 of Table 3 and, again, refers to the 20–22 °C experiments only. This represents the end of the juvenile period of body growth and the beginning of the mature phase in which reproductive as well as body growth occurs. It is an ecologically relevant temporal point in the life cycle for detecting any prolongation of the juvenile phase under food limitation. Table 3 shows that the prolongation of this phase is quite marked in severely limiting levels of $0.01-0.05$ mgC L^{-1} in all the species, from 1.5 to 3.3 times longer compared.

Fecundity

Associated with an increase in the age of the primipara female is a marked reduction in the size of her first brood, which is shown in column 7 of Table 3. However, a better assessment of the effect of low food concentrations upon the fecundity of cladoceran species is given by linear regression of eggs per brood against adult length. In the life cycle experiments, these could be calculated from the lengths and fecundities of the first three or four adult instars. An increasing fecundity with increase in body size of female has been reported by many workers (see review by Ivanova et al., 1987) and the regressions plotted in Fig. 2 for five species of Cladocera show very steep slopes in optimal non-limiting food concentrations. The slope of the regression flattens as the food level is lower until, at extreme food limitation, no significant regression can be calculated as all sizes of female are able to produce no more than one or two eggs in a brood. The steeper slopes of Moina are probably associated with her small size rather than high temperature since 32 °C is as high a temperature for tropical species as is 20 °C for temperate ones. These fecundity-length regressions are as good ecological indicators of the nutritive state of field populations of cladocerans as the length-weight regressions considered earlier.

Food limitation and life cycle studies in rotifers

Four periods in the rotifer life cycle are distinguished by King (1969): (1) the embryonic period from egg formation to egg hatching, (2) the pre-reproductive period from egg hatching to first reproduction, (3) the period of fecundity and (4) the post-reproductive period from cessation of reproduction to death. As in the cladocerans, the main periods in the rotifer life cycle which are vulnerable to the environmental stress of a sparse food supply are the juvenile phase (2) and the period of adult reproduction (3) although there is some evidence that egg size and so the duration of embryonic development may be affected by

18

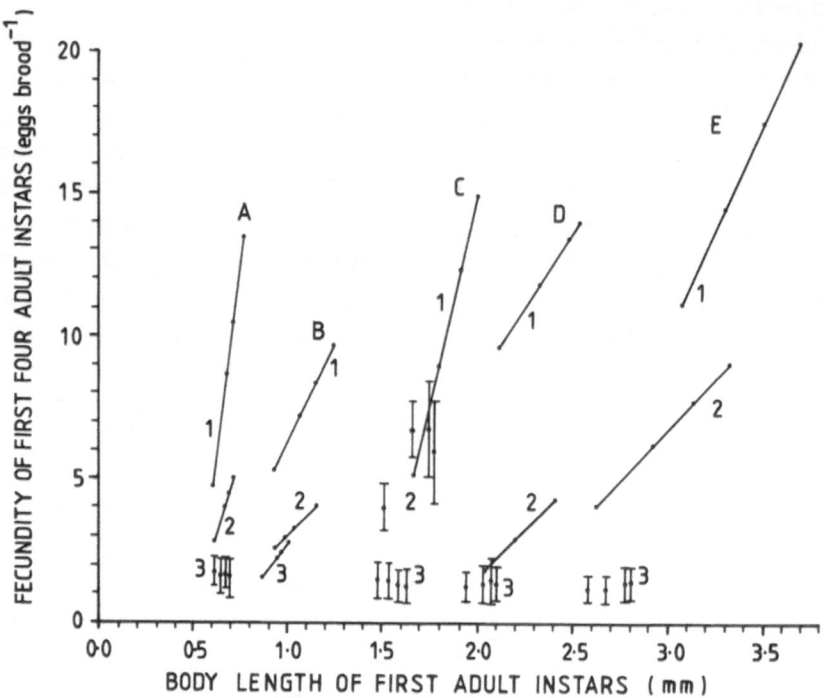

Fig. 2. Length-fecundity regressions for several species of temperate and tropical cladocerans which had been reared throughout their life cycle in different concentrations of food. Species: A – *Moina micrura* (32 °C); B – *Diaphanosoma excisum* (32 °C); C – *Daphnia thorata* (14 °C); D – *Daphnia pulicaria* (14 °C); E – *Daphnia magna* (20 °C). Food level: 1– 0.5–1.0 mgCl^{-1}; 2– 0.1 mgCl^{-1}; 3– 0.01–0.05 mgCl^{-1}

maternal diet (Yufera, 1987; see also Duncan, 1983, 1984). The postembryonic growth in rotifers is limited to expansion of cells or syncytial tissues since there is no proliferation of cells beyond the number fixed during the organ primordia (Starkweather, 1987). As Starkweather comments, this cytological constraint sets close bounds on adult body dimensions. Various authors have divided the juvenile period into an active feeding phase followed by a period of enlargement to an adult size by stretching and incorporation of assimilated material (King, 1967; Pourriott, 1973; Ruttner-Kolisko, 1974) but it is not clear whether the primiparous female is capable of body growth as well as egg production.

There are not many published papers on rotifer life cycle studies involving different food levels. The findings of King (1967) were that the growth rate in length of individual *Euchlanis dilatata* depends on food concentration but only in his three lower food levels from 1.64–16.4 µgDW

ml^{-1}. The rates were more or less similar in 16.4 and 49.2 µgDW ml^{-1}. His Fig. 2 shows that the primipara female became both older and smaller in food below 16.4 µgDW ml^{-1} so that, for this species at 22 °C, food appears to become limiting below this concentration. Figures 3, 4 and 10 in Robertson & Salt (1981) illustrate a reduced net reproductive rate, a prolongation of the age of first reproduction and a reduction in the body volume of the primiparous female of *Asplanchna girodi* when fed low numbers of *Paramecium* per individual per day.

There is only one published series of experiments which studied the influence of a wide range of food concentrations upon the life cycle characteristics of a rotifer species and which is comparable with the cladoceran experiments. This is the eco-physiological study of Pilarska (1977) on *Brachionus rubens* cultured throughout its life cycle at 20 °C and at ten concentrations of *Chlorella vulgaris*. Her food levels ranged from 0.12.10^6 to 10.10^6 cells ml^{-1} which, if one accepts

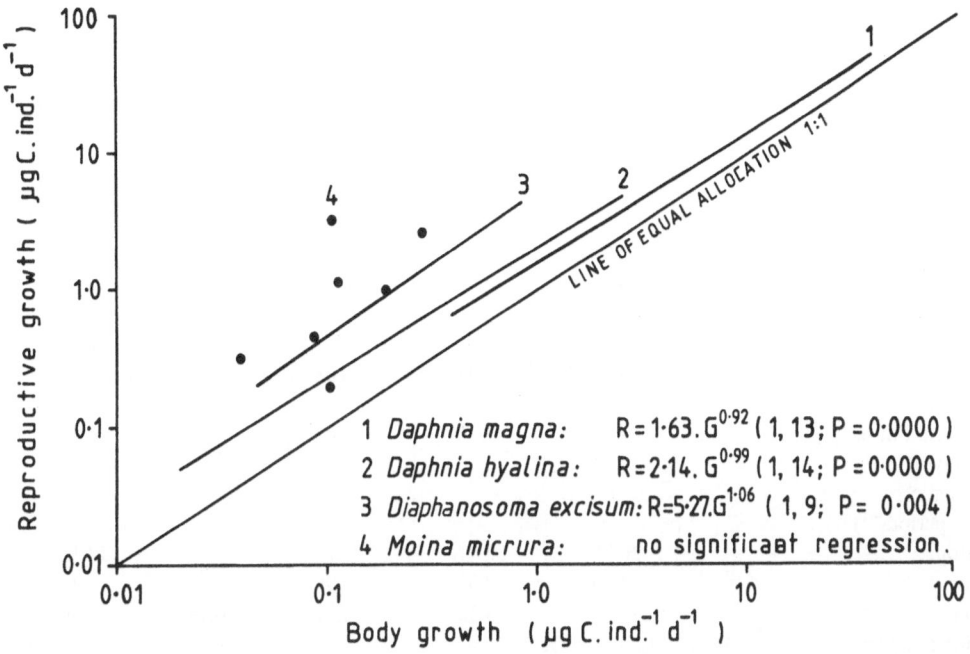

Fig. 3. The allocation of food resources to reproductive or body growth in adult animals of two temperate and two tropical species of cladocerans. The regressions quantify the daily growth rates of the adults under different temperature conditions and limiting and non-limiting food concentrations. The values in brackets are degrees of freedom and probability.

her very high cell weight for *C. vulgaris* (29.10^{-6} µgDW cell^{-1}), can be converted to 0.35 to 290 mgDW L^{-1} or to 0.18 to 145 gC L^{-1}, assuming carbon is 50% of dry weight. Similar published studies which used comparable batch culture rather than chemostat techniques either did not define the food concentration provided (Doohan, 1973) or used food levels which were too high to show food limitation effects (Leimeroth, 1980). Figure 3 in Starkweather (1987) shows that there is another unpublished study by Skrdla & Starkweather on the combined influences of temperature and food density on *Brachionus calyciflorus* fed on *Euglena gracilis* which used a wide range of food levels (0.1 to 100 µg ml^{-1}). This should provide a further comparison in future with both Pilarska's investigation and with the cladoceran ones.

Pilarska's methods and results on the life cycle energy balance for individual *B. rubens* have been published fully in a set of three papers (1977) and have been re-worked by Duncan (1984) to produce Table 4 which will form the basis for detecting any food effects on life cycle characteristics and for comparison with the cladoceran experi-

ments. Three food levels are compared which are characteristic of Pilarska's 'deficient', 'optimal' and 'excessive' diets; these are $0.012.10^6$, $1.0.10^6$ and 10.10^6 cells ml^{-1}.

Table 4 shows that the largest juvenile (less than 12 more than 24 hours old) and the largest amictic mature adult were those grown under optimal food conditions and that these larger females also produced the largest eggs. Her papers show that this is so in linear dimensions as well as in dry weight. In food deficient concentrations, the juveniles and the females were about 30% smaller by weight and the diploid eggs were smaller by 15–20%. There was a prolongation of the juvenile period to 2.8 days in 'deficient' food compared with 1.26 days in 'optimal' food. Pilarska also found a marked shortening of the period of egg production to 0.73 days in the lowest food level compared with 6.8 days in the 'optimal' food. Instantaneous rates of body growth in weight can be calculated from the sizes and duration of the juvenile phase; these were 0.21 day^{-1} and 0.35 day^{-1} for 'deficient' and 'optimal' food respectively. This work demonstrates that food-deficient juveniles grow slowly into small adults

Table 4. Energy budget of *Brachionus rubens* at 20 °C for the post-embryonic period and for the period of egg production and some resulting efficiencts, re-calculated from Pilarska (1977).

Food concentration	Period of growth			Period of egg production		
	Deficient	Optimal	Excessive	Deficient	Optimal	Excessive
Food concentration						
10^6 cell ml^{-1}	0.012	1.0	10.0	0.012	1.0	10.0
μgC ml^{-1} (a)	0.18	15.0	150.0	0.18	15.0	150.0
Body size (μgDW ind^{-1})						
Egg (2n)				0.04	0.06	0.06
Juvenile (young to old)	0.06–0.11	0.09–0.14	0.08–0.10			
Amictic female				0.13	0.17	0.14
Duration of period of body growth or egg production (days)	2.84	1.26	1.43	0.73	6.77	3.80
Energy parameters (cal 10^3 ind^{-1} period^{-1})						
Consumption	30.55	58.47	73.15	0.35	13.18	9.06
Production	4.26	8.00	3.96	0.05	2.34	0.32
Respiration	13.08	5.38	5.21	0.19	1.79	1.03
Efficiencies for the period						
Assimilation/consumption	0.57	0.23	0.13	0.68	0.31	0.15
Respiration/assimilation	0.75	0.40	0.57	0.80	0.43	0.77
Respiration/consumption	0.43	0.09	0.07	0.54	0.14	0.11

(a) Cell weight (29×10^{-6} μgDW cell^{-1} (Pilarska, 1977) and carbon weight is 50% of dry weight.

which (if they can) produce small eggs but only for a very limited period. The food concentration at which this occurred in *B. rubens* was 0.35 mgDW L^{-1} and this appears to be very near to the species' threshold for concentration for growth and reproduction. With a more usual cell weight of 2.5 pgC cell^{-1} for *Chlorella vulgaris* (Rocha & Duncan, 1985), the threshold concentration for the species would be lower (0.02–0.06 mgC L^{-1}). If we accept Pilarska's results, rotifers appear to respond phenotypically to limiting food stress similarly as cladocerans, by reducing the body size at first reproduction, prolonging development and reducing fecundity.

Because Pilarska measured the daily feeding rates and hourly respiratory rates of overigerous amictic females and three ages of juveniles at various food levels, she was able to calculate daily instantaneous energy budgets for each of these life cycle stages for deficient, optimal and excessive diets (her Table 1, page 345). However, the stress that food deficiency imposes upon the growth and reproductive periods of life cycle are better demonstrated by cumulative energy budgets, which she illustrated in her Fig. 3 but did not tabulate. These can be re-calculated for the two periods of growth and egg production and are presented in Table 4. This shows that, in 'deficient' food, a higher proportion of consumed food is required to cover the respiration of the juvenile phase than in 'optimal' food (due to its prolongation), despite a higher assimilation efficiency. It is therefore understandable why the oldest juvenile is small (0.1 μgDW compared with 0.14 μgDW). The energetic situation in cumulative terms is similar for the periods of egg production in the adults: in 'deficient' food, a greater proportion of consumed food is needed for respiration and less is available for egg production; this accounts for the shortness of the period of egg production and the smallness of the eggs and neonates. That all the life cycle stages measured were smaller in conditions of food deficiency is not surprising and nor is it surprising that food deficient juveniles and adults assimilate more efficiently (57% and 68%) than the same stages in optimal food (23% and 31%).

The value of Pilarska's research in the present context is to show that a high cost of cumulated respiration has to be paid in order to complete the life cycle at threshold food levels and that this appears to result from the food-limited prolongation of the rotifer juvenile growth phase. Another life cycle study on a species of benthic nematode demonstrates the same result (Schiemer et al., 1980). That the same bioenergetic pressures at threshold food concentrations may cause similar life cycle effects in cladocerans seems very likely, judging from the results of Bohrer & Lampert (1987) when they measured simultaneously the effect of food concentration on assimilation and respiration in adult *Daphnia magna*.

Allocation of food resources into body growth and reproductive growth

As the allocation of food resources into body growth or reproduction in different species is of ecological and evolutionary interest (Calow, 1978; Lynch, 1980, 1985; Sibly & Calow, 1986), an attempt was made to illustrate the investment allocation in a species of rotifer and in large and small, temperate and tropical species of Cladocera when supplied with non-limiting levels of food or when faced with continuously scarce food resource.

It proved possible to define the relationship between the daily rates of body and reproductive growth in adults of four of the planktonic species of cladocerans, using the rates calculated by Rocha (1983) for *Daphnia magna* and *Daphnia hyalina* and for *Diaphanosoma excisum* and *Moina micrura* by Jayatunga (1986). Rocha was able to fit the Chapman-Richards growth function (Richards, 1959) to the life cycle data of daily weights and ages for each temperature-food treatment and use the parameters of the curvilinear growth equation to calculate the relative growth rate of an animal which had the age of a primipara

female; knowing the body weight, her absolute daily growth rate could be computed. The daily reproductive rate was calculated from the number of eggs in the first brood, egg weight and the embryonic development duration for the experimental temperature involved. Rocha's growth rates, therefore, apply to the primiparous female. Jayatunga (1986) also found that the Richard's growth function, as modified by Schnute (1981), provided good fits to the life cycle growth patterns of her tropical species. However, she calculated the daily growth rates for all the first four adult instars so that her results are likely to demonstrate slightly lower body growth rates and slightly higher reproductive rates, because of the body size effect.

It is proved possible to obtain significant double logarithmic regressions for reproductive growth on body growth in three of the above four species and these are illustrated in Fig. 3. All three regressions lie above the line of equal allocation and more or less parallel to it; the slopes of the equations are close to one and are not significantly different ($P = 0.71$). As the regressions are based on all the replicates of all the food-temperature treatments that could provide growth and reproductive rates (Table 2), this result suggests that these three species are able to invest more food resource into reproduction than body growth at all the food-temperature conditions that permit them to complete their life cycle. The particular food-temperature condition merely determines the point along the regression line where the species population exists (low food-low temperature at its lower end, for example). The higher elevation of the tropical *Diaphanosoma excisum* may come from Jayatunga's inclusion of older adult instars in her calculations and this needs to be checked by a calculation for the primiparous female only.

A significant relationship between reproductive and body growth could not be obtained for *Moina micrura* but the points for the seven experimental treatments that produced positive body growth rates are plotted in Fig. 3. These lie at a higher elevation than the regression for *Diaphanosoma excisum*, the larger of the two tropical species. In only 12 of the 16 different experimental treatments to which *Moina micrura* was subjected in order to determine its threshold food concentration did the species attain a reproductive size and, in several of these food levels at high temperatures (32 °C), the very high reproductive rates were accompanied by negative body growth. At 32 °C, a possible water temperature in the tropics, *Moina* seems to have invested in reproduction at the expense of body growth.

Because of the nature of growth in rotifers, it is impossible to produce a comparable figure showing the allocation of food resources into growth or reproduction of adults. In Figure 4, a compromise was reached by comparing the daily growth rates of the oldest juvenile *Brachionus rubens* with the daily reproductive rate of adults when reared in ten different food concentrations (Pilarska, 1977); these food concentrations have been allocated in Fig. 4 into those representing a deficient,

optimal or excessive diet. Figure 4 shows that, unlike the cladoceran plot, the points for food-limited rotifers lies well below the line of equal allocation into juvenile growth and adult reproduction so that more of the scarce food is being put into enhancing juvenile body growth and much less into supporting reproductive growth. However, as the available food approaches the optimal level, the rotifers invest more equally into growth and reproduction. Within the group of optimal food concentrations, all the points lie well above the line of equal allocation; both rates are high and probably maximal so that they change little with food level. In excessively high food concentrations, the unfavourable circumstances are shown by the reduced body growth and a sharp decline in adult egg production.

Since the growth of the oldest juvenile under different food conditions is likely to be affected by its earlier growth, a comparison was made of the cumulated body growth for the whole juvenile period and the cumulated reproductive growth for the period of egg production under the different experimental conditions. This is plotted in Fig. 5. In this plot, the points for food-deficient rotifers lie on the line of equal allocation. When the food conditions are optimal, the cumulated growth for the juvenile phase is high, probably maximal but changes little with more food; an additional availability of food is chanelled directly into ever higher cumulated egg production. In the poor conditions of excessive food levels, the cumulated juvenile growth is low and changes very little with increase in food level; as the available food becomes ever more excessive, the cumulated egg production declines more and more.

Figures 3 and 4 show rather different daily allocations into growth and reproduction by adult cladocerans and near-adult *B. rubens*. They illustrate how the allocations differ in each taxonomic group when food resources span from non-limiting ('excessive' and 'optimal') to limiting ('deficient' and 'near-threshold') food levels. In *B. rubens* (Fig. 4), a more scarce resource is put into body growth than into egg production in deficient food conditions, whereas, once an optimal juvenile growth rate is achieved, additional

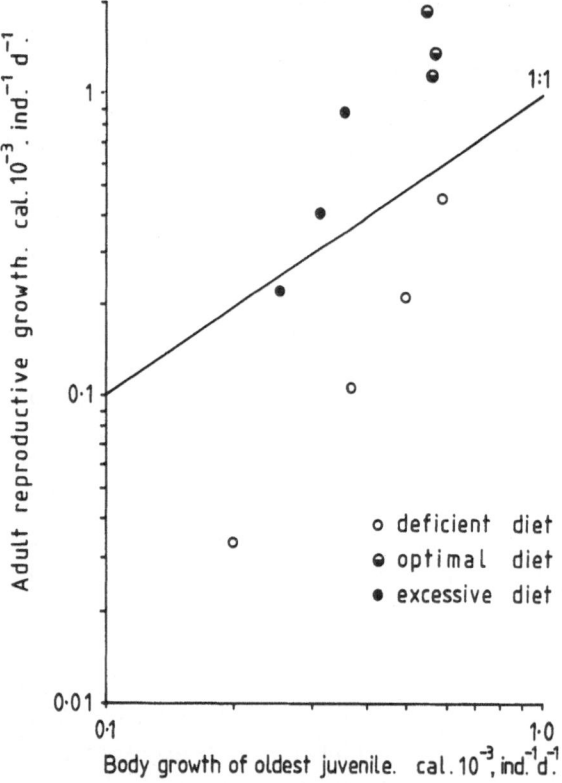

Fig. 4. The allocation of food resources to the body growth of the oldest juvenile *Brachionus rubens* and to the egg production of the adults under various conditions of food limitation: the daily picture. Data of Pilarska (1977).

23

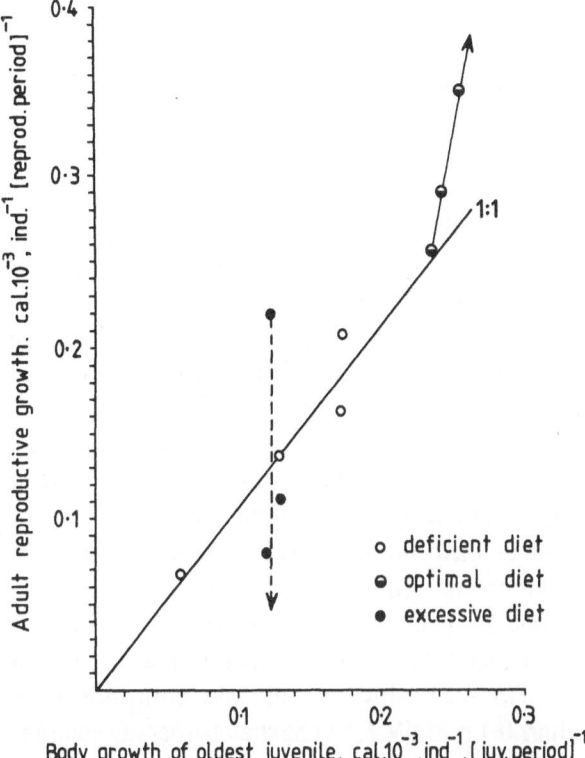

Fig. 5. The allocation of food resources to juvenile body growth and adult reproduction cumulated for the periods of body growth and egg production and when reared under various conditions of food limitation. *Brachionus rubens* (Pilarska, 1977).

food resource does not alter this and most goes to increase egg production. It seems that there are close bounds set on adult body dimensions of rotifers (Starkweather, 1987) but only in optimal food conditions and not in limiting or deficient food levels perhaps because maximal body dimensions were not achieved. In cladocerans (Fig. 3), the pattern of allocating slightly more to reproductive than body growth remains the same for the full span of non-limiting to limiting food resource. This is the implication of the slopes of the three significant regressions being near to one. When body growth is about 1 μgC ind^{-1} day^{-1} in good food conditions, the ratios of the reproductive ratios of the three species are 1.63 : 2.14 : 5.27 (or 1 : 1.3 : 3.2 for *D. magna*, *D. hyalina* and *D. excisum*). When the body growth rate is lower at about 0.5 μgC ind^{-1} day^{-1}, these ratios remain much the same (1 : 1.3 : 2.9). Irrespective of the level of food resource within the range from

above the threshold to optimal, the ratio of adult allocation into growth : reproduction seems to remain constant. This difference in allocation patterns seems to be a consequence of the fundamental difference in the nature of growth in rotifers and crustaceans, since the latter can continue to grow as adults by cell proliferation.

According to Snell & King (1977), a basic assumption within the area of trade-offs between life history characteristics is 'that of resource limitation: resources directed towards reproduction reduce those available for growth and maintenance. Thus one consequence of reproduction is presumed to be decreased survival.' In individual cladocerans reared in food concentrations that allow them to complete their life cycle and reproduce, the conflict is different for the juvenile and reproductive phases of the life cycle. In the juvenile phase, it lies between the highly cumulating metabolic costs due to food-limited prolongation of the phase and the body size at which it can mature. The latter characteristic is a species characteristic. In the reproductive phase, the conflict is between growth plus reproduction and maintenance, where the allocation of growth : reproduction is kept constant.

On the basis of results on cladocerans reared in undefined food conditions, Lynch (1980) grouped 'well-fed' animals into (1) the larger species which mature later in their life cycle at a size close to the maximal size and the channel most of their available food into reproduction and (2) small-sized species which mature early and continue to grow after the onset of maturity, thus channelling a larger proportion into body growth and a lesser proportion into reproduction. Although it is true that well-fed cladocerans of larger species mature at a later instar, older age and larger but not maximal body size, it is not at all established that a larger proportion of the food is channelled into reproduction in larger compared with smaller species. The two regressions for the larger *D. magna* and smaller *D. hyalina* in Fig. 3 are not significantly different, can be pooled and show a similar slightly higher investment into reproduction than body growth at all food-temperature treatments. The main difference lies in the smaller and larger absolute daily rates of the smaller and larger species, respectively. The higher investment into reproduction in the two smaller species which are also different in being tropical is interesting but not yet established and needs re-calculation for a comparable adult stage. If one takes well-fed rotifer individuals as representatives of even smaller zooplankton species, they allocate much more into reproduction once juvenile growth has been satisfied but the reverse is true in 'deficient' food levels (Figs. 4 and 5).

Comparison of threshold food concentrations in rotifers, cladocerans and ciliates

The series of papers by Stemberger & Gilbert (1985a, b; 1987) on rotifer threshold food levels, determined as the minimum food at which the population's instantaneous rate of increase (r) is more or less zero, permitted a comparison with the threshold concentrations for cladoceran which were determined experimentally and defined as the minimum food at which the life cycle could be completed. These two sets of data are plotted in Fig. 6, with food level on a log scale and in terms of dry weight as this was the measure used by Stemberger & Gilbert. Where the algal concentration was measured in carbon weight, the values were doubled, assuming that 0.5 dry weight is carbon.

The threshold food values tabulated in Stemberger & Gilbert (1987) for ten species of rotifers were determined using algal species whose cell sizes and types were most highly preferred or efficiently harvested by the rotifer species studied in order that the r-values would be comparable. They used *Cryptomonas erosa var. reflexa* (850 μm³; equivalent spherical diameter ESD of 6.5 μm) as food for the larger rotifer species and a series of smaller cryptomonads for the smaller species: *Cryptomonas ovata* (160 μM³; ESD 3.7 μm), *Cryptomonas* spp. (130 μm³; ESD 3.5 μM) and *Rhodomonas minuta* (85 μm; ESD 3.0 μm). The threshold concentrations ranged from 1.03 mgDW L^{-1} for *Keratella crassa* to 0.06 mgDW L^{-1} for *Keratella cochlearis*, a span

of 17 times, and revealed a strong relationship with the body weight of the rotifer species: larger species had higher threshold than smaller species. A data point has been added to Fig. 6 for Pilarska's (1977) *Brachionus rubens* reared at the lowest of the deficient food levels (0.35 mg DW L^{-1}) of *Chlorella vulgaris* (29 pgDW cell^{-1}) and this fits well into the series. When compared with the threshold values for temperate species of *Daphnia* (0.01–0.02 mgDW L^{-1}) and for tropical species of cladocerans (0.06–0.1 mgDW L^{-1} at 27 °C and 32 °C; 0.1–0.2 mgDW L^{-1} at 22 °C, which is a low temperature for tropical species), this series of threshold values for rotifers seem very high. Only a few of the smaller species of rotifers have threshold values as low as tropical cladocerans and none as low as temperate *Daphnia*. The threshold concentration cited by Schiemer (1985) for a temperate and a tropical species of freshwater calanoid copepods (less than 0.1 mgDW L^{-1}) fit the cladoceran picture.

These results appear to indicate a significant difference in the minimal food requirements between rotifers and planktonic crustaceans. It was fortunate that further data points for this diagram were provided during the rotifer symposium at Gargnano by K.O. Rothhaupt from his PhD dissertation on *Brachionus rubens* and *B. calyciflorus*, by P Starkweather from unpublished research with B Skrdla on *B. calyciflorus* and by T W Snell from recently published research on *B. plicatilis* (Snell & Boyer, 1988). The threshold value of 0.1 mgDW L^{-1} published in Starkweather (1987) for *B. calyciflorus* when reared on *Euglena gracilis* is lower than the value of 0.38 mgDW L^{-1} given by Stemberger & Gilbert (1987) with *Cryptomonas erosa* as the food organism. The threshold of 0.56 mgDW L^{-1} for amictic *B. plicatilis* cultured on *Dunaliella tertiolecta* (370 pgDW cell^{-1}) is higher than Dewey's value of 0.40 (Stemberger & Gilbert, 1987). Rothhaupt determined the food concentrations at which the instantaneous rate (r) is more or less zero for the two brachionid species when they were reared on a series of algal species of different cell and size and of different 'harvestability' efficiencies (which were tested). He defines the size

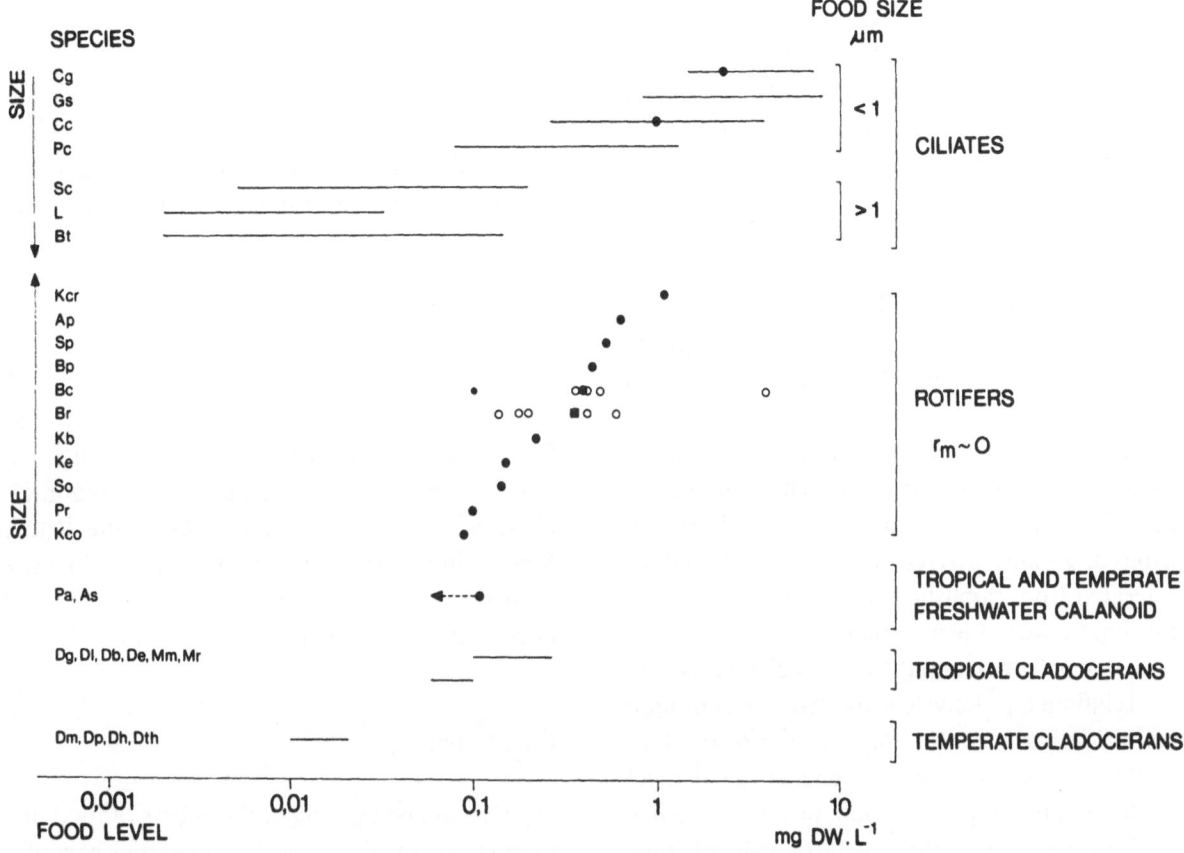

Fig. 6. A comparison of the threshold food concentrations of rotifers, planktonic cladocerans and copepods and ciliates. Key: Ciliates: Cg Cyclidium glaucoma, Gs Glaucoma scintillans, Cc Colpidium colpoda, Pc Paramecium caudatum, Sc Stentor coerulens, L Lembadion, Bt Bursaria truncatella. Rotifers: Kcr Keratella crasse, Ap Asplanchna priodonta, Sp Synchaeta pectinata, Bp Brachionus plicatilis, Bc Brachionus calyciflorus, Br Brachionus rubens, Kb Kellicottia bostoniensis, Ke Keratella earlinae, So Synchaeta oblonga, Kco Keratella cochlearis. Copepods: Pa Phyllodiaptomus annae, As Arctodiaptomus spinosus. Cladocerans: Dg Daphnia gessneri, Dl Daphnia lumholzi, Db Diaphanosoma brachyurum, De Diaphanosoma excisum, Mm Moina micrura, Mr Moina reticulata, Dm Daphnia magna, Dp Daphnia pulicaria, Dh Daphnia hyalina, Dth Daphnia thorata. Sources: Fenchel (1980); Stemberger & Gilbert (1987); Pilarska (1977); Starkweather (1987); Rothhaupt (pers. comm.); Schiemer (1985); Taylor (1978), Snell & Boyer (1988).

of his food cells as the equivalent spherical diameter (ESD). In the following list of algal species used by Rothhaupt, the values in brackets are first the cell ESD, then the threshold concentration in the original mgC L^{-1} for *Brachionus rubens* followed by that for *B. calyciflorus*: *Chlorella minutissima* (2 μm; 0.215 $\pm$ 0.014; 1.958 $\pm$ 0.217), *Monorhaphidium minutum* (3.5 μm; 0.071 $\pm$ 0.013; 0.236 $\pm$ 0.027), single-celled *Scenedesmus obliquus* (5.5 μm; 0.074 $\pm$ 0.009; 0.195 $\pm$ 0.011). *Chlamydomonas reinhardii* which was found to be nutritively poor (6 μm; 0.091 $\pm$ 0.011; not used for B.c), *Cyclotella meneghinianae* (8.5 μm; 0.107 $\pm$ 0.039; 0.193 $\pm$ 0.011) and *Chlamydomonas sphaeroides*

(12 μm; 0.31 $\pm$ 0.013; 0.185 $\pm$ 0.010). These data points, doubled to give dry weight, have been plotted in Figure 6, with Rothhaupt's permission, and show that a range of threshold concentrations exist for any one species according to the suitability of the food species. The most divergent points for *B. calyciflorus* are for the small-sized *Chlorella minutissima* (2 μm) which gave a very high threshold value and for *Euglena gracilis* whose threshold was the lowest; the values for other food cells are rather similar to that for *Cryptomonas erosa* (Stemberger & Gilbert, 1987). Rothhaupt's results gave both higher and lower threshold values for *B. rubens* compared with Pilarska's point with *Chlorella vulgaris*; the higher

threshold concentrations were given by both the smallest (*C. minutissima*, 2 μm) and the largest tested algal species (*Chlamydomonas sphaeroides*, 12 μm). An additional point of some significance is provided by Snell & Boyer (1988) who found that the threshold concentration for mictic reproduction in *B. plicatilis* was 5.7 mgDW L^{-1}, which is ten times the value for amictic animals; this point has not been plotted in Fig. 6.

For interest, threshold food concentrations for various kinds of ciliates have been included in Fig. 6. These come from Fenchel (1980), who calculated the minimum food particle concentration needed to cover metabolic costs. This comes from the volume of water cleared per ml of oxygen respired, a value derived from two relationships generated from published data, namely, the relationship between the maximal clearing rate at low food concentration and the ciliate cell volume and the relationship between oxygen consumption and cell volume. An assumption of a respiratory quotient of the one allowed Fenchel to calculate the minimum weight of food needed to cover respiration and, from the volume of water cleared, the concentration per unit volume of water. The threshold values are only very approximate because of the scatter in the published respiratory data and because of the existence of two different R/V regressions for ciliates. However, his calculations demonstrate that the larger ciliates feeding on larger food particles (1–5 μm) are similar to planktonic *Daphnia* in their low food threshold whereas the smaller ciliates which specialize on very small particles (0.2–1 μm) need much higher concentrations of suspended particles. The two points added for *Cyclidium glaucoma* and *Colpidium colpoda* come from Taylor (1978) and fall in the middle of Fenchel's ranges. The ecology of these two groups of ciliates differs; as Fenchel comments, the dependence of small-particle feeding bacterivorous ciliates upon the presence of high concentrations of suspended bacteria (10^9 cells ml^{-1}) precludes their colonisation of the open waters of lakes or the sea where bacteria are sparse (10^5 cells ml^{-1}). The planktonic ciliates (like tintinnids and oligotrich ciliates) feed on large particles and have lower thresholds. These insights into ciliate thresholds and trophic ecology might help in the interpretation of the rotifer thresholds shown in Figure 6. Are the smaller rotifer species with low thresholds planktonic forms, like the cladocerans, and are the larger species with high threshold inhabitants of particle-rich waters such as ponds and activated sludge?

Acknowledgements

I am grateful to Dr. O. Rocha, Dr. Y.N.A. Jayatunga, Dr. E.R. Hardy and Mr. L.C. dos Santos for use of their data without which this paper would not have been possible. I thank Dr. C. Ricci for splendid hospitality at the Rotifer Symposium in Gargnano and Royal Holloway & Bedford New College for some technical assistance with the production of this paper.

Conclusions

(1) Our understanding of the effects of food limitation upon the life cycle characteristics of rotifers and planktonic cladocerans is dependent upon the results of experiments in which the food concentrations are quantified and do not vary too much and in which something is known about the size, ingestibility and food value of the food organism offered. The results from many published studies are incomparable because this kind of information is not available.

(2) Although rotifers and cladocerans differ in their adult body sizes, there is an overlap between the two groups, particularly with the small, rarely studied tropical cladocerans. The elevation of the length-carbon weight regression of a cladoceran species reared in near-threshold food concentrations (0.01–0.05 mgC L^{-1}) is significantly lower than that for animals cultured in optimal concentrations. This occurs in both large and small cladoceran species and can be used as a useful ecological indicator of the nutritive state of field populations.

(3) Evidence from one study and on one species of rotifer showed that the weight of eggs, juveniles and adults were smaller, that the duration of the juvenile phase was longer and that the

period of egg production was shorter when it was reared in near-threshold concentrations of food compared with optimal food levels. There is a need to confirm these results with other life cycle experiments on rotifers under defined conditions of food limitation. Similar effects were demonstrated for six species of temperate and tropical cladocerans which varied greatly in their species size. In all these species, the primiparous female was smaller in size, appeared at a later instar, was older and had a reduced fecundity when reared under food limiting conditions. There is a conflict within the juvenile phase under resource limitation between the high cumulated respiratory costs caused by a prolongation of the phase and attaining a body size at which the juvenile can mature.

(4) Allocation of food resource into growth and reproduction was compared in the two taxonomic groups, using the primiparous female in the cladocerans and the oldest juvenile-adult female in the rotifer species studied. The pattern of allocation differed. In the rotifer, scarce food resource was invested more into body growth than egg production whereas most of an optimal food resource went to increase egg production. In cladocerans, the pattern of allocating slightly more to reproduction than to body growth remained the same for all conditions of food resource. Moreover, the ratio of adult allocation into growth : reproduction remained constant. It is suggested that these differences are the consequence of the fundamentally different nature of growth in the two taxonomic groups.

(5) Threshold food concentrations for eleven species of rotifers show that, for the group, they are much higher than for temperate planktonic cladocerans and that only the thresholds for smaller rotifers overlap with the small tropical species of cladocerans. Larger rotifer species have as high threshold food concentrations as Fenchel's small-particle feeding bacterivorous ciliates, and, like them, must be confined ecologically to non-pelagic particle-rich environments. Planktonic ciliates feed on larger particles and have lower threshold food concentrations. The work of K.O. Rothhaupt demonstrated that the threshold food concentration is not a constant for a species but varies with the suitability of the algal food species.

References

Bottrell, H. H., A. Duncan, Z. M. Gliwicz, E. Grygierek, A. Herzig, A. Hillbricht-Ilkowska, H. Kurasawa, P. Larsson & T. Weglenska, 1976. A review of some problems in zooplankton production studies. Norw. J. Zool. 24: 419–456.

Bohrer, R. N. & W. Lampert, 1988. Simultaneous measurement of the effect of food concentration on assimilation and respiration in *Daphnia magna*. Functional Ecology. 2: 463–471.

Calow, P., 1978. Life cycles. Chapman & Hall. 164 pp.

Doohan, M. & V. Rainbow, 1971. Determination of dry weights of small Aschelminthes ($< 0.1 \mu g$). Oecologia (Berl.). 6: 380–383.

Doohan, M., 1973. An energy budget for adult *Brachionus plicatilis* Muller (Rotatoria). Oecologia (Berl.). 13: 351–362.

Duncan, A., 1983. The influence of temperature upon the duration of embryonic development of tropical *Brachionus* species (Rotifera). IN: Schiemer, F. (ed.), Limnology of Parakrama Samudra – Sri Lanka. Junk. 107–115.

Duncan, A., 1984. Assessment of factors influencing the composition, body size and turnover rate of zooplankton in Parakrama Samudra, an irrigation reservoir in Sri Lanka. Hydrobiologia. 113: 201–215.

Duncan, A., 1985. Body carbon in daphnids as an indicator of the food concentration available in the field. Arch. Hydrobiol. Beih. Ergbn. Limnol. 21: 81–90.

Duncan, A., 1985b. Carbon weight on length regressions of *Daphnia* spp. grown at threshold food concentrations. Verh. Int. Ver. Limnol. 22: 3109–15.

Fenchel, T., 1980. Suspension feeding in ciliate Protozoa: feeding rates and their ecological significance. Microb. Ecol. 6: 13–25.

Gilbert, J.J., 1985. Competition between rotifers and Daphnia. Ecology 66(6): 1943–1950.

Gilbert, J. J. & R. S. Stemberger, 1984. Spine development in the rotifer *Keratella cochlearis*: induction by cyclopoid copepods and *Asplanchna*. Freshwat. Biol. 14: 639–647.

Hardy, E. R., 1989. Effect of temperature, food concentration and turbidity on the life cycle characteristics of planktonic cladocerans in a tropical lake, Central Amazon: a field and experimental study. PhD Thesis. University of London (RHBNC). 337 pp.

Hrabackova, M., 1971. The size of primipara and neonates of *Daphnia hyalina* Leydig (Crustacea: Cladocera) under natural and enriched food conditions. Vest. Ces. Spol. Zool. 38(2): 98–105.

Hrabackova, M. & J. Hrbacek, 1979. Rates of post-embryonic development in several populations of the group of species *Daphnia hyalina* Leydig at various concentrations of food. Vest. Ces. Spol. Zool. 43(4): 253–259.

Herzig, A., 1983a. The ecological significance of the relationship between temperature and duration of embryonic development in planktonic freshwater copepods. Hydrobiologia. 100: 65–91.

Herzig, A., 1983b. The ecological significance of the relationship between temperature and duration of embryonic development of rotifers. Hydrobiologia. 104: 237–46.

Ivanova, M. B. & S. V. Vasilenko, 1987. Relationship between number of eggs, brood weight and female body weight in Crustacea. Int. Revue ges. Hydrobiol. 72(2): 147–169.

Jayatunga, Y. N. A., 1986. The influence of food and temperature on the life cycle characteristics of tropical cladoceran species from Kalawewa Reservoir, Sri Lanka. PhD Thesis. University of London (RHBNC) 410 pp.

King, C. E., 1967. Food, age and the dynamics of a laboratory population of rotifers. Ecology. 48(1): 111–128.

King, C. E., 1969. Genetics of reproduction, variation and adaptation in rotifers. Arch. Hydrobiol. Beih. Ergebn. Limnol. 8: 187–201.

Lampert, W., 1977. Studies on the carbon balance of *Daphnia pulex* De Geer as related to environmental conditions. IV. Arch. Hydrobiol. Suppl. 48: 361–8.

Leimeroth, N., 1980. Respiration of different stages and energy budget of juvenile *Brachionus calyciflorus*. Hydrobiologia. 73: 195–197.

Lubzens, E., 1987. Raising rotifers for use in aquaculture. Hydrobiologia. 147: 245–255.

Lynch, M., 1980. Predation, enrichment and the evolution of cladoceran life histories: a theoretical approach. IN: W. C. Kerfoot (ed.). Evolution and ecology of zooplankton communities. 367–376. Univ. Press, New Hampshire.

Lynch, M., 1985. Elements of a mechanistic theory for the life history consequences of food limitation. Arch. Hydrobiol. Beih. Ergebn. Limnol. 21: 351–362.

Pejler, B., 1980. Variation in the genus *Keratella*. Hydrobiologia. 73: 207–213.

Pilarska, J., 1977. Eco-physiological studies on *Brachionus rubens* Ehrbg. (Rotatoria). Pol. Arch. Hydrobiol. 24: 319–354.

Pourriott, R., 1973. Rapports entre la temperature, la taille des adultes, la longuer des œufs et le taux de developpement embryonnaire chez *Brachionus calyciflorus*. Pallas (Rotifere). Ann. Hydrobiol. 4: 103–115.

Richards, F. J., 1959. A flexible growth function for empiracle use. J. exp. Botany. 10: 290–300.

Robertson, J. R. & Salt, G. W., 1981. Responses in growth, mortality and reproduction to variable food levels by the rotifer *Asplanchna girodi*. Ecology. 62(6): 1585–1596.

Rocha, O., 1983. The influence of food-temperature combinations on the duration of development, body size, growth and fecundity of *Daphnia* species. PhD Thesis, Royal Holloway College, University of London. 337 pp.

Rocha, O. & Duncan, A., 1985. The relationship between cell carbon and cell volume in freshwater algal species used in zooplanktonic studies. J. Plankton. Res. 7(2): 279–294.

Ruttner-Kolisko, A., 1974. Plankton Rotifers. Biology and Taxonomy. Die Binnengewasser. 26: 1–146.

Salonen, K., 1979. A versatile method for the rapid and accurate determination of carbon by high temperature combustion. Limnol. Oceanogr. 24: 177–183.

Santos, L. C. dos., 1989. The effects of food limitation on the population dynamics, production and biological interactions of three *Daphnia* species, co-existing in a London reservoir. PhD Thesis. University of London (RHBNC). 236 pp.

Schiemer, F., 1985. Bioenergetic niche differentiation of aquatic invertebrates. Verh. Int. Ver. Limnol. 22: 3014–3018.

Schiemer, F., A. Duncan & R. Z. Klekowski, 1980. A bioenergetic study of a benthic nematode, *Plectus palustris* de Man 1880, throughout its life cycle. Oecologia (Berl.). 44: 205–212.

Schnute, J., 1981. A versatile growth model with statistically stable parameters. Can. J. Fish. Aquat. Sci. 38: 1128–1140.

Serra, M. & M. R. Miracle, 1987. Biometric variation in three strains of *Brachionus plicatilis* as a direct response to abiotic variables. Hydrobiologia 147: 83–89.

Sibly, R. M. & P. Calow, 1986. Physiological Ecology of Animals. Balckwell Sci. Publ. 179 pp.

Snell, T. W., 1977. Clonal selection, competition among clones. Arch. Hydrobiol. Beih. Ergbn. Limnol. 8: 202–204.

Snell, T. W. & E. M. Boyer, 1988. Thresholds for mictic female production in the rotifer Brachionus plicatilis (Muller). J. exp. Mar. Biol. Ecol. 124: 73–85.

Snell, T. W. & K. Carrillo, 1984. Body size variations among strains of the rotifer Brachionus plicatilis. Aquaculture. 37: 359–367.

Snell, T. W. & C. E. King, 1977. Lifespan and fecundity patterns in rotifers: the cost of reproduction. Evolution. 31(4): 882–890.

Starkweather, P. L., 1987. Rotifera. chap. 5. IN: Pandian & Vernberg (eds.) Animal Energetics. 1: 159–183.

Stemberger, R. S. & J. J. Gilbert, 1985a. Body size, food concentration and population growth in planktonic rotifers. Ecology. 66(4): 1151–1159.

Stemberger, R. S. & J. J. Gilbert, 1985b. Assessment of threshold food levels and population growth in planktonic rotifers. Arch. Hydrobiol. Beih. Ergebn. Limnol. 21: 269–275.

Stemberger, R. S. & J. J. Gilbert, 1987. Rotifer threshold food concentrations and the size efficiency hypothesis. Ecology. 68(1): 181–187.

Taylor, W. D., 1978. Growth response of ciliate Protozoa to the abundance of their bacterial prey. Microb. Ecol. 4: 207–214.

Walz, N., 1983. Continuous culture of the pelagic rotifers, *Keratella cochlearis* and *Brachionus angularis*. Arch. Hydrobiol. 98(1): 70–92.

Weglenska, T., 1971. The influence of various food concentrations of natural food on the development, fecundity and production of planktonic crustacean filtrators. Ekol. Pol. 19(30): 427–473.

Yufera, M., 1987. Effect of algal diet and temperature on the embryonic development time of the rotifer *Brachionus plicatilis*. Hydrobiologia. 147: 319–322.

Hydrobiologia **186/187**: 29–34, 1989.
C. Ricci, T. W. Snell and C. E. King (eds), Rotifer Symposium V.
© 1989 Kluwer Academic Publishers.

Is food availability the main factor controlling the abundance of *Euchlanis dilatata lucksiana* Hauer in a shallow, hypertrophic lake?

J. Ejsmont-Karabin,[1] R. D. Gulati[2] & J. Rooth[2]
[1] *Polish Academy of Sciences, Institute of Ecology, Hydrobiological Station, ul. Leśna 13,
11-730 Mikołajki, Poland*; [2] *Limnological Institute, Vijverhof Laboratory, Rijksstraatweg 6,
3631 AC Nieuwersluis, The Netherlands*

Key words: *Euchlanis dilatata lucksiana*, blue-green algae, threshold concentrations of food, trophic conditions, predation

Abstract

Visual observations and experiments on food preference of *Euchlanis dilatata lucksiana* show that this euchlanid can feed on blue-green algae not consumed by the most planktonic animals. Nevertheless, even in lakes with blooms of blue-green algae, *E. d. lucksiana* occur infrequently and generally in low numbers. The paper is an attempt to explore into the causes for the rare occurrence of *Euchlanis* in the pelagial. A comparison of threshold food concentrations calculated from N and P excretion rates (Gulati *et al.*, this volume) with the concentrations of seston in the Lake Loosdrecht shows that the latter were several times higher during study period in 1984. This implies that the food requirements of *Euchlanis* were always satisfied in this lake. The time needed for the consumption of the total food fraction in a liter of lake water by a concentration of 50 *Euchlanis* l^{-1} was also calculated. This time varied from 70 to 200 days, so a *Euchlanis* population even at its maximum density will not cause major changes in blue-green algae biomass by grazing. Thus, food limitation cannot be viewed as a factor controlling the *Euchlanis* densities in Loosdrecht Lakes. There is some evidence that *Euchlanis* is heavily predated in Loosdrecht Lakes, losses in its biomass accounting for 126% of the production. Adaptation of this species to the littoral zone, as expressed by the deposition of eggs on plants, can also limit the occurrence of the *lucksiana* form to water bodies with blooms of blue-green algae.

Introduction

Unlike other forms of *Euchlanis dilatata* which are littoral, *E. dilatata lucksiana* is a pelagic form (Kutikova, 1970; Koste, 1978). In fact, this euchlanid is rarely observed also in pelagial and usually in small numbers. For example, Karabin (1985a), who studied summer communities of Rotifera in 64 lakes of north-eastern Poland, most of which had blue-green algae (Spodniewska, 1983), did not find *Euchlanis dilatata* in abun-

dance. This species may, however, rarely dominate in the pelagial like in Glubokoe Lake (Matveeva, 1986), coinciding with blooms of the blue-green alga *Aphanizomenon*. Also, Carlin (1943) noted a high correlation between the occurrence of *Euchlanis* and blue-green algae (*Oscillatoria* and others). Even if observations of this kind can be viewed as sufficient to say that larger appearances of *E. d. lucksiana* in the pelagic zone occur only during blooms of filamentous algae, there is some evidence that the reverse statement is not true. Blooms of blue-green algae are rarely combined with the appearance of *Euchlanis*.

The present paper is an attempt to explore the factors responsible for the rare occurrence of *E. d. lucksiana* in pelagial of lakes.

Material and methods

Information used in this paper comes from 3 main sources: 1) literature data, 2) unpublished results of field studies, and 3) partly published results (Rooth, 1985; Gulati *et al.*, 1987) of laboratory experiments on the rate of food assimilation and the rate of phosphorus and nitrogen excretion by *E. d. lucksiana*.

The study was conducted in Lake Loosdrecht, a shallow hypertrophic lake with a mean depth of 1.85 m. Except in early spring, the phytoplankton of this lake was dominated by filamentous blue-green algae (mainly *Oscillatoria redekei*, *O. agardhii* and *Aphanizomenon flos-aquae*) (Bosewinkel-de Bruyn *et al.*, 1986).

Zooplankton samples were collected every four weeks or fortnightly from January to October, from a station in the lake. Data on carbon, phosphorus and nitrogen concentrations in the seston fraction $< 3\,\mu$m taken concurrently with the zooplankton samples were provided by Gulati (in press).

The females of *E. d. lucksiana* were obtained from a culture developed from ten individuals taken from Loosdrecht Lakes. Cultured rotifers were fed on the lake seston fraction $< 33\,\mu$m, i.e. in lake water filtered using $33\,\mu$m mesh. This fraction comprised mostly blue-green algae which were the main food of the cultured animals. The optimum rotifer density maintained at a constant level in the culture was about 40 ind. ml^{-1}. Light in the laboratory corresponded to natural conditions prevailing during the same period.

Food requirements of *Euchlanis*, both consumption and assimilation rates, in relation to the available food supply, have been analysed in the laboratory using C^{14} technique (Gulati *et al.*, 1982; Gulati *et al.*, 1987). Threshold food requirements of *Euchlanis* were determined from the

rates of nitrogen and phosphorus excretion (Gulati *et al.*, this volume).

The formula given by Rooth (1985) was used to calculate a relationship between the rate of food assimilation by *E. d. lucksiana* and food concentration:

$$A = 0.98 - \frac{0.51}{FC}, \qquad (1)$$

where: A = assimilation in μgC ind.$^{-1}$ d^{-1}
FC = food concentration in μgC ml^{-1}.

The minimum metabolism of *Euchlanis* was determined by using two indices: the rate of phosphorus or nitrogen excretion by starved animals, assuming that the rate of food assimilation corresponding to this excretion rate is the minimum that *Euchlanis* requires to survive. Q_{10} values of 2.6 for phosphorus and 2.4 for nitrogen (Ejsmont-Karabin, 1984) were used to relate the rate of P and N excretion to ambient temperature. Using the rate of nutrient excretion, the corresponding rate of food assimilation, E_{PorN} (C:P or C:N), was calculated. This takes into account the C:P or C:N ratio by weight in the food fraction. Thus, the threshold concentration was calculated from the formula:

$$TFC = \frac{0.51}{0.98 - E_{PorN}/C:P \text{ or } C:N}, \qquad (2)$$

where: TFC = threshold concentration of food in μgC ml^{-1}
E_{PorN} = rate of phosphorus or nitrogen excretion in μg ind.$^{-1}$ d^{-1}
C:P or C:N = carbon to phosphorus or nitrogen ratio by weight in food.

The time needed for the consumption of the total content of carbon contained in 1 ml of lake water of the seston fraction $< 33\,\mu$m by one individual was calculated using the equation given by Rooth (1985):

$$CON = 1.17 - \frac{0.92}{FC}, \qquad (3)$$

where: CON = consumption in μgC ind.$^{-1}$ d^{-1}
　　　　FC = as in formula 1.

Results and discussion

The rotifer community of Loosdrecht Lakes was dominated by small species, mostly detritophagous *Anuraeopsis*, *Keratella* (Fig. 1). The highest densities occurred from June to early October. *Euchlanis dilatata lucksiana* appeared in small numbers (1 ind. l^{-1}) in April, but reached about 50 ind. l^{-1} in mid-August and September.

These densities are very low, however, compared with those of the dominant rotifers like *Anuraeopsis fissa* (Gosse), with which the *Euchlanis* maxima coincided.

Trophic conditions. Food availability and abundance are viewed as among the most important factors determining numbers and the structure of zooplankton (Benndorf & Horn, 1985; Karabin, 1985b). However, visual observations and experiments on food preference (Gulati *et al.*, 1987) show that *E. dilatata lucksiana* can feed on blue-green algae not consumed by most planktonic animals (Dumont, 1977). Nevertheless, even in

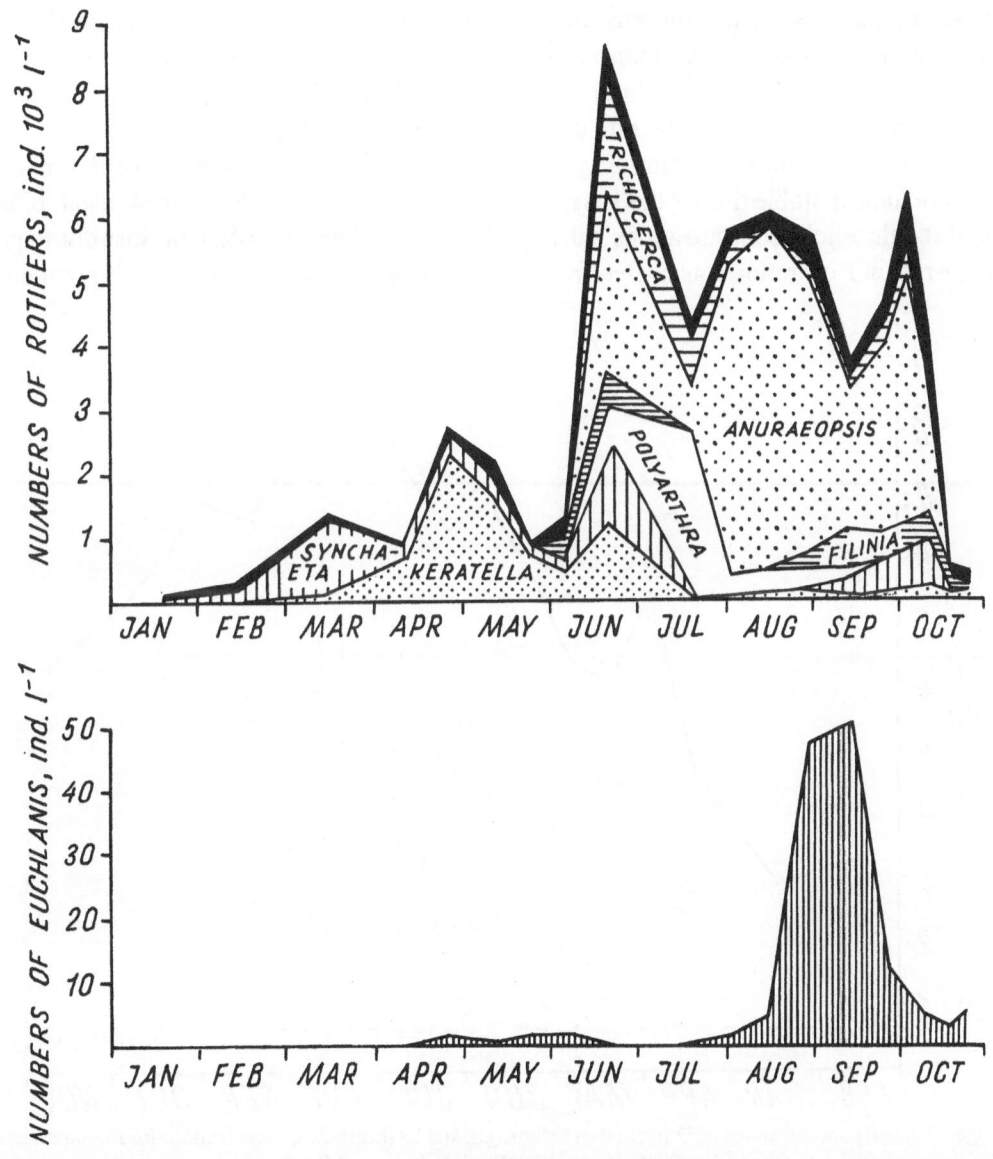

Fig. 1. Changes in numbers of the main genera of rotifers and *Euchlanis dilatata lucksiana* in Loosdrecht Lakes in 1984.

lakes with blooms of blue-green algae, *E. d. lucksiana* occur infrequently and generally in low numbers. Is such a high density of algae insufficient to meet the food requirements of this rotifer? To test this hypothesis, food consumption of *Euchlanis* was compared with food supply in water in an attempt to determine a critical food concentration (equations 1 & 2) below which *Euchlanis* would not be able to meet their metabolic needs.

The ratio by weight of the nutrients excreted by *Euchlanis* was C:N:P = 25:4:1 (Gulati *et al.*, this volume). Seasonal mean ratio of these nutrients in seston < 3 μm was 113:17:1, respectively. Thus, compared with phosphorus there was relatively more nitrogen in food than in the excretion products. Therefore, the threshold concentrations of food calculated from the rate of N excretion are lower than from P excretion (Fig. 2). Also, they were almost stable during the season. The range of the threshold concentrations calculated from the rate of P excretion was higher (from 0.53 to 0.73 μgC ml^{-1}) than for N excretion. A comparison of threshold concentrations with the actual concentration of seston in Loosdrecht Lakes (Fig. 2) shows that the latter were several times higher during study period. This implies that in 1984 there was no situation in which food requirements of *Euchlanis* could not be satisfied in this lake.

The effect of *Euchlanis* on the phytoplankton depends on rotifer densities. The time needed for consumption of the total carbon content of 1 ml lake water of the seston fraction < 33 μm by one individual was calculated using the equation 3. The calculated turnover time varied from 3 to 6 days in spring and autumn (Table 1) and 7 to 10 days in summer. Thus, a *Euchlanis* population comprising 1000 ind. l^{-1} (a concentration 2 orders of magnitude higher than normally encountered in the lake) would not cause any major change in blue-green algal biomass by foraging. The 10–20% daily loss due to grazing by *Euchlanis* could have been easily compensated for

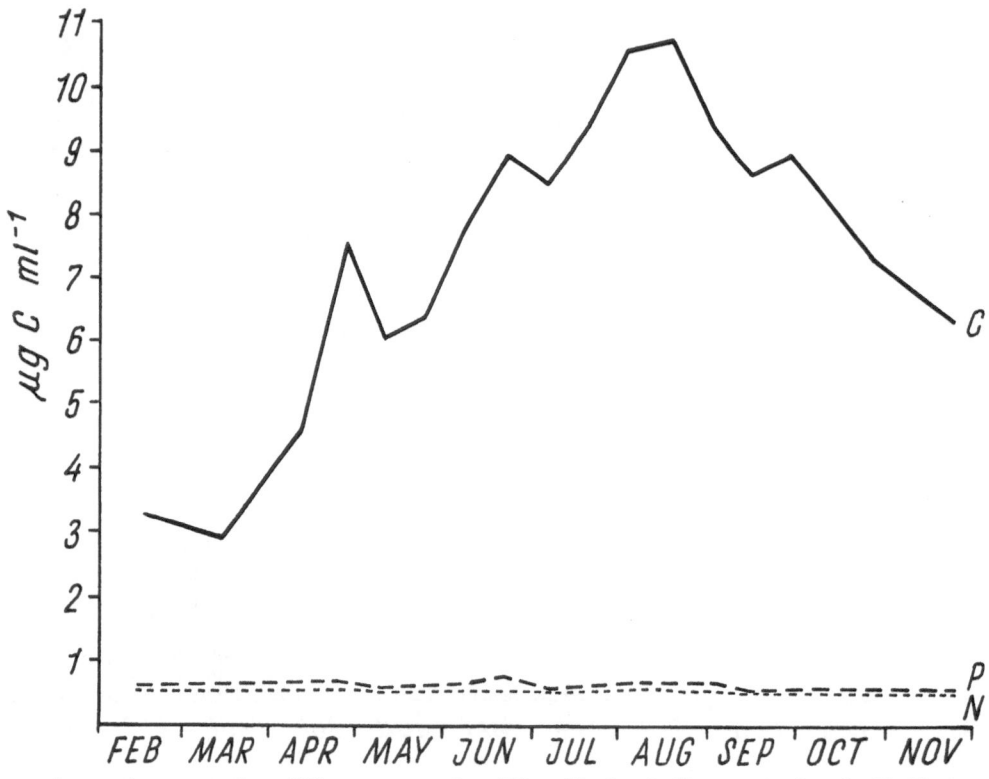

Fig. 2. Changes in actual concentrations of 33 μm seston carbon (C), and its threshold concentrations for *Euchlanis*, as calculated from the rate of phosphorus (P) and nitrogen (N) excretion.

Table 1. Seasonal changes in the concentration of 33 μm seston, consumption rate of algae by *Euchlanis dilatata lucksiana* and turnover time of the seston C as a result of *Euchlanis* foraging.

Date	Seston μgC ml^{-1}	Consumption rate μgC ind.$^{-1}$ d^{-1}	Turnover time (days)
84.02.15	3.24	0.88	4
03.14	2.85	0.85	3
04.11	4.63	0.97	5
04.24	7.48	1.05	7
05.09	6.03	1.02	6
06.06	7.85	1.05	7
06.20	8.98	1.07	8
07.04	8.51	1.06	8
07.18	9.51	1.07	9
08.01	10.66	1.08	10
08.15	10.77	1.08	10
08.29	9.56	1.07	9
09.12	8.69	1.06	8
09.26	8.98	1.07	8
10.24	7.36	1.06	7
11.21	6.36	1.03	6

due to primary production. Therefore, food availability cannot be the factor limiting *Euchlanis* densities in Loosdrecht Lakes.

Other factors

Temperature. The fact that peak numbers of *Euchlanis* in Loosdrecht Lakes were noted in late summer suggests that ambient temperature should be considered among the factors controlling the occurrence of this rotifer. *E. dilatata* can occur over a large range of ambient temperatures, from 0.1 to 31 °C, and is considered as an eurythermic species (Lair, 1980; Bazilevitsch, 1985; Kosova, 1985). There is no information of this kind on its pelagic form. Its clear peaks in late summer in both Glubokoe Lake (Matveeva, 1986) and Lake Loosdrecht suggest that this may be a thermophilous form. This suggestion is not confirmed, however, by the analysis of the occurrence of *Euchlanis* in Lake Loosdrecht in the annual cycle. Presence of this species in the lake was observed in April and the range of temperature at which it occurred was 12 to 21 °C, with two maxima at 13 °C and 21 °C. Its low numbers at intermediate temperatures (14–20 °C) in late spring and early summer cannot be explained.

Predation. The fact that in the cultures *Euchlanis* reached very high densities means that in laboratory the factors limiting the occurrence of these animals in the lake were relieved. The absence of predators, such as fish, may contribute to their high densities in the cultures.

E. d. lucksiana is a relatively large rotifer, which reaches the maximum 400 μm in length in Lake Loosdrecht. Their size is thus within the range of sizes of *Bosmina coregoni* Baird and cyclopoids, which are heavily predated by bream in this lake (Van Densen *et al.*, 1986). Smaller individuals of *Euchlanis* may also be predated by invertebrate predators, such as *Leptodora kindtii* (Focke) (Dumont, 1977).

To test this hypothesis, a total loss of rotifers (D) in the water column was estimated using the method described by Duncan & Gulati (1983), which is based on the formula:

$$D = B - \frac{\Delta N}{\Delta t},$$

where: B = recruitment rate

$\frac{\Delta N}{\Delta t}$ = change in standing crop.

In a field study lasting 29 h on 11-12 September, *Euchlanis* was found to be among the most heavily predated rotifers; only for the genus *Polyarthra* the elimination rates were higher. Losses in *Euchlanis* biomass accounted for 126% of its production. This result seems to support the hypothesis that the *Euchlanis* population is under heavy predation pressure.

Conclusion

The adaptation of *Euchlanis* to living in the littoral zone, as expressed by the deposition of eggs on plants, can limit the occurrence of its *lucksiana* form to the bodies of water with abundant blooms of blue-green algae. This form lay eggs on tight clusters of filamentous blue-green algae. However, *Euchlanis* females raised in aquaria did not use blue-green algae as a substratum, but attached their eggs to the aquarium walls. It is expected that there is a threshold of the abundance of filamentous algae below which searching for dense clusters of algae and the deposition of eggs on them cannot be worth the energy invested.

The dominance of blue-green algae was noted mostly in hypertrophic lakes dominated by small forms of zooplankton. *Euchlanis*, which is relatively large and conspicuous, is likely to be subject to a high predation pressure and, because of its size, it can be attacked by invertebrate predators as well as by early developmental stages of fish.

Due to their ability to feed on blue-green algae, which are not often consumed by other zooplankters, *Euchlanis* can occasionally reach densities of the order of several ten individuals per liter (as in Lake Loosdrecht) or even several hundred individuals (as in Glubokoe Lake; Matveeva, 1986), although it is not well adapted to the life in the pelagic zone.

References

Bazilevitsch, V. M., 1985. Rotifera of some canals of southern Ukraine. In L. A. Kutikova (ed.), Rotatoria, Proceedings of the National Rotifer Symposium. Nauka, Leningrad: 205–208. (in Russian).

Benndorf, J. & W. Horn, 1985. Theoretical considerations on the relative importance of food limitation and predation in structuring zooplankton communities. Arch. Hydrobiol. Beih. Ergebn. Limnol. 21: 383–396.

Boesewinkel-de Bruyn, P. J., L. van Liere & B. Z. Salome, 1984. Phytoplankton species composition. In S. Parma & R. D. Gulati (eds), Limnological Institute, Progress Report 1983. Verh. Kon. Ned. Akad. Wet., Afd. Nat., Tweede Reeks, 82.

Carlin, B., 1943. Die Planktonrotatorien des Motalaström. Medd. Lunds Univ. Limnol. Inst. 5: 260 pp.

Dumont, H. J., 1977. Biotic factors in the population dynamics of rotifers. Arch. Hydrobiol. Beih. Ergebn. Limnol. 8: 98–122.

Duncan, A. & R. D. Gulati, 1983. Feeding studies with natural food particles on tropical species of planktonic rotifers. In F. Schiemer (ed.), Limnology of Parakrama Samudra – Sri Lanka: a case study of an ancient man-made lake in the tropics. Developments in Hydrobiology 12: 117–125.

Ejsmont-Karabin, J., 1984. Phosphorus and nitrogen excretion by lake zooplankton (rotifers and crustaceans) in relationship to individual body weights of the animals, ambient temperature and presence or absence of food. Ekol. pol. 32: 3–42.

Gulati, R. D. (in press). Zooplankton structure in Loosdrecht Lakes in relation to the trophic status and the recent restoration measures. Hydrobiologia.

Gulati, R. D., J. Rooth & J. Ejsmont-Karabin, 1987. A laboratory study of feeding and assimilation in *Euchlanis dilatata lucksiana*. Hydrobiologia 147: 289–296.

Gulati, R. D., K. Siewertsen & G. Postema, 1982. The zooplankton: its community structure, food and feeding and its role in the ecosystem of Lake Vechten. Hydrobiologia 95: 127–163.

Gulati, R. D., J. Ejsmont-Karabin, J. Rooth & K. Siewertsen, 1989. A laboratory study of phosphorus and nitrogen excretion of *Euchlanis dilatata lucksiana*. this volume.

Karabin, A., 1985a. Pelagic zooplankton (Rotatoria + Crustacea) variation in the process of lake eutrophication. I. Structural and quantitative features. Ekol. pol. 33: 567–616.

Karabin, A., 1985b. Pelagic zooplankton (Rotatoria + Crustacea) variation in the process of lake eutrophication. II. Modifying effect of biotic agents. Ekol. pol. 33: 617–644.

Kosova, A. A., 1985. Ecological characteristics of Rotifera of reservoirs of the Volga Delta. In L. A. Kutikova (ed.), Rotatoria, Proceedings of the National Rotifer Symposium, Nauka, Leningrad: 199–204. (in Russian).

Koste, W., 1978. Rotatoria, Monogononta. Die Rädertiere Mitteleuropas. Gebrüder Borntraeger, Stuttgart, Vols 1 & 2, 673 pp., 234 pls.

Kutikova, L. A., 1970. Rotifer Fauna USSR. Fauna USSR. 104. Acad. Nauk. SSSR, Leningrad, 744 pp. (in Russian).

Lair, N., 1980. The rotifer fauna of the river Loire (France), at the level of the nuclear power plants. Hydrobiologia 73: 153–162.

Matveeva, L. K., 1986. Pelagic rotifers of Lake Glubokoe from 1897 to 1984. Hydrobiologia 141: 45–54.

Rooth, J., 1985. Een laboratoriumstudie van het metabolisme van de rotiferen – soort *Euchlanis dilatata lucksiana* uit de Loosdrechtse Plassen. Limnologisch Instituut Nieuwersluis/Oosterzee, Studentenverslag nr. 1985-5, 50 pp. (in Dutch).

Spodniewska, I., 1983. Ecological characteristics of lakes in north-eastern Poland versus their trophic gradient. VI. The phytoplankton of 43 lakes. Ekol. pol. 31: 353–382.

Van Densen, W. L. T., C. Dijkers & R. Veerman, 1986. The fish community of the Loosdrecht Lakes and the perspective for biomanipulation. Hydrobiol. Bull. 20: 147–163.

Hydrobiologia **186/187**: 35–38, 1989.
C. Ricci, T. W. Snell and C. E. King (eds), Rotifer Symposium V.
© 1989 *Kluwer Academic Publishers.*

Influence of cyanobacterial diet on *Asplanchna* predation risk in *Brachionus calyciflorus*

Peter L. Starkweather & Elizabeth J. Walsh[1]
Department of Biological Sciences, University of Nevada, Las Vegas, Las Vegas, Nevada 89154 U.S.A.;
[1]*Department of Zoology, Oregon State University, Corvallis, Oregon 97331 U.S.A.*

Key words: Anabaena, Asplanchna, Brachionus, rotifer, cyanobacteria, feeding behavior, predator-prey interaction

Abstract

Asplanchna sylvestrii does not discriminate between groups of *Brachionus calyciflorus* fed either the cyanobacterium *Anabaena flos-aquae* or a control diet of *Euglena gracilis*. We based our analysis on the observed probabilities of attack, capture and ingestion during encounters between predator and prey. While *A. sylvestrii* was very sensitive to brachionid size, we found no significant affects of prey diet on predatory behavior. Thus, cyanobacterial diet did not influence the short-term predation risk of *B. calyciflorus* exposed to an effective predator. On the other hand, matched cohorts of *A. sylvestrii* fed *B. calyciflorus* cultured on the cyanobacterium reproduced more slowly than those fed the same prey cultured on the control food. With prolonged sympatry, therefore, the long-term risk of *Asplanchna* predation may be reduced for *Brachionus* by the latter's consumption of cyanobacteria.

Introduction

The monogonont rotifer *Brachionus calyciflorus* consumes several species of cyanobacteria, with particularly well-documented benefit derived from the filamentous *Anabaena flos-aquae* (Starkweather, 1981; Starkweather & Kellar, 1983). Thus, in freshwater systems in which this rotifer or ecologically similar taxa are present, a fraction of cyanobacterial primary production may be directed into conventional grazing food webs. The more traditional view that cyanobacterial biomass is largely relegated to detrital energetic pathways may, therefore, be subject to some modification. These findings support the conclusions of field studies (cf. Lewis, 1979) which have shown correlations between cyanobacterial biomass and the productivity of several suspension-feeding rotifers.

On the other hand, if grazers on cyanobacteria were avoided by predatory species, the energetic impact of 'blue-greens' would be restricted. They would contribute to secondary production (of brachionid rotifers, for instance), but would represent a less-than-proportional contribution to the energetics of higher trophic levels.

In this paper we report the first experiments to determine whether or not a prey diet of cyanobacteria may deter immediate predation for individuals of a suspension feeding species or diminish predator effectiveness in the longer term. For rotifer predator-prey systems, both short and long term dietary influences are important. In the former case, single prey may benefit from reduced individual predation risk; in the latter, prey (clonal) survival is enhanced by inhibition of predator population growth. The intent of the work is both to examine the behavioral implications of predator functional response as influenced by prey diet and to determine if a moderately toxic prey diet may have a negative impact on predator population growth.

Methods and materials

We cultured the *Brachionus calyciflorus* used in this study on pure cultures of either *Euglena gracilis* (UTEX-753) or *Anabaena flos-aquae* (NRC-44-1). Both foods allow sustained population growth for the rotifer (Starkweather & Kellar, 1983), the latter despite its persistent toxicity (as measured in mammalian bioassays; W. Carmichael, pers. commun.). Before use in predation trials, we selected individual animals and washed them free of residual food materials and culture medium. The predator in this system, *Asplanchna sylvestrii*, had been previously isolated from Soda Lake, Nevada, U.S.A.; we cultured individuals and their immediate progeny in tissue culture well plates using excess numbers of *Euglena*-fed *B. calyciflorus* as their only food. For each experiment, we isolated a single *Asplanchna* and placed her in fresh rotifer medium. Predators were maintained without food for between 15 min to 4 h; however, the measurements we report here were drawn largely from observations of 85 animals held without food for less than 2.5 h. For each obser-

vation we pipeted several *Brachionus* into the vessel containing a single *Asplanchna*. We observed and recorded the following predator-prey interactions: encounter, attack, capture and ingestion, in each instance noting the success or failure of sequential steps in the functional process. To avoid any influence of satiation, we truncated observation of each predator after ingestion of three prey.

To determine the influence of prey diet on predator population dynamics we established matched cohorts of *A. sylvestrii* neonates drawn from cultures fed *Brachionus* grown on *Euglena*. We supplied each *Asplanchna* with 10 prey per day, a rate above the threshold for sustained population growth, but well below saturation (Stemberger & Gilbert, 1984). We observed each animal in each cohort (12-18 animals per cohort, n = 3 for each of two treatments) daily and recorded individual survival and reproduction. All steps in this study, from food culture through direct observation, were performed at 20 °C.

Results

Our first observations of predator-prey interactions determined that prey reproductive state, perhaps reflecting age and certainly body size, has a substantial influence on *Brachionus* susceptibility to predation by these *Asplanchna*. Table 1 shows that smaller non-ovigerous (and likely juvenile) *B. calyciflorus* have both significantly greater likelihood of attack after encounter and

Table 1. Influence of prey (*Brachionus calyciflorus* fed *Euglena gracilis*) reproductive condition on predatory responses of adult *Asplanchna sylvestrii*. Values shown are mean probabilities $\pm$ s.d. for all observed predator-prey encounters; parenthetical values indicate ranges for individual animals. Differences between groups evaluated using the Kruskal-Wallis ranks test (* = p < 0.05).

Prey condition	Attack after encounter	Capture after attack	Ingestion after capture	Ingestion after encounter
Non-ovigerous	0.80 ± 0.32 (0.19–1.00)	0.71 ± 0.28 (0.20–1.00)	0.69 ± 0.30 (0.20–1.00)	0.44 ± 0.37
Ovigerous	0.56 ± 0.24 (0–1.00)	0.56 ± 0.27 (0–1.00)	0.39 ± 0.31 (0–1.00)	0.14 ± 0.16
p(K.-W.)	*	n.s.	*	*

Table 2. Influence of prey diet on predator responses of adult *A. sylvestrii* to ovigerous *B. calyciflorus* cultured on dits of either *E. gracilis* or *Anabaena flos-aquae*. Numerical values and statistical treatment as in Tab. 1.

Prey diet	Attack after encounter	Capture after attack	Ingestion after capture	Ingestion after encounter
Euglena	0.56 ± 0.24 (0–1.00)	0.56 ± 0.27 (0–1.00)	0.39 ± 0.31 (0–1.00)	0.12 ± 0.16
Anabaena	0.58 ± 0.31 (0.19–1.00)	0.66 ± 0.35 (0.14–1.00)	0.37 ± 0.43 (0–1.00)	0.14 ± 0.38
p(K.-W.)	n.s.	n.s.	n.s.	n.s.

ingestion after capture than do ovigerous adults. The two reproductive classes do not differ in their probability of capture after attack. The combined probabilities for the entire functional sequence (ingestion after encounter) indicates that, in a given interaction, ovigerous animals are at least three times less likely to be ingested than the non-ovigerous and smaller individuals.

These results led us to use only ovigerous (and therefore somewhat larger and older) prey when examining the effect of prey diet on susceptibility to *Asplanchna* predation. We wanted to maximize the opportunity for the *Brachionus* to accumulate any chemical 'signal' which might influence *A. sylvestrii* behavior and to extend the period of handling of prey (due to their larger size) by individual predators – again, to maximize the possibility of the predators detecting and responding to any such factor.

Table 2 shows that in these experiments prey diet (*Euglena* vs. *Anabaena*) had no significant affect on *B. calyciflorus* vulnerability to predation by *A. sylvestrii*. In neither composite (ingestion after encounter) nor subsidiary behaviors was there any indication that the predator detected or responded to a factor correlated with prey diet. It is also important to note here that we observed no differences in prey size, shape or behavior based on their original diets. Thus, the lack of predator response was not due to confounding factors which might mask an influence based on prey diet.

Asplanchna sylvestrii cohorts fed *Brachionus* cultured on *Euglena* have instantaneous population growth rates approximately twice those of matched groups fed *Anabaena*-cultured prey (Table 3). This result appears to be due almost exclusively to the differential in net fecundity (R_o), with no significant differences between the two groups in median survivorship ($0.5\,l_x$) or calculated generation time (T).

Table 2. Influence of prey diet on predator responses of adult *A. sylvestrii* to ovigerous *B. calyciflorus* cultured on dits of either *E. gracilis* or *Anabaena flos-aquae*. Numerical values and statistical treatment as in Tab. 1.

Prey diet	Attack after encounter	Capture after attack	Ingestion after capture	Ingestion after encounter
Euglena	0.56 ± 0.24 (0–1.00)	0.56 ± 0.27 (0–1.00)	0.39 ± 0.31 (0–1.00)	0.12 ± 0.16
Anabaena	0.58 ± 0.31 (0.19–1.00)	0.66 ± 0.35 (0.14–1.00)	0.37 ± 0.43 (0–1.00)	0.14 ± 0.38
p(K.-W.)	n.s.	n.s.	n.s.	n.s.

38

Summary and conclusions

In these experiments, cyanobacterial diet had no effect on the short-term susceptibility of prey (*B. calyciflorus*) to predation by an expectedly selective predator (*A. sylvestrii*). The expectation here was based on literature reports which have shown species of *Asplanchna* to be extraordinarily sensitive to the nature of their prey, distinguishing even between alternate conspecific clones (see Gilbert, 1980). Accordingly, we conclude that cyanobacterial production in a pelagic system containing appropriate rotifers can be converted to animal biomass throughout the grazing food web.

As noted earlier, *Brachionus calyciflorus* can utilize the current strain of *A. flos-aquae* as a sole or complementary food source, despite its toxic properties (Starkweather & Kellar, 1983). While the cyanobacterial diet does not protect the rotifers from individual predator attack, including *A. flos-aquae* in the diet may have a beneficial influence beyond nutrition. The results described show that long-term population growth of a predatory species may be reduced when prey diet includes 'blue-greens'. This type of effect may be particularly significant in zooplankton predator-prey systems involving two rotifers, since generation times are short (2-5 d) and overlapping. It remains to be seen if such processes operate in natural pelagic communities; we would predict that the influences described here would depend upon the species of rotifer present, the type of cyanobacteria available in suspension and the period of sympatry of all involved taxa.

Acknowledgements

We extend sincere appreciation Penelope E. Kellar for expert technical assistance and to Dr. Wayne Carmichael for providing the cyanobacterial culture. Supported by the National Science Foundation (U.S.) DEB-8105724.

References

Gilbert, J. J., 1980. Feeding in the rotifer *Asplanchna*: Behavior, cannibalism, selectivity, prey defenses and impact on rotifer communities. pp. 158–172. In: W. C. Kerfoot (ed.). Evolution and Ecology of Zooplankton Communities. University Press of New England. Hanover, NH.

Lewis, W. M. Jr., 1979. Zooplankton Community Analysis. Springer Verlag. New York.

Starkweather, P. L., 1981. Trophic relationships between the rotifer *Brachionus calyciflorus* and the blue-green alga *Anabaena flos-aquae*. Verh. int. Ver. Limnol. 21: 1507–1514.

Starkweather, P. L. & P. E. Kellar, 1983. Utilization of cyanobacteria by *Brachionus calyciflorus*: *Anabaena flos-aquae* (NRC-44-1) as a sole or complementary food source. Hydrobiologia 104: 373–377.

Stemberger, R. S. & J. J. Gilbert, 1984. Body size, ration level and population growth in *Asplanchna*. Oecologia. 64: 355–359.

Hydrobiologia **186/187**: 39–42, 1989.
C. Ricci, T. W. Snell and C. E. King (eds), Rotifer Symposium V.
© *1989 Kluwer Academic Publishers.*

Nutritional effect of freshwater Chlorella on growth of the rotifer *Brachionus plicatilis*

K. Hirayama[1], I. Maruyama[1,2] & T. Maeda[2]
[1]*Graduate School of Marine Science and Engineering, Nagasaki University, Nagasaki, Nagasaki, 852 Japan*; [2]*Chlorella Industry Co., Ltd., Chikugo, Fukuoka, 833 Japan*

Key words: Freshwater Chlorella, nutritional effect, *Brachionus plicatilis*, Vitamin B_{12}, supplementary food

Abstract

Mass production of *Brachionus plicatilis* is usually accomplished by feeding so-called marine Chlorella (*Nannochloropsis oculata*) to the rotifers in marine fish hatcheries. If the marine Chlorella are in short supply, baker's yeast is usually used as a supplementary food. Recently, a condensed suspension of freshwater Chlorella (*Chlorella vulgaris*, k-22) was commercially developed as another supplementary food. We have evaluated the dietary value of this freshwater Chlorella for growth of the rotifer by means of individual and batch cultures. Rotifers cultured with the freshwater Chlorella suspension under almost bacteria-free conditions, showed very suppressed growth. However if the Chlorella was supplemented with vitamin B_{12} by adding the vitamin solution into the suspension or by culturing the Chlorella in a medium containing vitamin B_{12}, the nutritional value of freshwater Chlorella was greatly improved and almost at the same level as that of marine Chlorella. Condensed Chlorella may therefore be effective as a supplementary food if vitamin B_{12} is supplied.

Introduction

In Japan, mass production of the rotifer, *Brachionus plicatilis* is usually conducted by feeding so-called 'marine Chlorella', which have been taxonomically transferred from *Chlorella* to *Nannochloropsis oculata* (Maruyama *et al.*, 1986). In marine hatcheries, a much greater area is provided for mass production of marine Chlorella than for the rotifer. At times, especially in the summer, rotifer production exceeds the required supply of marine Chlorella for food. At that time, baker's yeast is usually supplied as a supplementary food for the rotifer. However, baker's yeast is nutritionally incomplete for rotifer growth

(Hirayama & Funamoto, 1983). There have been many attempts to develop another supplementary food or substitute for marine Chlorella. Condensed suspensions of freshwater Chlorella have been developed as commercially available supplementary foods. However, some culturists have complained that Chlorella suspensions sometimes produce suppressed or unstable growth. In this study, we evaluate the nutritional effect of freshwater Chlorella on rotifer growth under almost bacteria-free conditions. We report that freshwater Chlorella is nutritionally deficient for rotifer growth unless the food suspension contains vitamin B_{12}.

Materials and methods

A freshwater Chlorella (*Chlorella vulgaris* k-22) was cultured in 100 ml of a medium which does not contain vitamin B_{12} (Table 1), by using 500 ml glass flasks with continuous shaking (25 °C, 4000 lux). After harvesting at $(5-10) \times 10^8$ cells ml^{-1}, we washed this suspension with sterile water by centrifugation and then resuspended it at 10×10^8 cells ml^{-1}. This treatment made the suspension almost bacteria free. Marine Chlorella (*Nannochloropsis oculata*) used as a control was cultured under bacteria-free conditions in Erdschreiber medium supplemented with vitamin B_{12} at $0.5 \, \mu g$ l^{-1}. Additionally, freshwater Chlorella was cultured in medium supplemented with vitamin B_{12} at 50 μg l^{-1}. Baker's yeast was cultured under aeration in Yeast Nitrogen Base of Wako Company plus saccharose. Rotifer eggs used for the experiments were sterilized by using an antibiotic mixture (AM9) (Provasoli *et al.*, 1959). Throughout our experiments, we used bacteria-free techniques so as to avoid the excess contamination.

In order to evaluate the nutritional effects of foods or supplementary nutrients, the first-laid eggs collected from an actively growing culture of rotifers were divided into several groups. These groups were cultured in the suspension to be tested and in control food suspensions, by means of batch culture or individual culture. For the batch culture method, offspring hatched from first-laid eggs were placed in the food suspension to be tested and after 4-7 days the number of individuals was compared with that in control food suspensions. In the individual cultures, the first-laid eggs were cultured individually in the test suspension. We followed the growth of each individual by observation of daily survival and number of eggs laid. These data were used to calculate net reproduction rate (R_o) and intrinsic rate of population increase (r) on the basis of Birch's computational method (Birch, 1948). Details of the individual culture method are explained elsewhere (Hirayama & Funamoto, 1983).

Results and discussion

Results of these experiments are presented in Fig. 1. Growth of the rotifer on freshwater Chlorella was suppressed and did not change with increase of Chlorella density. Both r and R_o were

Table 1. Composition of culture medium for freshwater Chlorella.

Composition	Amount
Glucose	10 g
$(NH_2)_2CO$	1.2 g
KH_2PO_4	1.2 g
$MgSO_4 \cdot 7H_2O$	0.6 g
Fe.EDTA	15 mg
A5 solution	1 mg
Tap water	1000 ml

pH 6.6

A5 solution contains metals of Mn, Zn, Cu, Mo and B.

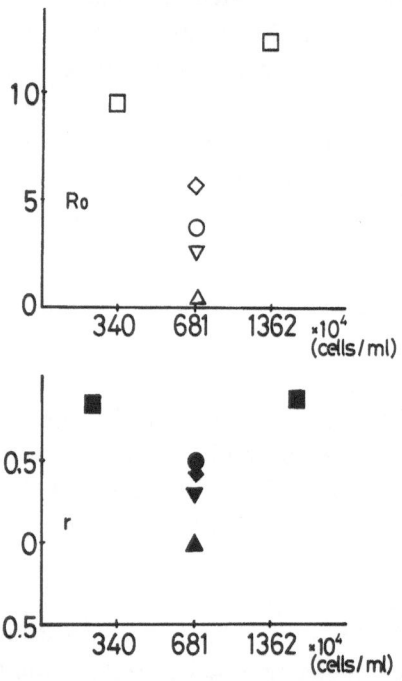

Fig. 1. Net reproduction rates (R_o) and intrinsic rates of population increase (r) of the rotifers cultured by individual cultures method in freshwater Chlorella suspensions at various cell densities. The same symbol is used to represent values from the same experiment.

quite variable. These observations suggest that the nutritional value of freshwater Chlorella is usually very low and that different lots of freshwater Chlorella varied in nutritional value even under similar conditions. We suggest that this may be due to the growth phase at harvest of the Chlorella culture.

Figure 2 presents the nutritional effects of marine and freshwater Chlorella on growth of the rotifer and also the effects of supplementation with baker's yeast. Freshwater Chlorella had very low nutritional value compared with marine Chlorella. Supplementation of low densities of both Chlorellas with baker's yeast had little effect on growth of the rotifer.

Figure 3 gives the results of 4 day batch cultures in the freshwater Chlorella suspensions with and without supplementation by vitamin B_{12} solution at 1.4 μg ml^{-1} and baker's yeast. Growth on a marine Chlorella suspension is presented as a control. Vitamin B_{12} substantially improved the nutritional value of the freshwater Chlorella suspension. Chlorella cultured in a medium containing vitamin B_{12} produced almost as many rotifers as either direct vitamin B_{12} supplemen-

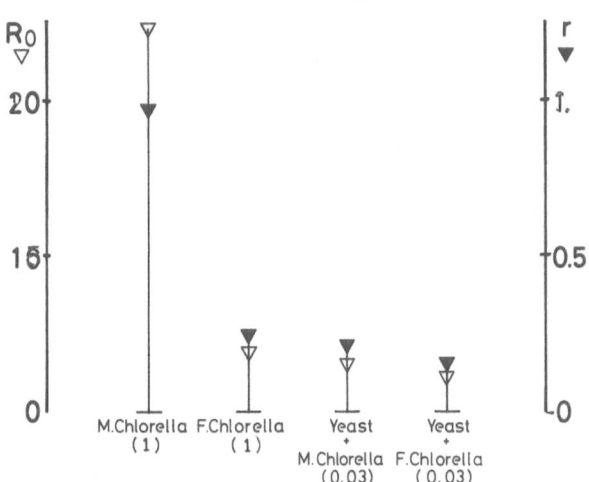

Fig. 2. Nutritional effects of marine and freshwater Chlorella on rotifer growth and effects of supplementation of both Chlorella with baker's yeast. The number in parentheses presents the cell density of Chlorella. Number one is used to indicate a density of 681 × 10^4 cells ml^{-1}. The density of baker's yeast is 200 μg ml^{-1}.

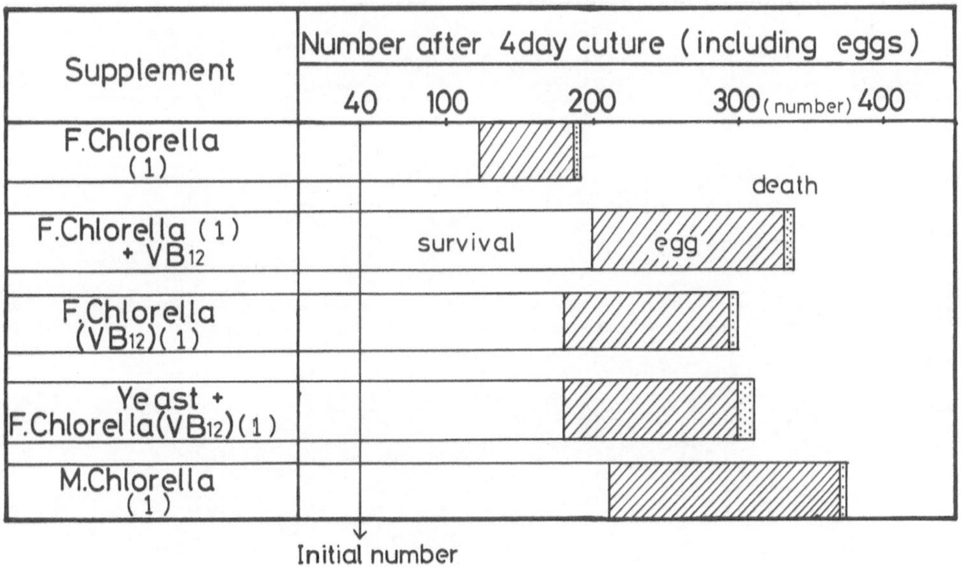

tation of the rotifer medium or use of marine Chlorella. Effects of supplementation by vitamin B_{12} using the two methods with individual cultures are presented in Fig. 4. These results con-

Supplement	Number after 4day cuture (including eggs)				
	40	100	200	300 (number)	400
F.Chlorella (1)				death	
F.Chlorella (1) + VB$_{12}$		survival	egg		
F.Chlorella (VB$_{12}$)(1)					
Yeast + F.Chlorella(VB$_{12}$)(1)					
M.Chlorella (1)					

Initial number

Fig. 3. Numbers of individuals after 4 day batch cultures of the rotifer in marine and freshwater Chlorella suspensions and in vitamin B_{12} supplemented freshwater Chlorella suspensions. Explanations of the numbers in parentheses and yeast density are in Fig. 2. F. Chlorella (V B_{12}) indicate freshwater Chlorella cultured in media containing vitamin B_{12} at 50 μg l^{-1}. Chlorella + V B_{12} indicates addition of vitamin B_{12} to the rotifer medium at 1.4 μg ml^{-1}.

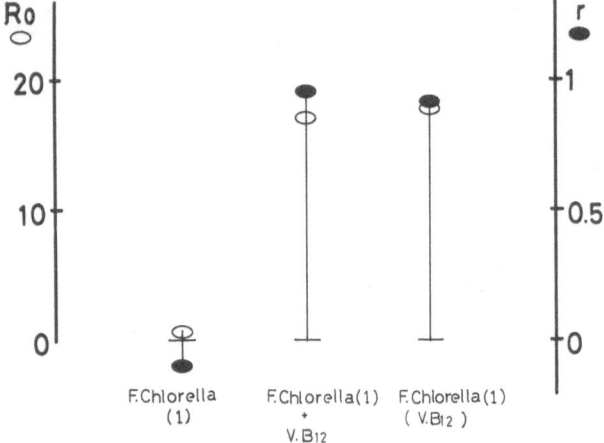

Fig. 4. Effects on *r* and R_o of vitamin B_{12} supplementation. Explanations of abscissa labels are in Fig. 3.

firm effects reported from our batch culture experiments.

Conclusion

Vitamin B_{12} is an essential nutrient for rotifer growth (Scott, 1981). The nutritional deficiency of baker's yeast suspension is mainly due to vitamin B_{12} (Hirayama & Funamoto, 1983). However, even if baker's yeast is supplemented with vitamin B_{12}, its nutritional value is still much lower than that of marine Chlorella. Freshwater Chlorella also has low nutritional value for rotifer growth. However, if vitamin B_{12} is used to supplement the freshwater Chlorella suspension by either addition to the algal or to the rotifer medium, rotifer growth is improved to the level of marine Chlorella which is considered as the best food for the rotifer.

References

Birch, L. C., 1948. The intrinsic rate of natural increase of an insect population. J. Anim. Ecol. 17: 15–26.

Hirayama, K. & H. Funamoto, 1983. Supplementary effect of several nutrients on nutritive deficiency of baker's yeast for population growth of the rotifer Brachionus plicatilis. Bull. Japan. Soc. Sci. Fish. 49: 505–510.

Maruyama, I., T. Nakamura, T. Matsubayashi & T. Maeda, 1986. Identification of the alga known as 'marine Chlorella' as a member of the Eustigmatophyceae. Jap. J. Phycol. 34: 319–325.

Provasoli, L., K. Shiraishi & J. R. Lance, 1959. Nutritional idiosyncrasies of artemia and tigriopus in monoxenic culture. Ann. N. Y. Acad. Sci. 77: 250–261.

Scott, J. M., 1981. The vitamin B_{12} requirement of the marine rotifer Brachionus plicatilis. J. mar biol. Ass. U.K. 61: 983–994.

Hydrobiologia **186/187**: 43–50, 1989.
C. Ricci, T. W. Snell and C. E. King (eds), Rotifer Symposium V.
© 1989 *Kluwer Academic Publishers.*

Life history characteristics of *Brachionus plicatilis* (rotifera) fed different algae

J. Korstad,[1] Y. Olsen & O. Vadstein
SINTEF Div. of Applied Chemistry, Aquaculture Group, 7034 Trondheim, Norway; [1]*Present Address:*
Department of Biology, Oral Roberts University, Tulsa, Oklahoma 74171, USA

Key words: Brachionus plicatilis, Rotifera, life history, food

Abstract

A detailed study of the life history of the rotifer *Brachionus plicatilis* was done at 20 °C, 20 ppt salinity, and 90 mg C l^{-1} food concentration. Rotifers were grown individually in culture plate wells (150 μl culture volume) and fed *Isochrysis galbana* Tahiti, *Tetraselmis* sp., *Nannochloris atomus*, or a 1 : 1 mixture (weight) of two of the algae. Observations were made every 2–8 hr and rotifers were sized and transferred to new food daily. A total of 19 different parameters were compared. Rotifers fed *Isochrysis* averaged 21 offspring per female, a 6.7 day reproductive period, a lifespan of 10.5 days and a mean length of 234 μm. After *Isochrysis*, the foods giving the highest growth, survival, and reproduction in decreasing order were *Isochrysis + Nannochloris, Nannochloris, Isochrysis + Tetraselmis, Tetraselmis + Nannochloris,* and *Tetraselmis*. Although the small volume culture system used in this study seems appropriate for studying life history of *B. plicatilis*, the results cannot always be directly applied to larger cultures.

Introduction

Brachionus plicatilis Müller is a euryhaline rotifer commonly used as initial food for a variety of marine fish larvae. Several reviews have discussed its biological characteristics and its importance in aquaculture (Walker, 1981; Lubzens, 1987, 1989).

Although much is known about the biology of *B. plicatilis*, relatively few researchers have investigated the life history of this rotifer. The only detailed studies of the life history of individually cultured *B. plicatilis* were done by Ruttner-Kolisko (1972), who tested the influence of temperature and salinity on several important life history parameters, and King & Miracle (1980), who tested the influence of temperature on rates of survivorship and reproduction in three clones of

B. plicatilis. Most other studies of various aspects of the life history of *B. plicatilis* have investigated the influence of food and temperature on population growth rates (Snell *et al.*, 1983; Yufera, 1987).

More information on the growth, survival, and reproduction of *B. plicatilis* fed different diets is needed before optimization of the culture of this rotifer can take place. Mathematical models of the production process of live feed for marine fish larvae (e.g., Slagstad *et al.*, 1987) might also be refined by this information.

This study examines the size at birth, size of first egg, size at first newborn, maximum size, juvenile period, time to last egg, time between egg-laying, egg development time, total number of eggs, total number of eggs hatching, percent eggs hatching, maximum number of eggs carried at one time, time to first newborn, time to last newborn, time between newborn, reproductive period, lifespan, post-reproductive period, and percent of lifespan post-reproductive of individually cultured *B. plicatilis* fed single and mixed diets of *Isochrysis galbana* Tahiti, *Tetraselmis* sp., and *Nannochloris atomus*.

Methods

Brachionus plicatilis were obtained from the Institute of Marine Research in Bergen, Norway. Separate cultures were maintained in 4 l beakers with continuous aeration at 20 °C and 20 ppt salinity. Sea water was obtained from direct pipeline supply to the laboratory from the nearby fjord, diluted to the desired concentration, and filtered through a Whatman GF/C glass-fiber filter. Rotifer cultures were daily fed one of the following algae: *Isochrysis galbana* Tahiti, *Tetraselmis* sp., and *Nannochloris atomus* (hereafter called Iso, Tetra and Nanno, respectively), or a 1 : 1 mixture (weight) of two of the algae: Iso + Tetra, Iso + Nanno, and Tetra + Nanno. Algae were obtained from the culture collections at the Institute of Microbiology and Plant Physiology in Bergen (Iso and Nanno) and the Laboratory of Biotechnology, University of Trondheim (local strain of Tetra; species unknown).

Algal cultures were grown in semi-continuous cultures diluted at 0.33 day^{-1} using f media (Guillard 1975) and constant light (160 μEin m^{-2} sec^{-1}). About 500 ml (13%) of the water in the rotifer cultures was removed each day and replenished by new sea water with algae. Rotifer cultures were physiologically adapted to the experimental conditions for a minimum of 10 days prior to the life history experiments to assure complete acclimation to the respective food (Korstad *et al.*, 1989). In another life history study, we determined that there are no significant differences among all life history parameters measured for four generations (F_1, F_2, F_3, and F_4 generations), indicating that 10 days acclimation is sufficient.

Twelve egg-bearing females were randomly selected from each stock culture and individually micro-pipetted into 150 μl tissue culture plate wells (Nunclon) containing the same algal culture at a concentration of 90 mg C l^{-1}. These concentrations were determined spectrophotometrically using an established carbon versus absorption relationship. These rotifers were observed every 1–2 hr under a Wild dissecting microscope at 25 × magnification, and as soon as neonates appeared, the mother was removed. Thereafter, observations were made every 2–8 hr for the appearance of eggs and neonates which were counted and removed. Rotifers were sized (under 50 × magnification) and transferred to new food daily. Between observations, the culture plates were kept inside a large beaker immersed in a water bath at 20 °C. Light intensity was approximately 6 μEin m^{-2} sec^{-1} (12L : 12D).

A separate experiment was done to study the low survival and reproduction of *B. plicatilis* which were fed *Tetraselmis*. Individual rotifers were cultured under the same conditions as the first study. Four groups of six rotifers each were fed Iso diluted with different solutions to a final concentration of 60 mg C l^{-1}. One group was diluted with filtrate from dense cultures of Tetra (Tetra grown in stock culture at a concentration of 240 mg C l^{-1} and filtered through a Whatman GF/C filter). Another group was diluted with Tetra filtrate at a concentration of 90 mg C l^{-1} filtered as above and a third group was diluted with f media to check if some compound in the media not metabolized by Tetra was toxic to *Brachionus*. A fourth group was diluted with 20 ppt sea water (control). All rotifers were observed and transferred to new food daily as in the other study.

Differences between means for each life history parameter were analyzed with one-way ANOVA. Significant differences ($p < 0.05$) were further analyzed with the Least Significant Difference multiple comparisons test (Statgraphics, Statistical Graphics Corp.).

Results

A total of 19 different parameters were determined for each rotifer (Table 1). Size at birth, time to first newborn, and post-productive period were not significantly different between foods. Other life history traits showing significant dietary effects included size at first egg, size at first newborn, and maximum size, all of which were larger for rotifers fed Iso and Iso + Nanno. Rotifers were smallest when fed Tetra. Rotifers fed Nanno and Iso + Nanno reached reproductive maturity in the shortest time, followed by animals fed Iso, Tetra + Nanno, Tetra, and Iso + Tetra. Time to last egg and last newborn, reproductive period, and lifespan were longest for rotifers fed Iso and shortest for those fed Tetra and Tetra + Nanno. Percent of lifespan post-reproductive was lowest for animals fed Iso, Iso + Tetra, and Iso + Nanno and highest for those fed Tetra, Nanno, and Tetra + Nanno.

Rotifers fed Iso had the highest reproductive

Table 1. Life history parameters for *Brachionus plicatilis* fed different algae. When one-way ANOVA for diet effects had significance values of $p < 0.05$, a Least Significance Difference (LSD) multiple comparison test was done. Numbers under Multiple Comparison indicate sample means which are similar (same number) or different (different number). A. Size parameters.

Food species		Size at birth (μm)	Size at first egg (μm)	Size at first newborn (μm)	Max. size (μm)
Iso:	N =	12	12	12	12
	mean =	130.00	190.00	207.50	234.17
	SE =	1.18	0.00	1.25	4.32
Tetra:	N =	12	5	0	11
	mean =	128.18	174.00	–	165.45
	SE =	1.73	2.19	–	4.52
Nanno:	N =	12	12	10	12
	mean =	128.33	189.17	202.00	208.33
	SE =	1.08	0.80	1.90	3.50
Iso + Tetra:	N =	9*	4	4	8
	mean =	126.67	187.50	197.50	183.75
	SE =	1.57	2.17	6.50	7.06
Iso + Nanno:	N =	12	12	12	6
	mean =	128.33	194.17	210.00	230.00
	SE =	1.08	1.42	2.04	6.24
Tetra + Nanno:	N =	12	7	0	8
	mean =	125.83	184.29	–	180.00
	SE =	1.42	2.75	–	6.12
Anova:	p =	0.37	<0.001	0.014	<0.001
LSD multiple comparison:					
Iso		–	1	1	1
Tetra		–	2	–	2
Nanno		–	1	2	3
Iso + Tetra		–	1	3	4
Iso + Nanno		–	3	1	1
Tetra + Nanno		–	4	–	4

* Only 9 newborn were obtained.

Table 1. (continued). B. Egg parameters.

Food species		Time to first egg (juvenile period) (days)	Time to Last egg (days)	Time between egg-laying (days)	Egg devel. time (days)	Total number eggs	Total number hatching (R_0)	Percent hatching	Max. no. eggs carried at one time
Iso:	N =	12	12	12	12	12	12	12	12
	mean =	1.39	8.11	0.33	0.41	22.42	21.17	0.93	4.00
	SE =	0.07	0.32	0.02	0.04	1.46	1.66	0.03	0.26
Tetra:	N =	5	5	1	0	12	12	12	0.12
	mean =	1.80	2.21	2.04	–	0.50	0	0	0.50
	SE =	0.09	0.29	–	–	0.19	0	0	0.19
Nanno:	N =	12	12	11	10	12	12	12	12
	mean =	1.23	4.92	0.79	1.97	5.75	3.08	0.41	2.50
	SE =	0.09	0.98	0.10	0.32	1.28	1.12	0.07	0.22
Iso + Tetra:	N =	4	4	4	4	9	9	9	9
	mean =	1.84	4.26	0.39	0.78	3.22	3.00	0.39	1.56
	SE =	0.19	0.44	0.02	0.21	1.39	1.37	0.15	0.59
Iso + Nanno:	N =	12	11	12	11	11	11	11	12
	mean =	1.19	6.73	0.42	0.42	17.36	16.00	0.92	4.33
	SE =	0.04	0.57	0.07	0.02	1.62	1.56	0.02	0.27
Tetra + Nanno:	N =	7	7	4	0	12	12	12	12
	mean =	1.57	2.20	0.89	–	1.00	0	0	0.92
	SE =	0.09	0.26	0.07	–	0.29	0	0	0.25
Anova:	p =	<0.001	<0.001	<0.001	<0.001	<0.001	<0.001	<0.001	<0.001
LSD multiple comparison:									
Iso		1	1	1	1	1	1	1	1
Tetra		2	2	2	–	2	2	2	2
Nanno		3	3	3	2	3	3	3	3
Iso + Tetra		2	3	1	3	4	3	3	4
Iso + Nanno		3	4	1	1	5	4	1	1
Tetra + Nanno		4	2	3	–	2	2	2	2

output both in terms of eggs and newborn. Those fed Iso and Iso + Nanno had the highest percentage of viable eggs and carried the most eggs at one time. Rotifers fed Tetra and Tetra + Nanno produced a few eggs but none were viable. Time between egg-laying, egg development time, and time between newborn were shortest for animals fed Iso, Iso + Nanno, and Iso + Tetra.

Percent survival was highest for rotifers fed Iso, followed by Iso + Nanno, Nanno, Iso + Tetra, Tetra, and Tetra + Nanno (Fig. 1). Rotifers fed Iso lived a minimum of 8.5 days and a maximum of 12.5 days, whereas rotifers fed Tetra and Tetra + Nanno lived a minimum of about 1.0 day and a maximum of 5.4 days.

The effects of Tetra filtrate on reproductive rate are presented in Table 2. The highest reproduction was achieved with rotifers fed a diet of Iso cells and Tetra filtrate and lowest for the control (sea water) and f media dilutions. Differences between the dilute and concentrated Tetra filtrate were not statistically significant.

Table 1. (continued).   C. Newborn, reproduction, lifespan, and post-reproductive parameters.

Food species		Time to first newborn (days)	Time to last newborn (days)	Time between newborn (days)	Repro. period (days)	Life-span (days)	Post-repro. period (days)	Percent of lifespan post-repro.
Iso:	N =	12	12	12	12	12	12	12
	mean =	2.25	8.88	0.36	6.72	10.46	2.35	0.22
	SE =	0.07	0.39	0.04	0.34	0.41	0.39	0.03
Tetra:	N =	0	0	0	12	12	5	4
	mean =	–	–	–	0.17	2.47	1.65	0.39
	SE =	–	–	–	0.16	0.40	0.46	0.08
Nanno:	N =	10	10	7	12	12	12	12
	mean =	2.66	5.47	1.06	3.70	7.92	3.00	0.39
	SE =	0.42	1.08	0.20	0.99	1.11	0.70	0.05
Iso + Tetra:	N =	4	4	4	9	9	4	4
	mean =	2.81	5.07	0.33	1.08	4.40	1.52	0.24
	SE =	0.31	0.44	0.11	0.48	0.60	0.60	0.07
Iso + Nanno:	N =	12	11	11	11	11	12	11
	mean =	1.96	7.21	0.36	5.54	8.87	1.96	0.20
	SE =	0.05	0.59	0.02	0.56	0.86	0.76	0.05
Tetra + Nanno:	N =	0	0	0	12	12	7	7
	mean =	–	–	–	0.37	2.71	2.71	1.29
	SE =	–	–	–	0.16	0.32	0.32	0.28
Anova:	p =	0.12	0.006	< 0.001	< 0.001	< 0.001	0.58	0.049
LSD multiple comparison:								
Iso		–	1	1	1	1	–	1
Tetra		–	–	–	2	2	–	2
Nanno		–	2	2	3	3	–	2
Iso + Tetra		–	2	1	4	4	–	1
Iso + Nanno		–	3	1	5	3	–	1
Tetra + Nanno		–	–	–	2,4	2	–	2

Discussion

The life history parameters of rotifers fed Iso in our study are similar to those for *B. plicatilis* fed excess *Dunaliella* at 20 °C reported by Ruttner-Kolisko (1972), indicating that the food value of these two algae is similar. Walker (1981) analyzed Ruttner-Kolisko's (1972) data and found that the rotifers spent approximately one-third of their lifespan in the post-reproductive stage at all temperatures tested. In our study, however, we observed a wider range, averaging about 22 percent for rotifers fed Iso, Iso + Tetra, and Iso + Nanno and about 38 percent for rotifers fed the other algal combinations.

The three clones of *B. plicatilis* used by King & Miracle (1980) had a wider range of mean lifespan at 20 °C than those used in our study. Clone SP had the longest lifespan (16.5 days), while rotifers fed Iso, which had the longest lifespan in our study, lived 11 days. Clones LA and MC in King & Miracle's study had fairly short lifespans (6–8.5 days), which were similar to rotifers fed all other algal types in our study (2.5–8.9 days).

Yufera (1987) measured the effect of algal diet and temperature on egg development time in two

48

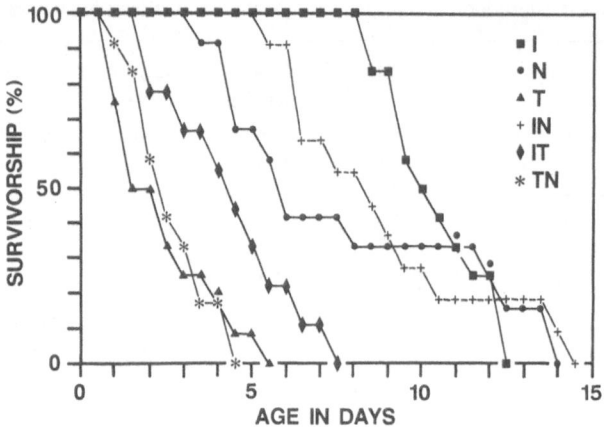

Fig. 1. Survivorship (fraction surviving over time) of *Brachionus plicatilis* fed Iso (I), Tetra (T), Nanno (N), Iso + Tetra (IT), Iso + Nanno (IN), and Tetra + Nanno (TN).

strains of *B. plicatilis*. He found the slowest development time in rotifers fed *Nannochloropsis gaditana*, which was most similar to rotifers fed *Nannochloris atomus* in our study.

Snell *et al.* (1983) studied the effects of unialgal and mixed diets of *Chlorella* sp., *Dunaliella tertiolecta*, and *Schizothrix calcicola* on the reproductive rate of *B. plicatilis*. They found that the reproductive rate was an average of 2.7 times

higher on the mixed diet than on either unialgal diet. Other researchers have reported similar findings with yeast and algae as food (Hirayama & Watanabe, 1973; Yufera & Pascual, 1980). Yufera and Pascual, however, found highest growth rates on rotifers fed *Tetraselmis suecica*. The results from our study are quite different from those reported above. Rotifers fed Tetra and Tetra + Nanno had the lowest reproductive

Table 2. Reproductive rate of *Brachionus plicatilis* fed a diet of Iso diluted to 60 mg C l^{-1} with 20 ppt sea water (Control), f Media, dilute Tetra filtrate (DTF), and concentrated Tetra filtrate (CTF). See text for details. Numbers represent mean ± (SE); n = 6 for each treatment. Description of statistical analyses same as for Table 1.

Dilution	Total no. eggs	Newborn R_0	Percent hatching
Control	7.50 (1.45)	3.33 (1.10)	38 (14)
f Media	6.17 (1.04)	2.33 (0.38)	38 (3)
DTF	16.00 (1.65)	13.00 (1.43)	81 (2)
CTF	17.33 (3.04)	15.17 (2.69)	89 (2)
Anova: p =	0.002	<0.001	<0.001
LSD multiple comparison:			
Control	1	1	1
f Media	1	1	1
DTF	2	2	2
CTF	2	2	2

rates, while those fed only Iso had the highest reproductive rate.

The results of the experiment testing the effect of Tetra filtrate on rotifers suggest that Tetra cells, but not filtrate, have an inhibitory effect on rotifers. However, we have also maintained larger volume aerated batch cultures of *B. plicatilis* fed the same algae used in this study (Korstad *et al.*, in prep.). Rotifers had somewhat similar growth rates on each food, indicating that the effects of Tetra are different in the culture plate wells than in the larger volume batch cultures. Tetra cells are motile but tend to readily attach to culture vessel surfaces, while Iso cells tend to remain motile (personal observation). Relatively more Tetra cells may therefore have been attached to the walls of the culture plate wells than in the batch culture vessels because of the larger surface-to-volume ratio in the smaller containers. This might account for the lower life history values for *B. plicatilis* fed Tetra in our study.

We have observed *B. plicatilis* feeding on Iso and Tetra. They seem to be able to ingest Iso cells more readily than Tetra. Whether this is because Iso cells move rapidly in the water while Tetra cells tend to adhere to surfaces is not clear. The Iso cells are also smaller than Tetra (approximately 11 and 250 μm^3, respectively), which may affect ingestion. Another possibility is that Tetra may be assimilated with lower efficiency than Iso. It's also probable that different species of Tetra may have differences in affinity to attachment, nutritional quality, ingestion by *Brachionus*, and other characteristics. Tetra has been commonly used as a nutritious food for *B. plicatilis* by many researchers (Trotta, 1983; Okauchi & Fukusho, 1984; Fukusho *et al.*, 1985).

Snell *et al.* (1983) reported that the enhancement of the reproductive rate of *B. plicatilis* fed a combination of *Chlorella* end *Schizothrix* was not dependent on ingestion of the *Schizothrix* cells. The reproductive rate of *B. plicatilis* increased in direct proportion to the amount of the *Schizothrix* filtrate added to the *Chlorella*. Further investigation revealed that the enhancing factor was a heat labile substance deactivated at 100 °C. Although we did not investigate the effects of

heating on the Tetra filtrate in our study, our results were different. Reproductive rate of *Brachionus* did not significantly increase from the dilute to the concentrated Tetra filtrate. There was, however, a nearly five-fold increase in R_0 between rotifers fed Iso in control (sea water) and *f* media treatments and those in the two Tetra filtrate treatments. Rotifers may possibly have obtained extra nutritional benefits from the chemicals released by the Tetra cells and present in the filtrate.

Acknowledgements

This study was supported by the Royal Norwegian Council for Scientific and Industrial Research (NTNF), and forms part of a research project on live feed for marine fish larvae financed by NTNF. The first author also gratefully acknowledges support from the Royal Norwegian Marshall Fund and Oral Roberts University.

References

Fukusho, K., M. Okauchi, H. Tanaka, S. I. Wahyuni, P. Kraisingdecha & T. Watanabe, 1985. Food value of a rotifer *Brachionus plicatilis*, cultured with *Tetraselmis tetrathele* for larvae of a flounder *Paralichthys olivaceus*. Bull. Natl. Res. Inst. Aquaculture 7: 29–36.

Guillard, R. R. L., 1975. Culture of phytoplankton for feeding marine invertebrates. In Smith, W. L. & Chanley, M. H. (eds) Culture of Marine Invertebrate Animals. Plenum Press, New York: 29–60.

Hirayama, K. & K. Watanabe, 1973. Fundamental studies on physiology of rotifer for its mass culture. IV. Nutritional effect of yeast on population growth of rotifer. Bull. Japan. Soc. Sci. Fish. 39: 1129–1133.

King, C. E. & M. R. Miracle, 1980. A perspective on aging in rotifers. Hydrobiologia, 73: 13–19.

Korstad, J., O. Vadstein & Y. Olsen, 1989. Feeding kinetics of *Brachionus plicatilis* fed *Isochrysis galbana*. Hydrobiologia, this volume.

Lubzens, E., 1987. Raising rotifers for use in aquaculture. Hydrobiologia, 147: 245–255.

Lubzens, E., 1989. Rotifers as food in aquaculture. Hydrobiologia, this volume.

Okauchi, M. & K. Fukusho, 1984. Food value of a minute alga, *Tetraselmis tetrathele*, for the rotifer *Brachionus plicatilis* culture. I. Population growth with batch culture. Bull. Natl. Res. Inst. Aquaculture 5: 13–18.

Ruttner-Kolisko, A., 1972. The metabolism of *Brachionus plicatilis* (Rotatoria) as related to temperature and chemical environment. Dt. Zool. Ges. 65: 89–95.

Slagstad, D., Y. Olsen & S. Tilseth, 1987. A model based system for control of live feed level for larval fish. Modeling, Identification and Control 8: 51–60.

Snell, T. W., C. J. Bieberich & R. Fuerst, 1983. The effects of green and blue-green algal diets on the reproductive rate of the rotifer *Brachionus plicatilis*. Aquaculture 31: 21–30.

Trotta, P., 1983. An indoor solution for mass production of the marine rotifer *Brachionus plicatilis* Müller fed on the marine microalga *Tetraselmis suecica* Butcher. Aquacult. Engin. 2: 93–100.

Walker, K. F., 1981. A synopsis of ecological information on the saline lake rotifer *Brachionus plicatilis* Müller, 1786. Hydrobiologia 81: 159–167.

Yufera, M., 1987. Effect of algal diet and temperature on the embryonic time of the rotifer *Brachionus plicatilis* in culture. Hydrobiologia 147: 319–322.

Yufera, M. & E. Pascual, 1980. Estudio del rendimiento de cultivos del rotifero *Brachionus plicatilis* O. F. Müller alimentados con levadura de panificación. Inves. Pesq. 44: 361–368.

Hydrobiologia **186/187**: 51–57, 1989.
C. Ricci, T. W. Snell and C. E. King (eds), Rotifer Symposium V.
© 1989 *Kluwer Academic Publishers.*

Feeding kinetics of *Brachionus plicatilis* fed *Isochrysis galbana*

John Korstad,[1] Ólav Vadstein & Yngvar Ólsen
SINTEF, Division of Applied Chemistry, Aquaculture Group, N-7034 Trondheim, Norway; [1] *Present Address: Department of Biology, Oral Roberts, University, Tulsa, Oklahoma 74171 USA*

Key words: *Brachionus plicatilis*, feeding kinetics, clearance rates, ingestion rates, *Isochrysis*

Abstract

Clearance and ingestion rates of *Brachionus plicatilis* were measured using ^{14}C-labeled *Isochrysis galbana* Tahiti. Experiments were conducted at 20–22 °C, 20 ppt salinity, and algal concentrations ranging from 0.13–64 mg C l^{-1}. Clearance rates were constant and maximal at concentrations <2 mg C l^{-1}, with maximum rates ranging from 3.4–6.9 μl ind.$^{-1}$ hr^{-1}. The ingestion rate varied with food concentration, and was described by a rectilinear model. The maximum ingestion rate varied considerably, and was dependent on the growth rate of the rotifers. Depending on the pre-conditions, *B. plicatilis* ingested about 0.5 to 2 times its body carbon per day at saturating food concentrations.

Introduction

A wealth of information has been published on feeding in the rotifer genus *Brachionus*, primarily in *B. calyciflorus* (see Starkweather, 1980). Although many feeding studies have also been done on *B. plicatilis*, most have employed methodology requiring several hours duration which can result in spurious results (cf. Dewey, 1976; Yufera & Pascual, 1985; Yamasaki & Hirata, 1986). Doohan (1973) applied the more sensitive radiotracer technique to measure clearance and ingestion rates in this species. Her results, however, were quite variable and the ingestion rates were probably underestimated because of the long incubation time used. Thus, knowledge of the feeding kinetics of *B. plicatilis* is incomplete. Because *B. plicatilis* is cultured as start food for marine fish larvae (Howell, 1973; Gatesoupe, 1987; Robin, 1987; Theilacker, 1987), this information would ultimately aid aquaculturists.

This paper describes a series of experiments using radio-actively labeled algae to measure clearance and ingestion rates of *B. plicatilis*. The results demonstrate the effect of food concentration and growth rate on rotifer feeding kinetics.

Materials and methods

Brachionus plicatilis was obtained from stock cultures at SINTEF Aquaculture Center in Trondheim, and maintained on a diet of *Isochrysis galbana* (Tahiti) for at least one week before experiments commenced. Stock rotifer cultures were continuously aerated and kept at 20 °C and 20 ppt salinity. The average length of adults of this strain is about 150 μm. *Isochrysis* cultures were grown in semi-continuous culture diluted at 0.33 day^{-1} using f media (Guillard, 1975). Radioactively labeled algae were prepared as follows: 20 μCi NaH[14]CO$_3$, unlabeled NaHCO$_3$ corresponding to 50 mg C l^{-1}, and about 25 ml new f media were added to approximately 25 ml *Iso-*

chrysis from the stock culture (ca. 120 mg C l^{-1} before dilution) in a 125 ml glass-stoppered bottle. The bottle was placed in constant light at about 60 μEin m^{-2} sec^{-1}.

B. plicatilis used for the feeding experiments was taken from the stock culture, placed in a 50 μm sieve in a beaker, and concentrated by carefully pipeting most of the excess water out of the beaker. The rotifers were pre-acclimated with non-radio-active *Isochrysis* at the desired concentration for 1–2 hours. All experiments were conducted at 20–22 °C, 20 ppt salinity, and about 15 μEin m^{-2} sec^{-1} light intensity. Sea water used for dilution and rinsing had been previously filtered through a Whatman GF/C filter, autoclaved, cooled to experimental temperature, and aerated.

After the pre-feeding incubation, the rotifers were concentrated as before and added to radioactive algae in flasks. Volume and incubation time depended upon the type of experiment and the flasks were agitated periodically during incubation (Schlosser & Anger, 1982). At the end of the experiment, replicate 3–5 ml samples were taken and filtered by gravity through 50 μm Nitex netting in 2.5 cm diameter Millipore filtering units, and rinsed with 50 ml 20 ppt sea water. The rotifers were maintained in about 3 ml of water throughout the rinsing. The Nitex netting was then placed in a scintillation vial. It took about 2 min from taking the sample from the flask to placing the Nitex netting in the scintillation vial.

Replicate samples were taken for determination of specific activity (3 × 100 μl) and for rotifer counts (5 × 200 μl). One hundred μl H$_2$O$_2$ was added to each vial to remove color. After 2 hr, 20 μl 0.5 N HCl was added to the specific activity samples to remove inorganic ^{14}C and 5 ml scintillation cocktail (Optifluor, Packard) was thereafter added to each vial. Samples were counted with a Nuclear Chicago Isocap 300 liquid scintillation counter for 10 min using the external standard ratio. Control experiments revealed similar counts and counting efficiencies of rotifer samples treated as above with the inclusion of adding tissue solubilizer (200 μl Soluene-350, Packard). To estimate background radioactivity on the screens we used the radioactive algal feeding

suspensions without added animals and processed samples as above. These controls were not significantly different from zero time samples with animals. Heat-killing of the rotifers before filtration yielded irreproducible results because the algae stuck to rotifers and netting.

Three different types of experiments were conducted to describe the feeding kinetics of *B. plicatilis*. A time course experiment was conducted at two food concentrations (5 and 15 mg C l^{-1}) to determine the gut passage time of the rotifer. Twenty measurements between 0 and 60 min were taken. Following the results of this experiment, subsequent feeding studies were conducted for 15–20 min. Two functional response experiments were done to determine the clearance and ingestion rates of the rotifers at different food concentrations. Rotifer stock cultures used in these experiments were fed daily in the first experiment and every other day in the second experiment. Finally, maximum ingestion rates were determined at 30 mg C l^{-1} throughout a batch culture experiment to observe variability in the ingestion rates following different long term preacclimation times. To obtain rotifers with close to zero growth rate, they were starved for 3 days prior to starting this experiment.

Results

The results of the time series experiment (Fig. 1) revealed that the gut passage time of *B. plicatilis* was dependent on food concentration. As indicated by the abrupt change in the slope of the regression lines, the gut passage time was approximately 30 and 45 min at 5 and 35 mg C l^{-1}, respectively. We therefore chose a feeding interval of 15–20 min for the other experiments.

The clearance rates for both functional response experiments (Fig. 2) were highest and constant at food concentrations below 1–2 mg C l^{-1} and decreased with increasing food concentration. Ingestion rates for both functional response experiments increased linearly with increasing food concentration up to a maximum and thereafter remained constant at higher food

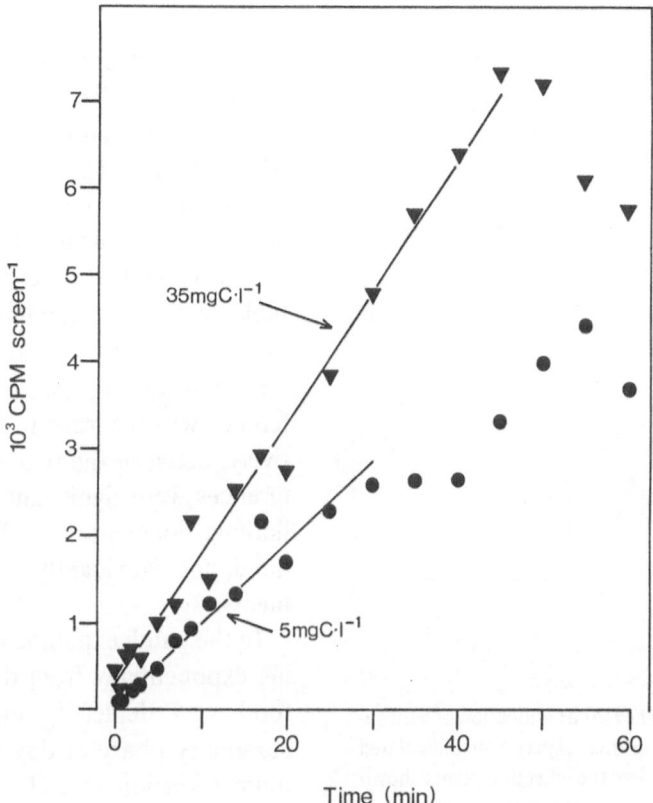

Fig. 1. Accumulation of radioactive food (^{14}C *I. galbana*) by *B. plicatilis* over time at two food concentrations. Gross CPM per Nitex screen is given on the ordinate.

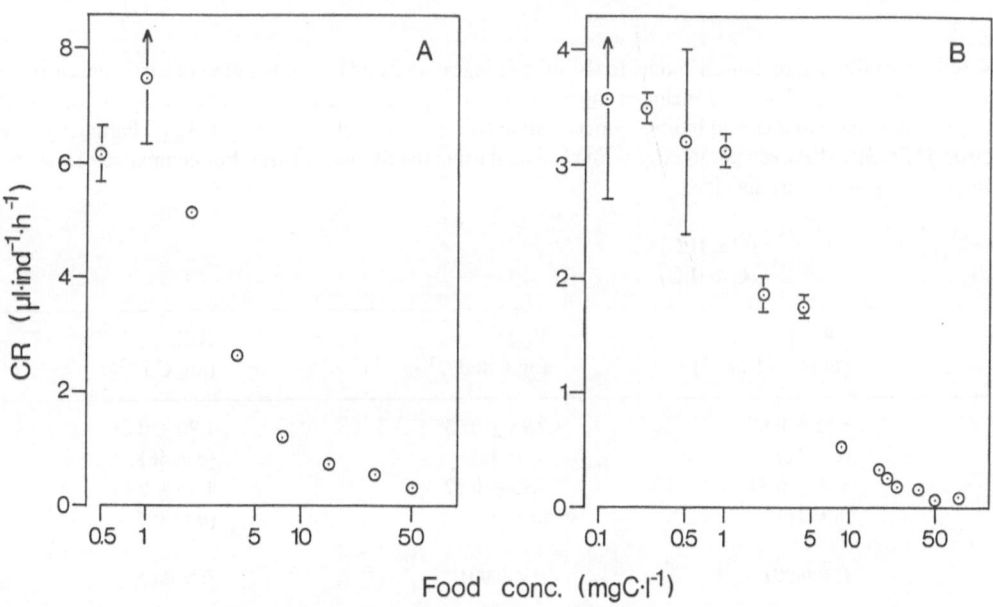

Fig. 2. Clearance rates (CR) of *B. plicatilis* for functional response experiment 1 (A) and 2 (B). Standard errors indicated when larger than symbol.

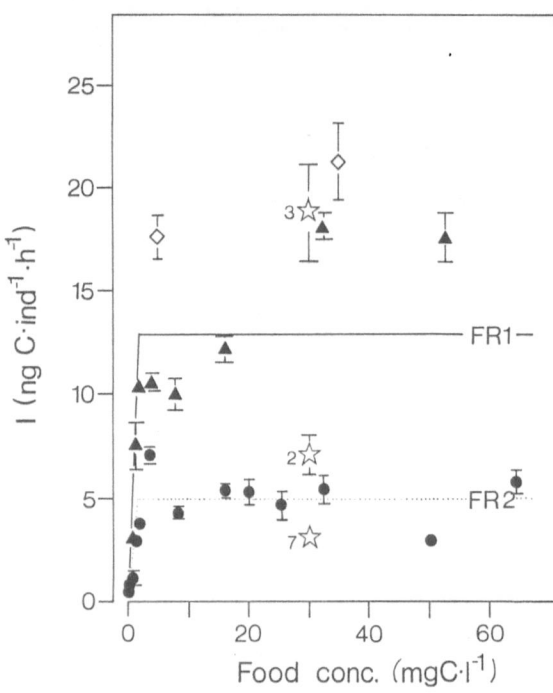

Fig. 3. Ingestion rates of *B. plicatilis* as a function of concentrations of *I. galbana*: Time Series experiment (calculated from the slope of the curve before the inflection points shown in Fig. 1) (◇), both Functional Response experiments (FR 1 ▲ and FR 2 ●), and the Batch experiment (☆). Day number indicated for the batch experiment. Error bars indicate ± SE. Lines represent model curves for the two functional response experiments (cf. Table 1).

concentrations (Fig. 3). Ingestion and clearance rates were higher in the first experiment than in the second experiment at all food concentrations. The ingestion data could be described by rectilinear, Ivlev and Michaelis-Menten models ($P < 0.001$). In Table 1 results from the rectilinear regressions are presented. We have chosen this model in the further treatment because of its simplicity (zero order approximation), and the biologically relevant parameters in the model – the maximum clearance rate (CR_{max}) and maximum ingestion rate (I_{max}). A 2.6 and a 2.0 times difference was recorded in I_{max} and CR_{max}, respectively, between the two experiments and the differences were significant (Table 1). The incipient limiting concentration (ILC) was, on the other hand, not significantly different in the two experiments.

In the batch experiment the rotifers were growing exponentially from day 1 to day 4 when the food was depleted, and the culture reached stationary phase at day 5–6 (Fig. 4). The maximum ingestion rate (I_{max}) was intermediate on day 2, increased by a factor of 2.7 on day 3, and decreased dramatically upon entering stationary phase (Fig. 3). Overall, a 6 fold difference was recorded in I_{max} during this experiment. No sig-

Table 1. Results from fitting a rectilinear model to the data of ingestion rate (I) as a function of food concentration (C) in the two functional response experiments. Maximum ingestion rate (I_{max}) and maximum clearance rate (CR_{max}) was fitted by zero-order approximation, and incipient limiting concentration (ILC) was calculated as I_{max}/CR_{max}. Parameters given with one standard error. Differences between parameters were evaluated using the Student's t-test. For comparison Dewey's (1976) data for *B. plicatilis* fed *I. galbana*, are included.

$$I = CR_{max} \cdot C \qquad (C \leq ILC)$$
$$I = I_{max} \qquad (C \geq ILC)$$

Exp.	CR_{max} (μl ind.$^{-1}$ hr^{-1})	I_{max} (ng C ind.$^{-1}$ hr^{-1})	ILC (mg C l^{-1})	P
1	6.81 ± 0.68 (*n* = 12)	12.93 ± 0.63 (*n* = 36)	1.90 ± 0.21 (*n* = 46)	<0.001
2	3.38 ± 0.31 (*n* = 18)	4.93 ± 0.22 (*n* = 54)	1.46 ± 0.15 (*n* = 70)	<0.001
	$P < 0.001$	$P < 0.001$	$P > 0.05$	
Dewey	6.01 ± 0.38 (*n* = 10)	19.1 ± 1.9 (*n* = 6)	3.18 ± 0.37 (*n* = 16)	

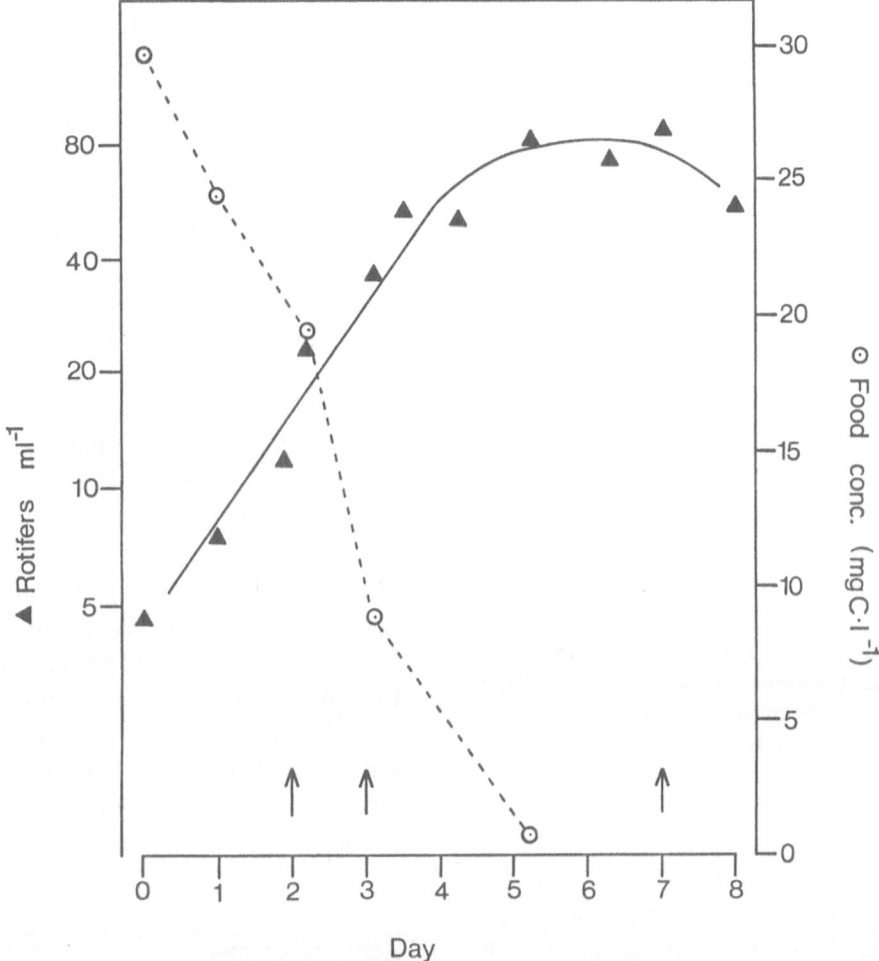

Fig. 4. Number of rotifers and *I. galbana* concentration versus day for the batch experiment. Arrows indicate days on which ingestion rates were measured.

nificant differences were recorded in the length of the animals, that averaged $155 \pm 4 \ \mu m$ ($\pm$ SD).

Discussion

The gut passage time is known to vary with species, temperature, food concentrations, and the nutritional state of the animals (Starkweather & Gilbert, 1977b). It probably also varies for foods with different assimilation efficiencies. Reported gut passage times in other species of rotifers under varying conditions ranges from 2–20 min (cf. Resvoi, 1926; Starkweather & Gil-

bert, 1977b). Our results with *B. plicatilis* are in agreement with the higher values of this range.

It has been suggested that filter feeders would benefit by reducing their filtering rate at very low food concentrations to ensure energy optimization (Mullin *et al.*, 1975; Lehman, 1976). No reduction in clearance rate was found in the present study for food concentrations down to $0.13 \ \mathrm{mg \ C \ l^{-1}}$. Our results are in accordance with most other studies on rotifer feeding (e.g. Erman 1956; Dewey, 1976; Starkweather & Gilbert, 1977a; Chotiyaputta & Hirayama, 1978). The maximum clearance rates reported by Dewey (1976) for *B. plicatilis* fed *Isochrysis galbana* were

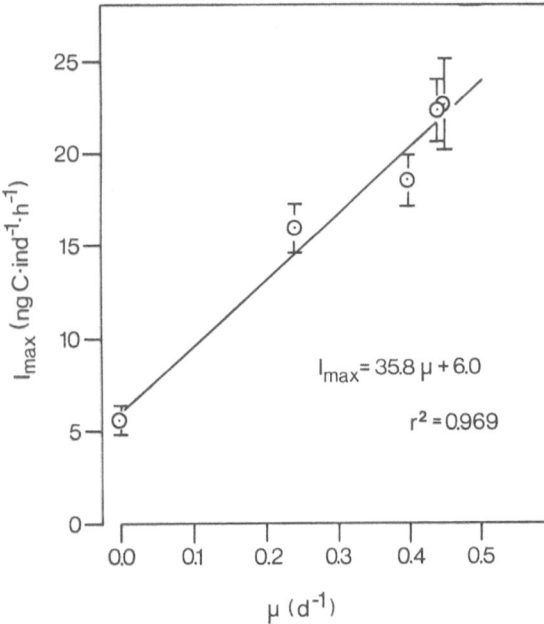

$$I_{max} = 35.8\,\mu + 6.0$$

$$r^2 = 0.969$$

Fig. 5. Maximum ingestion rate (I_{max}) at saturating food concentrations as a function of specific growth rate (μ) for *B. plicatilis* fed *Dunaliella tertiolecta*, assuming $\mu = 0$ for starved animals. Data are from Dewey (1976) and 95% CI are given for I_{max}.

intermediate to those estimated for the two functional response experiments (Table 1). The ingestion rates obtained on days 2 and 7 in the batch culture experiment were similar to I_{max} of functional response experiment 2, whereas those obtained on day 3 of the batch experiment and for the two time series experiment were comparable to that obtained in functional response experiment 1. The results presented in Fig. 3 indicate that individual *B. plicatilis* can ingest 0.5–2 times their body carbon per day at saturating food concentrations (assuming 0.23 μg C rotifer^{-1}).

The results obtained in the batch culture experiment suggest that the maximum ingestion rates were related to the physiological status of the rotifers. Starved rotifers exhibited 6 times lower I_{max} than well fed rotifers, although the starved rotifers were allowed to equilibrate their gut before the experiment (see methods). At least three days at saturating food concentrations are needed before the rotifers are capable of maintaining the highest I_{max}. This also agrees with the 2.6 times

difference in I_{max} which was recorded in the two functional response experiments. The rotifers used in experiment 2 were fed every second day, which caused the rotifers to experience large oscillations in the food concentration and periods of starvation. We found no significant correlations between maximum ingestion rates and size, swimming speed, egg ratio, or the abundance of rotifers in the experimental containers.

Dewey (1976) has reported similar variability in I_{max} and attributed it to size differences in the rotifers. Her results with *B. plicatilis* fed *Dunaliella tertiolecta* can be re-interpreted, and in Fig. 5 we have plotted I_{max} as a function of the specific growth rate. There is a significant linear relation between the two parameters ($P < 0.005$), which supports our hypothesis that the feeding kinetics of *B. plicatilis* is affected by the physiological status of the animals. We therefore conclude that the feeding rate of *B. plicatilis* is not only a function of the food concentration, but also of the growth rate of the rotifer. Further research into the feeding biology of *B. plicatilis* and other rotifers should take the long term (days) pre-feeding history of the animals into account.

Acknowledgements

This study was supported by the Royal Norwegian Council for Scientific and Industrial Research (NTNF), and forms part of a research project on live feed for marine fish larvae financed by NTNF. The first author also gratefully acknowledges support from the Royal Norwegian Marshall Fund and Oral Roberts University.

References

Chotiyaputta, C. & K. Hirayama, 1978. Food selectivity of the rotifer *Brachionus plicatilis* feeding on phytoplankton. Mar. Biol. (Berl.) 45: 105–110.

Dewey, J. M., 1976. Rates of feeding, respiration and growth of the rotifer *Brachionus plicatilis* and the dinoflagellate *Noctiluca miliaris* in the laboratory. Ph. D. Thesis Univ. Wash. 117 pp.

Doohan, M., 1973. An energy budget for adult *Brachionus plicatilis* Müller (Rotatoria). Oecologia 13: 351–362.

Erman, L. A., 1956. Quantitative aspects of the feeding of rotifers. Zool. Zh. 35: 965–971. (In Russian).

Gatesoupe, F. J., 1987. Further advances in the nutritional and antibacterial treatment of rotifers as food for turbot larvae, *Scophthalmus maximus* L. Aquaculture Europe '87 International Conference (in press).

Guillard, R. R. L., 1975. Culture of phytoplankton for feeding marine invertebrates. In Smith, W. L. & Chanley, M. H. (eds) Culture of Marine Invertebrate Animals. Plenum Press, New York: 29–60.

Howell, B. R., 1973. Marine fish culture in Britain. VIII. A marine rotifer, *Brachionus plicatilis* Müller, and the larvae of the mussel, *Mytilus edulis* L., as foods for larval flatfish. Cons. int. Explor. Mer 35: 1–6.

Lehman, J. T., 1976. The filter-feeder as an optimal forager, and the predicted shape of feeding curves. Limnol. Oceanogr. 21: 501–516.

Mullin, M. M., E. F. Stewart & F. J. Fuglister, 1975. Ingestion by planktonic grazers as a function of concentration of food. Limnol. Oceanogr. 20: 259–262.

Resvoi, P., 1926. Observations on the feeding of rotifers. Trav. Soc. Nat. Lenin. 56: 73–89. (In Russian; English translation provided by Dr. John Gilbert).

Robin, J. H., 1987. The quality of living preys for fish larval culture: preliminary results of mineral supplementation.

Aquaculture Europe '87 International Conference (in press).

Schlosser, J. J. & K. Anger, 1982. The significance of some methodological effects of filtration and ingestion rates on the rotifer *Brachionus plicatilis*. Heloglander wiss. Meersunters. 35: 215–225.

Starkweather, P. L., 1980. Aspects of the feeding behavior and trophic ecology of suspension-feeding rotifers. Hydrobiologia 73: 63–72.

Starkweather, P. L. & J. J. Gilbert, 1977a. Feeding in the rotifer *Brachionus calyciflorus*. II. Effect of food density on feeding rates using *Euglena glutinis*. Oecologia 28: 133–139.

Starkweather, P. L. & J. J. Gilbert, 1977b. Radiotracer determination of feeding in *Brachionus calyciflorus*. The importance of gut passage times. Arch. Hydrobiol. Beih. Ergebn. Limnol. 8: 261–263.

Theilacker, G. H., 1987. Feeding ecology and growth energetics of larval Northern Anchovy, *Engraulis mordax*. Fish. Bull. 85: 213–228.

Yamasaki, S. & H. Hirata, 1986. Food consumption rates of two types of rotifer *Brachionus plicatilis*. The Aquiculture 34: 137–140.

Yufera, M. & E. Pascual, 1985. Effects of algal food concentration on feeding and ingestion rates of *Brachionus plicatilis* in mass culture. Hydrobiologia 122: 181–187.

Hydrobiologia **186/187**: 59–67, 1989.
C. Ricci, T. W. Snell and C. E. King (eds), Rotifer Symposium V.
© 1989 *Kluwer Academic Publishers.*

Epizoic and parasitic rotifers

Linda May
*Institute of Freshwater Ecology, Edinburgh Research Station, Bush Estate, Penicuik,
Midlothian EH26 OQB, Scotland, UK*

Key words: epizoic, parasitic, rotifers

Abstract

Many rotifer species live in close association with plants or other animals. Most of these associations are of a commensal or synoecious nature, some rotifer species having lost the ability to live independently. Few rotifers are true parasites, actually harming their hosts.

The Seisonidae, Monogononta and Bdelloidea include epizoic and parasitic species. The most widely known are probably the parasites of colonial and filamentous algae (e.g. *Volvox, Vaucheria*). However, rotifers are also found on a wide range of invertebrates: colonial, sessile Protozoa; Porifera; Rotifera; Annelida; Bryozoa; Echinodermata; Mollusca, especially on the shells and egg masses of aquatic gastropods; Crustacea, including the lower forms (e.g. *Daphnia, Asellus, Gammarus*) and in the gill chambers of *Astacus* and *Chasmagnathus*; the aquatic larvae of insects. There appear to be few records of epizoic or parasitic rotifers among vertebrates, apart from *Encentrum kozminskii* on carp, *Limnias ceratophylli* on the Amazonian crocodile, *Melanosuchus niger*, and an unidentified Bdelloid apparently living as a pathogenic rotifer in Man.

Introduction

Rotifers have long been known to live in close association with other organisms, but little is known of the types of relationships involved. It is difficult to judge from the literature whether these associations are parasitic, symbiotic, commensal, epiphytic or epizoic. Many of the records are simply anecdotal and, while some authors have guessed at the type of relationship involved, few have based their guesses on careful biological observation. For the purposes of this paper, a parasitic rotifer is considered to be one which lives in or on a plant or animal species, feeding on its host and completing its life cycle in that environment. In contrast, an epizoic rotifer is defined as a rotifer which lives on another animal for all or part of its life, but does not feed on its host.

This paper reviews the literature which describes parasitic or epizoic associations between rotifers and other organisms. Much of this literature was published in the late 1800's and early 1900's, when naturalists spent many hours observing the behaviour of live specimens. Many of the species names given by the original authors are now out-of-date; these have been updated, as far as possible, according to Harring (1913), Bartos (1951), Kutikova (1970) and Koste (1978).

This review is not exhaustive and, in many ways, adds little to the information given by Budde (1925) because the topic has received so little attention since that time. Hopefully, however, the details given below, organised according to the type of host, will renew interest in this much neglected area of rotifer research. The text is divided into sections according to the kingdom and, in some cases, the phylum (after Barnes 1984) to which the various 'host' organisms belong.

Parasitic and epizoic associations

Protista and Monera

Rotifer parasites are common among colonies of *Volvox*, especially *V. globator* (Williams, 1852; Gosse, 1852; Hood, 1895; Galliford, 1946), *V. aureus* (Rousselet, 1911; Sauer, 1978), and *V. tertius* (Ganf *et al.*, 1983). These rotifers have been identified as *Proales parasita* (Ehr.) (Williams, 1852; Thompson, 1892; Rousselet, 1911, 1914; Harring & Myers, 1922; Rich & Pocock, 1933; Hollowday, 1949; Wulfert, 1960; Sauer, 1978), *Ascomorphella volvocicola* (Plate) (Gosse 1852; Hood, 1895; Murray, 1906; Galliford, 1946; Ganf *et al.*, 1983) and, more rarely, *Cephalodella catellina volvocicola* (Zawadovsky) (Edmondson & Hutchinson, 1934). Parasitism in *Ascomorphella volvocicola* seems to be specific to *Volvox* species, as this rotifer has never been observed on other phytoplankton species, or living freely (Ganf *et al.*, 1983). The mature rotifer enters the *Volvox* colony and feeds on the cells causing extensive damage (Hood, 1895; Hollowday, 1949; Ganf *et al.*, 1983). Whether this actually causes the *Volvox* population to decline is uncertain (Ganf *et al.*, 1983).

Rotifers have also been recorded as parasites on other colonial algae, *Ptygura melicerta* (Ehrb.) on *Gloeotrichia* (Edmondson, 1940; Koste, 1978), *Proales parasita* on *Uroglenopsis americana* (Carlin, 1939), *Proales uroglena* and *P. parasita* on *Uroglena* (Koste, 1978). Little is known of the relationships between these species.

Rotifers have long been known to parasitise *Vaucheria* filaments (Lister, 1884; Thompson, 1892; Brain, 1894; Wollny, 1879), although these early authors were unsure of the identification of the rotifer species involved. Later descriptions showed that *Vaucheria* filaments are usually parasitised by *Proales werneckii* (Ehrb.) (Jennings, 1894; Harring & Myers, 1922; Davis & Gworek, 1973; Ott, 1977; Christensen, 1987). Parasitism

by *P. werneckii* has been recorded in a variety of *Vaucheria* species, including *V. prona*, *V. geminata*, *V. dillwynii*, *V. erythrospora*, *V. ?racemosa* and *V. canalicularis* (Davis & Gworek, 1973; Ott, 1977; Christensen, 1987). In general, the rotifer enters the developing gametophore of the *Vaucheria* filament where it induces gall formation. Here it feeds on the cytoplasm of the host by breaking the tonoplast and ingesting the cytoplasmic organelles. The rotifer deposits its eggs within the galls; these hatch and the emergent rotifers escape to parasitise more filaments (Davis & Gworek, 1973; Ott, 1977).

Many rotifers are thought to be parasitic on protozoa, including *Dicranophorus difflugarium* (Penard) on *Difflugia acuminata* var. *inflata* (Penard, 1914; Koste, 1978), *Brachionus rubens* Ehrb. on colonies of *Carchesium*, *Proales parasita* and *Pleurotrocha petromyzon* on colonies of *Ophridium*, and *Albertia vermisculus* Dujardin on *Limax* species (Koste, 1978). In contrast, *Eosphora gibba* Garner seems to have a commensal association with colonies of *Carchesium* and *Vorticella*, the rotifer benefiting from the food carried towards it in the strong water currents generated by the protozoa (Hollowday, 1949).

Fungi

Garner (1937) records *Macrotrachela fungicola* Garner living in the gelatinous, orange-yellow fungus *Dacrymyces deliquescens* on decayed wood, in Britain. *M. fungicola* is probably an ectoparasite of this fungus, but the exact relationship between the species is unclear.

Plantae

Bryophyta

Although bdelloid rotifers are often found living among mosses, few could be thought of as parasites. *Habrotrocha roeperi* (Milne) and *Habrotrocha reclusa* (Milne) may be the exceptions, as they actually live inside the outer cells of submerged branches of *Sphagnum* (Milne, 1888; Bartos,

1951). Whether they simply live inside the empty cells, filtering food from the surrounding water (Milne, 1888), or whether they are true parasites, is unknown.

Animalia

Porifera
Several rotifer species have been found living in close association with the freshwater sponge *Spongilla lacustris* in lakes of southern Sweden (Berzins, 1950). The first, *Ptygura melicerta* (Ehrb.) does not seem to injure the sponge in any way and appears to have a commensal association with its host. This species is invariably found around the pores and oscula of the sponge, taking advantage of the increased food supply carried in the water currents generated by its host. However, *Lecane clara* (Bryce) and *Lepadella triba* Myers, do appear to be true parasites. They graze on the sponge itself, leaving furrows up to 50 μm deep in the surface tissues. Neither of these species appears to be an obligate parasite, as both are also found swimming freely and have no particular adaptation to a parasitic way of life.

Rotifera
Records of rotifers parasitising other rotifer species are rare. However, it has been noted that *Acyclus inquietus* Leidy is parasitic on colonies of *Sinatherina socialis* (Linne) (Koste, 1968), while *Proales decipiens* (Ehrb.) attacks *Stephanoceros fimbriatus* (Goldfusz) (Stevens, 1907). In the former case, the parasite feeds on the newly hatched young of its host. In the latter, *P. decipiens*, only one tenth the size of *Stephanoceros*, enters the protective tube and feeds on the adult and its developing eggs. The parasite lays its own eggs within the tube and these subsequently hatch, and leave in search of new hosts.

Annelida
Segmented worms and leeches have both ecto- and endo-parasitic rotifers. The ecto-parasites live on the epidermis of the host, feeding on the epidermal cells and piercing the skin with their trophi to suck out the body fluids. Among these parasites are *Cephalodella parasitica* (Jennings), found on a variety of freshwater oligochaetes (e.g. *Vejdovkyella comata*, *Stylaria lacustris*, *Chaetogaster* spp., *Nais* spp.) (Jennings, 1894; Beauchamp, 1905; Koste, 1972, 1978), and *Drilophaga bucephalus* Vejdovsky, found on oligochaetes (e.g. *Lumbriculus variegatus*, *Rhynchelmis* sp., *Stylodrilus* sp., *Nais elinguis*) (Vejdovsky, 1883; Koste, 1978) and the body and posterior sucker of leeches (e.g. *Herpobdella octoculata*, *H. nigricollis*, *H. testacea* and *Hirudo medicinalis*) (Pawlowski, 1935; Koste, 1978). Murray (1906) also records *Proales daphnicola* as living epizoically on oligochaetes. However, his identification of the rotifer species in thought to be incorrect (Dr W Koste, pers. comm.).

Endoparasitic rotifers of worms were originally discovered by Dujardin (1838) who found *Albertia vermisculus* in the expressed body fluids of earthworms and slugs. The parasites live in the gut of their host and most belong to the genera *Albertia* (*A. vermisculus*, *A. crystallina* Schültze, *A. naidis* Bousfield, *A. reichelti* Koste, *A. bernhardi* Hlava) and *Balatro* (*B. calvus* Claparède, *B. fridericiae* Kunst, *B. anguiformis* Issel, *B. aciliatus* (Radkevitsch)). The hosts of *Albertia* species include the freshwater oligochaetes *Stylaria* spp. (e.g. *S. proboscidae*, *S. lacustris*) (Levander, 1894; Hlava, 1905; Murray, 1906; Koste, 1969, 1970), *Nais* species (Hudson & Gosse, 1886; Bilfinger, 1894; Koste, 1969, 1978), *Paranais litoralis*, *Ripistis parasita* (Koste, 1978), and the earthworm *Allolobophora caliginosa* (Rees, 1960), while *Balatro* species are more commonly found in the guts of *Henlea ventriculosa*, *H. perpusila*, *Fridericia bulbosa*, *F. perrieri*, *Trichodrilus* sp., *Buchholzia appendiculata*, *Enchytraeus buchholzi* and *E. albidus* (Claparède, 1868; Hudson & Gosse, 1886; Koste, 1978). The rotifers attach themselves firmly to the intestine of their host by grasping, with their trophi, the small papillae of the intestinal mucosa (Rees, 1960). A firm attachment is essential to prevent the rotifer being dislodged by the peristaltic movements of the intestine, and the pressure of food moving along the gut. A detailed study of *Albertia naidis* in the gut of *Nais elinguis*

62

has shown that the rotifer feeds on cells from the intestine wall of its host and does not ingest any of the algal remains passing through the worms digestive tract (Coineau & Kunst, 1964).

Mollusca

One of the first records of rotifers parasitising freshwater snails was published by Glascott (1893), who observed the rotifer *Notommata gigantea* Glascott, inside the eggs of the common water snail. Stevens (1912) also observed this rotifer in the eggs of *Limnaea auricularia* L. and *Paludina vivipara* L., although he concluded that it belonged to another genus and renamed it *Proales gigantea* (Glascott). Stevens (1912) described the rotifer in detail, including an interesting account of its life history. The adult nibbles through the outer shell of the snail egg, wriggles through the hole and enters the shell. It then begins to feed on the fluid surrounding the snail embryo, killing the embryo within a few days. The adult rotifer lays up to 13 eggs, one at a time, inside the snail egg. She then moves on to parasitise a new egg, or dies and is eaten by her progeny. The developing eggs hatch and the first to hatch feed on the remains of the snail egg before escaping to parasitise another egg. Those which hatch later tend to starve. *P. gigantea* seems to be a true parasite, spending most of its life cycle inside the snail egg. It was only found outside an egg for short periods as it travelled through the jelly-like substance of the egg mass to parasitise another egg.

Similar observations were made by Giard (1908) & Nekrassow (1928), who also noticed that the eggs of several other snail species (e.g. *Lymnaea stagnalis* L., *L. auricularia* L., *L. ovata* Drap., *Myxas glutinosa* Müll., and *Physa fontinalis* L.) were similarly parasitised by *P. gigantea*. Nekrassow (1928) also noted that *L. stagnalis* eggs were parasitised less often than those of other species, and suggested that this may be because the outer shell of its eggs is tough and difficult to penetrate.

These early observations on *Proales gigantea* appear to be in contrast to the later findings of Boray (1964), from ecological studies of *Lymnaea*

tomentosa (Pfeiffer) in Australia. This author concluded that the many rotifer species found in the natural habitats, and in laboratory cultures, of these snails had no harmful effects. Had he looked more closely at the egg masses, his conclusions may well have been different.

Freshwater snails (e.g. *Lymnaea tomentosa*, *Biomphalaria alexandrina*, *B. galbrata*, *B. pfeifferi*, *Bulinus truncatus*) are of particular interest to researchers because of their associations with parasitic infestations such as schistosomiasis, bilharzia, and liver flukes. Many workers in this area have complained of infestations of rotifers in their experimental cultures which have caused a variety of problems (Stirewalt & Lewis, 1980; Hassan *et al.*, 1985; J.M. Jewsbury, pers. comm.; G.J. Greer, pers. comm.). Stirewalt & Lewis (1980), for example, describe how *Rotaria rotatoria* Pallas and *Philodina acuticornis* Murray colonize the snail shells, beginning at the centre of the whorl and developing over the entire surface. A serious infestation can result in snails looking as though they are 'wearing fur coats' (J.M. Jewsbury, pers. comm.). These rotifers seem to live commensally on the adult snails, but, like *Proales gigantea*, their main effect is on the egg masses. Hassan *et al.* (1985) have recorded up to 100% mortality among young snails (up to the blastula stage) in rotifer-infested cultures. The reason for this is unclear, but the authors suggest that it may be due to an irritating effect of the rotifer, or to the effect of its excretory products on the ootheca of the egg mass and on the embryo it contains. These results have led the authors to propose that rotifers could be used in the biological control of these snails.

A further effect of parasitic rotifers on disease-spreading snails is described by Stirewalt & Lewis (1980). They found that, in their studies of *Biomphalaria galbrata* infected with *Schistosoma mansoni*, cercariae production fell from 3123 cercariae/snail/day to 591 cercariae/snail/day when their cultures became infested with *Rotaria rotatoria* and *Philodina acuticornis*. They also found that the cercariae which did emerge showed reduced motility and infectivity. This effect occurred not only in the presence of rotifers, but also in rotifer-conditioned water from which the rotifers themselves had been removed. The swimming activity (speed and distance) of the cercariae was greatly depressed and their ability to penetrate the skin of their potential hosts (mice) was reduced by 50%, compared to the penetration rate of 'non-rotifer cercariae'.

Bryozoa

Philodina megalotrocha Ehrb. is the only rotifer which has been recorded as living on Bryozoa. This species was found in clusters around the upper parts of *Lophopus crystallinus* by Pittock (1894). The authors notes that even newly hatched young polypides were heavily infested. Colonies of this rotifer seemed to establish themselves on the shoulders of the bryozoan in order to take advantage of the plentiful food supply drawn into the area by the bryozoan's feeding activity; a colony of *Lophopus* creates a strong vortex in the surrounding water which brings food particles towards it. There was no evidence that *Philodina* was actually parasitic on the *Lophopus* itself, but the rotifer seemed to gain considerable benefit from the association (Pittock, 1894).

Crustacea

Epizoic and parasitic rotifers have often been found on marine and freshwater Crustacea. They are particularly common among freshwater Cladocera, where some rotifer species occur on a wide variety of hosts. *Brachionus sessilis* Varga has been found on *Diaphonosoma excisum* Sars in India and Sri Lanka (Chengelath *et al.*, 1973; Sharma, 1979), on *Diaphanosoma sarsii* Riahrd in India (Sharma, 1979) and on *Diaphanosoma brachyurum* Lieven in Hungary (Varga, 1931); this rotifer species has always been found on the same genus of cladoceran over a wide geographical area (Chengelath *et al.*, 1973). In contrast, *Brachionus rubens* is more widely distributed among the Cladocera. Although mostly found on *Daphnia* species (e.g. *D. magna*, *D. pulex*, *D. longispina*) (Bryce, 1924; Galliford, 1946; Viaud, 1947; Halbach, 1973; Koste, 1978; Schlüter, 1984), it has also been recorded on the carapace of *Moina rectirostris* and *Polyphemus pediculus* (Koste, 1978). The rotifer is attracted towards a *Daphnia* by the

water currents generated during swimming and filter feeding (Viaud, 1947). The rotifer attaches itself by its foot to the carapace of its host, and apparently benefits from this association by feeding on the excrement of the cladoceran (Schlüter, 1984).

Another rotifer which is often found on the carapace of *Daphnia* is *Proales daphnicola* (Thompson, 1892; Harring & Myers, 1922; Beauchamp, 1923; Galliford, 1946; Hollowday, 1949; Koste, 1968). This rotifer attaches itself firmly to its host by means of its pedal cement glands; it also fixes its eggs to the carapace. *P. daphnicola* appears to gain nothing from the association, apart from transportation (Hollowday, 1949). Other rotifers which are also epizoic on Cladocera, but occur less frequently, include *Testudinella epicopta* Myers on *Acantholeberis curvirostris* (Myers, 1934), *Brachionus variabilis* (Hempel) on *Daphnia longispina* and *Ceriodaphnia longispina* (Hempel, 1896; Ahlstrom, 1940), *Collotheca volutata* var. *sessilis* Sebestyen on *Monospilus dispar* (Sebestyen, 1937) and *Brachionus novaezealandiae* (Morris) on *Pseudomonia lemnae*, *Daphnia carinata* and *Ceriodaphnia* species.

Rotifers also occur as ectoparasites on species of the marine opossum shrimp *Nebalia*. Koste (1975) found *Seison annulatus* Claus among the thoracic appendages which the shrimp uses to filter organic detritus. The rotifer also attaches its eggs to the base of the gills of its host. *Paraseison nudus* Plate, *P. proboscideus* Plate and *P. ciliatus* Plate have also been found attached to the branchial lamellae of *Nebalia* species (Plate, 1888). These rotifers attach themselves by means of mucus secreted from their pedal gland.

Rotifers have been found on the thoracic and abdominal appendages, and the branchial plates, of the freshwater louse *Asellus aquaticus* (Giglioli, 1863; Hudson & Gosse, 1886; Bilfinger, 1894; Hood, 1895; Rousselet, 1898; Murray, 1906; Hofsten, 1909; Galliford, 1946; Hollowday, 1949; Bartos, 1951). Most of these rotifers belong to 3 main genera, namely *Rotaria* (*R. socialis* Kellicott, *R. magna-calcarata* Parsons), *Testudinella* (*T. elliptica* (Ehr.), *T. truncata* (Gosse), *T. caeca* (Parsons)) and *Embata* (*E. parasitica*

(Giglioli), *E. laticeps* (Murray), *E. commensalis* (Western)). These rotifers do not appear to cause any harm to their host and are generally considered to be epizoans, gaining nothing but transportation from their association with *Asellus*.

The branchial plates and appendages of *Gammarus pulex* are another common site for epizoic rotifers (Gigliolo, 1863; Hudson & Gosse, 1886; Plate, 1886; Hood, 1895; Murray, 1906; Varga, 1931; Bartos, 1951; Koste, 1968, 1978). These are mostly of the genus *Embata* (*E. parasitica* (Giglioli), *E. laticeps* (Murray), *E. hamata*), although *Proales daphnicola*, *Philodina convergens* Murray, *Dicranophorus hauerianus* var. *siedleckii* Wiszniewski and *Encentrum grande* (Western) have also been found. Again, none of these species seems to have adverse effects on the host.

A variety of rotifer species are commonly found in the branchial cavities of the freshwater crayfish, *Astacus* species (Carlin, 1939; Koste, 1978). These include *Lepadella astacicola* Hauer, *L. branchiola* Hauer, *L. parasitica* Hauer, *L. borealis* Harring, L. Wiszniewski, *Cephalodella crassipes* (Lord) and several varieties of *Dicranophorus hauerianus* Wiszniewski (Carlin, 1939; Koste, 1978). *L. borealis*, *L. astacicola*, *L. lata* and *D. hauerianus* have also been found in the gill cavity of the crayfish *Cambarus affinis* (Carlin, 1939; Wulfert, 1957; Koste, 1978). Little is known of the relationship between these species, but the rotifers may well be true parasites of these Crustacea.

Rotifers have also been found in the branchial chambers of both freshwater and marine crabs. Piovanelli (1903) describes two bdelloids, *Macrotrachela cancrophila* (Piovanelli) and *Anomopus telphusae* (Piovanelli), from the branchial chamber of *Telphusa fluviatilis*, while Mané-Garzon and Montero (1973) record another, *Anomopus chasmagnathi* (Mané-Garzon & Montero), from the branchial chamber of *Chasmagnathus granulata*. In addition, *Proales paguri* Thane-Fenchel was found on the gills of the hermit crab *Pagurus bernhardus* (L.) (Thane-Fenchel, 1966). In the latter case, the rotifers were seen browsing on the gills and sucking the epithelium tissue of their host. When lightly compressed under a cover-glass, the rotifers often expelled epithelial cells ingested from the hosts gills. These observations indicated a parasitic association between the species. The type of association between the species mentioned earlier is not known.

Uniramia

Many rotifers live epizoically on the aquatic larvae of insects, especially those of the dragonfly (*Rotaria rotatoria*, *Brachionus rubens*, *B. caudatus* Barrois) (Chandra & Kameswara, 1976; Sharma, 1979), damselfly (*Lepadella ovalis* Müller, *B. caudatus*) (Chandra & Kameswara, 1976; Sharma, 1979) and mayfly (*Philodina convergens*, Murray) (Bartos, 1951). Epizoic rotifers have also been found on the cases of caddis larvae (*Rotaria tardigrada* Ehr., *Embata parasitica*, *Habrotrocha collaris* Ehrb.) (Scherren, 1897; Bartos, 1951) and attached to the body and legs of the waterboatman, *Corixa* sp. (*Albertia naidis* Bousfield, *B. rubens*) (Varga, 1931; Koste, 1978). Little is known of the relationship between these rotifers and their hosts, but the associations are not thought to be truly parasitic.

Echinodermata

Zelinkiella synaptae (Zelinka) has been found in great abundance living in the body cavities of the echinoderms *Synapta inhoerens* and *S. digitata* (Lankester, 1868 (?); Zelinka, 1888; Cuènot, 1892). No detailed information of this relationship is given, but Cuènot (1892) suggests that *Zelinkiella* is commensal on the integument of the Echinoderm, and is unlikely to be a true parasite.

Chordata

An extensive search of the literature revealed only one record of a rotifer which was either parasitic or epizoic on fish. This was described by Wiszniewski (1946), who found the cold stenothermal species *Encentrum kozminskii* Wiszniewski on the skin and gills of carp from fish culture ponds. The author describes the species as a true parasite which feeds on the mucus, and probably the epithelium, of its host. He concludes that such an infestation could be harmful to the fish, but this has not been confirmed.

Rotifers have rarely been found living in close association with reptiles. However, Magnusson (1985) found the giant (up to 1500 μm) rotifer *Limnias ceratophylli* Schrank living on the trunk, limbs, tail and jaws of the Amazonian crocodile, *Melanosuchus niger*. The author suggests that this association is species specific, as the rotifer was not found on other crocodile species in the area (*Caiman crocodilus*, *Paleosuchus trigonatus*, *P. palpebrosus*). *M. niger* did not seem to suffer any ill effects from the association, indicating that the rotifer was simply epizoic on the crocodile. However, the author did express the concern that heavy infestations could reduce the value of the crocodile hide.

Giesen (1934) published what appears to be the only recorded case of a parasitic rotifer in mammals. He found a considerable number of bdelloids in urine samples from a seriously ill woman whose symptoms included sore throat, aching joints, stomach disorders, exhaustion, semi-coma and an enormous increase in blood leucocytes. Although Giesen (1934) suggests that this is 'an apparent case of a pathogenic rotifer in man', this has never been confirmed.

Discussion

Many rotifers live in close association with plants and other animals, but few are true parasites and actually harm their hosts. Most simply take advantage of increased food supplies and transportation provided by their host. However, some species are truly parasitic, feeding on the tissues, mucus or body fluids of their host. This is especially true of the parasites of worms and leeches, but is also occurs among algae, sponges, crustaceans, and fish. The damage to these hosts is rarely fatal, in contrast to the situation in snails and rotifers. Few attempts have been made to quantify effects of parasitic rotifers on host populations, although these effects could be considerable.

Epizoic and parasitic rotifers are found in freshwater and marine environments, and in sufficiently humid terrestrial environments. They belong to a variety of genera, both loricate and illoricate, within the Bdelloidea, Monogononta and Seisonidae. The relationships between these rotifers and their hosts may be specific or non-specific, obligate or opportunistic. Their method of attachment varies – some are attached by the foot, others by the mouthparts. However, little is known of the relationships between the species involved. Clearly, this interesting but neglected area of rotifer research warrants further attention.

Acknowledgements

I wish to thank Dr. A. E. Bailey-Watts for reading and improving the manuscript, Mrs. L. A. Dickson for help in obtaining the literature and Mrs. M. J. Ferguson who kindly typed the manuscript.

References

Ahlstrom, E. H., 1940. A revision of the Rotatorian genera *Brachionus* and *Platyias* with descriptions of one new species and two new varieties. Bull. amer. Mus. nat. Hist. 77: 148–184.

Barnes, R. S. K., 1984. A Synoptic Classification of Living Organisms. Blackwell Scientific Publications, Oxford, 273 pp.

Bartos, E., 1951. The Czechoslovak Rotifers of the order Bdelloidea. Vestnik. Cs. zool. spol. 15: 241–500.

Beauchamp, P. M. de, 1905. Remarques sur deux Rotifères parasites. Bull. Soc. zool. Fr. 30: 117–124.

Beauchamp, P. de, 1923. Courtes notes sur les Rotifères. Ann. Biol. lac. 12: 221–228.

Berzins, B., 1950. Observations on rotifers on sponges. Trans. am. microsc. Soc. 69: 189–193.

Bilfinger, L., 1894. Zur Rotatorien fauna Württembergs. 2. Beitrag. Jahresh. d. Verf. f. vaterländ. Naturk. Württembergs 50: 35–65.

Boray, J. C., 1964. Studies on the ecology of *Lymnaea tomentosa*, the intermediate host of *Fasciola hepatica*. I. History, geographical distribution and environment. Aust. J. Zool. 12: 217–230.

Brain, J. L., 1894. An inhabitant of *Vaucheria*. Science Gossip 1: 201–202.

Bryce, D., 1924. The Rotifera and Gastrotricha of Devil's and Stump Lakes, North Dakota, U.S.A. J. Quekett microsc. Club, Ser. 2, 15: 81–108.

Budde, E., 1925. Die parasitischen Rädertiere mit besonderer Berücksichtigung der in der umgegend von minden I.W. beobachteten Arten. Z. Morph. u. Ökol. Tiere 3: 706–785.

66

Carlin, B., 1939. Über die Rotatorien einiger Seen bei Aneboda. Medd. Lunds Univ. Limnol. Inst. 2: 3–68.

Chandra, M. P. & R. R. Kameswara, 1976. Epizoic rotifers observed on Odonata nymphs from Visakhapatnam. Science Cult. 42: 527–528.

Chengalath, R., C. H. Fernando & W. Koste, 1973. Rotifers from Sri Lanka (Ceylon). II. Further studies on the eurotatoria including new records. Bull. Fish. Res. Stat. Ceylon 24: 29–62.

Christensen, T., 1987. Some collections of Vaucheria (Tribophyceae) from south-eastern Australia. Aust. J. Bot. 35: 617–629.

Claparède, M., 1868. A new rotifer Balatro calvus. Quart. J. Microsc. Science 8: 170–172.

Cuènot, L., 1892. Commensaux et parasites des Echinodermes. Revue Biol. Nord-France, 5.

Coineau, Y. & M. Kunst, 1964. Une nouvelle espèce de rotifère parasite d'oligochète: Albertia soyeri n.sp. Vie Milieu 15: 1007–1015.

Davis, J. S. & W. F. Gworek, 1973. A rotifer parasitizing Vaucheria in a Florida spring. Trans. am. microsc. Soc. 92: 135–140.

Dujardin, F., 1838. Mémoire sur un ver parasite constituant un nouveau genre voisin des rotifères, sur le tardigrade, et sur les systolides on rotateurs en général. Ann. sci. natur., ser. 2, 10, Zool.: 175–191.

Edmondson, W. T., 1940. Sessile Rotatoria of Wisconsin. Trans. amer. Microsc. Soc. 59: 433–559.

Edmondson, W. T. & G. E. Hutchinson, 1934. Report on Rotatoria Yale North India Expedition. Art. IX. Mem. Connecticut Acad. Arts. Sci. 10: 153–186.

Galliford, A. L., 1946. A contribution to the Rotifer fauna of the Liverpool Area. Proc. Liverpool. Natur. Field Club 1945.

Ganf, G. G., R. J. Shiel & C. J. Merrick, 1983. Parasitism: The possible cause of the collapse of a Volvox population in Mount Bold Reservoirs, South Australia. Aust. J. mar. Freshwat. Res. 34: 489–494.

Garner, W. E., 1937. Two new species of Rotatoria. J. Quekett Microsc. Club, Series 3, 1.

Giard, A., 1908. Un nouveau rotifère, parasite des pontes de mollusques d'eau dance. Feuiles des jeunes naturalistes 38: 184.

Giesen, J., 1934. An apparent case of a pathogenic rotifer (Order Bdelloidea) in man. J. Parasitol. 20: 133.

Giglioli, H., 1863. On the genus Callidina (Ehr.); with the description and anatomy of a new species. Quart. J. Micr. Science London, New Ser., Vol. 3: 237–242.

Glascott, L. S., 1893. A list of some of the Rotifera of Ireland. Sciproc. Roy. Dublin Soc. B: 29–86.

Gosse, P. H., 1852. On the Notommata parasita Ehrbg., a rotiferous animal inhabiting the spheres of Volvox globator. Trans. microsc. Soc. London 3:

Halbach, U., 1973. Quantitative studies of rotifer associations in ponds. Arch. Hydrobiol. 71: 233–254.

Harring, H. K., 1913. Synopsis of the Rotatoria. Bull., U.S. Nat. Mus., Washington, 81: 7–226.

Harring, H. K. & F. J. Myers, 1922. The rotifer Fauna of Wisconsin. Trans. Wis. Acad. Sci. Arts Lett. 20: 553–662.

Hassan, A. A., A. M. S. El-Ridi & L. A. A. Magd, 1985. An approach to biological control of snails by rotifers. J. Egypt. Soc. Parasitol. 15: 553–558.

Hempel, A., 1896. Description of new species of Rotifers and Protozoa from the Illinois River and adjacent waters. Bull. Illinois State Lab. Nat. Hist. 4: 310–317.

Hlava, S., 1905. Über eine neue Rädertierart aus der Gattung Albertia. Zool. Anz. 28: 365–368.

Hofsten, von N., 1909. Rotatorien aus den Mästermyr (Gottland) und einigen andern Schwedischen Binnengewässern. Ark. Zool. 6, 125 pp.

Holloway, E. D., 1949. Introduction to the study of the Rotifera – IX Proales daphnicola Thompson: with reference to commensal and parasitic habits. The Microscope 6: 1–7.

Hood, J., 1895. On the Rotifera of the County Mayo. Proc. Roy. Irish Acad., ser. 3: 664–706.

Hudson, C. T. & P. H. Gosse, 1886. The Rotifera or Wheel-Animalcules, both British and Foreign. Longmans, London. I: I-VI & 1–128; II: 1–144.

Jennings, H. S., 1894. A list of the Rotatoria of the Great Lakes and some of the inland lakes of Michigan. Bull. Michigan Fish. Comm., Landsing 3: 1–34.

Koste, W., 1968. Über Proales sigmoidea (Skorikow) 1896 (eine für Mitteleuropa neue Rotatorienart) und Proales daphnicola (Thompson) 1892. Arch. Hydrobiol. 65: 240–245.

Koste, W., 1969. Parasitic Rotifera Albertia naidis. Mikrokosmos 58: 212–216.

Koste, W., 1970. Über eine parasitische Rotatorienart Albertia reichelti nov. spec. Zool. Anz. 184: 428–434.

Koste, W., 1972. Portrait of Rotifera a rare external parasite on freshwater Oligochaeta Cephalodella parasitica. Mikrokosmos 61: 10–12.

Koste, W., 1975. Seison annulatus, ein Ektoparasit des marinen Krebses Nebalia. Mikrokosmos 64: 341–347.

Koste, W., 1978. Rotatoria. Borntraeger, Berlin, 2 vol.: 673 pp., 234 plates.

Kutikova, L. A., 1970. Rädertierfauna der USSR. Fauna USSR, 104, Akad. Nauk. SSSR, Leningrad: 1–744 (In Russian).

Lankester, E. R., 1868. Note on the Synapte of Guernsey and a new parasitic rotifer. Quart. J. microsc. science 8: 53–55.

Levander, K. M., 1894. Materialien zur Kenntnis der Wasserfauna in der Umgebung von Helsingfors. Acta. soc. fauna et flora fennica. Helsingfors 12: 1–72.

Lister, 1884. On the parasitism of rotifers in cysts on Vaucheria. Proc. Essex Nat. Field Club 3: 45–48.

Magnusson, W. E., 1985. Habitat selection parasites and injuries in Amazonian Crocodilians. Amazoniana 9: 193–204.

Mané-Garzon, F. & R. Montero, 1973. A new species of Rotifers Bdelloidea Anomopus chasmagnathi (n.sp.) from the branchial chamber of the tidal crab Chasmagnathus granulata Decapoda Brachyura. Rev. Biol. Urug. 1: 139–144.

Milne, E. 1888. Rotifer as a parasite or tube dweller. Proc. Phil. Soc. Glasgow 20: 48–53.

Murray, J., 1906. The Rotifera of Scottish Lochs. Trans. r. Soc. Edinb. 45: 151–193.

Myers, F. J., 1934. The distribution of Rotifers on Mount Desert Island. VII. Amer. Mus. Mov. 761: 1–8.

Nekrassow, A. D., 1928. Vergleichende morphologie der Laiche von süsswasser gastropoden. Z. Morph. Okol. Tiere 12: 1–35.

Ott, D. W., 1977. Ultrastructural observations of parasitism of Vaucheria prona by Poales werneckii. J. Phycol. 13 (suppl.).

Pawlowski, L. K., 1935. Beiträge zur Anatomie und Biologie von Drilophaga delagei de Beauchamp. Arch. Hydrobiol. Rybactwa, Suwalki 9: 1–30.

Penard, E., 1914. A propos de Rotifères. Rev. Suisse 22: 1–25.

Piovanelli, S., 1903. Two new Bdelloidea commensal in the branchial cavity of Telphusa fluviatilis. J. Quekett Microsc. Club 8: 521–522.

Pittock, G. M., 1894. Rotifer-hunting in Minster Marshes, Thanet. Science Gossip 1: 173–175.

Plate, L. H., 1886. Untersuchungen einiger an der Kiemenblättern des Gammarus pulex lebenden Ektoparasiten. Z. wiss. Zool. 43: 175–241.

Plate, L. H., 1888. On some ectoparasitic Rotatoria of the Bay of Naples. Ann. Mag. Nat. Hist., London, ser. 6, 2: 86–112.

Rees, B., 1960. Albertia vermisculus (Rotifer) parasitic in the earthworm Allolobophora caliginosa. Parasitol. 50: 61–65.

Rich, F. & M. A. Pocock 1933. Observations on the genus Volvox in Africa. Ann. s. afr. Mus. 16: 427–471.

Rousselet, C. F., 1898. Notes on some little-known species of Pterodina. J. Quekett microsc. Club, Ser. 2, 7: 24–30.

Rousselet, C. F., 1911. Rotifera (excluding Bdelloidea). Proc. Royal Irish Acad. 31, No. 51, 10 pp.

Rousselet, C. F., 1914. Intelligence in parasitic rotifers. Knowledge 37: 191, 270–273.

Sauer, F., 1978. A rotifera as parasite in Volvox. Mikrokosmos 67: 110–111.

Scherren, H., 1897. Rotifers commensal with Caddis worms. Nature 56: 224.

Schlüter, M., 1984. Kontinuierliche Massenkultur des planktischen Rotators Brachionus rubens. Ber. Kernforschungsanlage Julich 1959, 195 pp.

Sebestyen, O., 1937. Investigations on an epizoic Collotheca (Rotifera). Ann. Inst. biol. Tihany 24: 183–192.

Sharma, B. K., 1979. On some epizoic rotifers from West Bengal. Bull. zool. Surv. India 2: 109–110.

Stevens, J., 1907. The Rotifera of the Exeter district. Proc. Coll. Field Club and Nat. Hist. Soc., Exeter, 1907: 30–52.

Stevens, J., 1912. Notes on Proales (Notommata) gigantea Glascott, a rotifer parasitic in the egg of the water-snail. J. Quekett. microsc. Club, Ser. 2, 11: 481–486.

Stirewalt, M. & F. A. Lewis, 1980. Schistosoma mansoni: Effects of rotifers on cercarial output, mortality and infectivity. Int. J. Parasitol., 11: 301–303.

Thane-Fenchel, A., 1966. Proales paguri sp. nov., a rotifer living on the gills of the hermit crab, Pagurus bernhardus. Ophelia 3: 93–97.

Thompson, P. G., 1892. Notes on the parasitic tendency of rotifers of the genus Proales, with an account of a new species. Science Gossip 28: 219–221.

Varga, L., 1931. Beiträge zur Rotatorienfauna Südschwedens. Zool. Anz. 96: 285–292.

Vejdovsky, F., 1883. Über Drilophaga bucephalus, ein parasitisches Rädertier Sitz. – Ber. K. Böhm. Ges. Wiss. Prag. (1882): 390–397.

Viaud, G., 1947. Recherches expérimentales sur les tropismes des Rotifères. L'oscillorhéotropisme des Brachionus rubens Ehrenberg, cause de la fixation de ce Rotifère phorétique sur les Daphnies et autres crustacés d'eau douce. Ann. Sci. nat. Zool. 9: 39–62.

Williams, J., 1852. On the occurrence of parasitic Rotifera in Volvox globator. Trans. microsc. Soc. London, 3: 129–131.

Wiszniewski, J., 1946. Sur un rotifère, parasite des carpes. Zool. pol. 4: 7–10.

Wollny, R., 1879. Parasitism of Notommata on Vaucheria. J. r. microsc. Soc. 2: 291.

Wulfert, K., 1957. Ein neues Rädertiere aus der Kiemenhöhle von Cambarus affinis. Zool. Anz. 158: 26–30.

Wulfert, K., 1960. Die Rädertiere Saurer Gewässer der Dubener Heide. II Die Rotatorien des Krebsscherentümpels Bei Winkelmühle. Arch. Hydrobiol. 56: 311–333.

Zelinka, C., 1888. Studien über Rädertiere. II. Raumparasitismus und Anatomie von Discopus synaptae. Z. f. wiss. Zool. 47.

Hydrobiologia **186/187**: 69–73, 1989.
C. Ricci, T. W. Snell and C. E. King (eds), Rotifer Symposium V.
© 1989 *Kluwer Academic Publishers.*

Interrelations of rotifers with predatory and herbivorous Cladocera: a review of Russian works

Lilian K. Matveeva

A.N. Severtsov Institute of Animal Evolutionary Morphology and Ecology, USSR Academy of Sciences, Leninsky prospekt 33, Moscow 117071, USSR

Key words: rotifers, cladocerans, epibionts, predation, *Polyphemus*

Abstract

Publications in Russian on the influence of epibiontic rotifers on cladocerans as well as on predator-prey interrelations between cladocerans and rotifers are reviewed. *Proales daphnicola* and *Brachionus rubens* are common epibionts of Cladocera. The biology of these species is described including the choice of host, feeding, reproduction and impact on the host. Representatives of the families Daphniidae and Moinidae are most readily colonized. At high densities of epibiontic rotifers, a high percentage of young Cladocera die. Predators consuming cladocerans (mainly Bosminidae) belong to the family Asplanchnidae. Rotifers are consumed by the predatory cladocerans *Leptodora* and *Polyphemus*. The results of functional response experiments by the author with *P. pediculus* feeding on *Synchaeta pectinata*, *Asplanchna priodonta*, colonial and solitary *Conochilus unicornis*, and *Platyias patulus* are given. In the range of prey densities of $100–1600 \ 1^{-1}$ a functional response was found in all the rotifers except *P. patulus*. Colonies of *C. unicornis* were not consumed. The highest level of feeding rate saturation was observed in *Synchaeta pectinata*. Various adaptations in prey morphology prevented effective predation: coloniality, large size of the prey and hard lorica with spines.

Introduction

Interrelationships between rotifers and cladocerans have drawn increasing attention in recent years. There are several investigations on this topic in Russian that are poorly known abroad because of linguistic difficulties. The aim of this paper is to review these works. I also present results of my own experiments on the feeding of *Polyphemus pediculus* on rotifers.

Rotifers as epibionts of Cladocera

Two species of epibiotic rotifers living on Cladocera are known from the Soviet literature,

these being *Proales daphnicola* and *Brachionus rubens*. *P. daphnicola* is an obligatory commensal of Cladocera (Vlastov, 1953a, b). It occurs most often on *Daphnia pulex* and *D. longispina* and, much more rarely, on *Scapholeberis*, *Ceriodaphnia* and *Polyphemus*. *Proales* feeds, breeds and develops on cladocerans. According to Vlastov (1956b), it feeds on the unicellular euglenacean alga *Colacium vesiculosum* common in epibioses of cladocerans and, to a lesser extent, on bacteria. The rotifers cannot live long or propagate in the absence of actively swimming cladocerans. Most eggs (65.5%) laid by *Proales* are attached to the head of cladocerans on their dorsal side. At a temperature of 21–22 °C, eggs developed in

20–21 hours. The mortality of *Daphnia* caused by *Proales* in nature was not estimated; however, movements of a young *Daphnia* are disrupted by as few as two attached rotifers.

B. rubens exploits crustaceans only as a substratum; it collects its food in mid-water. *Brachionus* firmly adheres to the carapace of crustaceans using an adhesive secretion produced by its two pedal glands. This secretion collects detritus particles, in 3–5 hours becoming as thick as 135 µm. Other *Brachionus* specimens do not settle onto the surface covered with detritus. Having molted, an old crustacean swims freely for some minutes and then again becomes covered with rotifers. *Brachionus* are most numerous on *Daphnia* one hour after molting and grow less numerous in 5 hours (Rivier & Markevich, 1978).

Selectivity of *Brachionus* colonizing substrata has been studied in nature and under laboratory conditions (Markevich, 1978). In a pond they definitely prefer *D. pulex* and *Moina brachiata*. The entire surface of their shells becomes covered with rotifers. On *Simocephalus vetulus* and *Scapholeberis mucronata*, *B. rubens* occurred as single individuals. In an experiment, rotifers were offered 36 species of Crustacea, plants and inanimate objects. At increasing densities of *Brachionus* their selectivity decreased. At a density up to $5 \times 10^4 \, 1^{-1}$ they settled only on *Daphnia* and *Moina*. At over $16 \times 10^4 \, 1^{-1}$ a slime sleeve is formed (threads of rotifers sitting on cladocerans hung down from cladocerans bodies and were used for settlement by other *B. rubens*). At $19 \times 10^4 \, 1^{-1}$, rotifers settled on other Cladocera such as *Ceriodaphnia*, *Polyphemus*, *Bythotrephes*, *Leptodora*, *Sida*, *Diaphanosoma*, *Scapholeberis*, *Simocephalus* and *Bosmina*. Copepoda and Chydoridae, with the exception of *Eurycercus lamellatus*, were not colonized. At higher densities, rotifers settled also on plants and dead objects. The intensity of settling on *D. pulex* depended directly on the density of *Brachionus* in the environment and, as it increased from 5×10^4 to $43 \times 10^4 \, 1^{-1}$, their quantity on one *Daphnia* of 2 mm increased from 16 to 301. In one case of a sleeve formation, up to 700 rotifers were recorded on one *Daphnia* (Markevich & Rivier, 1975). With intensive settling, the biomass of rotifers on an adult daphnid was equal to the weight of the host. On young daphnids it exceeded the weight of the host by 5–17 times, rendering the crustacean motionless. The impact of rotifers on a crustacean depended on its size. Large daphnids (larger than 2 mm) even fully covered with rotifers and bearing a sleeve, could remain swimming for 32 hours. Even less burdened juveniles could not swim after 1.5–3 hours; newborn daphnids were rendered motionless by 3–4 individuals of *Brachionus* (Markevich & Rivier, 1978). Therefore, populations of *Daphnia* affected by *B. rubens* lacked newborn individuals. Even in large daphnids (2.0–2.5 mm) covered with *Brachionus* the rate of movement dropped by 2 times and the character of movement was modified (fewer jumps, shorter races, more variable direction). Dead crustaceans were rapidly abandoned by rotifers; in 1.5 hours the rotifer number decreased by 5–7 times.

The joint cultivation of *Daphnia* and *Brachionus* demonstrated that in pond water *Daphnia* survival rate, growth and fecundity were lower in the presence of epibionts (Markevich & Rivier, 1975; Rivier & Markevich, 1978). If fresh organic material and detritus were added, the situation reversed. The filtering activity of daphnids apparently was not enough to suppress bacterial growth and to prevent deterioration of environmental conditions.

Rotifera as predators of Cladocera

Most species of the family Asplanchnidae can consume small Bosminidae. *Bosmina* remains are found in stomachs of *A. girodi*, *A. herricki*, *A. sieboldi* and *A. priodonta* (Tribush, 1960, Ghilarov, 1977, Timokhina, 1983). In experimental conditions, *A. priodonta* and *A. herricki* (0.4–0.8 mm long) consumed 3–4 individuals of *B. longirostris* daily. Food assimilation was 15–20% and the assimilation efficiency reached 34–36% (Sorokin & Mordukhai-Boltovskaya, 1962, Monakov & Sorokin, 1972). Adult females of *A. herricki* digested *Bosmina* in 6–6.5 hours at 18 °C and in about 4 hours at 21–22 °C

(Kutikova, 1970). I have noted up to 4 *Bosmina* in the stomach of *A. herricki* simultaneously.

Cladocera as predators of Rotifera

Predaceous cladocerans, such as *Polyphemus* and *Leptodora*, are known to consume rotifers. In experimental conditions, young and adult *L. kindti* readily consumed *A. herricki*, *A. priodonta*, *E. dilatata* and, in lesser quantities, *Conochilus* sp., *K. cochlearis*, *K. quadrata*. However rotifers were not a satisfactory food for *Leptodora* which lived for only 2–12 days when fed a mixture of *A. herricki*, *C. unicornis* and *Volvox* instead of 1.5 months when consuming a better diet (Mordukhai-Boltovskaya, 1960). A study of the gut contents of 200 specimens of *Leptodora* demonstrated that, in the Rybinsk reservoir on May, 8.5% fed on rotifers. In the gut of an adult *Leptodora* collected in nature, *K. cochlearis* was found (Mordukhai-Boltovskaya, 1958). Under experimental conditions, juveniles of *Leptodora* 3.5 mm long consumed 30–40 *B. calyciflorus* daily at a prey density of 800–1300 l^{-1} (L. V. Polishchuk, personal communication).

Rotifers and protozoans are the main food of newborn and adult males of *P. pediculus*. Adult females also prefer ciliates and soft-bodied rotifers rather than crustaceans: prefered prey were *Asplanchna*, *Synchaeta*, *Polyarthra* and in somewhat lesser quantities *Conochilus* colonies (Butorina & Sorokin, 1966; Butorina, 1970). Abrupt jumps of *Filinia* interfere with its consumption by predators, female *Polyphemus* catch this rotifer about half as frequently as *Polyarthra*. Among loricate rotifers, they prefer *Keratella*, consuming it at prey densities of more than 50 l^{-1} (Butorina, 1969, 1970). Nevertheless, for egg-bearing females, larger than 0.4 mm, rotifers are not a completely suitable food. Culturing on this food decreased size and fecundity but did not modify longevity. The maximum assimilation efficiency of *Keratella* at an experimental density of $35 \times 10^4 l^{-1}$ was only 3.6%, while that of newborn Copepoda and Cladocera was 30–45% (Butorina, 1971).

Functional response experiments on Polyphemus

Because most studies of the influence of *Polyphemus* on rotifers have been qualitative, I have conducted several quantitative experiments. The efficiency of feeding of invertebrate predators on rotifers is known to be affected by such prey characteristics as coloniality, body size, the presence of a hard lorica and spines (Williamson, 1983, Stemberger & Gilbert, 1987). Therefore, I chose several species of rotifers as prey which had one or the other of these characters. Females of *Polyphemus* (0.7–1.1 mm in length and with well developed but empty brood chambers) were caught in June-August 1988 in the littoral zone of Lake Glubokoe (Smirnov, 1986) and transfered to clean water for 2–5 hours. Then, they were added to 100 ml vessels filled with 50 μm-filtered lake water containing one of the following prey species: *Synchaeta pectinata*, *Platyias patulus*, solitary or colonial *Conochilus unicornis* or *Asplanchna priodonta*. One predatory animal was added to the first three prey types, 5–7 to the two latter. The rotifers were collected from the littoral zone which they successively dominated during summer. The experiments lasted 3 hours at 21–24 °C. During this period there was no rotifer mortality in the control vessels which lacked predator. *Polyphemus* feeding rate was estimated by the difference in rotifer numbers at the beginning and the end of the experiment. Because the sizes of the polyphemids used in these experiments varied, feeding rates were calculated per unit of dry weight (Dumont *et al.*, 1975).

The increase of feeding rate as a function of prey density was significant for *S. pectinata* ($F(5,10) = 13.495$, $P = 0.0004$), solitary *C. unicornis* ($F(5,7) = 7.084$, $P = 0.0115$) and *A. priodonta* ($F(2,6) = 10.474$, $P = 0.0110$), but insignificant in *P. patulus* ($F(3,8) = 0.097$, $P = 0.959$) (Fig. 1). The functional response curve of *P. pediculus* reached a plateau at a density of about 800 l^{-1} for *S. pectinata* and 300 l^{-1} for *C. unicornis*. No such saturation was found in *A. priodonta*. The low feeding rate for *A. priodonta* as prey (Fig. 1) was probably due to its large size preventing easy capture by the predators. For

72

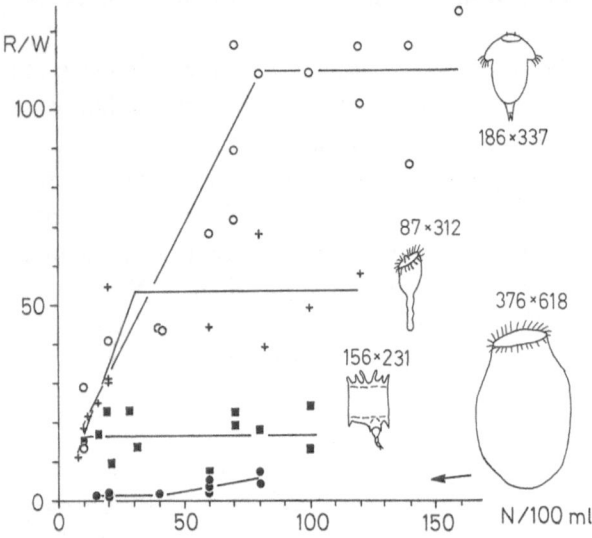

Fig. 1. Functional responses of *Polyphemus pediculus* for different prey. Ordinate: predator feeding rate per unit dry weight (R/W, prey μg^{-1} hr^{-1}); abscissa: density of prey (N, animals/100 ml). Open circles = *Synchaeta pectinata*, filled circles = *Asplanchna priodonta*, squares = *Platyias patulus*, crosses = solitary *Conochilus unicornis*. The averages for two major dimensions of body size (μm) ($n = 20$) are also shown.

example, several experiments with smaller *A. priodonta* (average dimensions 259×446 μm) yielded a 1.5–2 fold increase in the feeding rate of the predator. The plateau was highest in *S. pectinata*, intermediate in solitary *C. unicornis* and lowest in *P. patulus* (Fig. 1). The relatively low density feeding rate saturation in solitary *C. unicornis* was probably due to the fact that these animals either were not a familiar food for the crustacean or were distasteful because of gelatin remnants. Not a single colony of *C. unicornis* was consumed by *P. pediculus* at a density of 10–40 colonies per 100 ml. Poor feeding on *P. patulus* seemed to be due to its hard lorica and well developed spines.

The absolute maximum feeding rates of *P. pediculus* were observed at a prey density of $1–1.2 \times 10^3 l^{-1}$ and were as high as 7 *S. pectinata* $hour^{-1}$ $predator^{-1}$, or 4.7 *C. unicornis* or 1.5 *P. patulus*. Notwithstanding the high rate of *Synchaeta* consumption, *P. pediculus* significantly lowered its intensity of feeding on the rotifer in the presence of equal amounts of an alternative prey

(*Bosmina longirostris*, mean length 232 μm) in the range of densities of 50–1000 prey l^{-1} ($t = 2.355$, $P < 0.05$). Thus, as in other invertebrate predators, the successful feeding of *P. pediculus* was dependent upon the size of prey, the formation of colonies and the development of a hard exoskeleton with spines. Probably in nature this predator most efficiently preys upon populations of soft-bodied rotifers of about 300 μm in length.

Acknowledgements

This work has been done at the hydrobiological Station 'Lake Glubokoe'. I am grateful to Dr. C. Ricci for the arrangements concerning my participation in the 5th International Rotifer Symposium, and to my husband, Dr. V. Matveev, for various help throughout this study. I would like to thank Dr. P. Starkweather for language revision of the manuscript.

References

Butorina, L. G., 1969. O prichinakh obrazovaniya stay u *Polyphemus pediculus* (L.). Informatsionniy bulleten, Institut biologii vnutrennikh vod 3: 68–71 (On causes of swarm formation in *Polyphemus pediculus* (L.). In Russian).

Butorina, L. G., 1970. Ob izbiratelnosti pitaniya *Polyphemus pediculus* (L.). Informatsionniy bulleten, Institut biologii vnutrennikh vod 7: 46–50 (On feeding selectivity of *Polyphemus pediculus* (L.). In Russian).

Butorina, L. G., 1971. Intensivnost pitaniya *Polyphemus pediculus* (L.) v zavisimosti ot kontsentratsii korma. Informatsionniy bulleten, Institut biologii vnutrennikh vod 10: 35–39 (Feeding intensity of *Polyphemus pediculus* (L.) in relation to food concentration. In Russian).

Butorina, L. G. & Yu I. Sorokin, 1966. O pitanii *Polyphemus pediculus* (L.). Trudy Instituta biologii vnutrennikh vod 12(15): 170–174 (On feeding of *Polyphemus pediculus* (L.). In Russian).

Dumont, H. J., I. Van de Velde & S. Dumont, 1975. The dry weight estimate of biomass in a selection of Cladocera, Copepoda and Rotifera from the plankton, periphyton and benthos of continental waters. Oecologia 19: 75–97.

Ghilarov, A. M., 1977. Nablyudeniya nad sostavom pishchi kolovratok roda *Asplanchna*. Zoologicheskiy zhurnal 56, 12: 1874–1876 (Observations on food composition of rotifers of the genus *Asplanchna*. In Russian).

Kutikova, L. A., 1970. Kolovratki fauny SSSR. Nauka Publ.,

Leningrad: 744 pp. (Rotifers in the Fauna of the USSR. In Russian).

Markevich, G. I., 1978. Ob izbiratelnosti *Brachionus rubens* pri zaselenii substrata. Informatsionniy bulleten, Institut biologii vnutrennikh vod 39: 37–40 (On selective colonization of substrata by *Brachionus rubens*. In Russian).

Markevich, G. I. & I. K. Rivier, 1975. Vliyanie epibiontnykh bespozvonochnykh na kopepod i kladotser. In: Povedenie vodnykh bespozvonochnykh, Borok: 49–52 (The effect of epibiotic invertebrates on copepods and cladocerans. In Russian).

Markevich, G. I. & I. K. Rivier, 1978. Vliyanie *Brachionus rubens* na dvigatelnuyu aktivnost nekotorykh *Cladocera*. Informatsionniy bulleten, Institut biologii vnutrennikh vod 39: 45–48 (Influence of *Brachionus rubens* on motor activity of some *Cladocera*. In Russian).

Monakov, A. V. & Yu. I. Sorokin, 1972. Some results on investigations on nutrition of water-animals. In Z. Kajak & A. Hillbricht-Ilkowska (eds), Productivity problems of freshwaters. Warszawa-Krakow: 765–773.

Mordukhai-Boltovskaya, E. D., 1958. Predvaritelnye dannye po pitaniyu khishchnykh kladotser *Leptodora kindti* i *Bythotrephes*. Doklady AN SSSR 122, 4: 723–726 (Preliminary data on feeding of predatory cladocerans *Leptodora kindti* and *Bythotrephes*. In Russian).

Mordukhai-Boltovskaya, E. D., 1960. O pitanii khishchnykh kladotser *Leptodora* i *Bythotrephes*. Bulleten Instituta biologii vodokhranilishch 6: 21–22 (On the feeding of predatory cladocerans *Leptodora* and *Bythotrephes*. In Russian).

Rivier, I. K. & G. I. Markevich, 1978. Vliyanie *Brachionus rubens* na biologicheskie pokazateli nekotorykh *Cladocera* pri sovmestnom obitanii. Informatsionniy bulleten, Institut biologii vnutrennikh vod 39: 41–44 (Influence of *Brachionus rubens* on biological parameters of some *Cladocera* in case of cohabitation. In Russian).

Smirnov, N. N., 1986. Lake Glubokoe. Hydrobiologia 141, 1/2.

Sorokin, Yu. I. & E. D. Mordukhai-Boltovskaya, 1962. Izuchenie pitaniya kolovratok *Asplanchna* s pomoshchyu C^{14}. Bulleten Instituta biologii vodokhranilishch 12: 17–20 (Feeding of *Asplanchna* studied using C^{14}. In Russian).

Stemberger, R. S. & J. J. Gilbert, 1987. Defenses of planktonic rotifers against predators. In W. C. Kerfoot & A. Sih (eds), Predation: Direct and indirect impacts on aquatic communities. Univ. Press of New England, Hanover & London: 227–239.

Timokhina, A. F., 1983. Pitaniye kolovratok roda *Asplanchna* (*Ploimida, Asplanchnidae*). In G. G. Winberg (ed.), Troficheskie svyazi i ikh rol v produktivnosti prirodnykh vodoyemov. Leningrad: 81–84 (Feeding of rotifers of the genus *Asplanchna*. In Russian).

Tribush, T. M., 1960. Nekotorye nablyudeniya nad kolovratkami sem. *Asplanchnidae* Rybinskogo vodokhranilishcha. Bulleten Instituta biologii vodokhranilishch 6: 18–19 (Some observations on rotifers of the family *Asplanchnidae* in Rybinsk reservoir. In Russian).

Vlastov, B. V., 1953a. Evropeyskie i severo-amerikanskie kolovratki iz sem. Notommatid – sozhiteli *Cladocera* i ikh vidovaya prinadlezhnost. Zoologicheskiy zhurnal 32, 6: 1110–1114 (European and North-American rotifers of the Notommatidae family – cohabitants of *Cladocera* and their species attribution. In Russian).

Vlastov, B. V., 1953b. Vzaimootnosheniya mezhdu *Cladocera* i zhivushchimi na nikh kolovratkami iz roda *Proales*. Trudy Vsesoyuznogo hydrobiologicheskogo obshchestva 5: 299–317 (Interrelations between *Cladocera* and rotifers of the genus *Proales* living on them. In Russian).

Williamson, C. E., 1983. Invertebrate predation on planktonic rotifers. Hydrobiologia 104: 385–396.

Hydrobiologia **186/187**: 75–80, 1989.
C. Ricci, T. W. Snell and C. E. King (eds), Rotifer Symposium V.
© 1989 *Kluwer Academic Publishers.*

Competitive interactions between the rotifer *Synchaeta oblonga* and the cladoceran *Scapholeberis kingi* Sars

John J. Gilbert

Department of Biological Sciences, Dartmouth College, Hanover, New Hampshire 03755, USA

Key words: competition, interference, *Scapholeberis, Synchaeta,* zooplankton

Abstract

Competition experiments showed that the small cladoceran *Scapholeberis kingi* rapidly excluded the rotifer *Synchaeta oblonga* from mixed-species cultures, but was itself unaffected by the presence of *S. oblonga.* Short-term experiments testing the effect of *S. kingi* on the survivorship and reproduction of *S. oblonga* showed that the former imposed a high mortality on the latter, even though shared food resources were abundant. These results indicate that adult *S. kingi* mechanically interferes with *S. oblonga* either by ingesting, or by rejecting in a damaged condition, individuals swept into its branchial chamber. In contrast to many other small species of cladocerans, and like large species of *Daphnia, S. kingi* has the potential to markedly suppress populations of some rotifer species through a combination of interference and exploitative competition.

Introduction

Considerable direct and circumstantial evidence from field and laboratory studies shows or indicates that rotifers are strongly suppressed by large *Daphnia* – those with body lengths (head to base of tailspine) greater than about 1.2 mm (Gilbert, 1988a). Such *Daphnia* can monopolize shared food resources and also can mechanically interfere with (damage or eat) some small or soft-bodied rotifers swept into their branchial chambers (Gilbert, 1985, 1988b; Gilbert & Stemberger, 1985; Burns & Gilbert, 1986a, b).

In contrast, rotifers seem to be much less suppressed by small cladocerans and, in fact, often abound with them in natural zooplankton communities (Gilbert, 1988a). These cladocerans are much less able than large *Daphnia* to sequester food resources and to interfere with rotifers. For example, *Daphnia* and species of four other

cladoceran genera (*Bosmina longirostris, Ceriodaphnia dubia, Holopedium gibberum, Diaphanosoma brachyurum*) with body lengths less than 1.2 mm interfered slightly, if at all, with one of the most susceptible rotifer species (*K. cochlearis*) Burns & Gilbert, 1986a, b; Gilbert & MacIsaac, 1989), and one of these cladocerans (*C. dubia*) only interfered with newborn individuals of this rotifer (Gilbert & MacIsaac, 1989). Furthermore, laboratory competition experiments using a variety of different-sized cladoceran species and three rotifer species (*Keratella testudo, K. cochlearis, Synchaeta oblonga*) showed that one or more of these rotifers could coexist for many weeks with, and sometimes competitively suppress, two small cladoceran species (*B. longirostris, Daphnia ambigua*) but were rapidly excluded by three large *Daphnia* species (*D. galeata mendotae, D. pulex, D. magna*) (Gilbert & Stemberger, 1985; MacIsaac & Gilbert, in press).

In the present study I examine competitive interactions between the small, soft-bodied rotifer *Synchaeta oblonga* and the small cladoceran *Scapholeberis kingi* Sars — probably *S. armata armata* (Herrik) according to the revision of Dumont and Pensaert (1983). Unlike previous studies with other small cladoceran species, this study surprisingly shows that *S. kingi* rapidly excludes *S. oblonga* from mixed-species cultures and can impose a high mortality rate on *S. oblonga* through mechanical interference.

Materials and methods

Synchaeta oblonga and *Scapholeberis kingi* were isolated from water bodies within 30 km of Dartmouth College, cloned, and maintained on *Cryptomonas* sp. in glass-fiber-filtered lake water as described elsewhere (Gilbert, 1985, 1988b). Adult body lengths of female *S. oblonga* and *S. kingi* were about 150 μm and 0.4 to 0.6 mm, respectively. All experiments were conducted at 20 °C in a photoperiod (LD 16 : 8) with dim light ($\sim$ 300 lux).

Competition experiments were conducted in covered, 50-ml glass beakers containing 40 ml of a *Cryptomonas* suspension (10^4 cells ml^{-1}) in a 9 : 1 (v/v) mixture of lake water and the *Cryptomonas* growth medium (Stemberger, 1981). Initial population sizes were 20 randomly-selected *S. oblonga* in the single- and mixed-species cultures and 3 adult (experiment 1) or juvenile (experiment 2) *S. kingi* in the single- and mixed-species cultures. Three replicate cultures were set up for each treatment. All individuals in all populations of the two species were counted and transferred by pipette to fresh *Cryptomonas* suspension every 2 days until one of the species went to extinction. Oviposited eggs of *S. oblonga* were transferred along with the swimming individuals, but were not counted. Population growth trajectories for each species with and without the other were analyzed using univariate, repeated-measures ANOVA and Student-Newman-Keuls (SNK) tests (SAS Institute, 1985) on population-size data from time zero.

A short-term experiment was conducted to determine the effect of *S. kingi* on the survivorship of *S. oblonga* in the 2 cm-diameter concavities of glass, 9-spot plates. Each concavity contained 1 ml of a dense *Cryptomonas* suspension (5–10 × 10^4 cells ml^{-1}), 2 randomly-selected *S. oblonga*, and either no or 1 adult *S. kingi*. Nine replicate cultures of each treatment were in alternating concavities of each of two 9-spot plates. Each 9-spot plate was placed in a separate moist chamber. After 16 hours, the numbers of surviving *S. oblonga* and *S. kingi* in the cultures were determined. The pooled frequencies of living and dead (missing or eaten) *S. oblonga* in the cultures with and without *S. kingi* were compared with Fisher's exact test (Sokal & Rohlf, 1981).

Another short-term experiment was conducted to determine the effect of *S. kingi* on the reproductive rate of *S. oblonga* in covered, 20-ml, glass beakers. Each beaker contained 10 ml of a dense *Cryptomonas* suspension (5–10 × 10^4 cells ml^{-1}), 5 randomly-selected *S. oblonga*, and either no or 2 adult *S. kingi*. Six replicate cultures with and without *S. kingi* were organized as 6 matched pairs. After 23.3 to 24.2 hours, the numbers of living *S. oblonga* and *S. kingi* in each culture were determined. Six replicate death rates of *S. oblonga* due to the presence of *S. kingi* (d_s) were determined as described elsewhere (Gilbert, 1988b) from the reproductive rates of the *S. oblonga* in the 6 pairs of cultures with and without *S. kingi*.

Results

In both of the competition experiments, *S. oblonga* was significantly suppressed by *S. kingi* while *S. kingi* was unaffected by *S. oblonga* (Fig. 1, Tables 1 and 2). In experiments 1 and 2, the *S. kingi* excluded *S. oblonga* from mixed-species cultures after 4 and 16 days, respectively. The *S. kingi* populations in experiment 2 were initiated with juveniles and did not begin to noticeably affect the *S. oblonga* populations until day 6, when they contained adults. The *S. kingi* populations in experiment 1 were initiated with adults and markedly depressed *S. oblonga* population sizes by day

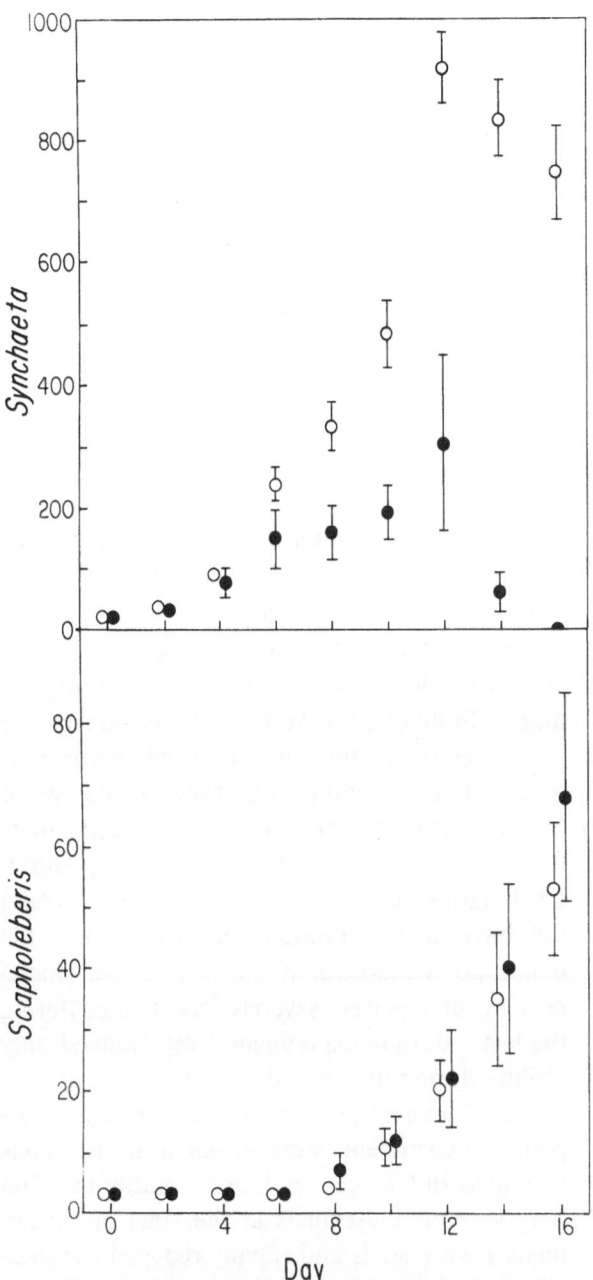

Fig. 1. Changes in population sizes over time for *Synchaeta oblonga* and *Scapholeberis kingi* in 40-ml, single-species (open circles) and mixed-species (closed circles) cultures. Experiment 2. Culture conditions: see Table 1. *Scapholeberis* populations initiated with juvenile females. Values are means of 3 replicates ± 1 SE (drawn when greater than circle).

Table 1. Changes in population sizes over time for *Synchaeta oblonga* and *Scapholeberis kingi* in 40-ml, single-species and mixed-species cultures. Experiment 1. *Scapholeberis* populations initiated with gravid adult females. Values are means of 3 replicates ± 1 SE.

Day	Culture			
	Synchaeta		*Scapholeberis*	
	Single species	Mixed species	Single species	Mixed species
0	20 ± 0	20 ± 0	3 ± 0	3 ± 0
2	39 ± 7	10 ± 2	20 ± 1	21 ± 1
4	88 ± 23	1 ± 1	59 ± 4	57 ± 2

over time (10.3 vs. 49.1 and 110.6 vs. 410.2 for experiments 1 and 2, respectively; SNK tests, $P < 0.05$) (Table 2). In neither experiment did the presence of *S. oblonga* significantly affect the population growth trajectories or the mean population sizes of *S. kingi* (treatment × time effects, $P > 0.05$; SNK tests, $P > 0.05$) (Table 2). Visual inspection of the cultures throughout the competition experiments showed that *Cryptomonas* cell densities were greatly depleted or negligible before renewal in cultures with *S. kingi* at days 2 and 4 in experiment 1 and in all cultures by day 8 and thereafter in experiment 2.

The results of the short-term survivorship experiment showed that the presence of *S. kingi* significantly increased the mortality of *S. oblonga*. The numbers of surviving *S. oblonga* with and without *S. kingi* were 9 and 17 out of 18, respectively (Fisher's exact test, $P = 0.007$). *Cryptomonas* cells were present in all cultures at the end of the experiment.

The results of the short-term reproductive rate experiment showed that *S. kingi* imposed a mean death rate (d_S) of 1.2 day^{-1} on the *S. oblonga* populations (Table 3). The mean reproductive rate of *S. oblonga* populations was significantly lower with *S. kingi* present (-1.1 day^{-1}) than with it absent (0.08 day^{-1}) (Student's $t = 4.4$ with 10 df, $P < 0.01$, 2-tailed). *Cryptomonas* cell densities remained high in all cultures at the end of the experiment.

2. In both experiments, the population sizes of *S. oblonga* with and without *S. kingi* diverged significantly over time (treatment × time effects, $P < 0.001$) and had significantly different means

Table 2. Analysis of results of competition experiments with *Scapholeberis kingi* (*S.k.*) and *Synchaeta oblonga* (*S.o.*) (See Table 1 and Fig. 1). Effects of each species on the other analyzed with repeated-measures ANOVA (type III model) and Student-Newman-Keuls (SNK) tests. Significant levels of $P > 0.05$, < 0.05, < 0.01, and < 0.001 indicated by NS, *, **, and ***, respectively.

Treatment	Experiment	ANOVA (F-value)				SNK test
		Treatemnt	Replicate	Time	Treatment × time	
S.k. on S.o.	1	2.3^{NS}	0.4^{NS}	6.1*	19.0***	*
	2	7.1**	1.8^{NS}	86.8***	65.6***	*
S.o. on S.k.	1	1.1^{NS}	0.8^{NS}	251.1***	0^{NS}	NS
	2	5.0*	3.2*	74.1***	0.9^{NS}	NS

Table 3. Short-term experiment on the effect of *Scapholeberis kingi* on the reproductive and death rates of *Synchaeta oblonga*. Experimental conditions: 10-ml cultures initiated with 5 *S. oblonga*, with and without 2 adult *S. kingi* females, and terminated after 0.97–1.01 days; 5–10×10^4 cells ml^{-1} *Cryptomonas* sp; 20 °C; photoperiod (LD 16 : 8, $\sim$300 lux). Symbols for rates: r_c and r_s (reproductive rates of *S. oblonga* in control cultures without *S. kingi* and in cultures with *S. kingi*), d_s (death rate of *S. oblonga* due to presence of *S. kingi*, equals r_c-r_s).

Replicate and statistic	Rate		
	r_c day^{-1}	r_s day^{-1}	d_s day^{-1}
1	0.19	− 0.23	0.42
2	0.19	− 1.64*	1.83*
3	− 0.23	− 1.63	1.40
4	0	− 0.93	0.93
5	0.34	− 0.51	0.85
6	0	− 0.59*	1.59*
Mean	0.08	− 1.09	1.17
1 SD	0.20	0.62	0.53

* Underestimate of rate; all *S. oblonga* killed or eaten, but 1 individual assumed to be alive at end of experiment so that r_s could be calculated from logarithmic formula.

Discussion

The combined results of the different experiments show than *Synchaeta oblonga* was rapidly excluded by *Scapholeberis kingi* from mixed-species cultures and that this exclusion probably was primarily due to direct, mechanical, interference competition from adult *S. kingi*. The high mortality rates imposed on *S. oblonga* by *S. kingi* in the short-term experiments can only be interpreted by some kind of interference, as food was still present in the cultures at the end of the experiments. Similarly, the very rapid declines of the *S. oblonga* populations in the mixed-species cultures of the competition experiments when *S. kingi* adults were present strongly suggest interference rather than exploitative competition. Exploitative competition alone probably would not have such a pronounced effect, since the dense populations of *S. oblonga* in the single-species cultures were severely food-limited during the last 6 days of experiment 2 yet declined only slightly during this period (Fig. 1).

It is of interest that the results of the two competition experiments were so similar despite considerable differences in initial conditions. The *Scapholeberis* individuals at the start of experiment 1 were adult and rapidly reduced the densities of the *Synchaeta* populations from their initial level (0.5 individual ml^{-1}). In contrast, the *Scapholeberis* individuals at the start of experiment 2 were juvenile and permitted the *Synchaeta* populations to increase ten-fold (from 0.5 to $\sim$5 individuals ml^{-1}) over a 12-day period before they became adult and rapidly suppressed the rotifers. Thus, the outcome of competition between these two species was the same at relatively low and high rotifer densities.

The mechanism by which *S. kingi* interferes with *S. oblonga* is not known. It could be entirely predation or some combination of predation and damage to rejected individuals. The high mortality rates attributed to interference in all of the experiments certainly was due to the very high population densities of the *Scapholeberis* (up to ~1 individual ml^{-1}). Preliminary observations of encounters between *S. kingi* and *S. oblonga* failed to detect either type of interference. There appears to be no information on the food niche of any species of *Scapholeberis*, and so the potential of these cladocerans to eat large food items is not known.

Of all the small (< 1.2 mm body length) cladoceran species tested thus far, *S. kingi* (0.6-0.8 mm adult body length) is the only one to markedly interfere with a rotifer. *Ceriodaphnia dubia* (~0.7 mm adult body length) interfered with juvenile but not older *Keratella cochlearis*, and *Daphnia ambigua* (~1.0 mm adult body length) interfered only to a limited extent, if at all, with *K. cochlearis* (Burns & Gilbert, 1986a, b; Gilbert & MacIsaac, 1989). *Bosmina longirostris* (~0.4 mm adult body length), *Diaphanosoma brachyurum* (~0.8 mm adult body length), and *Holopedium gibberum* (0.7-0.8 mm adult body length) did not interfere with *K. cochlearis* (Gilbert & MacIsaac, 1989).

The susceptibility of *S. oblonga* to cladoceran interference is very similar to that of *K. cochlearis*. This was shown for post-juvenile individuals of these two species in experiments with *Daphnia pulex* (Gilbert, 1988b). Juvenile *S. oblonga* was more vulnerable to *D. pulex* than older *K. cochlearis* (Gilbert, 1988b), but the relative susceptibilities of juvenile individuals of the two rotifer species have not been compared. At any rate, the small cladocerans that did not interfere with *K. cochlearis* probably would not interfere with *S. oblonga*, and, conversely, *K. cochlearis* should be as susceptible to interference from *Scapholeberis kingi* as *S. oblonga* was in the present study.

The ability of a cladoceran species to interfere with a rotifer species is positively correlated with its ability to competitively exclude that species from mixed-species cultures. Three species of *Daphnia* with adult body lengths greater than 2 mm strongly interfered with *K. cochlearis* and also rapidly drove this rotifer to extinction in competition experiments (Gilbert 1985; Gilbert & Stemberger, 1985; Burns & Gilbert, 1986a, b; MacIsaac & Gilbert, in press). The same was true for *S. kingi* with *S. oblonga* in the present study. *Ceriodaphnia dubia*, which interfered only with juvenile *K. cochlearis*, strongly suppressed this rotifer but did not completely exclude it from mixed-species cultures even after 8 weeks (MacIsaac and Gilbert, in press). *Bosmina longirostris*, which did not interfere with *K. cochlearis*, only weakly suppressed both *K. cochlearis* and *S. oblonga* and, in turn, was weakly suppressed by them (MacIsaac & Gilbert, in press). Finally, *Daphnia ambigua*, which interfered slightly or not at all with *K cochlearis*, only weakly suppressed *K. cochlearis* and *K. testudo*, and in one experiment it was excluded by *K. cochlearis* (Kirk & Gilbert, unpublished; MacIsaac & Gilbert, in press).

Cladoceran species that strongly interfere with a rotifer species have not been significantly suppressed by that rotifer species in competition experiments. This was true for the three large (> 2 mm adult body length) species of *Daphnia* (Gilbert, 1985; MacIsaac & Gilbert, in press) and also for the *S. kingi* in the present study.

Another cladoceran known to interfere with rotifers is *Moina hutchinsoni*. Females of this species attain body lengths of 1.4 to 1.6 mm (Goulden, 1968) and can eat the soft-bodied rotifer *Hexarthra* (Edmondson, 1974). Examination of the guts of 108 *M. hutchinsoni* from Soap Lake, Washington over 7 dates in 1974 revealed one set of *H. fennica* trophi in each of 3 individuals (N. G. Hairston, Jr., pers. comm.). While these *Moina* may not have preyed extensively on the *Hexarthra* they may have damaged and killed a high proportion of individuals swept into and then rejected from their branchial chambers. It is noteworthy that one field study provides strong correlational evidence for the competitive suppression of rotifers by a much smaller species of *Moina*, *M. micrura* (Gasith & Perry, 1985).

Acknowledgements

This research was supported by National Science Foundation research grants BSR-8415024 and BSR-8717074. I am very grateful to H.J. MacIsaac for performing the ANOVA and for improving the manuscript.

References

Burns, C. W. & J. J. Gilbert, 1986a. Effects of daphnid size and density on interference between *Daphnia* and *Keratella cochlearis*. Limnol. Oceanogr. 31: 848–858.

Burns, C. W. & J. J. Gilbert, 1986b. Direct observations of the mechanisms of interference between *Daphnia* and *Keratella cochlearis*. Limnol. Oceanogr. 31: 859–866.

Dumont, H. J. & J. Pensaert, 1983. A revision of the Scapholeberinae (Crustacea: Cladocera). Hydrobiologia 100: 3–45.

Edmondson, W. T., 1974. Secondary production. Mitt. int. Ver. Limnol. 20: 229–272.

Gasith, A. & A. S. Perry, 1985. Use of limnocorrals for pesticide toxicity studies: effect on zooplankton composition and dynamics. Verh. int. Ver. Limnol. 22: 2432–2436.

Gilbert, J. J., 1985. Competition between rotifers and *Daphnia*. Ecology 66: 1943–1950.

Gilbert, J. J., 1988a. Suppression of rotifer populations by *Daphnia*: a review of the evidence, the mechanisms, and the effects on zooplankton community structure. Limnol. Oceanogr. 33: 1286–1303.

Gilbert, J. J., 1988b. Susceptibilities of ten rotifer species to interference from *Daphnia pulex*. Ecology. 69: 1826–1838.

Gilbert, J. J. & R. S. Stemberger, 1985. Control of *Keratella* populations by interference competition from *Daphnia*. Limnol. Oceanogr. 30: 180–188.

Gilbert, J. J. & H. J. MacIsaac, 1989. The susceptibility of *Keratella cochlearis* to interference from small cladocerans. Freshwat. Biol. 22: in press.

Goulden, C. E., 1968. The systematics and evolution of the Moinidae. Trans. am. phil. Soc. NS 58(6): 1–98.

MacIsaac, H. J. & J. J. Gilbert. Competition between rotifers and cladocerans of different body sizes. Oecologia (Berlin). In press.

SAS (Statistical Analysis System) Institute, 1985. User's guide: statistics ed. 5. Cary, North Carolina, USA.

Sokal, R. R. & F. J. Rohlf, 1981. Biometry. W. H. Freeman and Co., San Francisco, California, USA.

Stemberger, R. S., 1981. A general approach to the culture of planktonic rotifers. Can. J. Fish. aquat. Sci. 38: 721–724.

Hydrobiologia **186/187**: 81–102, 1989.
C. Ricci, T. W. Snell and C. E. King (eds), Rotifer Symposium V.
© 1989 *Kluwer Academic Publishers.*

Salinity and temperature influence in rotifer life history characteristics

María R. Miracle & Manuel Serra
Dep. de Ecología, Fac. de C. Biológicas, Univ. de Valencia, 46100 Burjassot (Valencia), Spain

Key words: rotifers, rate of population increase, life span, temperature, salinity

Abstract

A review of temperature and salinity effects on rotifer population dynamics is presented together with original data of these effects for three clones of *Brachionus plicatilis*. There is a clear relationship between temperature and the intrinsic rate of increase, *r*: an increase of temperature – within the natural environmental range – produces an exponential increase of *r*, and the slope of the response depends on the genotype. The effect of salinity is also genetically dependent; the highest *r* for each clone is observed at the salinity close to that of its environmental origin. The response of *r* to temperature is mainly a consequence of the response of the individual rates of development and reproductive timing. The effect of temperature on fecundity (number of descendents per individual life time) is negligible when temperature values are within the normal habitat ranges. On the other hand, salinity seems to affect primarily fecundity. The interaction salinity-temperature may be important in clones or species living in fluctuating environments with positive response to the more frequent combinations found in the corresponding habitats.

Introduction

Extended studies of aquatic populations usually confirm a community typology that matches environmental heterogeneity. The main factors determining the community typology are most frequently (1) temperature, (2) salinity and the parameters associated with it, and (3) trophic status of the waters. For instance, principal components analysis, for data on the species abundance in zooplanctonic communities from a wide range of conditions, usually identify, as the first principal components, variables that are highly correlated with the above mentioned factors (e.g., Miracle, 1974; Armengol, 1978; Miracle *et al.*, 1987; Nogrady, 1988). Temperature, salinity and food are also the most frequently studied factors in life history analysis.

This paper reviews results of laboratory experiments on the relationships of temperature and salinity to the intrinsic rate of population increase, *r*, which is assumed to be the best measure of fitness in non-saturated environments. Rotifers in strongly fluctuating planctonic environments, which have to adapt to regular colonizing events, may frequently be subject to selection for high *r*. The paper is divided into two main sections. In the first, a brief bibliographic review of the effects of temperature and salinity on population dynamics parameters is given. Physiologic effects on individual rotifers will also be summarized to help interpret the effects on populations. In the second, original data combining the two factors from a series of experiments will be discussed. Our intention is to evaluate the interaction between temperature and salinity, *i.e.*, the dependence of the effect of one factor on the level of the other.

Temperature: effects on individuals

The direct positive effect of temperature on physiological rates in individual rotifers and on rotifer populations have been clearly demonstrated. Several of these studies deal with the relationship between temperature and duration of embryonic development. Vinberg & Galkovskaya (1979), Herzig (1983), Duncan (1983) and Galkovskaya (1987) have summarized data on the subject. The duration of embryonic development (D) in rotifers is a curvilinear function of temperature (T), which can be adequately described by Belehradek's equation $1/D = a(T - b)^c$ (a,b,c, fitted parameters) or Arrhenius' equation $1/D = ae^{-b/T}$ (Waltz, 1983) as in other organisms (Herzig, 1983). However other equations have also been proposed as best fit (*i.e.*, $1/D = a + bT + cT^2$, Galkovskaya, 1987; $1/D = ae^{bT}$, Duncan, 1983). Furthermore, when homogeneous data from one species, studied by the same author, in a limited range of temperatures, are used, the relationship between $1/D$ and T is nearly linear (Edmondson, 1965; Pourriot & Deluzarches, 1971). The slope of the curve and the absolute rate values are dependent on genotype. Cold-water adapted species (Galkovskaya,

1987) or clones (Pourriot & Deluzarches, 1971) have lower slopes and globally smaller rates than warm-water adapted ones. Therefore, in many cases, cold-adapted species show faster low-temperature development but much slower high-temperature development than warm-adapted species and vice versa. However, other factors such as the *r-K* strategy of the species have to be considered. The effect of temperature is also dependent on its stability. If temperature fluctuates around a mean value, the egg development time deviates from that at the constant mean (Ruttner-Kolisko, 1975, 1978); it is shortened, as would be expected, by the effect of oscillating temperature on the metabolism. Figure 1 presents idealized responses to temperature change of activities and functions discussed in this section.

It is often assumed that duration of embryonic development is only dependent upon the temperature (Herzig, 1983). However, relations with other parameters have also been described, mainly with egg volume (Pourriot & Deluzarches, 1971), which is in turn dependent on temperature (Pourriot, 1973a) or on the feeding conditions of the mother (Yúfera, 1987).

Duration of the pre-reproductive period is also a clear function of temperature. Its variation

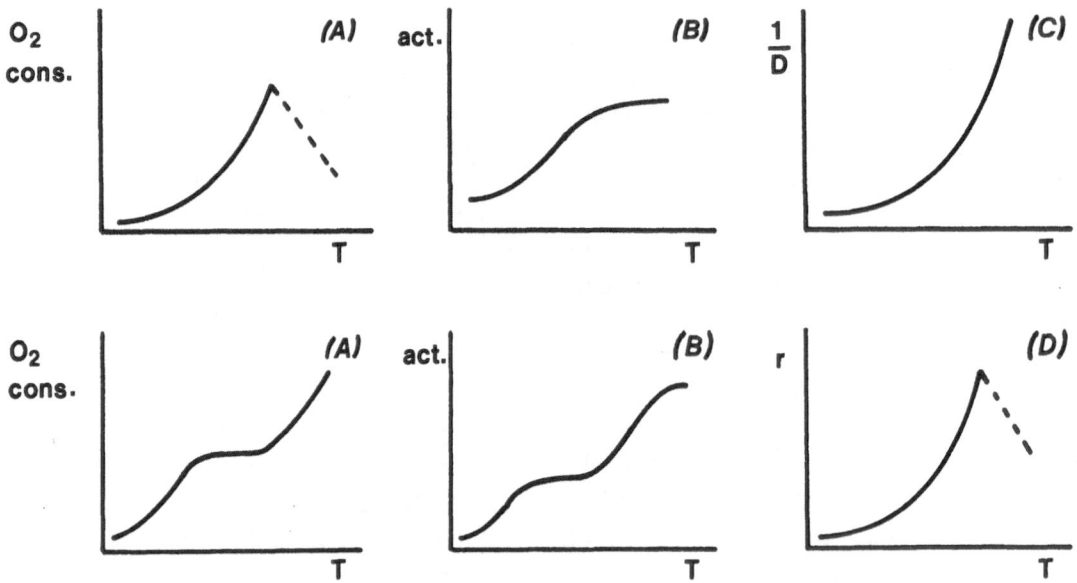

Fig. 1. Ideal curves showing relationships between temperature and (A) oxygen consumption, (B) activity (filtration rate), (C) inverse of development rate, and (D) intrinsic rate of growth.

parallels that of egg development. Thus, under the same feeding conditions the ratio postembryonic/-embryonic development time remains quite constant (Pourriot & Deluzarches, 1971; Ruttner-Kolisko, 1975). However, when rotifers are subjected to fluctuating temperature, the above mentioned ratio could be very different because there is some evidence (Ruttner-Kolisko, 1978) that the duration of the pre-reproductive period does not deviate significantly from that obtained at the corresponding mean temperature.

Post-embryonic growth is also highly dependent on temperature. Growth in length in rotifers can be described by the Bertalanffy equation (Lebedeva & Gerasimova, 1985; Carmona et al., this volume). Growth is almost totally restricted to the juvenile period (defined as lasting until the laying of the first egg): the growth rate undergoes a sharp decrease coincident with the beginning of reproduction and from then on, matter and energy are directed towards reproduction instead of growth. The somatic growth rate in juveniles is a clear function of temperature as well as the duration of the growth period. Consequently, the calculated Q_{10} for both somatic growth (during the pre-reproductive period) and egg development are similar, ranging approximately from 2 to 3 (Lebedeva & Gerasimova, 1985, Pourriot & Deluzarches, 1971, Duncan, 1983).

Very few studies exist on the effect of temperature on rotifer metabolism, and they have variable results. Pourriot (1973b) and Galkovskaya et al. (1987) find an exponential increase in oxygen consumption with temperature. On the other hand, Epp & Lewis (1980), studying a clone of B. plicatilis, suggest that there is a plateau in oxygen consumption, within the range of environmental temperatures in which the rotifer is most frequently found. The species studied by Pourriot (1973b) and Galkovskaya et al. (1987) were Brachionus calyciflorus and Rinoglena frontalis. Q_{10} was higher for B. calyciflorus (Q_{10} = 2.3–2.4) and much lower for the cold stenotherm R. frontalis (Q_{10} = 1.4). Thus, the oxygen consumption for R. frontalis is higher than that of B. calyciflorus at low temperatures, but much lower at medium and high temperatures. This

behaviour of cold vs. warm water species is similar to the trends described for egg development time.

Galkoskaya (1987) also reports data on O_2 consumption in natural populations of Brachionus calyciflorus and Hexarthra mira; the same exponential function was found but with smaller slope than the cultivated individuals. Another difference was that, after a maximum, a sharp decline occurred when temperatures were too high. H. mira, usually occurring in lower temperatures than B. calyciflorus, also showed lower O_2 consumption. Other metabolic rates have rarely been studied. Nitrogen excretion rates, investigated for B. calyciflorus by Galkovskaya et al. (1987), showed the same exponential function with temperature as O_2 consumption and a quite similar Q_{10} (= 2.2).

There is also only fragmentary information on rotifer activities, but from this, it seems that there exists a clear response to temperature. The observed response, when a wide range of temperatures is tested, could be interpreted either as a sigmoid curve with a decline at unsuitably high temperatures, or as two sigmoid functions with a plateau, at a specific optimum range of temperatures, in between. Tested activities are mainly filtration and ingestion rates (Hirayama & Ogawa, 1972; Galkovskaya, 1987), and locomotion (Epp & Lewis, 1984, Snell et al., 1987), with interpretation of the obtained results varying amongst the authors. Rates of beat of vibratile flames and of output from the contractile vesicles in Asplanchna have been also investigated (Pontin, 1966). Up to 24 °C, the relationship between the activity of vibratile flames and temperature is a exponential function.

Temperature: effects on populations

The intrinsic rate of increase

From comparison of the maximum rates of increase of zooplanktonic organisms (Allan, 1976, 1980) it can be inferred that (1) rotifers show amongst the highest r values of all zooplankton species, and (2) that the crucial

84

Table 1. The intrinsec rate of increase (r, days^{-1}) of some rotifer species growing at different temperature (if not preadapted, the pre-experimental temperature is indicated in bracket).

Rotifer species	Temperature (°C)																		Reference
	5	10	15	16	17	19	20	22	24	25	27	30	32	35	36	37	40	43	
Brachionus plicatilis																			
[1]							0.16			0.38		0.62							King & Miracle, 1980
[2]							0.12			0.32		0.50							
[3]							0.30			0.36		0.44							
plicatilis [4]		0.12		0.39		0.43		0.65		0.74		0.49							Hirayama & Kusano, 1972
plicatilis [5]							0.5			0.7		1.10	0.68		0.4		0.5		Snell, 1986
plicatilis [6]			<0		<0		0.54			0.74		0.98		1.35			1.07	<0	Pascual & Yúfera, 1983
plicatilis [7]							0.13		0.25										Yúfera & Pascual, 1980
plicatilis [8]		0.12					0.53												Nagata, 1985
plicatilis [9]			0.36				0.57			0.96									Ruttner-Kolisko, 1972
Brachionus calyciflorus [10]							0.23					2.18				0.38 [28 °C]	0.34 [35 °C]		Galkovskaya, 1983
calyciflorus [11]												1.56				2.83	2.18		Galkovskaya, 1987
[12]												2.18				2.95	1.94		
[13]										1.2			1.7						
calyciflorus [14]			0.34				0.48 / 0.60 [*T* variable]			0.82									Halbach, 1973
calyciflorus [15]			<0				0.1			<0		<0							In: Starkweather, 1987
[16]			0.2				0.6			1.7		0.9							
[17]			0.3				0.7			1.9		2.5							
[18]			0.4				2.2			1.9		3.7							
Brachionus dimidiatus																			
[19]							0.53			0.70		1.04							Pourriot & Rougier, 1975
[20]							0.38			0.43		1.22							
[21]							0.46			0.74									
[22]							0.47			0.52									
[23]							0.30			0.42									

Species						Reference
Euchlanis dilatata						King, 1972
[24]	0.81				1.78	
[25]	0.68				2.10	
[26]	0.61				2.13	
Asplanchna sieboldi						Gilbert, 1976
[27]		0.60		0.66	1.08	[Saccatte morphotype]
[28]		0.72		0.85	1.22	[Campanulate morphotype]
Brachionus angularis						Walz, 1987
[29]	−0.005	0.073	0.103	0.352	0.154	
Keratella cochlearis						Walz, 1983
[29]	0.004	0.034	0.098	0.082	0.076	

Food conditions and other observations:

[1]–[3] *Chlorella* (10^6 cell/ml); the three sets of values corresponds to three clones. Clone [3] coming from cold waters than clones [1] and [2].

[4] *Chlorella* (2.27×10^6 cell/ml).

[5] *Chlorella vulgaris* (10 μgC/ml).

[6] *Nannochloris oculata* (45–60×10^6 cell/ml).

[7] *Nannochloris* (6×10^6 cell/ml) plus *Saccharomyces cerevisiae* (0.35 g/l per day).

[8] *Chlorella saccharophila* (1–2×10^6 cell/ml).

[9] *Dunaliella* sp.

[10] *Chlorella* (5×10^6 cell/ml).

[11]–[13] *Chlorella*: [11] [12] [13]
                 5×10^6 50×10^6 240×10^6 cell/ml.

In [12] the experimental conditions were different than that in [10] and [11].

[14] *Kirchneriella lunaris* (10^6 cell/ml).

[15]–[18] *Euglena gracilis*, original data of B. Skrdla and P. L. Starkweather:
                 [15] [16] [17] [18]
                 0.1 1.0 10 100 μg dry weight/ml.

[19] *Dunaliella salina.*

[20] *Diogenes* sp.

[21] *Synechococcus cedrorum.*

[22] *Synechocystis* sp.

[23] *Saccharomyces cerevisiae.*

[24]–[26] *Chlamidomonas reinhardti* (0.5×10^6 cell/ml); the different values are average *r* from groups of clones collected at different times of the year:
                 [24] [25] [26]
                 May 1   June 5   June 12

[27] *Paramecium aurelia* (in excess); the different values are average *r* from groups of replicates.

[28] *Asplanchna brightwelli* (in excess); the different values are average *r* from groups of replicates.

[29] *Stichococcus bacillarius* (75×10^6 cell/ml).

advantage of one species against another, under a particular set of conditions, may lie in how r responds to any one of the changing conditions. Moreover, r summarizes all life tables parameters, because it combines survival, fecundity and the timing of development and reproduction. For this reason, we will center the review mainly on this parameter, and secondarily on life span as a measure of the stability or replacement time of the generations.

Table 1 compiles literature data on intrinsic rates of increase measured at different temperatures. Because of the important influence of other factors on r, mainly genotypes and food quantity and quality, we have only included data sets from experiments conducted at different temperatures by the same author, so that all other conditions would remain the same. In addition, it should also be considered that the different authors estimate r values by different methods; they may be obtained from individual, batch or continuous cultures, and by means of different mathematical approaches.

Very few species of rotifers have been studied to determine the relationship between r and temperature under laboratory controlled conditions. *Brachionus plicatilis* and *B. calyciflorus* are the most studied. Different clones of these species have maxima of r at different temperatures. This may be attributed to (1) selection at the environmental temperatures of origin, and (2) the interaction of food with temperature. This interaction is a complicated balance between the effect of temperature on filtering rates, low when the temperature is low, and the accelerated consumption of food for growth and reproduction with increasing temperatures. Thus, at low temperatures, if the food level is low, the reduction of food intake rate may not be compensated for by the corresponding decrease of the metabolic rate. On the other hand, at high temperatures the accelerated metabolism could only be satisfied if there were high quantities of food in the medium. This is clearly illustrated by the data between the interaction of food density and temperature given by Starkweather (1987, Table 1). However, it has also been observed, especially at high temperatures, that an exaggerated excess of food could be detrimental as a result of the decomposition of the surplus food (Stemberger & Gilbert, 1985; Galkovskaya, 1987).

From Table 1, three kinds of response can be discriminated. The first 5 species may be considered warm water adapted; they show their highest or maximum r values at temperatures over 27 °C. These maxima range most frequently from 1 to 2. Exceptional r values of 2.8–2.9 have also been found for clones with a high food supply, acclimatized to high temperatures (37 °C). *Keratella cochlearis* can be considered as a cold water species, the studied clone has a very low maximum r (<0.1) localized at 15 °C, which is the result of a low fecundity and slow development, thus, its longer life insures the maintenance of its population. The studied clone of *Brachionus angularis* has an intermediate position; it reaches a rather low maximum r (0.35) at 20 °C.

In the first group of species, the averaged relative increase of r, when the temperature rises from 20 to 25 °C, is around 10% per degree of temperature increase, although it varies from 2 to 35%, which correspond to ratios $r_{25 °C}/r_{20 °C}$ from 1.1 to 2.6 (mean around 1.5). Using the Q_{10} concept for r,

$$\ln Q_{10} = 10 (\ln r_2 - \ln r_1)/(T_2 - T_1),$$

we obtain values from 1.2 to 6.7 (mean around 2.3). Slightly smaller values of these parameters were found for increases from 25 to 30 °C. In most cases within clones, the relationship between r and temperature is almost linear from 20 to 30 °C.

On the other hand, the studied strain of *Keratella cochlearis* showed much higher relative increases: around 120% from 5 to 10 °C and 36% from 10 to 15 °C, with corresponding ratios $r_{10 °C}/r_{5 °C} = 7$ ($Q_{10} = 50$) and $r_{15 °C}/r_{10 °C} = 2.8$ ($Q_{10} = 8$). *Brachionus angularis* again had an intermediate position.

Although data are very fragmentary, it can be seen that if, for comparison, we plot (Fig. 2) the three types of response of r with temperature, standardizing the data of each clone to their corresponding maximum r value, r_{max}, and their optimum temperature (*i.e.*, temperature at which r is maximum, $T_{r_{max}}$), a family of similar curves is obtained. They are composite, highly asymmetric,

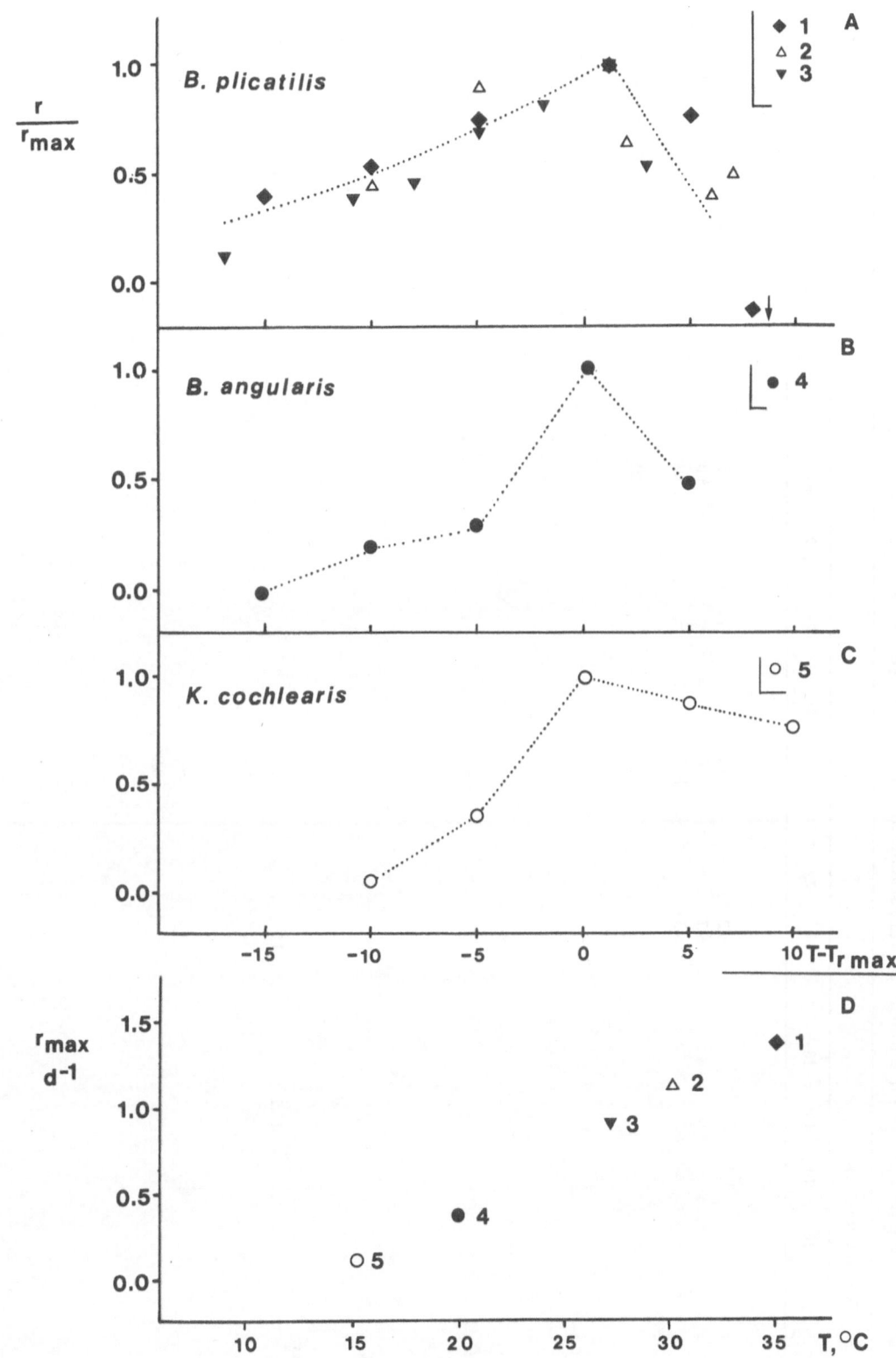

Fig. 2. A, B, and C. Relative r values (r/r_{max} on the ordinate) as a function of the distance from the optimum temperature $(T - T_{r_{max}})$ on the abcissa. D. Maximum r (r_{max}) and its corresponding temperature ($T_{r_{max}}$) for the indicated rotifers. (1: Hirayama & Kusano, 1980; 2: Snell, 1986; 3: Pascual & Yúfera, 1983; 4: Walz, 1987; 5: Walz, 1983).

Table 2. Duration of life (in days) of some rotifers growing at different temperatures (LT 50, age for a survival probability of 0.5).

Rotifer species	Temperature (°C)																			Reference
	5	10	14	15	18	19	20	22	24	25	26	27	28	30	32	35	37	38	40	
Brachionus plicatilis																				
[1]							6.2			6.2				6.8						King & Miracle, 1980
[2]							6.7			6.0				6.0						
[3]							16.5			13.5				11.0						
plicatilis [4]		16.2				19.2				8.0				2.9					[LT 50]	Hirayama & Kusano, 1972
plicatilis [5]					9.9								6.4					3.3		Shell, 1986
plicatilis [8]		11.5					26.5													Nagata, 1985
plicatilis [9]				15			10			7										Ruttner-Kolisko, 1972
Brachionus calyciflorus																				
[12]														2.61			2.20		1.5	Galkovskaya, 1987
[13]														2.72			1.75		1.16	
calyciflorus [14]				17.2			11			5									[LT 50]	Halbach, 1973
Brachionus dimidiatus																				
[19]							9.5			7.7				3.5						Pourriot & Rougier, 1975
[21]							9			8.9				3.6						
Euchlanis dilatata																				
[24]						3.3						2.5								King, 1972
[25]						3.1						3.2								
[26]						3.4						3.76								
Asplanchna sieboldi																				
[27]								6.1		4			2.5							Gilbert, 1976
[28]								6.9		5.26			2.9							

[Saccatte morphotype]
[Campanulate morphotype]

Species						Reference
Asplanchna brightwelli [30]			4.7	3.2	2.6	Snell & King, 1977
Brachionus angularis [29]	19.4	14.9	4.1	4.5	2.8	[LT 50] Walz, 1987
Keratella cochlearis [29]	27.0	21.8	14.9	8.8	6.3	Walz, 1983
Hexarthra fennica [31]		12.5		6.3	4	Ruttner-Kolisko, 1975
Philodina roseola [32]	35	47	(42)	15	5 3	Lebedeva & Gerasimova, 1987

Food conditions and other observations:

From [1] to [29], see the footnote of Table 1.

[30] *Paramecium aurelia.*
[31] *Chlorella vulgaris* (10^6 cell/ml).
[32] *Chlorella vulgaris* (11×10^6 cell/ml); the value in parentheses was included by the authors from Ricci, 1976.

curves divided by a sharp maximum. The left slope is higher than the right slope in the cold-adapted species and the opposite occurs in warm-adapted species. Below this maximum, r is an exponential function of temperature, but in a homogeneous experiment and for a specific range of temperature this function could be almost linear. This response to temperature reflects metabolic, embryonic and post-embryonic development rates, and could be fitted to Belehradek's or Arrhenius' equation. The second part of the curve is most probably a disequilibrium between reproduction and survival; reproduction at a too early stage, as a consequence of high temperature, could diminish further survival (Snell & King, 1977). This is most dramatic when the set of environmental conditions is not quite suitable for the rotifer at the experimental temperature.

In most cases, in the first group of warm-water species, net reproduction, R_0, varies little (Hirayama & Ogawa, 1972; King & Miracle, 1980; Pourriot & Rougier, 1975; Halbach, 1970; Ruttner-Kolisko, 1972), within a certain temperature range. The effect of temperature on r is mainly due to its effect on metabolism and development if the rotifer remains under a suitable set of conditions. R_0 is only low at the most extreme temperatures.

For species in colder temperatures, R_0 is also variable (Walz, 1983, 1987) and has an important effect on r. In this case, values of R_0 are low owing to the fact that reproduction starts very late, and then the juvenile period constitutes a high percentage of the mean life span (approx. 50% at low temperatures versus approx. 30% at the optimum temperature). Under these conditions the reproductive period is short. This delay in initial reproduction reduces r by increasing the generation time. At very high temperatures, gross reproduction may not vary, but low survival decreases R_0 as we mentioned before. At extreme values of temperature gross reproduction diminishes also.

The life span

Table 2 compiles all the data found on mean life span at different temperatures. Most littoral-

benthic rotifers (especially bdelloids: Ricci, 1978; Lebedeva & Gerasimova, 1987, but also *Lecane tenuiseta*, Hummon & Bevelhymer, 1979) have much longer life spans than planktonic rotifers, as expected by their occupation of more stable environments. However, there are exceptions, such as *Euchlanis dilatata* which showed the shortest of all the reviewed life spans. Apart from the littoral-benthic species, the mean life span varies, most frequently, from 3 to 20 days, and the ratio between life spans for a 5 °C increase is most frequently from 1.2 to 2, which indicates diminutions between 4 and 20% (averages 1.6, 11%).

Within a species, clones show longer life spans according to size and their adaptation to colder temperatures (King & Miracle, 1981). Also the life span of cold-adapted *Keratella cochlearis* is somewhat longer than that of the other studied planktonic species.

Life span decreases with a rise of temperature, the relationship again being similar to the duration of embryonic and post-embryonic development, intervals between egg depositions, etc. Life span matches the corresponding acceleration of reproductive effort (descendants per female per day) with temperature, and an inverse relationship exists between the two responses. As Snell & King (1977) pointed out, reproduction decreases the probability of future survival. However, in this case the physiological basis for an inverse relationship between reproduction and survival is the acceleration of development, the organisms more or less attain their potential number of descendents during their life time, but over a longer or shorter period with different timings.

Salinity effects on individuals

Total dissolved salts and relative specific ionic concentrations are important factors conditioning rotifer distribution (Ruttner-Kolisko, 1971; Miracle *et al.*, 1987). Several rotifer genera have halobiont species living in a very wide range of salinities. *Brachionus, Hexathra, Notholca, Synchaeta* are the most relevant. Moreover a number of essentially freshwater forms can also be salt tolerant to a certain extent. Nevertheless, studies on salinity effects on rotifer individuals or populations are extremely scarce.

Osmotic regulation has seldom been investigated. However, there is good evidence to suggest that highly evolved rotifers can regulate the salt concentration of the pseudocoelomic fluid by means of their flame cells and contractile vesicle. This has been confirmed for *Asplachna* (Pontin, 1964, 1966; Braun *et al.*, 1966) in which larger species or forms have a higher number of flame bulbs and in which the vibratile flames show an activity inversely related to the concentration of the medium. In addition, when the animal is placed in a more dilute medium, the protonephridium responds by sodium conservation, increased water excretion and decreased total solute excretion. Kabai & Gilbert (1978) relate the osmoregulatory capacity of these viviparous organisms to the independence of the embryos from the external environment, supported by the negligible effects that very severe osmotic decreases (10–100 fold dilutions of the normal medium) had on the response of body wall-outgrowth to tocopherol, in *A. sieboldi*. At extreme dilutions, *A. sieboldi* fecundity was drastically reduced. There is also evidence, from electron microscopy studies of protonephridia (Clément, 1968), that *Notommata copeus* also

osmoregulates. It seems however that both (*Asplanchna* and *Notommata*) are incapable of hypoosmotic regulation. These rotifers, as well as other rotifers, may behave as a group of 'essentially freshwater species that are salt tolerant' as described in Bayly (1972). These animals are hyperosmotic regulators until the salinity of the external medium more or less coincides with that of the body fluids (Fig. 3, A). At this point, many species reach their upper limit of tolerance, which may be widened by acclimatization through increases of the isosmotic point. Other species of this group which are somewhat more salt tolerant, behave as osmoconformers for salinities over isosmoticy. Many rotifer species could be placed in this group although little or no information exists on this subject.

Of the typical halobiont species which can tolerate a wide range of salinities, only *Brachionus plicatilis* has been studied. According to Epp & Winston (1977), *B. plicatilis* showed a very close correspondence between internal and external osmolarity within the range 41 to 957 mOsmol l⁻¹ with slight but consistent trend towards hyperosmolarity, which was more pronounced as the osmotic pressure of the external medium decreased. This could be represented as Fig. 3 (B). *B. plicatilis*, together with *H. fennica* and *H. jenkinae* were cited by Bayly (1972) as belong-

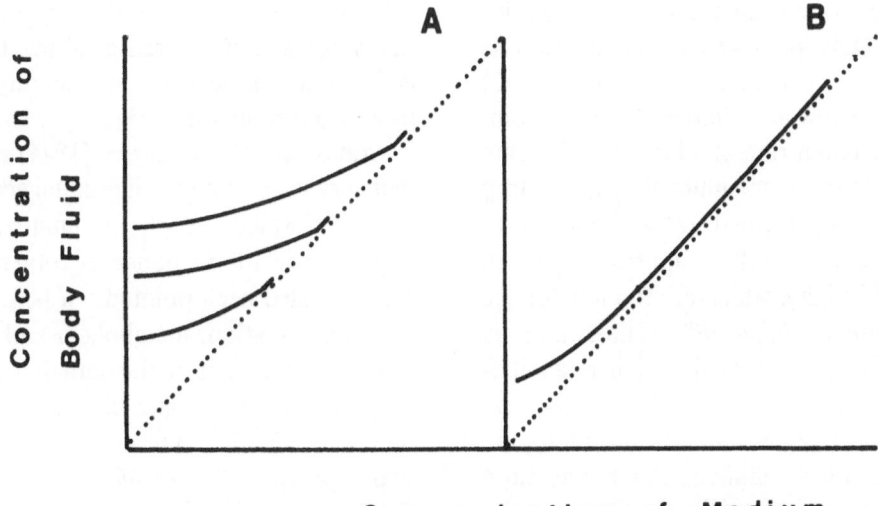

Fig. 3. Ideal relationships expected in rotifers among those proposed for other invertebrates (Bayly, 1972). (A) Hyperosmotic regulation till isosmotic point with different acclimations in freshwater species; (B) euryhaline halobiont species.

ing to the group of 'halobiont but entirely non-marine forms which osmoconform'. He considers that osmotic regulation of extracellular fluids is not necessary for successful colonization of saline waters. All the body cells of the animal may be able to tolerate or regulate their osmolarity instead of having a group of specialized cells of a specific organ regulating the internal fluids. Copepods of these groups, according to Bayly (1972), are capable of slight hyperosmotic regulation, which can be improved with previous acclimatization history, but they are incapable of any hyposmotic regulation.

Little or no information exists on osmoregulation in others rotifers, although genera such as *Synchaeta* and *Notholca* are subdivided in an array of species occupying almost all types of waters, including estuarine and marine environments.

There is some evidence that metabolic and activity responses to salinity correspond to an optimum curve with a more or less extensive plateau. Both the shape and position of the maximum with respect to the salinity axis, is strongly dependent on genotype. Each species or clone has a particular optimum curve and no general trend of a direct or inverse response to salinity has been found. Work in rotifers on activity response to salinity is restricted to *B. plicatilis*. An optimum curve can be extrapolated from the data of Hirayama & Ogawa (1972) on filtration rates, with a maximum occurring at moderate salinities ($= 12 \text{ g l}^{-1}$). Data on oxygen consumption is very fragmentary but Ruttner-Kolisko (1972) found rates to be low at salinities of 1 g l^{-1} compared to a maximum rate at 17 g l^{-1} and a rate only slightly below the maximum at 35 g l^{-1}. Epp & Winston (1978) studied rates from 10 to 100 mOsmol, corresponding approximately to $0.35–3.5 \text{ g l}^{-1}$ salinity, which are too low for the normal behavior of *B. plicatilis*. These authors found that rotifers, preadapted to higher salinities (487 mOsmol), showed a reduction in O_2 consumption and activity when transferred to the above mentioned low salinities, which was more pronounced the lower the salinity. These results were interpreted by the authors as being a consequence of tissue hydration and the dilution of active ions and enzymes, which follows the fall of

the external concentration, in osmo-conformers. They relate this to the assumption of Potts & Perry (1964) that osmo-conformers respond to any change in the accustomed osmotic concentration by reducing their metabolism. This may be true for concentrations well below the optimum, such as those described above, when the body cells are beyond their upper limit of tolerance or regulation. Epp & Winston (1978) also found that activity and O_2 consumption were less reduced in previously acclimatized rotifers, especially at the higher of the above mentioned concentrations.

Salinity effects on populations

The direct effects of salinity on life history traits are strongly dependent on the species and genotype. Raising rotifers at two salinities, 0.2 and 0.4 g l^{-1}, Lansing (1942) found that the mean life span of *Rotifer vulgaris* (= *Rotaria rotatoria*) decreases with salinity, without any significant variation in fecundity, while *Proales* sp. shows a small life span increase, but a reduction in the number of eggs laid per day. According to Aranovich & Spektorova (1974), *Brachionus calyciflorus* decreases its survival as the salinity rises from 2 to 10 g l^{-1}, the negative slope being progressively steeped. In addition, fecundity decreased with salinity, but levelled off at the highest salinities assayed. Thus the decrease in growth rate when salinity increases above the optimum, is due first to a decrease in fecundity and second to a decrease in survival.

Aranovich & Spektorova (1974) also observed that survival was higher if gradual acclimatization was performed. Such acclimatization seems important in the response of rotifers to salinity. This has also been pointed out before in relation to osmoregulation, metabolism and activity.

Table 3 compiles all the data that we could find on parthenogenetic rates of increase in rotifers, at different salinities. These studies are limited to two species of *Brachionus*, *B. plicatilis* and *B. dimidiatus*. The halobiont species *B. plicatilis* grows in a very broad range of salinities, but data on r, for most *B. plicatilis* clones, can be interpreted as an optimum curve with a maximum or

Table 3. The intrinsec rate of increase (*r*, days^{-1}) of the indicated species growing at different salinities.

Salinity (g/l)

Rotifer species	0	1	2	3	4	5	7	9	10	13	16	17	19	20	21	23	24	25	27	29	30	35	36	38	40	42	44	48	50	Reference
Brachionus plicatilis																														
acclimatization																														
3 d			0.64		0.82				0.91					0.59							0.42			0.32	0.21			0.34	0.13	Lubzens *et al.* 1985 [1]
8 d			0.54		0.34				0.89					0.40							0.42			0.33	0.18			0.01	0.03	
14 d			0.30		0.36				0.80					0.68							0.48			0.42	0.21			0.08	0.05	
16 h														0.70							0.57			0.40						
without	<0			<0					0.95					0.92									0.70						0.35	Pascual & Yúfera, 1983 [2]
with		<0	0.86	0.85					0.75					0.78									0.70							
food level, ×10⁶ cells/ml																														
0.1									0.05			<0								<0				<0						Lubzens, 1981 [3]
0.5									0.44			0.45								0.44				<0						
1.0									0.66			0.76								0.43					0.35					
8.0									0.95			0.81								0.64					0.50					
water type																														
sea water			0.46					0.62				0.98					= 0.72					0.56								Ruttner-Kolisko, 1972 [4]
NaHCO$_3$ solution			0.88					0.72				0.62			= 0												<0			
						0.85		0.83		0.94	0.84		0.69	0.69			0.5	0.66	0.4		0.77				0.9*					from Ito, 1966 [5] / Snell, 1986 [6]

Salinity (g/l)

	0.5	1.1	2.2	4.5	9
Brachionus dimidiatus	0.429	0.417	0.419	0.424	0.417	Pourriot & Rougier, 1975 [7]

Food, temperature conditions, and other observations:

[1] *Chlorella stigmatophora* (3 × 10⁶ cell/ml); 25 °C; pre-experimental salinity, 40 g/l; 16th acclimatization, values were calculated in mass cultures.
[2] *Nannochloris oculata* (40–60 × 10⁶ cell/ml); pre-experimental salinity, 36 g/l; 24 °C; growth was not observed at 60 and 80 g/l.
[3] *Chlorella stigmatophora*; 17–23 °C.
[4] *Dunaliella* sp.
[5] *Chlamydomonas* sp.; 1.2–2.9 × 10⁶ cell/ml.
[6] *Chlorella vulgaris* (10 µgC/ml); 25 °C; high sexual reproduction at 20 g/l; *: delayed growth.
[7] *Synechococcus cedrorum*; 20 °C.

a plateau located at moderate salinities between $10–20 \, g \, l^{-1}$. Some clones originally from southern latitudes and high salinity show rather high rates at high salinities (Pascual & Yúfera, 1983; Snell, 1986; however, in Snell, 1986, *r* was estimated from batch culture growth curves, measured at different time intervals and with a high percentage of mixis in the intermediate salinities).

Increased food increases *r* and broadens the range of tolerance, but, in contrast to the response to temperature, the maximum is maintained at about the same position on the salinity axis (Table 3, Lubzens, 1981).

It has been proposed that measures of fitness of rotifer populations are better based on the production of sexual resting eggs, rather than on parthenogenetic growth rates. Several studies have been carried out to determine the importance of sexual reproduction in relation to temperature (Hino & Hirano, 1984; Snell, 1985) and salinity (Ito, 1960; Lubzens *et al.*, 1980; Lubzens, 1981; Hino & Hirano, 1988; Snell, 1985; Lubzens *et al.*,

1985). The reproductive response curve of sexual females to an environmental gradient is more constrained and peaked than that of asexual females, with maxima located at the optimal conditions which are probably the same for both amictic and mictic growth. However, the sexual or asexual reproductive responses to temperature are very different, because sexual reproduction is much diminished at extreme temperatures, and male rates of increase do not follow the same patterns as those of amictic females. By contrast, in a salinity gradient, the maximal sexual reproductive effort is more or less coincident with the optimum salinity for parthenogenetic growth.

Interaction between temperature and salinity

In order to establish the combined effect of temperature and salinity on the intrinsic rate of increase, *r*, we followed individual cultures of three clones of *Brachionus plicatilis* (CU, SPO, and FCA; Serra & Miracle, 1983, 1985, 1987;

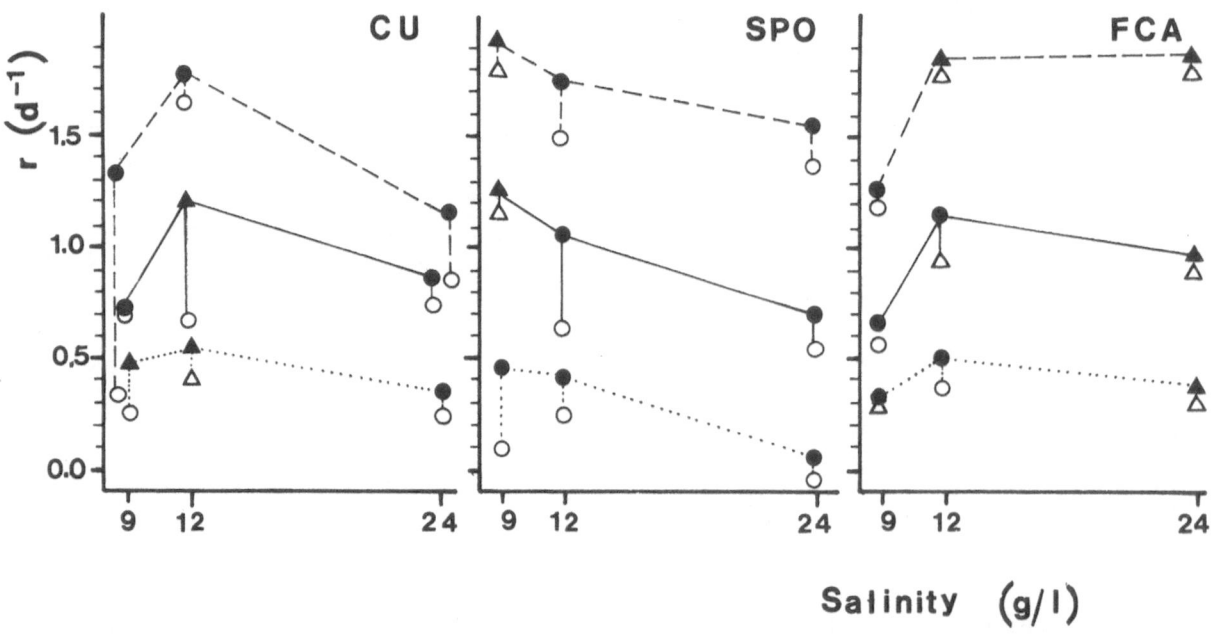

Fig. 4. *r* values of three clones of *Brachionus plicatilis* growing at the indicated salinities and at 20 °C (dotted lines), 25 °C (solid lines), and 30 °C (dashed lines). Closed symbols represent the values calculated for amictic females; open symbols correspond to the total number of females (mictic plus amictic). The triangles indicate which clone shows a maximum *r* at each temperature-salinity combination.

Serra, 1987), at three salinities (9, 12, and 24 g l^{-1}; diluted sea water), and at three temperatures (20, 25, and 30 °C). The r values were calculated from daily survival and fecundity schedules from sets of 50 newborn females grown at each of the nine experimental conditions and observed until their death. They were derived from parents coming from log-phase cultures pre-adapted at the corresponding experimental combinations. The cultures were grown at light and the medium containing 0.5 × 10^6 cell ml^{-1} of *Tetraselmis* sp. was renewed daily.

Two types of r values were calculated. The first type considered only the amictic females – disregarding the mictic ones – and assumed that all their daughters were amictic females. Thus, it estimates the intrinsic rate of increase, assuming the absence of mixis. The second type included all individuals either mictic or amictic. Both r values were computed using an integrated solution of the Lotka's integral equation, assuming a negative exponential interpolation between the observed survival schedules (Serra, 1987).

Figure 4 shows the effects of genotype, temperature and salinity on r. The r values calculated from the subset of amictic females have a clearer pattern in relation to these factors; because of this, we will focus on it. The main feature is an increase of r as the temperature rises, showing most frequently a linear relationship for our data points. These r values, within a temperature, showed great variability due to clone and salinity, but they never overlapped with those at other temperatures. The slope of the r-temperature relationship varies for the different clones. Clone CU, bigger in size and, according to its origin, adapted to lower temperatures shows a much smaller slope than the other two clones (Fig. 5). The same comments made before for the duration of embryonic development (in different species, Pourriot & Rouger, 1975, or clones, Amrén, 1964) or oxygen consumption (Pourriot, 1973b) could be applied here. At low temperature, cold-adapted genotypes have relatively higher r than more thermophilic ones, while the latter have relatively higher rates at higher temperatures. The same results are obtained in other studies when different *B. plicatilis* clones are compared. In Fig. 5

we have plotted together our results on the response of r to temperature and those calculated from data of King & Miracle (1981), where also a bigger clone (SP) was compared with two smaller and more thermophilic clones.

The effect of salinity on the intrinsic rate of increase is less pronounced than that of temperature. It is conditioned mainly by clone, but has a high interaction with temperature. Thus, for clone CU the highest value of r has been found at middle salinity, while SPO shows a decrease of r as the salinity rises. The FCA clone shows a trend to maintain high r values at high salinity, most notably when temperature is high.

Figure 4 indicates which clone would hypothetically be dominant in each condition (triangle vs. circles). According to these results, CU would have advantage in relatively cold and low salinity waters, SPO in warm waters with relatively low salinities, and FCA in warm waters with high salinities. FCA shows also a good response to conditions where salinity and temperature are positively correlated.

This behaviour of clones is quite closely adjusted to the one expected if they were adapted to their habitats. Thus, the origin of FCA is a coastal marsh with highly variable salinity, reaching more than 35 g l^{-1}. We obtained this clone from the Institute of Fisheries Research of Castellón, where it was cultured in sea water and at approximately 26 °C. SPO was isolated by us from a relatively stable coastal lagoon, in where salinity was 13 g l^{-1} and temperature, 26.6 °C. Finally, CU was isolated from a endorreic athalasic lagoon, which, at the time of collection, had a salinity of 25 g l^{-1} and a temperature of 17 °C.

The response of r to some combinations of temperature and salinity suggests a synergestic effect between these parameters, which has been explored studying their additive interaction. The interaction ($r_{T \cdot S}$, Fig. 6) has been calculated as the difference between the observed value of r, at each condition combination, and the expected value of r assuming an additive effect of salinity and temperature. The expected r is calculated as the addition of the following three terms: (1) grand mean for the clone, (2) the deviation to this grand mean of the averaged *r* for the specific temperature, and (3) the deviation from this grand mean of the averaged *r* for the specific salinity. The results are shown in Fig. 6.

The most interesting response is that of FCA. It shows a strong interaction between the studied factors. At the high salinity-high temperature, at middle salinity-middle temperature, and at low salinity-low temperature FCA has a higher *r* than would be expected without interaction. This can be related to the fact that a positive correlation between temperature and salinity is frequent in the natural waters of Mediterranean region, where in summer high temperature and dryness causes an increase of water salinity. Thus, FCA, the clone isolated from a variable marsh, seems to be adapted to the fluctuations of natural conditions.

For clones CU, the interaction between tem-

perature and salinity is more complex. This clone shows a relative advantage at low and middle levels of both salinity and temperature, and if only these conditions are considered, the pattern of CU is similar to that of FCA; the *r* is above the expected value at both low-low and middle-middle combinations. Finally, clone SPO presents a 'smooth relief', pointing out a poor response to specific combinations of temperature and salinity.

To investigate the underlying causes of the obtained variation of *r*, we have explored the relationships between *r* and other life history traits, being particularly easy for global – age independent – life history traits. An approximation of *r* could be obtained from the net reproductive rate, R_0, and the cohort generation time, T_c, *i.e.*, $r_c = \ln(R_0)/T_c$. In turn, R_0 can be decomposed in

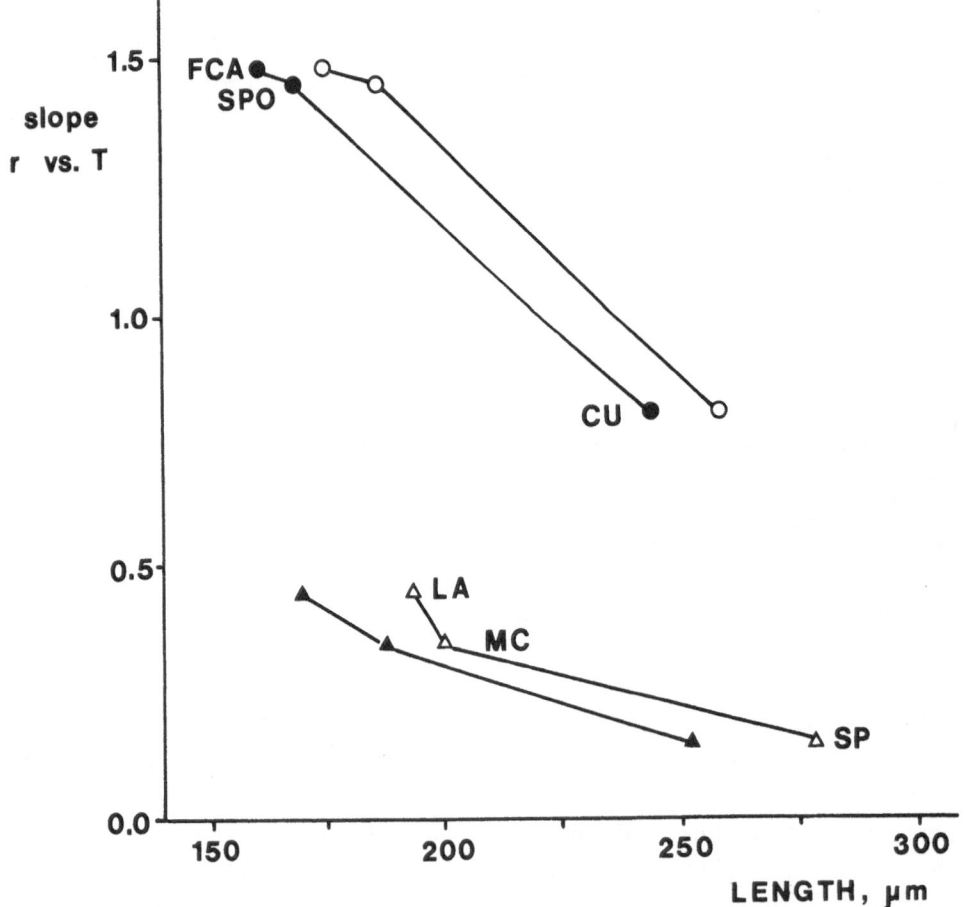

Fig. 5. The slope of the relationship between *r* and temperature as a function of the mean body length of four *Brachionus plicatilis* strains, measured at two fixed temperatures (data for LA, MC, and SP strains were obtained from King & Miracle, 1980).

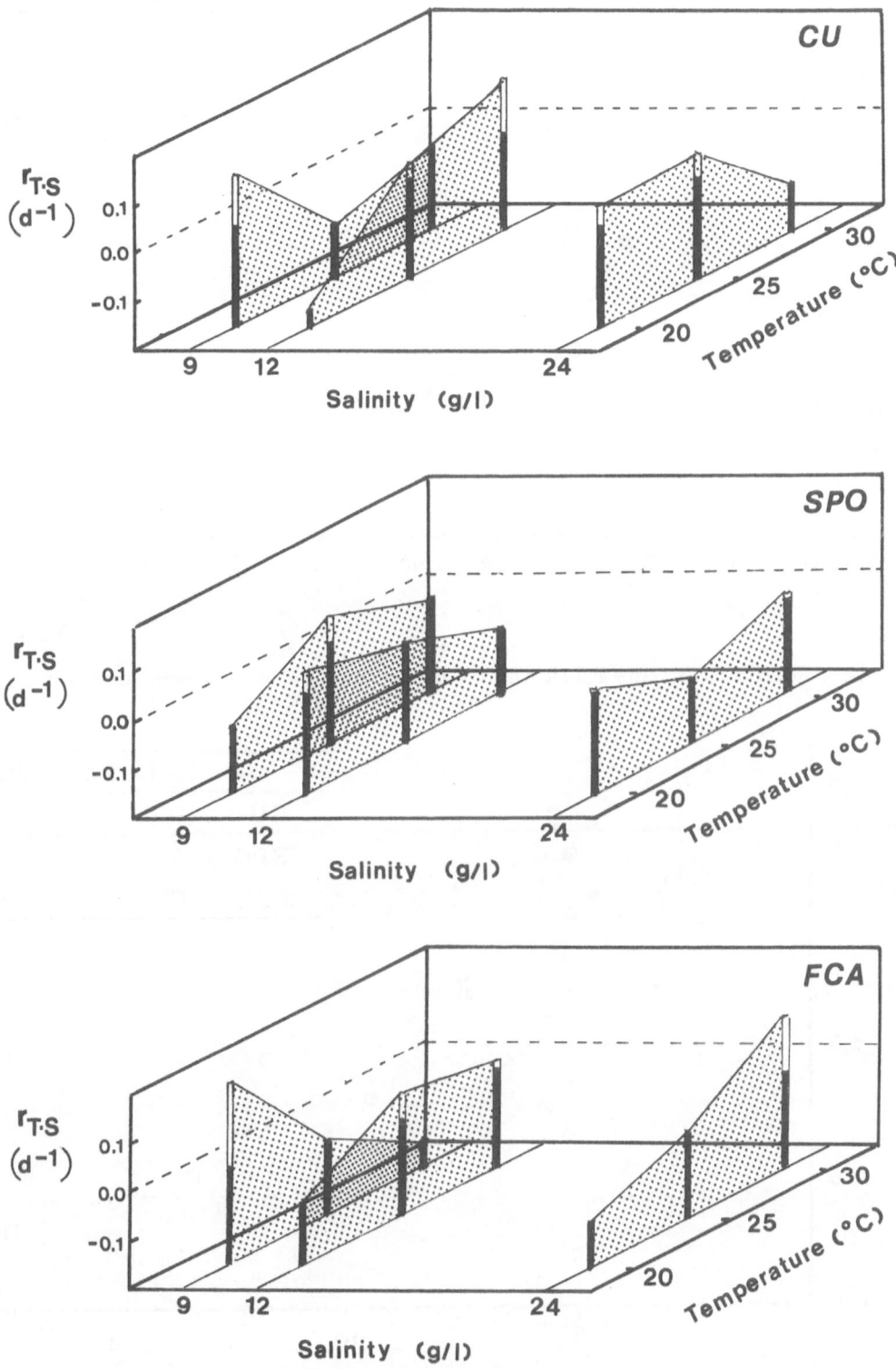

Fig. 6. Salinity-temperature interaction, $r_{T \cdot S}$ (days^{-1}, see text) for three clones of *Brachionus plicatilis* cultured at the indicated conditions. The bars represent that part of r which is due to interaction, and are solid until the interaction equals zero and empty when it is positive.

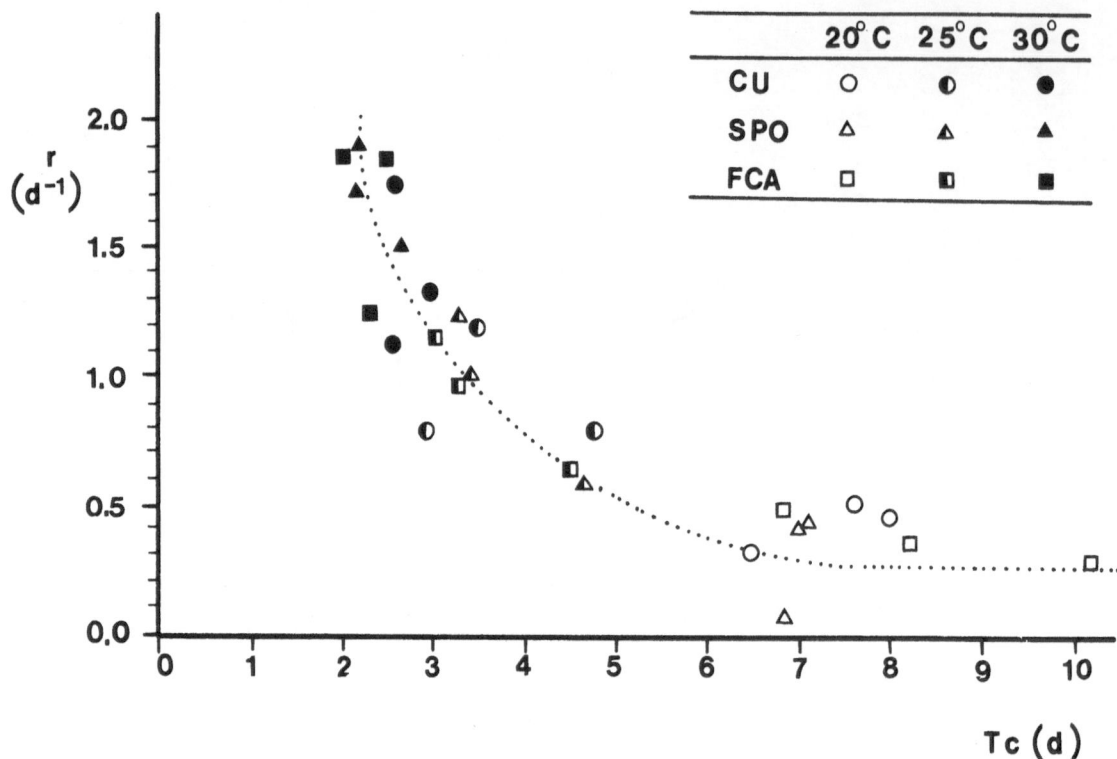

Fig. 7. Relationship between *r* and the cohort generation time, T_c (days), in *Brachionus plicatilis*.

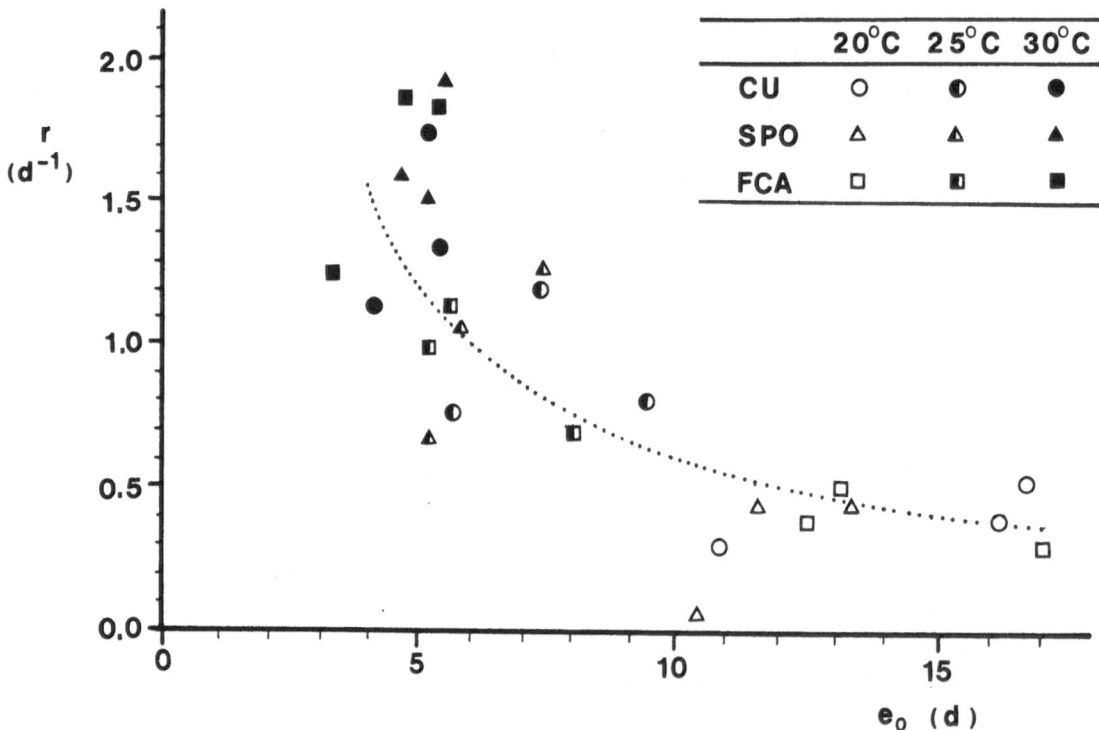

Fig. 8. Relationship between *r* and the mean life expectancy, e_0, in *Brachionus plicatilis*. The dotted line has been calculated as $r \cdot e_0 = K$, *K* being the average of the $r_i \cdot e_{0i}$ for the 27 experiments, that is $K = 6.3$. This equation that assumes the potential population growth throughout the life of an individual is the same for all conditions, *i.e.*, assumes that the intrinsic rate of increase relative to lifespan is constant.

the product between the duration of life, measured as the mean expectancy of life, e_0, and the average number of daughters per day: $F_d = R_0/e_0$.

As expected, r and T_c (Fig. 7) are related by an asynthotic inverse curve. The values are mainly arranged according to temperature, and there are not clear distinctions between values belonging to different clones or salinities. Looking at the slope of this relationship, it is observed that the covariation between r and T_c is high at 25 °C, lower at 30 °C, and nearly zero at 20 °C.

The mean expectancy of life, e_0, presents a pattern similar to T_c in its relationship with r (Fig. 8). This similarity could be expected as a consequence of the relationship between T_c and the duration of life.

The relationship between r and F_d is also strongly dependent on temperature (Fig. 9), and within the same temperature, a linear correlation between r and F_d has been found. When a functional regression analysis between these parameters was performed, lines with similar slopes were obtained.

Our results show that temperature not only affects the r values, but also the relationships between r and the other demographic parameters analyzed. In contrast, genotype and salinity do not cause special relationships between the studied traits.

Two important conclusions can be deduced from our results: (1) At changing temperatures the rate of growth and life span or generation time are adaptively and physiologically adjusted, so the number of descendants per female (R_0) remains constant. If we plot (Fig. 8) the relation between r_c and T_c by inferring r_c from the potential R_0 (the maximum R_0 found in our results; i.e., 24) by $r_c = \ln24/T_c$, we obtain a line which matches almost exactly that of expected r assuming constancy in the r relative to life span (dotted line in Fig. 8). (2) Both salinity and genotype have important effects on fecundity, which can be seen in the relationship between r and F_d at different temperatures. On the other hand, their effect on life span is much less apparent.

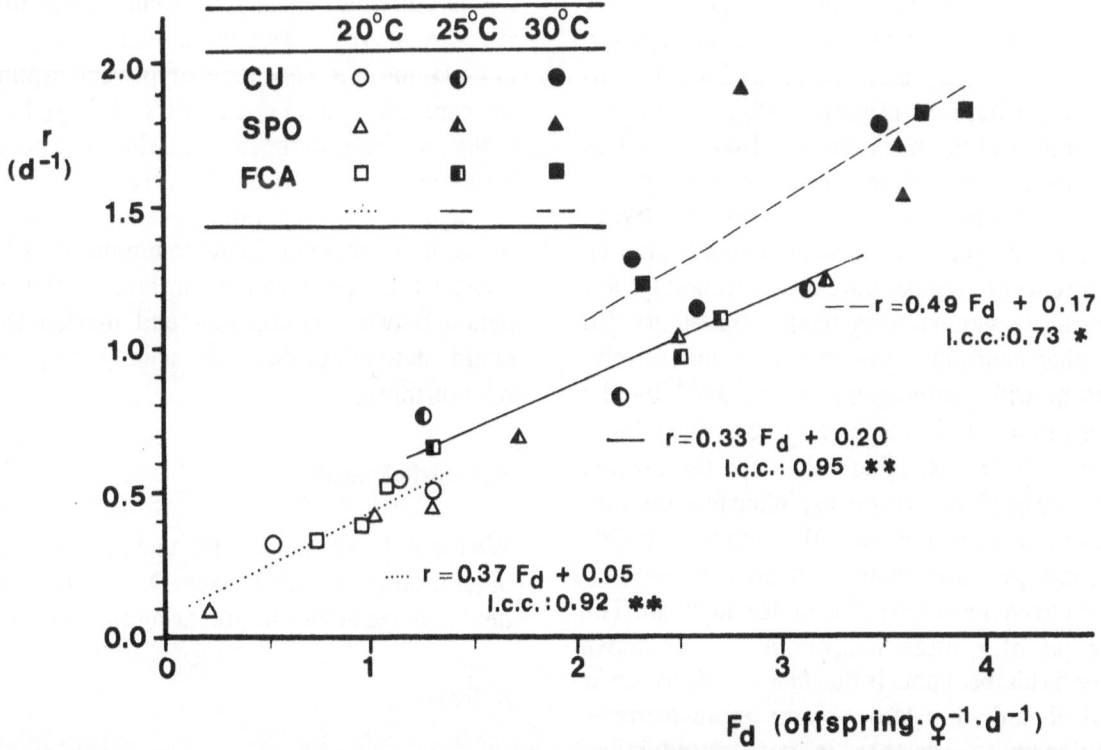

Fig. 9. Relationship between r and the reproductive effort, F_d, average offspring per female and day of life) in *Brachionus plicatilis*. The relationship is plotted for each temperature, indicating the corresponding regression equation with its significance (l.c.c., linear correlation coefficient).

General considerations

On the basis of laboratory studies we observe that temperature has a direct effect on r. This influence is mainly derived from the response direct of developmental rates or metabolic activities to increasing temperature, and the resultant r values reach a maximum. Beyond this point, r sharply decreases (Fig. 2). The temperature of the maximum is the upper limit for the normal functioning of the animal in the particular set of remaining conditions. The slope of the exponential curve and the temperature at which the maximum occurs depend on genotype (Fig. 2), and cold- or warm-adapted genotypes are clearly differentiated. Cold-adapted species or clones have in general, smaller slopes than warm-adapted ones (Fig. 2). This confirms the general observation that cold-adapted species have relatively higher rates of increase at lower temperatures than warm-adapted ones, and the latter have relatively higher rates at higher temperatures. It is interesting to note also that within species cold-adapted genotypes have a larger size (Fig. 5). This trend is coincident with the direct effect of temperature on the size of genotypically identical individuals, which have a bigger size when reared at low temperatures (Serra & Miracle, 1987). Slope and temperature of the maximum are also more or less influenced by environmental factors. Food in particular determines the temperature limits between which the response of r is exponential (Table 1).

The effect of temperature on net fecundity, R_0, is negligible within normal temperature limits. On the other hand, life span shows an inverse relationship with temperature, as expected by the acceleration of development rates. Thus, temperature (within suitable values for the normal functioning of the organisms) influences the timing, but not fertility (i.e., the number of descendents per individual), and an increase of r results from the acceleration of development. The decrease of r when temperature values move beyond suitable limits is due first to a decrease in survival and, if values become more extreme, additionally to a decrease in gross reproduction.

On the other hand, the direct effect of salinity on r depends on the genotype; the same findings can be observed in other invertebrates (Fava et al., 1983). The genotype is adapted to an optimum salinity in which r is maximum. Then, r decreases as salinity conditions move away from this optimum. The decrease of r is due first to a decrease of fertility and, if salinity moves still farther from the optimum, a decrease of survival.

In the response of rotifer rate of increase, the interaction of temperature-salinity is important. In our studies of Brachionus plicatilis clones from habitats with greatly fluctuating temperature and salinity, a positive interaction or beneficial effect is found for the low/low and high/high temperature/salinity combinations within the normal limits of those factors in the habitats. Kinne in several papers (reviewed in Kinne, 1970; Alderdice, 1972) found similar results for other invertebrates and fishes. However, this is not general and Dorgelo (1976) compiles other beneficial combinations for survival in crustaceans, such as high/low temperature/salinity combinations in the tropics. The responses of species or ecotypes seem to match their habitats of origin. In temperate climates a positive correlation between temperature and salinity is usually found during the annual cycle. Temperature-salinity positive interactions in B. plicatilis ecotypes correspond to the expected annual covariation of those factors in their habitats. In the tropics, where temperature is more constant and salinity varies with the wet-dry seasons (corresponding to summer and winter) other factors may be more important. Also in coastal brackish water environments the interaction between continental and marine waters could determine different salinity-temperature relationships.

Acknowledgments

We thank L. Serrano for her technical assistance in laboratory. We also thank Dr. C. Dawson for her language advice in preparing the manuscript.

References

Alderdice, D. F., 1972. Responses of marine poikilotherms to environmental factors acting in concert. In: O. Kinne (ed.), Marine ecology, vol. 1. Wiley, London: 1659–1722.

Allan, J. D., 1976. Life history patterns in zooplankton. Am. Nat. 110: 165–180.

Allan, J. D. & C. E. Goulden, 1980. Some aspects of reproductive variation among freshwater zooplankton. In: W. C. Kerfoot (ed.), Evolution and ecology of zooplankton communities. University Press of New England, Hanover: 388–410.

Amrén, H., 1964. Ecological studies of zooplankton populations in some ponds on Spitsbergen. Zool. Bidr. Upps. 36: 162–191.

Armengol, J., 1978. Zooplankton crustaceans in Spanish reservoirs. Verh. int. Ver. Limnol. 20: 1652–1656.

Aronovich, T. M. & L. V. Spektorova, 1974. Survival and fecundity of Brachionus calyciflorus in water of different salinities. Hydrobiol. J. 10: 71–74.

Bayly, I. A. E., 1972. Salinity tolerance and osmotic behavior of animals in athalassic saline and marine waters. Annu. Rev. Ecol. Syst. 3: 233–268.

Braun, G., G. Kummel & J. A. Mangos, 1966. Studies on the ultrastructure and function of a primitive excretory organ, the protonephridium of the rotifer Asplanchna priodonta. Pflügers Arch. 189: 141–154.

Clément, P., 1968. Ultrastructures d'un rotifère, Notommata copeus. I. La cellule-flamme. Hypothèses physiologiques. Z. Zellforsch. 89: 478–498.

Dorgelo, J., 1976. Salt tolerance in crustacea and the effect of temperature upon it. Biol. Rev. 51: 255–290.

Duncan, A., 1983. The influence of temperature upon the duration of embryonic development of tropical Brachionus species (Rotifera). In: F. Schiemer (ed.), Limnology of Parakrama Samudra Sri-Lanka. Dr. W. Junk, The Hague.

Edmondson, W. T., 1965. Reproductive rate of planktonic rotifers as related to food and temperature in nature. Ecol. Monogr. 35: 61–111.

Epp, R. W. & P. W. Winston, 1977. Osmotic regulation in brackish-water rotifer Brachionus plicatilis (Müller). J. Exp. Biol. 68: 151–156.

Epp, R. W. & P. W. Winston, 1978. The effect of salinity and pH on the activity and oxygen consumption of Brachionus plicatilis (Rotatoria). Comp. Biochem. Physiol. 59A: 9–12.

Epp, R. W. & W. M. Lewis, 1980. Metabolic uniformity over the environmental temperature range in Brachionus plicatilis (Rotifera). Hydrobiologia 73: 145–147.

Epp, R. W. & W. M. Lewis, 1984. Cost and speed of locomotion for rotifers. Oecologia (Berl.) 61: 289–292.

Fava, G., I. Lazzaretto & E. Martini, 1983. Effetti della riduzione di salinità in due diverse popolazioni lagunari di Tisbe clodiensis (Copepoda, Harpacticoida). Atti Ist. Veneto Sci. 141: 105–120.

Galkovskaya, G. A., 1983. On temperature acclimation in a experimental population of Brachionus calyciflorus. Hydrobiologia 104: 225–227.

Galkovskaya, G. A., 1987. Planktonic rotifers and temperature. Hydrobiologia 147: 307–317.

Galkovskaya, G. A., J. Ejsmont-Karabin & V. N. Evdokimov, 1987. Relative protein metabolism in rotifer Brachionus calyciflorus Pallas, in relation to temperature. Int. Rev. Gesamten Hydrobiol. 72: 59–69.

Gilbert, J. J., 1976. Polymorphism in the rotifer Asplanchna sieboldi: biomass, growth, and reproductive rates of the saccate and campanulate morphotypes. Ecology 57: 542–551.

Ito, T., 1960. On the culture of mixohaline rotifer Brachionus plicatilis O. F. Müller in sea water. Rep. Fac. Fish., Prefec. Univ. Mie. 3: 708–740.

Halbach, U., 1973. Life table data and population dynamics of the rotifer Brachionus calyciflorus Pallas as influenced by periodically oscillating temperature. In: W. Wieser (ed.), Effects of temperature in ectothermic organism. Springer-Verlag. Berlin: 217–228.

Herzig, A., 1983. Comparative studies on the relationship between temperature and duration of embryonic development of rotifers. Hydrobiologia 104: 237–246.

Hirayama, K. & S. Ogawa, 1972. Fundamental studies on physiology of rotifer for its mass culture. I. Filter Feeding of Rotifer. Bull. Jpn. Soc. Sci. Fish. 38: 1207–1214.

Hirayama, K. & T. Kusano, 1972. Fundamental studies on physiology of rotifer for its mass culture. II. Influence of water temperature on population growth of rotifer. Bull. Jap. Soc. Sci. Fish. 38: 1357–1363.

Hino, A. & R. Hirano, 1984. Relationship between water temperature and bisexual reproduction rate in the rotifer Brachionus plicatilis. Bull. Jap. Soc. Sci. Fish. 50: 1481–1485.

Hino, A. & R. Hirano, 1988. Relationship between water chlorinity and bisexual reproduction rate in the rotifer Brachionus plicatilis. Nippon Suasan Gakkaishi 54: 1329–1332.

Hummon, W. D. & D. P. Bevelhymer, 1980. Life table demography of the rotifer Lecane tenuiseta under culture conditions and various age distributions. Hydrobiologia 70: 25–28.

Kabay, M. E. & J. J. Gilbert, 1978. Polymorphism in the rotifer Asplanchna sieboldi: Insensitivity of the body-wall-outgrowth response to temperature, food density, pH and osmolarity differences. Arch. Hydrobiol. 83: 377–390.

Kinne, O., 1970. Temperature: Animals-Invertebrates. In: O. Kinne (ed.), Marine ecology, vol. 1. Wiley, London: 407–514.

King, C. E., 1972. Adaptation of rotifers to seasonal variation. Ecology 53: 408–418.

King, C. E. & M. R. Miracle, 1980. A perspective on aging in rotifers; Hydrobiologia 73, 13–19.

Lansing, A. I., 1942. Some effects of hydrogen ion concentration, total salt concentration, calcium and citrate on longevity and fecundity of rotifer. J. Exp. Zool. 91: 195–211.

Lebedeva, L. I. & T. N. Gerasimova, 1985. Peculiarities of Philodina roseola (Ehrbg.) (Rotatoria Bdelloida). Growth and reproduction under various temperature conditions. Int. Rev. Gesamten Hydrobiol. 70: 509–525.

Lebedeva, L. I. & T. N. Gerasimova, 1987. Survival and reproduction potential of Philodina roseola (Ehrenberg) (Rotatoria Bdelloida) under various temperature conditions. Int. Rev. Gesamten Hydrobiol. 72: 695–707.

Lubzens, E., 1981. Rotifer resting eggs and their application to marine aquaculture. Eur. Maricult. Soc. Spec. Publ. 6: 163–180.

Lubzens, E., R. Fishler & V. Berdugo-White, 1980. Induction of sexual reproduction and resting egg production in *Brachionus plicatilis* reared in sea water, Hydrobiologia 73: 55–58.

Lubzens, E., G. Minkoff & S. Marom, 1985. Salinity dependence of sexual and asexual reproduction in the rotifer *Brachionus plicatilis*. Mar. Biol. (Berl.) 85: 123–126.

Miracle, M. R., 1974. Niche structure in freshwater zooplancton: a principal components approach. Ecology 55: 1306–1317.

Miracle, M. R., M. Serra, E. Vicente & C. Blanco, 1987. Distribution of *Brachionus* species in Spanish mediterranean wetlands. Hydrobiologia 147: 75–81.

Nagata, W. D., 1985. Long-term acclimation of a parthenogenetic Strain of *Brachionus plicatilis* to subnormal temperatures. I. Influence on size, growth, and reproduction. Bull. mar. Sci. 37: 716–725.

Nogrady, T., 1988. The littoral rotifers plankton of the Bay Quinte (Lake Ontario) and its horizontal distribution as indicators of trophy. I. A full season study. Arch. Hydrobiol. Suppl. 79: 145–156.

Pascual, E. & M. Yúfera, 1983. Crecimiento en cultivo de una cepa de *Brachionus plicatilis* O. F. Müller en función de la temperatura y la salinidad. Invest. Pesq. 47: 151–159.

Pontin, R. M., 1964. A comparative account of the protonephridia of *Asplanchna* (Rotifera) with special reference to the flame bulbs. Proc. zool. Soc. Lond. 142: 511–525.

Pontin, R. M., 1966. The osmoregulatory function on the vibratile flames and the contractile vesicles of *Asplanchna* (Rotifera). Comp. Biochem. Physiol. 17: 1111–1126.

Potts, W. T. & G. Parry, 1964. Osmotic and ionic regulation in animals. Pergamon Press, Oxford.

Pourriot, R., 1973a. Rapports entre la température, la taille des adultes, la longueur des œufs en le taux de développement embryonnaire chez *Brachionus calyciflorus* Pallas (Rotifère). Ann. Hydrobiol. 4: 103–115.

Pourriot, R., 1973b. Influence de la teneur en protéines, de la température et du jeûne sur la respiration de Rotifères héléoplanctonctoniques. Verh. int. Ver. Limnol. 18: 1429–1433.

Pourriot, R. & M. Deluzarches, 1971. Recherches sur la biologie des rotifères. II. Influence de la température sur la durée du développement embryonaire et post-embryonaire. Ann. Limnol. 7: 25–52.

Pourriot, R. & C. Rougier, 1975. Dynamique d'une population expérimentale de *Brachionus dimidiatus* (Bryce) (Rotifère) en fonction de la nourriture et de la température. Ibidem, 11: 125–143.

Ricci, C., 1976. Note preliminare sull'allevamento di un rotifero Bdelloidea. Atti. Soc. Ital. Sci. Nat. Museo Civ. Stor. Nat. Milano 117: 144–148.

Ricci, C., 1978. Some aspects of the biology of *Philodina roseola* (Rotifera). Me. Ist. ital. Idrobiol. 36: 109–116.

Ruttner-Kolisko, A., 1971. Rotatorien als Indikatoren für den Chemismus von Binnensalzgewässern. Sitz-Ber. Österr. Akad. Wiss. Math. Nat. Kl. Abt. I. 179: 283–298.

Ruttner-Kolisko, A., 1972. Der Einfluß von Temperatur und Salzgehalt des Mediums auf Stoffwechsel- und Vemehrungsintensität von *Brachionus plicatilis* (Rotatoria). Dt. Zool. Ges. 65: 89–95.

Ruttner-Kolisko, A., 1975. The influence of fluctuating temperature of plankton rotifers. A graphical model based on live data of *Hexarthra fennica* from Neusiedlersee, Austria. Symp. Biol. Hung. 15: 197–204.

Ruttner-Kolisko, A., 1978. Influence of fluctuating temperature of plankton rotifers. II. Laboratory experiments. Verh. int. Ver. Limnol. 20: 2400–2405.

Serra, M., 1987. Variación morfométrica, isoenzimática y demográfica en poblaciones de *Brachionus plicatilis*. Diferenciación genética y plasticidad fenotípica. Ph. D. Thesis, Universitat de Valencia, Valencia.

Serra, M. & M. R. Miracle, 1983. Biometric analysis of *Brachionus plicatilis* ecotypes from Spanish lagoons. Hydrobiologia 104: 279–291.

Serra, M. & M. R. Miracle, 1985. Enzyme polymorphism in *Brachionus plicatilis* populations from several Spanish lagoons. Verh. int. Ver. Limnol. 22: 2991–2996.

Serra M. & M. R. Miracle, 1987. Biometric variation in three strains of *Brachionus plicatilis*. Hydrobiologia 147: 83–89.

Stemberger, R. S. & J. J. Gilbert, 1985. Body size, food concentration, and population growth in planktonic rotifers. Ecology 66: 1151–1159.

Snell, T. W., 1986. Effect of temperature, salinity and food level on sexual and asexual reproduction in *Brachionus plicatilis* (Rotifera). Mar. Biol. 92: 157–162.

Snell, T. W., M. J. Childress, E. M. Boyer & F. H. Hoff, 1987. Assessing the status of rotifer mass cultures. J. World Aquacult. Soc. 18: 270–277.

Snell, T. W. & C. E. King, 1977. Lifespan and fecundity patterns in rotifers: the cost of reproduction. Evolution 31: 882–890.

Starkweather, P. L., 1987. Rotifera. In: Animal Energetics, Vol. 1. Academic Press, N.Y.

Vinberg, G. G. & G. A. Galkovskaya, 1979. Relationship between the development rate of rotifers and temperature. V kn. Obschyie osnovyiznchieniya vodnykh ekosistiem. L, Nauka: 149–155.

Walz, N., 1983. Individual culture and experimental population dynamics of *Keratella cochlearis* (Rotatoria). Hydrobiologia 107: 35–45.

Walz, N., 1987. Comparative population dynamics of the rotifers *Brachionus angularis* and *Keratella cochlearis*. Hydrobiologia 147: 209–213.

Yúfera, M., 1987. Effect of algal diet and temperature on the embryonic development time of the rotifer *Brachionus plicatilis* in culture. Hydrobiologia 147: 319–322.

Yúfera, M. & E. Pascual, 1980. Estudio del rendimiento de cultivos del rotífero *Brachionus plicatilis* O. F. Müller alimentados con levadura de panificación. Invest. Pesq. 44: 361–368.

Hydrobiologia **186/187**: 103–108, 1989.
C. Ricci, T. W. Snell and C. E. King (eds), Rotifer Symposium V.
© 1989 *Kluwer Academic Publishers.*

Empirical evidence for a complex diurnal movement in *Hexarthra bulgarica* from an oligotrophic high mountain lake (La Caldera, Spain)[1]

P. Carrillo, L. Cruz-Pizarro & R. Morales-Baquero
Dpto. Biología Animal, Ecología y Genética. Fac. Ciencias. Universidad de Granada, 18701 Granada, Spain

Key words: Hexarthra bulgarica, diurnal vertical movement, horizontal distribution

Abstract

A detailed 24 hour sampling program has been carried out at 26 depths of 6 stations located along the two main transects of lake La Caldera. The resultant data has allowed us to define for *H. bulgarica* a general daily trend of movement which couples a typical nocturnal vertical migration with an 'horizontal' one that is particularly conspicuous at dawn and dusk when the population seemed to moved toward or away from the shore, respectively.

These results confirm our previous data and suggest that light is responsible for these complex movements.

Introduction

Data on rotifer spatial distribution and movement of populations come, basically, from two main approaches. Diurnal vertical migration is a common phenomenon, but is variable and far less important than in crustaceans in terms of amplitude and velocity of migration. Horizontal distribution also varies and the resultant patchiness may reflect shore avoidance processes.

Although the underlying mechanisms involved in both phenomena as well as their adaptive value still remain uncertain, light has an important role as a releasing (and directing) stimulus (Ringelberg, 1980; Preissler, 1977a, b). Our previous data (Cruz-Pizarro, 1978, 1981) suggest that these movements are important, thus we have measured a 24-hour cycle of movement for *H. bulgarica* in lake La Caldera. As far as we know, no previous attempts have been made to obtain field data coupling vertical and horizontal movements for an entire day.

Material and methods

Selected site and species

La Caldera is an oligotrophic high mountain lake in southern Spain. It is a typical 'winter-kill' lake of glacial origin which remains frozen for 8 to 9 months a year. Martinez (1975, 1980) gives data on its morphometric and physico-chemical characteristics and the papers from Martinez (1977), Cruz-Pizarro (1978, 1981), Morales (1985, 1988) and Sanchez-Castillo *et al.* (1988) consider different aspects of its plankton composition and ecology. Of particular note is its small size, the absence of littoral rooted vegetation and of visible inlet and outlets, and the lack of a true thermocline and oxycline.

---

[1] Research supported by CAICYT Project No. 3069/83.

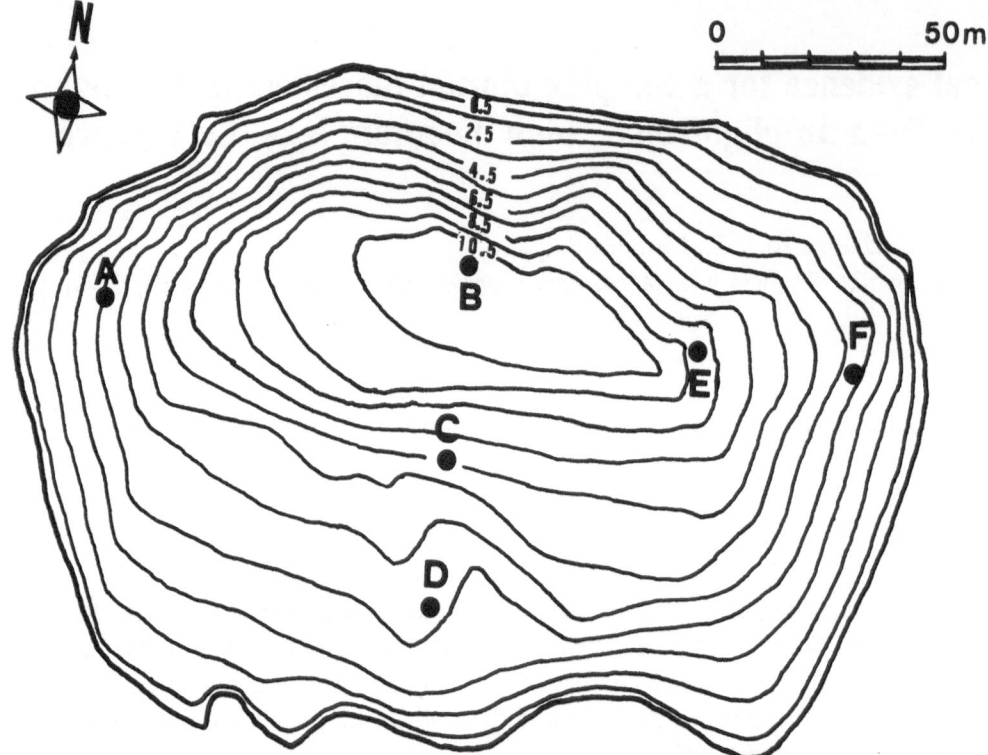

Fig. 1. Bathymetric map of lake La Caldera and location of sampling stations.

Hexarthra bulgarica, is the dominant rotifer species in the lake, both in number of individuals and in biomass. Data from Cruz-Pizarro (1978, 1981) reveal a definite nocturnal vertical migration and suggest a simultaneous horizontal movement of the population.

Sampling and counting

To entend and explain these previous observations, we carried out in August 1986, during the peak of population density for the species, a 24 h sampling program which included different sampling stations and depths along our two main transects of the lake (Fig. 1).

The samples were taken from two small boats working simultaneously (starting from different points) so that each sampling series required less than 25 minutes. A double Van Dorn sampler of 8 l capacity was used. Plankton was removed with a 45 μm mesh net and immediately preserved in 4% formaldehyde. Counts were made with an inverted microscope at 100 × magnification. Surface light measurements (Luxmeter) were recorded every two hours at the B station.

Results

Data obtained from all stations and depths for the daily cycle are shown in Fig. 2. At 15 hours, the first sampling carried out, individuals are located, predominantly, at the deepest layers of stations C and D and, to a lesser extent, at B and E also. The relative low densities are noteworthy.

Results from the 18 hour sampling suggest an offshore movement has taken place giving rise to the great differences observed between pelagic and littoral catches. This pointed movement seems to continue such that at 21 hour most individuals are found close to the eastern shore which remains longer in the light.

The situation at midnight reveals a change in

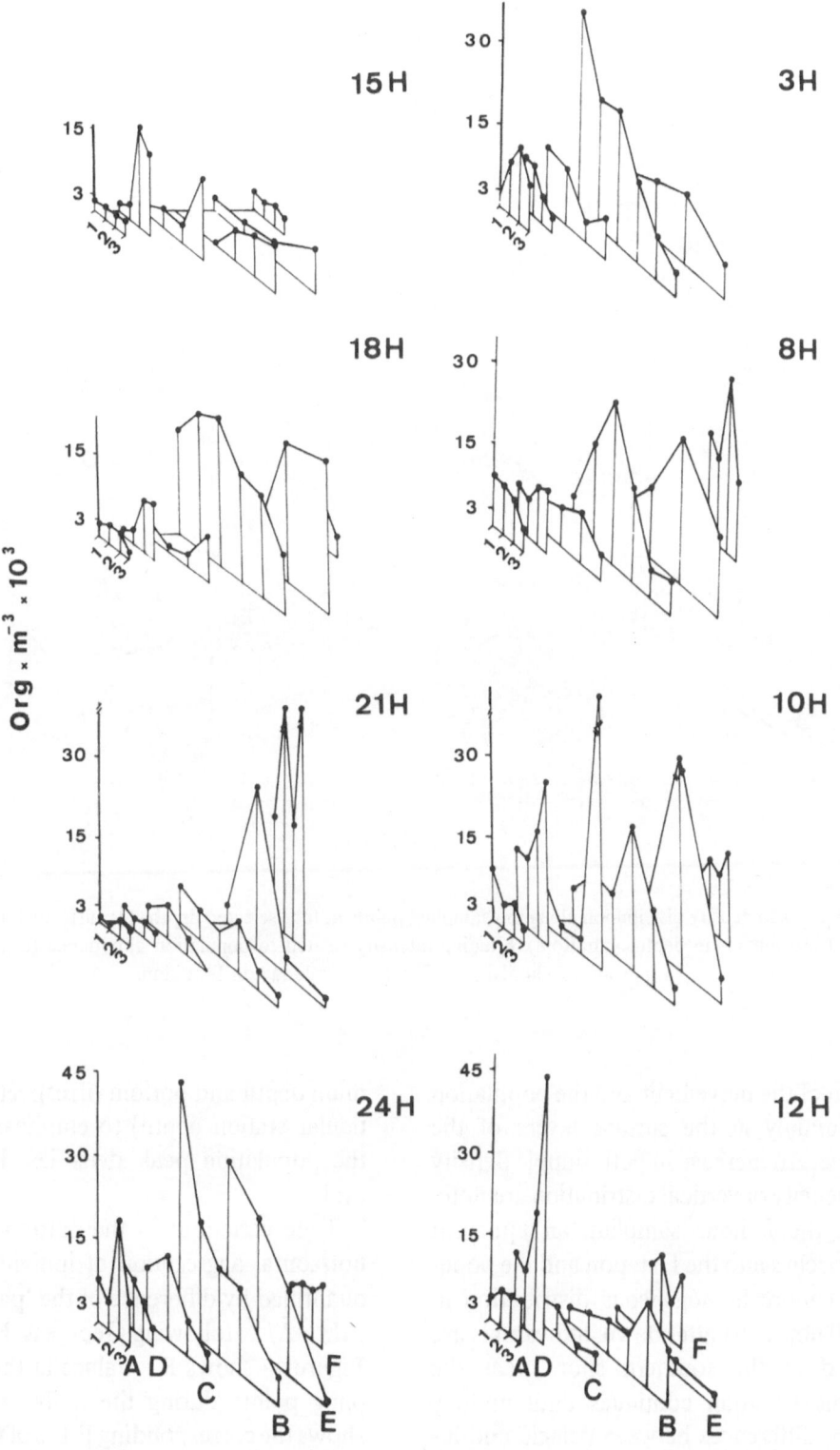

Fig. 2. Population density along the daily cycle at the different sampling stations and depths.

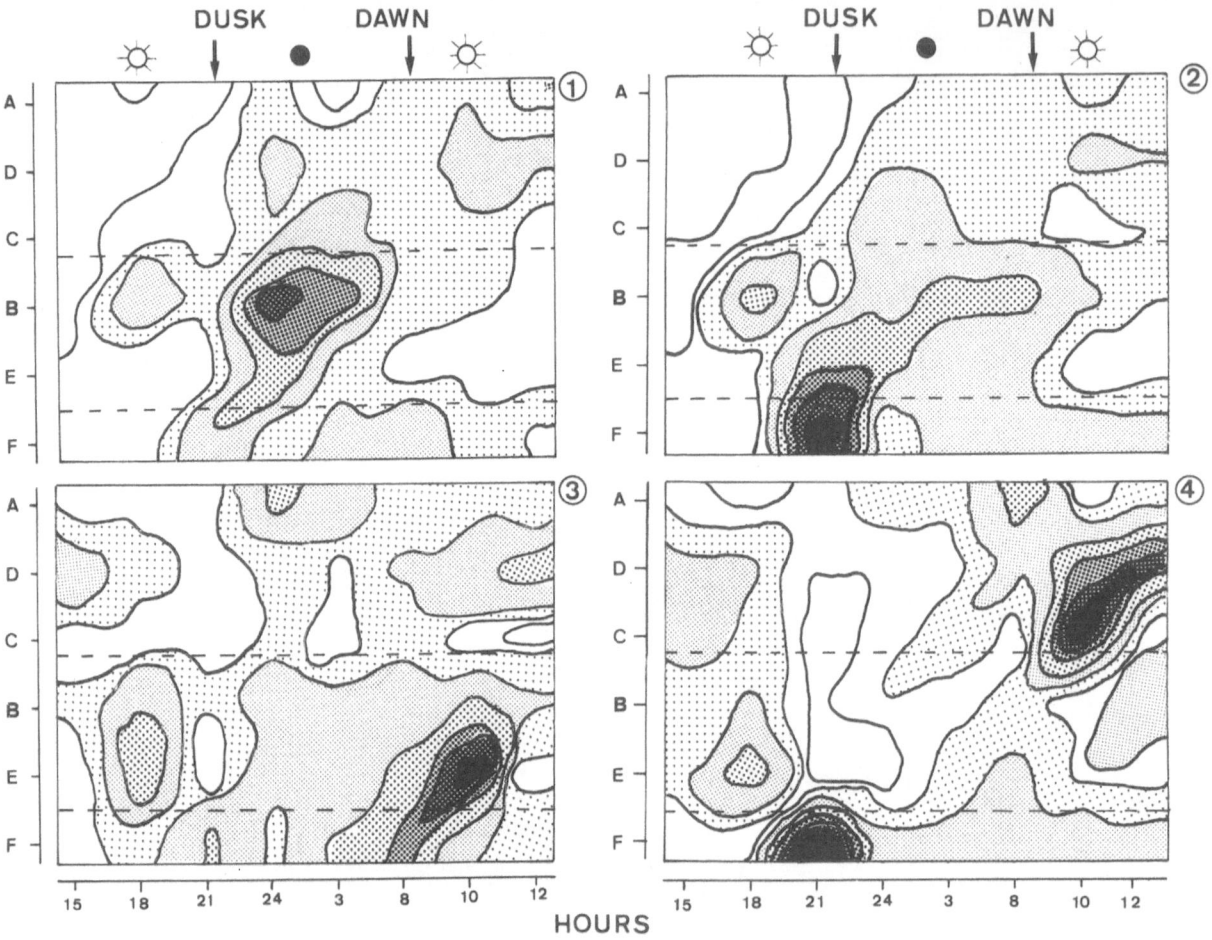

Fig. 3. Diel changes in density distribution along the sampling points at four selected depths (1: surface; 2: one-third maximum depth; 3: two-third maximum depth; 4: bottom). Shading intensity related to population abundance (org/m³). Dashed lines indicate secchi disk visibility (m) at midday on B station.

the direction of the movement and the population is located mainly at the surface layers of the limnetic zone. An increase in both sample density and heterogeneity of vertical distribution are noteworthy. At the 8 hour sampling an apparent movement occurs into the F station and the population has a more homogeneous distribution in space. At 10 hour, just after dawn, individuals are concentrated at the southern shore near the bottom. This situation continues until midday when, again, differences between pelagic and littoral catches are quite evident.

In Fig. 3, we have selected four depths: surface; one-third maximum depth; two-thirds maxi-

mum depth and bottom (irrespective of each particular station depth) to emphasize positions of the population peak densities during the daily cycle.

Time variations in the pattern of vertical and horizontal aggregation of individuals have been quantified by difference of the 'patchiness index', P.I. = $\dot{X}/\overline{X}$, following George & Edwards (1976). Figure 4a shows P.I. values at the different sampling points during the daily cycle and fig. 4b shows the corresponding P.I.'s of the four selected 'horizontal' planes.

Deviations from a random distribution are the lowest during nighttime, particularly near the

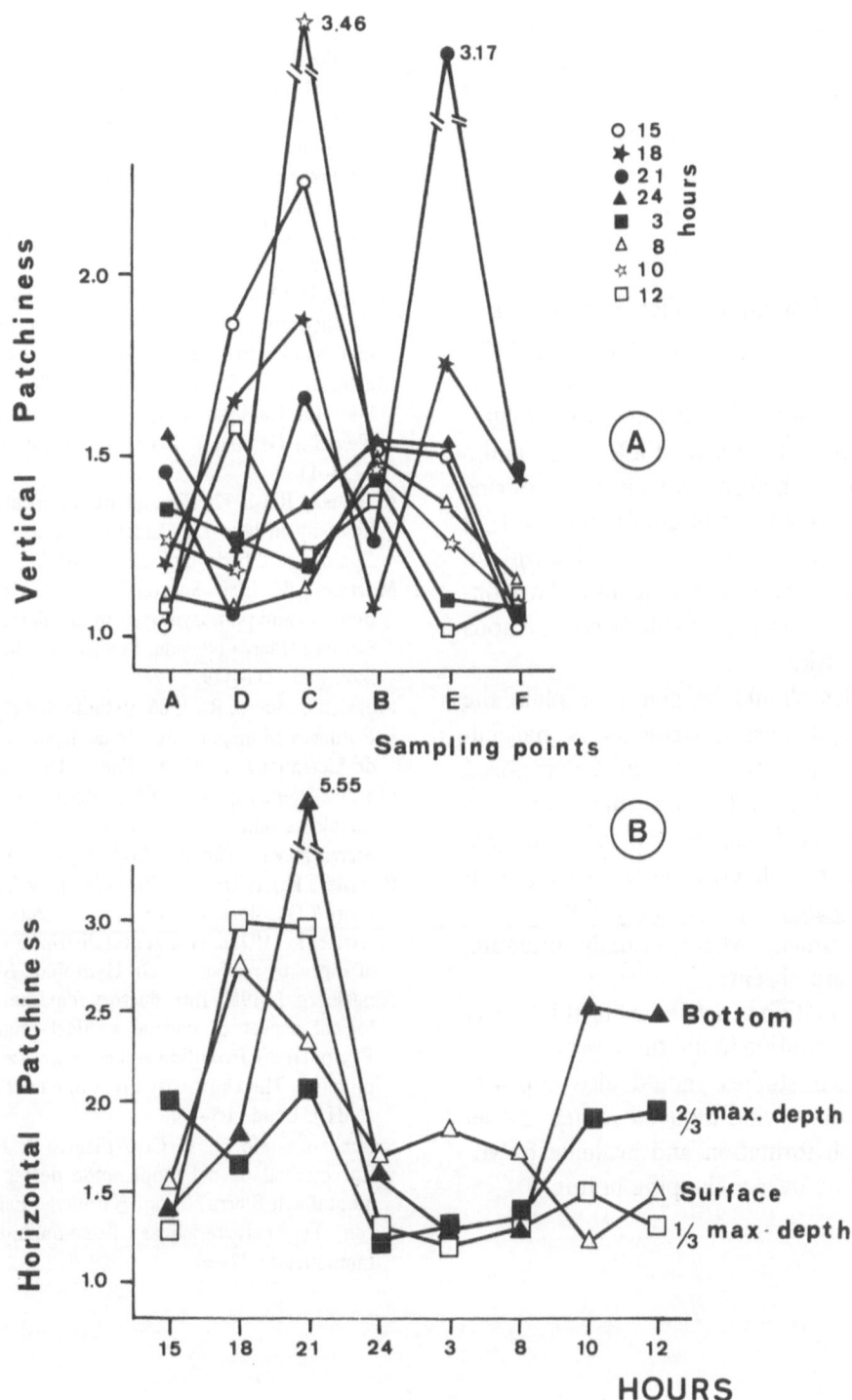

Fig. 4. Time variations of vertical (A) and horizontal (B) patchiness index (see text) at the different sampling stations and selected depths.

bottom, whereas the greatest patchiness measures were obtained around dusk and dawn in the upper and deepest layers, respectively. The extremely high aggregation value of the population at 21 hour at maximum depths is noteworthy.

Discussion

Approving some limitations inherent in the lack of sample replicates, these results depict a general daily trend of movement for *H. bulgarica*. The vertical component is a typical nocturnal migration with an upward movement after dusk and a downward one just before dawn gives rise, during nighttime, to the highest values of vertical P.I. at pelagic stations. This seems to be coupled with an 'horizontal' migration (highest values of horizontal P.I.) that is, likewise, particularly conspicuous at dawn and dusk.

Light changes should be able to explain the adaptive value of these movements as, particularly for vertical migration, other major proposed hypothesis (energetic and demographic, predator-avoidance, increased chance for mating and genetic exchange) are unlikely to hold for this, basically parthenogenetic species, in a virtually iso-thermic environment where visually orienting planktivorous are absent.

Further information is needed on light intensity and angular distribution in the shore region of the lake. Subsequent studies should also consider advective factors (wind-induced water movements) in patch formation and evaluate downward movements over a slopping bottom.

References

Cruz-Pizarro, L., 1978. Comparative vertical zonation and diurnal migration among Crustacea and Rotifera in the small high mountain lake La Caldera (Granada, Spain). Verh. int. Ver. Limnol. 20: 1026–1032.

Cruz-Pizarro, L., 1981. Estudio de la comunidad zoo-planctónica de un lago de alta montaña (La Caldera, Sierra Nevada, Granada). Ph. D. Thesis. Univ. Granada. 186 pp.

George, D. G. & R. W. Edwards, 1976. The effect of wind on the distribution of chlorophyll a and crustacean plankton in a shallow eutrophic reservoir. J. appl. Ecol. 13: 667–690.

Martinez, R., 1975. First report on the limnology of the alpine lake La Caldera, in the Penibetic Mountains (Sierra Nevada, Granada, Spain). Verh. int. Ver. Limnol. 19: 1133–1139.

Martinez, R., 1977. Phytoplankton species, biomass and diversity in lake La Caldera (Sierra Nevada, Granada, Spain). Acta hydrobiol. 19: 95–107.

Martinez, R., 1980. Seasonal variations of phytoplankton biomass and photosynthesis in the high mountain lake La Caldera (Sierra Nevada, Spain). Development in Hydro-biology 3: 111–119.

Morales-Baquero, R., 1985. Estudio de las comunidades de Rotiferos Monogonontes de las lagunas de alta montaña de Sierra Nevada. Ph. D. Thesis. Univ. Granada. 296 pp.

Morales-Baquero, R., 1988. Body size variability of Euchlanis dilatata Ehrenberg in high mountain lakes of Sierra Nevada (Spain). Arch. Hydrobiol. 112: 597–609.

Preissler, K., 1977a. Do Rotifers show 'Avoidance of the shore'? Oecologia (Berl.). 27: 253–260.

Preissler, K., 1977b. Horizontal distribution and 'avoidance of shore' by rotifers. Arch. Hydrobiol. Beih. 8: 43–46.

Ringelberg, J., 1980. Introductory remarks: causal and teleo-logical aspect of diurnal vertical migration. In W. C. Kerfoot (ed.), Evolution and Ecology of zooplankton Communities. The University Press of New England, Hanover (N.H.); Lond.: 65–68.

Sanchez-Castillo, P., L. Cruz-Pizarro & P. Carrillo, 1989. Caracterización del fitoplancton de las lagunas de alta montaña de Sierra Nevada (Granada, España) en relación con las características físico-químicas del medio. Limnetica 5: 37–50.

Hydrobiologia **186/187**: 109–117, 1989.
C. Ricci, T. W. Snell and C. E. King (eds), Rotifer Symposium V.
© 1989 *Kluwer Academic Publishers.*

Morphological variation in *Kellicottia longispina*

W. T. Edmondson & Arni H. Litt
Department of Zoology, NJ-15, University of Washington, Seattle, Washington 98195 U.S.A.

Key words: zooplankton, Rotifera, biogeography, arctic limnology

Abstract

The lengths of the body, the posterior spine and the three longest anterior spines were measured for 25 specimens of *Kellicottia longispina* from each of the eight lakes distributed from Imikpuk at Point Barrow, Alaska (latitude 71° 15′) to Lake Washington (latitude 47° 38′). Collections were available for more than two dates from six of the lakes. Temperature ranged from 1.2° to 18 °C. Mean lengths and ratios were examined in relation to latitude and temperature. Each population differed from the others in some aspect of absolute size, variability, or shape as expressed by the ratios of the dimensions. The population from Point Barrow is similar but not identical to Olofsson's var. *heterospina*.

Introduction

Kellicottia longispina is a widespread, common planktonic rotifer that can become very abundant. It is strikingly different in appearance from all other rotifers, even from its only congener, *K. bostoniensis*. Interestingly, both species were described from North America. *Kellicottia bostoniensis* was confined to that continent until about 1943 when it was first noticed in Sweden (Arnemo *et al.*, 1968). It subsequently spread to Finland (Dr. Pertti Eloranta, personal communication).

Kellicottia longispina was found in 'Niagara water' near Buffalo, N.Y. and assigned to *Anuraea* by Kellicott in 1879 (Fig. 1-a). Following Hudson & Gosse (1886) it was known as *Notholca* until 1938 when Ahlstrom created *Kellicottia* to recognize its unusual structure. The species was noticed in England and Europe soon after its discovery in North America. Hood's *Ertemias tetrathrix* (1888) probably was this species (Fig. 1-b). Some

species described as *Ertemia* may have been rhizopod shells into which a rotifer had crept (see Hudson & Gosse).

Although *K. longispina* is easily recognized, even causal examination shows that there are differences in size and shape among different populations. A form from arctic Norway is so different that it was given the varietal name *heterospina* by Olofsson (1917) (Fig. 1-c). A similar form was relatively abundant in samples collected by the late G.W. Comita from Imikpuk (Freshwater Lake) near Point Barrow, Alaska in 1951. Its occurrence in this small arctic lake raised questions about its distribution in the Arctic and the occurrence of forms resembling the ones in temperate regions. Therefore we examined collections from other lakes in northern Alaska and material from five lakes at lower latitudes (Fig. 2, Table 1). We measured major anatomical structures in an attempt to define objectively what made the populations look so different (Fig. 1-d, Fig. 3, Table 2). We also examined correlations

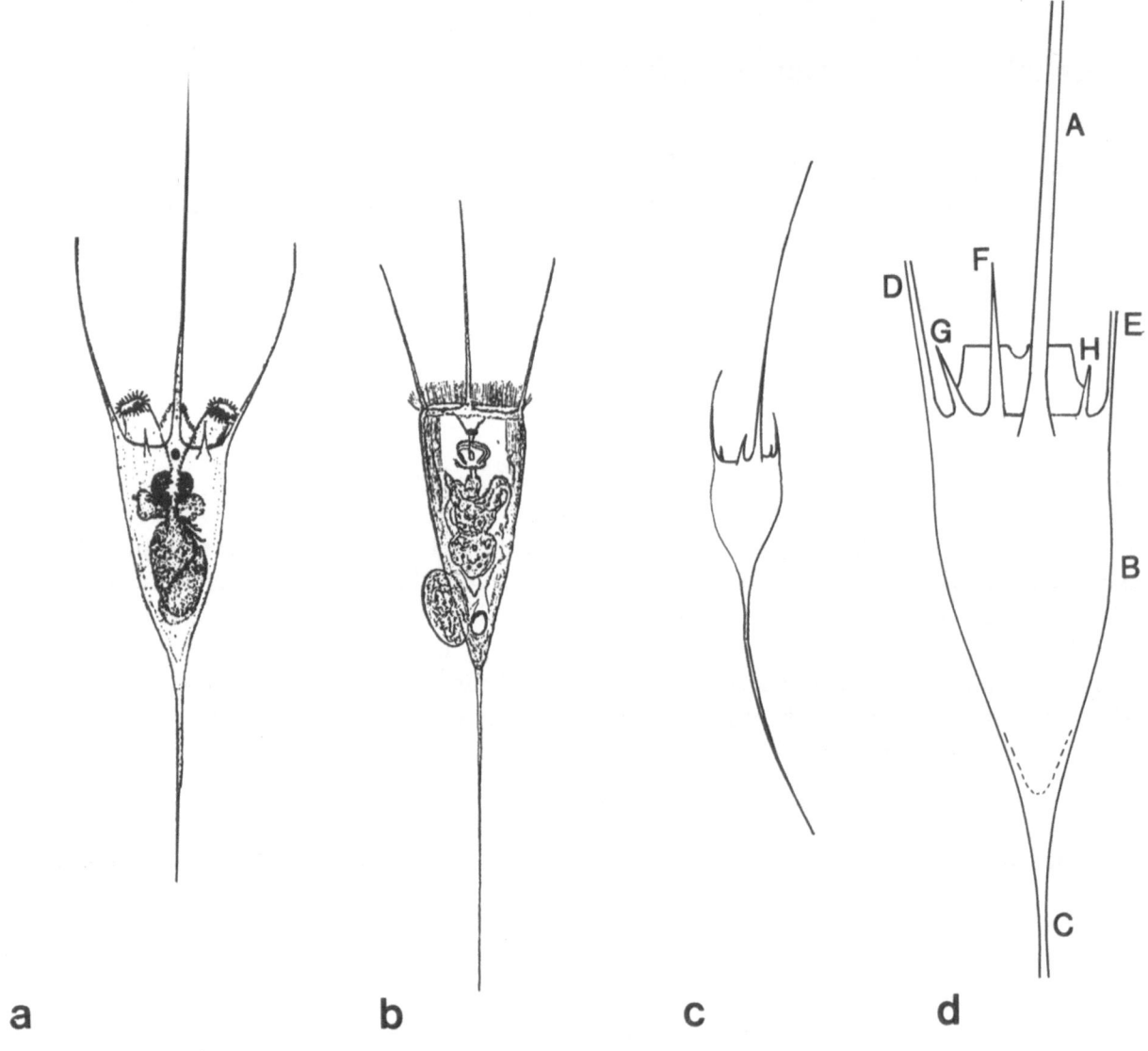

Fig. 1. Four drawings of *Kellicottia longispina*. a. *Anuraea longispina*, from original drawing by Kellicott (1879). b. *Ertemias tetrathrix* from Hood (1888). c. *K. longispina* var. *heterospina* after Olofsson (1917). d. Diagram indicating symbols used throughout the text.

among the various dimensions and between morphometric features and the location and environmental conditions of the lakes.

Materials and methods

Collections from two arctic lakes in addition to Imikpuk were made by Comita in 1951. Imikpuk was studied more in detail in 1952 (Comita, 1956;

Edmondson, 1955). For reasons not understood, *Kellicottia* was extremely scarce in that year. In addition to our own material from Bare Lake, Hall Lake and Lake Washington, we were kindly provided with material from Great Slave by the late Donald S. Rawson and from Lake Pend Oreille by Raymond A. Stross.

The three arctic lakes are small tundra ponds. Great Slave Lake is very large and oligotrophic (Rawson, 1956). Bare Lake is small and oligo-

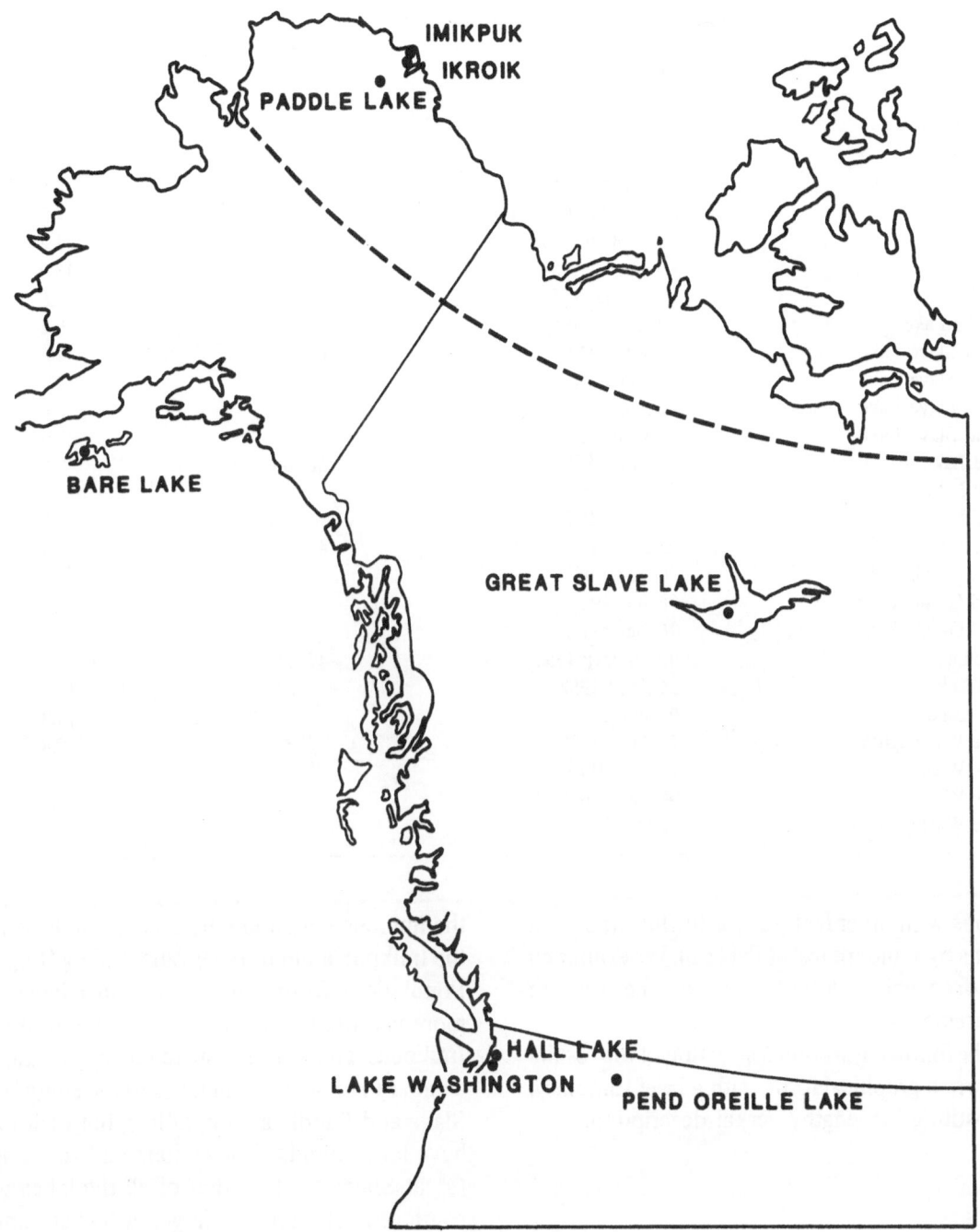

Fig. 2. Location map.

trophic, but was fertilized on June 11, 1952 (Nelson & Edmondson, 1955). Lake Pend Oreille is large and mesotrophic. Hall Lake is small and slightly dystrophic. Lake Washington is moderately large and mesotrophic (Edmondson, 1963).

Twenty-five specimens from each collection were measured under light compression.

Measurements were made of the three longest anterior spines (A,D,E), the posterior spine (C), and the length (B) and width (W) of the body (Fig. 1-d). After a preliminary graphical analysis of the dimensions and their ratios to each other, we selected dimensions or proportions that showed distinct relation to location or strong cor-

Table 1. Data on collections used for measurements of *Kellicottia longispina.*

Name	Date	Latitude	Temperature ° C
1 Imikpuk	09 August 1951	71° 20′	12.5
2 Imikpuk	18 August 1951		6.6
3 Imikpuk	06 September 1951		4.2
4 Imikpuk	18 September 1951		1.2
5 Ikroik	01 August 1951	71° 20′	13.0
6 Ikroik	30 August 1951		5.6
7 Paddle Lake	22 August 1951	71° 15′	9.2
8 Great Slave Lake	03 July 1949	62° 00′	4.0
9 Great Slave Lake	11 August 1949		12.0
10 Great Slave Lake	22 August 1947		8.8
11 Great Slave Lake	21 September 1947		7.0
12 Bare Lake	03 June 1952	58° 00′	8.0
13 Bare Lake	17 June 1952		11.1
14 Bare Lake	25 June 1952		14.5
15 Pend Oreille Lake	05 April 1953	48° 25′	4.3
16 Pend Oreille Lake	20 June 1953		13.8
17 Pend Oreille Lake	06 July 1953		18.2
18 Pend Oreille Lake	26 October 1953		12.0
19 Hall Lake	02 January 1952	47° 50′	4.4
20 Hall Lake	24 April 1952		11.5
21 Hall Lake	07 July 1952		14.2
22 Lake Washington	23 April 1980	47° 38′	9.8
23 Lake Washington	27 May 1981		14.2
24 Lake Washington	04 June 1980		13.6
25 Lake Washington	09 October 1981		13.7

relations with other features for further statistical analysis by standard tests (Table 2). We examined length-frequency plots for all of the sets of measureménts.

In the following section a selection of the results is shown in graphical form, with a brief summary, to substitute for lengthy verbal descriptions.

Results

The clearest generalization that we can make is that the population of each lake was different from all the others in at least one feature. Some of the features varied among samples taken from a lake at different times. We looked to see whether the variation was related to time or temperature, as might be expected of cyclomorphosis.

The following comments call attention to facts shown by Fig. 4 and Table 2. The total length of the animals varied greatly among the lakes, with the Imikpuk animals being outstanding (Fig. 4-T). Populations from some of the other lakes show more variation with temperature than those from Imikpuk. There is a clear tendency for the total length (T) to decrease in lakes to the south. Great Slave and Paddle are out of line, but their means have large standard deviations and the range of Great Slave overlaps that of all the lakes to the south of it. Rotifers are larger at higher temperature in three of the lakes and smaller in two, but only the greatest differences in size are significant.

The variation among lakes is more easily seen in plots of the range of all the samples for each lake (Fig. 4). Of all the features, the length of spine A shows the clearest geographical pattern (Fig. 4-A). The population of Ikroik is out of line and shows few similarities to that of the other two arctic lakes.

Differences in the proportions of spines and

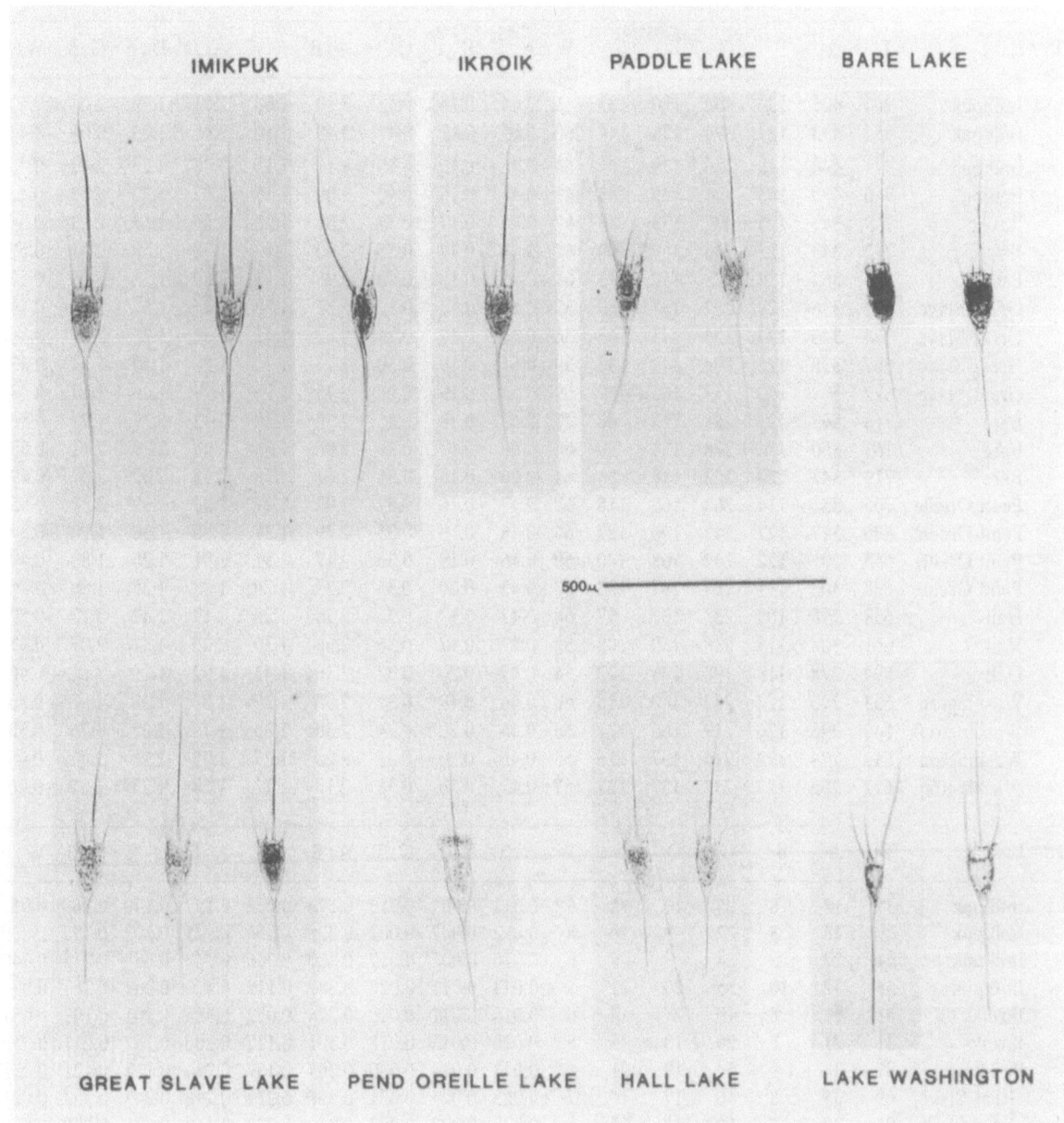

Fig. 3. Photograph of representative individuals from each lake.

body determine the different appearances of the population. The way the total length is shared by the spines and body is shown by ratios A : T, B : T and C : T. The ratio of A : C relates the two most conspicuous features. In four of the eight lakes values of A : C vary from 1.0 to 1.4. In Imikpuk they are nearly equal, varying from 1.0 to 1.1, but in Great Salve the ratio is 1.7 to 1.9, as is consistent with the shorter posterior spine. Within its range of variation in each lake, A : C is not correlated with body size. The ratio A : B has a clear relation to latitude as did the absolute size of spine A, but the ranges of the ratio are proportionately much larger. Body length does not vary as much among the lakes as the spines, and its variations have less effect on appearance.

Table 2. Upper. Mean values and ratios of some of the features measured. See Fig. 1-d for symbols. T is total length (A + B + C). W is width of body at widest point. Note: The ratios are means of ratios of individual measurements; the values may be slightly different from the ratios of the tabulated mean lengths. Lower. Standard deviations of the means, $n = 25$. All measurements are in micrometers. Smallest unit of measure for lakes 1 to 7 was 12 μ, for lake 8 was 4 μ.

OBS	Lake	T	A	B	C	D	E	W	A:T	B:T	C:T	A:B	A:C	A:D	D:E	C:B	W:B
1	Imikpuk	964	427	134	402	170	133	64	0.44	0.14	0.42	3.20	1.06	2.51	1.29	3.04	0.47
2	Imikpuk	954	433	127	394	176	144	60	0.45	0.13	0.41	3.40	1.10	2.46	1.23	3.09	0.48
3	Imikpuk	971	444	140	386	179	144	64	0.46	0.15	0.40	3.17	1.15	2.48	1.25	2.73	0.45
4	Imikpuk	940	422	143	374	169	143	65	0.45	0.15	0.40	3.00	1.15	2.57	1.22	2.63	0.45
5	Ikroik	749	347	125	277	133	88	67	0.46	0.17	0.37	2.80	1.25	2.66	1.53	2.23	0.54
6	Ikroik	712	343	122	246	150	100	64	0.48	0.17	0.35	2.80	1.40	2.30	1.53	2.01	0.52
7	Paddle	815	364	124	328	150	113	61	0.45	0.15	0.40	2.96	1.13	2.60	1.34	2.66	0.50
8	Great Slave	673	336	120	217	137	89	55	0.50	0.18	0.32	2.82	1.58	2.47	1.57	1.81	0.46
9	Great Slave	724	346	124	254	173	143	56	0.48	0.17	0.35	2.82	1.41	2.00	1.22	2.07	0.46
10	Great Slave	667	338	125	204	172	133	59	0.51	0.19	0.30	2.73	1.72	1.98	1.30	1.64	0.47
11	Great Slave	622	329	120	173	166	130	59	0.53	0.19	0.28	2.75	1.95	1.99	1.29	1.44	0.49
12	Bare	718	342	127	248	112	48	71	0.48	0.18	0.35	2.72	1.38	3.12	2.39	1.97	0.56
13	Bare	763	350	134	278	138	70	66	0.48	0.17	0.35	2.90	1.38	2.98	2.29	2.11	0.53
14	Bare	719	347	120	252	118	53	64	0.46	0.18	0.36	2.61	1.26	2.55	2.02	2.07	0.49
15	Pend Oreille	706	331	114	260	162	118	62	0.47	0.16	0.37	2.92	1.27	2.06	1.39	2.30	0.55
16	Pend Oreille	686	317	127	242	170	132	64	0.46	0.19	0.35	2.49	1.31	1.86	1.30	1.90	0.50
17	Pend Oreille	655	301	122	232	168	140	60	0.46	0.19	0.35	2.47	1.31	1.81	1.20	1.89	0.49
18	Pend Oreille	668	302	131	235	167	133	60	0.45	0.20	0.35	2.32	1.29	1.82	1.25	1.80	0.46
19	Hall	638	298	108	233	137	97	54	0.47	0.17	0.36	2.76	1.28	2.19	1.42	2.15	0.50
20	Hall	661	308	115	238	150	108	54	0.47	0.17	0.36	2.69	1.30	2.06	1.41	2.07	0.47
21	Hall	594	278	118	198	145	122	54	0.47	0.20	0.33	2.36	1.41	1.92	1.19	1.68	0.46
22	Washington	653	290	122	242	160	115	60	0.44	0.19	0.37	2.38	1.20	1.83	1.38	1.98	0.50
23	Washington	640	296	126	219	166	131	63	0.46	0.20	0.34	2.36	1.35	1.79	1.27	1.75	0.50
24	Washington	612	274	122	216	157	126	55	0.45	0.20	0.35	2.25	1.27	1.75	1.25	1.77	0.45
25	Washington	622	262	123	237	157	127	57	0.42	0.20	0.38	2.13	1.12	1.68	1.23	1.92	0.46

OBS	Lake	T	A	B	C	D	E	W	A:T	B:T	C:T	A:B	A:C	A:D	D:E	C:B	W:B
1	Imikpuk	37	12	8	25	10	12	5	0.011	0.007	0.013	0.173	0.062	0.127	0.140	0.209	0.051
2	Imikpuk	39	18	8	22	9	10	2	0.012	0.007	0.002	0.225	0.056	0.125	0.102	0.195	0.039
3	Imikpuk	41	22	5	28	9	9	6	0.016	0.008	0.017	0.220	0.086	0.147	0.089	0.213	0.040
4	Imikpuk	58	32	10	35	20	21	5	0.023	0.011	0.022	0.282	0.119	0.429	0.261	0.267	0.035
5	Ikroik	41	21	8	19	21	14	6	0.010	0.011	0.010	0.234	0.052	0.386	0.201	0.193	0.051
6	Ikroik	31	21	8	20	13	15	5	0.020	0.014	0.018	0.301	0.117	0.206	0.167	0.211	0.058
7	Paddle	87	36	9	54	19	21	4	0.023	0.014	0.030	0.267	0.137	0.294	0.118	0.421	0.053
8	Great Slave	69	35	8	40	17	10	7	0.025	0.021	0.031	0.350	0.221	0.296	0.243	0.352	0.061
9	Great Slave	81	30	7	58	11	12	5	0.029	0.022	0.043	0.310	0.248	0.158	0.075	0.470	0.050
10	Great Slave	84	41	13	49	20	19	3	0.034	0.022	0.040	0.370	0.316	0.234	0.137	0.360	0.004
11	Great Slave	47	25	8	31	18	13	4	0.026	0.018	0.032	0.282	0.292	0.167	0.091	0.270	0.045
12	Bare	16	12	11	11	13	8	3	0.157	0.015	0.012	0.308	0.078	0.370	0.410	0.208	0.055
13	Bare	29	13	7	17	12	10	4	0.011	0.010	0.014	0.199	0.077	0.313	0.573	0.186	0.047
14	Bare	23	10	7	18	14	13	4	0.013	0.009	0.015	0.166	0.087	0.269	0.304	0.176	0.033
15	Pend Oreille	36	19	7	20	15	12	4	0.016	0.010	0.017	0.233	0.097	0.164	0.104	0.218	0.051
16	Pend Oreille	51	25	9	25	14	11	4	0.018	0.012	0.016	0.204	0.105	0.111	0.103	0.153	0.056
17	Pend Oreille	28	15	7	19	10	13	4	0.016	0.013	0.018	0.199	0.106	0.131	0.089	0.191	0.040
18	Pend Oreille	41	21	8	22	10	9	2	0.013	0.016	0.016	0.202	0.078	0.136	0.075	0.192	0.030
19	Hall	17	10	5	14	15	13	5	0.015	0.008	0.015	0.173	0.089	0.210	0.107	0.144	0.039
20	Hall	30	15	8	15	8	11	5	0.009	0.010	0.013	0.168	0.066	0.127	0.133	0.175	0.052
21	Hall	24	14	6	16	11	12	5	0.015	0.012	0.018	0.166	0.109	0.118	0.093	0.178	0.050
22	Washington	38	26	5	17	20	11	4	0.022	0.009	0.019	0.195	0.113	0.273	0.193	0.137	0.037
23	Washington	28	19	4	12	12	7	3	0.016	0.008	0.013	0.141	0.092	0.088	0.087	0.096	0.024
24	Washington	34	22	8	17	14	13	4	0.017	0.015	0.018	0.225	0.100	0.141	0.099	0.200	0.038
25	Washington	38	18	3	27	14	12	4	0.018	0.014	0.023	0.168	0.111	0.141	0.081	0.227	0.038

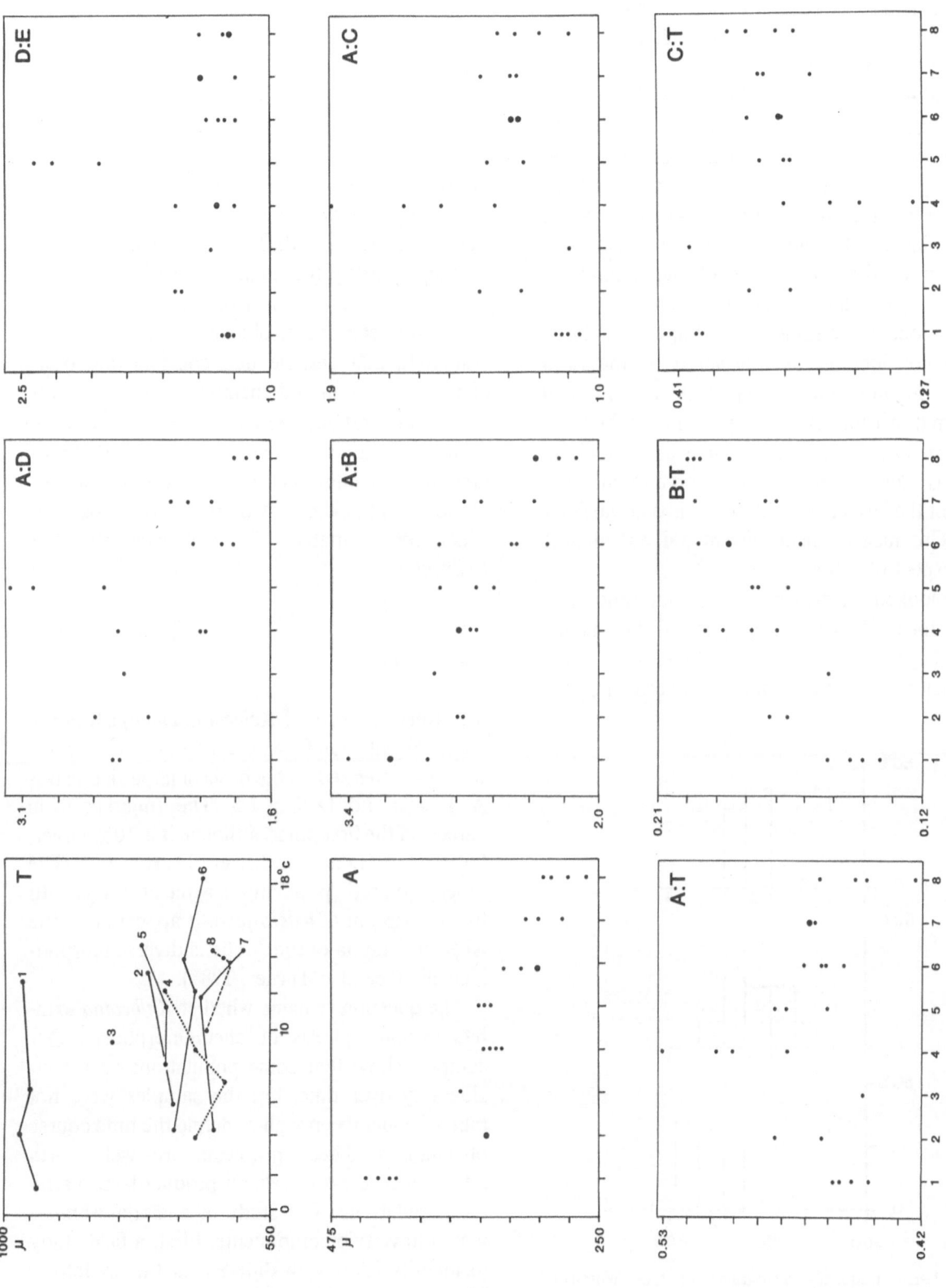

Fig. 4. Mean lengths of all features measured and a selection of ratios for all lakes, arranged by latitude and lake number. See Fig. 1-d for symbols. Large dots represent overlapping measurements. Numbers on curves represent lakes: 1-Imikpuk, 2-Ikroik, 3-Paddle, 4-Great Slave, 5-Bare, 6-Pend Oreille, 7-Hall, 8-Washington. Other panels show the individual values arranged in a vertical series for each lake. Lake are numbered at the bottom, arranged in order of latitude.

Two general patterns emerge, one a trend from north to south, the other humped or depressed in the middle because of the special features of the Great Slave population.

Another feature that catches the eye is the position of the body along the total length, determined by the length of the two spines A and C. Great Slave specimens have the body further back on the axis because the C spine is shorter than in other lakes (Fig. 4-C:T). One's subjective reaction to the appearance of the animals is affected by the shape of the body as well as by the proportions of the spines. The shape of the body varies considerably, as measured by the ratio W:B, but not with a very clear distributional pattern (see Table 2). The appearance of the body is affected by the fact that in some, as in Imikpuk, the maximum width of the body is well behind the front of the lorica, while in others it is at the front end. Our measurement system did not include that aspect of shape.

We looked for evidence of cyclomorphosis or other form changes in graphs of dimensions plotted against temperature and time. Some of the dimensions and ratios change between sampling dates in some of the lakes, but most of the differences are small and statistically insignificant (*t*-test, $\alpha = 0.05$). We examined the samples that showed the biggest differences by plotting length against frequency. The relative frequency of different stages is affected by changes in birth and death rates. An increase in mean value can be caused either by an increase in the frequency of the largest size or by a decrease in the frequency of the smallest, or both (Fig. 5). However, we may be seeing simply the replacement of a cohort of large animals by a cohort of different size.

Any time the lengths of spines of a rotifer show systematic changes, we must consider the possibility of predation and chemomorphosis, as with *Brachionus* and *Asplanchna*. As far as we know each of these lakes has a tactile invertebrate predator. In addition to the predatory *Limnocalanus*, Imikpuk had two species of Anostraca which are capable of swallowing the big *Kellicottia*.

Conclusions

The Imikpuk form of *Kellicottia longispina* is not identical with Olofsson's var. *heterospina* which was characterized on the basis of large size, ratios A:D > 3 and D:E > 1.5. The Imikpuk form agrees on the first point, although it is 10% larger, but ratio A:D is only 2.5 and D:E is 1.25. The other two arctic populations agree even less with the description of *heterospina*. It appears that the *Kellicottia* fauna of the Arctic is diverse morphologically (see also Turner, 1987).

The question remains whether *Kellicottia* exhibits cyclomorphosis or chemomorphosis. Our samples show that some populations vary considerably over time, but the samples were not taken frequently enough to define the time course of changes. These problems are well worth pursuing in lakes that reliably produce large, variable populations. The study of chemomorphosis would have two components. First, a field study to identify lakes with different tactile predators. Then one would do laboratory experiments. These will be difficult because *Kellicottia* is hard

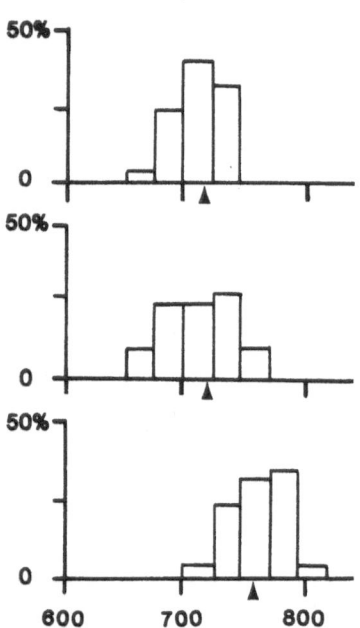

Fig. 5. Length-frequency histograms of total length of *K. longispina* in Bare Lake. Above – 03 June 1952, Middle – 17 June 1952, Below – 25 June 1952.

to culture, largely because it sicks to the surface of the water so easily. Possibly some success could be achieved by tethering individuals by the posterior spine to the bottom of a culture dish on a microblob of petroleum jelly. We know that the spines are tightly curled around the body in the egg and rapidly straighten after hatching, but nothing seems to be known about the subsequent rate and pattern of growth. It is surprising that an animal as conspicuous as *Kellicottia longispina* is not better known.

Acknowledgements

The work at Point Barrow was supported by the United States Office of Naval Research. Preparation of this paper was made possible by the Andrew W. Mellon Foundation.

References

Arnemo, R., B. Berzins, B. Grönberg & I. Mellgren, 1968. Dispersal in Swedish waters of *Kellicottia bostoniensis* (Rousselet) (Rotatoria). Oikos 19: 351–358.

Comita, G. W., 1956. A study of a calanoid copepod population in an arctic lake. Ecology 37: 576–595.

Edmondson, W. T., 1955. Seasonal life history of *Daphnia* in an arctic lake. Ecology 36: 439–455.

Edmondson, W. T., 1963. Pacific Coast and Great Basin. pp. 371–392 in D. G. Frey, ed. Limnology in North America. University of Wisconsin Press.

Hood, J., 1888. Chats about rotifers (*Ertemias tetrathrix*). Hardwickes' Science Gossip. 24: 27–28.

Hudson, C. T & P. H. Gosse, 1886. The Rotifera; or wheel-animalcules. Longmans, Green & Co. London.

Kellicott, D. S., 1879. A new rotifer. Am. J. Microsc. Pop. Sci. 4: 19–20.

Nelson, P. R. & W. T. Edmondson, 1955. Limnological effects of fertilizing Bare Lake, Alaska. U.S. Fish and Wildlife Service, Fish. Bull. 102: 413–436.

Olofsson, O., 1917. Süsswasser-Entomostraken und Rotatorien von der Murmankuste und aus den nordlichsten Norwegen. Zool. Bidrag Uppsala, 5: 259–294.

Pejler, B., 1977. On the global distribution of the family Brachionidae (Rotatoria). Arch. Hydrobiol./Suppl. 53. 2: 255–306.

Rawson, D. S., 1956. The net plankton of Great Slave Lake. J. Fish Res. Bd Can. 13: 53–127.

Turner, P. N., 1987. Some rotifers from Alaska, U.S.A. with a note to researchers. Microscopy 35: 541–548.

Hydrobiologia **186/187**: 119–128, 1989.
C. Ricci, T. W. Snell and C. E. King (eds), Rotifer Symposium V.
© 1989 *Kluwer Academic Publishers.*

Morphological structure and functional patterns of *Keratella cochlearis* (Gosse) populations in stratified lakes

Galina A. Galkovskaya & Inessa F. Mityanina
Institute of Zoology, Byelorussian Academy of Sciences, Academycheskaya 27, Minsk 220072, USSR

Key words: *Keratella cochlearis*, diurnal vertical distribution, morphological structure, population and energetic patterns

Abstract

Vertical distribution of the rotifer *Keratella cochlearis* in stratified water columns of mesotrophic and eutrophic lakes during summer stagnation has been studied. Coexisting morphs *K. cochlearis hispida* (Lauterborn, 1898), *K. c. tecta* (Gosse, 1851) and *K. c. cochlearis* (Gosse, 1851) inhabit different layers in the water column and are vertically subdivided. The distribution of morph abundance and reproductive potential indicate that substitution of morphs within the vertical water column may be due to trophic conditions. The maximum population productivity is observed at the epi-metalimnion border. The maximum density zone lies below the zone of the highest productivity. The principle of 'sliced functioning' is used to explain the adaptive significance of the morphological structure of populations under heterogeneous environmental conditions.

Introduction

The rotifer *Keratella cochlearis*, like many other species of planktonic rotifers, is characterized by high spatial and temporal heterogeneity of species populations. This is especially true in stratified lakes. Variability in morphotype, lorica length, and posterior spine length occurs seasonally as well as in response to lake trophy (Carlin, 1943; Gallagher, 1957; Green, 1981; Hofmann, 1980, 1983, 1987; Hillbricht-Ilkowska, 1972). In these studies, the short-spined morph of *K. cochlearis* is generally found in the epilimnion and the long-spined morph is found in the hypolimnion. However, there are no data on eco-physiological or production parameters of *K. cochlearis* populations pertaining specifically to affects of environ-

mental spatial heterogeneity. These data are necessary to understand how these populations form functional units.

The purpose of this investigation is to study population functioning of *K. cochlearis* in stratified lakes with consideration of morphological structure and spatial distribution.

Methods

Zooplankton samples were collected in June-July 1985 and July, 1987 from oligo-mesotrophic Lake South Volos, Byelorussia (area 1.2 km, max. depth 40.4 m (Yakushko, 1971)) and in June 1986 from eutrophic Lake Mikolajki, Poland (area 4.6 km, max. depth 27.8 m (Hillbricht-Ilkowska

120

et al., 1971)). Samples were collected from three stations of Lake South Volos in the vertical water column from surface to bottom in 5 m increments. In addition, daily collections were made every 3 hours using the same increments at the station with the maximum depth. In June 1986 and July 1987, samples were taken from the stations with the maximum depth at 1 m intervals from both lakes at midday and midnight. Rotifer and chlorophyl l samples were taken simultaneously.

Water samples were taken with a 2 l bathometer and fixed with 40% formalin, so as to attain a 2% concentration. Zooplankton samples (1 l) were sedimented for 3–4 days, then the overlying liquid was decanted. Each sample was diluted to 100 ml and a 20 ml aliquot from each was completely counted.

For calculating population and energetic parameters, morphometric measurements, embryonic development rates, and oxygen consumption rates were calculated for each population. Wet weight of rotifers was calculated relative to volume, dry mass was taken as 10% of wet weight, and the caloric value of the dry weight was 4.7 cal mg^{-1}.

Birth rate (b) was calculated according the formula (Paloheimo, 1974):

$$b = 1/D_e \cdot \ln (1 + E/N),$$

where D_e = duration of embryogenesis, day^{-1}; E = number of parthenogenetic eggs, eggs l^{-1}; N = density of parthenogenetic females, ind l^{-1}. The E/N ration was named the egg ratio or rela-

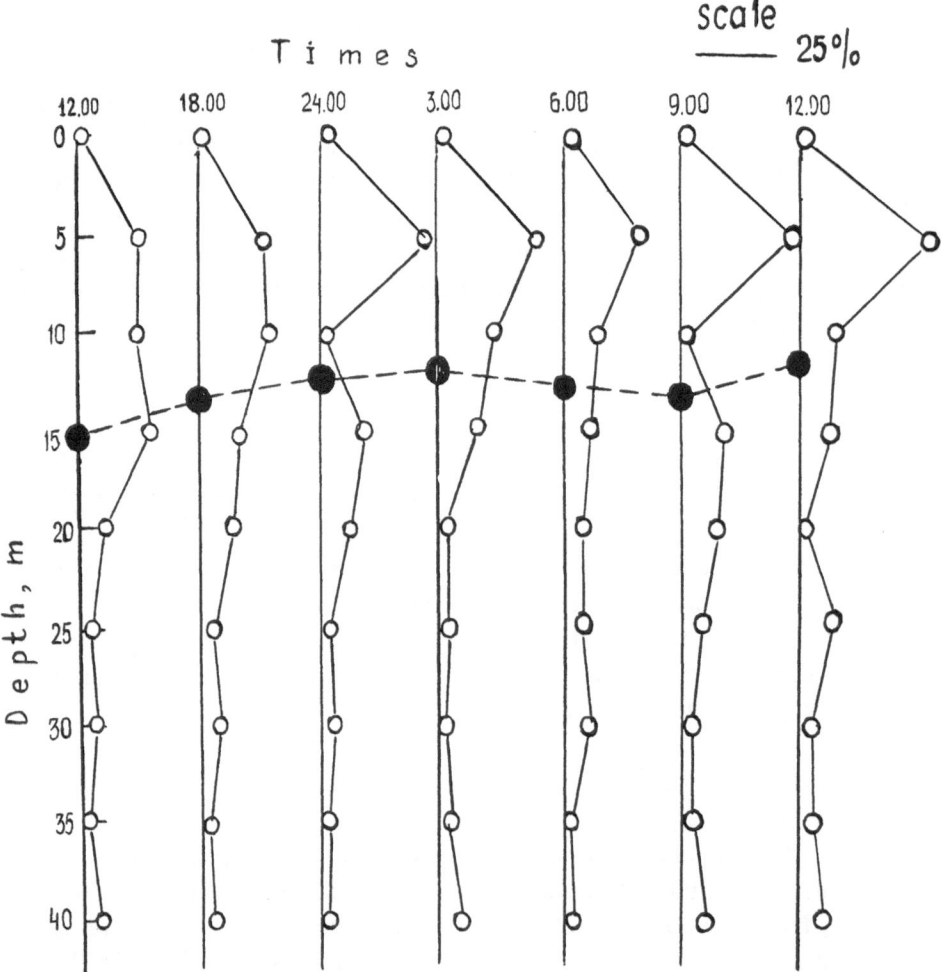

Fig. 1. Diurnal vertical distribution of *Keratella cochlearis* in S. Volos Lake, June 24–25, 1985. Distribution of relative density (density in each layer relative to total density at all layers in %). Black circles indicate mean depth of the population 'center' site.

tive population fecundity. Potential production (P) was determined as a product of ($e^b - 1$) and the density or biomass was expressed as ind l^{-1} day^{-1} or j^{-1} day^{-1}. The net growth efficiency was calculated as the production divided by the sum of production and respiration.

Chlorophyl l concentration in Lake Mikolajki was determined by Dr. Irena Kufel (Hydrobiological station, Institute of Ecology PAN). All other data were obtained by the authors.

Vertical distribution of populations of *K. cochlearis*

In Lake S. Volos, June 1985, diurnal vertical migration was not observed (Fig. 1). At the surface, the density was minimal throughout the day and was generally not more than 10 ind l^{-1}. The maximum density was observed at 5 m. In the 20–40 m layer, the number of individuals was evenly distributed and formed not more than 15% of the total density of *K. cochlearis* in the water column. More than 60% of the total number of *K. cochlearis* were found between the surface and 10 m layer. The absence of vertical migration of *K. cochlearis* and in other planktonic rotifers in S. Volos Lake has also been reported by Galkovskaya (1984).

Examination of population vertical density profiles of *K. cochlearis* at three stations (Fig. 2), and at the same station on different days (Fig. 3), shows the stability of the vertical distribution of this species in both space and time. The surface density is low (50 ind l^{-1}), density then increases up to 150–200 ind l^{-1} at a depth of 10 m, and decreases abruptly at depths of 15 m and greater.

Increased sampling intensity of the vertical water column up to 1 m shows a clearer picture of the vertical profile of the *K. cochlearis* population at midday and midnight (Fig. 4). The absence of vertical migration of *K. cochlearis* was observed in

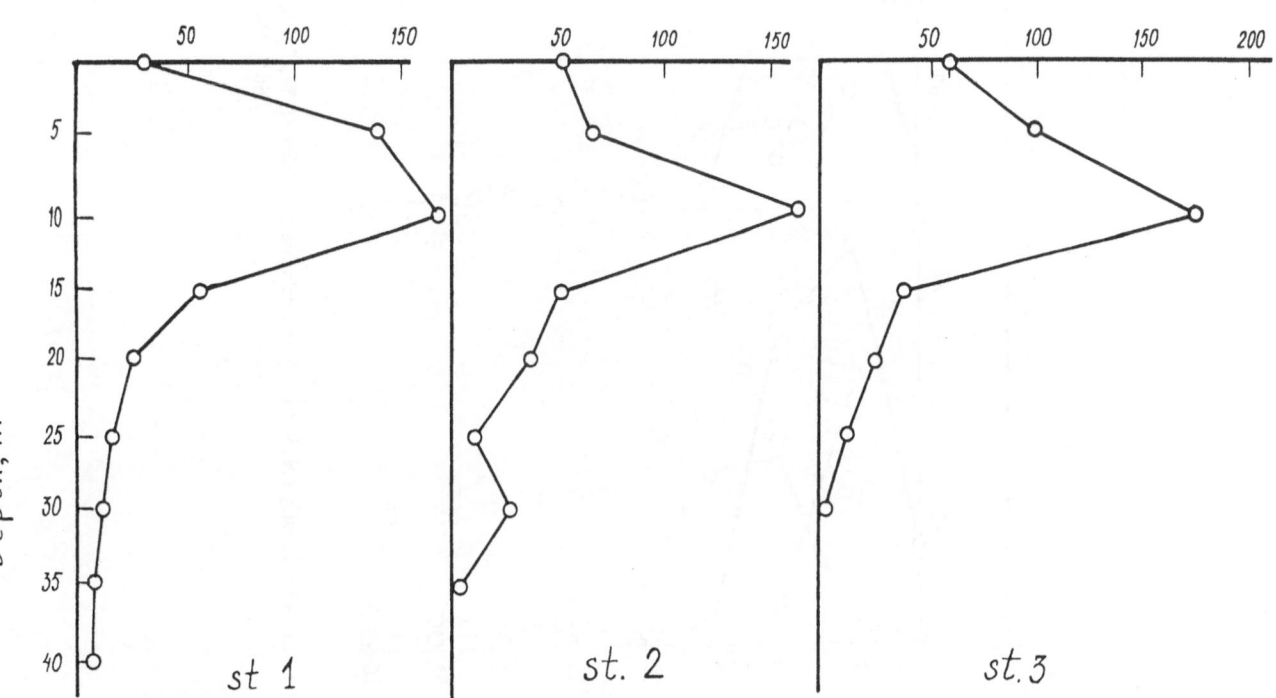

Fig. 2. Vertical distribution of the *Keratella cochlearis* population (1 – N, ind l^{-1}) in water column of S. Volos Lake at three stations. July 1985.

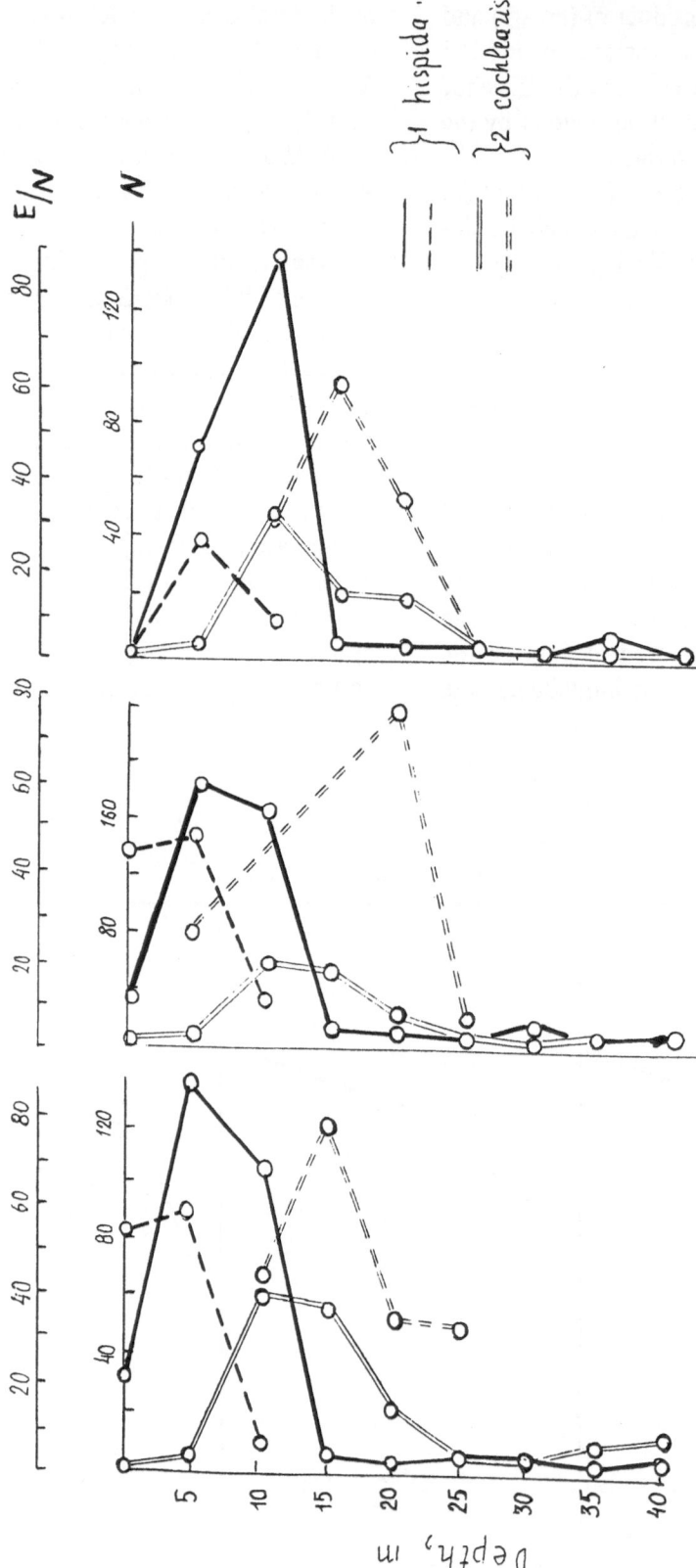

Fig. 3. Variations in density (N, ind l⁻¹) and relative fecundity (E/N, %) for *Keratella cochlearis hispida* (1) and *Keratella cochlearis cochlearis* (2) in water column of S. Volos Lake, July 1985.

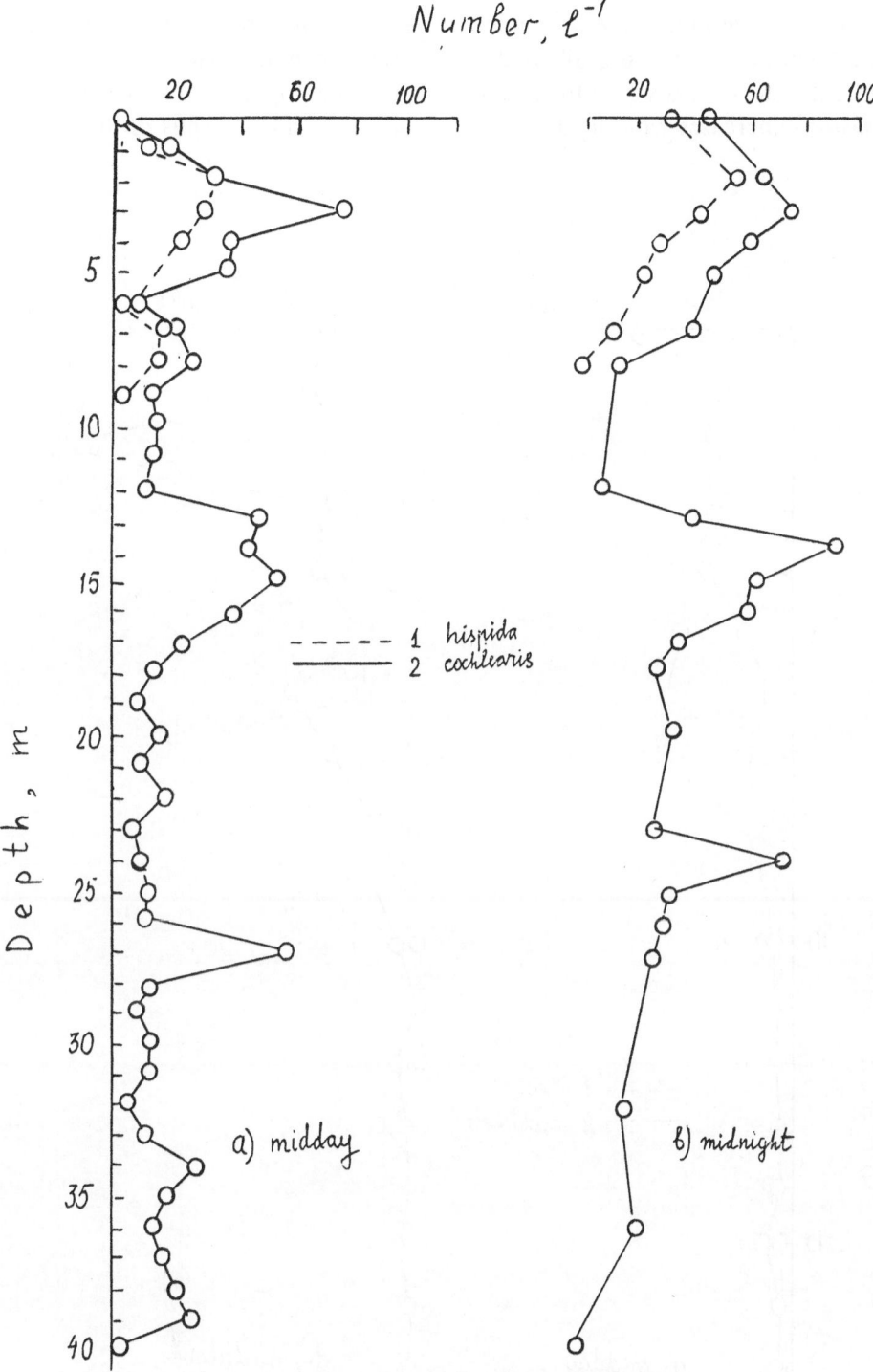

Number, ℓ^{-1}

Depth, m

1 hispida
2 cochlearis

a) midday

b) midnight

Fig. 4. Vertical distribution of densities (N, ind l^{-1}) for *Keratella cochlearis hispida* (1) and *Keratella cochlearis cochlearis* (2) in water column of S. Volos Lake at midday (a) and midnight (b). July, 17 (24.00) – 18 (12.00), 1986.

eutrophic lake Mikolajki as well (Fig. 5). In both lakes, smaller morphs – *tecta* in Lake Mikolajki and *hispida* in Lake S. Volos – prevail in the epilimnion while the relative density of the larger morph – *cochlearis* – increases with depth.

A positive correlation between the density of *K. cochlearis* and chlorophyl *l* concentration was found in both lakes (Fig. 6), thus indicating a potential trophic factor contributing to the spatial distribution of morphs of this species.

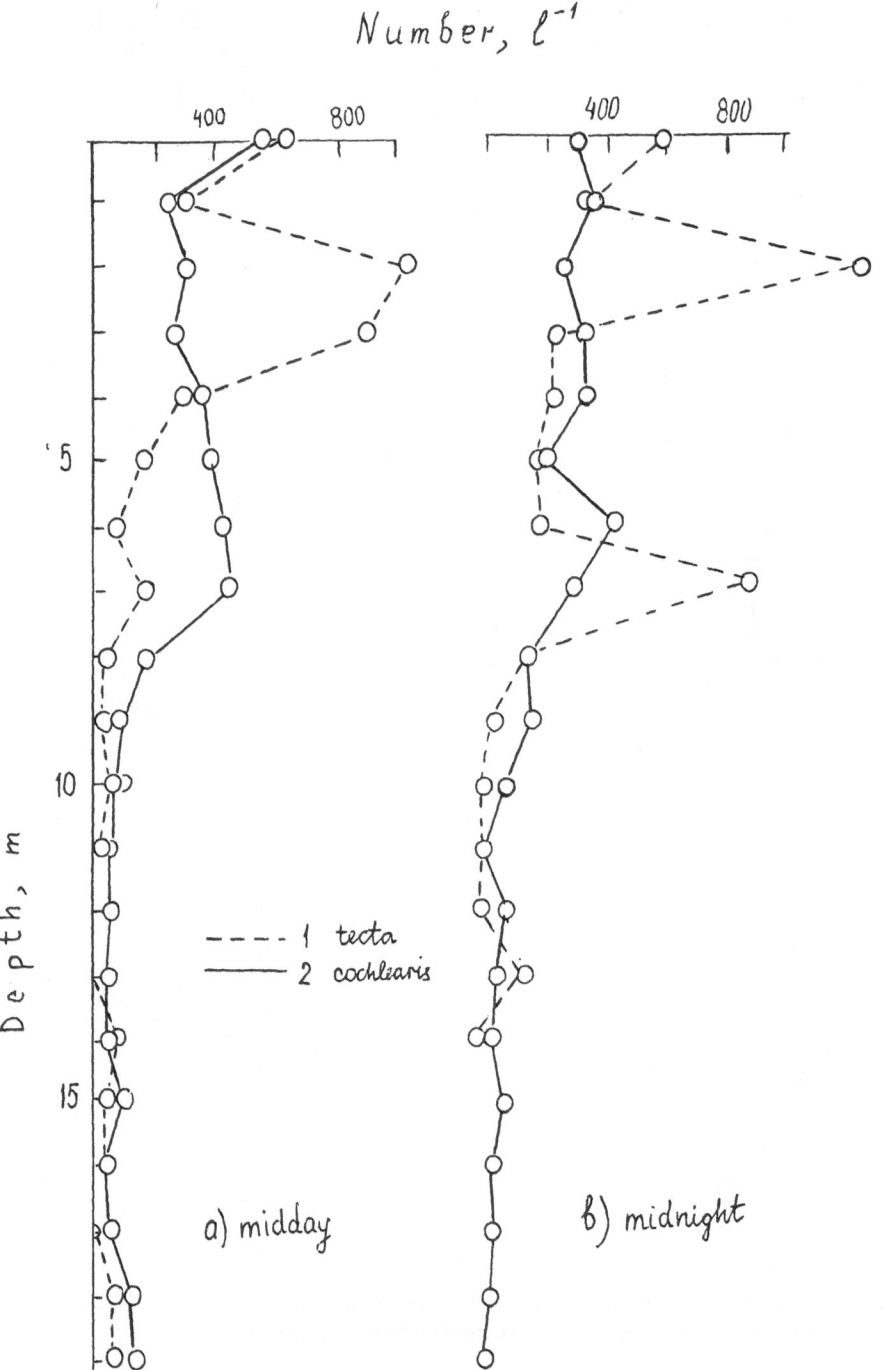

Fig. 5. Vertical distribution of densities (N, ind l⁻¹) for *Keratella cochlearis tecta* (1) and *Keratella cochlearis cochlearis* (2) in water column of Mikolajskie Lake at midday (a) and midnight (b). June, 24 (12.00) – 25 (24.00), 1986.

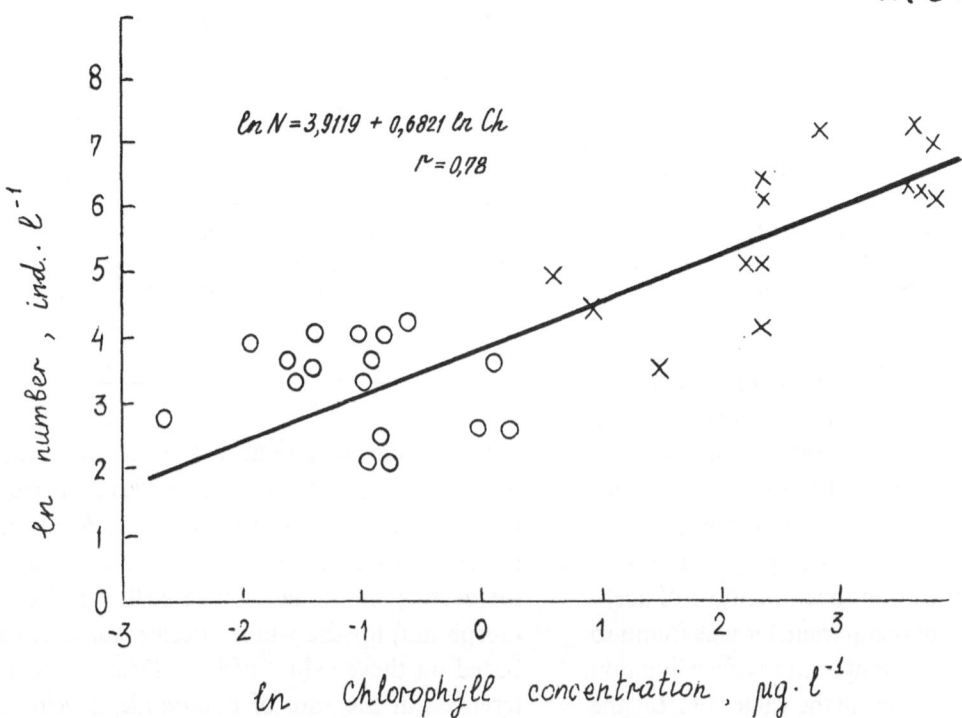

Fig. 6. Relationship between density of *Keratella cochlearis* (N, ind l^{-1}) and chlorophyll concentration (Ch, $\mu g \, l^{-1}$) based on data from concurrent sampling of water column layers in Lakes S. Volos (1) and Mikolajskie (2).

Some population and energetic characteristics of *K. cochlearis*

In July, 1985, the population size (*N*) and the *E/N* ratio for two morphs of *K. cochlearis* in the vertical water column at the same station were observed continuously for several days (Fig. 3). Maximum density of *K. c. hispida* (140 ind l^{-1}) was observed in the 5–10 m layer. *K. c. cochlearis* had a maximum density of 60 ind l^{-1} in the 10–15 m layer. This morph was not found in the 0–5 m layer.

The vertical profile of the parameters *N* and *E/N* from the surface to the bottom showed a maximal *E/N* value for *K. c. hispida* above or in the layer with maximal density. In lower layers there was an abrupt decrease in the value of this parameter. The sudden decrease of the *E/N* ratio along with the increased density of *K. c. hispida* allows us to suggest that there may be trophic deficiencies under these conditions.

A different pattern was observed in *K. c. cochlearis*. The density of this morph increased gradually with depth. Maximum *E/N* values were observed below the maximial density zone. The maximum occurred in the zone of 10–15 m.

The minimum egg ratio and maximum density of the species population were observed at the meta-hypolimnion border. The apparent density effect is probably a result of trophic limitation. It is apparent that the density of species population is determined by *K. c. hispida* except under trophically limiting conditions where it is unable to reproduce itself and is gradually supplanted by the larger morph *K. c. cochlearis*.

Embryonic development rates (l/D_e) in Lake S. Volos were obtained for rotifers selected from water layers with temperatures of 6, 12, and 18 °C both under natural conditions (Galkovskaya & Mityanina, 1986) and under abruptly changed temperature conditions as well (for example, ani-

Table 1. Empryonic development rate ($1/D_e$, day^{-1}) and embryonic growth intensity (W/D_e, 10^{-5} mg wet mass day^{-1}) for *Keratella cochlearis*. S. Volos Lake, June 1984.

Temperature of adapta-tion, °C	Experimental temperature, °C					
	18		12		6	
	$1/D_e$	W/D_e	$1/D_e$	W/D_e	$1/D_e$	W/D_e
18	0.96	3.08	0.66	2.12	0.50	1.60
12	0.75	3.29	0.63	2.77	0.36	1.58
6	0.63	3.48	0.38	2.10	0.29	1.60

Table 2. Variations in relative populational fecundity (E/N, %), embryonic development rate ($1/D_e$, day^{-1}) and birth rate (b, day^{-1}) for two *Keratella cochlearis* morphs in the water column of S. Volos Lake. July 1985.

Depth, m	*Keratella cochlearis* hispida			*Keratella cochlearis* cochlearis		
	E/N	$1/D_e$	b	E/N	$1/D_e$	b
0	22	1.02	0.202	-	-	-
5	24	1.08	0.232	33	0.88	0.250
10	6	0.95	0.055	24	0.62	0.134
15	-	-	-	21	0.39	0.075
20	-	-	-	37	0.36	0.112

mals collected from 6 °C water temperature were tested for embryonic development times at 12 °C and 18 °C, etc.). The volume-specific egg wet weight was also determined during these experiments. Using these data we introduce the parameter W/D_e the embryonic development intensity, and consider it in relation to temperature of acclimation. The value of this parameter was found to be independent of the temperature of acclimation (Table 1). When comparing the values of $1/D_e$, the effect of preliminary temperature adaptation is due to egg mass differences. Using the values of W/D_e, the Q_{10} is 1.7 at temperatures ranging from 6 °C to 18 °C.

Using the empirical curve of the relation between W/D_e and temperature, we reconstructed the expected values of $1/D_e$ for *K. cochlearis* from the values of the egg mass and the temperature of the corresponding layer. The birth rate for each morph and for the whole species was calculated based on these values of $1/D_e$. Despite great differences in the rate of embryonic development between the morphs, the birth rate of both morphs was equal. Maximum birth rates were observed in the epilimnion at a depth of 5 m (Table 2). The maximum daily population production in the

Table 3. Some energetic characteristics of the Keratella cochlearis population (1 – *K. cochlearis hispida*, 2 – *K. cochlearis cochlearis*) from S. Volos Lake. July 20, 1987.

Depth	Tem-pera-ture °C	N, ind·l^{-1}		E/N		B, 10^{-2} J·l^{-1}		P, 10^{-3} J·l^{-1} day^{-1}	R, 10^{-2} J·l^{-1} day^{-1}
		1	2	1	2	1	2		
0	20.6	33	-	0.27	-	0.40	-	0.96	4.94
5	19.3	139	-	0.29	-	2.06	-	5.50	19.45
10	8.1	106	61	0.02	0.21	1.27	1.10	0.95	8.06
15	5.7	-	55	-	0.38	-	0.10	0.08	1.59
20	5.0	-	19	-	0.16	-	0.34	0.12	0.46
25	4.9	-	6	-	0	-	0.11	0	0.15
30	4.8	8	3	0.25	0.33	0.10	0.05	0.09	0.27
35	4.8	1	7	0	0	0.01	0.13	0	0.19
40	4.6	6	8	0	0	0.07	0.14	0	0.34

Mean values for the lake: $B = $ 1.51 10^{-2} J·l^{-1}.
$P = $ 0.124 10^{-2} J·l^{-1}·day^{-1}.
$R = $ 0.319 10^{-2} J·l^{-1}·day^{-1}.
$P/B = $ 0.08.
$K_2 = $ 0.28.

Table 4. Some energetic characteristics of the *Keratella cochlearis* population (1 – *Keratella cochlearis tecta*, 2 – *Keratella cochlearis cochlearis*) from Mikolajskie Lake. June 24–25, 1986.

Depth	Temperature, °C	N, ind·l^{-1}		E/N		B, 10^{-2} J·l^{-1}		P, 10^{-3} J·l^{-1} day^{-1}	R, 10^{-2} J·l^{-1} day^{-1}
		1	2	1	2	1	2		
0	17.0	653	375	0.23	0.15	3.76	2.22	1.10	10.06
5	15.1	171	300	0.28	0.16	0.99	2.58	0.54	4.65
10	11.0	35	55	0.06	0	0.61	0.47	0.02	0.60
15	7.5	46	76	0.26	0	0.21	0.65	0.03	0.62
19	6.8	32	66	0.28	0	0.18	0.57	0.02	0.47

Mean values for the lake: $B = 3.09\ 10^{-2}$J·l^{-1}.
$P = 0.42\ 10^{-2}$ J·l^{-1}·day^{-1}.
$P/B = 0.135$.
$K_2 = 0.10$.

water column was observed at 5 m as well. Potential production decreases abruptly at depths of 10 m and greater despite stable, and even increasing, relative fecundities.

Figure 6 shows that the rotifer density is closely correlated with chlorophyl *l* concentration. This may explain why there is a change of morphs between layers; i.e., the production of each morph is likely to be dependent upon trophic conditions. The population production rate (P/B) of *K. cochlearis* is higher in the eutrophic lake in comparison with the mesotrophic lake (0.135 and 0.08 respectively), the net growth efficiency of the populations, on the contrary, decreases abruptly in the eutrophic lake in comparison with the mesotrophic lake (0.10 and 0.28 respectively) (Tables 3 and 4).

Discussion

There were no active vertical diurnal movements of *K. cochlearis* populations at least between the epi-, meta-, and hypolimnion in Lake S. Volos or Mikolajki during the summer stagnation period.

The absence of diurnal vertical migration and the relative temporal stability of the population vertical profile suggest eco-physiological patterns of processes taking place in the epi-, meta-, and hypolimnion. In particular, it becomes apparent while examining the affects of temperature adaptation on embryonic development.

Cases of simultaneous presence of several morphs of the rotifer *K. cochlearis* in lakes are known (Green, 1981; Hillbricht-Ilkowska, 1972; Hofmann, 1987). Pejler (1980) explained morphological variability of *K. cochlearis* within one water body as a purely physiological adaptation. As a basis for his explanation, he used data on the absence of sexual reproduction in *K. cochlearis* during one year and some successive years (Carlin, 1943; Ruttner-Kolisko, 1949) and hence found no reason for applying King's (1977) hypothesis of genetic population interruption to *K. cochlearis*. Protein polymophism in natural rotifer populations has been demonstrated by King (1977); i.e., population genetic heterogeneity has been shown. Only selection characteristics can be discussed until it is proven that mutation accumulation takes place during parthenogenetic reproduction within a season.

Presently there are not enough data to explain morphological variability as a type of morpho-physiological adaptation occurring within the limits of the genotype, nor to characterize morphs from the point of total genetic interruption as genotypes adapted to different lake conditions. Further investigations in this area are necessary.

If morphs are the phenotypic realization of

128

genotypes, strict division of regions inhabited by individual morphs may be facilitated in stratified lakes. The absence of shifting condition gradients, i.e., correspondence of the definite part of population to epi-, meta-, hypoliminion can promote selection even during successive parthenogenetic generations. The relative spatio-temporal isolation of population components may create a 'sliced functioning' of population as a whole.

Discussing *K. cochlearis* populations of deepwater mesotrophic and eutrophic lakes in terms of a heterogeneous medium, we find the realization of optimal population productivity may be due to the formation of morphotypic-distinct functional subunits.

High correlations between rotifer densities and chlorophyll concentration permits one to consider spatial morphotypic shifting as a component of a population's adaptive ability in response to variability in trophic conditions.

Acknowledgements

The authors are grateful to Dr. I. Kufel who has provided unpublished data. Also to Dr. J. Ejsmont-Karabin and Dr. A. Karabin for their help in sampling Mikolajskie Lake and Dr. V. Vezhnovets for his help in sampling Lake S. Volos.

References

Bosch, F. V. D. & J. Ringelberg, 1985. Seasonal succession and population dynamics of *Keratella cochlearis* (Gosse) and *Kellicottia longispina* (Kellicott) in lake Maarsseveen 1 (Netherlands). Arch. Hydrobiol. 103: 273–290.

Carlin, B., 1943. Die planktonrotatorien des Motalastrom. Lund: Carl Bloms Bortryckeri, 255 pp.

Djokosetiyanto, D. & N. Lair, 1983. Strategies d'utilisation des ressources dans un lac meso-oligotrophe: Migration verticales. Ann. Stat. biol. Besse- en Chandess. N 17: 65–91.

Galkovskaya, G. A., 1984. Spatial distribution of planktonic rotifers and their trophic relations in lake ecosystems. V kn. Limnologia gornykh vodoemov. Yerevan: 54 (In Russian).

Galkovskaya, G. A. & I. F. Mityanina, 1986. Birth rate characteristics of natural population of the rotifer *Keratella cochlearis* (Gosse). DAN BSSR 30: 568–570 (In Russian).

Gallagher, J. J., 1957. Cyclomorphosis in the rotifer *Keratella cochlearis* (Gosse). Trans. am. micr. Soc. 76: 197–203.

Green, J., 1981. Altitude and seasonal polymorphism of *Keratella cochlearis* (Rotifera) in lakes of Auvergne, Central France. Biol. J. limnean Soc. 16: 55–61.

Hillbricht-Ilkowska, A., 1972. Morphological variation of the *Keratella cochlearis* (Gosse) (Rotatoria) in several Masurian lakes of different trophic level. Pol. Arch. Hydrobiol. 19: 253–264.

Hillbricht-Ilkowska, A., E. Pieczynska & E. Pieczynski, 1971. Studies on the productivity of several Mazurian lakes and on the effect of fish on lake biocenosis. Wiadomosci ekologiczne 17: 124–146.

Hofmann, W., 1980. On morphological variation in *Keratella cochlearis* population from Holstein lakes (Northern Germany). Hydrobiologia 73: 255–258.

Hofmann, W., 1983. On temporal variation in the rotifer *Keratella cochlearis* (Gosse): the question of 'Lauterborncycles'. Hydrobiologia 101: 247–254.

Hofmann, W., 1987. Population dynamics of hypolimnetic rotifers in the Pluss-see (North Germany). Hydrobiologia 147: 197–201.

King, C. E., 1977. Genetics of reproduction, variation and adaptation in rotifers. Arch. Hydrobiol. 8: 187–201.

King, C. E., 1980. The genetic structure of zooplankton populations. In W. C. Kerfoot (ed.), Evolution and Ecology of Zooplankton Communities. Special Sumposium 3. Univ. Press New England: 315–329.

Kutikova, L. A., 1970. Rotifer fauna of the USSR. Fauna USSR 104. Akad. Nauk. SSR, Leningrad, 744 pp. (In Russian).

Paloheimo, J. E., 1974. Calculation of instantaneous birth rate. Limnol. Oceanogr. 19: 692–694.

Pejler, B., 1980. Variation in the genus *Keratella*. Hydrobiologia 73: 207–213.

Ruttner-Kolisko, A., 1949. Zum Formwechsel und Artproblem von *Anuraea aculeata* (*Keratella quadrata*). Hydrobiologia 1: 425–468.

Snell, T., 1977. Clonal selection: Competition among clones. Arch. Hydrobiol. 8: 202–204.

Yakushko, O. F., 1971. Byelorussian Poozerje. Historical development and present situation of lakes of Northern Byelorussia. Minsk, 336 pp. (In Russian).

Hydrobiologia **186/187**: 129–136, 1989.
C. Ricci, T. W. Snell and C. E. King (eds), Rotifer Symposium V.
© 1989 *Kluwer Academic Publishers.*

The development of *Hexarthra* spp. in a shallow alkaline lake

Alois Herzig & Walter Koste
*Biologische Station Neusiedler See, A-7142 Illmitz, Austria and Ludwig-Brill-Straße 5, D-4570
Quakenbrück, FRG*

Key words: *Hexarthra*, salinity, alkalinity, wind action, temperature

Abstract

In Neusiedler See, a shallow alkaline lake with fluctuating water level and salinity, four species of *Hexarthra* occur: *H. mira*, *H. fennica*, *H. jenkinae* (occasional) and *H. polyodonta*. The analysis of long-term data reveals a general phenological pattern which does not change from year to year. They first occur in May, develop a maximum in June/July, sometimes a second one in August/September and disappear in October. But the species succession is different in the various years, occasionally only one species (*H. mira* or *H. polyodonta*) being present. There is a fairly consistent relation between the chemical conditions and the prevalent species; an increase in salinity favours the development of *H. polyodonta*. Low temperature and wind generated suspended particles have a negative influence on the development of the *Hexarthra* populations. Smaller populations of *Hexarthra* are in a relation to the occurrence of *Leptodora* indicating predation pressure of the latter species. In Neusiedler See the *Hexarthra* populations seem to be controlled to a great extent by abiotic factors, but predation by *Leptodora* and most probably by young fish seems to play an important role too.

Introduction

Morphological and ecological characteristics of the genus *Hexarthra* Schmarda 1854 make these rotifers challenging objects for field and laboratory investigations. These animals have complex and subtle appendages which produce a skipping motion. They inhabit the pelagic zone of fresh and inland saline waters as well as the sea. Some of the species tolerate a wide range of salinities. A preference for high temperatures is obvious (Ruttner-Kolisko, 1974; Koste, 1978).

For the past 20 years (1968–1988) four species have occurred in Neusiedler See: *Hexarthra mira* (Hudson, 1871), *Hexarthra fennica* (Levander, 1892), *Hexarthra jenkinae* (De Beauchamp, 1932)

and *Hexarthra polyodonta* (Hauer, 1957). In earlier papers on the plankton of this lake *H. mira*, *H. fennica* and *H. intermedia* Wiszniewski 1929 are mentioned (Varga, 1926, 1934; Zakovsek, 1961). One to three species may be detected within a year and all of them develop throughout the warmer season (May-October). During summer 8–61% of the rotifer plankton may consist of *Hexarthra* specimens.

The present paper gives a description of the phenology of these species over the years and a discussion of the possible influences of abiotic factors (e.g. salinity, temperature, wind) and biological interactions (e.g. predator-prey) on the population development.

Materials and methods

The lake

Neusiedler See is a shallow ($\bar{z} = 1.3$ m) well-mixed lake which is characterized chemically by its high alkalinity (7.5–14.6 meq l^{-1}) and a conductivity of 1150–2800 μS (18 °C). The surface area (321 km^2) is divided into two different habitat types. The open water zone (143 km^2) is characterized by a high concentration of suspended solids which are stirred up from the bottom by wind action. The marginal areas are densely covered by *Phragmites communis* (178 km^2) and have a clear water colored by humic substances. The low altitude (115 m a.s.l.), the influence of the Pannonian climate, large surface area, shallow water and continuous mixing result in high water temperatures during the summer months (maximum >28 °C) and extremely low ones throughout winter. Since 1970 the lake has become increasingly eutrophic. A detailed description of the physiography and ecological aspects may be found in Löffler (1979).

Taxonomy

The identification of the various species was done on preserved material only and was based on the trophi of the mastax. *Hexarthra* is characterized by a mastax with malleoramate trophi and the number of teeth on the uncus is used in taxonomy. According to Ruttner-Kolisko (1974) the various morphological characteristics (e.g. size, relation of the ventral arm/body length, number of bristles, number of teeth on the uncus) by which the species of this genus are distinguished show considerable variation in different environmental conditions; there seems to be a tendency towards an increase in the number of teeth on the uncus as salinity increases. Nevertheless, according to Koste (1978), the four species can be identified on the basis of the number of teeth and display clear ecological differences. Drawings of the trophi of the various species are given in Fig. 1.

Methods

In 1968 and 1969, samples were taken with a Ruttner bottle at 8–16 stations at three different depths; in 1970–1974 integrated samples were taken with horizontal tows using an Isaacs-Kidd high-speed sampler. The mesh size of the filters used was 50 μm. Since 1975 samples have been collected with vertical net hauls (30 μm mesh size). The sample interval has varied between 6–15 days in summer, 2 weeks in spring and autumn and 3–5 weeks during winter. In order to obtain a reasonable estimate of the population in the lake 8–20 separate stations have been sampled and from these counted samples a mean for the whole lake was calculated (for details see Herzig, 1974, 1979).

Data on chemical analysis is taken from Neuhuber (1971), Ruttner-Kolisko (1971), Berger & Neuhuber (1979), Löffler & Newrkla (1985) and unpublished reports of the Biological Station Neusiedler See.

Results and discussion

In the quantitative plankton samples from Neusiedler See, *Hexarthra* spp. consistently appear in the summer plankton (Zakovsek, 1961; Herzig, 1979; Ruttner-Kolisko pers. comm.). During 1951–1979 the highest relative contribution of species in this genus to the rotifer community varied between 8 and 61% and the summer mean relative abundance varied between 5 and 18%. The abundance of *Hexarthra* spp. for several years is shown in Fig. 2. In general the first specimens appear in May (rarely in April) at water temperatures of 8–10 °C; subsequently an exponential growth phase takes place (T °C: 17–20) and maximum abundance is reached in June/July or August when water temperatures exceed 21 °C; *Hexarthra* quickly disappears in October, as soon as water temperature drops below 15 °C.

Both qualitative and quantitative changes in seasonal abundance are seen in Fig. 2. The populations of 1968–1970 are dominated by *H. mira*

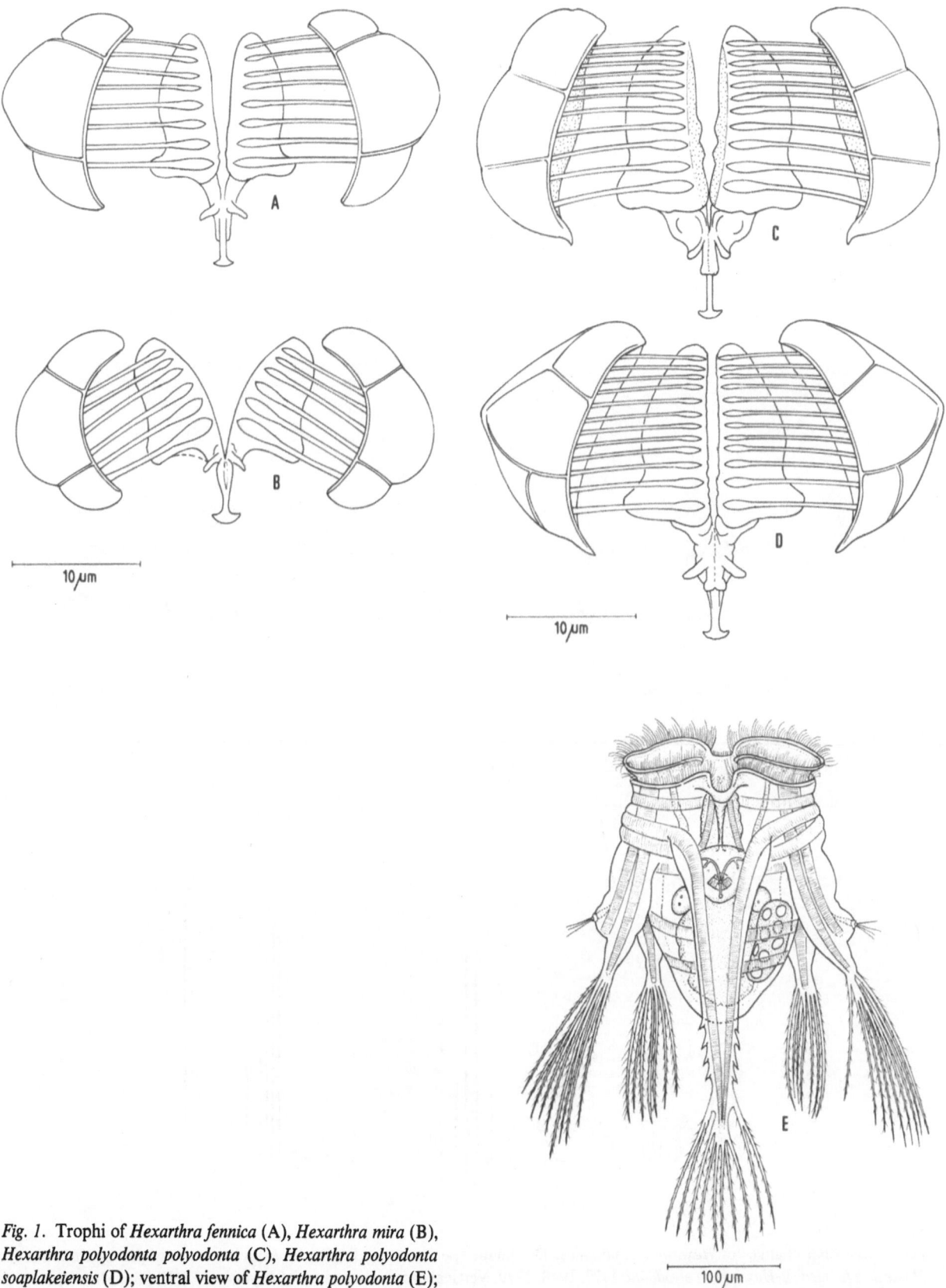

Fig. 1. Trophi of *Hexarthra fennica* (A), *Hexarthra mira* (B), *Hexarthra polyodonta polyodonta* (C), *Hexarthra polyodonta soaplakeiensis* (D); ventral view of *Hexarthra polyodonta* (E); drawings by W. Koste.

10 μm

10 μm

100 μm

132

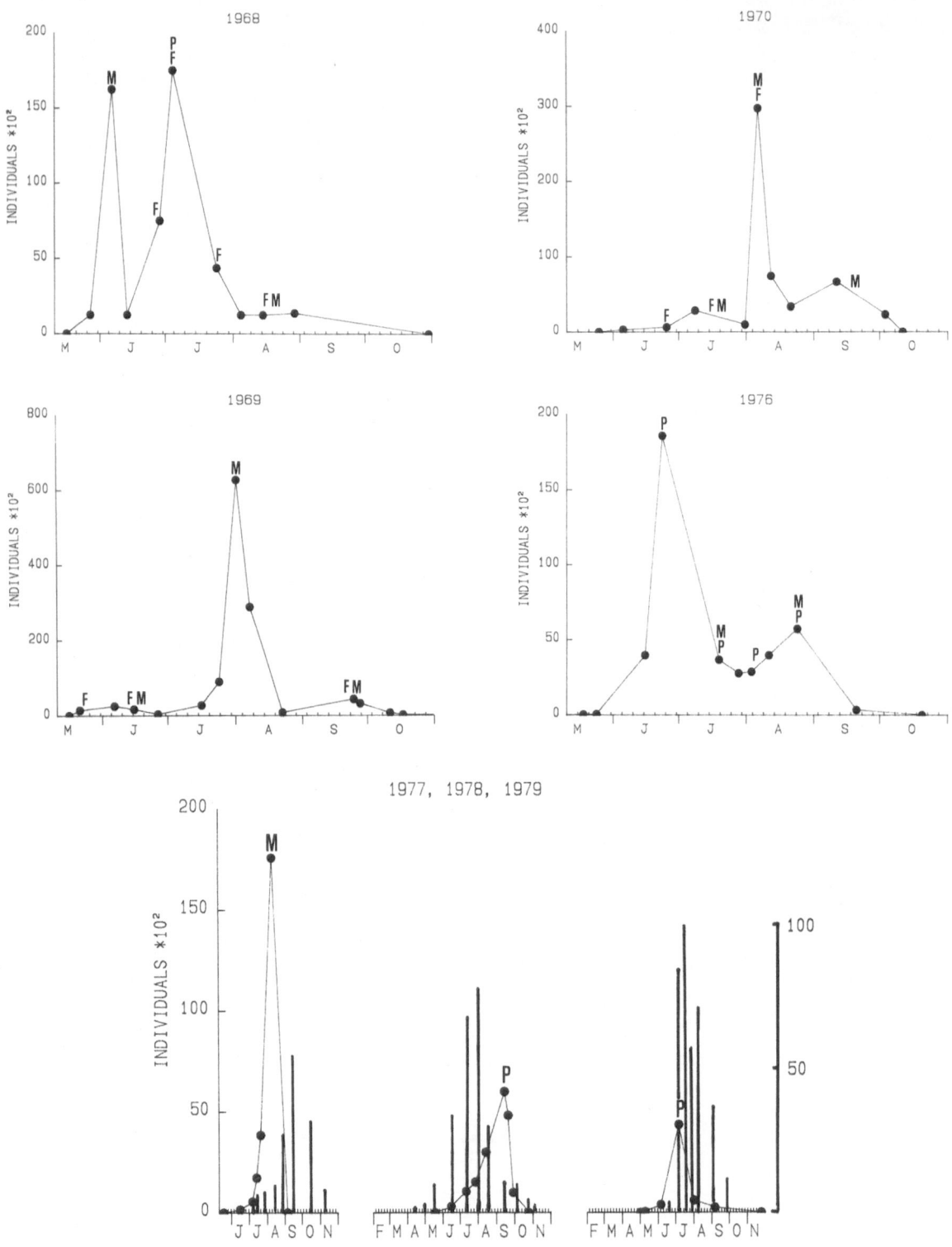

Fig. 2. Seasonal changes of *Hexarthra* abundance. Dominant species are indicated by capital letters. M-*Hexarthra mira*, F-*Hexarthra fennica*, P-*Hexarthra polyodonta*. 1977, 1978, 1979: Vertical bars – abundance of *Leptodora kindti*. (abundance per m³).

and *H. fennica*. Within this period *H. jenkinae* and *H. polyodonta* were found only once (beginning of June 1969 and beginning of July 1968 respectively). In 1976 *H. polyodonta* occurred together with *H. mira*. In 1977 *H. mira* was the dominant species, but *H. polyodonta* was present in low numbers. Since 1978 *H. polyodonta* has frequently been the only *Hexarthra* species found in the summer rotifer plankton.

The various *Hexarthra* populations hatch from resting eggs. Resting egg production was observed in Neusiedler See from late August until the disappearance of the animals. In addition a continuous input of *Hexarthra* resting eggs may occur from the various shallow lakes ('Seewinkellacken') or gravel pits east of Neusiedler See. Some of these water bodies chemically resemble Neusiedler See, others are more alkaline, and in some chlorides and/or sulphates prevail. Accordingly *H. jenkinae* and *H. polyodonta* may be observed in lakes containing soda; *H. fennica* occurs in water bodies with a smaller amount of soda and in waters characterized by chlorides and sulphates (Löffler, 1957, 1959; Ruttner-Kolisko, 1971). *H. mira* is frequently found in waters of low salinity and in gravel pits (Herzig, unpublished). The majority of these shallow lakes dry up completely during summer or autumn and resting eggs of various *Hexarthra* populations may be introduced by wind into Neusiedler See. Passive transfer of the eggs may also take place by water fowl which frequently migrate between these smaller water bodies and Neusiedler See.

Variation in the occurrence of these species may reflect variation in the number of viable resting eggs. Alternatively, the chemical conditions of Neusiedler See may not suit all species and/or populations. According to Ruttner-Kolisko (1971) *H. jenkinae*, which was found only once, mainly inhabits waters with a higher salt content (> 5 g 1^{-1}) and higher alkalinity (> 50 meq 1^{-1}). These values are far above the levels reached in Neusiedler See.

Nevertheless there is a fairly consistent relation between chemical conditions and the prevalent species (Fig. 3, Table 1). Between 1968 and 1979 moderate, but significant, increases in electrical

Table 1. Mean values ($\pm$ SD) of electrical conductance and alkalinity for the period of occurrence of the various species.

	Electrical conductance (μS)	Alkalinity meq $\cdot 1^{-1}$
Hexarthra mira	1327 ± 225	8.48 ± 0.9
Hexarthra fennica	1228 ± 87	8.18 ± 0.72
Hexarthra polyodonta	1846 ± 220	10.29 ± 0.95

Table 2. Mean values ($\pm$ SD) of electrical conductance and alkalinity for various years.

Years	Electrical conductance (μS)	Alkalinity (meq $\cdot 1^{-1}$)
1968/69/70	1269 ($\pm$ 103)	8.20 ($\pm$ 0.63)
1975/76	1836 ($\pm$ 172)	10.19 ($\pm$ 0.41)
1977/78/79	1867 ($\pm$ 232)	10.35 ($\pm$ 0.73)

conductance and alkalinity were observed (Table 2). The relation between alkalies (A) and alkaline earth compounds (E) (E-A (meq 1^{-1}): $-1.22 - +1.63$) and pH (7,8–9.0) remained relatively stable (Berger & Neuhuber, 1979; Brossmann in Löffler & Newrkla, 1985). The higher the values of electrical conductance and alkalinity, the more likely is the prevalence of *H. polyodonta* (Fig. 3). An analysis of variance (one factor ANOVA for repeated measures) reveals clear differences (significant at 95%) between *H. polyodonta* and *H. fennica* and *H. mira*. No significant differences exist between *H. fennica* and *H. mira*. This is consistent with the results of Hutchinson (1932) who was not able to detect any great contrast in the occurrence of the two species. This overlap in occurrence seems to happen as long as the salt content remains rather low (< 2 g 1^{-1}). At higher concentrations, predominantly in chloride dominated inland waters, only *H. fennica* will occur (Ruttner-Kolisko, 1971, 1974).

The results of the investigation so far suggest that the increase in salinity did favour the development of *H. polyodonta*. None the less more comparable field studies and especially laboratory experiments are needed to get a better knowledge about how the various *Hexarthra* species

134

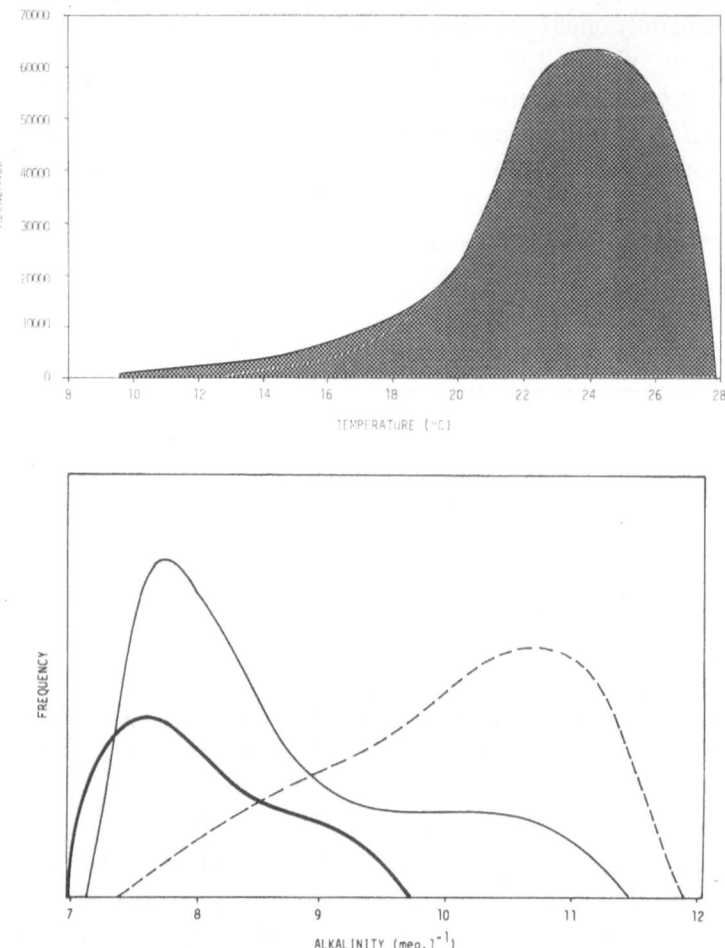

Fig. 3. Occurrence of *Hexarthra* spp. in relation to temperature, electrical conductance and alkalinity in Neusiedlersee.

cope with a changing chemical environment. Besides the alteration in species composition well expressed quantitative changes took place within the years 1968–1979 resulting from the impact of abiotic factors and biological interactions. Irrespective of the species, the population development of *Hexarthra* is mainly influenced by temperature and wind action. A correlation analysis revealed highly significant positive correlations between abundance and temperature ($r = 0.575$), as well as between population growth rate and mean temperature for the period considered ($r = 0.579$). The correlation between population growth rate and mean wind speed was highly significant and negative ($r = -0.455$).

In summer, temperatures above 22 °C and several days of calm weather (mean wind speed < 12 km h^{-1}) lead to rapid population growth (daily growth rate > 0.2). But even at these temperatures high wind speed (mean > 18 km h^{-1}, maximum > 40 km h^{-1}) produces suspended inorganic particles which mechanically interfere with the animals (cf. *Diaphanosoma brachyurum*, Herzig, 1975, 1979) and may have a negative effect on the population development (e.g. Fig. 2, August 1970). Frequently strong summer winds are associated with a rapid temperature decline which may last for about one to two days. This combination of abiotic factors has a detrimental effect on the *Hexarthra* populations (e.g. Fig. 2, June and July 1968, August 1969).

As mentioned earlier, hatching of resting eggs

is followed by a period of exponential growth (e.g. Fig. 2, 1968, 1976). This exponential growth phase can be delayed (e.g. Fig. 2, 1969, 1970). As temperature and food conditions (Dokulil, 1984) were comparable for all years, but heavy winds prevailed (mean wind speed > 20 km h^{-1}) during these periods in 1969 and 1970, the growth delay may again be attributed to mechanical interference by suspended particles.

These results are consistent with the idea that low temperature and high amounts of wind-generated suspended particles can negatively influence the development of Hexarthra populations. Nevertheless a multiple regression analysis based on the data from the summer months (y = population growth rate; x_1 = mean temperature; x_2 = mean wind speed for the same period) demonstrates that only 25.4% of the total variance is explained by these two factors. One therefore has to search for an explanation of the 74.6% residual variance.

There is evidence from long-term analysis of the phytoplankton that the food conditions improved throughout the years 1968–1979 (Dokulil, 1979, 1984) resulting in an increase of the zooplankton (Herzig, 1980). The increased abundance of zooplankton favored the development of planktivorous fish. The current idea is that planktivorous fish and the still unknown quantities of fish fry and fish that are less than one year old (mainly cyprinids) to some extent control the zooplankton populations (especially during the summer months). So far we do not know whether these fishes have an impact on the rotifer populations and in particular on Hexarthra spp.

In any case, the Hexarthra populations show quite a variation in their quantitative appearance; peak values range from 15 l^{-1} to more than 60 l^{-1} (see Fig. 2). But in 1978 and 1979 the maximum numbers remain below 10 or even below 5 animals per litre. These much smaller populations are in a fairly consistent relation to the occurrence of Leptodora kindti which was detected for the first time in mid June 1977 and since then has become more and more abundant (Fig. 2, 1977, 1978, 1989). Preliminary estimates of birth rates of Hexarthra (applying embryonic

development data derived for H. fennica, Ruttner-Kolisko, 1975) gave values of 0.2–0.42 for the summer months and lead to the suggestion that, without interference by Leptodora, Hexarthra could have developed a much larger population.

There is little evidence from the literature that Leptodora is preying upon rotifers; only two references can be found in the review paper on 'Invertebrate predators of planktonic rotifers' by Williamson (1983). Recently Edmondson & Litt (1987) demonstrated a relation between Conochilus hippocrepis and Leptodora kindti; in three years the decrease in Conochilus came during the peak of Leptodora. No further information on feeding behavior or feeding rates is available but at least the authors could identify trophi of Conochilus in the Leptodora stomach.

So far we have neither information on stomach contents nor experimental work that demonstrate this predator-prey relationship. But it seems likely that the size (up to 400 μm), the reasonable soft body, and the skipping movement of Hexarthra species may make them suitable prey items for Leptodora kindti.

References

Berger, F. & F. Neuhuber, 1979. The hydrochemical problem. In H. Löffler (ed.), Neusiedler See. The limnology of a shallow lake in central Europe. Dr. W. Junk bv Publ., The Hague, Boston, London. Monogr. Biol. 37: 89–99.

Dokulil, M., 1979. Seasonal pattern of phytoplankton. In H. Löffler (ed.), Neusiedler See. The limnology of a shallow lake in central Europe. Dr. W. Junk bv Publ., The Hague, Boston, London. Monogr. Biol. 37: 203–234.

Dokulil, M., 1984. Assessment of components controlling phytoplankton photosynthesis and bacterioplankton production in a shallow, alkaline, turbid lake (Neusiedler See, Austria). Int. Revue ges. Hydrobiol. 69: 679–727.

Edmondson, W. T. & A. H. Litt, 1987. Conochilus in Lake Washington. Hydrobiologia 147: 157–162.

Herzig, A., 1974. Some population characteristics of planktonic crustaceans in Neusiedler See. Oecologia 15: 127–141.

Herzig, A., 1975. Der Neusiedler See – charakteristische Eigenschaften und deren Auswirkungen auf das Zooplankton. Verh. Ges. Ökologie, Wien 1975: 189–196.

Herzig, A., 1979. The zooplankton of the open lake. In H. Löffler (ed.), Neusiedler See. The limnology of a shallow lake in central Europe. Dr. W. Junk bv Publ., The Hague, Boston, London. Monogr. Biol. 37: 281–335.

Herzig, A., 1980. Effects of food, predation and competition in the plankton community of a shallow lake (Neusiedler See, Austria). In M. Dokulil, H. Metz & D. Jewson (eds), Shallow lakes. Contribution to their limnology. Dr. W. Junk bv Publ., The Hague, Boston, London. Developments in Hydrobiology 3: 45–51.

Herzig, A., 1987. The analysis of planktonic rotifer populations: A plea for long-term investigations. Hydrobiologia 147: 163–180.

Hutchinson, G. E., G. E. Pickford & J. F. M. Schuurman, 1932. A contribution to the hydrobiology of pans and other inland waters of South-Africa. Arch. Hydrobiol. 24: 1–154.

Koste, W., 1978. Rotatoria. Die Rädertiere Mitteleuropas. Monogononta. Gebrüder Borntraeger, Berlin, Stuttgart, Bd. 1, 673 pp.

Löffler, H., 1957. Vergleichende limnologische Untersuchungen and den Gewässern des Seewinkels (Burgenland). Verh. Zoo.-Bot. Ges. Wien, 97: 27–52.

Löffler, H., 1959. Zur Limnologie, Entomostraken- und Rotatorienfauna des Seewinkelgebietes (Burgenland, Österreich). Sitz. ber. Öst. Akad. Wiss., mathem.-naturwiss. Kl., Abt. I, 168: 315–362.

Löffler, H. (ed.), 1979. Neusiedler See. The limnology of a shallow lake in central Europe. Dr. W. Junk bv Publ., The Hague, Boston, London. Monogr. Biol. 37: 543 pp.

Löffler, H. & P. Newrkla, 1985. Der Einfluß des diffusen und punktuellen Nährstoffeintrages auf die Eutrophierung von Seen. Teil 2: Neusiedler See, Attersee, Universitätsverlag Wagner, Innsbruck, Veröff. Öst. MaB Progr. 8: 121 pp.

Neuhuber, F., 1971. Ein Beitrag zum Chemismus des Neusiedler Sees. Sitz. ber. Öst. Akad. Wiss., mathem.-naturwiss. Kl., Abt. I, 179: 225–298.

Ruttner-Kolisko, A., 1971. Rotatorien als Indikatoren für den Chemismus von Binnensalzgewässern. Sitz. ber. Öst. Akad. Wiss., mathem.-naturwiss. Kl., Abt. I, 179: 283–298.

Ruttner-Kolisko, A., 1974. Plankton Rotifers. Biology and Taxonomy. E. Schweizerbart'sche Verlagsbuchhandlung, Stuttgart. Die Binnengewässer XXVI/1, Suppl., 146 pp.

Ruttner-Kolisko, A., 1975. The influence of fluctuating temperature on plankton rotifers. A graphical model based on life data of *Hexarthra fennica* from Neusiedler See, Austria. Symp. Biol. Hung. 15: 197–204.

Varga, L., 1926. Die Rotatorien des Fertö (Neusiedler See). Arch. Balat. 1: 181–225.

Varga, L., 1934. Neue Beiträge zur Kenntnis der Rotatorienfauna des Neusiedler Sees. Allatani Közlemenyek 31: 139–150.

Williamson, C. E., 1983. Invertebrate predation on planktonic rotifers. Hydrobiologia 104: 385–396.

Zakovsek. G., 1961. Jahreszyklische Untersuchungen am Zooplankton des Neusiedler Sees. Wiss. Arb. Bgld. 27: 1–85.

Hydrobiologia **186/187**: 137–144, 1989.
C. Ricci, T. W. Snell and C. E. King (eds), Rotifer Symposium V.
© 1989 *Kluwer Academic Publishers.*

On choice of substrate and habitat in brachionid rotifers

Birger Pejler & Bruno Bērziņš
Institute of Limnology, P.O. Box 557, 751 22 Uppsala, Sweden

Key words: rotifers, Brachionidae, ecology, substrate, habitat

Abstract

Information on the distribution of 28 rotifers of the family Brachionidae from diverse waters in south and central Sweden was analyzed to reveal their relationships to substrate and habitat. Some brachionids are preferably planktic, others periphytic and/or benthic. Some non-planktic habitats are utilized more than others, but there is no evidence of a chemical attraction from any substrate. Instead, some substrates seem to be avoided, possibly depending on a poorer flora of periphytic algae. Besides substrate type, the following factors are found to be important for creating separate ecological niches in the brachionid family: temperature, oxygen content, trophic degree, chemical environment, food choice and sensitivity to predation. It is possible to delineate separate ecological niches for all brachionid rotifers, implying that Hutchinson's ideas about the 'plankton paradox' are contradicted. Some species are specialists, other are generalists, the latter being characterized by a great morphological variation. The species are adapted in different ways to their preferential habitats, as regards foot, egg-carrying, protrusions and other lorical structures etc. Longer spines, for instance, are generally found in more transparent water, being a supposed protection against visual predators.

Introduction

Most ecological studies of rotifers have been on planktic forms. Good studies have also been made on sessile rotifers (e.g. Edmondson 1944, 1945; Wallace, 1980; Francez, 1984), as well as on psammic ones (e.g. Ruttner-Kolisko, 1956; Tzschaschel, 1983). As regards some other non-planktic environments, important information may be obtained in, e.g., Meuche (1939), Wulfert (1951), Donner (1964, 1970 and 1972), Martin (1976 and 1977), Bateman & Davis (1980), and Francez & Dévaux (1985). However, a comprehensive study comparing major substrates has never been made before. This has now been possible by utilizing the large amount of data collected by Bruno Bērziņš during the years 1945–1982 from various types of water bodies in southern and central Sweden. His data include information on several abiotic parameters (see Bērziņš & Pejler 1987 and in press) and substrate, as well as the occurrence of different microzoans (planktic, periphytic and benthic). The material has now been computerized and will be published in several steps. Data on substrate exist for protozoans and rotifers and they will be reported separately for different taxonomic groups. In the present paper only one rotifer family is discussed, the Brachionidae.

Bruno Bērziņš is solely responsible for the collection of data and the taxonomic identification. Both authors participated in the computerization. Soon after the computer treatment was finished Bruno Bērziņš unfortunately died (Pejler, 1987), and thus Birger Pejler wrote the present paper (as well as the others).

Material and methods

The entire study comprises about 350 lakes, 50 ponds, 20 pools, 150 running water localities, and 15 mires. However, in most cases only planktic samples were investigated, together with physical and chemical analyses. This is reflected in Table 1, where the number of samples from different substrates is denoted. Most material derives from southern Sweden, especially the provinces of Småland and Skåne (= Scania). The waters in this part of Sweden exhibit a very broad range of varying character, from decidedly oligotrophic to hypertrophic, from oligohumic to polyhumic environments, etc.

Zooplankton material was collected with a water sampler (1 or 5 l). The samples were filtered through a 20 μm mesh net, and the concentrates were fixed with formalin and counted. In most cases net samples were collected as well; they were studied alive, focusing on species which are difficult to identify in a fixed condition.

Periphyton was collected in the following way: moist macrophyte material of at least 0.5 l was squeezed into a bottle, dipped again and squeezed. This was repeated several times. The sample was investigated alive. Subsamples were sucked up several times and analyzed microscopically until no additional species were encountered. Densities were estimated on a scale from 1 to 10, where the Figure 1 designated stray individuals, 6 a common occurrence and 10 abundant occurrence. Periphytic and benthic species were seldom found in high abundance. By investigating the epifauna of stones and logs an area of 100–200 cm^2 was scraped off. The material was sucked up into a bottle directly from the knife, being kept at the substrate (air was slowly let out from the bottle, causing an influx of water and scraped-off material).

For acquiring benthos a small pot was slowly moved over the bottom, whereby a layer of bottom material, some millimeters thick, was removed. An area of c. 100–200 cm^2 was sampled. The material was analysed alive, 20–30 subsamples per sample. The estimation scale from 1 to 10 was used in this case as well.

Results

As in the other papers dealing with the present material, a drastic data reduction was made. The entire data set is contained in computer sheets and tapes, archived at the Limnological Institutes in Lund and Uppsala.

The reduction was made in different respects: 1) Only substrates from which at least 50 samples were taken are included in Table 1. *Chara* constitutes an exception, depending on its richness of species. 2) Abundance is not reported in Table 1, only occurrence. 3) In the computer sheets, lakes, ponds, running waters and mires are separated, but in Table 1 they are united. However, in the discussion the additional information obtained from the computer sheets will be considered and sometimes referred to in the text.

In the table the planktic environments are mentioned first (also the mire hollows might be regarded as an open-water environment in miniature). Then the benthic substrates are brought up, followed by submersed vegetation, floating-leaved plants and helophytes. An artificial substrate, cotton, is given at the end of the series. The cotton was dipped into the water for some days and then removed for investigation.

The figure '1' in the table means that the species has been found in 1–10% of the samples collected from the substrate in question, as against 10–50% for figure '2' and > 50% for figure '3'. The latter figure was obtained only in plankton and only for one brachionid, *Keratella c. cochlearis*, perhaps the most common freshwater metazoan in the world. Interpretation of results will focus on rank 2- and 3-occurrences, but some attention will also be paid to the patterns of distribution on various substrates.

As regards the sum of species, the planktic environment dominates strongly over other 'substrates'. This may partly depend on the very high number of plankton samples taken, but certainly also upon a generally higher tendency to a free-swimming life in brachionids than in most other rotifer families. The following species were usually found in the plankton of lakes and/or running waters: *Brachionus angularis, Kellicottia longispina,*

Table 1. Occurrence of brachionid rotifers on different substrates. Only substrates with at least 50 samples are included (exception: Chara). The figure '1' means that the species was found in 1–10% of the samples from the respective substrate, while '2' corresponds to 10–50% and '3' to >50%.

	Plankton	Mire hollows	Minerogenous bottom	Mixed bottom	Thick organogenous sediment	Tree trunks, branches	Leaves of birch	Filamentous algae	Chara	Fontinalis	Lobelia, Isoëtes	Myriophyllum alterniflorum	Sphagnum	Sphaerotilus	Utricularia	Lemna minor	Nymphaea, Nuphar	Potamogeton natans	Alisma	Alopecurus	Carex spp.	Equisetum	Glyceria	Juncus	Phragmites	Scirpus lacustris	Sparganium spp.	Typha	Cotton
Number of samples	4838	50	215	336	273	53	59	176	27	122	61	94	194	78	60	77	125	136	67	161	234	72	58	74	67	52	130	97	50
Anuraeopsis fissa (Gosse)	1								1							1	1				1				1		1	1	1
Brachionus a. angularis (Gosse)	1															1	1						1				1	1	1
Brachionus budapestinensis (Daday)	1															1													1
Brachionus calyciflorus (Pallas)	1								1							1							1						
Brachionus l. leydigi (Cohn)																							1			1			
Brachionus patulus (Müll.)	1																												
Brachionus q. quadridentatus (Hermann)								1		1		1				1	1	1	1				1	1			1	1	2
Brachionus q. brevispinus (Ehrbg)																1			1										2
Brachionus u. urceolaris (L.)	1		1	1				1	1								1		1				1						1
Brachionus u. rubens (Ehrbg)	1		1	1	1			1		1		1					1	1				1	1			1		1	2
Kellicottia longispina (Kellicott)	2		1	1	1	1											1	1				1		1		1			1
Keratella c. cochlearis (Gosse)	3		1	1	1	2		1		1	1	1					1	1				1	1	1			1	1	
Keratella c. hispida (Lauterborn)								1																					
Keratella c. macracantha (Lauterborn)	2																												
Keratella c. robusta (Lauterborn)	2																												
Keratella c. tecta (Gosse)	1					1																						1	1
Keratella hiemalis Carlin	1																												
Keratella i. irregularis (Lauterborn)	1					1																	1						
Keratella i. wartmanni (Asper and Heuscher)	1																												
Keratella paludosa (Lucks)		2											1																
Keratella q. quadrata (Müll.)	1	2		1		1				1						2							1					1	1
Keratella serrulata (Ehrbg)	1	1			1								1		2	1	1	1				1	1					1	
Keratella ticinensis (Callerio)	1						1				1																		
Keratella valga (Ehrbg)	1																												
Notholca acuminata (Ehrbg)	1	1	1				1					1																1	
Notholca caudata Carlin																													
Notholca squamula (Müll.)	1	1		1	1																								
Platyias quadricornis (Ehrbg)												1											1					1	
Sum of species	22	4	2	3	4	5	2	3	6	3	2	4	2	0	1	8	6	6	3	4	5	5	9	3	1	2	3	10	10

Keratella cochlearis (with its different forms), *K. hiemalis, K. irregularis, K. quadrata* and *Notholca caudata*. In contrast, a pronounced heleoplanktic (pond-preferring) occurrence was found for *Anuraeopsis fissa, Brachionus patulus, B. quadridentatus, Keratella serrulata, K. ticinensis, K. valga, Notholca squamula* and *Platyias quadricornis*.

The most typical cold-water eulimnoplankters, viz. *Keratella cochlearis macracantha, K. hiemalis* and *Notholca caudata*, were not found in any benthic or periphytic habitat, probably because no vegetation was investigated during winter. On the other hand, brachionids tolerating warm water often appeared in non-planktic habitats as well, but then preferably in environments containing large areas of open water (floating-leaved vegetation, etc.), in which the rotifers may have been swimming freely among the plants. (The method used makes it impossible to distinguish between different degrees of connection to the plants or bottom substrates.)

When studying the different sums of species for different substrates, the total number of samples from the respective substrates has to be considered. However, even if doing so, it is rather evident that some substrates and habitats are more attractive than others. *Sphaerotilus*, for instance, is avoided by all brachionids, as well as by the members of most other rotifer families. This is in contrast to the case for ciliates, where some species even occur in > 50% of the samples (to be published in future papers). Only *Keratella serrulata* was found together with *Utricularia*, exclusively in most samples from mires. The low preference for this substrate by brachionids is surprising with regard to its attractiveness for many other rotifers (Wallace, 1980; Pejler & Bērziņš, in press). In general, the benthic substrates do not seem to be very popular among brachionids, and for the periphytic habitats the power of attraction varies strongly, even between macrophytes living in about the same way, as between *Phragmites* and *Typha*. The high occupation frequency of the artificial substrate, cotton, is remarkable.

Different macrophytes prefer different types of waters; e.g., isoëtids (*Lobelia, Isoëtes*) prefer oligotrophic lakes, *Fontinalis* running waters and *Sphaerotilus* strongly polluted environments. Thus an inclination of a rotifer for a certain substrate often also reflects its preference for the habitat as a whole.

Discussion

Relation to substrate and habitat

There is no reason to suspect a chemical confinement to a certain substrate for any brachionid similar to what has been found for certain sessile rotifers (Wallace, 1980). The high occupation frequency of cotton taken in concert with the absence of brachionids on some of the available plants, instead suggests an avoidance of certain natural substrates. It is impossible to decide if such an avoidance depends on chemical repellants or on an indirect influence. The nature of the periphytic algal growth is probably of importance, but was not investigated in the present connection. It may be supposed that the physical structure of a substrate plays a role, and this could be one reason for the attractiveness of the cotton.

Based on results presented earlier it is evident that different species prefer different habitats, e.g. lakes vs. ponds or mire hollows. However, species showing a very similar pattern in Table 1 may be found to differ strongly when subjected to more detailed examination. Thus, in the south-Swedish mire Åkhultsmyren, *Keratella paludosa* and *K. serrulata* were very clearly segregated, the former almost exclusively occurring in stands of *Carex rostrata* and among *Sphagnum apiculatum*, the latter close to sediments and, especially, in the free water of mire hollows and soligenous fen soaks ('dråg' in Swedish) as well as among *Utricularia* (Bērziņš, unpublished). On the whole, *K. serrulata* tends more to a planktic life, e.g. also occurring in the zooplankton of acidified lakes (Morling & Pejler in press). Hauer (1935) designates *K. paludosa* as a typical bog species ('Leitform im engeren Sinne'), while *K. serrulata* is considered to have a broader ecological range.

Other niche-creating factors

Reasoning from the competitive exclusion principle, each species should have its own ecological niche, and the question arises as to which forces are important for this diversification. It is evident from the above that the choices of substrate and macrohabitat are far from sufficient for this explanation, and this might be the proper place for a discussion of other important factors.

The relation to *temperature* divides the brachionids into separate groups. *Keratella hiemalis* and all *Notholca* species are cold stenothermal; the other species are eurythermal or warm stenothermal (Pejler & Bērziņš, in press a). Similarly, some species evidently can tolerate lower levels of *oxygen concentration*, whereby especially *Keratella ticinensis*, *K. valga*, *K. hiemalis*, *Platyias quadricornis*, *Brachionus patulus* and *Anuraeopsis fissa* ought to be mentioned (Pejler & Bērziņš, in press b). For both temperature and oxygen the spatial and temporal segregation of the separate species is much more pronounced within the same localities than when comparing different waters as done in the regional material presented here (cf. Bērziņš & Pejler, 1989).

Also with respect to *trophic degree*, there are obvious differences existing between different species. Thus, *Brachionus angularis*, *B. calyciflorus*, *Keratella irregularis* and *Anuraeopsis fissa* may be regarded as indicators of eutrophy, *Kellicottia longispina* rather the reverse (Pejler, 1965, 1983; Pejler & Bērziņš, in press c). In some cases different forms of the same species are good indicators of different trophic degree; *K. cochlearis macracantha* is found only in oligotrophic lakes during summer (but during winter in eutrophic lakes as well), *K. cochlearis tecta* only occurring during summer in eutrophic environments (Pejler, 1962b). As regards *K. quadrata*, the results are contradictory (Pejler, 1965, p. 478; Pejler & Bērziņš, in press c). Probably different ecological races occur, the oligotrophic form belonging to the 'frenzeli-group' (Carlin, 1943).

A different tolerance to *deviating chemical environments* certainly also plays an important niche-creating role. Thus, some brachionids can occur at rather low pH-values, especially *Keratella serrulata* but also *K. ticinensis* and *K. valga* (Bērziņš & Pejler, 1987). *K. paludosa* surely belongs to this category as well, but it did not reach the '50-sample block' needed for inclusion in the diagrams of the referred paper. Another brachionid, *Brachionus urceolaris sericus* (Rousselet), is a specialist on very acid conditions (Morling & Pejler, in press) but it was not included in the present material because of the lack of such localities in southern and central Sweden. Humic material may have a varying toxicity depending on the pH-level (e.g., Petersen & Persson, 1987). From Pejler & Bērziņš (in press d) it appears that the following brachionids have an unusually good ability to withstand a high concentration of humolimnic substances; *Keratella serrulata*, *K. ticinensis*, *K. valga*, *Brachionus patulus*, *B. quadridentatus* and *Anuraeopsis fissa*.

For co-occurring species a differential choice of *food* may help to prevent competition. Thus, all *Notholca* species seem to prefer diatoms, in contrast to all other brachionids (Thane-Fenchel, 1968; Pourriot, 1977; May, 1980a, b; Laxhuber & Hartmann, 1988). There are also differences between other genera and species although perhaps not equally clear-cut (Pourriot, 1977; Pejler, 1977; Starkweather, 1980; Bogdan *et al.*, 1980). When offered the same food, *Keratella cochlearis* and *Brachionus angularis* were influenced in different ways, the former behaving as a K-strategist, the latter as a r-strategist (Walz, 1983). Finally, a varying sensitivity to *predation* may be decisive for the environment where a species will be able to live (Williamson, 1983).

Concluding remarks

In conclusion, it seems possible with present knowledge to delineate separate ecological niches for all brachionid rotifers. This also agrees with results and opinions put forward by Makarewicz & Likens (1975) and Miracle (1977). Consequently, our results contradict ideas of a 'plankton paradox' suggested by Hutchinson (1961) and supported by Ghilarov (1984). In fact, these ideas are also contradicted by more recent results on *Daphnia cucullata* and *D. galeata*, two coexisting species used by Ghilarov in support of his position. These two species show a pronounced difference in their vertical distribution in the Polish lakes studied by Pijanowska & Dawidowicz (1987) and are therefore likely to have a different vulnerability to predation.

Some species, like *Keratella paludosa* and *Notholca caudata*, seem to be relatively restricted in their ecology, while others are very euryecious. The most extreme example of the latter category is *Keratella cochlearis*, which shows an enormous morphological variation, and is composed of several forms (Pejler, 1980). The morphological variation is certainly underlaid by genetical and physiological differences which may explain why different forms exhibit different ecological requirements.

Adaptations to different environments

The brachionids show a smooth transition from a non-planktic via a semiplanktic and heleoplanktic to a eulimnoplanktic way of life. One adaptation of true eulimnoplankters, discussed by de Beauchamp (1909), is the disappearance of the *foot*. However, a foot is also lacking in some species dwelling in very small open waters, like mire hollows, and in close connection to plants and sediments. Thus, all *Keratella* species are devoid of a foot, even the semi- or non-planktic *K. paludosa*, *K. serrulata*, *K. ticinensis* and *K. valga*. Certainly these species must differ from coexisting forms of *Brachionus* and *Platyias* which may be fixed to a substrate when feeding. On the other hand, some species with feet are able to invade the pelagic zone. *B. angularis* may become planktic during blooms of cyanobacteria ('blue-green algae'), when they use these phytoplankters as a substrate, as observed by Wesenberg-Lund (1930, p. 119). *B. urceolaris rubens*, on the other hand, is mainly found on *Daphnia* and other cladocerans (Koste, 1978). The possession or lack of an adhesive organ surely contributes to determining the ecological niche of a species.

Egg-carrying is another phenomenon considered to constitute an adaptation for a planktic life. According to the general formulations by Koste (1978) and Sudzuki (1957), all brachionids possess this habit irrespective of whether they are planktic or not. This is in contrast to the condition in some other rotifer families, where only true planktic species (and not even all of these) show this characteristic. In the genus *Notholca*, eggs are carried for a very short time (cf. Pejler, 1962a).

Another feature typical of planktic organisms is the possession of varying types of *protrusions* (spines, etc.). These have been interpreted as means of increasing the buoyancy. However, in brachionids their variation in relation to viscosity is the reverse of what is expected, the longest extensions being found in cold water, when viscosity is highest (Pejler, 1962b). In addition, the spines are much longer in oligotrophic than in eutrophic lakes during summer at the same temperature (*op. cit.*). Thus the most long-spined forms, *Kellicottia longispina*, *Keratella cochlearis macracantha* and f. *frenzeli* of *K. quadrata*, are found in cold water and/or oligotrophic waters, while the short-spined *Brachionus angularis* and *Keratella cochlearis tecta* are typical of eutrophic waters during summer (Pejler & Bērziņš in press c). Long-spined forms are characteristic of transparent water and may provide protection against visual predators (cf. Nilsson & Pejler, 1973, pp. 69–70). Evidently, non-visual predators like cyclopoids and *Asplanchna* also avoid the long-spined forms as suggested by Stemberger & Evans (1984) and by Stemberger & Gilbert (1984, 1987).

In addition, soft-bodied forms are more readily ingested than loricated rotifers (Anderson, 1970; Gilbert & Williamson, 1978; Stemberger, 1982; Williamson, 1983). It is a striking fact that the species with the longest spines, *Kellicottia longispina*, has a weaker lorica than most *Keratella* forms. These two adaptive traits might be conceived to compensate for each other. Forms living in acid waters often have an unusually thick lorica with diverse structures. Thus, the extremely acidobiontic form *Brachionus urceolaris sericus* is distinguished from *B. u. urceolaris* through its strongly sculptured lorica (Koste 1978), and the same is characteristic of *Keratella paludosa* (Hauer, 1935; Wulfert, 1956) and *K. serrulata* (Hendelberg *et al.*, 1979), as well as in several mire-living rotifers of other families. This might be explained by the fact that fishes are more or less lacking in such waters, whereby their role as predators is taken over by invertebrates (e.g., Nyberg, 1984). These invertebrate predators generally prey upon smaller animals than fishes do, and consequently rotifers need a stronger lorica as a protection. Possibly a thick lorica may also serve as a protection against plants feeding on small animals, and it ought to be pointed out that *K. serrulata* is the only brachionid found among *Utricularia*, where it occurs frequently (see Table 1). A pustulation is often found, not only in mire-living rotifers, as part of an interspecific or infraspecific variation. Attempts at explaining its adaptive value (which it certainly has) are almost never found in the literature. Only Ruttner-Kolisko (1972, p. 173) links the occurrence of *Keratella cochlearis hispida* (a pustulated form-series) with the turbulence of the water; these rotifers being said to occur mainly in shallow and small waters (an observation which agrees fairly well with the authors' experience from south and central Sweden). However, with regard to the fact that biotic relations have been increasingly found to be important in ecology during recent decades, it may be suspected that these structures function as protections against predators as well.

Acknowledgements

Data analysis and subsequent routine work was supported by the National Environmental Protection Board of Sweden. Among the persons engaged in this treatment we especially wish to mention Jan Bertilsson.

References

Anderson, R. S., 1970. Predator-prey relationships and predation rates for crustacean zooplankters from some lakes in western Canada. Ca. J. Zool. 48: 1229–1240.

Bateman, L. & C. Davis, 1980. The Rotifera of hummock-hollow formations in a poor (mesotrophic) fen in Newfoundland. Int. Revue Hydrobiol. 65: 127–153.

Beauchamp, P. M. de, 1909. Recherches sur les Rotifères: les formations tégumentaires et l'appareil digestif. Årch. Zool. exp. gén., ser. 4, 10: 1–410.

Bērziņš, B. & B. Pejler, 1987. Rotifer occurrence in relation to pH. Hydrobiologia 147: 107–116.

Bogdan, K., J. Gilbert & P. Starkweather, 1980. In situ clearance rates of planktonic rotifers. Hydrobiologia 73: 73–77.

Carlin, B., 1943. Die Planktonrotatorien des Motalaström. Zur Taxonomie und Ökologie der Planktonrotatorien. Medd. Lunds Univ. limnol. Instn. 5: 1–256.

Donner, J., 1964. Die Rotatorien-Synusien submerser Makrophyten der Donau bei Wien und mehrerer Alpenbäche. Arch. Hydrobiol. Suppl. 27: 227–324.

Donner, J., 1970. Die Rädertierbestände submerser Moose der Salzach und anderer Wasser-Biotope des Flussgebietes. Ibid. 36: 109–254.

Donner, J., 1972. Die Rädertierbestände submerser Moose und weiterer Merotope im Bereich der Stauräume der Donau an der deutsch-österreichischen Landesgrenze. Ibid. 44: 49–114.

Edmondson, W. T., 1944. Ecological studies of sessile Rotatoria. Part I. Factors affecting distribution. Ecol. Monogr. 14: 31–66.

Edmondson, W. T., 1945. Ecological studies of sessile Rotatoria. Part II. Dynamics of populations and social structures. Ibid. 15: 141–172.

Francez, A.-J., 1984. Rotifères sessiles observés en Auvergne. Cahiers Naturalistes, Bull. nat. paris., N.S. 40: 73–80.

Francez, A.-J. & J. Dévaux, 1985. Répartition des rotifères dans deux lacs-tourbieres du Massif Central (France). Hydrobiologia 128: 265–276.

Ghilarov, A. M., 1984. The paradox of the plankton reconsidered; or, why do species coexist? Oikos 43: 46–52.

Gilbert, J. & C. Williamson, 1978. Predator-prey behavior and its effect on rotifer survival and associations of Mesocyclops edax, Asplanchna girodi, Polyarthra vulgaris, and Keratella cochlearis. Oecologia 37: 13–22.

Hauer, J., 1935. Rotatorien aus dem Schluchseemoor und seiner Umgebung. Verh. naturw. Ver. Karlsruhe 29: 47–130.

Hendelberg, M., G. Morling & B. Pejler, 1979. The ultrastructure of the lorica of the rotifer Keratella serrulata (Ehrbg). Zoon 7: 49–54.

Hutchinson, G. E., 1961. The paradox of the plankton. Am. Nat. 95: 137–145.

Koste, W., 1978. Rotatoria. Die Rädertiere Mitteleuropas. Ein Bestimmungswerk begr. von Max Voigt. Überordning Monogononta. Vol. 1-2. 673 pp. + 234 pl.

Laxhuber, R. & U. Hartmann, 1988. The influence of temperature on the life cycle of the cold-stenothermal rotifer Notholca caudata Carlin. Verh. int. Ver. Limnol. 23: 2016–2018.

144

Makarewicz, J. & G. Likens, 1975. Niche analysis of a zooplankton community. Science 190: 1000–1003.

Martin, L. V., 1976. Rotifers in the Sphagnum pools on Thursley Common. Microscopy 33: 90–93.

Martin, L. V., 1977. Rotifers in the Sphagnum pools on Thursley Common. Part 2. Ibid. 33: 236–241.

May, L., 1980a. Studies on the grazing rate of Notholca squamula Müller on Asterionella formosa Hass. at different temperatures. Hydrobiologia 73: 79–81.

May, L., 1980b. On the ecology of Notholca squamula Müller in Loch Leven, Kinross, Scotland. Ibid. 73: 177–180.

Meuche, A., 1939. Die Fauna im Algenbewuchs. Nach Untersuchungen im Litoral ostholsteinischer Seen. Arch. Hydrobiol. 34: 349–520.

Miracle, M. R., 1977. Migration, patchiness, and distribution in time and space of planktonic rotifers. Arch. Hydrobiol. Beih. 8: 19–37.

Morling, G. & B. Pejler, in press. Acidification and zooplankton development in some west-Swedish lakes 1966-1983. Limnologica 21.

Nilsson, N.-A. & B. Pejler, 1973. On the relation between fish fauna and zooplankton composition in North Swedish lakes. Rep. Inst. Freshw. Res. Drottningholm 53: 51–77.

Nyberg, P., 1984. Impact of Chaoborus predation on planktonic crustacean communities in some acidified and limed forest lakes in Sweden. Ibid. 61: 154–166.

Pejler, B., 1962a. Morphological studies on the genera Notholca, Kellicottia and Keratella (Rotatoria). Zool. Bidr. Uppsala 33: 295–309.

Pejler, B., 1962b. On the variation of the rotifer Keratella cochlearis (Gosse). Ibid. 35: 1–17.

Pejler, B., 1965. Regional-ecological studies of Swedish fresh-water zooplankton. Ibid. 36: 407–515.

Pejler, B., 1977. On the global distribution of the family Brachionidae (Rotatoria). Arch. Hydrobiol. Suppl. 53: 255–306.

Pejler, B., 1980. Variation in the genus Keratella. Hydrobiologia 73: 207–213.

Pejler, B., 1987. Bruno Bērziņš in memoriam. 1909-1985. Ibid. 147: 1–2.

Pejler, B. & B. Bērziņš, 1989. Rotifer occurrence in relation to temperature. Ibid. 175: 223–231.

Pejler, B. & B. Bērziņš, 1989. Rotifer occurrence in relation to oxygen content. Ibid. 183: 165–172

Pejler, B. & B. Bērziņš, 1989. Rotifer occurrence and trophic degree. Ibid. 172: 171–180

Pejler, B. & B. Bērziņš, 1989. Rotifer occurrence in relation to water colour. Ibid. 184: 23–28

Petersen, R. & U. Persson, 1987. Comparison of the biological effects of humic materials under acidified conditions. Sci. Tot. Envir. 62: 387–398.

Pijanowska, J. & P. Dawidowicz, 1987. The lack of vertical migration in Daphnia: the effect of homogenously distributed food. Hydrobiologia 148: 175–181.

Pourriot, R., 1977. Food and feeding habits of Rotifera. Arch. Hydrobiol. Beih. 8: 243–260.

Ruttner-Kolisko, A., 1956. Der Lebensraum des Limnopsammals. Verh. dtsch. zool. Ges. Hamburg 1956: 421–427.

Ruttner-Kolisko, A., 1972. Rotatoria. Binnengewässer 26: 99–234.

Stemberger, R., 1982. Mechanisms controlling selection and rates of predation on rotifers in Cyclops bicuspidatus thomasi. Ph. D. Thesis, Univ. Mich. 95 pp.

Stemberger, R. & M. Evans, 1984. Rotifer seasonal succession and copepod predation in Lake Michigan. J. Great Lakes Res. 10: 417–428.

Stemberger, R. & J. Gilbert, 1984. Spine development in the rotifer Keratella cochlearis: induction by cyclopoid copepods and Asplanchna. Freshwat. Biol. 14: 639–647.

Stemberger, R. & J. Gilbert, 1987. Defenses of planktonic rotifers against predators. In: W. C. Kerfoot and A. Sih (eds.). Predation. Direct and indirect impacts on aquatic communities. 386 pp.

Sudzuki, M., 1957. Studies on the egg-carrying types in Rotifera. II. Genera Brachionus and Keratella. Zool. Mag., Tokyo 66: 11–20. (Japanese, with English summary.)

Thane-Fenchel, A., 1968. Distribution and ecology of non-planktonic brackish-water rotifers from Scandinavian waters. Ophelia 5: 273–297.

Tzschaschel, G., 1983. Seasonal abundance of psammon rotifers. Hydrobiologia 104: 275–278.

Wallace, R., 1980. Ecology of sessile rotifers. Hydrobiologia 73: 181–193.

Walz, N., 1983. Continuous culture of the pelagic rotifers Keratella cochlearis and Brachionus angularis. Arch. Hydrobiol. 98: 70–92.

Wesenberg-Lund, C., 1930. Contributions to the biology of the Rotifera. II. The periodicity and sexual periods. K. danske vidensk. Selsk., nat.-math. Afd., Raekke 9, Bd 2, no. 1. 230 pp.

Williamson, C., 1983. Invertebrate predation on planktonic rotifers. Hydrobiologia 104: 385–396.

Wulfert, K., 1951. Das Naturschutzgebiet auf dem Glatzer Schneeberg. Die Rädertiere des Naturschutzgebietes. Arch. Hydrobiol. 44: 441–471.

Wulfert, K., 1956. Die Rädertiere des Teufelssees bei Friedrichshagen. Ibid. 51: 457–495.

Hydrobiologia **186/187**: 145–152, 1989.
C. Ricci, T. W. Snell and C. E. King (eds), Rotifer Symposium V.
© *1989 Kluwer Academic Publishers.*

Temporal analysis of clonal structure in a moss bdelloid population

Claudia Ricci[1], Manuela Pagani[2] & Anna Maria Bolzern[2]
[1] *Dipartimento di Biologia Animale, via Accademia Albertina 17, Torino, Italy*; [2] *Dipartimento di Biologia, via Celoria 26, Milano, Italy*

Key words: electrophoresis, isozymes, moss population, bdelloid rotifers

Abstract

Clonal structure of a population of *Macrotrachela quadricornifera* (Rotifera, Bdelloidea) from a terrestrial moss in Northern Italy, was investigated over a 16 month period. Every month, 40-60 specimens of *M. quadricornifera* were collected from about 0.1 m² of moss. The individual animals were homogenized and their isozyme phenotypes analyzed by electrophoresis on vertical polyacrylamide gel. One enzyme, phosphoglucose isomerase (PGI), was used as a marker to distinguish the different clones present in the sample. A few clones were established from the rotifers sampled and patterns of esterases α and β, and malic enzyme were studied. Nine electrophoretic patterns for PGI were seen. One was dominant, a second was almost always present, but in lesser amounts. The remainder were present occasionally.

There seemed to be no seasonal replacement of the clones and the composition of the population appeared to be unaffected by variations in temperature. Relative humidity seemed to be the more important factor in regulating the number of electromorphs of the rotifer population.

Introduction

Much attention has been paid to the seasonal structure of parthenogenetic populations of cladocerans and rotifers in freshwater plankton. Here, several clones coexist in the same environment, with temporal changes in their relative frequencies (e.g. Hebert, 1977; Young, 1979; King & Zhao, 1987; Hebert *et al.*, 1988). In contrast to the competitive exclusion principle, the coexistence of clones with partially or totally overlapping niches can be maintained by changes in the environment which they inhabit. It is generally assumed that different clones compete for the same resources with different competitive abilities and that varying the conditions of the environment may shift the selective advantage from one clone to another over time. Changes in environment are commonly recognized as the major factors affecting the clonal dynamics of parthenogenetic zooplankton (Hebert, 1977; Korpelainen, 1986).

Lakes and ponds may be considered predictable environments that vary through the year without dramatic changes. The situation faced by animals inhabiting terrestrial mosses, where conditions change dramatically and rapidly, is different.

The present study concerns the clonal ecology of a bdelloid rotifer, *Macrotrachela quadricornifera*, in a terrestrial moss. Bdelloid rotifers reproduce through obligatory apomictic thelytoky and, although they are aquatic animals, they can inhabit water bodies and terrestrial mosses because they can respond to desiccation by entering an anhydrobiotic state. When anhydrobiotic, any bdelloid can be a propagule, able to colonize a new environment and give rise to a clonal population: its success in establishing such a population will depend on its ability to survive the anhydrobiotic state and on its growth rate. It seems reasonable to assume that competition among different clones in the same habitat will affect the clonal composition of a population throughout the different seasons.

At present, there is no experimental evidence that any species of bdelloid rotifers is composed of different clones, but most of parthenogenetic morphospecies investigated so far, have been found to be composed of several clones (Parker, 1979). Moreover, a recent investigation of electrophoretic patterns of six enzymes from 5 strains of the same morphospecies *M. quadricornifera* (unpublished) has revealed that the different strains can be distinguished by their enzymatic patterns. Therefore, as is will known for other parthenogenetic species (e.g. Snell, 1979; King, 1980; Hebert & Crease, 1983; Lynch, 1984), bdelloid species also consist of many different clones.

Material and methods

Specimens of *Macrotrachela quadricornifera* were collected from moss on a wall near Milan, Northern Italy, between February 1987 and June 1988. Each moss sample was approximately 0.1 m². Samples were taken every 45 days from February to October 1987 and monthly from November 1987 to June 1988. On some occasion no *M. quadricornifera* were found in the moss.

The moss had not been disturbed, for instance by pesticide treatments. No attempt was made to classify it botanically since the presence of bdelloid species is not thought to be related to moss species (Dobers, 1915; Bartos, 1951).

The moss was brought to the laboratory and the rotifers were extracted as soon as possible, usually on the day of sampling. For extraction, the moss was hydrated, suspended in a beaker with some water, immersed into an ice bath and heated from above with a ordinary incandescent lamp. Within a couple of hours, rotifers and other moss inhabitants moved downward to escape the light and heat. The contents of the beaker were

examined under a stereo-microscope and the rotifers collected with a Pasteur pipette. Each rotifer was identified under the light microscope. *M. quadricornifera* was easily recognized by the two appendages on its foot, dorsal to the spurs, and by its mastax.

Macrotrachela quadricornifera Milne 1886 is a cosmopolitan species-complex which lives in ponds, lakes and terrestrial mosses. Despite the many varieties described (Donner, 1965), the differences in morphology of the specimens concerned in this study were considered unimportant and the animals were classified as *M. quadricornifera sensu strictu*.

To obtain a total of about 50 specimens, more moss than usual had to be collected on some occasion. When the number of rotifers were excessive, some animals were isolated and cultured under laboratory conditions. The animals were fed on a mixture of algae (*Chlamydomonas reinhardti*) and bacteria (*E. coli*), sometimes with yeast (*Saccharomyces cerevisiae*) added.

Shortly after identification, the animals were processed for electrophoretic investigation. Extracts for electrophoresis were prepared by homogenizing each bdelloid in 5 μl sample buffer (0.05 M Tris HCl, pH 8, + 10% sucrose + 0.02% bromophenol). The entire extract was loaded into a sample slot. Electrophoretic separation was performed in 0.8 mm thick 7.5% polyacrylamide gel with 4.5% of N,N'-methylenebisacrylamide, using a mini slab vertical gel apparatus (Scientific Idea, Corvallis, U.S.A.) in a continuous system of 0.082 M Tris-borate buffer, pH 8.5. Since the size of the rotifers enabled analysis of a single locus system only, phosphoglucose isomerase, PGI (E.C. 5.3.1.9) was used to distinguish electromorphs. Previous experience with bdelloid electrophoresis (unpublished) showed that PGI has good staining resolution and is polymorphic in this species, therefore, although insufficient, it is effective for detecting at least part of the variability of the population. Two additional enzymes, malic enzyme (ME) and esterases (Est α and β), were tested in the clonal populations when possible. The staining technique of Harris & Hopkinson (1976) was used for the three enzymes, with minor modifications (unpublished). Since the data were mostly for a single locus, only PGI electromorphs are considered and are listed according to speed of migration through the gel, with 1 being slowest and 6 being fastest.

Physical parameters of the sampling site, such as rain, relative humidity and average daily air temperature, were are also related to the changes in clone diversity over the investigation period. Rain, humidity and temperature were obtained from the Brera Metereological Station (Milan).

The Shannon index, the Evenness or, simply, the number of electromorphs were used to express clonal diversity in each sample. All metereological parameters were calculated over 10, 15 and 20 days prior to sampling.

Results

550 morphologically classified individuals of *M. quadricornifera* were screened, and 9 electrophoretic patterns for the enzyme PGI were distinguished (Fig. 1). The term 'clone' is used as a synonym for electromorph, and does not necessarily refer to a single lineage.

The distribution of the clones throughout the

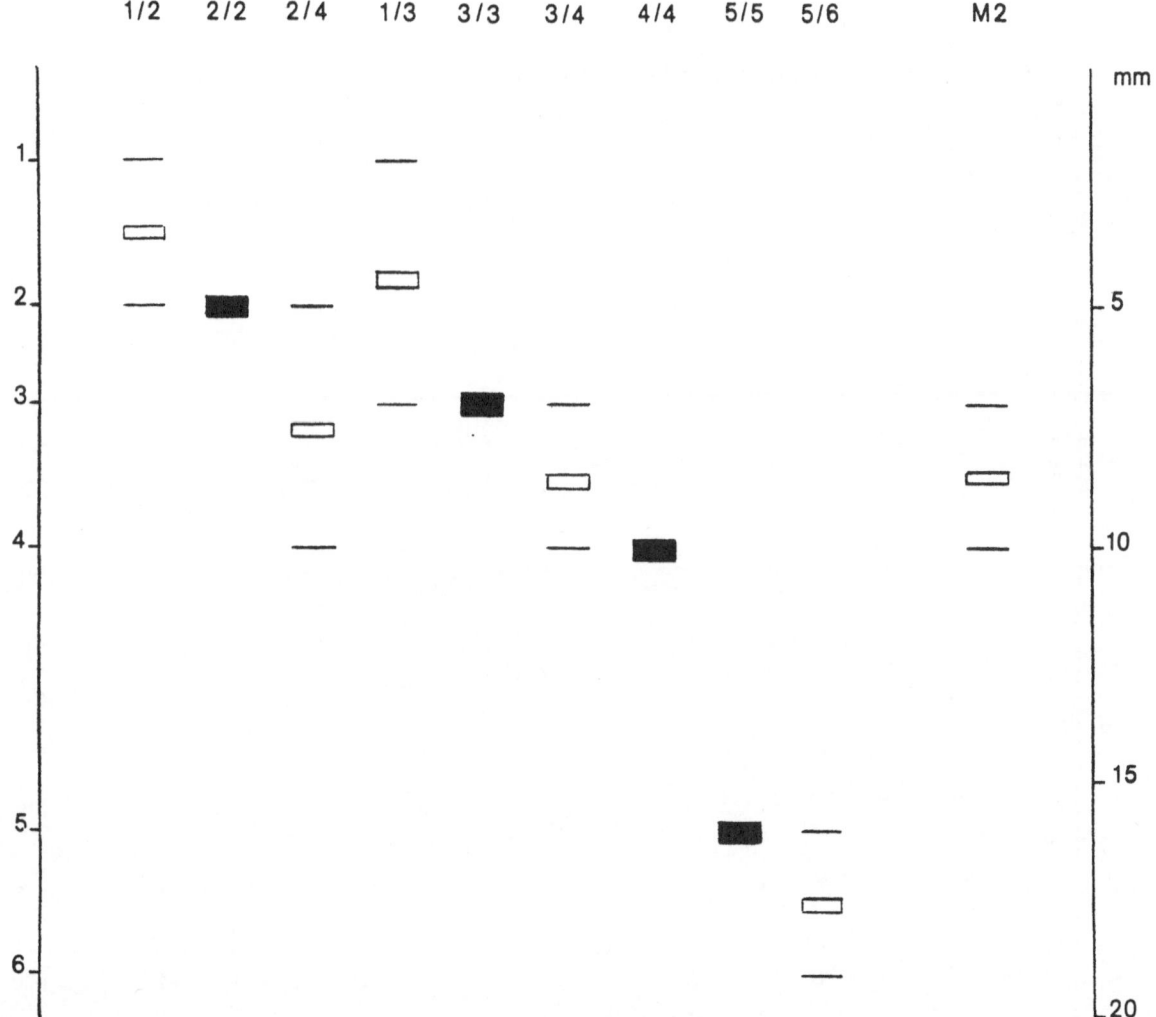

Fig. 1. Patterns for phosphoglucose isomerase (PGI) of the population of *M. quadricornifera.* M2 is a laboratory clonal population used as marker. Migration distances are expressed in mm (right axis). Banding patterns are numbered (1-6) according to speed of migration through the gel.

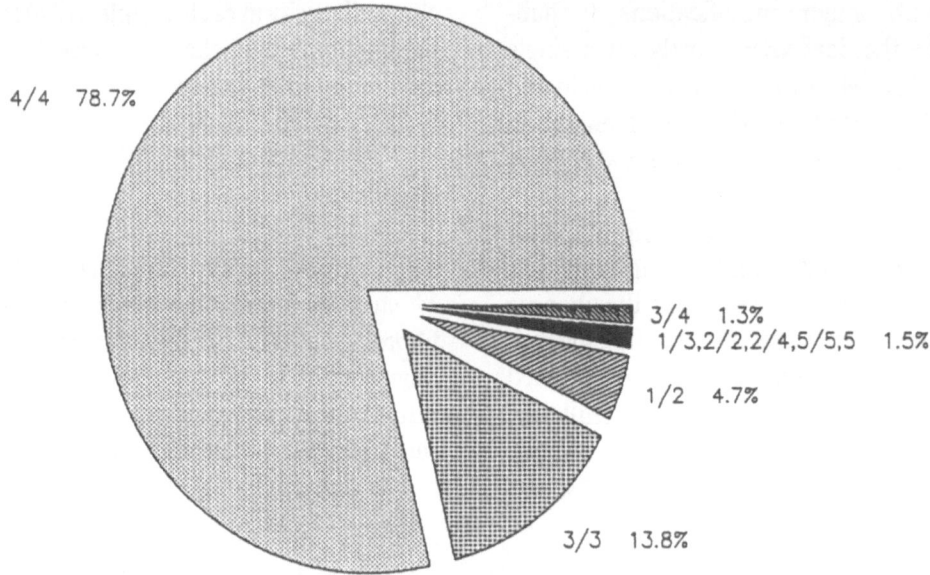

Fig. 2. Percentages of the electromorphs throughout the study period. See text for explanation.

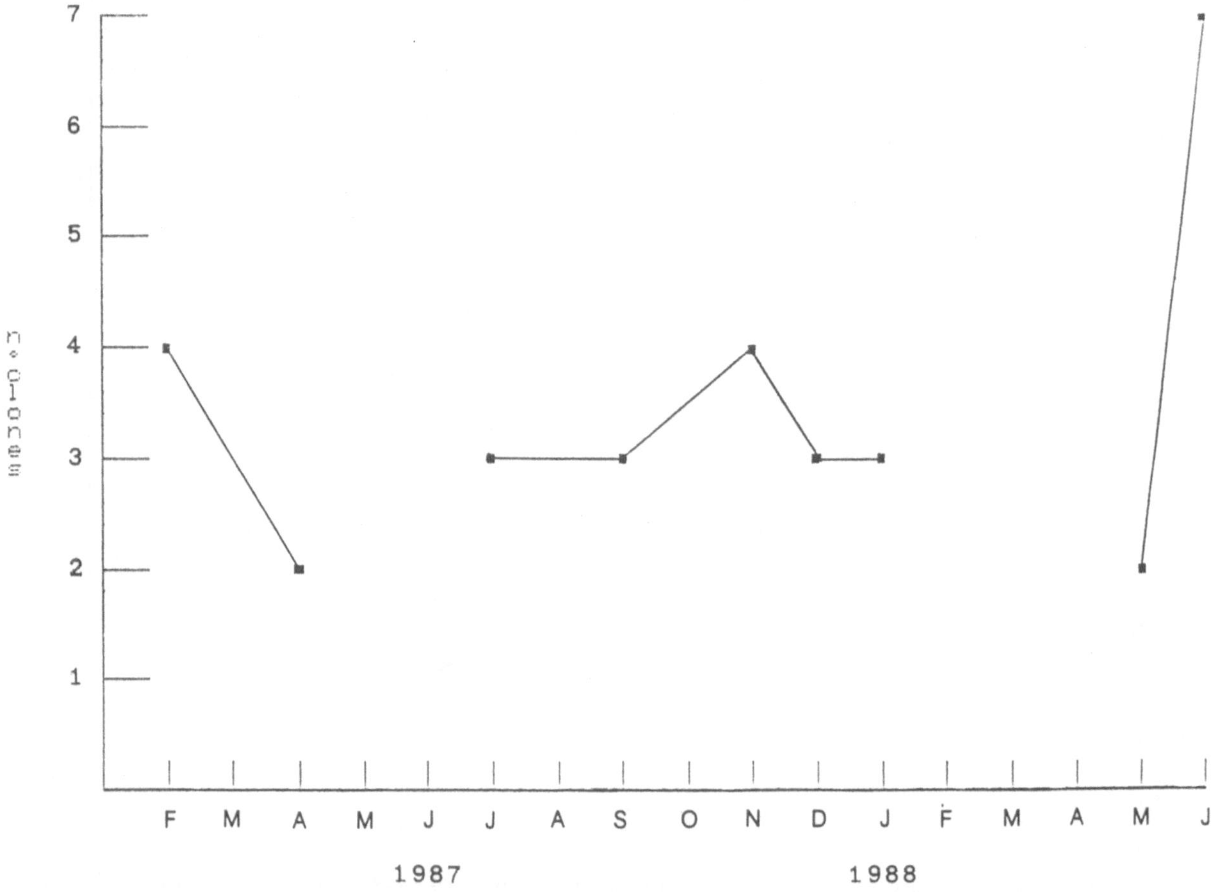

Fig. 3. Temporal variation of the number of electromorphs from each collection.

Table 1. Number of *M. quadricornifera*, Shannon index, number and percentage of the electromorphs from each collection.

Sample date	N M.q.	Shannon index	Electromorphe																
			1/2		2/2		2/4		1/3		3/3		3/4		4/4		5/5		5/6	
			N	%	N	%	N	%	N	%	N	%	N	%	N	%	N	%	N	%
Feb '87	41	0.60	1	2.5	–		–		–		2	5	1	2.5	37	90	–		–	
Apr '87	62	0.11	1	2	–		–		–		–		–		61	98	–		–	
May '87	0	–	–		–		–		–		–		–		–		–		–	
Jun '87	0	–	–		–		–		–		–		–		–		–		–	
Jul '87	42	1.35	–		–		–		–		21	50	4	10	17	40	–		–	
Sep '87	74	0.20	1	1.4	–		–		–		–		1	1.4	72	97.2	–		–	
Nov '87	59	1.20	–		–		1	1.7	–		32	54.2	–		25	42.4	–		1	1.7
Dec '87	74	0.34	1	1.4	–		–		–		3	4	–		70	94.6	–		–	
Jan '88	43	0.84	1	2.4	–		–		–		8	18.6	–		34	79	–		–	
Feb '88	0	–	–		–		–		–		–		–		–		–		–	
Mar '88	0	–	–		–		–		–		–		–		–		–		–	
Apr '88	0	–	–		–		–		–		–		–		–		–		–	
May '88	57	0.21	2	3.5	–		–		–		–		–		55	96.5	–		–	
Jun '88	98	1.61	19	19.4	2	2	–		3	3	10	10.2	1	1	62	63.4	1	1	–	

investigation period was partitioned as shown in Fig. 2: a very common electromorph (4/4) represented approximately 79% of all rotifers, a second (3/3) about 14%, a third (1/2) approximately 5%, and a fourth (3/4) 1%. Up to 5 less frequent electromorphs, occurring as less than 0.5%, are grouped into a single zone containing electromorphs 2/4 (0.2%), 5/5 (0.2%), 5/6 (0.2%), 2/2 (0.4%) and 1/3 (0.5%).

Clonal diversity, the number of clones recorded on each sampling occasion, is shown in Fig. 3. No bdelloid rotifers were extracted from moss samples taken in May or June 1987. Both of these collections were made after very heavy rain, and a washing out of the moss by the rain was suspected. The ground below the wall was sampled without success. In February, March and April 1988 the weather was exceptionally dry and again the extraction was unsuccessful. However, the most frequent clone (electromorph 4/4) was collected before and after both climatic disturbances, suggesting continuity of occurrence in the bdelloid population.

Table 1 lists the numbers and percentages of the electromorphs present on the sampling dates. Clone 4/4 was almost always dominant, but decreased to less than 50% on some occasions. The minor clones occurred sporadically. There was no evidence of seasonal succession of any clone, as the dominant clone was present throughout the year and no replacement of the others was apparent.

Temporal variation of clones was analysed in relation to physical parameters. Relative humidity ($r = 0.94$, $P \ll 0.001$), calculated as the average value over 10 days prior to sampling (Fig. 4) was most closely related to the number of clones. No relationship was detected between number of clones and mean temperature over 20 ($r = -0.20$, $P > 0.5$) or 10 days prior to sampling ($r = -0.22$, $P > 0.5$). The same was true of rainfall (average over 10 days before sampling) and number of clones ($r = 0.52$, $P > 0.1$). In addition, we attempted to relate rainfall in the 5 days prior to sampling to the number of clones, suspecting a washing effect on the moss population, but no relationship was found ($r = 0.25$, $P > 0.1$).

In the cultivation of clones established from single animals, we obtained contrasting results. Some lineages reproduced promptly when fed *E. coli* and *Chlamydomonas*, while others did not. The latter increased after addition of *S. cerevisiae*

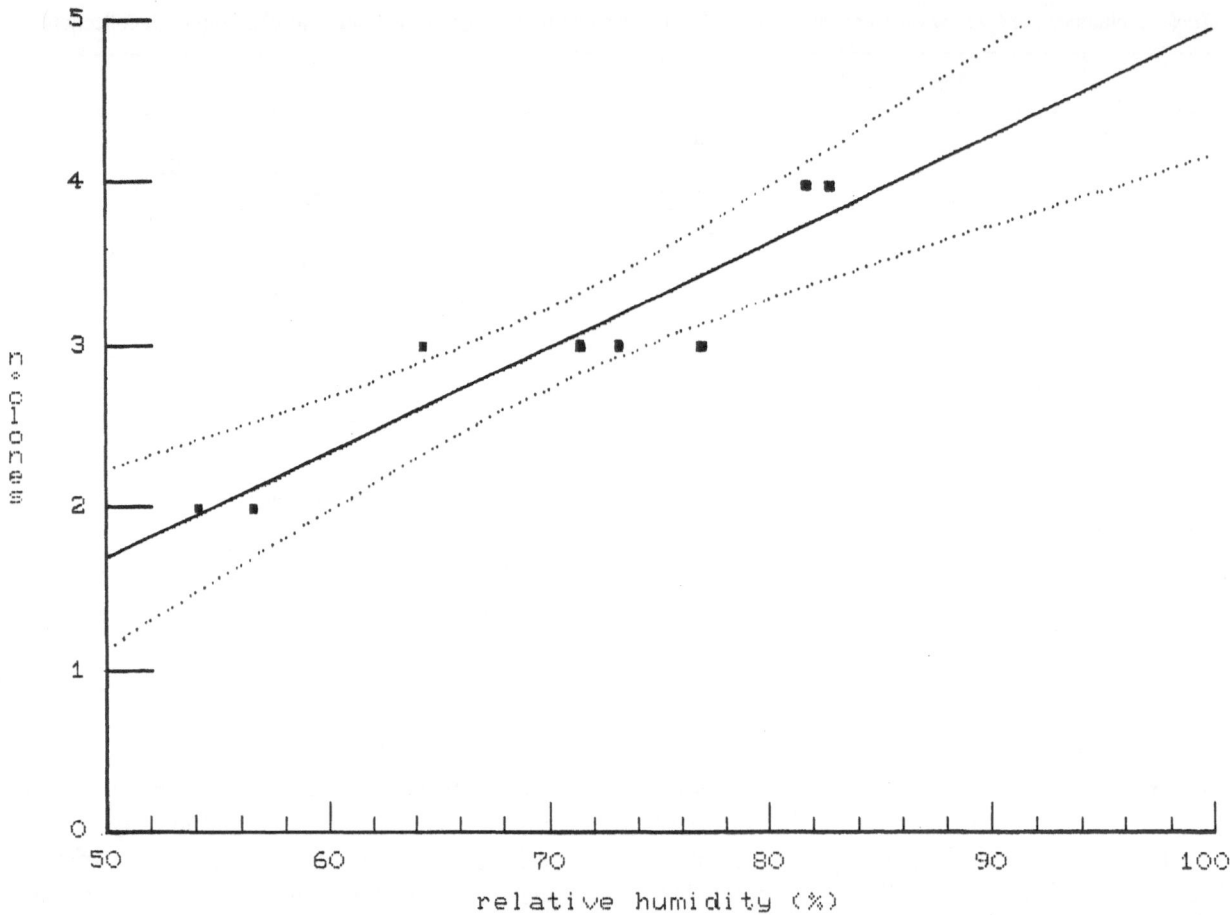

Fig. 4. Correlation between number of electromorphs and relative humidity of the air, calculated as the average value over 10 days prior to sampling. $r = 0.94$, $P \ll 0.001$.

to the food mixture. These clones, when analyzed for PGI enzymatic patterns, were the 4/4 electromorph. The other loci investigated in laboratory clones confirmed the results of PGI staining: PGI electromorphs gave similar ME and EST electromorphs, but no composite electromorph could be identified because of the paucity of the cultured clones.

Discussion

The most serious limitation of this study is the analysis of a single locus system, used as a simple marker. This approach underestimates the number of clones, since each electrophoretic clone probably contains several genotypes due to differentiation at other loci. An alternative approach

would have been to rear animals in order to establish clonal cultures in which to screen more loci. However, we thought it more important to get information about the whole population of a single species, also, maintenance of clones established from collected animals is often difficult. One could easily miss a number of clones only because of difficulties of culturing under laboratory conditions.

The amount of polymorphism at the PGI locus in *M. quadricornifera* is very high, with as many as 6 alleles arranged into 9 different 'genotypes'. If more loci had been examined, the number would probably have increased further, implying that this population consists of many distinct clones. Moreover, the sample size is not always large enough to obtain a satisfactory number of bdelloids. A greater variability might have been detected if a larger number of animals had been studied.

The results suggest high stability of clonal structure among the rotifers investigated. One clone seemed to be dominant throughout the entire period, winter and summer, sharing its leadership with a second common electromorph (3/3). The sporadic electromorphs may represent unsuccessful colonization events in the moss. However, bearing in mind how unstable and unpredictable terrestrial moss is and how easily a bdelloid rotifer should in theory colonize a habitat, the population stability was unexpected.

Temperature and water content are the major variables affecting the life of the moss inhabitants. In any terrestrial moss, rapid fluctuations in temperature occur daily. A field study of the population dynamics of moss tardigrades showed that humidity and rainfall ten to twenty days prior to sampling affect the population (Morgan, 1977). Here, the water content has not been investigated, but relative humidity of the air might also be related to it. When humidity increased, a larger number of electromorphs were detected suggesting that bdelloid colonization takes place when environmental conditions offer more resources. On the other hand, when the water content is too low to sustain life, rotifers survive as anhydrobionts. However, the new colonization is ineffective and the few well established clones continue to dominate the habitat. Clonal competition cannot be ruled out, of course, since fluctuations in electromorph frequencies are found. However, the clonal frequencies do not vary greatly and clonal replacement does not appear to take place.

The differences in food requirements of the clonal populations under laboratory conditions suggest that at least some clones may use different food resources in nature. This supports the hypothesis that different clones can co-occur because they occupy separate niches (Loaring & Hebert, 1981).

Generally, the results of this study are in contrast to the situation in parthenogenetic invertebrates living in plankton, where seasonal clonal succession is commonly observed in response to changing temperature (Carvalho & Crisp, 1987;

Lynch, 1984; Rossi & Menozzi, in press). However, variations in temperature in terrestrial mosses are so dramatic and sudden that no specialization is possible. In fact, there is not even adaptation to summer or winter seasons.

Competition among clones leading to the exclusion of all but a few genotypes in planktonic rotifers (Snell, 1977; King, 1980) has been hypothesized, while King & Zhao (1987), studying a population of *Brachionus plicatilis*, found that several clones 'may coexist over an extensive period', supporting King's (1977) model of incomplete genetic discontinuity. All authors (e.g. Loaring & Hebert, 1981; Hebert & Crease, 1980; Weider *et al.*, 1987) agree that clonal diversity in natural habitats is maintained by environmental disturbance and temporal heterogeneity, and that clones are not ecological analogues. Their differences in competitive ability and intrinsic rates of increase, coupled with habitat heterogeneity, shape the clonal composition of the environment. Our results, if reliable, are discordant with this idea, since we found a single dominant clone in a very unstable habitat. However, because the terrestrial moss dries out completely from time to time, the ability to withstand anhydrobiosis may be the determining factor for successful maintenance of a clone. No studies of differences in anhydrobiotic capability among clones have been performed, but this may be worth future investigation. Moreover, the founder effect (Hebert & Moran, 1980) may be important in preventing other genotypes from colonizing the same habitat.

In conclusion, the results of this far from exhaustive study show unexpected dynamics of clones of a strictly parthenogenetic animal in an unstable and unpredictable environment. The dominance of single clone in the moss through the study period can be attributed to the founder effect, competitive superiority and anhydrobiotic viability, all of which are important factors. Only further investigation, involving field and laboratory studies, can reveal which biological capabilities of such a parthenogenetic population account for its unusual dynamics.

152

Acknowledgements

We particularly thank Giancarlo Fava and Paolo Bisol. Very helpful suggestions on the extracting procedure were given by Umberto Parenti. Linda May read and provided excellent comments on earlier draft of the manuscript. The study has been supported by grant CT 84.01201.11 from CNR and by MPI 40% fund to C.R.

References

Bartos, E., 1951. The Czechoslovak Rotatoria of the Order Bdelloidea. Vestnik Cs. Zool. spol. 15: 241–500.

Carvalho, G. R., 1987. The clonal ecology of Daphnia magna (Crustacea: Cladocera). I. Temporal changes in the clonal structure of a natural population. J. anim. Ecol. 56: 453–468.

Dobers, E., 1915. Uber die Biologie der Bdelloidea. Int. Revue ges. Hydrobiol. Hydrol. Suppl. zu Bd. 7: 1–128.

Donner, J., 1965. Ordnung Bdelloidea. Akademie Verlag. 297 pp.

Harris, H. & D. A. Hopkinson, 1976. Handbook of enzyme electrophoresis in human genetics. North-Holland, Amsterdam, Neth.

Hebert, P. D. N., 1977. Niche overlap among species in the Daphnia carinata complex. J. anim. Ecol. 46: 399–409.

Hebert, P. D. N. & T. J. Crease, 1980. Clonal coexistence in Daphnia pulex (Leydig): another planktonic paradox. Science 207: 1363–1365.

Hebert, P. D. N. & T. J. Crease, 1983. Clonal diversity in populations of Daphnia pulex reproducing by obligate parthenogenesis. Heredity 51: 353–369.

Hebert, P. D. N. & C. Moran, 1981. Enzyme variability in natural populations of Daphnia carinata King. Heredity 45: 313–321.

King, C. E., 1977. Genetics of reproduction, variation, and adaptation in rotifers. Arch. Hydrobiol. Ergeb. Limnol. 8: 187–201.

King, C. E., 1980. The genetic structure of zooplankton populations, p. 315–328. In: W. C. Kerfoot (ed.), Evolution and ecology of zooplankton communities. Univ. Press New England, Hanover.

King, C. E. & Y. Zhao, 1987. Coexistence of rotifer (Brachionus plicatilis) clones in Soda Lake, Nevada. Hydrobiologia 147: 57–64.

Korpelainen, H., 1986. Competition between clones: an experimental study in a natural population of Daphnia magna. Hereditas 105: 29–35.

Loaring, J. M. & P. D. N. Hebert, 1981. Ecological differences among clones of Daphnia pulex Leydig. Oecologia 51: 162–168.

Lynch, M., 1984. The genetic structure of a cyclical parthenogen. Evolution 38: 186–203.

Morgan, C. I., 1977. Population dynamics of two species of Tardigrada, Macrobiotus hufelandii (Schultze) and Echiniscus (Echiniscus) testudo (Doyere), in roof moss from Swansea. J. anim. Ecol. 46: 263–279.

Parker, E. D., Jr., 1979. Ecological implications of clonal diversity in parthenogenetic morphospecies. Am. Zool. 19: 753–762.

Rossi, V. & Menozzi P. The clonal ecology of Heterocypris incongruens (Crustacea: Ostracoda). Ecology, in press.

Snell, T. W., 1977. Clonal selection: competition among clones. Arch. Hydrobiol. Ergeb. Limnol. 8: 202–204.

Snell, T. W., 1979. Interspecific competition and population structure in rotifers. Ecology 60: 494–502.

Weider, L. J., M. J. Beaton & P. D. N. Hebert, 1987. Clonal diversity in High-Artic populations of Daphnia pulex, a polyploid apomictic complex. Evolution 41: 1335–1346.

Young, J. P. W., 1979. Enzyme polymorphism and cyclic parthenogenesis in Daphnia magna. I. Selection and clonal

Hydrobiologia **186/187**: 153–156, 1989.
C. Ricci, T. W. Snell and C. E. King (eds), Rotifer Symposium V.
© 1989 *Kluwer Academic Publishers.*

Horizontal distribution of the plankton rotifers *Keratella cochlearis* (Bory de St Vincent) and *Polyarthra vulgaris* (Carlin) in a small eutrophic lake

A.P. Saunders-Davies
33 Park Road, Esher, Surrey KT10 8NP, UK

Key words: horizontal distribution *Keratella cochlearis, Polyarthra vulgaris*, lux

Abstract

The plankton rotifers *Keratella cochlearis* and *Polyarthra vulgaris* were sampled at 10 cm below the surface at different distances from two dissimilar shores and in the centre of a small eutrophic lake. Light and depth were measured at each sampling point. In each case the numbers of rotifers per liter increased with distance from the shore. There was a significant correlation between the numbers for the two species for the two shores, but none in the centre. In the case of one shore there was a strong correlation between rotifer numbers and supra-surface ambient light.

Introduction

The vertical distribution of plankton rotifers has been well-studied (George, 1970; Stewart & George, 1987), but there have been fewer investigations of the horizontal distribution. Plankton rotifers are known to be strongly phototactic. This paper describes the variation of rotifer numbers with distance from two shores and relates this to the mean supra-surface light levels.

Site studied

Brooklands lake is a small (6 hectare) eutrophic lake with an average depth of 1 m, and a maximum of about 1.5 m. It is approximately L-shaped. Part of the North East (NE) shore is a re-entrant curve exposed to the prevailing West and South West wind. It is comparatively free of macrophytes. It supports a vigorous phyto- and zooplankton community. At the times of sampling the water was uniformly turbid with a Secchi depth of 10–20 cm. The South West (SW) shore shelves much more gently and a number of the surrounding trees of alder (*Alnus*) and willow (*Salix*) have shed branches which have re-rooted and formed long 'gullies' or inlets. The height of the fallen branches is about 1–2 m.

Methods

Traverses were made on 6 occasions of a gully on the SW shore together with similar traverses of the opposite NE shore sampling at 2.5 m and 5 m intervals. Sampling was carried out 10 cm below the surface with closing bottles of 125 ml capacity. Three samples were taken at each site. The samples were then decanted into specimen tubes (4 × 15 cm), fixed with a 10% formalin solution and allowed to sediment for a minimum of 24 hr. The supernatant was carefully siphoned off leaving the bottom 4 ml undisturbed. These 4 ml were

154

then vigorously stirred and transferred to a perspex chamber and the numbers of rotifers were counted.

The light was measured with a Gossen Lunasix meter fitted with an integrating cone which recorded the light over a 180° hemisphere. Days on which the sky was overcast were chosen to minimise the effects of direct sunlight.

The temperature was measured at each sampling point and found not to vary by more than 0.2 °C between sites, except on one occasion when the temperature in the centre was 2.1 °C higher than in the shaded areas.

Results

My major interest was the horizontal distribution of the rotifers. However the relative numbers of each species varied considerably, and the numbers per liter varied by as much as a factor of 10 during the 2 months of the study. Therefore the numbers in each sample were expressed as a percentage of the total numbers for each traverse. This avoids any weighting by occasions when the total numbers were high. The figures are shown in the Table together with mean light levels, depth and mean numbers liter^{-1} for each site. The numbers per sample increased with progressive distance from the shore. In the case of the NE shore both species show a significant correlation against light levels, with values for the coefficient of correlation of 0.96 for *Keratella*, and 0.97 for *Polyarthra*. This is significant at the 0.2% level for *Polyarthra* and at the 1% level for *Keratella*. The slopes of the regression lines (and intercepts) were remarkably similar, with values of 26.98% per log lux for *Polyarthra* and 25.05% for *Keratella*. As might be expected from the foregoing, the correlation between the two species was also significant, in fact at the 0.2% level.

In the case of the SW shore there was also a

Table. Rotifer numbers as a percentage of the total numbers for the traverse, together with distances from the shores, mean log lux, standard deviation, mean numbers per litre and depth. Percentages shown represent the mean of 6 traverses.

Table 1A. Rotifer numbers, average light values and depth for the SW shore

Distance ⇒ (m)	2.5	5	10	15	20	25	30	35
Depth (m)		0.84	0.90	1.00	1.05	1.40	1.40	1.37
Log mean lux	8.51	9.20	9.41	9.70	9.83	9.93	9.89	9.89
Polyarthra % total			7.58	4.21	10.90	16.11	45.78	43.84
Std dev for 6 runs			3.83	3.13	6.69	8.92	9.86	6.50
Mean no/liter			57	27	55	142	389	296
K. cochlearis % total			3.36	6.05	5.88	12.32	34.83	37.57
Std dev for 6 runs			1.57	6.12	4.11	9.15	9.83	8.03
Mean no/liter			488	510	749	1481	5245	6190

Table 1B. Rotifer numbers, average light values and depth for the NE shore

Distance ⇒ (m)	2.5	5	10	15	20	25
Depth (m)	0.90	1.20	1.40	1.38	1.50	1.48
Log mean lux	9.29	9.55	9.72	9.88	9.93	9.93
Polyarthra % total	7.84	10.71	18.51	26.63	26.60	21.19
Sd for 6 runs	6.74	2.96	7.49	8.49	5.83	6.63
Mean no/liter	1,058	1,520	3,153	4,154	4,132	3,240
K. cochlearis % total	9.29	12.20	18.27	24.71	25.55	21.72
Std for 6 runs	7.80	6.75	3.18	5.84	4.09	6.03
Mean no/liter	2,595	4,934	6,536	9,784	6,970	7,880

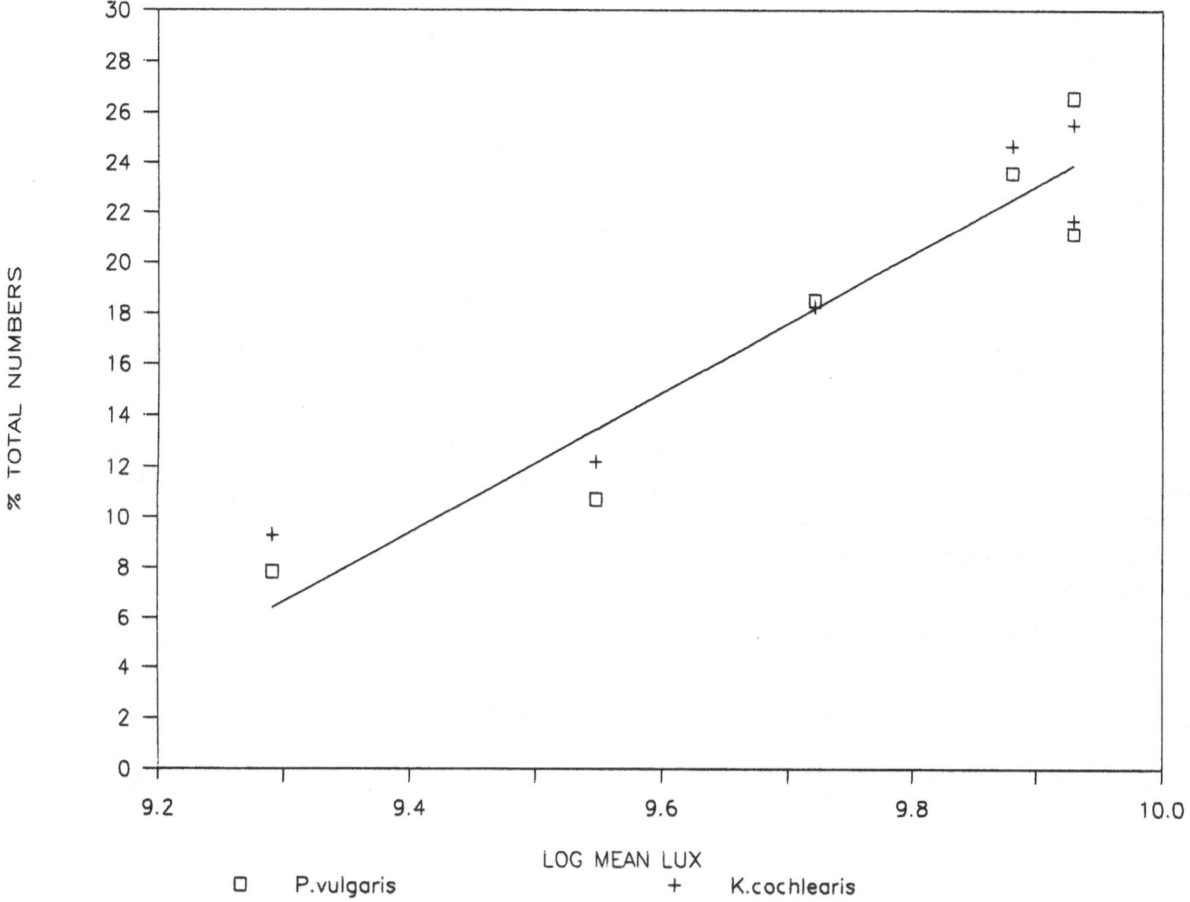

Fig. 1. Numbers of rotifers plotted as a percentage of the total for the traverse for the NE shore as a function of mean light level in log lux. Points represent the mean of 6 traverses. (Cross = K. cochlearis, square = P. vulgaris). Regression line shown is for both species.

marked rise in numbers per liter as the distance increased. These were plotted in a similar way to those of the NE shore (Fig. 1). The correlation between the two species was again highly significant with a correlation coefficient of 0.99. In this case, however there was no significant correlation between light levels and the rotifer numbers. For both species there was a slow increase during the length of the gully and then a sharp rise when the entrance to the gully was cleared.

In the centre of the lake the light levels were constant over the length of the traverses. In the data from 6 traverses of 20 m each there was no correlation between the species, and the distribution appeared to be completely random and independent for each species.

Discussion

The profiles of the relative distribution as a function of distance from the shore indicates that planktonic rotifers follow the pattern established for pelagic crustaceans (Burckhardt, 1910; Ruttner, 1974). Similar profiles for plankton rotifers were described by Preissler (1977), although the species present in his samples were different.

The strong and significant correlation, coupled with the known phototaxic response of the plankton rotifers concerned, argues for a causal relation between light levels and rotifer numbers for the distribution off the NE shore. The slight drop in numbers for the last (25 m) point at first sight appears anomalous, particularly since the point on the graph represents the mean of 6 samples

156

and thus is not a single aberrant data point. The swimming speeds of *K. cochlearis* over a range of temperatures averages 0.5 mm sec^{-1} with a maximum of 0.8 mm sec^{-1}, while *Polyarthra* averaged 0.6 mm^{-1} with a maximum of 0.9 mm sec^{-1}. (Saunders-Davies – in preparation). A surface wind will generate currents at the surface of 1–3% of the wind speed (Harvey, 1974). A wind speed of less than 0.2 ms^{-1} is described as 'calm' (Pedgley, 1978) but this could generate a surface water current of 2 mm sec^{-1}. It would seem that in all but the most sheltered areas, water movements are an important factor in determining the distribution of rotifers in the horizontal plane. There will also be convection currents generated by insolation. The problems of measuring small water currents has been commented on by many writers, including Harvey (1976).

References

Burckhardt, G., 1910. Hypothesen und Beobachtungen uber die Bedcutung der vertikalen Planktowanderung. Int. Revue Hydrobiol. 3: 156–172.

George, M. G., 1970. Diurnal Migration in three species of rotifers in Sunfish Lake, Ontario. Limnol. Oceanogr. 15: 218–223.

Harvey, J. G., 1976. Atmosphere and Ocean. Artemis.

Pedgely, D. E., 1978. A Course in Elementary Meteorology. HMSO.

Preissler, K., 1977. Do Rotifers Show 'Avoidance of the Shore'? Oecologia (Berl.) 27: 253–260.

Ruttner, F., 1974. Fundamentals of Limnology. 150 pp.

Stewart, L. J. & M. G. George, 1987. Environmental factors influencing the vertical migration of planktonic rotifers in a hypereutrophic tarn. Hydrobiologia 147: 203–208.

Hydrobiologia **186/187**: 157–161, 1989.
C. Ricci, T. W. Snell and C. E. King (eds), Rotifer Symposium V.
© 1989 *Kluwer Academic Publishers.*

Oviposition behavior of the littoral rotifer *Euchlanis dilatata*

Elizabeth J. Walsh
Department of Zoology, Oregon State University, Corvallis, OR, 97331-2914, USA

Key words: oviposition behavior, *Euchlanis dilatata*

Abstract

Several components of egg-laying behavior in 3 clones of *E. dilatata* were investigated. Using nearest-neighbor analyses it was shown that individual animals lay their eggs randomly within culture dishes in the absence of an artificial substrate while eggs laid in the presence of a substrate are clumped in 2 of the 3 clones studied. Animals preferentially oviposited on artificial substrates containing eggs when given a choice between artificial substrates with and without eggs of conspecifics. When given a choice of natural substrates, animals consistently laid more eggs on the plant species from which they were collected. This preference persisted for 2 generations after animals had been removed from plants. In addition, observations of neonate hatching behavior demonstrated that neonates often spend relatively long times in the immediate vicinity of hatching.

Introduction

Oviposition behavior can directly influence female fitness by influencing offspring reproductive success (Thorpe, 1945; Chew & Robbins, 1984). Via (1986) has shown in leafminers that the choice of egg deposition site affects pupal weight of offspring. Butterflies have genotypic preferences for oviposition site both between and within host species which can affect larval performance (Rausher, 1979; 1984). Learning may play a role in determining butterfly egg-laying sites depending upon other conflicting constraints such as time (Papaj, 1986; Papaj & Rausher, 1987). In *Drosophila*, Del Solar (1968) has shown that egg deposition pattern is genetically determined. Theoretically, the pattern of egg deposition may also influence competitive interactions (Atkinson & Shorrocks, 1984) and subsequent sympatric speciation (Rice, 1984). Egg deposition pattern, choice of deposition site and the ecological and evolutionary consequences of these behaviors have not been explored in rotifers.

Littoral rotifers are ideal study organisms for investigations of oviposition behavior because they reproduce asexually and live in complex habitats offering a wide variety of potential egg deposition sites. During asexual reproduction, cloning generates a large number of genotypically identical individuals useful for a variety of behavioral tests.

This research investigates two questions relating to oviposition behavior in the rotifer *Euchlanis dilatata*. First, are there genetically-based differences in rotifer egg-laying behavior? The pattern of egg deposition in the presence and absence of artificial substrates is compared among three clones. Secondly, what factors influence oviposition site selection? Whether cues from egg/eggshells of clonemates are used in determining egg-laying site is tested and substrate preferences are investigated. Additionally, neonate hatching behavior was monitored because it may also play a role in site recognition and subsequent habitat preferences in adults.

Study organism

Euchlanis dilatata is a freshwater rotifer commonly found in the littoral zone of nutrient-rich ponds and lakes. This rotifer attaches it eggs directly to littoral vegetation (Edmondson, 1960). In laboratory mass cultures, eggs tend to be laid in clumps (King, 1967) but variables potentially affecting this behavior have not been studied. Because this rotifer reproduces by cyclical parthenogenesis (King, 1967), individual females isolated from natural populations produce genetically identical daughters (King, 1977). The original female and her daughters constitute a clone.

Clones used in laboratory experiments were derived from collections from three geographically isolated, natural populations of *E. dilatata* (DL = Devil's Lake, Lincoln City, OR; QP = Quarry Pond, East Corvallis, Oregon; and SC = Soap Creek Pond, North Corvallis, Oregon). Animals were maintained in a temperature controlled chamber (25 °C) with a 16l:8d photoperiod. Rotifers were cultured in Pourriot-Gilbert medium (Gilbert, 1968) at pH 7.5 and fed *Chlamydomonas reinhartii* (Univ. Texas culture collection 89) grown on Bristol's agar (Starr, 1978). Clones used in experiment 1-4 described below were kept under these conditions for at least 10 generations prior to experimentation.

Methods and results

Experiment 1

The pattern of egg deposition in the presence or absence of an artificial substrate (nylon monofilament line) was compared among the 3 clones. Females of each clone were individually placed in 1 cm diameter culture wells. Half of the wells contained a 1-cm piece of monofilament, half were empty. The pattern of egg deposition was recorded daily, using a grid system. Neonates were removed upon hatching.

Nearest neighbor analyses (Vandermeer, 1981) were performed to determine whether eggs were laid randomly, clumped or superdispersed within culture wells. For this test, an index of aggregation

Table 1. Egg deposition behavior in the presence and absence of artificial substrates by *E. dilatata*. R is the index of aggregation, where $R = 1$ represents a random distribution; $R < 1$, clumped; and $R > 1$, superdispersed. n represents the number of individuals tested from each clone. The number of eggs laid per individual varied between 3-7.

Artificial substrata						
	Present			Absent		
Clone	n	$\overline{X}(R)$	Range(R)	n	$\overline{X}(R)$	Range(R)
DL	21	0.396	0.089-0.949	13	1.098	0.313-4.75
QP	7	0.446	0.200-0.773	8	0.831	0.552-1.25
SC	28	1.050	0.155-1.940	9	2.563	0.259-6.05

(R) is calculated by measuring the distance between individual eggs. When $R = 1$ the egg distribution is considered random, when $R < 1$ the distribution is clumped and when $R > 1$ the pattern is said to be superdispersed. For 2 of the 3 clones tested, animals laid their eggs in clumps in the presence of artificial substrates and randomly in the absence of artificial substrates (Table 1). Females of the third clone, SC, deposited its eggs randomly both in wells containing artificial substrates and in those without them.

Experiment 2

To determine potential cues used in choosing an egg deposition site, females were given a choice between artificial substrates with and without the eggs of clone-mated attached. Two replicates of sixty ovigerous females from clone SC were placed into glass culture dishes (5.0 cm diameter) containing six 1-cm pieces of monofilament line: three lines contained 15-25 eggs and three were blank (i.e., without eggs). The number of additional eggs deposited was recorded after 12 h.

Animals laid significantly more eggs on those lines containing eggs than blanks (ANOVA, F test, $p < 0.001$; Fig. 1).

Experiment 3

An experiment was conducted to ascertain whether morphological cues (i.e., the shape of the eggs) or chemical cues from the eggs are used in

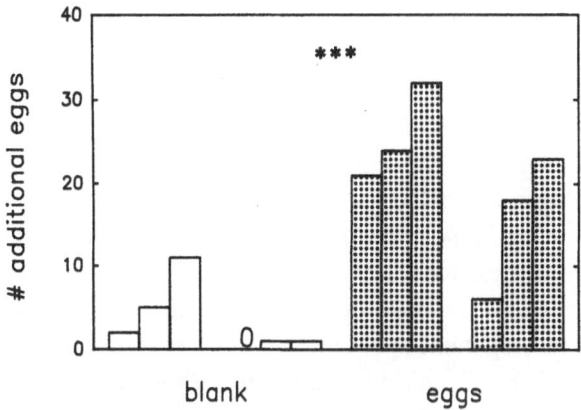

Fig. 1. Egg deposition behavior of clone SC on artificial substrates with and without eggs of clonemates. Bars represent the number of additional eggs laid per monofilament line. *$p < 0.05$; **$p < 0.01$; ***$p < 0.001$.

choosing an egg deposition site by *E. dilatata*. Animals were given the choice of laying eggs on a blank monofilament, one with eggshells attached, or one with intact eggs. The procedure described in experiment 4 was followed with these modifications:

(1) 150 instead of 60 females of clone SC were placed individually into the experimental culture dishes.

(2) monofilament lines containing approximately 80 eggshells were included along with those containing eggs and blanks.

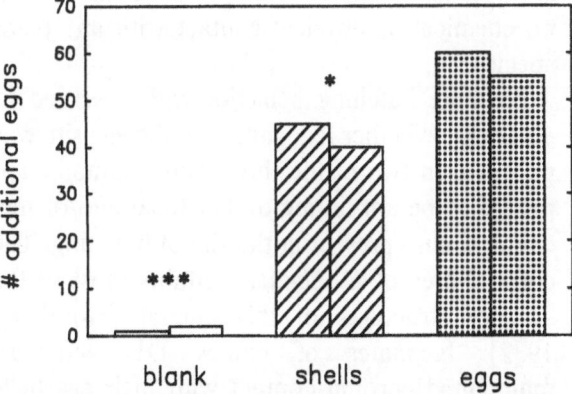

Fig. 2. Egg deposition behavior of clone SC on artificial substrates with eggs, with eggshells, or without either. Bars represent the number of additional eggs laid per monofilament line. *$p < 0.05$; **$p < 0.01$; ***$p < 0.001$.

The number of eggs laid on monofilament depended upon the presence or absence of eggs and eggshells (ANOVA, F test, $p < 0.001$). More eggs were laid in the presence of intact eggs of clonemates than on either blank monofilament or monofilament containing eggshells (Newman Keuls multiple comparison test, $p < 0.05$) (Fig. 2).

Experiments 4 and 5

The role of substrate preference in determining egg-laying site was further investigated in two experiments. Females were offered a choice among 3 natural substrates. Individuals of *Euchlanis dilatata* were isolated from a mass culture of millfoil (*Myriophyllum exalbescens*) collected from Devil's Lake. Fifty females were placed into each of 5 replicate 100 ml beakers containing 0.25 g (wet weight) of millfoil, *Elodea*, and coontail (*Ceratophyllum*) in 50 ml millipore filtered lake water. Before use, all plants were rinsed thoroughly in filtered lake water and inspected for eggs. No eggs were found on any plant. Plants were inspected for eggs and eggshells after 54 h.

For experiment 5, these modifications were made:

(1) all animals were obtained from a single clone collected from millfoil.

(2) half of the replicates contained animals grown in the presence of millfoil until the time of experimentation and half contained animals which had been removed from millfoil for 2 generations. Six beakers were set-up: 3 containing females grown in association with millfoil and 3 which had been grown in culture dishes for 2 generations.

Animals preferentially laid eggs on millfoil when given a choice among millfoil, *Elodea* and coontail. A one-tailed t-test based on the numerical difference in the number of eggs laid on substrate pairs (millfoil versus *Elodea* and millfoil versus coontail) was conducted. The overall probability that animals prefer millfoil over both coontail and *Elodea* is 0.9945 (1-tailed *t*-test). This preference for millfoil persisted even after animals have been removed from the plant for 2 generations (Table 2).

Table 2. Egg-laying pattern in the presence of three natural substrates of a single clone. d_{M-E} represents the mean difference in the number of eggs laid on millfoil versus *Elodea*. d_{M-C} represents the mean difference in the number of eggs laid on millfoil versus coontail. Significance levels for a one-tailed t-test are given for animals which were raised in contact with the natal substrate and those which had been removed from it for 2 generations.

	Millfoil vs *Elodea*		Millfoil vs coontail	
	$\bar{d}_{M-E}$	P	$\bar{d}_{M-C}$	P
Experiment 1	0.4252	***	0.4830	***
Experiment 2				
Raised w/millfoil	0.5093	*	0.4839	**
Raised w/o millfoil	0.6047	*	0.6195	*

Experiment 6

Neonate hatching behavior was compared among the 3 clones. One cm pieces of monofilament line were placed into mass cultures of DL, QP, and SC. Eggs were deposited onto the monofilament line. The line was then removed and placed into a small culture well. I recorded the amount of time individual neonates from the three clones spent in contact with the eggshell from which it hatched and nearby eggs as well as observations of posture and movement following hatching.

Neonate behavior was similar between individuals of clones DL and QP. Each clone DL individual ($n = 10$) remained in contact with the monofilament line for longer than 20 minutes. Similarly, individuals from clone QP ($n = 8$) spent an average of 24.0 ± 11.7 min in the immediate vicinity of hatching. Neonates from these clones slowly extended their coronas, bent over and then made coronal contact with their eggshell and nearby eggs for the durations given above. In contrast, neonates of clone SC ($n = 12$) spent very little time ($X = 0.97 \pm 0.81$ min) in contact with the monofilament line. They typically extended their coronas and immediately swam free of the monofilament line.

Discussion

Oviposition behavior varied among the three clones of *E. dilatata* tested. Choice of egg deposition site was affected by the presence of other eggs, eggshells, and substrate-type. Individuals of some *E. dilatata* clones lay their eggs in clumps in the presence of artificial substrates. This clumping pattern has also been observed in several other monogonont species (*Mytilina, Monostyla, E. triquetra, E. calypida*, pers. obs.; *Proales decipiens*, Noyes, 1922) and in one unidentified bdelloid rotifer (pers. obs.). Potential benefits of laying eggs in clumps for rotifers include exploitation of favorable physical conditions for embryonic development and subsequent resource utilization by neonates and avoidance of interclonal competition. A clumped distribution of eggs may also have consequences for the genetic structure of populations, particularly if genotype-specific clumps are formed and habitat preferences exist.

The results of experiment 3 demonstrate that these clones lay their eggs preferentially in the vicinity of other eggs and eggshells. This is also supported by observation of egg aggregations in laboratory mass cultures (King, 1967; pers. obs.). *E. dilatata* also chooses to deposit eggs on the plant species from which they were collected. It should be noted that in extensive collections of individuals of these three plant species, *E. dilatata* is most often found on millfoil (unpubl. data). This plant preference may have a genetic component because it persists over two generations with no chemical or physical contact with any plant species.

Neonate hatching behavior was observed to determine whether recognition of the egg site was possible. In two of the three clones animals appeared to be imprinting on chemical and/or mechanical cues present at the site of hatching. The coronal area of *E. dilatata* contains most of the sensory structures of this animal (Stoszberg, 1932). Neonates of clones DL and QP maintained coronal contact with their eggshells and nearby eggs for extended periods and did not appear to be feeding during this time. This behavior may be similar to that described during oviposition in *Notommata copeus* (Clement *et al.*, 1983). These data suggest that some form of imprinting may occur. However, neonates from clone SC did not display this behavior indicating that there is variability in this behavior among clones. A genetic basis for this is supported by the results of the egg-laying pattern experiment (1) reported here as well as significant differences in body size, embryonic development time, and life history traits among these clones (unpubl. data). While the data presented here are intriguing, the effect of oviposition behavior on the genetic structure of natural populations of rotifers is still unknown. In an attempt to assess these effects, research in progress explores genotypic associations between *E. dilatata* clones and the plants with which they are associated.

Acknowledgements

I thank Charles E. King, Annette M. Olson & Deanna Olson for improving the manuscript. Eric Atkinson, Michelle Schroeder and Chip Rutledge provided field and laboratory assistance. This research was supported by grants from ZORF and Sigma Xi. Travel funds were provided by Dan & Judy Jensen and Jim & Sally Nelson.

References

Atkinson, W. D. & B. Shorrocks, 1984. Aggregation of larval Diptera over discreet and ephemeral breeding sites: the implications for coexistence. Am. Nat. 124: 336–351.

Chew, F. S. & R. K. Robbins, 1984. Egg-laying in butterflies. Symposia of the Royal Entomological Society of London, No. 11. The Biology of Butterflies. R. I. Vane-Wright and P. R. Ackeny (eds.) pp. 65–79. Academic Press, London.

Clement, P., 1987. Movements in rotifers: correlations of ultrastructure and behavior. Hydrobiologia 147: 339–359.

Clement, P., E. Wurdak, & J. Amsellem, 1983. Behavior and ultrastructure of sensory organs in rotifers. Hydrobiologia 104: 83–129.

Del Solar, E., 1968. Selection for and against gregariousness in the choice of oviposition sites by *Drosophila pseudoobscura*. Genetics 58: 275–282.

Edmondson, W. T. 1960. Reproductive rates of rotifers in natural populations. Mem. Ist. ital. Idrobiol. 12: 21–77.

Garcia-Dorado, A., 1987. Polymorphism from environmental heterogeneity. Some features of genetically induced niche preference. Theor. Pop. Biol. 32: 66–75.

Gilbert, J. J., 1968. Dietary control of sexuality in the rotifer *Asplanchna brightwelli* Gosse. Physiol. Zool. 41: 14–43.

King, C. E., 1967. Food, age, and the dynamics of laboratory population of rotifers. Ecology 48: 111–128.

King, C. E. & T. W. Snell, 1977. Sexual recombination in rotifers. Heredity. 39: 357–360.

Noyes, B. 1922. Experimental studies on the life history of a rotifer reproducing parthenogenetically (*Proales dicipiens*). J. exp. Zool. 35: 225–255.

Papaj, D. R., 1986. Interpopulation differences in host preference and the evolution of learning in the butterfly, *Battus philenor*. Evolution 40: 518–530.

Rausher, M. D., 1979. Egg recognition: Its advantage to a butterfly. Anim. Behav. 27: 1034–1040.

Rausher, M. D., 1984. The evolution of habitat preference. II. Evolutionary genetic stability under soft selection. Theor. Pop. Biol. 31: 116–139.

Rausher, M. D. & D. R. Papaj, 1987. Genetic differences and phenotypic plasticity as causes of variation in oviposition preference in *Battus philenor*. Oecologia. 74: 24–30.

Rice, W. R., 1984. Disruptive selection on habitat preference and the evolution of reproductive isolation: A simulation study. Evolution 38: 1251–1260.

Starr, R. C., 1978. The culture collection of algae at the University of Texas at Austin. J. Phycol. Suppl. 14: 47–100.

Stoszberg, K., 1932. *Euchlanis, Brachionus* and *Rhinoglena*. Z. wiss. Zool. 142.

Thorpe, W. H., 1945. The evolutionary significance of habitat selection. J. anim. Ecol. 14: 67–70.

Via, S., 1986. Genetic covariance between oviposition preference and larval performance in an insect herbivore. Evolution 40: 778–785.

Vandermeer, J. H., 1981. Elementary Mathematical Ecology. Wiley, New York. 294 pp.

Hydrobiologia **186/187**: 163–165, 1989.
C. Ricci, T. W. Snell and C. E. King (eds), Rotifer Symposium V.
© 1989 *Kluwer Academic Publishers.*

Developmental times of *Synchaeta oblonga* eggs from the Danube (Austria)

Wolfgang Zoufal
Schweglerstr. 42/4, A-1150 Wien, Austria

Abstract

The relationship between temperature and development time of *Synchaeta oblonga* eggs from the Danube was investigated. Growth rates were also measured in populations established from the Danube.

Introduction

The Altenwörth run-of-river station, which is located at stream km 1979.8 at the Austrian part of the Danube, was the subject of a 'Man and Biosphere Ecosystem Study' from July 1985 to March 1988. Zooplankton samples were taken at monthly intervals. Compared with literature data from this part of the river (Schallgruber, 1944; Naidenow & Saiz, 1975; Naidenow, 1985) the observed zooplankton densities were much higher, showing maxima in summer of 100,000, 130,000 and 230,000 m^{-3} in the years 1985, 1987 and 1986 respectively. *Synchaeta* (mainly *Synchaeta oblonga*) was the most abundant genus followed by *Keratella* and *Brachionus*.

Field data from a river, even with short sampling-intervals, can never be interpreted as the result of local factors alone because they are, at least in part, the result of input from other environments. Information about the growth of populations may be obtained by comparing samples from different localities in the river. Although the backwater-area has a length of 27 kilometers even at low discharge no significant increase of densities towards the dam could be observed. Populations do not have enough time in the backwater-area (retention time about 10–15 hours) to show an increase in numbers or biomass. To get an idea about the dynamic aspect of the river-rotifer populations it was necessary to perform laboratory experiments.

Material and methods

All experiments were performed at the actual temperature of the Danube at the day of collection (kept constant at $\pm 0.1\,°C$). For the enclosure experiments river-water was collected with a 50 liter Schindler plankton trap at stream km 2000 and placed in 3 l plastic bottles and kept for up to 6 days in dark without adding food. Daily one of the bottles was filtered and the animals were preserved with 4% formalin. A 30 μm nylon netting was chosen to ensure that neither small rotifers nor rotifer eggs were lost. Even when counting at 100 × magnification it was impossible to distinguish species for all genera.

Low zooplankton densities and the small volume of the bottles allowed reliable calculation of population growth only for *Synchaeta oblonga*. Population growth (r) and birth (b) rates were calculated as follows:

$$r = \frac{\ln(N_{t+1}) - \ln(N_t)}{t_{t+1} - t_t},$$

where: N = number of individuals and t = time;

$$b = \ln(E + 1) * D^{-1},$$

164

where: E = eggs per female and D = egg-development time.

Development time of eggs from individuals collected in the river was determined. The animals were kept in 10 ml dishes. All eggs were isolated after laying and checked until hatching at 2-hour intervals.

Results

As no relationship between temperature and embryonic development time for *Synchaeta oblonga* was available from the literature, it was necessary to determine one from my own data. A Bèlehrádek equation was fitted by the method of least squares and is presented in Fig. 1. The derived parameters are: $a = 16.0932$, $b = -1.9$ and $c = 1.0796$.

Birth rates, calculated for the enclosures, are constant throughout the experiments, reaching values of about 0.4–0.8 (Fig. 2a). The population growth rates are in a range of -0.8 to 1.2 (Fig. 2b). No significant correlation between the maximum growth rates and temperature was found.

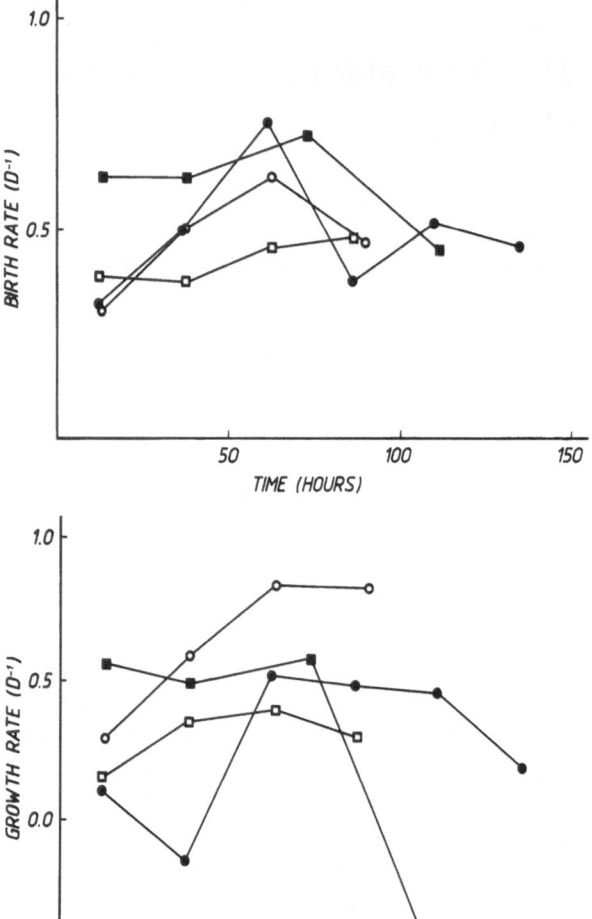

Fig. 2. Birth and growth rates of *Synchaeta oblonga* in the enclosure experiments at different temperatures: ○: 8.6 °C; ●: 9.8 °C; □: 11.4 °C; ■: 14.8 °C.

Discussion

The embryonic development time of *Synchaeta oblonga* is very short at low temperatures and can be compared with the cold-stenothermous *Rhinoglena fertoeensis* (Herzig, 1983). As the temperature of the Danube exceeds 20 °C only on some days in summer but drops below than 4 °C for several weeks during winter, this short development time can be regarded as an adaptation to the temperature conditions in the river.

Although no food was added during the experiments the rotifers showed high rates of increase.

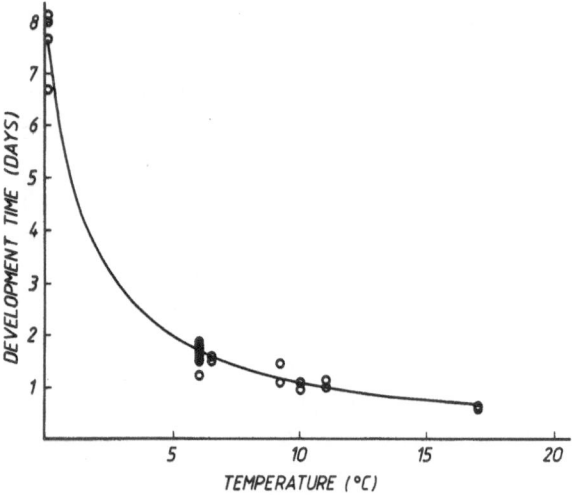

Fig. 1. Relationship between embryonic development time of *Synchaeta oblonga* and temperature.

Presumably, bacteria from the Danube was the food source used by the *S. oblonga*.

Acknowledgements

My sincere thanks are due to Univ. Doz. Dr. A. Herzig for his constructive criticism and help improving this manuscript. I am grateful to the Austrian Bundesministerium für Wissenschaft und Forschung and the Donaukraft (Donaukraftwerke AG) for financial assistance. This study was done in the scope of the 'Ecosystem study Donaustau Altenwörth'.

References

Herzig, A., 1983. Comparative studies on the relationship between temperature and duration of embryonic development of rotifers. Hydrobiologia 104: 237–246.

Koste, W., 1978. Die Rädertiere Mitteleuropas. Gebr. Borntäger, Berlin, Stuttgart, 673 pp.

Naidenow, W., 1985. Die Auswirkungen der Wasserbaulichen Maßnahmen und der Belastung auf das Plankton und das Benthos der Donau. Verlag der Bulgarischen Akademie der Wissenschaften: 72–102.

Naidenow, W. & D. Saiz, 1975. Der Einfluß des regulierten Abflusses auf das Plankton der Oberen Donau. In: Berichte der 18. Tagung der IAD. Regensburg 1975, I. Teil: 239–243.

Schallgruber, F., 1944. Das Plankton des Donaustromes bei Wien in qualitativer und quantitativer Hinsicht. Arch. Hydrobiol. 39: 665–689.

Hydrobiologia **186/187**: 167–179, 1989.
C. Ricci, T. W. Snell and C. E. King (eds), Rotifer Symposium V.
© 1989 *Kluwer Academic Publishers.*

Community structure and coexistence of the rotifers of an artificial crater lake

Geert Bogaert & Henri J. Dumont
Laboratorium voor ecologie der dieren, zoögeografie & natuurbehoud, State University of Ghent, K.L. Ledeganckstraat 35, B-9000 Ghent, Belgium

Key words: rotifers, succession, migration, abundance, distribution

Abstract

The community structure and dynamics of the rotifers of an abandoned and inundated limestone quarry are described. Zooplankton standing crop is low, but the community is rich in species. *Ploesoma hudsoni* and *Filinia hofmanni* are rare or unknown in a perimeter of several hundreds of kilometers around the lake. This illustrates that (passive?) dispersal across large stretches of land even occurs in rare species. Temporal and spatial distribution, together with specific diurnal vertical movements in different seasons facilitate the coexistence of numerous species in the lake. This complex ecology contrasts with the small size of the initial propagules that helped colonise the lake.

Introduction

Many studies deal with the dynamics of rotifers in a variety of lake types. In Belgium, numerous studies of stagnant waters focus on shallow artificial basins. These are nearly all under direct human influence and have recently undergone eutrophication. The present paper initiates the study of fairly unknown limnological sites in the country: the inundated quarries in the south. Their depth and trophic status, their remarkable plankton fauna and the unknown way in which they were colonised, prompted the given research.

In the site studied, quarrying ceased around 1948 and the cratershaped cavity was allowed to fill with groundwater. Filling-up presumably took not more than a year. The quarry and the resulting lake is not connected to any surface water, and thus has no in- or outlet. Even though no information on the colonisation of the lake is available, 40 years can be considered a short period for populating such an isolated lake. The objectives of the present paper were to identify the species assemblage living in the lake and to see how they compare to similar but natural lakes elsewhere. Also, because the lake's populations can be expected to have originated from a limited number of small initial propagules, each with little or no genetic variation, this narrow genetic basis might well be expressed in an unusual ecology.

Site description

The study site is a 47 meter deep, artificial lake (Lake Delwart), a former limestone quarry near Doornik, Province of Henegouwen, Belgium (50° 35′ 30″ N, 3° 24′ E). The lake is mesotrophic (Table 1). The shores are steep and a littoral zone is negligible: there is no *Phragmitetum* and no

Table 1. The morphometric characteristics of Lake Delwart.

Area	4.8 ha
Length	295 m
Breadth	250 m
Mean breadth	163 m
Maximum depth	47 m
Relative depth	19%
Shore line	880 m
Shore line development	1.13
Altitude above sea level	+ 35 m
Bathymetry	complex

submerged vegetation. Fish were introduced artificially and the ichthyofauna is dominated by perch (*Perca fluviatilis*) and carp (*Cyprinus carpio*).

Methods

The plankton was sampled biweekly between October 12, 1987 and October 8, 1988, at the deepest point of the lake. Duplicate samples were collected at 5 meter intervals, using a 2 liter Friedinger bottle (October-November 1987) with a 35 µm screen or a 35 liter plankton sampler (December 1987-October 1988) with a 50 µm screen. Plankton was preserved in 4% formaldehyde and counted in the laboratory using a dissecting microscope at 45 × magnification.

Temperature, pH, oxygen concentration and conductivity were also measured.

Diurnal vertical migration was studied over 24 hours in winter (20–21/12/1987) and in summer (4–5/7/1988). Samples were taken every 3 hours at 2.5 meter intervals in the upper 20 meters and in deeper waters every 10 (winter) or 5 meters (summer). Mean residence depth of a population was calculated as

$$\overline{D} = \Sigma N_i \times d_i / \Sigma N_i$$

with N_i = concentration of individuals at depth i, d_i = depth of ith sample.

Results

Abiotic environment

The lake is slightly alkaline with pH values between 6.32 and 8.70. It was warm and monomictic during the study period, with a winter low of 5.3 °C and a summer high of 21.5 °C. Oxygen stratification was clinograde in autumn, nearly uniform during turn-over, positive heterograde in spring, and clinograde again in autumn. At the deepest point (45 meter), the water was permanently anoxic, due to a sapropelium layer

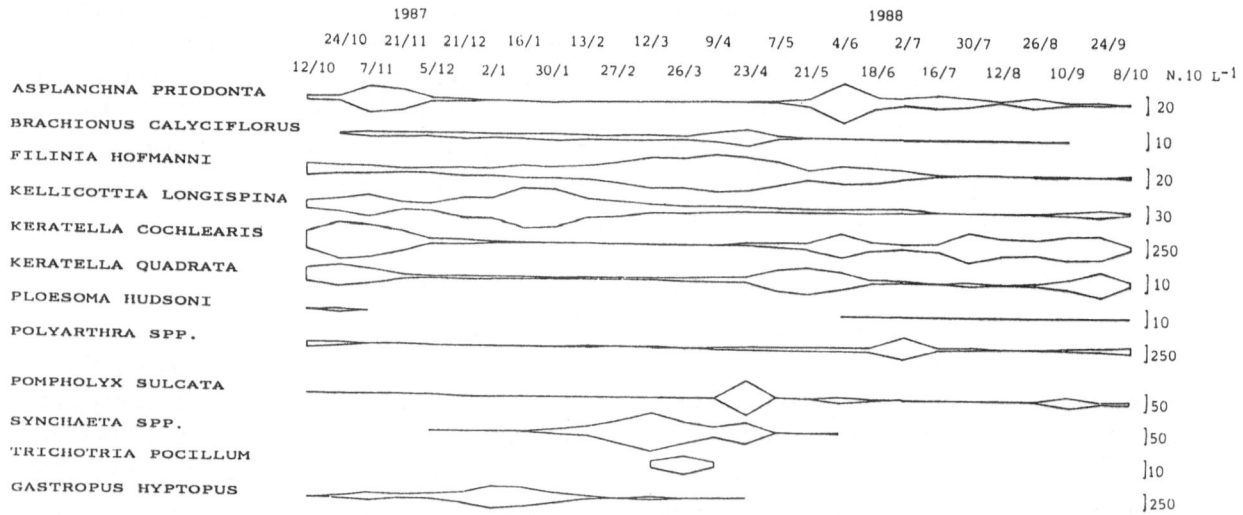

Fig. 1. Succession of the rotifers in Lake Delwart during the study period.

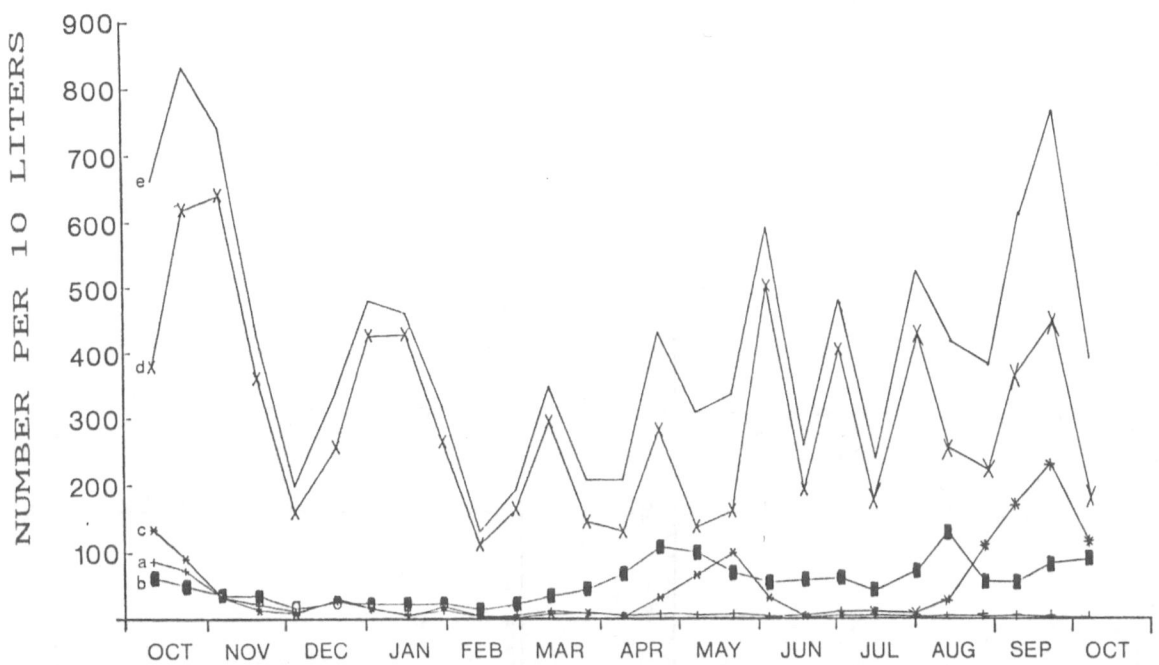

Fig. 2. Seasonal abundance of copepods (a), nauplii and copepodites (b), cladocerans (c), rotifers (d) and sum of these categories (e) in the lake in 1987–1988. Average over the water column, in numbers per 10 liters.

on the bottom. Conductivity values ranged between 560 and 840 μS cm^{-1}.

Community composition

The pelagic zooplankton community is dominated by rotifers: 18 species were found, including a few incursive littoral forms. The number and species composition is comparable to that of many natural deep lakes in the temperate region (Table 2). Two taxa among this assemblage stand out: the *Filinia* species present shows most of the morphological characteristics of *Filinia hofmanni*, as described by Koste (1980) and Schaber & Schrimpf (1984). Its ecology follows that reported by these authors: the species is found at temperatures from 5 to 10 °C and at oxygen contents from 1 to 14 mg l^{-1}. The highest abundance is at lower oxygen concentrations near the anoxic layer in the hypolimnion. The species has been reported from Norway, Austria and Germany. *Ploesoma hudsoni* has only been reported twice from Belgium (De

Ridder, 1973). Its nearest known occurrence is in France, at 250 kilometers from Lake Delwart (De Ridder, Pers. Com.).

Seasonal occurrence related to abiotic conditions

Many rotifer species are perennial (Fig. 1): the dominant *Keratella cochlearis*, with a broad maximum from spring to autumn; *Keratella quadrata*, with maxima in spring and autumn; *Polyarthra* spp., with dominance in autumn and summer. *Filinia hofmanni* and *Kellicottia longispina* also occur during the whole year, the former species with a broad maximum from winter to late spring, the latter abundant mostly in winter. *Brachionus calyciflorus* is absent only during a short period in autumn and reaches a weak spring maximum. There is a marked seasonal separation of the incidence of the raptorial forms: *Ploesoma hudsoni* (summer and autumn), *Asplanchna priodonta* (perennial, with highest densities late in autumn and late in spring) and *Synchaeta pectinata* (late

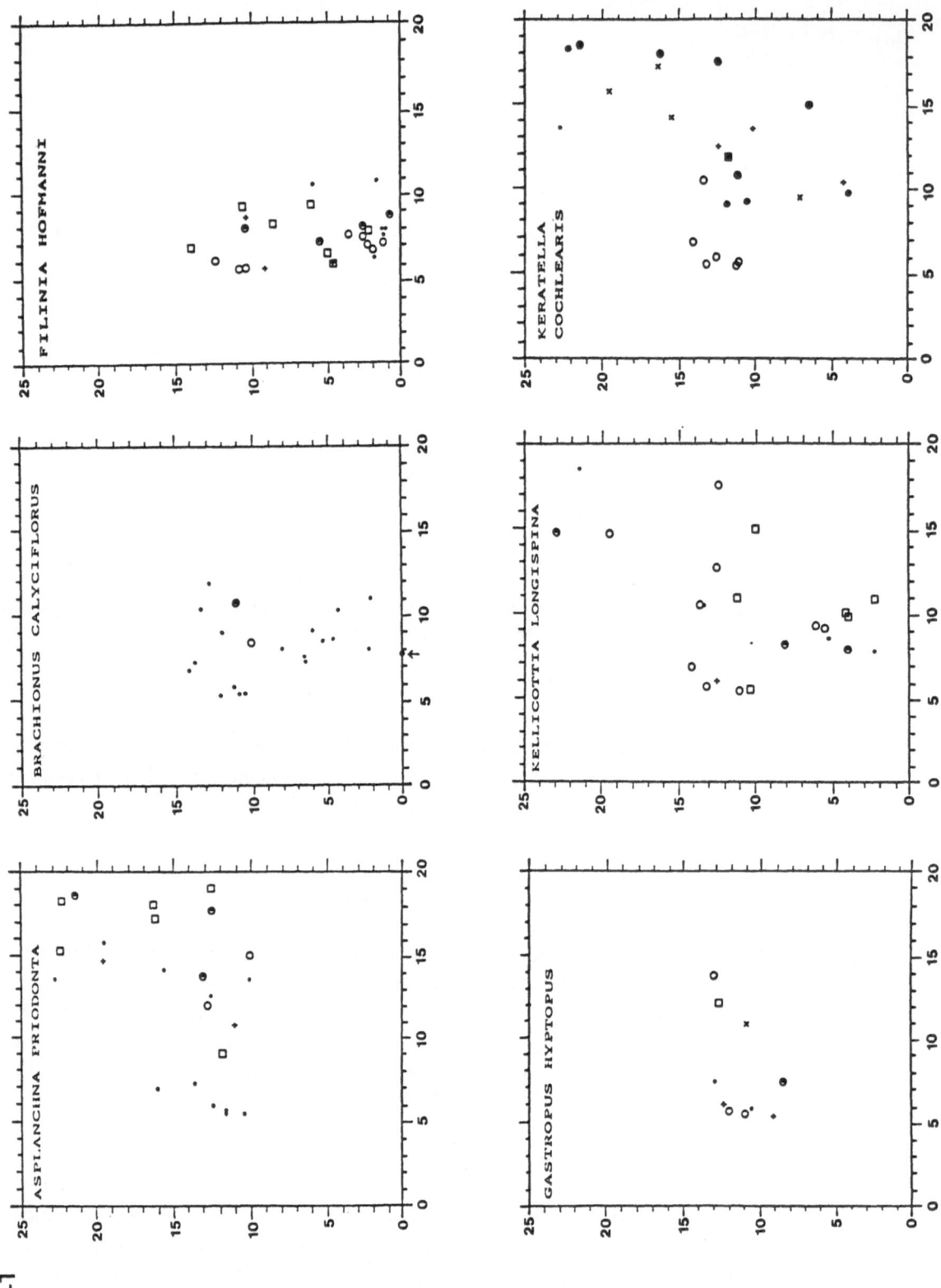

$O_2 mg.l^{-1}$

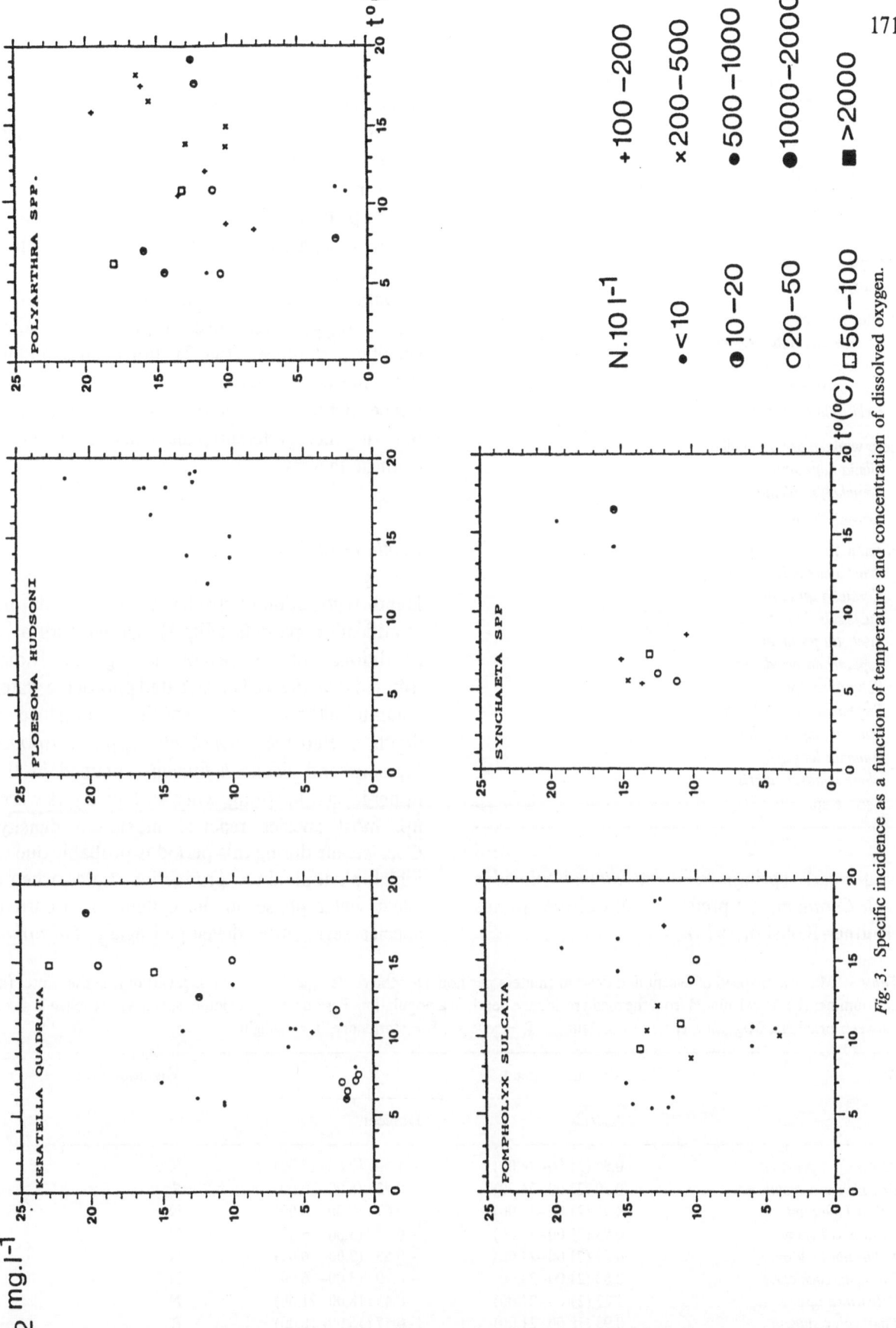

171

Fig. 3. Specific incidence as a function of temperature and concentration of dissolved oxygen.

172

Table 2. Pelagic species composition of Lake Delwart.

Summer/autumn species
  Cyclops vicinus vicinus
  Thermocyclops oithonoides
  Diaphanosoma brachyurum
  Ceriodaphnia quadrangula
  Ploesoma hudsoni
  Chydorus sphaericus
  Diacyclops bicuspidatus

Winter species
  Cyclops strenuus strenuus
  Eudiaptomus gracilis
  Gastropus hyptopus
  Kellicottia longispina

Late winter/spring species
  Filinia hofmanni
  Pompholyx sulcata
  Synchaeta spp.

Perennial and rare species
  Brachionus calyciflorus
  Keratella quadrata
  Euchlanis dilatata
  Trichotia pocillum
  Asplanchna priodonta
  Polyarthra spp.
  Stephanoceros fimbriatus
  Macrocyclops albidus
  Daphnia longispina
  Alone quadrangularis
  Lecane sp.

Gastropus hyptopus is a winter species. The littoral *Trichotria pocillum* occurs in the plankton only in late spring, but in relatively high densities, while *Pompholyx sulcata* is perennial with a maximum in summer.

The peak of the rotifer community taken as a whole, is separated in time from that of copepod nauplii and small cladocerans (Fig. 2).

Data on temperature and concentration of dissolved oxygen were used to determine some specific preferences (Fig. 3). For each sampling date, the combinations of the maximum abundance depth were plotted per species, summarizing their preferential incidence as a function of these factors.

Depth distribution

Depth segregation of species was marked during stratification periods (Fig. 4). Under turn-over conditions rotifers were more homogeneously distributed over the well oxygenated part of the water column, although some species maintained a depth preference: *Synchaeta* spp., *Gastropus hyptopus* and *Filinia hofmanni* remained in the upper layers. In spring, when surface layers warm up, most species reached maximum density. Coexistence during this period is probably due to a high phytoplankton production. Later, when a 'clear water phase' in the epilimnion occurred, species segregation developed again: *Polyarthra*

winter-early spring). *Ploesoma hudsoni* coincides with *Ceratium*, the preferred food of the species (Ruttner-Kolisko, 1974).

Table 3. Maximum speed of ascent and descent (meters per hour) for the rotifer species in Lake Delwart during the winter (a) and summer (b), as calculated from the mean residence depth of a population. Period when maximum speed was reached is given between brackets. Migration type: N = nocturnal, R = reverse, S = stationary, T = twilight.

A	Maximum speed		Migration type
	Ascent	Descent	
Asplanchna priodonta	0.80 (21.00–24.00)	− 1.60 (12.00–15.00)	R
Brachionus calyciflorus	0.50 (21.00–24.00)	− 0.43 (6.00– 9.00)	S
Filinia hofmanni	0.93 (21.00–24.00)	− 0.67 (3.00– 6.00)	N
Gastropus hyptopus	0.53 (21.00–24.00)	− 0.27 (3.00– 6.00)	N
Kellicottia longispina	0.57 (21.00–24.00)	− 0.63 (3.00– 6.00)	N
Keratella cochlearis	0.60 (21.00–24.00)	− 0.90 (3.00– 6.00)	N
Polyarthra spp.	1.23 (21.00–24.00)	− 0.43 (18.00–21.00)	N
Pompholyx sulcata	0.93 (18.00–21.00)	− 0.57 (15.00–18.00)	R

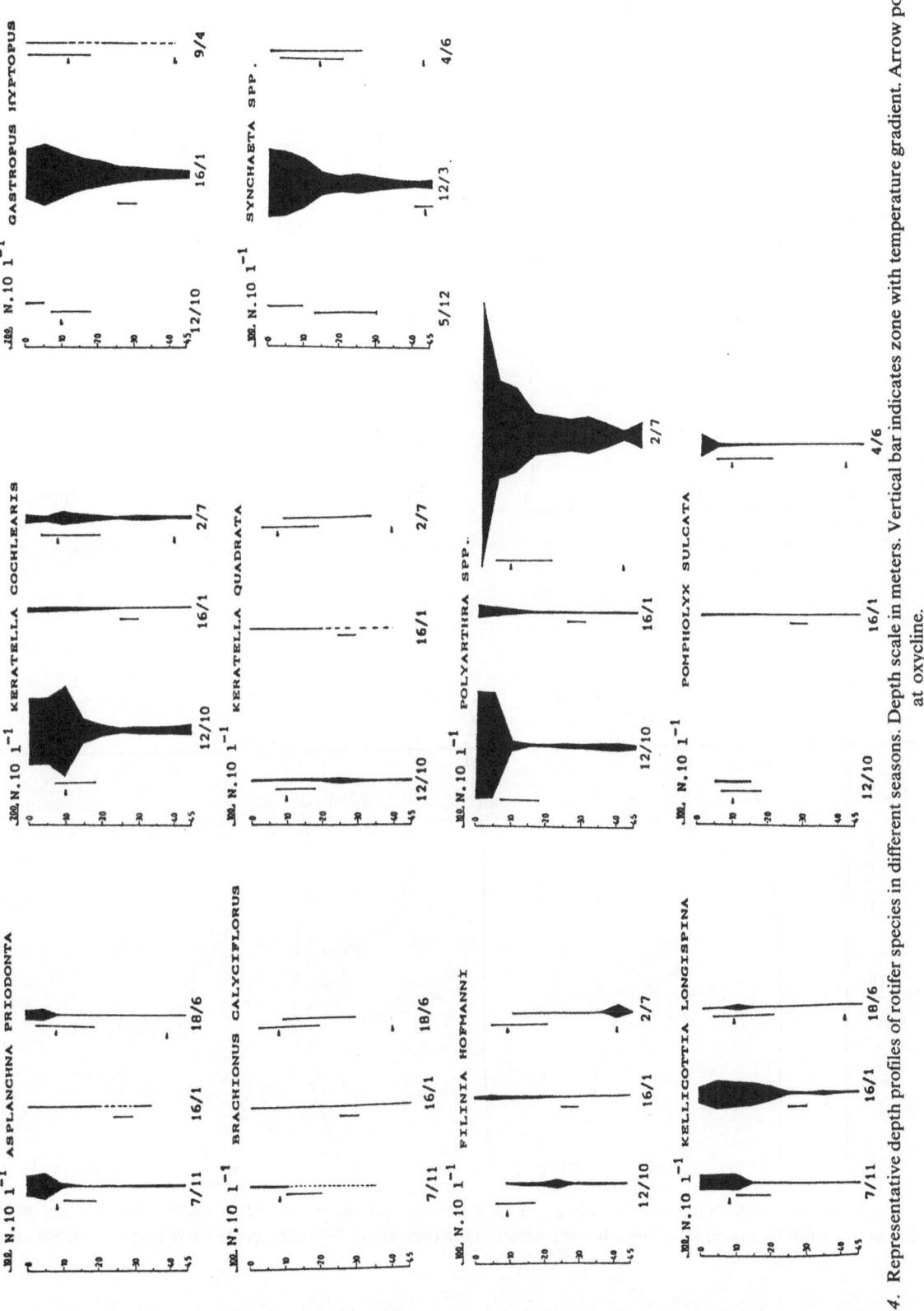

173

Fig. 4. Representative depth profiles of rotifer species in different seasons. Depth scale in meters. Vertical bar indicates zone with temperature gradient. Arrow points at oxycline.

174

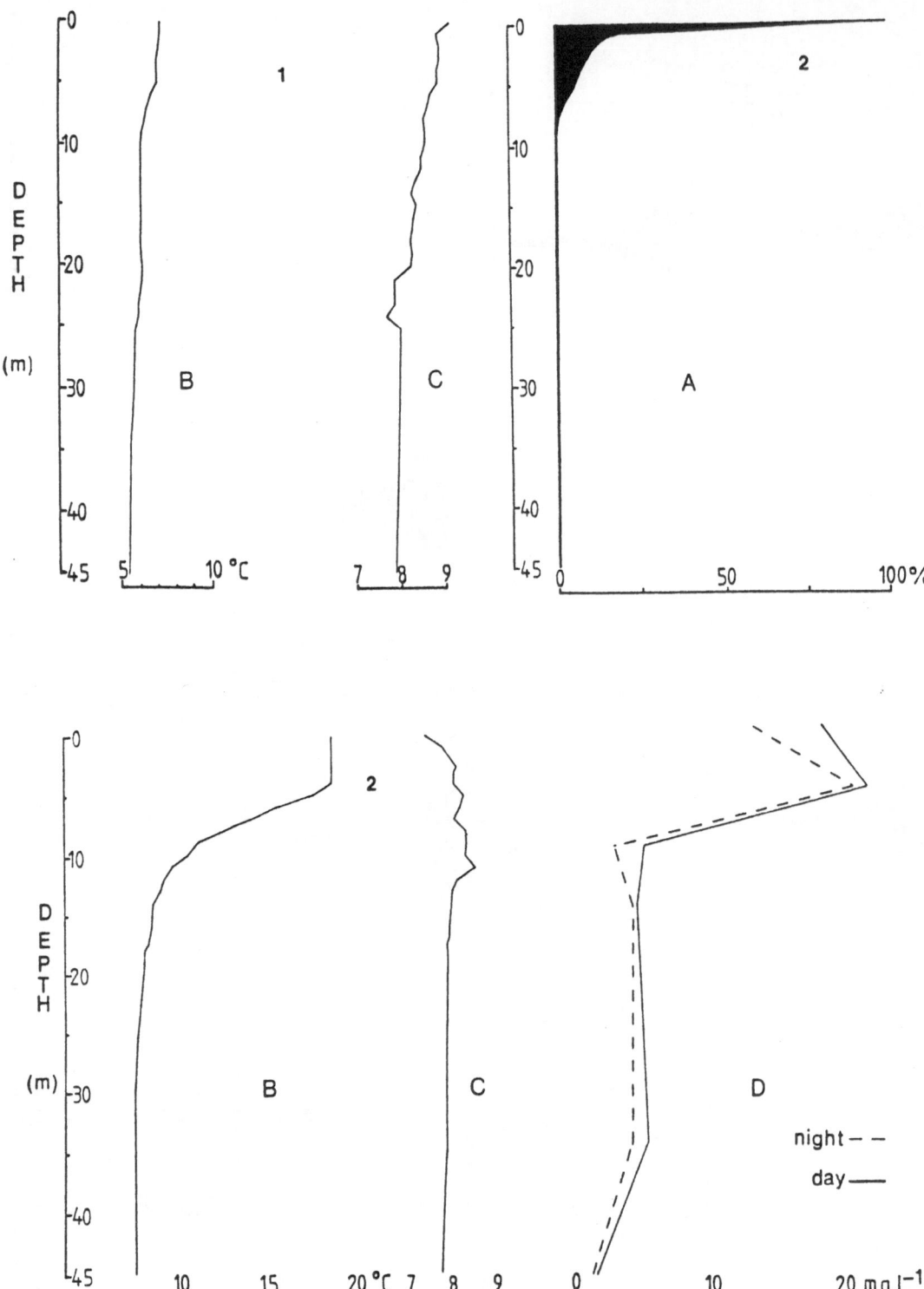

Fig. 5. Environmental conditions during the 24 hour sampling in winter (1) and summer (2). a: under water illumination in %
of that of the subsurface (average of day), b: temperature curve (average of 24 hours), c: pH curve (day), d: oxygen curve.

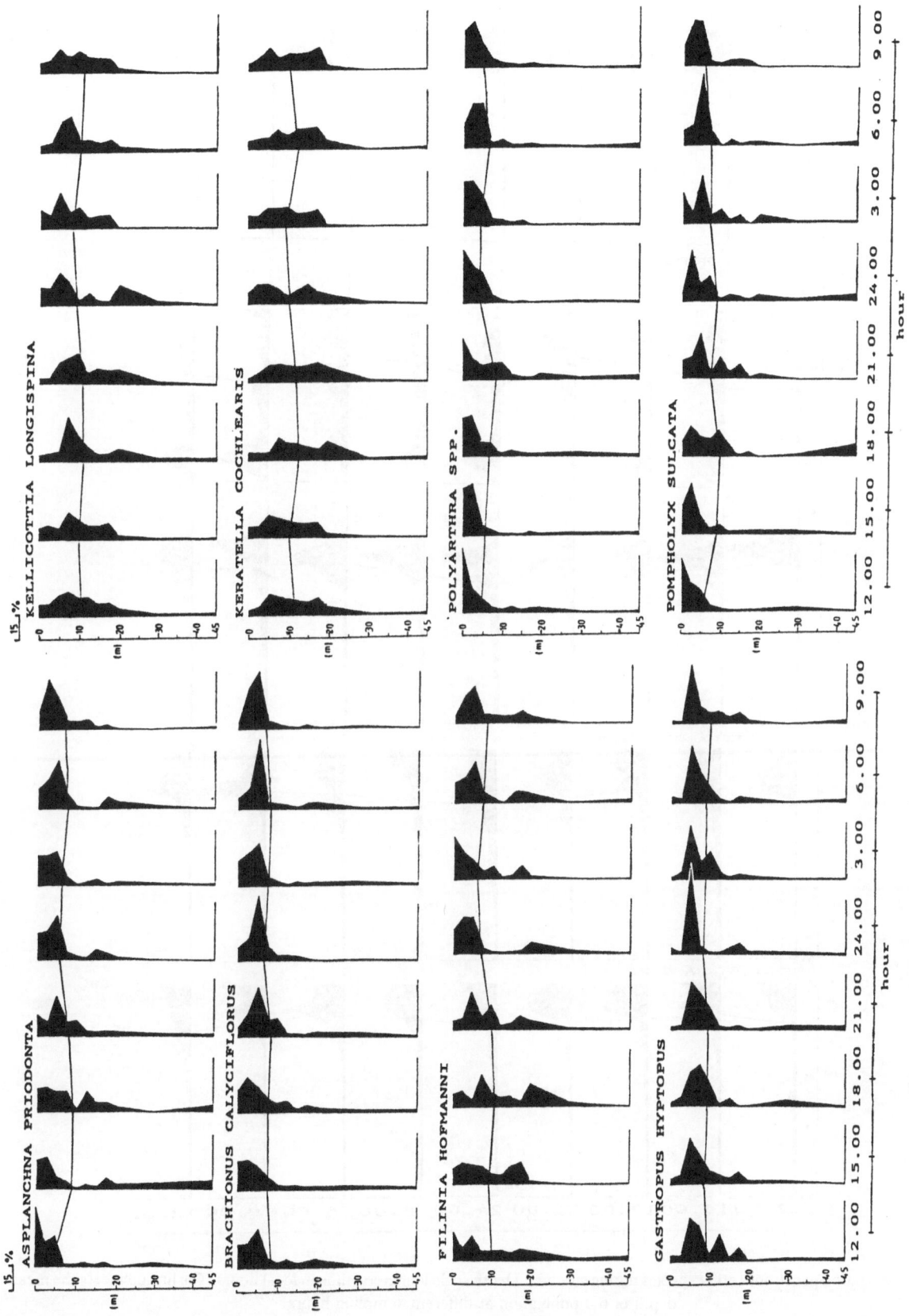

Fig. 6. Migration panels for rotifer species during winter. Depth scale in meters, time in local hours. The line connects the mean depth of the population at different sampling hours.

176

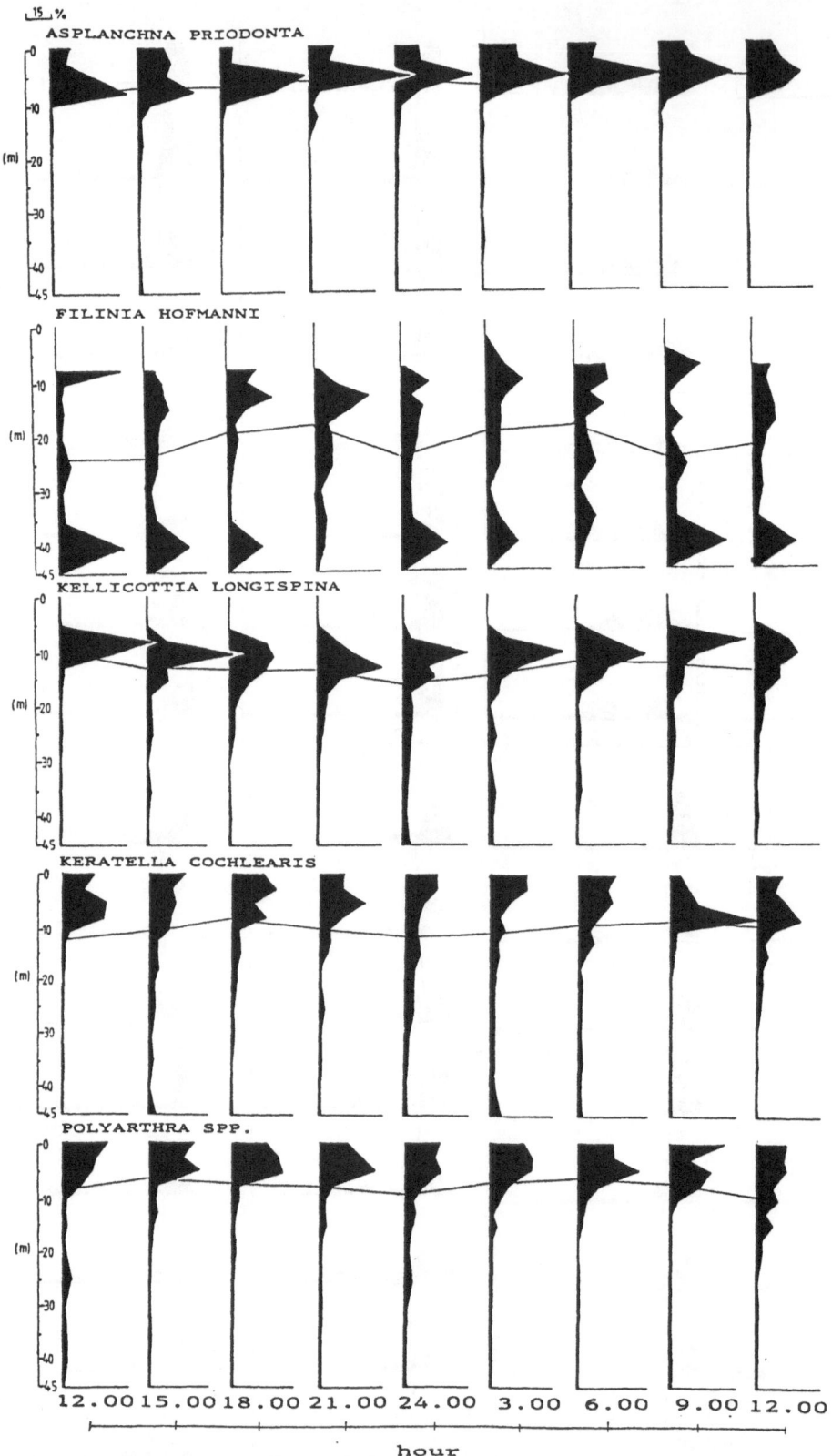

Fig. 7. Migration panels for rotifer species during summer. Depth scale in meters, time in local hours. The line connects the mean depth of the population at different sampling hours.

B	Maximum speed		Migration type
	Ascent	Descent	
Asplanchna priodonta	0.77 (3.00– 6.00)	– 0.23 (21.00–24.00)	S? or R?
Filinia hofmanni	1.77 (15.00–18.00)	– 2.23 (21.00–24.00)	T
Kellicottia longispina	0.53 (3.00– 6.00)	– 1.10 (12.00–15.00)	N
Keratella cochlearis	0.87 (3.00– 6.00)	– 0.87 (12.00–15.00)	R?
Polyarthra spp.	0.73 (24.00– 3.00)	– 0.87 (9.00–12.00)	R?

spp. and *Keratella cochlearis* remained in the epilimnion where they were slightly segregated. Most rotifer species accumulated near the thermocline, where bacteria, algae and other food particles concentrate. *Filinia hofmanni* and *Keratella quadrata* sank to the hypolimnion as soon as spatial overlap in the upper layers increased.

Diurnal vertical migration

Winter sampling (20–21/12/1987)
The weather was cloudy, with rain. Sunset was at 16.39 and sunrise at 8.43. A bloom of blue-green algae occurred in the upper 20 meters. Oxygen stratification was present with the oxycline between 20 and 25 meters (Fig. 5a).

The rotifers migrated over small distances (Fig. 6, Table 3), but depth segregation was diminished compared to the biweekly sampling. Copepods and cladocerans showed nocturnal migration behaviour.

Summer sampling (4–5/7/1988)
There was cloudy weather and rain. Sunset was at 21.58, sunrise at 5.36. Turbidity in the epilimnion was slight, with a bloom of bluegreen algae occurring at a depth of approximately 7 meters (Fig. 5b).

Different rotifers were located in and under the bloom layer and were segregated from species occurring nearer the surface. Data on migration type and speed are given in Fig. 7 and in Table 3. Copepods and cladocerans were concentrated in the bloom layer during the day and migrated upward at night.

Discussion

The diversity of the rotifer community indicates that relatively isolated lakes can get populated by a fairly 'complete' plankton zoocenosis in a remarkably short period of time. As stressed earlier, *Ploesoma hudsoni* is an epilimnetic species, *Filinia hofmanni* is a rare, largely hypolimnetic species. Further, their nearest known occurrence is about 250 kilometers away from Lake Delwart. While this illustrates the power of passive dispersal in rotifers, the mechanism through which this has occurred remains undocumented.

Environmental factors in a lake change in time and space. The way in which species relate to them in terms of abundance and depth distribution, is determined by their specific ecological tolerance, competitive interactions, the occurrence of food (Pourriot, 1965) and predation (Stemberger & Evans, 1984). Separate density peaks among the carnivorous species in Lake Delwart suggest an avoidance of competition through seasonal succession. Likewise, maxima of the herbivorous species are also segregated across the year. Cooccurrence with the filter-feeding nauplii and small cladocerans is reduced by a succession of their density peaks during bloom periods. Given these temporal segregations, differences in food selection (e.g. consumption of food particles from different size-class; Edmondson, 1965; Hutchinson, 1967; Peyler, 1957; Dumont, 1977; Starkweather, 1980), which can allow coexistence of even closely related species, probably do not play the most important role in the present study.

Apart from temporal segregation, overlap between species is reduced by differing depth-distributions: the marked segregations, seen on a small scale during the 24 hour sampling and the sinking of species like *Keratella quadrata* and *Filinia hofmanni* to the hypolimnion when densities in the upper layers increase, illustrate this point well. The slight segregation between *Polyarthra* spp. and *Keratella cochlearis* is probably due to light preferences (Peyler, 1957). Many species are abundant over a broad range of temperature and oxygen concentrations, but the absolute maximum of each species occurs at specific combinations of these factors. *Filinia hofmanni* behaves as a cold-stenotherm (5–10 °C), as does *Brachionus calyciflorus* to a lesser degree, while *Ploesoma hudsoni* is warm-stenotherm. *Gastropus hyptopus, Asplanchna priodonta, Ploesoma hudsoni* and *Synchaeta* spp. have specific oxygen preferences. The data also illustrate the ability of rotifers to tolerate low oxygen concentrations, as described by different authors (Larsson, 1971; Peyler, 1957; Elliot, 1977; Pourriot, 1977; Hofmann, 1977); *Keratella quadrata, Kellicottia longispina* and *Brachionus calyciflorus* survive in nearly anoxic layers.

The diurnal vertical movements of rotifers in the lake are also noteworthy. Migrations are present but not pronounced in winter, while they are more marked in summer. A relation with the homogeneous conditions during turn-over in winter and with the higher light levels, oxygen, temperature stratifications and food concentration in the trophogenic layer in summer seems obvious. Light regulation also seems important, as in many other planktonic organisms (Cushing, 1951; McNaught & Hasler, 1964; Hutchinson, 1967). Comparing winter and summer, we recorded significant variation in migration patterns. Stewart & George (1987) studied migration of rotifers in different months and found a similar variation.

The speed of ascent and descent, as calculated from $\overline{D}$ on different sampling hours is high. It is higher than in the (shallower) lakes studied by George & Fernando (1970) and by Stewart & George (1987). It is highest in summer, when migration amplitudes are larger.

The numerous cases of spatial segregation within the vertical migration event again suggests a role for biotic interactions in organizing the lake's community. The positioning and migration of rotifers is consistent with a response to predation pressure from copepods and competition with cladocerans and other rotifer species, i.e. it is a spatial and temporal adaptive strategy which facilitates co-occurrence of numerous species (Miracle, 1974, 1977; Makarewicz & Likens, 1975; Lane, 1975).

It follows that the ecology of the rotifer species of Lake Delwart is complex and comparable to that of other, older lakes. This indicates that even in a lake as young as Delwart, the (narrow?) gene pools of the populations present acquired sufficient variation to successfully respond to changing abiotic and biotic conditions.

Acknowledgements

We thank Guido Goos for practical help and Dr M. De Ridder for help in species identification and for providing unpublished data.

References

Cushing, D. H., 1951. The vertical migration of planktonic Crustacea. Biol. Rev. 26: 158–192.

De Ridder, M., 1973. Atlas provisoire des Rotifères de Belgique. J. Leclercq & C. Gaspar, eds, Gembloux.

Dumont, H. J., 1977. Biotic factors in the population dynamics of rotifers. Arch. Hydrobiol. Beih. 8: 98–122.

Elliot, J. E., 1977. Seasonal changes in the abundance and distribution of planktonic rotifers in Grasmere (English Lake District). Freshwat. Biol. 7: 147–166.

George, M. G. & C. H. Fernando, 1970. Diurnal migration in three species of Rotifers in Sunfish Lake, Ontario. Limnol. Oceanogr. 15: 218–223.

Hofmann, W., 1977. The influence of abiotic environmental factors on population dynamics in planktonic rotifers. Arch. Hydrobiol. Beih. 8: 77–83.

Hutchinson, G. E., 1967. A treatise on limnology. Vol. 2: Introduction to Lake biology and the Limnoplankton. Wiley, New York. 1155 pp.

Koste, W., 1980. Über zwei Plankton-Radertiertaxa *Filinia australiensis* n.sp. und *Filinia hofmanni* n.sp., mit Bemerkungen zur Taxonomy der *longiseta-terminalis*-Gruppe. Genus *Filinia* Bory de St. Vincent, 1824, Familie Filiniidae Bartos 1959, (Überordnung Monogononta). Arch. Hydrobiol. 90: 230–256.

Lane, P. A., 1975. The dynamics of aquatic ecosystems: a comparative study of the structure and dynamics of four zooplankton communities. Ecol. Monogr. 45: 307–336.

Larsson, P., 1971. Vertical distribution of planktonic rotifers in a meromictic lake; Blankvatn near Oslo, Norway. Norway. J. Zool. 19: 47–75.

Makarewicz, J. C. & G. E. Likens, 1975. Niche analysis of a zooplankton community. Science 190: 1000–1003.

Mc Naught, D. C. & A. D. Hasler, 1964. Rate of movement of populations of *Daphnia* in relation to changes in light intensity. J. Fish Res. Bd Can. 21: 291–316.

Miracle, M. R., 1974. Niche structure in freshwater zooplankton: a principal components approach. Ecology 55: 1306–1317.

Miracle, M. R., 1977. Migration, patchiness, and distribution in time and space of planktonic rotifers. Arch. Hydrobiol. Beih. 8: 19–37.

Pejler, B., 1957. Taxonomical and ecological studies on planktonic Rotatoria from northern Swedish Lapland. Kgl. Svenska Vetenskapskad. Handl. 4: 1–68.

Pourriot, R., 1965. Recherches sur l'écologie des Rotifères. Vie Milieu Suppl. 21. 224 pp.

Ruttner-Kolisko, A., 1974. Plankton rotifers: Biology and Taxonomy. Binnengewasser Suppl. 26. 146 pp.

Schaber, P. & A. Schrimpf, 1984. On morphology and ecology of the *Filinia-terminalis-longiseta*-group (Rotatoria) in Bavarian and Tyrolean lakes. Arch. Hydrobiol. 101: 247–257.

Starkweather, P. L., 1980. Aspects of the feeding behavior and trophic ecology of suspension-feeding rotifers. Hydrobiologia 73: 63–72.

Stemberger, R. S. & M. S. Evans, 1984. Rotifer seasonal succession and copepod predation in Lake Michigan. J. Great Lakes Res. 10: 417–428.

Stewart, L. G. & D. G. George, 1987. Environmental factors influencing the vertical migration of planktonic rotifers in a hypereutrophic tarn. Hydrobiologia 147: 203–208.

Hydrobiologia **186/187**: 181–189, 1989.
C. Ricci, T. W. Snell and C. E. King (eds), Rotifer Symposium V.
© 1989 *Kluwer Academic Publishers.*

The rotifer communities of acid-stressed lakes of Maine

Michael T. Brett
Institute of Limnology, Uppsala University, Uppsala, Sweden

Key words: rotifer communities, acidification, Maine

Abstract

The structure of the rotifer community in relation to lake pH, trophic status, the type of planktivore assemblage and the crustacean community was assessed in a survey of 23 lakes ranging in pH from 4.4 to 7.3, and in a study of two lakes – one acidic, the other circumneutral – during two summers. In both investigations the number of rotifer species encountered per sample was strongly reduced with pH. Although the reason for this is not clear acid-stress, the ultraoligotrophic nature of the acidic lakes, and competitive interactions with crustacean zooplankters may all have played a role. More importantly the ecological significance of this relationship is not known. The rotifer *Keratella taurocephala* was a principle species in the most acidic lakes, while several common rotifers were notably absent from these lakes. Although rotifer abundance was correlated with lake pH, the results of this study indicate that rotifer abundance is not a result of lake pH *per se*, but of lake trophic status and interactions with the crustacean community.

Introduction

A considerable amount of research has been carried out on acidification influences on zooplankton communities, however, the vast majority of this work has focused on the crustacean component of the zooplankton (Brett, 1989b). Of those studies dealing with rotifers the general findings include reductions in the number of species encountered per sample with pH (Almer *et al.*, 1974; Roff & Kwiatkowski, 1977; Hobæk & Raddum, 1980; Brezonik *et al.*, 1984; Chengalath *et al.*, 1984; Siegfried *et al.*, 1984, 1988; Yan & Geiling, 1985; Carter *et al.*, 1986; MacIssac *et al.*, 1987), and dominance of the communities of acidic temperate eastern Northern American lakes by *Keratella taurocephala* (Chengalath *et al.*, 1984; Siegfried *et al.*, 1984, 1988; Yan & Geiling, 1985, Carter *et al.*, 1986;

MacIssac *et al.*, 1987; Pinel-Alloul *et al.*, 1987; Schaffner, 1989).

Considerable confusion exists as to the relationship between rotifer abundance or biomass and lake pH. Yan & Geiling (1985) reported extensive rotifer communities, 15 to 51% of total zooplankton biomass, in two acidic metal contaminated lakes relative to six circumneutral lakes, where rotifers comprised no more than 2% of the total biomass. During the experimental acidification of lake 223 the importance of rotifers, as portion of total zooplankton biomass, increased slightly as pH was reduced from 6.8 to 5.0 (Schindler *et al.*, 1985). In apparent contradiction to these results Roff & Kwiatkowski (1977), Carter *et al.* (1986), and MacIssac *et al.* (1987) have all reported reduced rotifer abundance with pH. However, Pinel-Alloul *et al.* (1987) and Siegfried *et al.* (1987) observed that even across broad lake acidity gradients rotifer community abundance was more closely related to factors associated with lake trophic status, i.e. chlorophyll *a* and total phosphorus, than to lake pH.

This study examined the relationship between lake pH, trophic status, the planktivore community and the planktic crustacean and rotifer communities in a survey of 23 lakes across a pH gradient, and in a seasonal comparison of two lakes (one acidic and the other circumneutral) during two summers.

Methods

The 23 lakes investigated were all forest lakes located in isolated regions of Maine, northeastern United States; i.e. these lakes had little or no human development in their watersheds. Lake

selection was biased to include a broad range of pH values, from 4.45 to 7.20, with relatively little variation in lake morphology. These lakes were small, oligo- to mesohumic, with soft water, and were generally oligotrophic, see Table 1. Five of the lakes, the most acidic, were fishless and contained dense populations of limnetic hemipterans – especially *Buenoa macrotibialis*, while four contained allopatric populations of brook char (*Salvelinus fontinalis*), and the remaining 14 contained cyprinid fish, mainly golden shiners (*Notomigonus crysoleucas*) (Brett, 1989a).

Each of the lakes was sampled once at its deepest point for zooplankton and water chemistry in July or August of 1984. Zooplankton sampling was carried out using a 63 μm Wisconsin net drawn at a rate of 0.5 m sec^{-1} from 1 to 2 m above the sediment to the surface, except for Mud Pond which was sampled from 5 m above the

Table 1. Limnological characteristics of the 23 lakes surveyed, all values reported as averages for paired duplicate surface (0.5 m) and bottom (1 m above sediment surface) samples taken on the same dates as the zooplankton samples.

Lake	Elevation (m)	Surface area (ha)	Maximum depth (m)	pH	Alkalinity (μeq/l)	Conductivity (μs/cm)	Color (Pt units)	Total P (μg/l)	Planktivore type classification
Unnamed P. (WC)	73	2.8	5.2	4.45	− 29.1	18	35	7	Insect
Dead P.	48	4.9	10.8	4.50	− 30.4	14	30	6	Insect
Kerosene P.	67	2.8	5.2	4.50	− 19.1	17	45	7	Insect
Mud P.	97	2.2	14.0	4.60	− 26.5	25	20	2	Insect
Salmon P. (WC)	42	4.5	6.7	4.85	− 12.3	14	35	6	Insect
Lily L.	72	13.0	7.3	5.15	− 1.8	17	40	6	Cyprinid
Ledge P.	948	2.4	7.3	5.45	8.9	17	50	5	Salmonid
Anderson P.	65	5.0	6.0	5.45	2.8	16	5	5	Cyprinid
Eddy P.	793	3.6	6.7	5.65	10.8	16	10	6	Salmonid
Garrock P.	145	0.8	6.0	5.75	29.2	17	60	31	Cyprinid
Grenell P.	388	2.4	5.5	5.80	86.2	18	70	9	Salmonid
Hosea Pug L.	88	23.5	10.7	5.85	13.4	16	35	8	Cyprinid
Round P.	279	2.4	9.5	5.85	88.4	25	45	13	Cyprinid
Salmon P. (HC)	85	4.1	10.7	5.90	55.8	25	15	11	Cyprinid
Unnamed P. (PC)	380	6.8	2.0	5.90	43.0	16	85	27	Cyprinid
Secret P.	385	5.7	10.4	6.00	88.2	22	35	12	Cyprinid
Possum P.	80	12.1	5.5	6.10	7.9	14	35	8	Cyprinid
Harvey P.	291	4.1	2.4	6.15	54.8	26	10	11	Cyprinid
Daicey P.	329	15.4	7.9	6.15	6.28	18	10	8	Cyprinid
Trout L.	274	2.0	3.0	6.20	50.7	22	25	11	Cyprinid
Upper Basin P.	745	14.2	9.0	6.30	14.8	14	5	1	Salmonid
Grindstone P.	394	2.4	9.1	6.40	195.9	31	20	9	Cyprinid
Carry P.	900	6.9	3.7	7.20	534.4	63	20	13	Cyprinid

sediment. These samples were preserved in formalin.

Two of the lakes included in the survey, Mud Pond and Salmon Pond (HC), were chosen for more extensive investigations because they represented extremes in the lake types included in the lake survey – Mud Pond was acidic and fishless, while Salmon Pond (HC) was circumneutral with fish – and because they were morphologically similar and located approximately 100 m apart, see Table 2. These lakes were sampled on four dates in 1983 and six dates in 1984 at biweekly intervals during the mid-summer period. The sampling was carried out at eight stations in the littoral zone using a vertical tube sampler (1 m). Samples were filtered through a 64 μm filter apparatus and preserved in Lugol's solution. The results for the two-lake comparison are presented as pooled data for all stations on each date.

For both the lake survey and the two-lake comparison, rotifers and nauplii were counted by decanting known volumes of the samples, subsampling with a Henson-Stemple pipette, and enumerating with 1 ml Sedgwick-Rafter counting

cells until at least 400 individuals were counted. The entire sample was scanned for rarer species. The whole of each sample was counted for crustacean zooplankters. In order to generate biomass values for the lake survey the first 20 individuals of each crustacean group, generally genus, were measured and their lengths recorded. Then length weight regressions (McCaully, 1984) were used to calculated the average weight for the group or species. When possible, equations for the same species, or most common species of a genus, from the same region were used. Alternatively the same species from another region or a similar species regression was used. Rotifer weights were adapted from Makarewicz & Likens (1979). Species identifications were based on the taxonomic works of Edmondson (1959), Ruttner-Kolisko (1974), Chengalath & Mulamoothil (1975), Kiefer & Fryer (1978), Pennak (1978) and Stemberger (1979).

In order to rank the lakes surveyed according to their presumed predation pressure, the lakes were first divided into three categories based on the type of planktivore present, and then ranked within these groupings according to the capture rate of the dominant planktivore in that lake (Brett, 1989a). Thus the lake dominated by limnetic insects with the lowest capture rate, during sweep net sampling, received the lowest rank while the cyprinid-fish dominated lake with the highest capture rate received the highest ranking.

Table 2. Selected physical, biological, and chemical characteristics of Mud Pond and Salmon Pond (HC). All chemical data expressed as μeq L^{-1}, except as otherwise noted.

	Mund Pond	Salmon Pond (HC)
Surface area (ha)	1.6	2.4
Maximum depth (m)	14	11
Elevation (m.a.s.l.)	97	85
Fish	absent	present
Notonectidis	abundant	rare
Chaoborus	rare	present
pH	4.6	6.2
Alkalinity	− 26.5	55.8
Ca	28.3	72.4
Al	37.5	3.6
SO$_4$	104	52
Total phosphorus (ppb)	2	11
Color (Pt units), Na, K, Mg, Fe, Zn, Mn, NO$_3$, Cl	similar between the lakes	

Results

In the lake survey 29 rotifer species were collected and identified, of which only 12 were found in 40% or more of the lakes. Of these *Keratella cochlearis, Kellicottia bostoniensis, Kellicottia longispina, Polyarthra remata, Trichocerca multicrinis, Asplanchna priodonta* and *Collotheca mutabilis* were greatly reduced or absent from those lakes below pH 5.0. In fact these seven species had a 73% rate of occurrence in the lakes above pH 5.0., but only a 17% rate in lakes below this pH. In contrast the rotifer *Keratella taurocephala* was found in every lake investigated (Table 3).

Table 3. The rotifer community structure of the lakes surveyed, expressed as individuals per liter. Those species listed individually occurred in $\geq 40\%$ of the lakes surveyed, while those occurring in fewer lakes are listed all together under other rotifers.

	NRS	RA	Kt	Kc	Kb	Kl	Pv	Pr	Cn	Cu	Tm	Tc	Ap	Cm	oth
Unnamed P. (WC)	5	2	2												<1
Dead P.	7	6	1					4			<1	<1			1
Kerosene P.	7	19	5			1	11	1							1
Mud P.	5	8	7	<1	+		1					<1			
Salmon P. (WC)	5	132	1	<1						131					+
Lily L.	7	138	5	17	6	1	27		67	15					+
Ledge P.	9	132	33	<1	5		+	1	+	93	<1				<1
Anderson P.	7	86	25	30	+	8	22	1							+
Eddy P.	9	17	9	+			2	+	1	5		+		+	
Garcock P.	14	1920	2	318	1220		276	+	1	50	11	19	20	+	3
Grenell P.	10	124	6	28	<1	19	+	5	39	4					
Hosea Pug L.	14	80	4	27	1	12	28	2	3		2	1	1	+	+
Round P	14	100	4	44	8	13	31	9	24	2		1	1		+
Salmon P. (HC)	14	89	10	<1	6	2	31	2	1	20	1	13	1	3	2
Unnamed P. (PC)	8	471	348	43	1		71	+			1	+	6		
Secret P.	14	171	1	12	8	87	34	1	+	3	4		+	+	24
Possum P.	12	210	18	18	<1	11	106	7	4	35	6	5		+	
Harvey P.	14	92	46	10	1	1	5	5	+	23	<1	+		1	+
Daicey P.	9	128	+	3	3	8	11			103			+	+	+
Trout L.	9	176	3	1	+	167	5			+	+	1			+
Upper Basin P.	5	9	<1	<1	1			<1		8					
Grindstone P.	12	250	1	158	20	52	9	4		+		1	1		5
Carry P.	8	191	1	30	2	8	136	12					4		+

NRS = number of rotifer species per sample; RA = rotifer abundance per liter; Kt = *Keratella taurocephala*; Kc = *Keratella cochlearis* including *K.c. cochlearis*, *K.c. tecta* and *K.c.* ssp.; Kb = *Kellicottia bostoniensis*; Kl = *Kellicottia longispina*; Pv = *Polyarthra vulgaris*, including *P. dolichoptera* from one lake; Pr = *Polyarthra remata*; Cn = *Conochiloides natans*: Cu = *Conochilus unicornus*, including *C. hippocrepis* from one lake; Tm = *Trichocerca multicrinis*; Tc = *Trichocerca cylindrica*; Ap = *Asplanchna priodonta*; Cm = *Collotheca mutablis*; oth = other rotifer species, including *Lecane lunaris*, *L. luna*, *Synchaeta* spp., *Ascomorpha ecaudis*, *Keratella hiemalis*, *Trichocerca elongata*, *T. platessa*, *T. myersi*, *T. simillis*, *Polyarthra euryptera*, *Ploesoma truncatum*, *P. hudsoni*, *Filinia longiseta*, and *Gastropus stylifer*.

Table 4. A correlation matrix for various chemical and biological characteristics of the lakes surveyed. All statistically significant relationships are shown in bold type.

	pH											
Total phosphorus	**0.474** Total phosphorus											
Predation rank	**0.589**	**0.788** Predation rank										
Specific conductivity	0.396	**0.508**	0.304 Specific conductivity									
Color	−0.353	0.336	0.174	−0.114 Color								
Rotifer species	**0.482**	**0.704**	**0.634**	0.412	0.148 Rotifer species							
Rotifer abundance	**0.580**	**0.669**	**0.799**	0.261	0.357	**0.526** Rotifer abundance						
Rotifer biomass	**0.594**	**0.550**	**0.678**	0.090	−0.055	**0.548**	0.403 Rotifer biomass					
Crustacean species	0.169	**0.499**	0.303	0.326	0.122	**0.465**	0.054	0.273 Crustacean species				
Crustacean abundance	0.145	0.023	0.006	−0.019	−0.093	−0.117	0.257	−0.204	−0.386 Crustacean abundance			
Crustacean biomass	0.129	0.229	0.078	**0.413**	−0.198	−0.035	−0.088	0.031	0.158	−0.044 Crustacean biomass		
Total biomass	0.272	0.389	0.250	0.287	−0.117	0.293	−0.041	**0.536**	0.402	−0.317	**0.718**	

If $r > 0.412$ then $p < 0.05$.
If $r > 0.531$ then $p < 0.01$
If $r > 0.633$ then $p < 0.001$.

Table 5. Mean zooplankton community parameters when the lakes surveyed are divided into dominant planktivore catagories. Least-square means were calculated, then the data was ranked and compared for significant differences ($P < 0.05$).

Class	Limnetic insect	Salmonid	Cyprinid	Significant differences
$n =$	5	4	14	
pH	4.6	5.8	6.0	LI*SL; LI*CY
Total phosphorus 1	5	5	12	LI*CY; SL*CY
Rotifer species	5.9	8.3	11.2	LI*CY
Rotifer abundance 2	13.6	48.9	263.2	LI*CY; SL*CY
Rotifer biomass 1	0.9	4.5	21.9	LI*CY; SL*CY
Total biomass 1	66.2	72.6	115.4	
RA/CA 3	0.6	3.3	5.9	LI*CY
	Zooplankton community biomass proportional composition			
Rotifers	1	6	19	LI*CY; SL*CY
Nauplii	5	4	18	LI*CY; SL*CY
Bosmina	9	2	3	
Diaphanosoma	8	<1	2	LI*CY; LI*SL
Holopedium	29	32	6	LI*CY; SL*CY
Daphnia	<1	17	17	LI*CY
Diaptomus	45	32	24	
Cyclopida	3	2	11	LI*CY; SL*CY

1. in μg per liter.
2. in individuals per liter.
3. RA/CA = rotifer abundance/crustacean abundance.

Table 4 is a correlation matrix for various chemical and biological characteristics of the lakes surveyed. Notably this matrix shows a very high amount of interrelatedness between the predictor variables pH, total phosphorus, and predation rank. These variables, in turn, predicted nearly identical variation in the rotifer community parameters – rotifer species, rotifer abundance, and rotifer biomass.

By dividing the lakes into planktivore categories and analyzing for differences in zooplankton community composition between them, strong differences emerge. Namely, in both absolute terms (rotifer abundance and rotifer biomass) and relative terms (the ratio between rotifer abundance and crustacean abundance and the rotifer portion of the zooplankton biomass) rotifers were strongly favored by cyprinid predation relative to limnetic insect predation. Likewise, cyprinid predation was associated with enhanced nauplii and cyclopoid components of the zooplankton, whereas the relatively large and

less evasive cladocerans *Holopedium gibberum* and *Diaphanosoma birgei* were more important in the limnetic insect dominated lakes.

In the two-lake study 16 rotifer species were identified from the acidic lake, and 21 species were observed in samples from the circumneutral lake. The acidic lake averaged 4.5 ± 2.2 ($\pm$ sd) species per sample while the circumneutral lake averaged 10.4 ± 2.2. The rotifer community of the acidic lake was strongly dominated by *Keratella taurocephala* (98% of individuals observed) while the rotifer community of the circumneutral lake was composed of 3 to 5 species, with *Keratella taurocephala* again the dominant (77% of individuals observed), but with *Keratella cochlearis*, *Kellicottia longispina*, *Trichocerca cylindrica*, *Conochilus unicornis*, and *Collotheca mutabilis* periodically common.

Notably, the crustacean community of the acidic fishless lake was principally composed of the calanoid *Diaptomus minutus* and the cladocerans *Diaphanosoma birgei* and *Polyphemus pediculus*, while the crustacean community of the circumneutral fish-containing lake was again dominated by *D. minutus*, but also contained large cyclopoid and *Bosmina longirostris* components, Tables 6 and 7. Rotifer abundance was generally reduced in the acidic lake, relative to the crustacean community of that lake and to the rotifer community of the circumneutral lake, except for the month of August 1984 when rather high abundances were observed (Tables 6, 7). This rise in rotifer abundance coincided with a marked decline in the cladoceran abundance

Table 6. Zooplankton community structure in Mud Pond, expressed as individuals per liter.

	Date 1983				1984					
	21/6	7/7	21/7	8/8	12/6	28/6	12/7	26/7	9/8	27/8
Copepods 1	6	8	4	1	26	3	3	6	14	0
Nauplii	37	10	5	4	28	<1	<1	<1	<1	<1
Cladocerans 2	7	4	5	1	7	4	9	2	2	0
Keratella taurocephala	<1	9	7	8	6	1	8	37	229	287
Other rotifers	<1		<1	<1	<1	<1	<1	<1	<1	<1
Total rotifer abundance	<1	9	7	8	6	1	8	37	229	287
Number of rotifer species per sample	3	1	5	7	3	3	6	6	8	3

1. copepods = 96% *Diaptomus minutus*, 4% *Mesocyclops edax*.
2. cladocerans = 58% *Diaphanosoma bigrei*, 26% *Polyphemus pediculus*, 16% *Bosmina longirostris*.

Table 7. Zooplankton community structure in Salmon Pond (HC), expressed as individuals per liter.

	Date 1983				1984					
	21/6	7/7	21/7	8/8	12/6	28/6	12/7	26/7	9/8	27/8
Copepods 1	21	10	7	10	24	17	19	8	46	6
Nauplii	13	32	12	13	1	12	60	12	5	1
Cladocerans 2	32	9	5	8	11	5	6	9	20	2
Keratella taurocephala	27	36	36	63	49	93	120	50	77	11
Keratella cochlearis	20	2	1	26	6	8	4	3	12	4
Kellicottia longispina		<1	<1	<1	3	5	8	2	1	
Trichocerca cylindrica	<1		<1	3	<1	1		4	13	1
Conchilus unicornis		<1			10	1		1		
Collotheca mutabilis	<1					<1	2	6		
Asplanchna priodonta	1	<1	<1	2	3			<1	1	2
Polyarthra vulgaris	1	<1	4	<1	1	1	<1			
Other rotifers	<1	<1	<1	<1	<1	<1	<1	<1	<1	<1
Total rotifer abundance	48	38	38	97	72	109	135	67	103	19
Number of rotifer species per sample	8	7	13	12	12	13	8	11	9	11

1. copepods = 72% *Diaptomus minutus*, 18% *Tropocyclops prasinus mexicanus*, 10% *Mesocylops edax*.
2. cladocerans = 92% *Bosmina longirostris*, 7% *Diaphanosoma birgei*.

during this period. In samples collected at the same time, but from the pelagic as apposed to the littoral region of Mud Pond, crustacean zooplankters were much more abundant and the rotifers were greatly reduced (Table 3, 6).

Discussion

The most obvious result of this study, for both the lake survey and the two-lake comparison, was a reduction in the number of rotifer species per sample with pH. This observation has been made in several other studies (Almer et al., 1974, Roff & Kwiatkowski, 1977; Hobæk & Raddum, 1980; Brezonik et al., 1984; Chengalath et al., 1984; Siegfried et al., 1984, 1988; Yan & Geiling, 1985; Carter et al., 1986; MacIssac et al., 1987), however to what extent this is a direct result of lake acidity or the otherwise simplified nature of acidic lakes is poorly understood. In the present material the number of species per sample was also strongly, in fact more strongly, correlated to total phosphorus, lake predation rank, and rotifer density. It has been suggested (Pejler, 1983) that ultra-oligotrophic lakes, such as the most acidic lakes of the present study, are generally species depauperate relative to mesotrophic lakes, such as several of the circumneutral lakes of the present study. The strong relationship between rotifer species number and abundance may in fact be important, for instance reduced rotifer abundance might be indicative of conditions unfavorable for rotifers as a whole. One could also speculate that reduced predation pressure on the crustaceans – invertebrate versus fish, could act to increase competitive interactions and hence dominance of large cladocerans over rotifers as a group, thereby reducing 'niche space' for rotifers in the most acidic lakes. The fact that *Keratella taurocephala* was so strongly dominant in the most acidic of the two lakes compared, and that it was important in several of the other acidic lakes examined, as well as other acidic lakes in eastern North America (Chengalath et al., 1984; Siegfried et al., 1984, 1988; Yan & Geiling, 1985; Carter et al., 1986; MacIssac et al., 1987; Pinel-Alloul et al., 1987;

Schaffner, 1989) indicates that acid-stress, at least in influencing competitive interactions between species, might have an important influence on rotifer species number. Furthermore that several common species had greatly reduced rates of occurrence in lakes below pH 5.0 relative to lakes above this pH is also indicative of stress influences. Notably MacIssac *et al.* (1986) reported that when a severely acidic metal-contaminated lake partially recovered, pH change from 4.0 to 5.0, the rotifer community shifted from one dominated almost exclusively by *Keratella taurocephala* and *Synchaeta* spp., to one dominated by these species plus *Polyarthra* spp., *Trichocerca similis*, and *Conociloides natans*. A more fundamental problem is the ecological significance of the reduced species number in the acidic lakes. Is this observation ecologically irrelevant – but technically easy to generate, or is it of real importance to the functioning of acid-stressed ecosystems?

The abundance of rotifers in the present study was also correlated with pH as has been observed in several other studies (Roff & Kwiatkowski, 1977; Brezonik et al., 1984; Carter et al., 1986, MacIssac et al., 1987). However, rotifer abundance was more strongly correlated with total phosphorus. This in agreement with the results of several other studies which found zooplankton abundance in acid-stressed lakes to be more closely related to lake trophic status, and hence presumably the availability of edible algae and bacteria, than to pH *per se* (Brezonik et al., 1984; Pinel-Alloul et al., 1987; Kerekes et al., 1988; Siegfried et al., 1987). This finding is also consistent with several nutrient enrichment studies of acidic lakes (Yan et al., 1982; DeCosta et al., 1983; Yan & Lafrance, 1984) showing dramatic increases in zooplankton community abundance or biomass after enrichment, and is consistent with the relationship between nutrient levels and community standing crop typical for non-acidic lakes.

Analysis of zooplankton community structure across planktivore type categories suggests that planktivore type influenced the relative composition of the zooplankton. Cyprinid fish predation was associated with increased representation of the rotifer community, nauplii, and cyclopoids – small or highly evasive groups – at the expense of the much larger *Holopedium gibberum* and *Diaphanosoma birgei*, both in absolute, and more importantly relative terms, e.g. as regards portion of the zooplankton community biomass. This would indicate that fish predation, which is presumably most concentrated on larger cladocerans – for reasons of size and evasiveness, liberated resources or reduced competitive interactions with crustacean zooplankters thereby benefitting the rotifers. The extreme case of this can be exemplified by Garcock Pond which had an allopatric population of golden shiners, a very large population relative to lakes of this region, which reduced the crustacean community to less than one individual per liter, despite the fact that this lake had the highest total phosphorus concentration of the lakes examined. In this lake the rotifer abundance was estimated at 2 000 individuals per liter, an abundance 5 times higher than that recorded in any of the other lakes studied.

Interactions between the crustacean and rotifer communities was also indicated by the results from the acidic lake of the two-lake comparison, where the rotifer *K. taurocephala* increased dramatically after first the cladocerans and then the copepods were greatly reduced in the littoral zone. This pattern was not observed in the pelagic zone of this lake during the same period, presumably because the chief planktivore in this lake was not particularly abundant in the pelagic relative to the littoral zone. Other studies of acid-stressed zooplankton communities have also indicated the importance of interactions between the rotifer and crustacean components of the zooplankton, namely Yan & Geiling (1985) used the reduced crustacean biomass, relative to six circumneutral lakes, and hence reduced competition with rotifers, to explain the extremely high rotifer biomasses attained in their two acidic lakes. While Yan *et al.*, (1982) showed dramatic increases in *K. taurocephala* abundance of an acidic lake after predation by *Chaoborus* had virtually eliminated the crustacean community. Furthermore, several studies of non-acidic systems have

clearly shown the ability of crustaceans to depress rotifer populations (Gilbert, 1985), and have also shown that intense predation on the crustacean component of the zooplankton can greatly increase rotifer abundance (Neill, 1984).

Acknowledgements

I thank T. Haines who provided considerable assistance during the course of this study.

References

Almer, B., W. Dickson, C. Ekström, & E. Hörnström, 1974. Effects of acidification on Swedish lakes. Ambio 3: 30–36.

Brett, M. T., 1989a. The distribution of limnetic macroinvertebrates in acidic lakes of Maine: the Role of fish predation. Aque Fennica (in press).

Brett, M. T., 1989b. Zooplankton communities and acidification processes – a review. Wat. Air Soil Pollut. 44: 387–414.

Brezonik, P. L., T. L. Crisman & R. L. Schulze, 1984. Planktonic communities in Florida softwater lakes of varying pH. Ca. J. Fish. aquat. Sci. 41: 46–56.

Carter, J. C. H., W. D. Taylor, R. Chengalath & D. A. Scruton, 1986. Limnetic zooplankton assemblages in Atlantic Canada with special reference to acidification. Ca. J. Fish. aquat. Sci. 43: 444–456.

Chengalath, R., W. J. Bruce & D. A. Scruton, 1984. Rotifer and crustacean plankton of lakes in insular Newfoundland. Verh. int. Ver. Limnol. 22: 419–430.

Chengalath, R., & G. Mulamoottil, 1975. Littoral rotifera of Ontario – genus Trichocerca. Ca. J. Zool. 53: 1403–1411.

DeCosta, J., A. Janicki, G. Shellito & G. Wilcox, 1983. The effect of phosphorus additions in enclosures on the phytoplankton and zooplankton of an acid lake. Oikos 40: 283–294.

Edmondson, W. T., 1959. Freshwater-biology. John Wiley, New York.

Gilbert, J. J. 1985. Competition between rotifers and *Daphnia*. Ecology 66: 1943–1950.

Hobæk, A. & G. G. Raddum, 1980. Zooplankton communities in acidified regions of south Norway. SNSF-project. IR75/80.

Kerekes, J. J., A. C. Blouin & S. T. Beauchamp, 1988. Trophic response to phosphorus in acidic and non acidic lakes in Nova Scotia, Canada. P. Biro (ed.) Trophic relationships in inland waters symposium, Tihany, Hungary. Junk Publ. Co. (in press).

Kiefer, F. & G. Fryer, 1978. Das zooplankton def binnengewässer. Die binnengewässer. 26 (2).

MacIsaac, H. J., W. Keller, T. C. Hutchinson & N. D. Yan, 1986. Natural changes in the planktonic rotifera of a small acid lake near Sudbury, Ontario following water quality improvements. Wat. Air Soil Pollut. 31: 791–797.

MacIssac, H. J., T. C. Hutchinson & W. Keller, 1987. Analysis of planktonic rotifer assemblages from Sudbury, Ontario, area lakes of varying chemical composition. Ca. J. Fish. aquat. Sci. 44: 1692–1701.

Makarewicz, J. C. & G. E. Likens, 1979. Structure and the function of the zooplankton community of Mirror Lake, New Hampshire. Ecol. Monogr. 49: 109–127.
function of the zooplankton community of Mirror Lake, New Hampshire. Ecol. Monogr. 49: 109–127.

McCaully, E., 1984. The estimation of the abundance and biomass of zooplankton in samples. In Downing, J. A. and Rigler, F. H. (eds.), Secondary productivity in freshwaters. Blackwell Scientific Publications, Oxford. 228–265.

Neill, W. E., 1984. Regulation of rotifer densities by crustacean zooplankton in an oligotrophic montane lake in British Columbia. Oecologia 61: 175–181.

Pejler, B., 1983. Zooplanktic indicators of trophy and their food. Hydrobiologia 101: 111–114.

Pennak, R. W., 1978. Fresh-water invertebrates of the United States. John Wiley, New York.

Pinel-Alloul, B., G. Méthot & G. Codin-Blumer, 1987. Structure spatiale du zooplancton des lacs du Québec: relation avec l'acidité. Nat. can. 114: 295–305.

Roff, J. C. & R. E. Kwiatkowski, 1977. Zooplankton and zoobenthos communities of selected northern Ontario lakes of different acidities. Ca. J. Zool. 55: 899–911.

Ruttner-Kolisko, A., 1974. Plankton rotifers. Biology and taxonomy. Die binnengewässer. 26(1).

Schaffner, W. R. 1989. Effects of neutralization an the addition of brook trout (*Salvelinus fontinalis*) on the limnetic zooplankton communities of two acidic lakes. Ca. J. Fish. aquat. Sci. 46: 295–305.

Schindler, D. W., K. H. Mills, D. F. Malley, D. L. Findlay, J. A. Shearer, I. J. Davies, M. A. Turner, G. A. Linsey & D. R. Cruikshank, 1985. Long-term ecosystem stress: the effects of years of experimental acidification on a small lake. Science (Washington, DC) 228: 1395–1401.

Siegfried, C. A., J. A. Bloomfield & J. W. Sutherland, 1988. The planktonic rotifer community structure in Adirondack lakes in relation to acidity, trophic status, and related water quality characteristics. Hydrobiologia (in press).

Siegfried, C. A., J. W. Sutherland & J. A. Bloomfield, 1987b. Analysis of plankton community structure in Adirondack lakes in relation to acidification. In R. Perry, R. M. Harrison, J. N. B. Bell & J. N. Lester (eds.). Acid Rain: Scientific and Technical Advances. Selper Ltd., London. 445–450.

Siegfried, C. A., J. W. Sutherland, S. O. Quinn & J. A. Bloomfield, 1984. Lake acidification and the biology of Adirondack lakes: I. Rotifer communities. Verh. int. Ver. Limnol. 22: 549–558.

Stemberger, R. S., 1979. A guide to rotifers of the Laurentian Great Lakes. US EPA, Report EPA 600/4-79-021.

Yan, N. D. & W. Geiling, 1985. Elevated planktonic rotifer biomass in acidified metal-contaminated lakes near Sudbury, Ontario. Hydrobiologia 120: 199–205.

Yan, N. D. & C. Lafrance, 1984. Responses of acidic and neutralized lakes near Sudbury, Ontario, to nutrient enrichment. In J. Nriagu (ed.) Environmental impact of smelters. John Wiley and Sons, INC., New York. 457–521.

Yan, N. D., C. J. Lafrance & G. G. Hitchin, 1982. Planktonic fluctuations in a fertilized, acidic lake: the role of invertebrate predators. In R. E. Johnson (ed.) Acid Rain/Fisheries. Proceedings of a symposium on acidic rain and fishery impacts in northeastern North America. 137–154.

Hydrobiologia **186/187**: 191–200, 1989.
C. Ricci, T. W. Snell and C. E. King (eds), Rotifer Symposium V.
© 1989 *Kluwer Academic Publishers.*

Composition and distributional patterns in arctic rotifers

R. Chengalath & W. Koste

National Museum of Natural Sciences, P.O. Box 3443, Station 'D', Ottawa, Ontario, Canada K1P 6P4;
D 4570 Quakenbruck, Ludwig-Brill-Strsse 5, Federal Republic of Germany

Key words: arctic North America, rotifera, distribution, community composition

Abstract

Based on collections of rotifers from 212 localities in arctic North America, the patterns of distribution and composition are evaluated. An attempt is made to discern the dominant components of the rotifer community in arctic habitats. One hundred and sixty five species of rotifers are reported, three of which represent new records for North America. With increasing latitude and decreasing summer temperatures, a decline in species richness and change in species composition is observed. Some rotifers that previously were not adequately described are redescribed and illustrated. The significance of the dispersal capacity of rotifers in the arctic is discussed.

Introduction

Studies on the rotifers of arctic North America have been very few (Ahlstrom, 1940, 1943; Chengalath, 1978; Chengalath & Koste, 1987; Harring, 1921; Hooper, 1947; Moore, 1978.) and were generally limited to restricted areas. There are forty-eight publications, mainly dealing with fish and general zooplankton studies, that list rotifers from the North American arctic (Chengalath, 1984). In most of these studies rotifers were identified only to the generic level and no morphological or ecological data were provided. Comprehensive biogeographic studies are necessary for complete systematic and ecological understanding of any animal group. This paper is part of an ongoing investigation to examine extensively the occurrence, distribution, and species diversity of the Rotifera of Canada. We will also attempt to discern the dominant components of the rotifer communities in North American arctic, to redescribe and illustrate some species that previously were not adequately described and to assess the significance of dispersal capacity as well as overall distribution patterns.

Materials and methods

Qualitative sampling of lakes and ponds throughout Alaska (USA), and Yukon Territory, Northwest Territories, and Bathurst Island (Canadian arctic archipelago) were carried out during summer field expeditions. A total of 212 samples have so far been examined and are included in this study. The Bathurst Island samples were collected in 1974 and the rest in the summer of 1986 and 1987. At each of the five arctic regions (See Map, Table 1, 2), from 22 to 84 different aquatic habitats were sampled with a plankton net with a 64 μm aperture mesh. The waterbodies in the main North American land mass under

192

Map: The study area showing all the localities sampled.

Table 1. Distributional records for sub-arctic and arctic rotifers. (SYN = southern Yukon and Northwest Territories, SA = southern Alaska, NYN = northern Yukon and Northwest Territories, NA = northern Alaska, BI = Bathurst Island, Canadian arctic archipelago).

Rotifer taxon	SYN	SA	NYN	NA	BI
Ascomorpha ecaudis Perty	×	×	×		
A. saltans Bartsch	×				
Aspelta aper (Harring)	×	×		×	
A. circinator (Gosse)		×	×		
Asplanchna priodonta Gosse	×	×	×		
Asplanchnopus multiceps (Schrank)	×				
Bipalpus hudsoni (Imhoff)	×				
Cephalodella catellina (Muller)	×			×	
C. forficata (Ehrenberg)		×		×	
C. forficula (Ehrenberg)		×		×	
C. gibba (Ehrenberg)	×	×	×	×	×
C. globata (Gosse)	×				
C. gracilis (Ehrenberg)				×	
C. megalocephala (Glasscott)	×			×	
C. sterea (Gosse)	×		×	×	
C. ventripes (Dixon-Nuttall)	×	×		×	
Collotheca companulata (Dobie)	×				
C. edentata (Collins)	×		×	×	
C. libera (Zacharias)				×	
C. mutabilis (Hudson)	×	×			
C. ornata (Ehrenberg)	×			×	×
C. pelagica (Rousselet)	×				
Colurella adriatica Ehrenberg		×		×	
C. colurus (Ehrenberg)	×				
C. obtusa (Gosse)	×	×	×	×	
C. uncinata (Muller)	×	×	×	×	
C. dossuarius (Hudson)	×	×	×		
C. natans (Seligo)	×	×			
Conochilus hippocrepis (Schrank)	×	×	×	×	
C. unicornis (Rousselet)	×	×	×	×	
Dicranophorus epicharis Harring & Myers	×			×	
D. forcipatus (Muller)	×		×		
D. lutkeni (Bergendal)		×			
D. robustus Harring & Myers				×	
D. rostratus (Dixon-Nuttall & Freeman)				×	
Dipleuchlanis propatula (Gosse)	×				
Encentrum grande (Western)					×

Rotifer taxon	SYN	SA	NYN	NA	BI
Eothinia elongata (Ehrenberg)	×		×		
Epiphanes senta (Muller)			×		
Euchlanis alata Voronkov				×	
E. deflexa Gosse		×	×		
E. dilatata Ehrenberg	×	×	×	×	×
E. incisa Carlin	×	×	×	×	
E. lyra Hudson	×	×	×	×	
E. meneta Myers	×	×	×	×	×
E. oropha Gosse	×	×			
E. triquetra Ehrenberg	×	×	×	×	
Filinia longiseta Ehrenberg				×	
F. terminalis (Plate)	×		×		
Floscualria melicerta Ehrenberg			×		
F. ringens Linnaeus	×	×			
Gastropus stylifer Imhoff	×		×	×	
Kellicottia longispina (Kellicott)	×	×	×	×	
Keratella cochlearis (Gosse)	×		×	×	
K. faluta Ahlstrom	×				
K. hiemalis Carlin	×		×		×
K. quadrata (Muller)	×		×		
K. recurvispina (Jagerskiold)	×	×	×		
K. serrulata (Ehrenberg)	×	×	×		
Lecane acus (Harring)	×	×	×		
L. bulla (Gosse)	×	×	×	×	
L. closterocerca (Schmarda)	×	×	×	×	
L. cornuta (Muller)	×		×		
L. flexilis (Gosse)	×	×	×	×	×
L. furcata (Murray)	×	×		×	
L. hamata (Stokes)	×	×	×		
L. lauterborni Hauer			×		
L. levistyla (Olofsson)	×	×			
L. ligona (Dunlop)	×	×	×		
L. luna (Muller)	×	×	×	×	
L. lunaris (Ehrenberg)	×	×	×	×	×
L. mira (Murray)	×	×	×	×	
L. perplexa Ahlstrom		×	×	×	
L. pyriformis (Daday)			×		
L. quadridentata (Ehrenberg)	×				
L. rotundata (Olofsson)					×
L. scutata (Harring & Myers)	×		×	×	
L. stichaea Harring	×	×	×	×	
L. stokesi (Pell)	×	×	×		

Table 1. (continued).

Rotifer taxon	Region				
	SYN	SA	NYN	NA	BI
L. ungulata (Gosse)	×		×	×	
Lepadella acuminata (Ehrenberg)	×	×	×	×	×
L. ovalis (Muller)	×	×	×	×	×
L. patella (Muller)	×	×	×	×	×
L. quadricarinata (Stenroos)	×	×	×		
L. rhomboides (Gosse)	×				
L. triptera Ehrenberg	×	×		×	
Limnias melicerta Weiss	×				
Lophcharis salpina (Ehrenberg)	×	×	×	×	×
L. oxysternon (Gosse)	×			×	
Microcodon clavus Ehrenberg		×		×	
Monommata aeschyna Myers				×	
M. caudata Myers				×	
M. dentata Wulfert				×	
M. grandis Tessin	×	×			
M. phoxa Myers	×				
Mytilina bicarinata (Perty)					×
M. mucronata (Muller)	×	×	×	×	×
M. ventralis (Ehrenberg)	×	×	×	×	×
Notholca acuminata (Ehrenberg)	×	×	×		
N. foliacea (Ehrenberg)	×	×		×	
N. labis Gosse	×				
N. lapponica (Ruttner-Kolisko)	×	×			
N. squamula (Muller)	×	×	×	×	×
Notommata allantois Wulfert				×	
C. cerberus (Gosse	×		×		
N. copeus Ehrenberg	×	×	×	×	
N. cyrtopus (Gosse)	×		×		
N. doneta Harring & Myers		×		×	
N. glyphura Wulfert	×	×	×	×	
N. pachyura (Gosse)		×		×	
Playias qudrocornis (Ehrenberg)	×			×	
Pleurotrocha petromyzon Ehrenberg	×	×	×	×	
Ploesoma triacanthum (Bergendal)	×	×	×		
P. truncatum (Levander)	×	×	×	×	
Polyarthra dolichoptera Idelson	×	×	×	×	
P. euryptera Wierzejski	×				

Table 1. (continued).

Rotifer taxon	Region				
	SYN	SA	NYN	NA	BI
P. loniremis Carlin	×				
P. major Burckhardt	×		×	×	
P. minor Voigt		×		×	
P. remata Skorikov				×	
P. vulgaris Carlin	×	×			
Proales decipiens (Ehrenberg)	×				
P. fallacioca Wulfert	×	×		×	
P. theodora (Gosse)		×		×	
Ptygura brachiata (Hudson)		×			
P. crystllina (Ehrenberg)	×				
P. mucicola (Kellicott)				×	
Resticula melandocus (Gosse)			×		
Scaridium longicaudum (Muller)	×	×		×	
Squatinella rostrum (Schmarda)			×		
Stephanoceros fimbriatus (Goldfusz)			×		
Synchaeta grandis Zacharias	×	×			
S. longipes Gosse	×	×			
S. oblonga Ehrenberg			×	×	
S. pectinata Ehrenberg		×			
S. tremula (Muller)				×	
Taphrocampa selenura Gosse	×				
Testudinella emarginula (Stenroos)	×		×	×	
T. mucronata (Gosse)		×			
T. naumanni Carlin	×				
T. parva (Ternetz)	×	×	×	×	
T. patina (Hermann)	×	×	×	×	
Tetrasiphon hydrocora Ehrenberg	×	×	×		
Trichocerca bidens (Lucks)	×	×			
T. bicristata (Gosse)	×	×			
T. brachyura (Gosse)					×
T. cavia (Gosse)		×	×		×
T. collaris (Rousselet)		×			
T. elongata (Gosse)	×	×		×	
T. insignis (Herrick)	×	×			
T. intermedia (Stenroos)			×		
T. longiseta (Schrank)	×	×	×	×	
T. porcellus (Gosse)	×	×		×	
T. pusilla (Lauterborn)		×		×	
T. rattus (Muller)	×	×	×	×	
T. rosea (Stenroos)		×	×		
T. rousseleti (Voigt)	×				

Table 1. (continued).

Rotifer taxon	Region				
	SYN	SA	NYN	NA	BI
T. similis (Wierzejski)	×				
T. tenuoir (Gosse)	×	×		×	
T. tigris (Muller)			×		
T. uncinata (Voigt)				×	
T. weberi (Jennings)	×				
Trichotria pocillum (Muller)	×	×	×	×	
T. tetractis (Ehrenberg)	×	×	×	×	×
Wolga spinifera (Western)	×				

study vary from small, highly coloured, ponds that have a low content of electrolytes and profuse vegetation to alpine lakes surrounded by snow with clear blue waters and sandy bottom with little or no vegetation. The high arctic site, Bathurst Island, is one of the Parry group of islands situated in the central part of the Canadian arctic archipelago (see Map). The island has rather low elevation and is composed pre-dominantly of sedimentary rocks (Blake, 1964). The samples were collected from within a radius of three miles of a field station at 75° 43′ N, 98° 25′ W. The ponds sampled in this region were essentially shallow basins, with a diameter of 30 m, a maximum depth of 25 cm, and with a soft muddy bottom.

Results and discussion

A list of 165 rotifers found during this study is given in Table 1, including the three species that are new for North America. Except for the three common species of rotifers, *Keratella cochlearis*, *Keratella quadrata*, and *Kellicottia longispina*, all the other 110 species are recorded from Alaska for the first time. Bdelloid rotifers were present in most samples but could not be positively identified to species level and are, therefore, not included in this study. The total number of species encountered in the sample varied at each site from 0-22, with the majority of the samples containing 8-12 species. Twenty-one samples contained no rotifers at all. The rotifer community in the high arctic sites of Bathurst Island is composed of only 21 species (Tables 1, 2), and a high percentage of ponds lacked rotifers entirely. These ponds are characterised by extremely low allochthonous nutrient input and lower conductivity. However, at a number of low arctic sites in Alaska, Yukon, and Northwest Territories, ponds with low conductivity contained a rich rotifer fauna, indicating that lack of rotifers may be correlated more strongly with the lack of available nutrients for phytoplankton growth.

The size of the species pool gradually increased from the central arctic island southwards (Table 2), with the most dramatic change in the assemblage occurring upon moving inland onto continental North America. The species encountered in this study are generally widely distributed inhabitants of ponds and littoral zone of lakes (Koste, 1978). However, the occurrence of several

Table 2. Numbers of rotifer species inhabiting designated regions in North American arctic.

Region	N. lat.	W. long.	No. of localities sampled	Loricate rotifers	Illoricate rotifers
Southern Yukon & NWT (SYN)	60° 00′–66° 33′	112° 12′–142° 08′	84	88	30
Southern Alaska (SA)	61° 30′–66° 33′	112° 15′–152° 00′	30	66	23
Northern Yukon & NWT (NYN)	66° 33′–69° 10′	122° 10′–132° 15′	22	63	18
Northern Alaska (NA)	66° 33′–70° 30′	150° 00′–151° 10′	38	66	21
Bathurst Island, Canadian arctic archipelago, (BI)	75° 43′	98° 25′	38	19	2

species represents an extension of their known range. Members of the genera *Notholca*, *Mytilina*, *Lepadella*, *Lecane*, and *Trichotria* were the most widely distributed and present in all five regions (Table 1). Distribution pattern for individual species can provide useful qualitative data; however no attempt was made during this study to define the distribution of each species as a function of any one or combination of limnological parameters.

Three little known species of rotifers, *Encentrum grande*, *Lecane rotundata*, and *Mytilina bicarinata*, were found in samples from the Canadian arctic archipelago, and represent new records for North America. *Encentrum grande* (Figs. 1-5) was first recorded from the arctic by Murray (1908) from Spitzbergen living in moss. This species is mostly known from Europe where it is found in the psammon of shallow ponds and

ditches or living as epibionts of *Asellus aquaticus* and *Gammarus* sp. (Western, 1891; Koste, 1976). Specimens are usually contracted in preserved samples to about half their normal size and seldom reveal the coronal structure or the posterior end showing the toes. The trophi (Figs. 2-5) however is very characteristic and is seen invariably through the body. Another interesting rotifer, *Mytilina bicarinata*, (Figs. 10-12) has previously been reported from temporary waterbodies and from among vegetation in smaller ponds mainly from northern Europe (Olofsson, 1918; Pejler, 1962). *Lecane rotundata* (Fig. 6) was found in two samples from Bathurst Island and has hitherto been reported only from coastal waters of Novaja Semlja, Spitzbergen, and Swedish Lapland (Idelson, 1926; Olofsson, 1918; Pejler, 1962). The lorica of this species is almost spherical in outline and could easily be

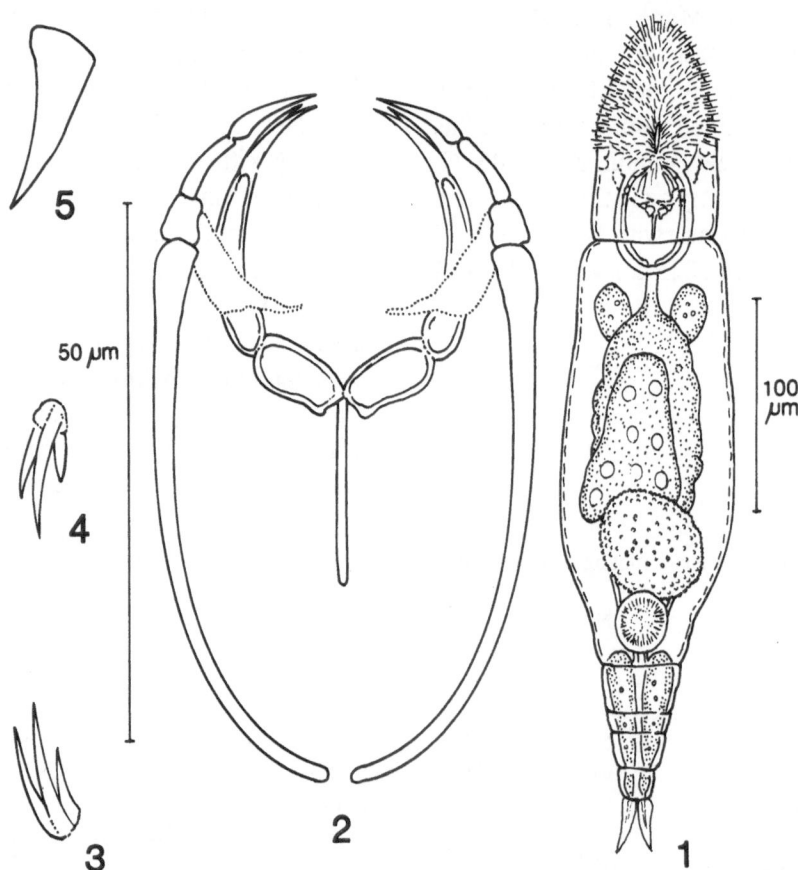

Figs. 1-5. Encentrum grande. 1. Ventral view. 2. Trophi. 3&4. Teeth on unci and rami, apical view. 5. Fulcrum, lateral view.

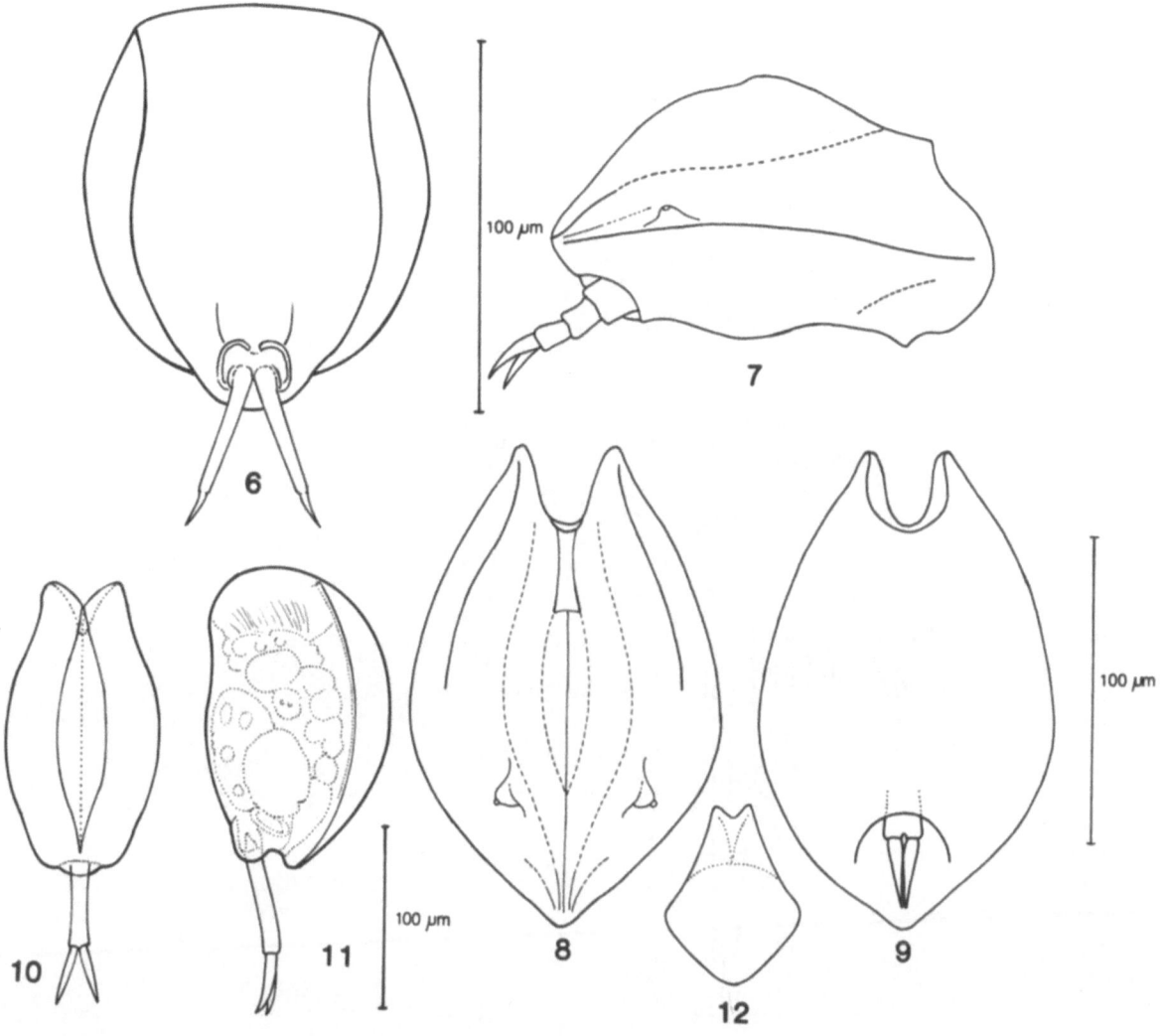

Figs. 6-12. 6. *Lecane rotundata*, ventral view. 7. *Lophocharis oxysternon*, lateral view. 8. *L. oxysternon*, dorsal view. 9. *L. oxysternon*, ventral view. 10. *Mytilina bicarinata*, dorsal view. 11. *M. bicarinata*, lateral view. 12. *M. bicarinata*, cross-section of lorica.

mistaken for a number of other *Lecane* species under low magnifications.

Polyarthra minor and *Lophocharis oxysternon* were found fairly commonly in northern Alaskan ponds. Both these species have been reported from various parts of the world (Bartos, 1951; Koste, 1978; Ruttner-Kolisko, 1974). Although *Polyarthra minor* is considered to be a rare acid water form, during the present study it was found in alkaline ponds with low conductivity. The morphology of this species has long been neglected mainly because of its very small size and transparency which make it very difficult to discern all the parts clearly. It is readily recognised by its asymmetry, the left dorsal appendages being almost twice as long and a little wider than the rest. Also, one of the appendages in the right ventral bundle is broader than the other two (Fig. 13). All the appendages have clearly visible mid-ribs and serrated edges. The more or less broadly oval body seldom exceeds a size of 90 μm. The vitellarium contains four distinct nuclei (Fig. 13). *Lophocharis oxysternon* is mentioned in the literature as a variable species (Bartos, 1959; Gosse, 1851; Koste, 1978; Kutikova, 1970) but the illustrations provided were far from satis-

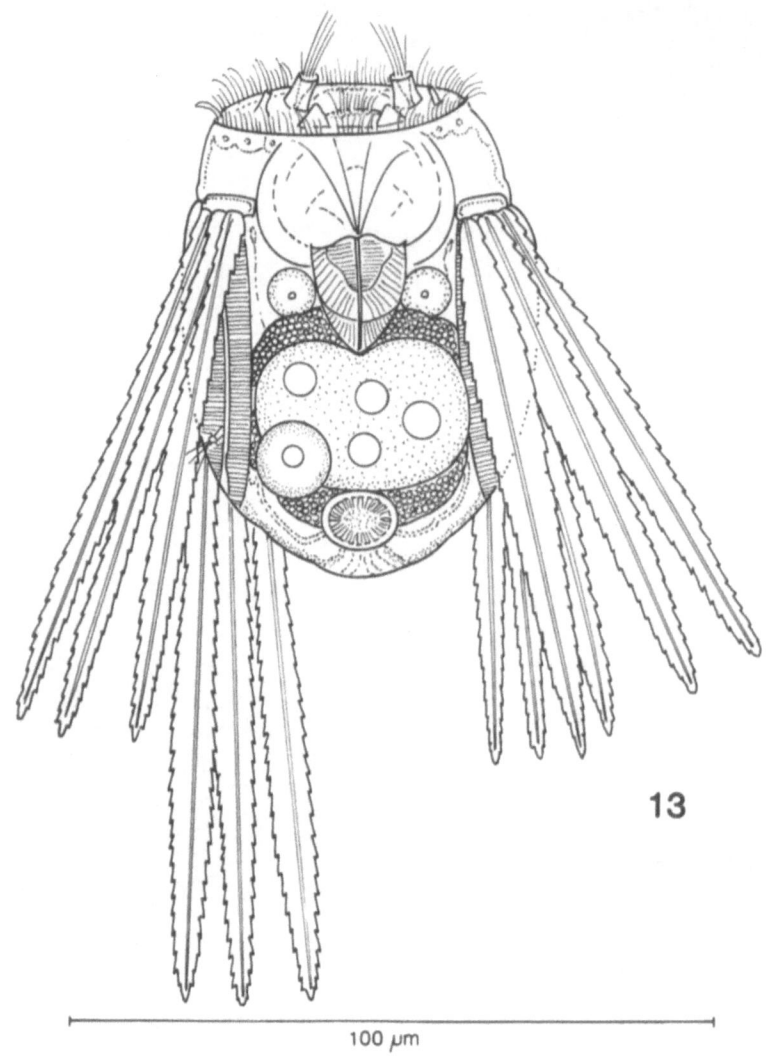

100 µm

Fig. 13. Polyarthra minor, ventral view.

factory. The lorica of this animal is oval shaped with distinct ridges along the dorsal side (Figs. 7-9). It is also distinguished from other members of this genus by its characteristic lateral view.

Species of the genera *Notholca* and *Mytilina* predominated the high arctic rotifer fauna, and the body sizes of the animals found during this study were much larger than the previously recorded forms. There are references to giant forms of rotifers in the literature (Chengalath & Koste, 1983; Koste & Shiel, 1980). The reasons for the occurrence of such rotifers may be related to life history patterns, favourable environmental conditions, or absence of predators. Several morphs of *Notholca acuminata* and *Notholca squamula* were found in the high arctic localities. Many of these forms are also found in the low arctic and have been described previously (Chengalath & Koste, 1987). In several samples in the arctic two or three morphs of a species with varying body sizes were present. For example, *Notholca acuminata* and *N. squamula* ranged from 180 µm to 383 µm in body size. The significance of such size differences in rotifers is not fully understood. Water temperatures are usually cited as a factor for variation in sizes. However, the body size variation among conspecifics cannot be attributed to temperature. Patterns of size separation are found to be common in copepod communities and it has been suggested that such separation minimises competition for resources among species (Cole, 1961; Hutchinson, 1967). A number of studies on calanoid copepods have indicated that food abundance is the primary factor determining their body sizes (Deevey, 1960; Harris & Paffenhoffer, 1976). It has also been shown that copepod body sizes increased greatly when population densities were reduced (Hebert, 1985). Perhaps these are also contributing factors to the increased body size of arctic rotifers.

Some patterns of rotifer community composition in the arctic emerge from this study. The waterbodies in the arctic archipelago support a very depauperate assemblage of species compared with habitats located to the south (Table 1, 2). The severity of the physical environment, low temperature and low primary production are clearly factors limiting size and composition of rotifer community in the high arctic habitats. In comparison to other regions, the Alaskan habitats have a high diversity of rotifers that may be attributable to the fact that much of the area escaped glaciation during the pleistocene (Hopkins, *et al.*, 1982). The possibility exists that the diversity patterns of rotifers reflect the effects of survival in glacial refugia. Most long distance movements of zooplankton is probably mediated through birds and opportunities for such dispersal are undoubtedly greater along the north-south axis. The high arctic localities in Bathurst Island have been colonised by several loricate rotifers, most notably by members of the genera *Notholca*, *Mytilina*, *Trichotria*, *Lecane* and *Lepadella*. However, other groups, especially illoricate rotifers, are conspicuous in their scarcity. This may suggest a very slow dispersal of some rotifers from areas in which they survive. Differences in life cycles among various rotifers may be involved in their differential colonisation success in the arctic. Most cladocerans require a minimum of two generations in order to produce resting eggs, although some species have become obligate par-

thenogens producing ephippial eggs asexually. While physiochemical factors may play a role in preventing successful colonisation in high arctic regions, dispersal difficulties including failure to produce resting eggs may be the controlling factor. Too little is known about the production, morphology and dispersal of rotifer resting eggs for any definite conclusions to be drawn. Although passive dispersal of resting eggs and ephippial eggs via birds and wind has been documented (Maguire, 1963), the dispersal capability and opportunity may be responsible for distributional patterns among animals. However, dispersal history alone does not restrict species distributions. Survival in a new habitat involves other factors such as avoidance of predators, availability of food, completion of the life cycle within the short favourable season including the production of resting eggs.

Acknowledgements

We thank Colin Eades, Chief, Zoology Division, National Museum of Natural Sciences, for his support and encouragement. Thanks are also due to Chang-tai Shih for critically reading the manuscript. The help of Maria Wing in assisting with the manuscript and that of Brett Purdy, Jerry Chemilowski, and Bonnie Smith, all of the Department of Biology, University of Calgary, in collecting the samples from Yukon, Northwest Territories, and Alaska is gratefully acknowledged. Ian Sutherland, Canadian Heritage Information Network, Department of Communications, collected the samples from Bathurst Island.

References

Ahlstrom, E. H., 1940. A revision of the rotatorian genera *Brachionus* and *Platyias* with descriptions of one new species and two new varieties. Bull. Am. Mus. nat. Hist. 77: 148–184.

Ahlstrom, E. H., 1943. A revision of the rotatorian genus *Keratella* with descriptions of three new species and five new varieties. Bull. am. Mus. nat. Hist. 80: 411–457.

Bartos, E., 1951. Klic k urcovani virniku rodu *Palyarthra* Ehrenberg. Csaop Narodn. musea. Odd. prirodved. 118–119: 82–91.

200

Bartos, E., 1959. Virnici- Rotatoria. Fauna CSR, Praha. 15: 1–969.

Blake Jr., W., 1964. Preliminary account of the glacial history of Bathurst Island, Arctic Archipelago. Geol. Surv. Can. Publ. 64–30: 1–8.

Chengalath, R., 1978. A new species of the genus Notholca Gosse, 1886 (Brachionidae: Rotifera), from Great Slave Lake, N.W.T. Ca. J. Zool. 56: 363–364.

Chengalath, R., 1984. Synopsis Speciorum: Rotifera. Bibliographia Invertebratorum Aquaticorum Canadensium 3. National Museum of Natural Sciences, National Museums of Canada, Ottawa. 102 pp.

Chengalath, R. & W. Koste, 1983. Rotifera from northeastern Quebec, Newfoundland and Labrador, Canada. Hydrobiologia 104: 49–56.

Chengalath, R. & W. Koste, 1987. Rotifera from northwestern Canada. Hydrobiologia 147: 49–56.

Cole, G. A., 1961. Some calanoid copepods from Arizona with notes on congeneric occurrences of Diaptomus species. Limnol. Oceanogr. 6: 432–442.

Deevey, G. B., 1960. Relative effects of temperature and food on seasonal variations in length of marine copepods in some eastern American and western European waters. Bull. Bingham Oceanogr. Collect. 17: 54–85.

Gosse, P. H., 1851. A catalogue of Rotifera found in Britain with descriptions of five new genera and thirty-two new species. Ann. Mag. nat. Hist. 2: 197–203.

Harring, H. K., 1921. The Rotifera of the Canadian Arctic expedition, 1913-1918. Rep. Can. Arct. Exped. 1913-1918, 8E: 1–23.

Harris, R. P. & G. A. Paffenhoffer, 1976. Feeding, growth and reproduction of the marine copepod *Temora longicornis* Muller. J. mar. biol. Ass. U.K. 56: 675–690.

Hebert, P. D. N., 1985. Ecology of the dominant copepod species at a Low Arctic site. Ca. J. Zool. 63: 1138–1147.

Hooper, F. F., 1947. Plankton collections from the Yukon and Mackenzie river systems. Trans. am. micros. Soc. 66: 74–84.

Hopkins, D. M., J. V. Mathews, Jr., C. E. Schweger & S. B. Young (Ed.) 1982. Paleoecology of Beringia. Academic Press. New York. NY.

Hutchinson, G. E., 1967. A treatise on limnology. Vol. II. Introduction to lake biology and the limnoplankton. John Wiley & sons, New York.

Idelson, M. S., 1926. Zur Erforschung der Rotatorianfauna der Gewasser auf Nowaja Semlja. Ber. Wissenschaft. Meeresinstitut. Moskau. 12: 77–99 (in Russian).

Koste, W., 1976. Uber die Radertierbestande. Rotatoria der oberen und mittleren Hase in den Jahren 1966-1969. Osnabrucker Naturwiss. Mitt. 4: 191–263.

Koste, W., 1978. Die Radertiere Mitteleuropas. Oberordnung Monogononta. Borntraeger, Stuttgart-Berlin, 2 Vols., 673 pp., 234 plates.

Koste, W. & R. J. Shiel, 1980. Preliminary remarks on the characteristics of the rotifer fauna of Australia (Notogea). Hydrobiologia 73: 221–227.

Kutikova, L. A., 1970. Rotifer fauna SSSR. Subclass Eurotatoria. Nauka, Leningrad. 742 pp. (In Russian).

Maguire, B., 1963. The passive dispersal of small aquatic organisms and their colonization of isolated bodies of water. Ecol. Monogr. 33: 161–185.

Moore, J. W., 1978. Composition and structure of zooplankton communities in eighteen arctic and subarctic lakes. Int. Revue ges. Hydrobiol. 63: 545–565.

Murray, J., 1908. Arctic rotifers collected by Dr. W. Bruce. Proc. Roy. Philosophical Soc. Edinburgh. 17: 121–127.

Olofsson, O., 1918. Studien uber dei susswasser fauna Spitzbergens. Beitr. z. Systematik, Biol. u. Tiergeogr. der Crustaceen u. Rotatorien. Zool. Bidr. Uppsala 6: 183–648.

Pejler, B., 1962. On the taxonomy and ecology of benthic and periphytic Rotatoria (Lapland). Zool. Bidr. Uppsala. 33: 327–422.

Ruttner-Kolisko, A., 1974. Plankton rotifers, biology and taxonomy. Suppl. Die Binnengewasser 26: 1–164.

Western, G., 1891. Notes on rotifers. J. Queckett Micorsc. Club ser. 2: 320–322.

Hydrobiologia **186/187**: 201–208, 1989.
C. Ricci, T. W. Snell and C. E. King (eds), Rotifer Symposium V.
© *1989 Kluwer Academic Publishers.*

Abundance and diversity of planktonic rotifers in the Po River

I. Ferrari[1], A. Farabegoli[1] & R. Mazzoni[2]
[1] *Istituto di Zoologia, Università, 44100 Ferrara, Italy*; [2] *Istituto di Ecologia, Università, 43100 Parma, Italy*

Key words: Po River, zooplankton, rotifers, diversity

Abstract

Zooplankton samples from the middle reach of the Po River were collected daily from 27 July to 24 August 1988 from a station located near Viadana. Changes in the biocoenosis structure were analyzed in relation to variations in flow rate. Rotifers accounted for more than 99% of the total zooplankton (protozoans excluded) in every sample. The dominant species were *Brachionus calyciflorus*, *Brachionus bennini*, *Brachionus budapestinensis* and *Epiphanes macrourus*. Under scanty flow conditions, the taxocoenosis showed marked stability. An increase in flow rate acts as a disturbance factor leading to a significant decrease in both total density and dominance.

Introduction

Considerable research has been carried out on zooplankton composition and variability in rivers of Europe and other continents (Williams, 1966; Hynes, 1970; Mordukhai-Boltovskoi, 1979; Lair, 1980; Pourriot *et al.*, 1982; Shiel, 1985; Davies & Walker, 1986; Guisande & Toja, 1988). Plankton studies of the Po River since the 1970's have chiefly involved the analysis of zooplankton samples collected with seasonal or monthly frequency in the delta (Ferrari *et al.*, 1983) and in the middle reach of the river (Ferrari & Benassi, 1984). These studies demonstrated that during the low flow summer periods the biocoenosis is characterized by very high densities and is dominated by rotifers.

Ferrari *et al.* (1987) examined a series of samples collected every other day during July-August 1985 from a station near Viadana (Fig. 1). A close relationship was found between zooplankton structure, hydrometric parameters and trophic state of the water. There was also a definite tendency of the biocoenosis to maintain its structure under reduced flow conditions and to reconstitute after disturbance by floods.

Studies on zooplankton variability were continued in July-August 1988 with another short-term sampling series at the same station near Viadana. Changes in zooplankton composition and density during this period were analyzed in relation to river flow variations. Particular attention was paid to trends of the rotifer taxocoenosis and results were compared to those obtained in 1985.

Materials and methods

Zooplankton samples were collected daily or every other day from 27 July to 24 August 1988 from a station near Viadana. Sampling was carried out from the surface to a depth of 1 m with a 15 l Patalas trap. Sixty l of water were taken for each of the 26 samples collected and filtered through a 50 μm mesh net. Daily data of river flow rate recorded at Viadana were kindly supplied by the Ufficio Idrografico del Po, Parma.

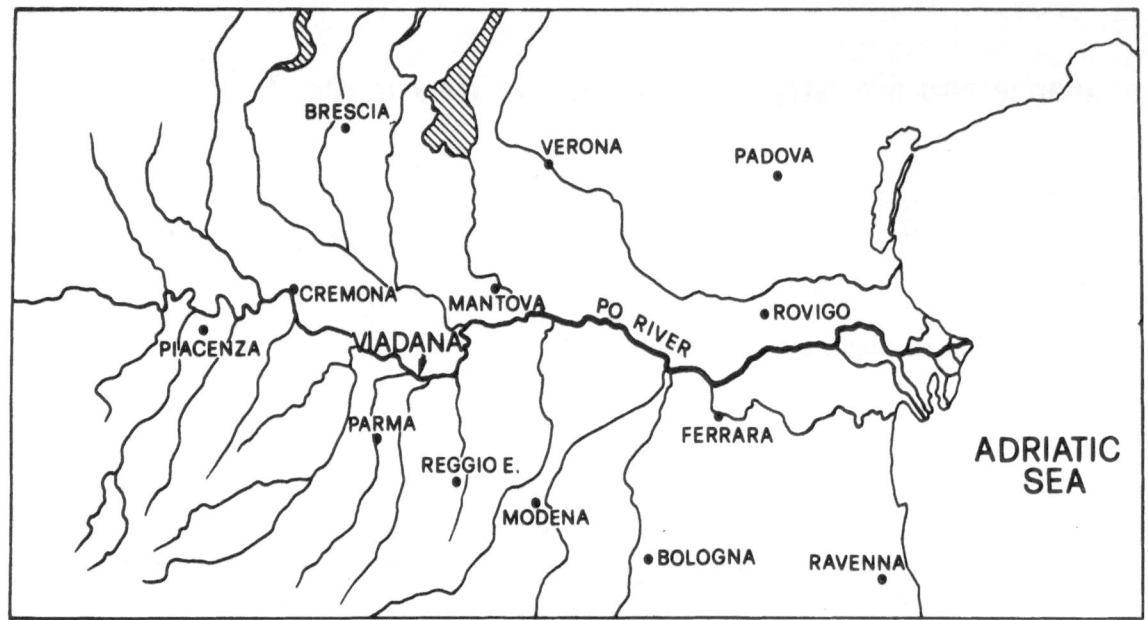

Fig. 1. Po River basin: the Viadana sampling station is indicated.

Table 1. List of zooplankton species found in the Po River at Viadana.

Ciliophora	*Dinamoeba mirabilis*	*L. luna*	Bdelloidea unidentif.
Actinobolina radians		*L. (Monostyla) bulla*	Rotifera unidentif.
Cinetochilum margaritaceum	Heliozoea	*L. (M.) closterocerca*	
Codonella cratera	*Acanthocystis* sp.	*L. (M.) hamata*	Cladocera
Coleps incurvus	*Actinosphaerium eichhorni*	*L. (M.) lunaris*	*Bosmina longirostris*
Cyclidium glaucoma		*L. (M.) quadridentata*	*Scapholeberis* sp.
Dileptus falciformis	Rotifera	*L. (Monostyla)* sp.	*Moina micrura*
Frontonia leucas	*Epiphanes macrourus*	*Scaridium longicaudum*	*Ilyocryptus agilis*
Mesodinium pulex	*E. brachionus spinosus*	*Monommata* sp.	*Macrothrix* sp.
Metacineta sp.	*Brachionus angularis*	*Cephalodella catellina*	*Pleuroxus* sp.
Nassula tumida	*B. bennini*	*Cephalodella* spp.	*P. aduncus*
Opercularia sp.	*B. budapestinensis*	*Trichocerca capucina*	*Chydorus sphaericus*
Paracineta möbiusi	*B. calyciflorus*	*T. elongata*	*Leydigia leydigi*
Phascodolon vorticella	*B. diversicornis*	*T. pusilla*	*L. acanthocercoides*
Plagiocampa rouxi	*B. falcatus*	*T. similis*	*Alona quadrangularis*
Podophrya sp.	*B. patulus*	*Trichocerca* sp.	*A. rectangula*
Stentor polymorphus	*B. quadridentatus*	*Synchaeta* spp.	*A. costata*
Strobilidium sp.	*B. urceolaris*	*Polyarthra dolich.-vulgaris*	
Tintinnidium fluviatile	*Keratella cochlearis*	*P. major-euryptera*	Copepoda
Tokophrya quadripartita	*K. cochlearis tecta*	*Asplanchna brightwelli*	Nauplii
Trachelophyllum pusillum	*K. quadrata*	*A. priodonta*	Harpacticoida copepodites
Urocentrum turbo	*K. tropica*	*Encentrum* sp.	Cyclopoida copepodites
Uronema nigricans	*Anuraeopsis fissa*	*Pompholyx* sp.	*Macrocyclops albidus*
Vorticella campanula	*Euchlanis* sp.	Flosculariacea unifentif.	*Paracyclops fimbriatus*
V. mayeri	*Mytilina ventralis*	*Testudinella patina*	*Acanthocyclops* sp.
Zoothamnium sp.	*Trichotria pocillum*	*Hexarthra mira*	*Cryptocyclops* sp.
	Colurella sp.	*Filinia cornuta-brachiata*	*Thermocyclops crassus*
Sarcomastigophora	*Lepadella* sp.	*F. longiseta*	
Rhizopoda	*Lecane elsa*	*F. opoliensis*	
Arcella sp.	*L. flexilis*	*Collotheca* sp.	

Identification and counting was completed for rotifers, cladocerans and copepods. Although abundant in a number of samples, protozoans were not counted. Other taxa, especially including drift fauna organisms (e.g. nematodes, oligochaetes, chironomid larvae), present with low abundances in some samples, were not taken into account.

The rotifer taxocoenosis was analyzed by computing the Shannon-Wiener diversity index for each sample. Density data of the rotifer taxa were processed by Principal Component Analysis (PCA) in order to obtain a coincise representation of both the association pattern among species and the degree of similarity among samples.

Results and discussion

During the first ten days of the sampling period, river flow rate was decreasing. On 27 July it was $526 \, m^3 \cdot s^{-1}$ and from 2 to 4 August it was $368 \, m^3 \cdot s^{-1}$. During the next few days, flow rate increased sharply to a maximum of $1150 \, m^3 \cdot s^{-1}$ on 7 August. It then gradually decreased to $394 \, m^3 \cdot s^{-1}$ on 20 August. During the final four

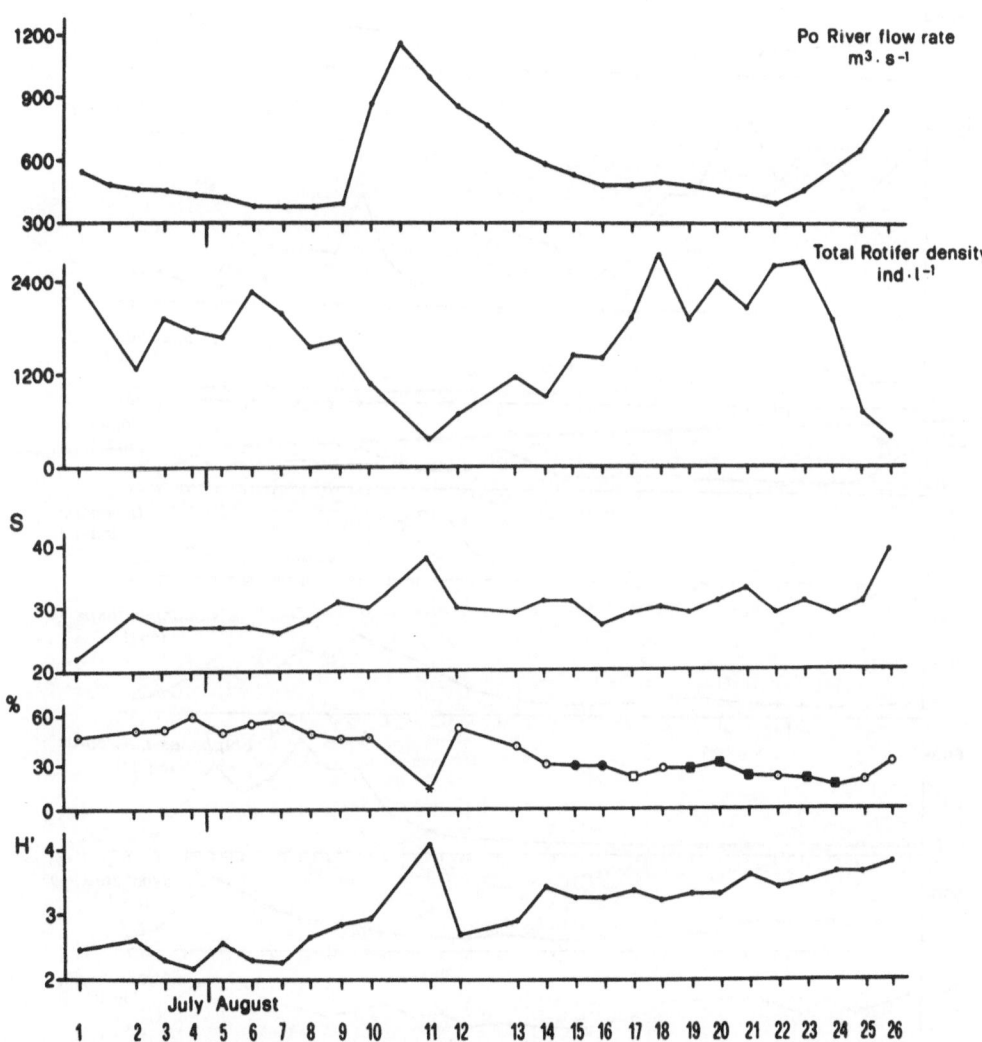

Fig. 2. Po River flow rate at Viadana and parameters of rotifer community structure: total density, number of species (S), percent incidence of the most abundant species in each sample (o *Brachionus calyciflorus*, ● *Brachionus bennini*, □ *Brachionus budapestinensis*, ■ *Epiphanes macrourus*, * *Keratella cochlearis tecta*) and Shannon-Wiener diversity index (H').

204

days of the sampling period flow rate again increased with a peak of 827 m³·s⁻¹ on 24 August.

Flow rate was, on average, lower than that recorded during the corresponding sampling period (23 July-30 August) in 1985 (Ferrari *et al.*, 1987). This is confirmed by flow rates recorded at Pontelagoscuro, a permanent station located in the terminal reach of the Po River, near the city of Ferrara, with recordings dating back to 1915.

During the first days of August 1988 the lowest flow rate at Pontelagoscuro was above 500 m³·s⁻¹, a value much higher than the all-time low (307 m³·s⁻¹) recorded at that station.

Water temperature shows fluctuations which are to a large extent related to flow rate variations. The highest temperature (27.5 °C) was recorded at the end of July, the lowest (22-23 °C) coincided with flood phases on 7 and 24 August.

Settling measurements carried out on

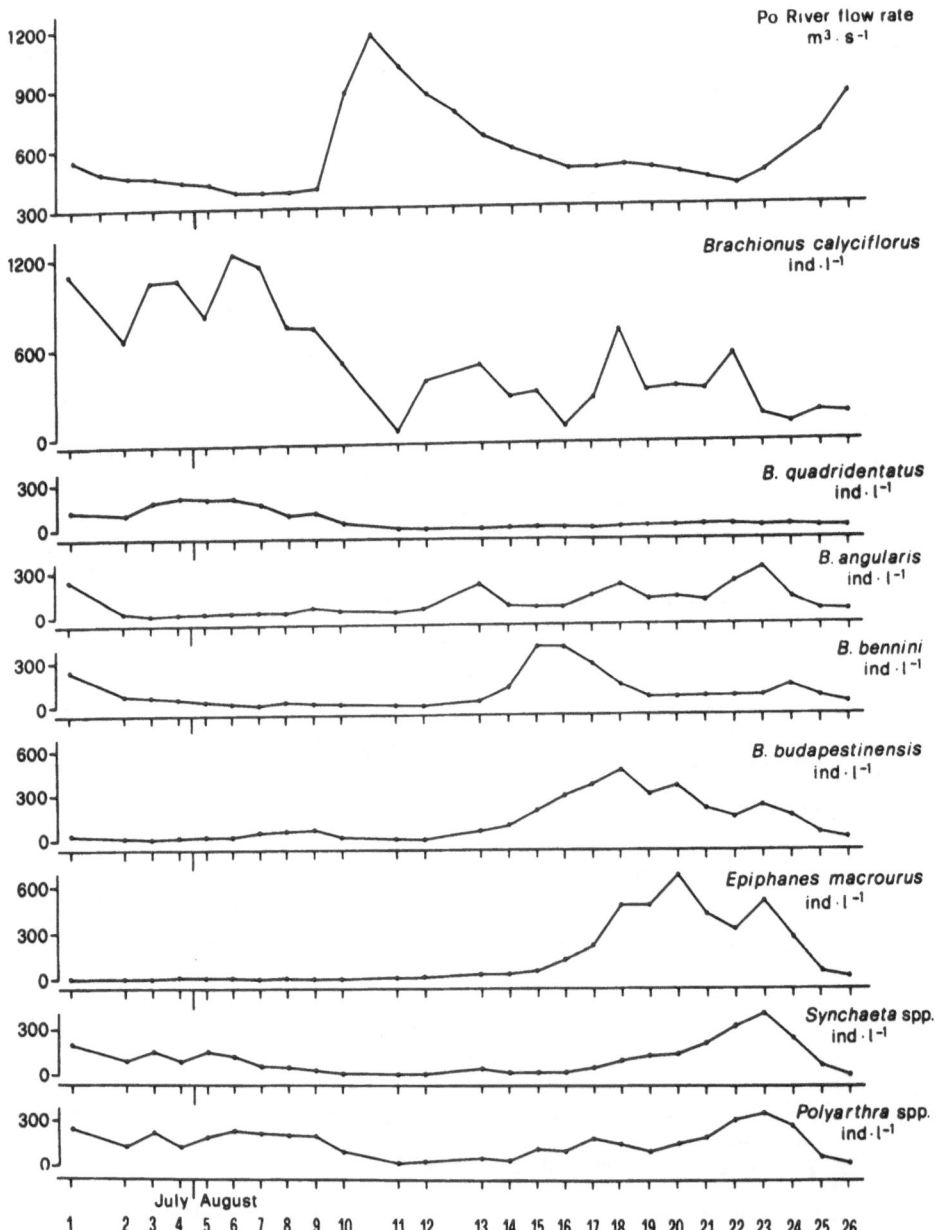

Fig. 3. Po River flow rate at Viadana and densities of the dominant rotifer species.

zooplankton samples have evidenced that the volume of the sandy fraction was greater than that of the plankton fraction. In the samples collected during the lowest flow phases of July-August 1985, on the contrary, the plankton fraction prevailed significantly over the sandy one.

Phytoplankton collected with the 50 μm mesh net were dominated by *Fragilaria crotonensis* in the first samples of the 1988 series. After the flow peak of 7 August, phytoplankton diversity increased, due particularly to the appearance of blue-green algae.

All the rotifer taxa found in 1988 are also present in 1985, except for some species of *Lecane* and *Trichocerca* (Table 1). Copepods are mainly represented by cyclopoid nauplii and copepodites and by *Acanthocyclops robustus* adults. Four clado-

ceran species appear: *Moina micrura, Bosmina longirostris, Chydorus sphaericus* and *Alona* sp. Copepods and cladocerans show very low densities; in all samples the two taxa together represent less than 1% of the total zooplankton (protozoans excluded) while over 99% is made up of rotifers.

The overall rotifer density was inversely correlated to river flow (Fig. 2). During the first ten days it always exceeded 1000 ind·l⁻¹ with a peak of 2350 ind·l⁻¹ on 27 July. On 8 August, when river flow abruptly increased, density fell to 350 ind·l⁻¹. Total density then gradually rose to a maximum of 2700 ind·l⁻¹ on 18 August; at the flow peak on 24 August it dropped to 400 ind·l⁻¹.

There is also a relationship between river flow and Shannon-Wiener index. This index was cal-

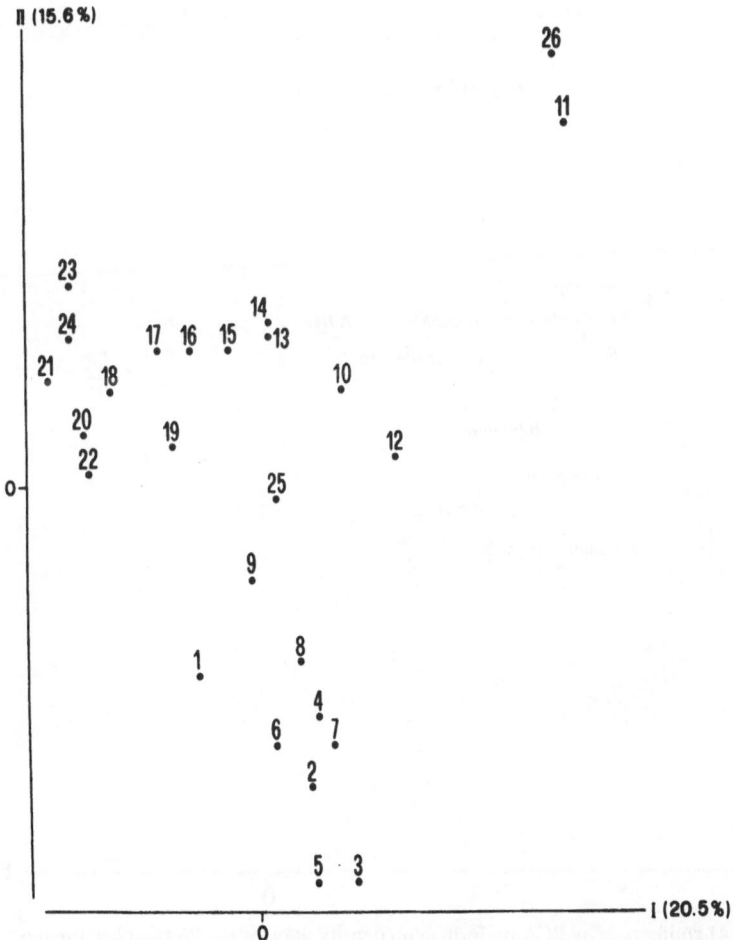

Fig. 4. Ordination of the 26 samples (see Fig. 2 and Fig. 3) by PCA performed on density data of 41 rotifer taxa.

culated for 41 taxa: some of them are represented by genera, like *Cephalodella*, *Trichocerca*, *Synchaeta* and *Polyarthra*.

During the first low water phase (from 27 July to 6 August) the dominant species is *Brachionus calyciflorus*, which accounts for about 50% of all rotifers. During this phase *Brachionus quadridentatus*, *Brachionus angularis*, *Brachionus bennini* and *Polyarthra* spp. also appear, with densities of more than 200 ind·l⁻¹ in at least one sample. On 8 August there is a marked increase in diversity due to an increased number of species, but a greater evenness in distribution of abundances. *Keratella cochlearis tecta* becomes the most abundant species, comprising 13% of the total rotifers. During the second low water phase *B. calyciflorus* is still abundant but does not return to its previous relative importance. Several species are abundant including *Epiphanes macrourus*, *B. bennini*, *Brachionus budapestinensis*, *Polyarthra* spp. and *Synchaeta* spp., and diversity remains high.

Density trends of the most important species are shown in Fig. 3. Other species appearing with high frequency are *Keratella cochlearis*, *Keratella cochlearis tecta*, *Anuraeopsis fissa*, *Lecane (Monostyla) closterocerca*, *Cephalodella catellina*, *Trichocerca pusilla*, *Asplanchna priodonta*, *Asplanchna brightwelli* and *Filinia longiseta*. Some of these species reach density maxima during the last week of August, in particular *K. cochlearis tecta* and *F. longiseta* with 123 ind·l⁻¹ and *Trichocerca* spp. with 159 ind·l⁻¹. The predatory species *A. brightwelli* also attains a relatively high density (60 ind·l⁻¹) during this period. The presence of males is generally negligible.

PCA was carried out on the densities of 41

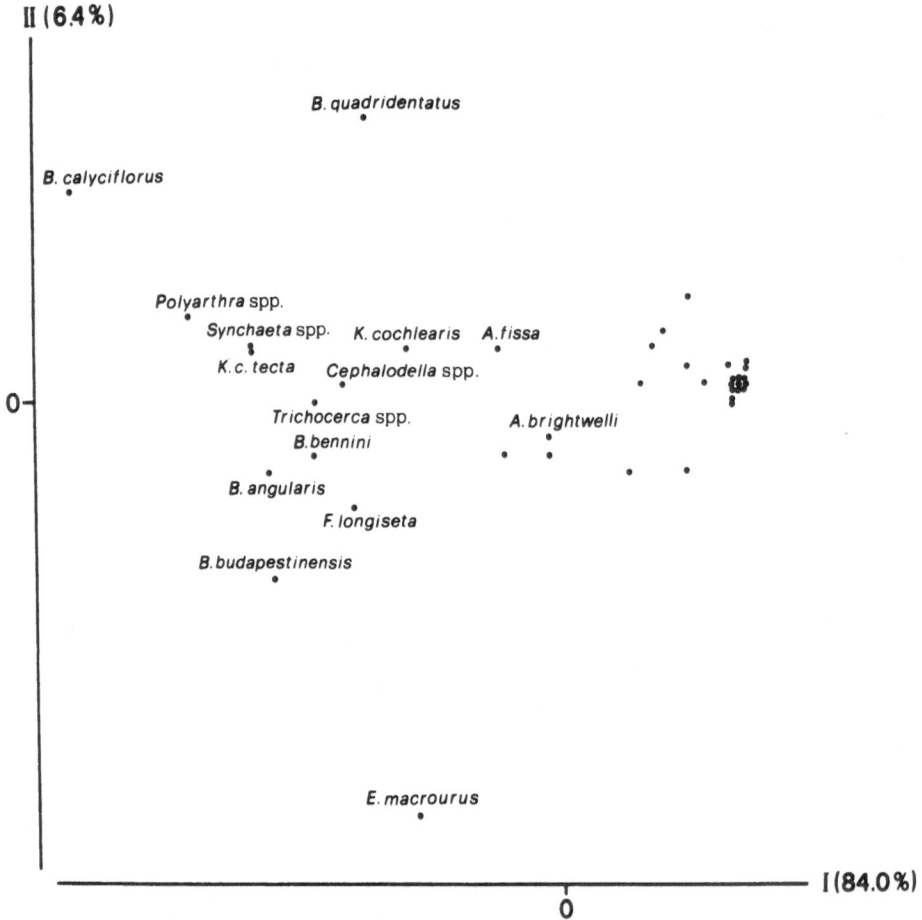

Fig. 5. Ordination of the 41 rotifer taxa by PCA performed on density data for the 26 samples: the most important taxa, in terms of both abundance and frequency, are indicated.

taxa, the same ones for which the diversity index was computed (Figs 4 and 5). Samples of the first low water phase (No 1-No 9) are distinctly separated from those of the second low water phase (No 13-No 24) (Fig. 4). Samples collected at the two flow peaks in August (No 11 and No 26) are closely associated with each other and sharply detached from all the others.

PCA on rotifer taxa reveals a density gradient from the most abundant (*Brachionus calyciflorus*) to the rarest species (Fig. 5). On the second component axis, species at their maxima during the first low flow phase (*B. calyciflorus* and *Brachionus quadridentatus*) are inversely correlated to species with the highest densities during the second low water phase, in particular *Filinia longiseta*, *Brachionus budapestinensis* and *Epiphanes macrourus*.

Conclusions

1. During the summer low flow phases, the Po River waters have a trophic structure comparable to that of shallow, highly productive bodies of water. Microzooplankton are dominant, especially rotifers which attain densities that are among the highest reported (Hynes, 1970; Pourriot *et al.*, 1982). High densities of the Po planktonic rotifers might be related to the increase in concentrations of nutrients (nitrogen and phosphorus) as documented by Marchetti *et al.* (1985) over the past 20 years.

2. Under low water conditions zooplankton tend to become a structured community, with a stable composition, abundance and diversity in time and space. During the summer, zooplankton composition in the Po Delta branches (Ferrari *et al.*, 1983; Ferrari & Mazzocchi, 1985) is very similar to what has been described here.

3. Interesting similarities and differences come to light when data from the 1985 and 1988 series are compared. The ranges of water temperature (22-28 °C) and flow rates are very similar for the two years. The floods observed in both August 1985 and August 1988 have a destabilizing effect on the rotifer community, leading to a sharp increase in diversity. This increase is due to the higher number of species (benthic and phytophilous forms dislodged by the current from the bottom and vegetation), but especially to the reduction in *Brachionus calyciflorus* density. On the other hand, in July-August 1985 total rotifer density was very high (up to 6600 ind $\cdot$ l^{-1}). *B. calyciflorus*, with a maximum density of 4450 ind $\cdot$ l^{-1}, remained the dominant species during both low water phases, thus contributing to the maintenance of low diversity. High production of *Brachionus* males, which is likely to be a density-dependent phenomenon, was also observed.

Diversity was higher in 1988, above all during the last two weeks of the sampling period. This might be attributed to the higher turbidity due to dredging activities upstream. Turbidity acts as a disturbing factor promoting diversity. The intermediate disturbance hypothesis (Connell, 1978; Huston, 1979; Colinvaux, 1986) suggests that moderate physical disturbances promote diversity.

4. The two distinct phases observed in the 1988 sampling series, the first characterized by the clear-cut dominance of *Brachionus calyciflorus* and the second by high rotifer diversity, could be interpreted as stages of succession (Margalef, 1974). In the final phase biotic interactions probably become important factors controlling zooplankton community. Predatory behavior of *Asplanchna brightwelli* could also play a significant role in structuring rotifer communities (Gilbert, 1980).

5. Further research should shed light on the mechanisms which allow the zooplankton community, under different hydrological and trophic conditions, to assemble, persist or modify its diversity. Parameters of the community structure seem to be valid descriptors of river ecosystem dynamics. In this context, studies on plankton can make useful contributions to expansion of the River Continuum Concept (Minshall *et al.*, 1985).

208

Acknowledgements

The authors wish to thank Mr. L. Cati and Mr. G. Allodi of the Ufficio Idrografico del Po, Parma, who provided us with river flow data and Mr. A. Carrieri of the Institute of Zoology, University of Ferrara, who carried out the statistical elaboration of the rotifer density data. Protozoan species listed in Table 1 were identified by Dr. P. Madoni of the Institute of Ecology, University of Parma. Thanks also go to Mrs. Nancy Birch Podini for her help in translating the manuscript.

References

Colinvaux, P., 1986. Ecology. J. Wiley & Sons, New York, 725 pp.

Connell, J. H., 1978. Diversity in tropical rain forests and coral reefs. Science 199: 1302–1310.

Davies, B. R. & K. F. Walker (Eds.) 1986. The ecology of river systems. Dr W. Junk Pbls, 793 pp.

Ferrari, I., A. Malice, M. G. Mazzocchi, & G. Matteucci, 1983. Struttura dello zooplancton dulcicolo nei rami terminali del Po e in una laguna del Delta. Atti V Congresso A.I.O.L., Stresa, 1982: 505–514.

Ferrari, I. & G. Benassi, 1984. Ricerche sullo zooplancton estivo nel tratto medio del Fiume Po. In: Moroni, A. (Ed.), Contributi di ecologia fondamentale. Zara, Parma: 27–41.

Ferrari, I. & M. G. Mazzocchi, 1985. Composizione, dinamica e ruolo trofico dello zooplancton nel Delta del Po. Nova Thalassia 7: 187–214.

Ferrari, I., R. Mazzoni & A. Solazzi, 1987. Il popolamento zooplanctonico del fiume Po nell'estate 1985. Atti VII Congresso A.I.O.L., Trieste, 1986: 261–266.

Gilbert, J. J., 1980. Feeding in the rotifer *Asplanchna*: behavior, cannibalism, selectivity, prey defenses and impact on rotifer communities. In: Kerfoot, W. C. (Ed.), Evolution and ecology of zooplankton communities. University Press of New England: 158–172.

Guisande, C. & J. Toja, 1988. The dynamics of various species of the genus *Brachionus* (Rotatoria) in the Guadalquivir River. Arch. Hydrobiol. 112: 579–595.

Huston, M., 1979. A general hypothesis of species diversity. Am. Nat. 113: 81–101.

Hynes, H. B. N., 1970. The ecology of running waters. Liverpool University Press, 555 pp.

Lair, N., 1980. The rotifer fauna of the River Loire (France), at the level of the nuclear power plants. Hydrobiologia 73: 153–160.

Marchetti, R., G. Pacchetti & A. Provini, 1985. Tendenze evolutive della qualità delle acque del Po. Nova Thalassia 7: 311–340.

Margalef, R., 1974. Ecologia. Ed. Omega, Barcelona, 951 pp.

Minshall, G. W., K. W. Cummins, R. C. Petersen, C. E. Cushing, D. A. Bruns, J. R. Sedell & R. L. Vannote, 1985. Developments in stream ecosystem theory. Ca. J. Fish. Aquat. Sci. 42: 1045–1055.

Mordukhai-Boltovskoi, Ph. D. (Ed.), 1978. The river Volga and its life. Dr W. Junk Pbls, The Hague, 473 pp.

Pourriot, R., D. Benest, P. Champ & C. Rougier, 1982. Influence de quelques facteurs du milieu sur la composition et la dynamique saisonnière du zooplancton de la Loire. Acta Oecol., Oecol. Gener. 3: 353–371.

Shiel, R. J., 1985. Zooplankton of the Darling River system, Australia. Verh. int. Ver. Limnol. 22: 2136–2140.

Williams, L. G., 1966. Dominant planktonic rotifers of major waterways of the United States. Limnol. Oceanogr. 11: 83–91.

Hydrobiologia **186/187**: 209–214, 1989.
C. Ricci, T. W. Snell and C. E. King (eds), Rotifer Symposium V.
© 1989 Kluwer Academic Publishers.

Rotifer distribution in relation to temperature and oxygen content

Ernst Mikschi
Goethegasse 4, A-2380 Perchtoldsdorf, Austria

Key words: rotifers, distribution, temperature, oxygen content

Abstract

Lunzer Obersee, a small lake located at an altitude of 1100 m above sea level, was investigated from July 1985 to October 1987. The rotifer community consists of 7 dominant species, 7 subdominant species and 34 species which occasionally occurred in the plankton. The dominant species show rather different demands in relation to temperature and oxygen content; e.g.: *Filinia hofmanni* was found at a wide range of oxygen concentrations (0.6-13.3 mg O_2 l^{-1}) and low temperatures (4-6 °C), living in the upper water layers (1-7 m) during spring and in the deeper, anoxic zone in summer. In contrast, *Asplanchna priodonta* was found at rather high oxygen contents (> 9 mg O_2 l^{-1}), and showed a wide range of temperature tolerance (4-15 °C).

On the basis of field data the temperature and oxygen requirements of several species are described and discussed.

Introduction

Temperature and oxygen content are key factors in restricting rotifer occurrence. Physiological and population parameters are influenced by temperature, and the population development of rotifers is limited by the combined effect of oxygen concentration and temperature (Herzig, 1987). The present study is based on field data and describes and discusses the requirements of different rotifer species in relation to these important physical parameters.

The lake

Lunzer Obersee is a small lake situated at an altitude of 1100 m above sea level. The area of the lake is 0,144 km², 44% of the surface being covered by quaking bogs. There are five basins. In the main basin the maximum depth is 15.5 m; the mean depth of the lake is about two meters.

Methods

From July 1985 to October 1987 samples were taken at the main basin from the surface to a depth of 14 m at one meter intervals using a 10 l Schindler sampler. Samples were concentrated by filtering through a 0.034 mm mesh size net. Using the Winkler method oxygen content and temperature were measured simultaneously at each sampling date. Except during the time of ice-cover, samples were taken biweekly. In winter the sampling interval was extended to about one month.

210

Temperature and oxygen content

After ice thaw, usually at the end of April or beginning of May, a rapid temperature increase occurs in the topmost water layers and inhibits complete mixing. The spring turnover reaches a water depth of only about 7 m and the chemocline becomes established in 9-10 m (Fig. 1).

In general the temperature of the lower stratum (10-14 m) never rose above 5 °C. The temperature of the middle stratum (5-10 m) always remained below 10 °C.

Rapid cooling in October and November resulted in a uniform temperature profile (4-5 °C). The intensity of the autumnal turnover primarily depends on the duration of the homothermous period and the wind intensity during this period.

Planktonic rotifers

The rotifer community consists of 7 dominant species, 7 subdominant species and 34 species which occasionally occurred in the plankton but were usually confined to the littoral or benthic zone.

Only *Keratella cochlearis* is a perennial and forms two maxima. The more significant one occurs in summer (July/August); the second one is in winter (January/March). All the other species are monocyclic. The seasonal occurrence and abundance of the different species is shown in Table 1.

Occurrence and abundance of different species in relation to temperature and oxygen content.

The analysis of the relationship between these two parameters and the occurrence of each species was restricted to samples containing more than 5 ind. l^{-1} (*Synchaeta pectinata*, *Asplanchna priodonta* and *Filinia hofmanni*) and 10 ind. l^{-1} (*Polyarthra dolichoptera*, *Ascomorpha ecaudis* and *Keratella hiemalis*), respectively. Because of the high densities of *Keratella cochlearis*, only samples containing more then 100 ind. l^{-1} were used for analysis.

F. hofmanni and *K. hiemalis* were most abundant at low temperatures (<7.5 °C) under a wide range of oxygen concentrations. Both species inhabited the upper layers (0-7 m) during spring and migrated to the deeper, deoxygenated zone in summer. *P. dolichoptera* and *S. pectinata* preferred waters with a high oxygen content but developed at a wide range of temperatures (>8.7 mg $O_2 l^{-1}$, 5.5-16.0 °C for *P. dolichoptera*; >9.1 mg $O_2 l^{-1}$, 5.1-12.7 °C for *S. pectinata*). Both species were restricted to the upper water layers (0-6 m). As the preceding two species, *A. priodonta* showed a wide range of temperatures being tolerated (6.8-15.4 °C) and – apart from a few samples – high numbers were found at oxygen contents

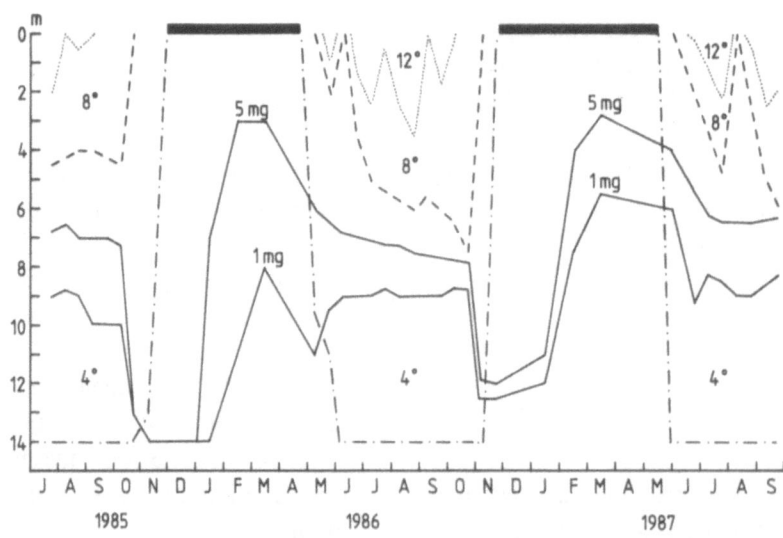

Fig. 1. Temperature and oxygen content in Lunzer Obersee during the years 1985-1987. The isopletes of 5 and 1 mg $O_2 l^{-1}$ and the isotherms of 4, 8 and 12 °C are shown. Black bars indicate ice-cover.

Table 1. Seasonal occurrence (–) and abundance (∗), maximal density found during the period of investigation, vertical distribution (stratum in which a species was most abundant: U = upper stratum (0-5 m), M = middle stratum (6-10 m), L = lower stratum (10-14 m), ranges of temperature and oxygen content within which the highest densities were attained, and type to which each species belongs (A, B or C) as described in the text.

Dominant species	Occurrence and abundance J F M A M J J A S O N D	Max. dens. (ind. l^{-1})	Vert. dist.	High densities °C	High densities mg O_2 l^{-1}	Type
Keratella cochlearis	∗∗∗∗∗∗∗––––––∗∗∗∗–––––––	7,690	UM	5.5–13.5	1.0–13.7	C
Keratella hiemalis	–∗∗∗∗––––––	450	ML	4.4– 8.4	1.4–15.0	A
Polyarthra dolichoptera	––∗∗∗∗––––––	660	U	5.5–16.0	8.7–18.4	B
Ascomorpha ecaudis	––∗∗∗∗––––	1,590	UM	5.0–16.0	2.4–18.1	C
Synchaeta pectinata	––∗∗∗∗–––––	110	U	5.1–12.7	9.1–13.7	B
Asplanchna priodonta	––∗∗∗∗∗–––––	200	U	6.8–15.4	8.0–14.4	B
Filinia hofmanni	–––––∗∗∗–––––––– ––	110	ML	4.0– 7.5	1.0–11.7	A
Subdominant species						
Notholca squamula	––––––––––	3	UM	5.1–11.9	6.7–13.5	–
Synchaeta lakowitziana	––––––––––	4	U	2.0–10.2	8.0–15.7	–
Trichotria tetractis	––––––	<1	M	–	–	–
Euchlanis dilatata	––––––––	<1	U	–	–	–
Synchaeta tremula	–––––––––	6	UM	6.5–14.3	7.8–18.2	–
Keratella quadrata	–––––	<1	UM	–	–	–
Synchaeta longipes	–––––––	22	U	5.9–15.0	6.9–15.2	–

exceeding 8 mg $O_2 l^{-1}$. This species inhabited the upper stratum (0-6 m) while the lake was stratified and was distributed throughout most of the water column during the autumnal turnover.

A. ecaudis was abundant over a wide range of temperatures (5.0-16.0 °C) and oxygen contents (2.4-18.7 mg $O_2 l^{-1}$), but high oxygen contents (>9 mg $O_2 l^{-1}$) obviously were preferred. *K. cochlearis* – well known as a generalist with respect to many ecological factors – was abundant over a wide range of temperatures

(5.5-13.5 °C) and oxygen contents (1.0-13.7 mg $O_2 l^{-1}$).

Discussion

Some results of this study on the dominant rotifers are summarized in Table 1. All species occurring in Lunzer Obersee tolerate low temperatures and high oxygen concentrations. Differences between species become obvious when comparing the lowest oxygen content and the highest temperature tolerated. According to their requirements, three different groups of animals can be distinguished (Fig. 2):

Type A: Cold stenothermous species tolerating a wide range of oxygen contents (i.e. *F. hofmanni* and *K. hiemalis*)

F. hofmanni was described by Koste (1980) as exhibiting the ecologial behavior of *F. terminalis*. Miracle & Vicente (1983) found *F. hofmanni* being clearly restricted to the chemocline and Schaber

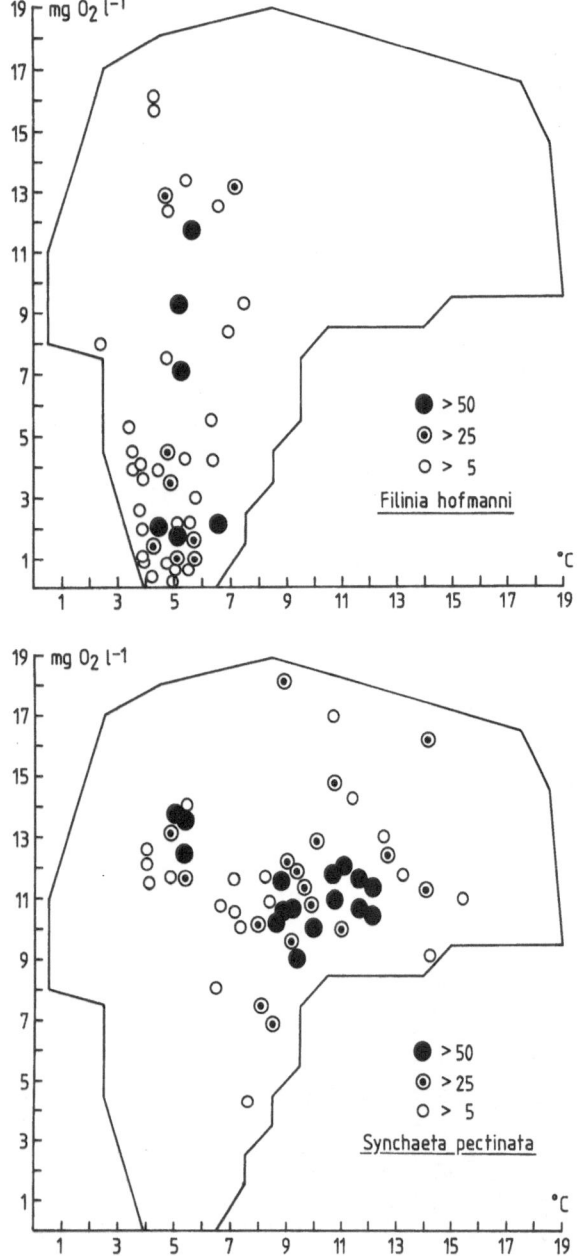

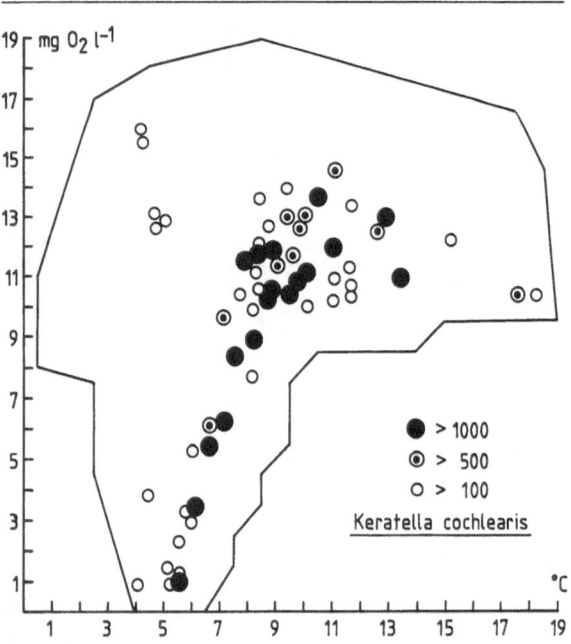

Fig. 2. Occurrence and abundance in relation to temperature (°C) and oxygen content (mg $O_2 l^{-1}$) for: (a) *Filinia hofmanni*, (b) *Syncheata pectinata*, (c) *Keratella cochlearis*. Numbers mean ind. l^{-1}, the continuous line includes all the combinations of temperature and oxygen content occurring in Lunzer Obersee during the present study.

& Schrimpf (1984) described *F. hofmanni* migrating along the 12 °C isotherm, disappearing at higher temperatures. Herzig (1987) notes that cold stenothermous species can be trapped between the warm epilimnion and the deoxygenated hypolimnion. Similar observations were made during the present study (Fig. 3a). According to other workers (Ruttner-Kolisko, 1975; Koste, 1978), *K. hiemalis* must be characterized as a cold stenothermous species.

Bosselmann (1981) reported a downward movement of this species similar to that found in Lunzer Obersee.

Type B: Species tolerating a wide range of temperatures but requiring high oxygen content (i.e. *S. pectinata*, *P. dolichoptera* and *A. priodonta*).

Ruttner-Kolisko (1974) described *S. pectinata* being a eurythermous species, showing high den-

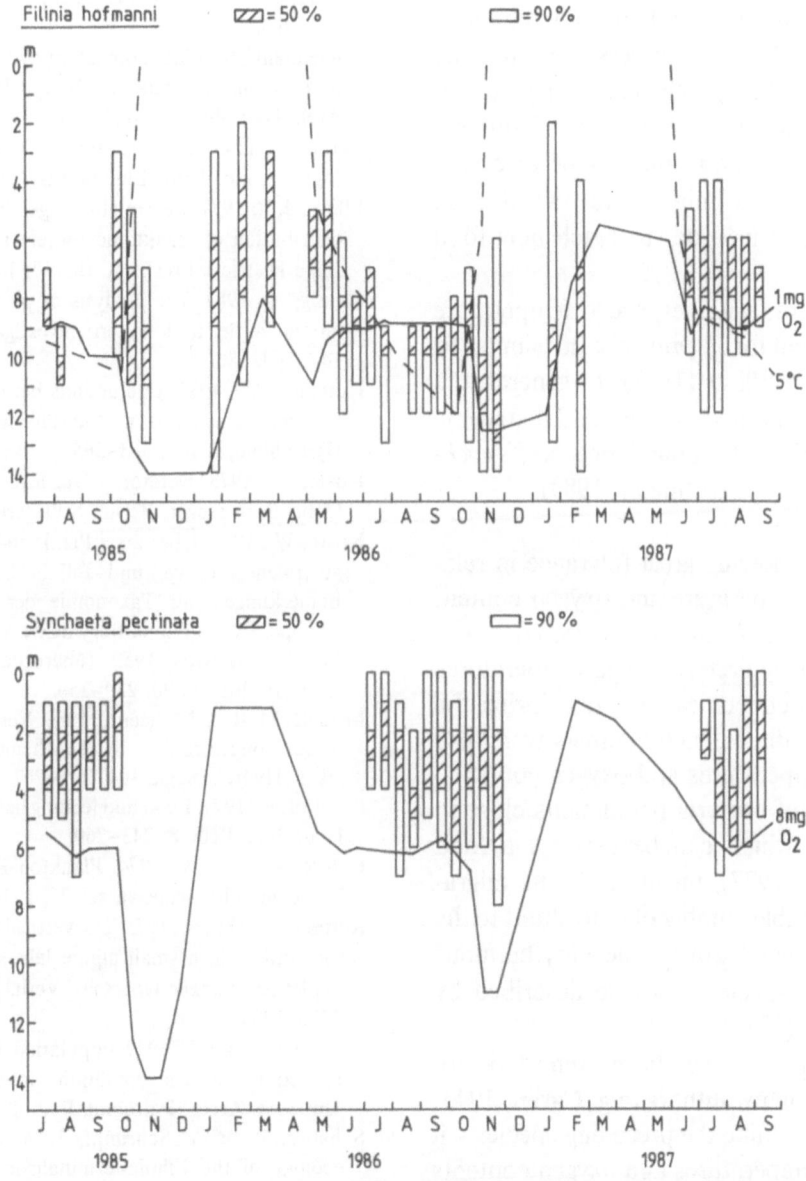

Fig. 3. Vertical distribution of: (a) *Filinia hofmanni* and (b) *Syncheata pectinata* during the years 1985-1987. The total number of individuals in the whole water column was taken as 100%; layers where 50% and 90% of the individuals are concentrated are indicated.

sities at low temperatures. Similar observations have been made by Elliott (1977). During the present study this species always inhabited the upper water layers (Fig. 3b).

While Ruttner-Kolisko (1975) found *P. dolichoptera* in Lunzer Obersee tolerated low oxygen contents and its distribution pattern was governed by temperature and food resources, this species was always found at high oxygen contents during the present study. The discrepancy might be due to the marked decrease in abundance of *P. dolichoptera* within the last ten years (values found during the present study usually remained below 10-20% of the numbers reported by Ruttner-Kolisko (1977)). When occurring at relatively low numbers, the availability of suitable food may be sufficient and it may not be necessary for *P. dolichoptera* to migrate to the deeper, deoxygenated zone in order to reach new food resources.

The requirements with respect to temperature and oxygen content for *A. priodonta* are similar to those reported by Elliott (1977). In general, the success of this species in Lunzer Obersee is obviously related to the abundance of *K. cochlearis* (Pourriot, 1977; Hofmann, 1983).

Type C: Species showing great tolerance in relation to temperature and oxygen content (i.e. *A. ecaudis* and *K. cochlearis*).

A. ecaudis obviously preferred high temperatures and high oxygen concentrations, but during the time of high abundance, high numbers were also found at low temperatures and oxygen contents. The appearance of bacteria populations close to the chemocline, which can be used as a food supply (Pourriot, 1977), might cause the migration of a considerable number of individuals to the deeper, deoxygenated zone. The eurythermous character of this species was also described by Ruttner-Kolisko (1974).

High tolerances have been reported for *K. cochlearis* by many authors (e.g. Carlin, 1943; Elliott, 1977), but – like the preceding species – it preferred high temperatures and oxygen contents and its distribution pattern seemed to be governed by food sources (Bosch & Ringelberg, 1985).

Acknowledgements

I wish to thank Dr. Alois Herzig for his helpful advice and for revision of the manuscript and Dr. Walter Koste for his aid in species identification.

References

Bosch, F. v.d. & J. Ringelberg, 1985. Seasonal succession and population dynamics of Keratella cochlearis (EHRB.) and Kellicottia longispina (KELLICOTT) in Lake Maarsseveen I (Netherlands). Arch. Hydrobiol. 103: 273–290.

Bosselmann, S., 1981. Population dynamics and production of Keratella hiemalis and K. quadrata in Lake Esrom. Arch. Hydrobiol. 90: 427–447.

Carlin, B., 1943. Die Planktonrotatorien des Motalaström. Medd. Lunds Univ. Limnol. Inst. 5: 256 pp.

Elliott, J. I., 1977. Seasonal changes in the abundance and distribution of planktonic rotifers in Grasmere (English Lake District). Freshwat. Biol. 7: 147–166.

Herzig, A., 1987. The analysis of planktonic rotifer populations: A plea for long-term investigations. Hydrobiologia 147: 163–180.

Hofmann, W., 1983. Interactions between Asplanchna and Keratella cochlearis in the Plußsee (north Germany). Hydrobiologia 104: 363–365.

Koste, W., 1978. Rotatoria. Die Rädertiere Mitteleuropas. Gebr. Borntraeger, Berlin, Stuttgart, 673 Seiten.

Koste, W., 1980. Über zwei Plankton-Rädertiertaxa Filinia australensis n. sp. und Filinia hofmanni n. sp., mit Bemerkungen zur Taxonomie der longiseta-terminalis-Gruppe. Genus Filinia Bory de St. Vincent, 1824, Familie Filiniidae Bartos 1959 (überordnung Monogononta). Arch. Hydrobiol. 90: 230–256.

Miracle, M. R. & E. Vicente, 1983. Vertical distribution and rotifer concentrations in the chemocline of meromictic lakes. Hydrobiologia 104: 259–267.

Pourriot, R., 1977. Food and feeding habits of Rotifera. Arch. Hydrobiol. Beih. 8: 243–260.

Ruttner-Kolisko, A., 1974. Plankton Rotifers; Biology and Taxonomy. Binnengewässer 26, 1, 146 pp.

Ruttner-Kolisko, A., 1975. The vertical distribution of plankton rotifers in a small alpine lake with a sharp oxygen depletion (Lunzer Obersee). Verh. int. Ver. Limnol. 19: 1286–1294.

Ruttner-Kolisko, A., 1977. Population dynamics of rotifers as related to climatic conditions in Lunzer Obersee and Untersee. Arch. Hydrobiol. Beih. 8: 88–83.

Schaber, P. & A. Schrimpf, 1984. On morphology and ecology of the Filinia-terminalis-longiseta-group (Rotatoria) in Bavarian and Tyrolean lakes. Arch. Hydrobiol. 101: 247–257.

Hydrobiologia **186/187**: 215–221, 1989.
C. Ricci, T. W. Snell and C. E. King (eds), Rotifer Symposium V.
© *1989 Kluwer Academic Publishers.*

Patterns in the composition of the rotifer communities from high mountain lakes and ponds in Sierra Nevada (Spain)*

R. Morales-Baquero, L. Cruz-Pizarro & P. Carrillo
*Departamento de Biología Animal, Ecología y Genética, Facultad de Ciencias, Universidad de Granada.
18071 Granada, Spain*

Key words: rotifer communities, high mountain lakes

Abstract

On the basis of periodic collections of rotifers from 29 lakes and ponds over 2500 m above sea level in the Sierra Nevada (Southern Spain), patterns of species richness, distribution and community composition were evaluated. Results allow us to distinguish communities which fall into two major lake types. One is defined by the presence of typically planktonic species as well as lower specific richness whereas the other includes communities of mainly benthic and periphytic species. Both lake types seem to be related to small differences in their physical and chemical characteristics. These relationships and the influence of littoral vegetation are discussed.

Introduction

The study of the composition of rotifer communities in different lakes in a specific geographical area provides valuable information on those factors which control species distribution and grouping (Tonolli & Tonolli, 1951; Pejler, 1965; Miracle, 1978; Green, 1981; Chengalath *et al.*, 1984; Siegfried *et al.*, 1984; Ebert & Balko, 1987). A series of small alpine lakes are found in the highest part of the Sierra Nevada mountains, where the action of glaciers was able to create appropriate conditions in this southern mountain range. Morales-Baquero (1987) has previously published a list of the species found in this area. In this paper we will attempt to discern overall distribution patterns as well as discussing some of the physical and chemical factors which control them.

Material and methods

Figure 1 shows the area of study. Some basic limnological characteristics of the lakes were described in earlier publications (Sanchez-Castillo, 1986; Morales-Baquero, 1988) and are summarized here. Lakes rarely exceed 100 m in length or 3 m in depth and have siliceous substrates. Littoral vegetation is sparse along the shores consisting of Bryophyta and filamentous green algae. There are no aquatic vertebrates and the lakes are ice-covered from October through May.

Samples were collected from 29 lakes located at altitudes ranging from 2800 to 3050 m above sea level. Lakes were sampled from 2 to 6 times (most frequently 4) during the ice free period in 1981. A conical plankton net (45 µm mesh) was swept repeatedly through the open water.

*) Research supported by CAICYT Project n° 3069/83

216

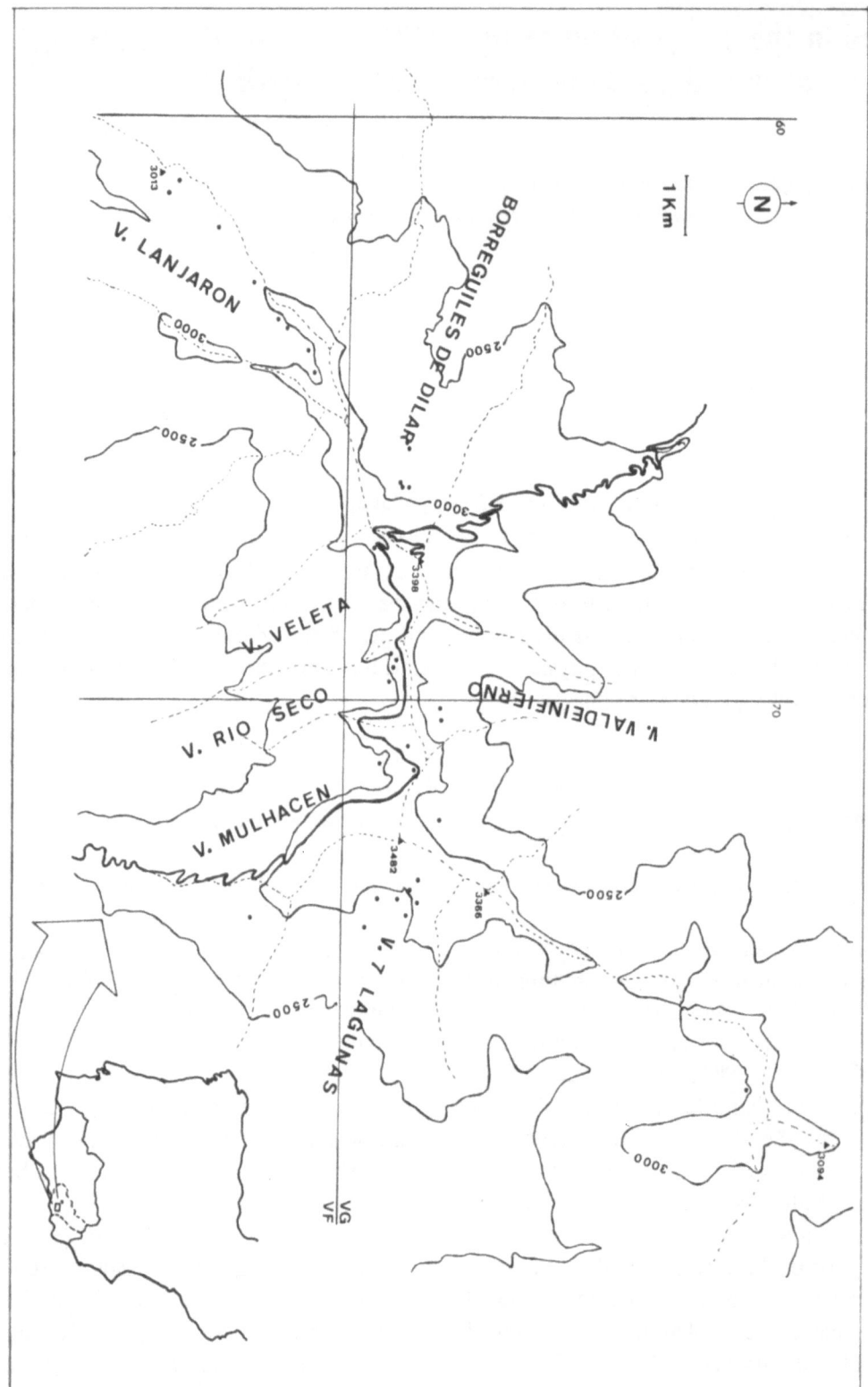

Fig. 1. Map of the upper region of Sierra Nevada. (UTM coordinates overimposed).

Samples were preserved in 4% formaldehyde solution immediately after collection. Environmental variables were measured with a Hydrolab 4041 for simultaneous reading of pH, conductivity, oxygen concentration and temperature.

Similarity values between lakes were obtained from presence-absence species data using the Sørensen index. Lakes were then clustered with the UPGMA method of Sneath & Sokal (1973). This was followed by a polar ordination analysis (Poole, 1974) performed by transforming the same Sorensen values into a disimilarity matrix.

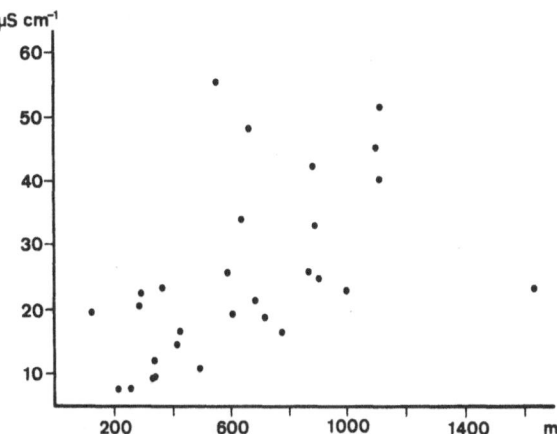

Fig. 2. Relationships between mean conductivity from study lakes and distance to the nearest crest line.

Results

Sampled lakes were all well oxygenated, with more than 5 mg l^{-1} dissolved oxygen. Most lakes were acidic and varied widely in range of mean temperatures (8.1° to 22.7 °C). Mean conductivity values were usually low and never exceeded 55.4 μS cm^{-1}. Conductivity generally tended to rise as the distance between the lake and the nearest crest line (as determined from maps) became greater (Fig. 2).

Table 1. Distribution of rotifer species in 29 lakes of Sierra Nevada (bdelloids and rare species omitted)

	Lakes																												
	1	2	3	4	5	6	7	8	9	10	11	12	13	14	15	16	17	18	19	20	21	22	23	24	25	26	27	28	29
Euchlanis dilatata	+	+	+	+	+	+	+	+	+		+	+		+	+	+	+	+	+	+	+	+	+	+	+	+	+	+	+
Hexarthra bulgarica									+		+	+		+	+	+	+	+	+	+	+	+	+	+	+				
Polyarthra dolichoptera					+												+	+											
Lepadella patella	+	+	+	+	+	+	+	+	+		+	+	+			+		+											
Lepadella quinquecostata nevadensis		+		+	+																								
Lepadella acuminata			+																			+							
Notholca squamula			+		+	+		+							+														
Trichocerca relicta	+	+	+	+	+	+	+	+	+	+	+	+	+	+	+	+	+	+	+	+		+	+	+	+	+	+	+	+
Trichocerca tenuior			+	+	+	+	+	+		+	+		+				+					+	+						
Trichocerca rattus	+	+	+			+				+												+							
Trichocerca bicristata			+		+		+	+	+							+		+	+	+	+	+							+
Trichocerca cavia						+			+																				
Trichotria tetractis	+	+	+	+	+			+	+	+	+	+											+						
Colurella obtusa	+	+	+	+			+			+					+	+	+				+	+							
Lecane lunaris		+	+	+						+											+	+	+						
Lecane flexilis				+	+	+				+		+									+				+				+
Lecane furcata	+	+	+	+	+					+	+						+						+	+	+	+	+	+	
Lecane closterocerca					+																	+				+	+	+	
Lecane kluchor		+	+																										+
Cephalodella gibba microdactyla		+	+	+	+	+	+	+	+	+						+	+												+

1: Mosca; 2: Aguas Verdes; 3: Virgen Inferior; 4: Virgen Superior; 5: Gemelas; 6: Siete Lagunas (4); 7: Siete Lagunas (7); 8: Siete Lagunas (5); 9: Siete Lagunas (6); 10: Dilar (1); 11: Dilar (2); 12: Siete Lagunas (2); 13: Rio Seco; 14: Laguneto Laguna Larga; 15: Peñón Negro; 16: Caldera; 17: Majano; 18: Rio Seco Inferior; 19: Lanjarón; 20: Peñón Colorao; 21: Lanjarón (4); 22: Rio Seco Superior; 23: Larga; 24: Dilar (3); 25: Caballo Inferior; 26: Cuadrada; 27: Caballo; 28: Caldereta; 29: Virgen Media.

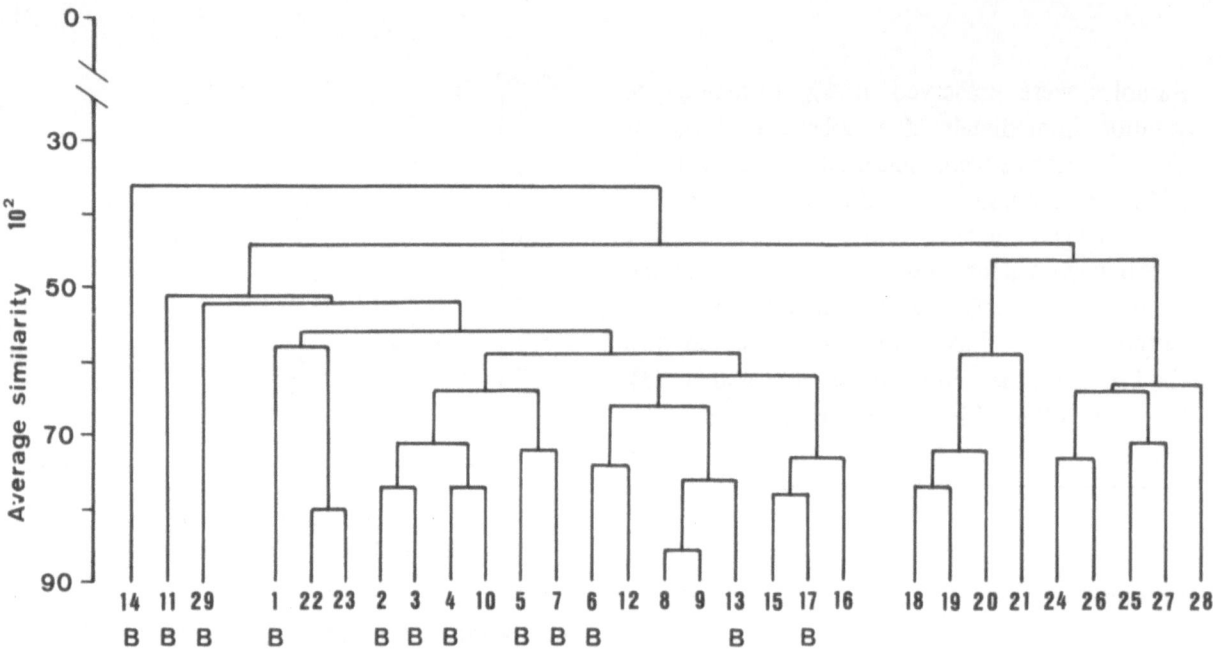

Fig. 3. Average Linkage Cluster Analysis (UPGMA) of communities from 29 lakes (numbers like in Table 1). Those with bryophytic vegetation in their shores are labeled with a B.

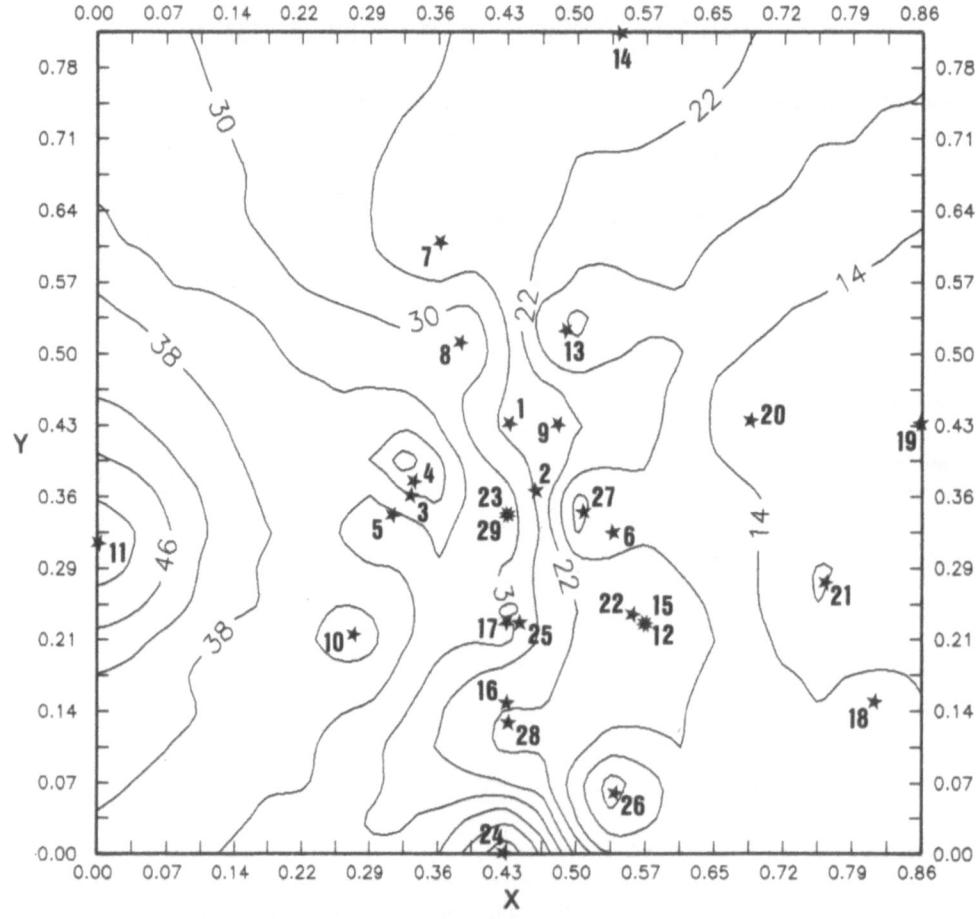

Fig. 4. Two dimensional Polar Ordination of rotifer communities in 29 lakes of Sierra Nevada (data from Table 1). Isolines of conductivity (μs cm^{-1}) has been overimposed in this plane according to the mean values of each one of the lakes. (X,Y: first and second ordination axis, respectively).

Table 1 summarizes the occurrence of the species of rotifers found. Most species are benthic-periphytic. Only 5 can be considered pelagic: *Hexarthra bulgarica*, *Polyarthra dolichoptera*, *Notholca squamula*, *Euchlanis dilatata* and *Trichocerca bicristata*.

Compositional trends are illustrated in the results of the cluster analysis (Fig. 3). Two main groups of lakes not necessarily near one another could be distinguished. The first group comprised 9 lakes containing few species. Their rotifer communities were dominated, with one exception, by *H. bulgarica*. The second group generally contained a greater variety of species, which were mainly benthic-periphytic, as well as more developed littoral vegetation. This group included all lakes with bryophytes in the shores (Fig. 3).

Polar ordination reflects the division between the two lake groups (Fig. 4). Ordination of the lakes on the basis of disimilarity in fauna consistently revealed differences in the degree of mineralization. The isolines of conductivity are plotted in the plane defined by the ordination model and are based on mean values for each lake. Conductivity rises steadily from right to left.

Both number of species per lake and mean density of individuals show a clear pattern (Fig. 5). Those lakes with fewer species and higher densities lie in the low conductivity portion of the graphs. In most of these lakes *H. bulgarica* alone is responsible for the majority of organisms; only *P. dolichoptera* reached similar figures in Lanjaron lake.

The remaining parameters failed to show such

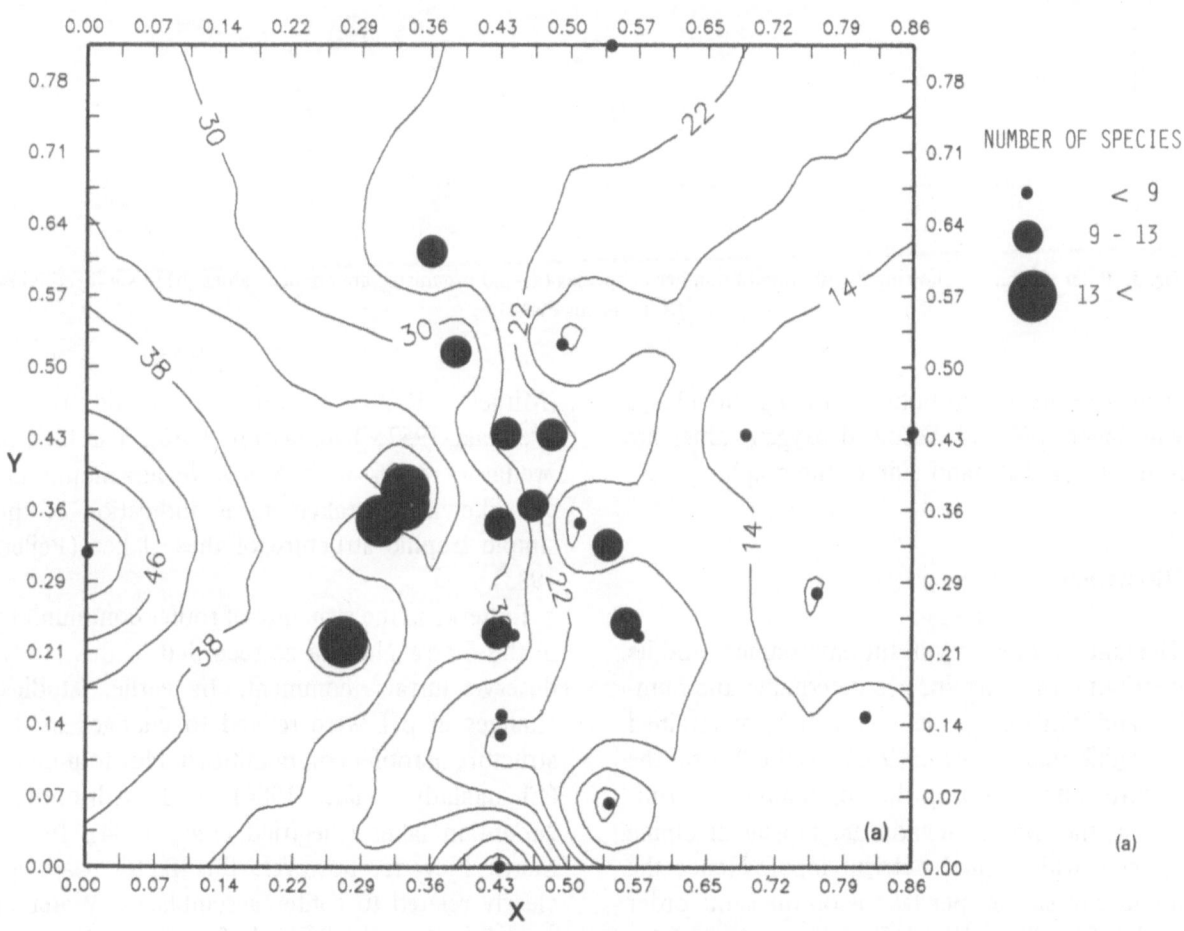

Fig. 5a. See p. 220 for caption.

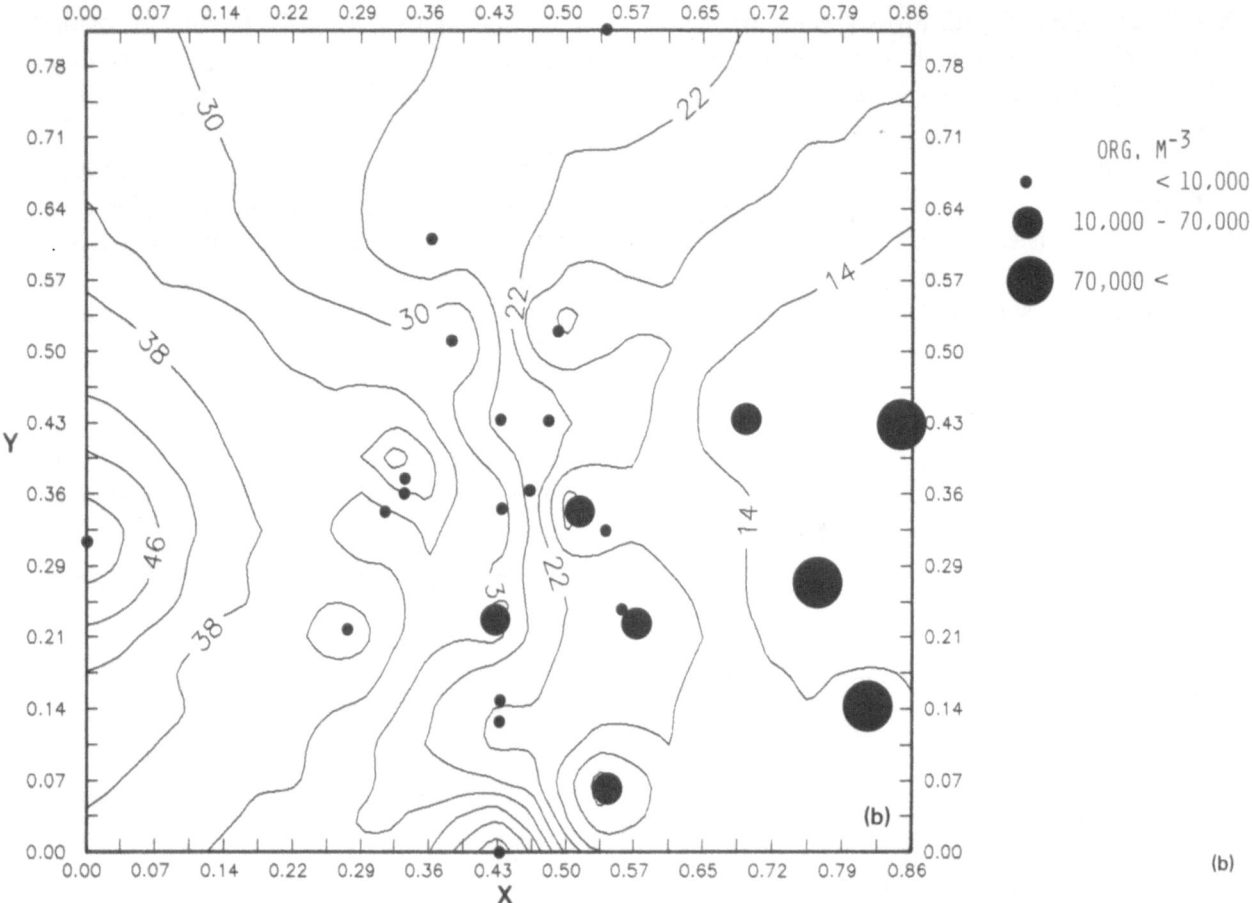

Fig. 5. Polar ordination showing distribution of number of species (a) and organisms abundance values (b) in surveyed lakes. (X,Y: as in Fig. 4).

clear patterns of distribution, although most lakes with higher pH and dissolved oxygen values are found in the left hand side of the graph.

Discussion

The amount of energy in the environment and its distribution among species determines the number and kinds of species that can be maintained in equilibrium (Brown, 1981). In this sense, the features characterizing the communities of rotifers in the Sierra Nevada are typical of alpine systems with limited nutrient input. Hence the number of species per lake is on the same order as that found in other alpine lakes such as those in the Alps (Tonolli & Tonolli, 1951), the Pyrenees

(Miracle, 1978), or other mountain ranges (Pennak, 1957; Thomasson, 1956). The lack of predatory rotifers in the Sierra Nevada mountains can likewise be taken as an indication of the simple trophic structure of these lakes (Pejler, 1983).

Patterns in the structure of rotifer communities in the Sierra Nevada as recorded in this study, deserve further comment. In earlier studies, changes in pH were related to changes in the structure of rotifer communities in Newfoundland (Chengalath *et al.*, 1984) and Adirondack Mountain lakes (Siegfried *et al.*, 1984). In the Sierra Nevada, however, this factor was not clearly related to rotifer assemblages. Water in our lakes was only slightly buffered and pH varied markedly during the thawing period within indi-

vidual lakes (Morales *et al.*, 1986), making it difficult to rely on mean pH values over the entire
sampling period. Another important factor in the
overall structure of rotifer communities is mineralization. Margalef *et al.* (1976) found changes in
the composition of rotifer communities in 102
Spanish reservoirs to be related with their degree
of mineralization. Green (1981) also observed a
drop in the diversity of such communities in
Australian crater lakes with conductivities ranging from 550 to 30 420 μS cm^{-1}. It is nonetheless
surprising that such slight variations in conductivity were so closely related to the ordination of
Sierra Nevada lakes based strictly on rotifer communities.

The finding that low conductivity lakes yielded
greater densities of typically planktonic species,
whereas high conductivity lakes contained predominantly benthic and periphytic species, suggests that this change is associated with the development of littoral vegetation in these small alpine
lakes. In the former, phytoplankton species are
probably favored by their greater surface to volume ratio and high turnover rate (Gliwicz, 1975),
which facilitates the development of filterfeeding
planktonic rotifers. As the difference in altitude
between lakes and the crest line become greater,
nutrient input should be more efficiently used in
the littoral zone. This gives rise to a community
of rotifers of more varied habitat. This also could
explain why some Sierra Nevada alpine lakes of
similar morphometric characteristics were found
to have markedly different rotifer communities.

References

Brown, J. H., 1981. Two decades of homage to Santa Rosalia: towards a general theory of diversity. Am. Zool. 21: 877–888.

Chengalath, R., J. W. Bruce & D. A. Scruton, 1984. Rotifers and crustacean zooplankton communities of lakes in cen-tral Newfoundland, Canada. Verh. int. Ver. Limnol. 22: 419–430.

Ebert, T. A. & M. L. Balko, 1987. Temporary pools as islands in space and in time: the biota of vernal pools in San Diego, Southern California, USA. Arch. Hydrobiol. 110 (1): 101–123.

Gliwicz, Z. M., 1975. Effect of zooplankton grazing on photo-synthetic activity and composition of phytoplankton. Verh. int. Ver. Limnol. 19: 1490–1497.

Green, J., 1981. Associations of rotifers in Australian crater lakes. J. Zool., Lond. 193: 469–486.

Margalef, R., D. Planas, J. Armengol, A. Vidal, N. Prat, A. Guiset, J. Toja & M. Estrada, 1976. Limnología de los embalses españoles. Ministerio de Obras Públicas. Madrid. 422 + 85 pp.

Miracle, M. R., 1978. Composición específica de las comuni-dades zooplanctónicas de 153 lagos de los Pirineos y su interés biogeográfico. Oecol. Aquat. 3: 167–212.

Morales-Baquero, R., 1987. Rotifer fauna of lakes and ponds over 2500 m.a.s.l. in the Sierra Nevada, Spain, with description of a new subspecies. Hydrobiologia 147: 97–101.

Morales-Baquero, R., 1988. Body size variability of *Euchlanis dilatata* HERENBERG in high mountain lakes of Sierra Nevada (Spain). Arch. Hydrobiol. 112 (4): 597–609.

Morales-Baquero, R., L. Cruz-Pizarro & P. Carrillo, 1986. Lagunas de alta montaña en Sierra Nevada: algunas características físicas y químicas. Actas II Simp. Agua en Andalucía. 1: 413–424.

Pejler, B., 1965. Regional-ecological studies of swedish fresh-water zooplankton. Zool. Bidr. Upps. 36 (4) 407–515.

Pejler, B., 1983. Zooplanktic indicatory of trophy and their food. Hydrobiologia. 101: 111–114.

Pennak, R. W., 1957. Species composition of limnetic zooplankton communities. Limnol. Oceanogr. 2: 222–232.

Poole, R. W., 1974. An Introduction to Quantitative Ecology. McGraw-Hill, New York. 532 pp.

Sanchez-Castillo, P., 1986. Estudio de las comunidades fito-planctónicas de las lagunas de alta montaña de Sierra Nevada. Unpl. Ph. D. Thesis, Universidad de Granada.

Siegfried, C. A., J. W. Sutherland, S. O. Quinn & J. A. Bloomfield, 1984. Lake acidification and the biology of Adirondack lakes: Rotifers communities. Verh. int. Ver. Limnol. 22: 549–558.

Sneath, P. H. A. & R. R. Sokal, 1973. Numerical taxonomy. W. H. Freeman, San Francisco. 573 pp.

Thomasson, K., 1956. Reflections on Artic and Alpine Lakes. Oikos. 7 (1): 117–143.

Tonolli, L. & V. Tonolli, 1951. Osservazioni sulla biologia ed ecologia di 170 popolamenti zooplactonici di laghi italiani di alta quota. Me. Ist. ital. Idrobiol. 6: 53–136.

Hydrobiologia **186/187**: 223–228, 1989.
C. Ricci, T. W. Snell and C. E. King (eds), Rotifer Symposium V.
© 1989 *Kluwer Academic Publishers.*

Rotifer associations of some wetlands in Ontario, Canada

Thomas Nogrady
Department of Biology, Queen's University, Kingston, Ontario, Canada, K7L 3N6

Key words: rotifers, wetlands, marsh, *Pleurotrochopsis anebodica*, *Trichocerca bicuspes*

Abstract

The rotifer assemblage of a small marsh-pond and one large marsh/swamp system in the Kingston, Ontario area was investigated for four years. The remarkably diverse (177 species) rotifer fauna contained many species considered very rare. Autecological and taxonomic description of some of these species is presented, and the wider implications of wetland investigations are discussed. The following species are new for Canada: *Collotheca trilobata, Diplois daviesiae, Encentrum tyrphos, Eosphora thoides, Itura chamadis, Notommata fasciola, Pleurotrochopsis anebodica, Trichocerca bicuspes, Trochosphaera solstitialis.*

Introduction

While there is an enormous literature on the synecology, autecology, population dynamics, seasonal distribution and taxonomy of planktonic rotifers, relatively little is known about the same aspects of rotifers occupying wetland habitats. There are probably numerous reasons for this. Quantitative and even qualitative sampling of these habitats is fraught with difficulties and problems (Downing & Cyr, 1985); differentiation of wetland niches is difficult; last but not least, the taxonomy of many species encountered (primarily the notommatids) is problematic. These considerations are also true of littoral associations of lakes, macrophyte beds (Pennak, 1966) and even bogs. Consequently, we cannot even begin to formulate coherent hypotheses of the synecology of rotifer associations inhabiting shallow water with abundant submerged and emergent macrophyte vegetation, regardless of the prevailing pH (see Francez & Devaux, 1985). Two relatively recent publications using modern ecological methods,

that of Anderson *et al.* (1977) dealing with an acidic swamp and the paper of Goddard & McDiffett (1985) describing the rotifer community of an alkaline marsh, underline rather than solve the numerous problems.

The present paper summarizes data from two marshes that were sampled during 1984-88 about every 10-14 days during ice-free periods. It also describes the habitats, rotifer associations, and the autecology and taxonomy of the two most noteworthy rotifer species; a future publication will attempt to evaluate and ordinate the associations using multivariate methodology.

Long-term investigations of littoral or wetland rotifer associations are even rarer than those of planktonic populations. Thus their species succession, composition of assemblages, their niches, and many autecological parameters are poorly known. This paucity of data is compounded by the vastly larger number of species, smaller number of individuals, the urgent need for taxonomic revision of some genera, and the methodological difficulties associated with littoral biotopes. The

advantages of long-term investigations and the obvious need to extend such work to the vast majority of rotifers living in the benthic/littoral zone are amply illustrated in the important and sophisticated paper of Herzig (1987). The present paper shows that extensive and frequent sampling can uncover an unexpected faunal diversity, yield species that are 'rare' or present for a very short time only, and contributes to the knowledge of the autecology of these species. It seems, however, that even more data and longer observations are needed to be able to attempt the elucidation of populations dynamics and interrelation of assemblages using methods of numerical ecology. Such research will also help to uncover the suspected groups of 'equivalent species' that probably occupy the same niches, and their presence or absence depends on the 'first come first served' principle of fortuitious occupancy of an available niche; thus any member of such group can occupy an available niche, but to the exclusion of other members of the 'equivalence group'. Recognition of such groupings may be able to extricate us from the seemingly insoluble and frustrating quandary of correlating environmental parameters and physiological properties of species with ecological phenomena (Ivanova, 1987). Thus, beside contributing to the knowledge of wetland rotifer

–	copeus	Sinantherina	semibullata
–	fasciola	–	socialis
–	tripus	Squatinella	longispinata
Platyias	quadricornis	–	m. mutica
Pleurotrocha	petromyzon	–	m. tridentata
Pleurotrochopsis	anebodica	150 –	rostrum
Ploesoma	lenticulare	Synchaeta	asymmetrica
120 –	truncatum	–	lakowitziana proloba
Polyarthra	dolichoptera	–	longipes
–	dolichoptera aptera	–	oblonga
–	major	–	pectinata
–	major proloba	–	stylata
–	remata	–	tremula
–	remata proloba	Taphrocampa	annulosa
–	vulgaris luminosa	Testudinella	patina
–	v. vulgaris	160 –	patina intermedia
Proales	decipiens	Trichocerca	bicristata
130 –	fallaciosa	–	bicuspes
–	reinhardti	–	elongata
–	sigmoidea	–	iernis
–	theodora	–	intermedia
–	vernecki	–	longiseta
Ptygura	furcillata	–	porcellus
–	p. pilula	–	pusilla
Resticula	gelida	–	rattus
–	nyssa	170 –	rattus carinata
–	plicata	–	similis
140 Rotaria	neptunia	–	tenuior
–	rotatoria	–	weberi
–	sordida	Trichotria	pocillum
–	tardigrada	–	tetractis
Scaridium	longicaudum	Trochosphaera	soltitialis

Fig. 1. Species list of Bayview Marsh and Little Cararaqui Marsh (Kingston, Ontario, Canada) in 1984-1987.

* The subgenera *Trichocerca* (*Diurella*) and *Lecane* (*Monostyla*) are shown as genera.

assemblages, more basic ecological principles could be elucidated by prolonged and frequent sampling in wetlands.

Description of study areas

Both marshes are in or the immediate vicinity of Kingston, Ontario. The first one, Little Cataraqui Marsh (Kingston Township), is a marsh pond about 100 × 50 meters in an area of *Carex* marsh, a few hundred meters from the Little Cataraqui Creek and across the road from a bay of Lake Ontario, but is not in direct contact with either. It is about 1-2 meters deep, but there is a 0.5-1 meter thick loose sediment layer at the bottom. In summer it is filled with submerged macrophytes. Its physicochemical parameters are very ordinary: pH 7.2-8.5, conductivity 400-1800 μmhos/cm, $CaCO_3/CaHCO_3$ type water chemistry, no H_2S during ice-free periods, but due to its immediate vicinity to a major road, it is somewhat polluted. The second study area, Bayview Marsh, is about 20 km west of Kingston in Ernestown Township. It is a depression area of about 1 by 2 km in calcareous till in a non-cultivated open forest-meadow area and is fed by precipitation, but has a small drainage creek. It is in the final stages of filling, and some parts are dry land most of the year. Two areas of the marsh, about 300 meters apart, were studied: one is a *Carex* marsh with a maximum of 30 cm deep water over about 1 m sediment, that dries out in July or August until filled again by fall rains. The other section is a swamp containing numerous dead treetrunks and only submerged macrophytes. There are small cyprinid fish and wild ducks in this part, because it contains water all year round. The water chemistry is as unremarkable as the physiography: pH 7.1–7.8, conductivity 400–700 μmhos/cm, and occasional faint H_2S smell if the thick sediment is disturbed. The free water layer can be up to 80 cm deep. Bayview Marsh is relatively undisturbed apart from occasional hunting.

Methods

Both marshes were sampled every 10-14 days with few exceptions, and data obtained between September 1984 and July 1988 are considered in this paper. Only qualitative samples were taken using a plankton net of 28 μm mesh, due to the unavoidable and well known difficulties of sampling among macrophytes (Downing & Cyr, 1985). Open water, sediment, and water squeezed from submerged macrophytes were investigated; macrophytes were scanned for sessile rotifers. Living as well as preserved material was examined, and percentage composition of the rotifer assemblage counted.

Results and discussion

The rotifer association

The list of rotifers in the two marshes is combined in Table 1, since there is considerable overlap and no discernible information could be obtained from separating the two lists. However, Little Cataraqui Marsh has a lower diversity, and contained, on occasion, very few rotifers; this was probably due to competition from *Bosmina longirostris* or *Chydorus sphaericus*, judging from the cladoceran population maxima at such occasions.

It was also most remarkable that the rotifer assemblages of the two areas of Bayview Marsh, only about 300 m apart, were often, if not always, very different qualitatively as well as quantitatively. This indicates that wetlands, where water movement is normally very restricted, must be sampled at as many locations as feasible.

The list of species is conspicuous for containing over 170 taxa from an area as small as these two marshes. However, about 30% of all species occurred only once, even if present in large numbers at that occasion. It was not rare to find over 30 species in a single sample. Varga (1939) describes a similarly rich rotifer fauna in his paper on a small wetland off Lake Balaton in Hungary,

Trichotria spp.	58	Lepadella patella	25
Synnchaeta spp.	48	Cephalodella gibba	24
Euchlanis incisa	48	Mytilina spp.	23
Euchlanis dilatata	44	Polyarthra vulgaris	22
Testudinelia patina	41	Monostyla quadricornis	21
Polyarthra remata	39	Lophocharis spp.	20
Trichocerca longiseta	38	Trichocerca rattus carinata	19
Cephalodella auriculata	34	Anureopsis fissa	18
Monostyla lunaris	30	Keratella cochlearis	17
Squatinella mutica	29	Platyias quadricornis	17
Ascomorpha ecaudis	29	Keratella serrulata curvicornis	15
Brachionus patulus	29	Trichocerca bicuspes	12
Scaridium longicaudum	27	Itura viridis	9
Gastropus hyptopus	27	Brachionus calyciflorus	5
Monommata spp.	26		

Fig. 2. Percent occurrence of most common rotifer species in all samples.

whileas the paper of Goddard and McDiffett (1983) lists only 30 species in a year.

Only 16% of rotifer species (or genera) encountered can be designated as common or dominant, and their list is shown in Table 2. Planktonic genera normally considered 'common' are very poorly represented (*Keratella, Brachionus*).

Taxonomic remarks

Probably due to the frequent and prolonged sampling a large number of rotifer species were found that are considered rare. In view of the sporadic sampling of Canadian waters (see Chengalath, 1984), this is not surprising. Two of these species are discussed and illustrated below, because they have been either poorly described or no figures are available in the modern literature.

Pleurotrochopsis anebodica Berzins 1973 (Fig. 3). The single specimen seen differs in many respects from the figure given by Berzins (1973) as well as from the figure of Fadeew (as shown in Koste, 1978), and one has to concur with Koste regarding the unsatisfactory nature of these figures. The animal seen among submerged macrophytes in Bayview Marsh had a distinctly segmented notommatid-type body, but not like shown by Berzins. The head is large and changes shape easily; corona extends down to the ventral side. There is a rigid rhinoceros-like pointed 'horn' on the dorsal side. The stout and flexible body carries a series of thorns beginning at the neck, thus resembling Fadeew's description. The number of these thorns is hard to establish. There are apparently four groups of three thorns ventrally, perhaps a few smaller thorns as well. At the very end of the trunk there is an especially prominent curved ventral thorn, and 5 more equidistant thorns circling the body. In addition there are two dorsolateral series of fine teeth at the distal end. The foot has two segments, the first one carries 5 or 6 thorns, the second one only 4. The lateral thorny lamellae described by Berzins could not be found. Toes are small, dagger-like.

The central ganglion is long and carries an eyespot with 4-5 additional red pigment spots. The stomach consists of large, round cells, the lines shown by Berzins were not seen. The

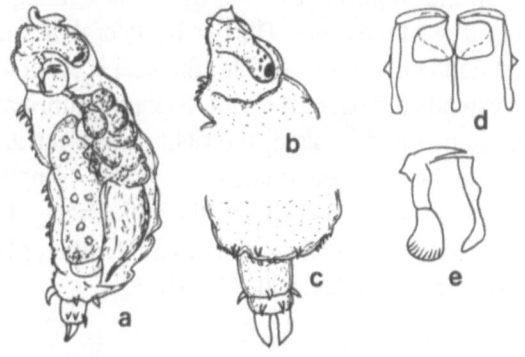

Fig. 3. Pleurotrochopsis anebodica Berzins 1973. *a.* from left, *b.* head from left, *c.* end of body and foot dorsally, *d.* trophi dorsally, *e.* trophy sideways.

228

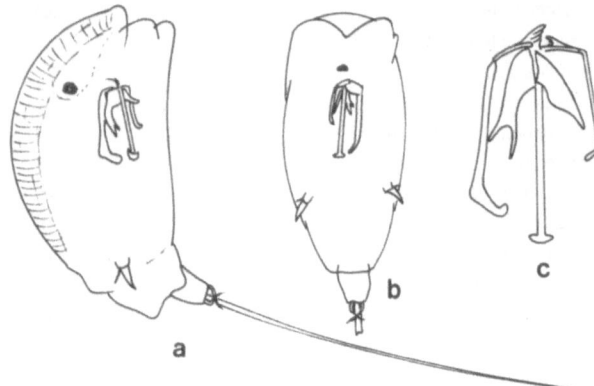

Fig. 4. Trichocerca bicuspes (Pell, 1890). *a.* sideways, *b.* ventrally, semiextended, *c.* trophi.

intestine is smooth. The mastax is small, rami square, manubria straight, spindle-shaped with a small triangular alula in the middle. The fulcrum is wide and paddle-shaped in side view. Measurements: total length 180 μm, toe 13 μm, terminal ventral thorn 5 μm. Found among submerged vegetation in September, temperature 17 °C, pH 8.2, conductivity 640 μS.

Trichocerca bicuspes (Pell, 1490) (Fig. 4) is another one of those species that appear mysteriously, not to be seen again for decades. After its description by Pell, it was properly figured by Jennings (1904). It was listed next by Harring (1921) from the Canadian Arctic, and listed again in Dismal Swamp, Virgina, by Anderson *et al.* (1977). It is an abundant and regular summer constituent of the Bayview Marsh, both in the open water as well as between submerged plants, seen every year. The species conforms to the figure of Jennings, but is drawn here again. The two big lateral thorns are unique and unmistakable; the keel is striated and extends almost to the foot. Right manubrium is L-shaped and hooked, the right ramus carries two alulae. Measurements: body length 90–110 μm, toe 120–140 μm, thorn 9 μm, trophi 43–45 μm. It occurs from May to September, pH 7.2–7.9, conductivity 400–650 μS.

References

Anderson, C. B., A. F. Benfield & A. L. Buikema Jr., 1977. Zooplankton of a swamp water ecosystem. Hydrobiologia 55: 177–185.

Berzins, B., 1973. *Pleurotrochopsis anebodica* n.g., n.sp. (Rotatoria) aus Schweden. Hydrobiologia 41: 449–451.

Chengalath, R., 1984. Synopsis Speciorum. Rotifera. Bibliographia Invertebratorum Aquaticorum Canadiensium Vol. 3. National Museum of Natural Sciences, National Museums of Canada., Ottawa. pp. 1–102.

Downing, J. A. & H. Cyr, 1985. Quantitative estimation of epiphytic invertebrate populations. Can. J. Fish. Aquat. Sci. 42: 1570–1579.

Francez, A.-J. & J. Dévaux, 1985. Répartition des rotiferes dans deux lac-tourbieres du massif Central (France). Hydrobiologia 128: 265–276.

Goddard, K. A. & W. F. McDiffett, 1983. Rotifer distribution, abundance and community structure in four habitats of a freshwater marsh. J. Freshw. Ecol. 2: 199–211.

Harring, H. K., 1921. The Rotatoria of the Canadian Arctic Expedition, 1913-1918. Rep. Can. Arctic Exp. Vol. 8 Pt E. p. 3–23.

Hauer, J., 1958. Rädertiere aus dem Sumpfe 'Grosse Seewiese' bei Kist. Nachr. Naturw. Mus. Aschaffenburg 60, pp. 1–19.

Herzig, A., 1987. The analysis of plankton rotifer populations: A plea for long-term investigations. Hydrobiologia 147: 163–180.

Ivanova, M. B. (1987). Relationship between zooplankton development and environmental conditions in different types of lakes in the zone of temperate climate. Int. Revue ges. Hydrobiol. 72: 669–684.

Jennings, H. S., 1904. Rotatoria of the United States II. A monograph of the Rattulidae. Bull. U.S. Fish Comission 22: 275–352.

Koste, W., 1978. Rotatoria. Die Rädertiere Mitteleuropas begründet von Max Voigt. Bornträger, Berlin.

Koste, W., R. J. Shiel & M. A. Brock, 1983. Rotifera from Western Australian wetlands with description of two new species. Hydrobiologia 104: 9–17.

Pell, A., 1890. Three new rotifers. The Microscope, Detroit, 10: 43–45.

Pennak, R. W., 1966. Structure of zooplankton populations in the littoral macrophyte zone of some Colorado lakes. Tr. Am. Micr. Soc. 85: 329–349.

Varga, L., 1939. Adatok a Balaton kerekesféreg-faunájának ismeretéhez. Az Aszoföi nàdas öböl kerekesférgei. (Contributions to the rotifer fauna of Lake Balaton. Rotifers of the Aszofö bay.) Magyar Biol. Kut. Int. Munkái 11: 316–371.

Hydrobiologia **186/187**: 229–234, 1989.
C. Ricci, T. W. Snell and C. E. King (eds), Rotifer Symposium V.
© 1989 *Kluwer Academic Publishers.*

Opportunist rotifers: colonising species of young ponds in Surrey, England

Rosalind M. Pontin
26 Hermitage Woods Crescent, St. Johns, Woking, Surrey, GU21 1UE, England

Key words: rotifers, new ponds, colonisers, herbivores, omnivores, reproduction

Abstract

Two young ponds of known age were investigated and their populations of rotifers and other organisms compared. The earliest colonisers appeared after a few days and included the rotifers *Keratella valga* (Ehrb.) and *Brachionus urceolaris* Müller, the cladoceran *Daphnia obtusa* Kurz and cyclopoid copepods. The rotifers quickly produced males and resting eggs, as well as large numbers of parthenogenetic females. These herbivorous species were followed by an omnivorous *Asplanchna* species. These species continued to flourish in the ponds several years after digging, in spite of the presence of additional species, although their season of occurrence changed. The greater growth of macrophytes in one pond may account for the greater variety of species in this community.

Introduction

Surrey is a heavily-populated area of Southern England. Numerous ponds which have silted up and have been dry for many years have been re-dug and allowed to fill with water naturally. Two of these ponds were compared to find which species can fill the roles of opportunist colonisers and whether they persist in the pond as it ages.

The ponds

Four Horseshoes Pond was dug out in November 1982 after it had been dry for nearly forty years. It varies from 1/2-1 1/2 m deep; in winter it is often flooded and in summer it may be nearly dry. Consequently its area is also variable, but is about 300 m². Initially the pH was 5-5.5, but now averages 6.5, while the conductivity averages 80 μS cm^{-1} at 16 °C. Macrophytes include two small beds of *Nymphaea* and a few clumps of *Iris* at the margins.

Bulhousen Farm Pond was dug out in February 1979 after it had been dry for at least twenty years, having been filled with wood waste from a sawmill (Pontin & Pontin, 1980). It varies from 1 1/2-2 1/2 m deep, the water level being highest in winter and falling in summer. Its area is about 120 m². Initially the pH was 5-5.5, but now averages 6.5 and conductivity is 160 μS cm^{-1} at 16 °C. Macrophytes include *Glyceria*, *Hottonia* and *Myriophyllum*, which grow thickly in spring and summer.

Methods

Qualitative samples were collected by throwing a plankton net on a rope from the banks. It is possible to reach any part of the ponds by this means. Three horizontal hauls, each of about 5 m and just below the surface, were taken in different direc-

230

tions on each sampling occasion. These were combined, examined live and also preserved in 5% formalin. The net pore size was $50 \times 50\ \mu m$. Quantitative samples were taken just below the surface, using a vessel of standard volume suspended on a rope. Three samples were taken on each occasion, at positions chosen at random. The samples were combined, making a total of 1050 ml, filtered through bolting silk of pore size $70 \times 70\ \mu m$, preserved in 5% formalin and counted in total.

Samples were taken weekly during March-September with extra samples every two or three days from Four Horseshoes Pond during late July and early August. Observations and counts of stomach contents were made on living animals.

Measurements of pH and conductivity were made in the field with Whatman microsensors.

Initial colonisers

The first colonisers appeared in Four Horseshoes Pond about ten days after digging. They were *Keratella valga* (Ehrb.), *Daphnia obtusa* Kurz and cyclopoid copepods including nauplii. They were still present after one month, but in Bulhousen Farm Pond, which was dug in winter rather than in late autumn, no animals were found during the first month.

After two months, males of *K. valga* had appeared in Four Horseshoes Pond, but *D. obtusa* had disappeared. *Brachionus urceolaris* Müller, both females and males, was now in Bulhousen Farm Pond, as well as cyclopoid copepods.

After three months, only *K. valga* females and copepods were in Four Horseshoes Pond, but Bulhousen Farm Pond also contained both sexes of *K. valga* and *B. urceolaris* and females of *Gastropus hyptopus* (Ehrb.).

After four months, Four Horseshoes Pond remained as it had been the previous month i.e. *K. valga* females and copepods. Bulhousen Farm Pond also had *K. valga* females plus copepods and males and females of *Asplanchna brightwelli* Gosse.

A relatively simple food chain can be con-

structed to represent this stage in the pond's development. Algae of different sizes may be taken by the herbivores *K. valga*, *B. urceolaris*, *G. hyptopus*, *D. obtusa* and copepods, either by filtering, piercing or seizing. The omnivore *Asplanchna*, however, takes *K. valga* as prey as well as the larger algae.

Quantitative study

A quantitative study was made in 1987 and the results are shown in Fig. 1 and 2. It can be seen that the herbivores make up the majority of the population in each pond. As well as those present

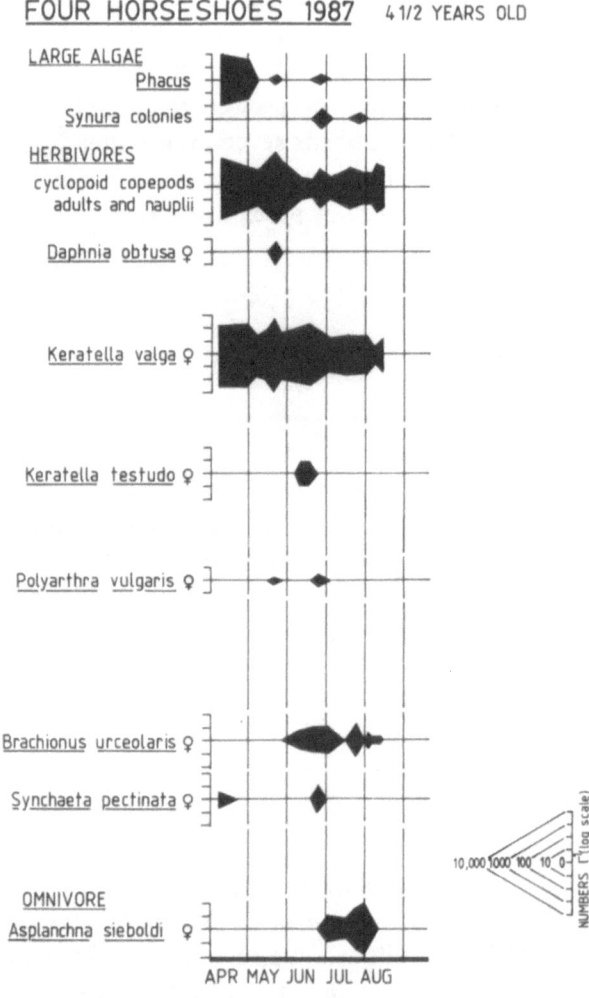

Fig. 1. Four Horseshoes Pond 1987, 4½ years after digging. Changes in log. numbers per litre with time.

BULHOUSEN FARM 1987 8 YEARS OLD

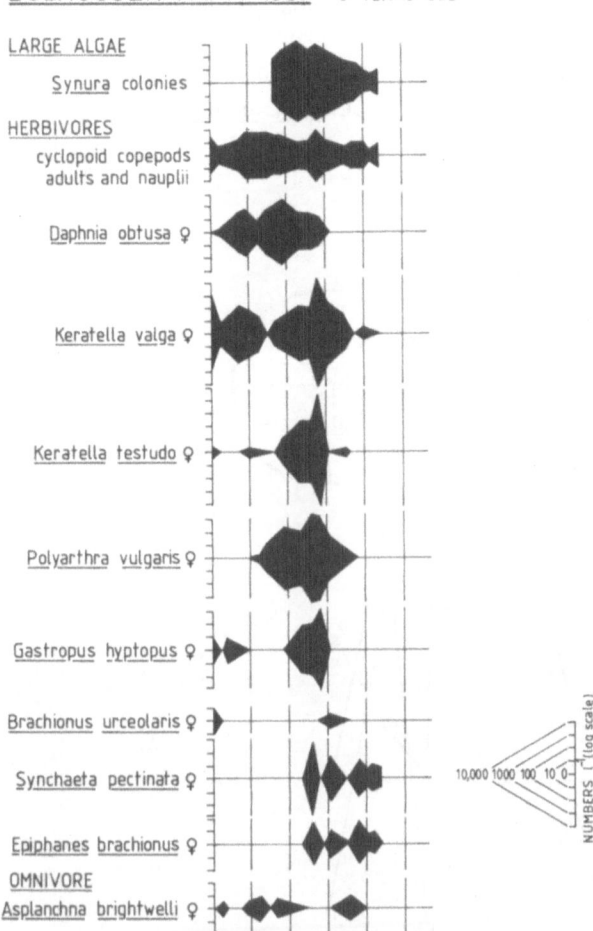

LARGE ALGAE
  Synura colonies

HERBIVORES
  cyclopoid copepods
  adults and nauplii

  Daphnia obtusa ♀

  Keratella valga ♀

  Keratella testudo ♀

  Polyarthra vulgaris ♀

  Gastropus hyptopus ♀

  Brachionus urceolaris ♀

  Synchaeta pectinata ♀

  Epiphanes brachionus ♀

OMNIVORE
  Asplanchna brightwelli ♀

APR MAY JUN JUL AUG

10,000 1000 100 10 0

NUMBERS (log scale)

Fig. 2. Bulhousen Farm Pond 1987, 8 years after digging. Changes in log. numbers per litre with time.

initially, the following appeared in each pond: *K. testudo* (Ehrb.), *Polyarthra vulgaris* Carlin and *Synchaeta pectinata* Ehrb., plus in Bulhousen Farm Pond only, *Epiphanes brachionus* (Ehrb.). Estimates of their potential food algae have so far been limited to counts of the larger species, *Phacus* and the colonial *Synura*, which may be eaten by piercers and suckers such as *S. pectinata* and *G. hyptopus*. However, an estimate of the numbers of small-sized algae is also needed. The stomachs of *K. valga* and *K. testudo* contain very small algal cells of about 2 μm in diameter, while some of about 10 μm diameter were observed in *S. pectinata*.

Each pond has an omnivorous species of

Asplanchna. The species present in Four Horseshoes Pond was very rapacious, eating *K. valga*, copepod nauplii and *B. urceolaris*. *Asplanchna* predation, therefore, had a considerable effect on its prey populations (Fig. 1 and 3). As *Asplanchna* density increased, it became increasingly cannibalistic (Fig. 3). Cannibalism, and its large toothed jaws, lobed nuclei in the vitelline gland and the presence of humps in the males, indicate that this was *A. sieboldi* (Leydig), although no humped females were seen.

A. brightwelli occurred in Bulhousen Farm Pond, but was present only in very small numbers. Both ponds also contain carnivorous planktonic larvae of the phantom midge *Chaoborus*. These may take considerable numbers of prey, but their effect has not yet been determined.

The 1987 survey showed all the original colonising species to be present in each pond and *K. valga* in very large numbers. Of the additional species, *K. testudo* and *P. vulgaris* also had very large maxima in Bulhousen Farm Pond, although they are represented by only small numbers in Four Horseshoes Pond. The production of resting eggs in *Keratella* and other genera precedes the maxima, beginning almost as soon as females are seen. By the time maximum numbers of females are reached, production of both sexual and asexual eggs per female is already falling (Fig. 4).

Bulhousen Farm Pond at 8 years old in 1987 had more species than did Four Horseshoes Pond at 4½ years. However, Bulhousen Farm Pond also had more species in its early days. Apart from age, other differences between the ponds include nutrient levels and macrophyte growth. Conductivity values are twice as high on average in Bulhousen Farm Pond, perhaps because of the debris originally remaining in the bottom or floating on the surface or perhaps because of run-off from the nearby farm. This in turn would promote greater algal and macrophyte growth, thus providing a greater potential food source and more microhabitats.

Adjustment of a species to a maturing pond community might involve a change in timing or extent of occurrence, as is suggested by Fig. 5, which compares occurrence time for species in

232

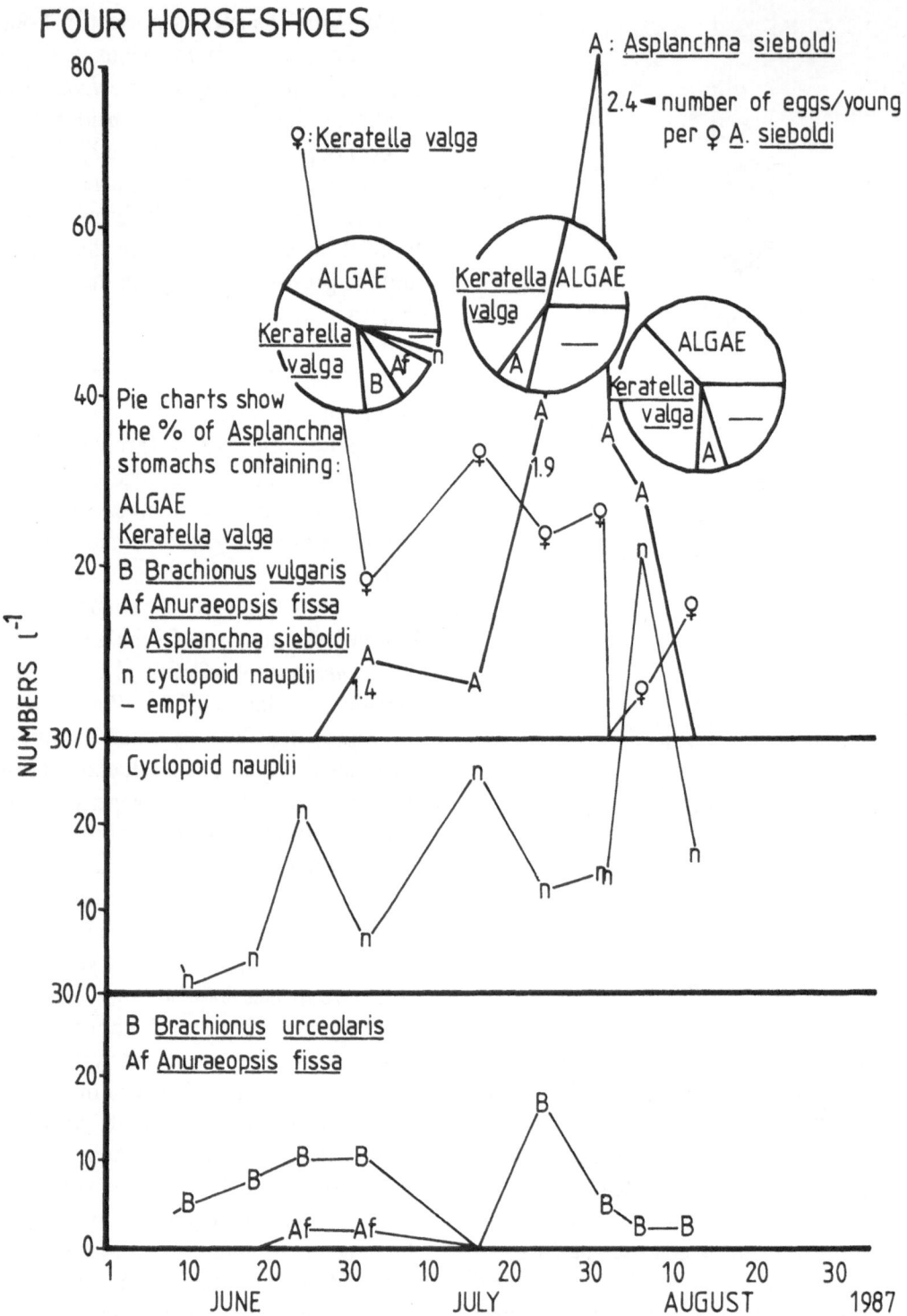

Fig. 3. Asplanchna diet in Four Horseshoes Pond 1987. Numbers per litre of *A. sieboldi* and its food organisms compared with its dietary choice. Pie charts indicate proportion of *A. sieboldi* stomachs containing different food organisms.

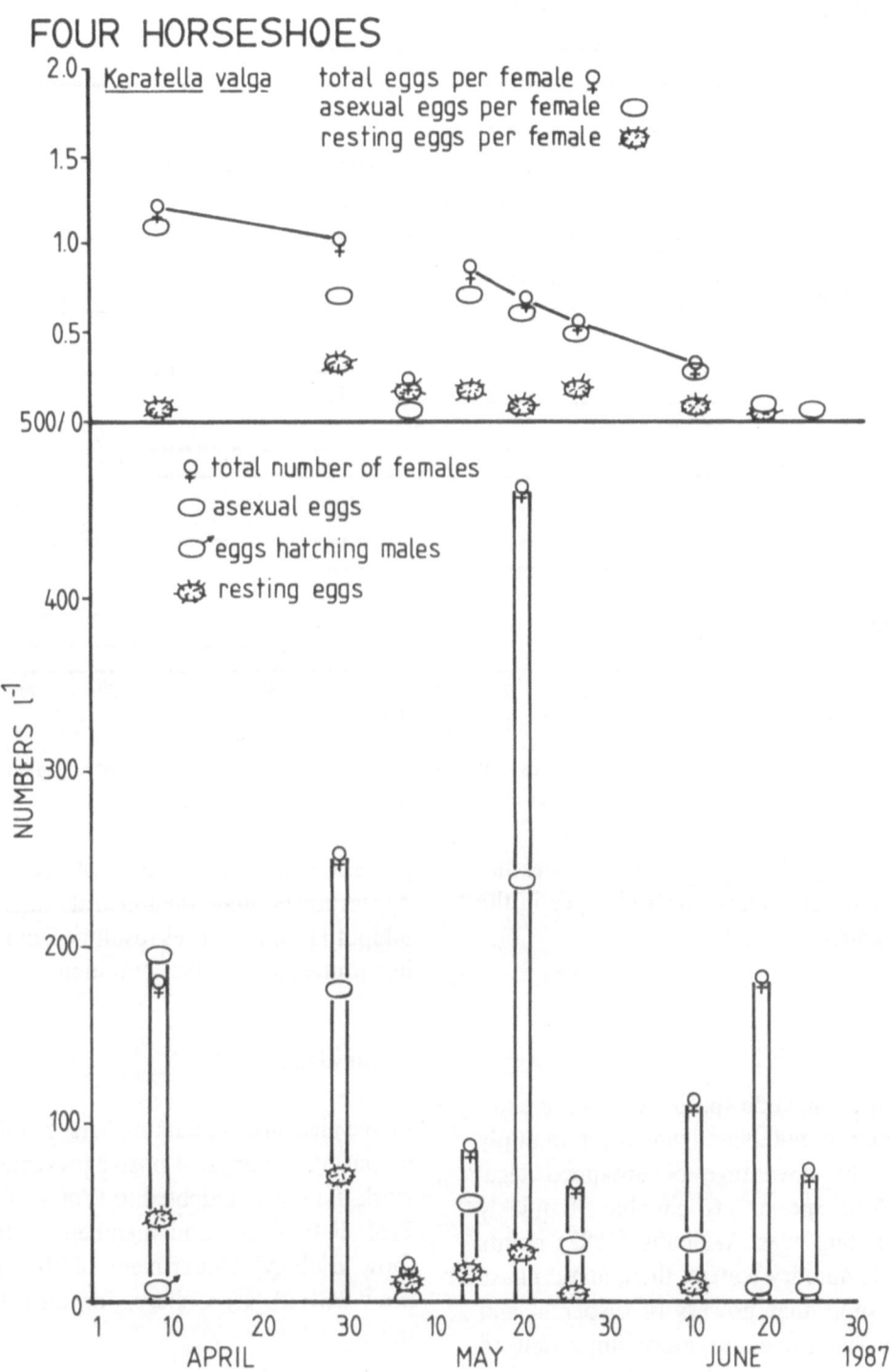

Fig. 4. Egg production in *Keratella valga* in Four Horseshoes Pond 1987. Below: numbers per litre *K. valga* females, asexual, sexual and male eggs with time. Above: ratio of eggs per female for the same period.

Bulhousen Farm Pond in its first year with 1987. *K. valga* and copepods, for example, appear to have extended their season, while *K. testudo* and *B. urceolaris* occurred at different times. These

234

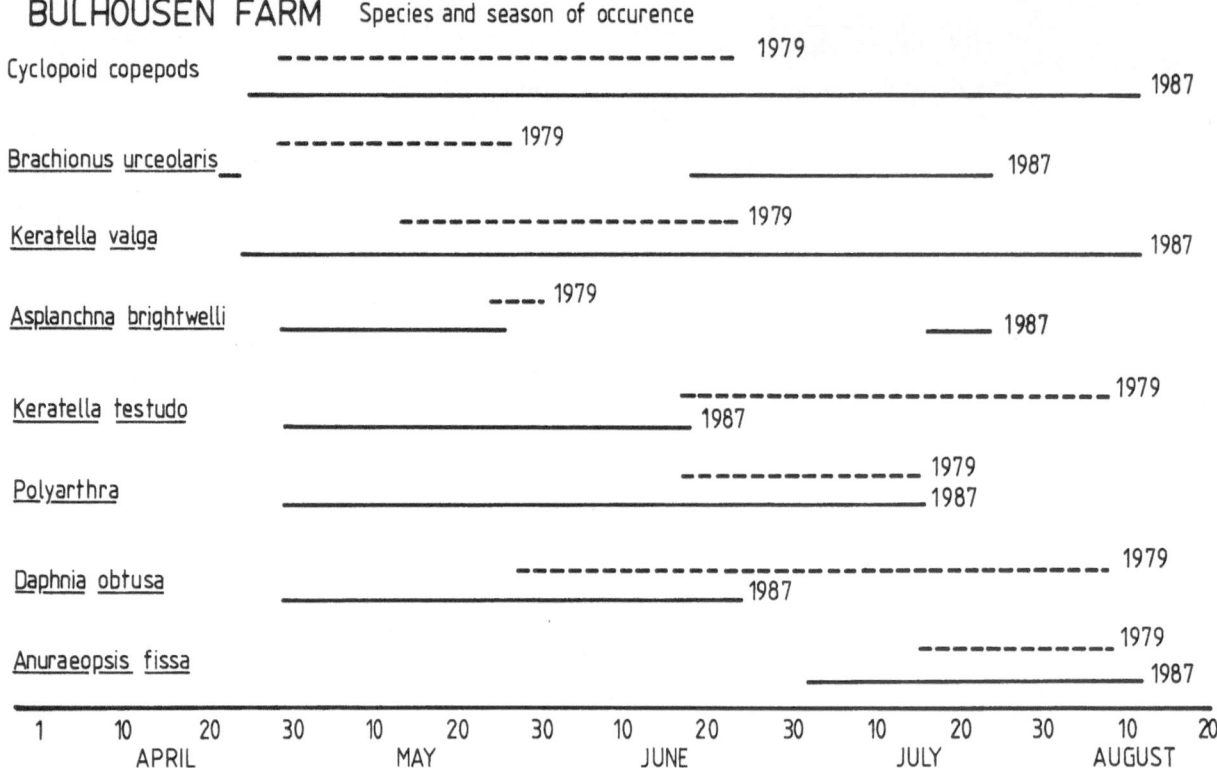

Fig. 5. Comparison of season of occurrence of species in Bulhousen Farm Pond in 1987 (8 years after digging) with the same species in 1979 (first year).

changes could be a consequence solely of weather conditions, but could also reflect changes in the total community.

Conclusions

Initial colonisers include species which are common in temporary ponds and which can multiply rapidly to take advantage of untapped algal supplies. These species are capable of quickly producing resting eggs. Williams (1987) points out that, in temporary waters, the aquatic phase may be so short that powers of dispersal and plasticity of life cycle are more important to success than competitive ability. In a persistent water body, however, competitive ability and avoidance of predation may become more significant. In these two ponds, the colonising species were still present after some years in spite of the arrival of additional species, including omnivores, which followed the populations of filterers and grazers. Changes in season of occurrence of the opportunists may demonstrate their ability to adapt to new pressures resulting from the increasing complexity of the community.

Acknowledgements

I have pleasure in thanking Mrs. Sarah Wroot for preparing the original poster presentation of this work. I am also indebted to Prof. C.T. Lewis and Prof. R.P. Dales and members of the Zoology (now Biology) Department of Royal Holloway (now R.H.B.N.) College, for their kind hospitality.

References

Pontin, A. J. & R. M. Pontin, 1980. Bullhousen Pond, Bisley, 1979. Surrey Trust for Nat. Conserv. Newsletter 55: 8–9.
Williams, D. D., 1987. The Ecology of Temporary Waters. Croom Helm, Beckenham & N. S. Wales, 205 pp.

Hydrobiologia **186/187**: 235–238, 1989.
C. Ricci, T. W. Snell and C. E. King (eds), Rotifer Symposium V.
© *1989 Kluwer Academic Publishers.*

Percentage of rotifers in spring zooplankton in lakes of different trophy

Stanisław Radwan & Barbara Popiołek
Academy of Agriculture, Department of Zoology and Hydrobiology, Akademicka 13, 20-934 Lublin, Poland

Key words: zooplankton community, rotifers, lake trophy, Poland

Abstract

Studies carried out on 8 lakes in the Łęczna-Włodawa Lakeland of eastern Poland indicated that the qualitative and quantitative structure of zooplankton was clearly correlated with the lake trophic state. In the spring zooplankton of lakes affected by gradual natural eutrophication were dominated by rotifers. In the zooplankton of lakes strongly affected by human activities, Cladocera dominated. With an increase in lake trophy there was an increase in the number of species that were indicators of eutrophy and a decrease in the number of indicators of mesotrophy. The total number of species in individual lakes tended to increase with an increase in trophy.

Introduction

Relationships between the qualitative and quantitative composition of the zooplankton community and the trophic state of lakes have been the object of numerous studies. These studies have focused on three issues: 1) determination of species composition and abundance of crustacean plankton in lakes of different trophy, 2) the relationship between the trophic state of lakes and the qualitative and quantitative structure of main zooplankton groups (Rotifera, Cladocera, Copepoda), 3) determination of groups of species that are bioindicators of lake trophy. In the present study special attention has been paid to the role of rotifers in the creation of the qualitative and quantitative structure of zooplankton in lakes of different trophic status.

Site and methods of studies

Studies have been carried out on 8 lakes in the Łęczna and Włodawa lakeland. The area of these lakes ranges from 12 to 85 ha and the maximum depth varies from 4,3 to 38,8 m. Most of the basic physical and chemical factors are at a low or medium level. Only two factors – electrolytic conductivity and calcium have high values in most of the investigated lakes. These lakes are affected by different forms of human activity. Almost all of them are under the influence of agricultural activities. Five deeper lakes with a low state of eutrophication are progressively affected by recreational activities and three lakes are affected by strong hydrotechnical activities, because they are connected by a system of canals with two rivers. The investigated lakes represent different trophic states, namely: mesotrophic, dystrophic – eutrophic and eutrophic.

In spring of 1987 in the pelagial zone of these lakes quantitative planktonic samples were collected. A 2 l apparatus of Patalas type was used in collecting samples (Patalas, 1954). The collected samples were processed according to the generally used method (Hillbricht-Ilkowska & Patalas, 1967). Three major groups of zooplankton (Rotifera, Cladocera, Copepoda) were subjected to detailed microscopical analysis.

Results and discussion

The spring zooplankton of the examined lakes was composed of 33 taxa of Rotifera, 10 taxa of Cladocera and 7 taxa of Copepoda. The qualitative composition of zooplankton in most of lakes was relatively uniform, because, as a rule, it was represented by 20 to 23 species. There were 28 species in one shallow and highly eutrophic lake. The species composition of zooplankton in the Łęczna and Włodawa lakeland was represented by many more species than in the Florida lakes of similar trophy. In each of the 8 lakes of the Łęczna-Włodawa lakeland there were 11 to 18

taxons, whereas in the Lake Conway system only 1 to 3 taxons were found (Blancher, 1984). The difference may result from the small area of the Polish lakes and from migration of benthic – periphytonic species from the littoral to the pelagial zone (Radwan, 1973; Kowalczyk & Radwan, 1982). In both lake systems rotifers dominated the zooplankton community.

In most of the lakes in the Łęczna-Włodawa lakeland undergoing gradual eutrophication the quantitative structure of spring zooplankton was similar. Rotifers predominated in these lakes. Their percentage share ranged from 60% to 94%, and was highest in the zooplankton of strongly eutrophic lakes. Eight taxa constituted the dominant group in the rotifer community. Both eurytopic species and lake trophy indicators predominated among the rotifers. In mesotrophic lakes *Conochilus unicornis* Rouss. and *Kellicottia longispina* (Kell.) preferring waters of low trophy were found among the dominants (Tab. 1, Fig. 1). However in eutrophic lakes the clear dominants were *Keratella cochlearis tecta* (Gosse) and *Keratella quadrata* (Mull.) – indicators of high trophic-status (Tab. 1, Fig. 1). A similar species

Table 1. Number of indicators of waters trophy in lakes of different trophism.

Indicator	Piaseczno	Rogóźno	Bialskie	Rotcze	Czarne Sosn.	Krzczeń	Głebokie Cyc.	Uścim.
EUTROPHIC WATERS:								
Bosmina longirostris (O.F.Mull.)		+		+	+	+	+	+
Branchionus angularis Gosse							+	+
Filinia longiseta (Ehrb.)					+		+	+
Keratella c. tecta (Goose)	+	+			+	+	+	+
Pompholyx sulcata (Huds.)		+	+		+	+	+	+
Trichocerca cylindrica (Imf.)								+
Keratella c. hispida (Laut.)	+			+	+	+	+	+
Keratella quadrata (Mull.)	+	+	+	+	+	+	+	+
Polyarthra euryptera Wierz.					+		+	
Total	3	4	2	3	7	5	8	8
MESOTROPHIC WATERS:								
Bosmina coregoni Baird.	+							
Conochilus unicornis Rouss.	+	+	+	+	+	+		+
Gastropus stylifer Imh.	+	+		+				
Kellicottia longispina (Kell.)	+	+	+	+	+	+		+
Total	4	3	2	3	2	2		2

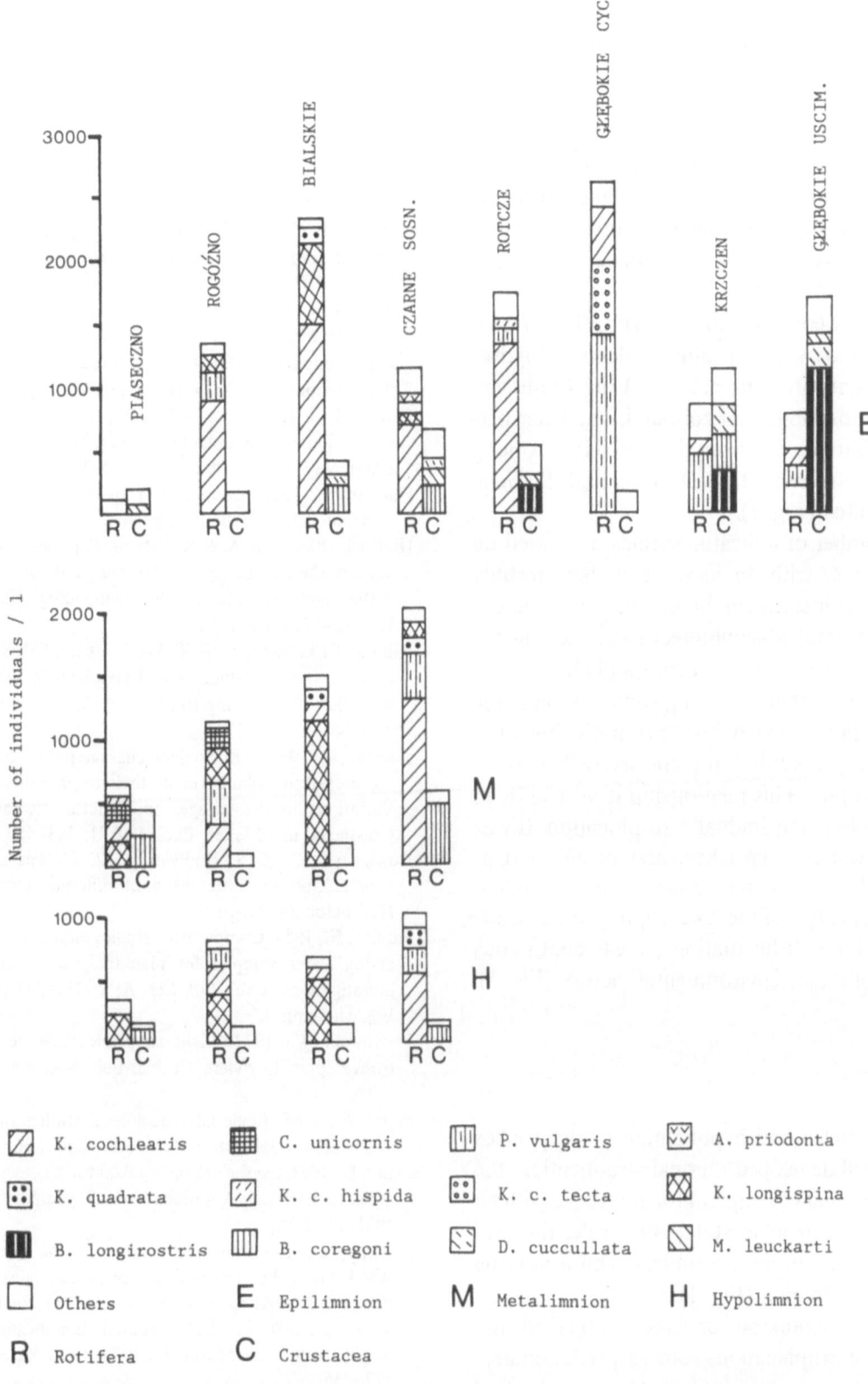

Fig. 1. Abundance structure of zooplankton in particular thermal zone in lakes of different trophy.

238

composition of dominants has been found in other Polish lakes (Hillbricht-Ilkowska & Węgleńska, 1970; Radwan, 1976; Karabin, 1983; Radwan *et al.*, 1984) and in other countries (Pejler, 1965, 1981; Ferrari & Rossi, 1970; Blancher, 1984).

The crustacean plankton in the examined lakes was scarce and of a lowest abundance in highly eutrophized lakes. In slightly eutrophic lakes Cladocera were the predominant crustacean plankton, but in highly eutrophic lakes the Copepoda were predominant (Fig. 1). In two lakes affected by strong human influences (hydro-technical and breeding activities) the Cladocera were the most abundant zooplankton. Their percentage share ranged from 51% to 54%. Among them small filter-feeding *Daphnia* and *Bosmina* predominated (Fig. 1).

The number of indicator species depended on trophic state, with an increase in lake fertility there was an increase in the number of indicators of eutrophy and a simultaneous decrease in the number of mesotrophy indicators (Tab. 1).

The total numbers of spring zooplankton in the examined lakes ranged from 576 individuals l^{-1} to 2,667 individuals l^{-1}. It increased with increasing lake trophy. This relationship is very clear in 6 lakes undergoing gradual eutrophication. It was the strongest in deep lakes, first of all, in their meta- and hypolimnion. However, in the epilimnion, irrespective of the lake trophic status, these were considerable fluctuations, due to continuous changes of major environmental factors (Fig. 1).

Conclusions

1. In the meta- and hypolimnion of deep lakes with well developed thermal stratification, the total number of zooplankton increases with an increase in trophic state. As a rule, the predominating species are rotifers. This is not true of the epilimnion (Fig. 1).
2. In the zooplankton of lakes subjected to gradual eutrophication, rotifers predominate, however, in the zooplankton of lakes subjected to intensive human activite cladocerans from the genera *Daphnia* and *Bosmina* predominate (Fig. 1).

3. With an increase in lake trophy there is an increase in the number of indicators of eutrophy and a decrease in the number of indicators of mesotrophy.

References

Blancher, E. C., 1984. Zooplankton – trophic state relationship in some north and central Florida lakes. Hydrobiologia 109: 251–263.

Brzęk, G., C. Kowalczyk, W. Lecewicz, S. Radwan, W. Wojciechowska & I. Wojciechowski, 1975. Influence of abiotic environmental factors on plankton in lakes of different trophy. Pol. Arch. Hydrobiol. 22: 123–139.

Ferrari, I. & O. Rossi, 1976. Compositions and seasonal succession of the zooplankton in lake Mergozzo (Northern Italy). Me. Ist. ital. Idrobiol. 33: 85–104.

Hakkari, I., 1972. Zooplankton species as indicators of environment. Aqua Fenn. pp. 46–54.

Hillbricht-Ilkowska, A. & K. Patalas, 1967. Methods of estimate of the production and biomass and some problems of the quantitative zooplankton methodology. Ekol. Pol. B, 13: 139–172 (in Polish).

Hillbricht-Ilkowska, A. & T. Węgleńska, 1970. Some relations between production and zooplankton structure of two lakes of a varying trophy. Pol. Arch. Hydrobiol. 17: 233–240.

Karabin, A., 1983. Ecological characteristics of lakes in north-eastern Poland versus their trophic gradient. VII. Variations in the pelagic zooplankton (Rotatoria and Crustacea) in 42 lakes. Ekol. Pol. 31: 383–409.

Kowalczyk, C. & S. Radwan, 1982. Groups of pelagic zooplankton in three lakes of different trophy. Acta Hydrobiol. 24: 39–51.

Patalas, K., 1954. Comparative studies on a new type of self acting water sampler for plankton and hydrochemical investigations. Ekol. Pol. Ser. A, 2: 231–242 (in Polish, English summ.).

Pawłowski, L., 1972. Liste supplementaire de Rotiferés trouvés dans la riviére Grabia. Bull. Soc. Sci. Łódź 21: 1–10.

Pejler, B., 1965. Regional – ecological studies of Swedish freshwater zooplankton. Zool. Bidr. Upps. 36: 407–515.

Pejler, B., 1981. On the use of zooplankton as environmental indicators. Some Approaches to Saprobiol. Problems 1981: 9–12.

Radwan, S., 1973. Pelagic rotifers of the Łęczna and Włodawa Lake District. Faunistic-ecological studies. Rozpr. Nauk. Akad. Roln. Lublin 8: 1–57, (in Polish).

Radwan, S., 1976. Planktonic rotifers as indicators of lake trophy. Ann. Univ. Mariae Curie-Skłodowska, Sec. C, 31: 227–235.

Radwan, S., C. Kowalczyk, W. Kowalik & W. Zwolski, 1984. Littoral fauna in two lakes of different trophy (Bialskie and Piaseczno, Eastern Poland). Verh. int. Ver. Limnol. 22: 991–995.

Hydrobiologia **186/187**: 239–245, 1989.
C. Ricci, T. W. Snell and C. E. King (eds), Rotifer Symposium V.
© *1989 Kluwer Academic Publishers.*

Tasmania revisited: rotifer communities and habitat heterogeneity

R.J. Shiel[1], W. Koste[2] & L.W. Tan[3]

[1] *Murray-Darling Freshwater Research Centre, P.O. Box 921 Albury, N.S.W. 2640, Australia*; [2] *Ludwig Brill Strasse 5, Quakenbrück D-4570, Federal Republic of Germany*; [3] *Western Mining Corporation, P.O. Box 409 Unley, S.A. 5061, Australia*

Key words: rotifers, Tasmania, dune lakes, endemism

Abstract

The results of four field surveys for Rotifera in Tasmania are summarized. Most new species and records in a 1987 survey were from acid waters (pH < 4.0) of dune lakes on the west coast (42° S). Marked intra- and interhabitat differences in rotifer communities of lakes and ponds were demonstrated by cluster analysis and related to habitat heterogeneity.

Introduction

Our first survey of Tasmanian natural lakes on the central plateau (1980) provided 62 new rotifer records (one n. sp., *Aspelta tilba*). In 1984 the survey was extended to impoundments and a range of smaller ponds and stock dams, with 62 further taxa (including n. spp. *Lecane tasmaniensis, Lepadella tyleri* and *Testudinella unicornuta*) added to the known Rotifera (Koste & Shiel, 1987a). In 1985, a third survey (100 sites, including all sites samples earlier), added 63 taxa (including *Brachionus lyratus tasmaniensis, Cephalodella lindamaya, Lepadella tana* and *Testudinella mucronata tasmaniensis*) (Koste & Shiel, 1986).

To provide seasonal information from the same 100 localities, a fourth survey was made in spring 1987, with the addition of eight new sites in a series of dune lakes on the west coast of the island. We had not investigated dune lakes previously; work on the mainland (e.g. Timms, 1973) suggested that dune lakes were not productive of Rotifera. Indeed, in a review of these habitats, Timms (1986: 425) noted that rotifers seem to be

excluded by acidity; 'several taxa, including... rotifers... are rare or absent in acidic dune lakes'. Our attention was directed to these lakes by Dr P.A. Tyler at the University of Tasmania, who had sampled them for phytoplankton.

The 1987 survey resulted in 159 identified taxa, 59 of these first records for the island (with *Lecane herzigi, Notommata tyleri, Trichotria buchneri* and *T. pseudocurta* new) (Koste *et al.*, 1988). Most of the new records came from the western plateau area (Fig. 1). Some 250 taxa of Rotifera are now known from the island, and it is likely that only a fraction of the rotifer fauna has been encountered. We summarize here the results of the four surveys to date, and discuss some of the peculiarities of the Tasmanian limnetic and littoral rotifer communities relative to those of the nearby mainland.

Sampling sites

All sites described in the earlier reports were resampled in 1987. Habitats ranged from shallow roadside pools, ponds, turbid stock dams, to large

240

Fig. 1. Tasmania, with nine lakes mentioned in text. Arrows indicate approximate locations of endemic species figured.

impoundments and natural lakes above 1000 m A.S.L. on the central plateau. A number of stream and river sites also were sampled. For convenience, the sites are grouped here into (a) lakes, (b) ponds, and (c) streams, however the division is arbitrary in view of the range of habitats within each category.

Species

Species lists are provided in each of the survey reports cited above, considerably extending the known distribution of most taxa recorded. New species apparently confined to Tasmania are shown in Fig. 1. Notably absent from Tasmania are *Brachionus* species known from the mainland (28 taxa) (Koste & Shiel, 1987b). Only *B. angularis* was common in Tasmanian waters, with occasional occurrences of *B. calyciflorus*, *B. quadridentatus* and a local variant of *B. lyratus* (Fig. 1). The most abundant taxa in limnetic assemblages were species of *Keratella*, with nine identified and four (*K. australis*, *K. cochlearis*, *K. procurva* and *K. slacki*) widely distributed. Also abundant were species of *Lecane* (28 spp.), with four taxa common in the limnetic regions of ponds and stock dams (*L. bulla*, *L. flexilis*, *L. hamata* and *L. lunaris*), and an endemic (*L. tasmaniensis*) occurring widely in acid waters in the southwest. *Trichocerca* species (26) were widespread, but only *T. similis* abundant.

Of 250 taxa recorded, 116 (48%) were found only once (or twice from the same locality in different surveys). Such patchiness is generally in accord with Australian mainland surveys (Shiel & Koste, 1986) and with studies reported elsewhere (Dumont, 1983). Approx. 20% (49 spp.) have not been reported from the mainland; in most cases these are acidophile species known previously from tropical regions of South America or South East Asia, e.g. *Testudinella ahlstromi*, *Trichocerca braziliensis* (Koste, 1978). About half the recorded taxa were of limited distribution (<20% of localities). Only 10 species (4%) occurred in >20% of samples (*n* = 267): *K. slacki* (38%), *T. similis* (35%), *K. cochlearis* (32%), *K. australis/L. flexilis* (30%), *L. lunaris* (28%), *L. bulla* (23%), *B. angularis/Polyarthra vulgaris* (21%) and *K. procurva* (20%).

Comparison of species distribution data from the Tasmanian series with those from a more intensive sampling effort reported by Hillman (1986) from the R. Murray catchment on the mainland suggests that the Tasmanian rotifer fauna is more diverse (Table 1), i.e. the R. Murray data represent >1500 samples from a 300 000 km² area, the Tasmanian data 267 samples from a 68 000 km² area. More species occur in Tasmanian standing waters, with a marked decrease in streams possibly reflecting the steeper gradients and greater turbulence at Tasmanian sites. R. Murray gradients are gentler,

Table 1. Summary of rotifer recorded from Tasmanian and Australian mainland habitats. *Lakes* includes artificial impoundments; *Ponds* includes turbid stock dams. (R. Murray data after Hillman, 1986).

	No. of species	
	Tasmania (240)	R. Murray (229)
1. Lakes	148	126
2. Ponds	189	174*
3. Streams	50	73
4. Lakes and ponds	59	85
5. Lakes and streams	3	45
6. Ponds and streams	10	54
7. Lakes. ponds and streams	41	39

* Billabongs

the flows slower, and a complex rotifer plankton is maintained.

A simple plot of species numbers against sampling frequency (Fig. 2) suggests that more rotifer species await discovery in Tasmania. Typically, such plots produce an asymptote as repeated surveys produce fewer unknown taxa. Our sample series has not reached an asymptote, although first records have declined (90%, 47%, 32%, 28% for respective surveys). Given that we have collected from only 100 of >4000 available sites, it is conceivable that more rotifer taxa ultimately will be recorded from Tasmania than presently are known from the mainland (about 600 species).

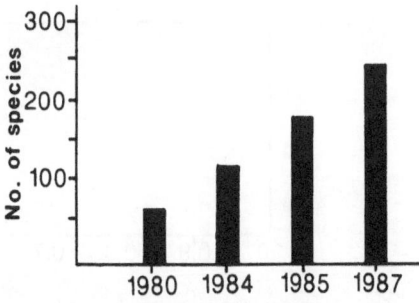

Fig. 2. Cumulative species numbers per sampling survey, Tasmania 1980-1987.

Community composition

Using Sokal's (1961) distance measure to compare species assemblages (number of taxa and proportion of each) across all sites in all years, the most striking feature of the rotifer assemblages was the marked dissimilarity in species composition between habitats, even within the same habitat category. The whole data set is too large to reproduce here, however two smaller subsets are used below as examples.

(a) Lakes

In summary, over the four surveys, the natural lakes on the central plateau had the most diverse rotifer assemblages, although not necessarily the most species. The rotifer assemblages (1987 survey) of ten sites (mean no.spp. site^{-1} = 10.4; mean diversity (H') = 2.31) are compared by average linkage cluster analysis in Fig. 3a. Relative proximity of the lakes is shown in Fig. 1.

Although communities of the ten lakes were ranked of low similarity (0-22% taxa shared) by Jaccard coefficient run concurrently with Sokal's niche distance, clustering grouped lakes in close proximity, e.g. Great Lake and Arthur's Lake, both dominated by *Polyarthra vulgaris*, and sharing *Gastropus minor* and *Filinia longiseta* from 16 and 10 species, respectively. Similarly, Plateau (16 spp.) and Augusta (11 spp.) were both domi-

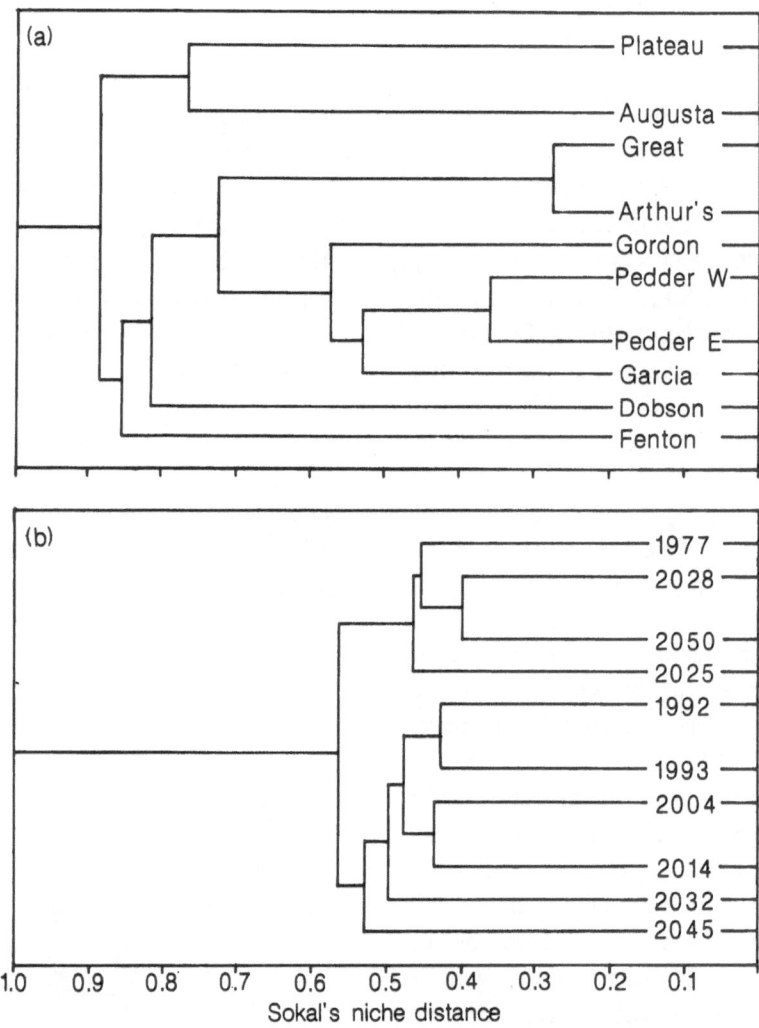

Fig. 3. Average linkage dendrograms (Sokal's niche distance) for (a) ten lake sites and (b) ten ponds/stock dams (indicated by site #). 1987 survey.

nated by *Synchaeta oblonga* and shared *Euchlanis meneta*, *Lepadella patella* and *Lecane flexilis*.

Generally, the widely distributed taxa (*K. cochlearis*, *P. vulgaris/dolichoptera*, *F. pejleri/longiseta* and *T. similis*) comprised the shared species; the remaining taxa were specific to each lake. No two lakes had the same combination of temperature, pH and conductivity (within the range 7–14 °C, pH 4.0–7.7, 9.0–33.0 μS cm^{-1}), suggesting other differences in chemistry which may account in part for the disparity of resident rotifer assemblages by their influences on the algal and bacterial communities.

Notably, a net tow from Lake Garcia, a dune lake first sampled in the 1987 survey (Fig. 1) (17.0 °C, pH 4.3, 98.3 μS cm^{-1}) had 35 rotifer species, including 8 new records (3 n. spp., with one (*Trichocerca*) yet to be described) and the highest diversity (*H'* = 4.37) yet recorded from Tasmania. The assemblage was a mixture of limnetic and littoral taxa indicative of shallow eutrophic waters, including *K. procurva* (dominant), 5 spp. of *Lecane* and 8 of *Trichocerca*.

Seasonal changes in species composition within each of the lakes were reflected in low similarities between surveys, with only 1-3 of the common species present on successive visits. In the larger lakes, e.g. L. Pedder, there were appreciable community differences at opposite ends of the lake on the same day. In L. Pedder in the 1987 series (cf. Fig. 3a), community dominants at the eastern end of the lake were *K. cochlearis/Synchaeta pectinata/Conochilus dossuarius*, while at the western end, 12 km away, dominants were *K. australis/F. pejleri/C. dossuarius* with only 20% of taxa shared.

(b) Ponds

Smaller standing waters commonly had 8-20 tychoplanktonic species, with dominants one or more species of *Keratella* (the most common in ponds and stock dams was *K. slacki*, followed by *K. australis* > *K. cochlearis* > *K. procurva* > *K. javana*, with incidental occurrences of *K. quadrata*, *K. valga*, *K. hispida* and *K. tropica*). Three sites had six co-occurring *Keratella* species. Other

abundant pond rotifers were species of *Polyarthra*, *Lecane*, *Filinia*, *Trichocerca* and *Synchaeta*.

As in the lake communities, there was little similarity in inter-site rotifer species composition on each survey. Fig. 3b shows clusters produced for ten unnamed pond sites (mean no. spp. site^{-1} = 17.3; mean *H'* = 3.25) selected from widely separated areas of the island i.e. 1977, 1992, 1993 are north coast, 2004 and 2014 east coast, 2025 near L. Pedder, 2050 near Garcia and so on. Most obviously, tighter clustering suggests that the pond rotifer communities are more uniform than those of the lakes, i.e. more species are shared. Again, the analysis grouped sites in closest geographical proximity, e.g. 1992/1993 and 2004/2014, with outliers (1977, 2032, 2045) the most isolated sites.

The most diverse pond rotifer assemblage (*H'* = 3.92) and greatest number of species (32) occurred in a small black-water humic roadside pool in the dune lakes area (site 2050, near L. Garcia, Fig. 1). This pool had the lowest pH (3.1) of the eight sites in the dune series (pH range 3.1–4.3). Dominant taxa were: *K. javana*, *Lecane tasmaniensis*, *Synchaeta oblonga*, *Cephalodella mucronata* and *Trichotria buchneri*. The high species diversity at this site accounts for the anomalous grouping with 2028 in Fig. 3b, which was not geographically close, but which also was dominated by the acidophile *K. javana*, and shared other acidophile taxa (incl. *L. flexilis* and *L. tasmaniensis*) of 15 spp. present.

The ten pond data set was run against the ten lake set to compare the respective assemblages. The dendrogram derived (Fig. 4) maintains the separation of all lakes except Garcia, which clustered with the pond assemblage, sharing five of the widely distributed taxa with site 1992, and supported our impression of a mixed limnetic/littoral assemblage from the shallow margin of this lake.

The dune lakes

The unexpectedly high species richness of the dune lake series is clearly in contrast with the

244

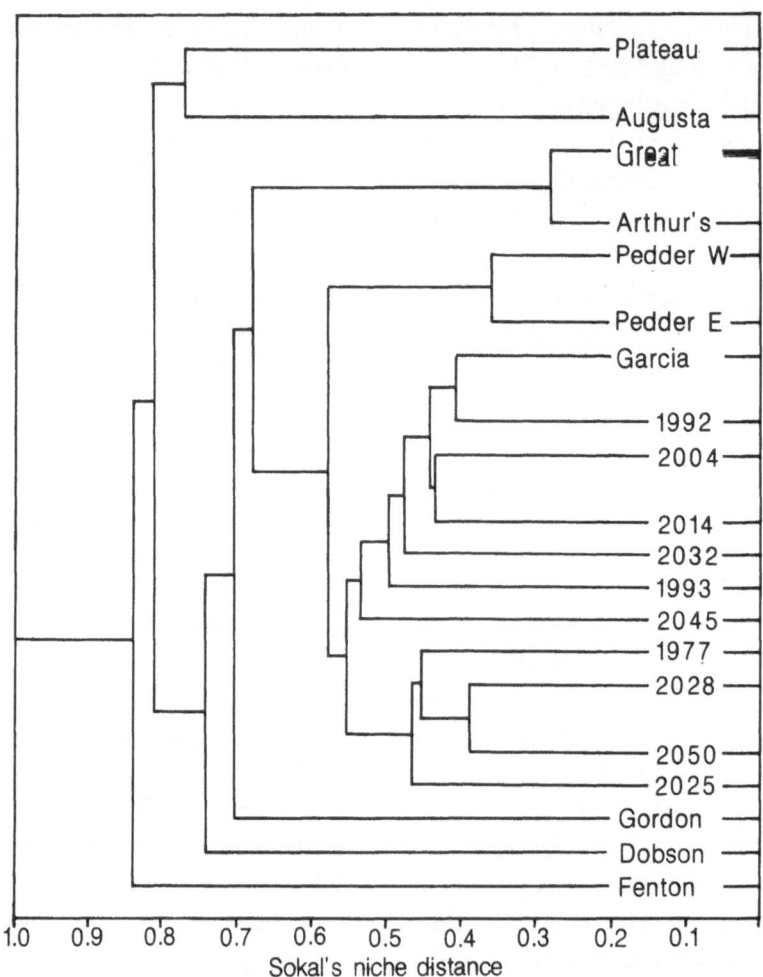

Fig. 4. Sokal's niche distance for combined ponds/lakes data sets. L. Garcia separated from lakes, clustered with ponds.

observations of Timms (1986), and may represent a real difference between the dune lakes of Tasmania and the mainland. The rotifer communities of these acid waters are among the richest yet recorded from Australia. A possible explanation is provided by recent geological evidence reviewed by De Deckker (1986), which suggests that these western Tasmanian waters have a long, undisturbed evolutionary history.

At the height of the last glaciation (>20 000 years B.P.), only part of Tasmania was covered by ice. It is likely that some of the many lakes, particularly in coastal areas, remained ice-free. Geological evidence points to the persistence of some of these lakes during an extremely arid climatic phase 18 000 years B.P., when their waters may have been refugia for aquatic species as the mainland dried (De Deckker, 1986; Koste &

Shiel, 1986). Many thousands of years isolation during periods of climatic change could account for the high species diversity and predominance of taxa now recorded primarily in the tropics, but which may have persisted in Tasmanian waters as relict populations from a time when Tasmania's climate, and that of southern Australia, was tropical. Notably, the west coast of Tasmania, directly in the path of prevailing westerlies, has the highest rainfall in southern Australia (ca. 5 m annually); even shallow pools are permanent.

Rotifer biogeography in Tasmania

Dumont's (1983) question 'Do rotifers have a biogeography?' was applied in the global context, with evidence that although most species were cosmopolitan or widely distributed, some were remarkably restricted. Our Tasmanian surveys must be regarded as preliminary, however it is apparent that Tasmanian rotifers have their own biogeography.

If an arc is drawn on Fig. 1 approximately from Great Lake southwest through *Aspelta tilba*, east of Lakes Dobson and Fenton to *B. lyratus tasmaniensis*, the eastern margin of the central plateau is demarcated. We have named this arc 'Tyler's Line'. It separates two broad rotifer assemblages: west of Tyler's Line, predominantly acidophiles, contains 11 of the 13 endemic taxa and most of the species which have not been recorded from the mainland. Waters are dark-black humic acid type, low in electrolytes, and presumably differ markedly in characteristics such as nutrients and phytoplankton, and rotifer composition from those east of the line. To the east, which is a rainshadow area, waters are less humic, some are alkaline, and electrolytes are higher, including some saline and hypersaline waters on the east coast, i.e. they reflect geological and climatic differences between the two areas.

The heterogeneity of rotifer assemblages across the island appears to be a response to local, even microhabitat, variations in their requirements.

Acknowledgements

In 1987 survey was funded by the Australian Biological Resources Study. We thank Peter Tyler for help with field work, and for hospitality and use of facilities at Hobart. Initial MS preparation was at the Botany Department, University of Adelaide; thanks to the then Chairman, George Ganf, for access to facilities. Particular thanks to Terry Hillman, MDFRC, for cluster analysis, and to Kim Jenkins for graphics.

References

De Deckker, P., 1986. What happened to the Australian aquatic biota 18 000 years ago? In P. De Deckker & W. D. Williams (eds) Limnology in Australia. CSIRO & Dr W. Junk BV, Melbourne and Dordrecht: 487–496.

Dumont, H. J., 1983. Biogeography of rotifers. Hydrobiologia 104: 19–30.

Hillman, T. J., 1986. Billabongs. In P. De Deckker & W. D. Williams (eds) Limnology in Australia. CSIRO & Dr W. Junk BV, Melbourne and Dordrecht: 457–470.

Koste, W., 1978. Rotatoria. Die Rädertiere Mitteleuropas. Bestimmungswerk begründet von Max Voigt. 2 vols. Borntraeger, Stuttgart.

Koste, W. & R. J. Shiel, 1986. New Rotifera (Aschelminthes) from Tasmania. Trans. r. Soc. S. Aust. 110: 93–109.

Koste, W. & R. J. Shiel, 1987a. Tasmanian Rotifera: affinities with the Australian fauna. Hydrobiologia 147: 31–43.

Koste, W. & R. J. Shiel, 1987b. Rotifera from Australian inland waters. II. Epiphanidae and Brachionidae (Rotifera: Monogononta). Inv. Taxon. 1: 949–1021.

Koste, W., R. J. Shiel & L. W. Tan, 1988. New rotifers (Rotifera) from Tasmania. Trans. r. Soc. S. Aust. 112 (in press).

Shiel, R. J. & W. Koste, 1986. Australian Rotifera: ecology and biogeography. In P. De Deckker & W. D. Williams (eds) Limnology in Australia. CSIRO & Dr W. Junk BV, Melbourne and Dordrecht: 142–150.

Sokal, R. R., 1961. Distance as a measure of taxonomic similarity. Syst. Zool. 10: 71–79.

Timms, B. V., 1973. A limnological survey of the freshwater coastal lakes of east Gippsland, Victoria. Aust. J. mar. Freshwat. Res. 24: 1–20.

Timms, B. V., 1986. The coastal dune lakes of eastern Australia. In P. De Deckker & W. D. Williams (eds) Limnology in Australia. CSIRO & Dr W. Junk BV, Melbourne and Dordrecht: 421–432.

Hydrobiologia **186/187**: 247–254, 1989.
C. Ricci, T. W. Snell and C. E. King (eds), Rotifer Symposium V.
© 1989 *Kluwer Academic Publishers.*

Occurrence of Rotifera in the field under natural and intentionally-changed conditions. II. Lake Numasawa

M. Sudzuki, T. Matsumoto[1] & K. Narita[1]
Biological Laboratory, Nihon Daigaku-University, Omiya-shi Saitama-ken, Japan; [1]*Naisuimen Fisheries Experimental Station, Inawashiro-cho, Fukushima-ken, Japan*

Key words: seasonal-vertical occurrences, power-water circulation, saprobic rotifers

Abstract

Seasonal and vertical occurrences of representative rotifer species were recorded together with such taxa as Cyanophycea, Phytomastigophorea, Bacillariophycea, Protozoa, Rotifera and Crustacea, from 1982 to 1986 at two sites S1 (natural) and S2 (nearby water circulated since 1952). 1) The following species were observed from S1: *K. hiemalis, C. ovalis, N. labis, L. patella* and *Anuraeopsis* sp., from S2: *B. urceolaris* and *A. ecaudis,* 2) *P. hudsoni* appeared in 1984 as a successor to *P. truncatum,* 3) *K. longispina* was negatively associated with *P. t. vulgaris,* 4) *Synchaeta* spp suddenly appeared at both sites in 1985, 5) *K. cochlearis, B. calyciflorus, Proalides* sp., *Diurella* sp., *N. labis, B. urceolaris* and bdelloids did not appear until 1986, 6) The following species decreased or disappeared: *A. p. herricki* and *Collotheca* sp. since 1982, *C. coenobasis, C. hippocrepis* and *K. hiemalis* since 1984, 7) A complicated relation was observed between rotifer population density and other plankton. 8) Occurrence of Rotifera seemed to be affected by circulation of an electric power plant.

Introduction

Seasonal and vertical occurrences of representative plankton in a lake, locally called Numasawa-numa, were investigated from 1982 to 1986 as a continuation of a 1980-81 project (Sudzuki *et al.*, 1983). In this paper, a correlation between some 10 rotifer species and 6 plankton groups is figured and tabulated.

Materials and methods

All the plankton were concentrated by pulling or towing a Rigosha's plankton net (diameter = 20 cm, mesh size = 47 μm, messenger = 500 gr) through 4 m of water. Samples were collected, at least once per season, except winter, from 0 to a 50 m depth at 7-11 different strata (0,5,10,15,20,30,40 m occasionally 25,35,45 and 50 m) at 2 sites of a lake. S1 is natural lake water and S2 is nearby water that has been circulated by an electric power plant. A total of 256 samples was examined.

Further detail for materials and methods is given in Sudzuki *et al.* (1983).

Results and discussion

Results are presented in Figs. 1-5 and Tables 1-2. We may make the following observations: 1) *Polyarthra t. vulgaris* is considered to be eurythermal and perennial, but prefers to concentrate in the

248

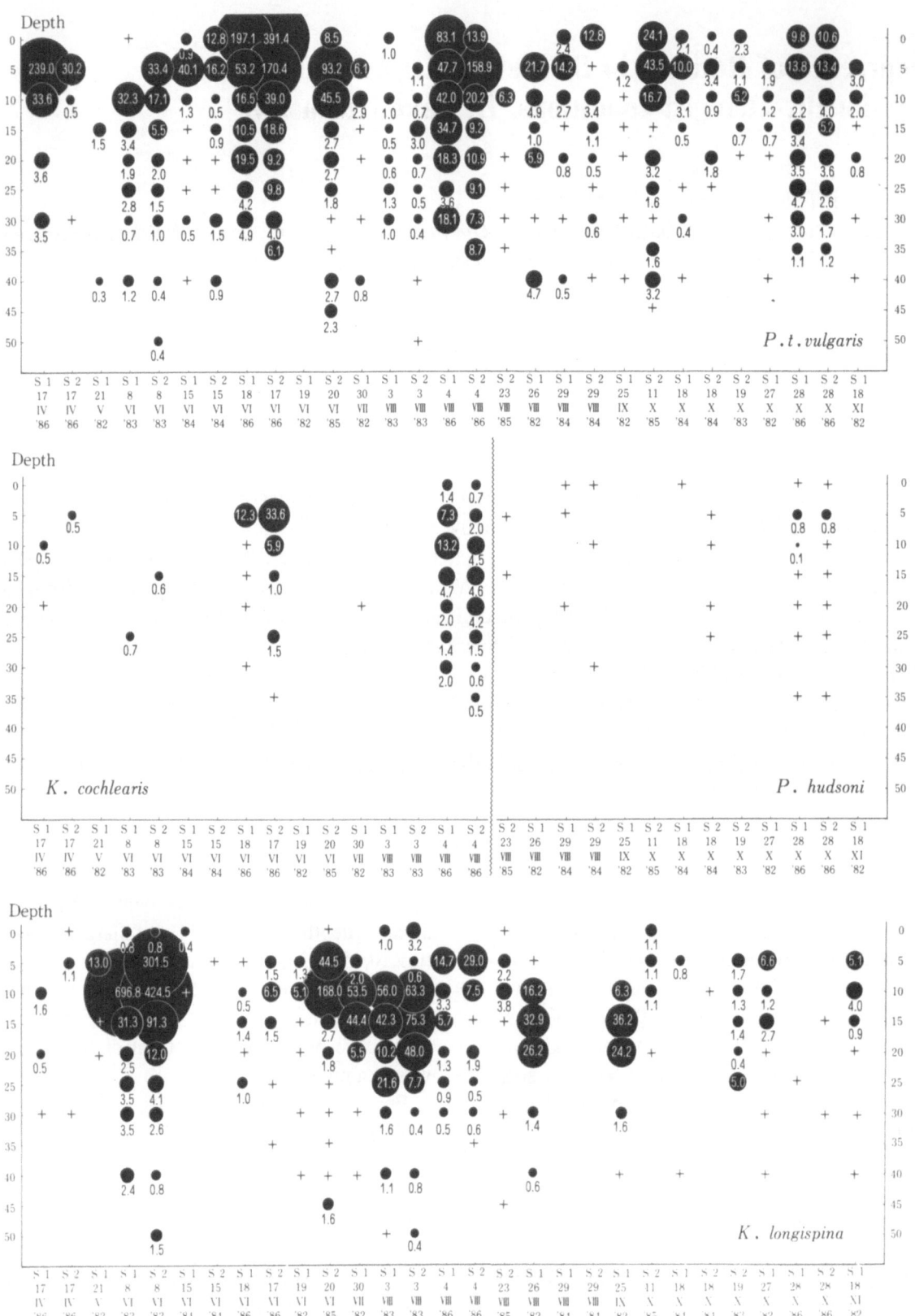

Fig. 1. Seasonal and vertical occurrences of 4 rotifers in lake, Numasawa during 1982-86. S1 = natural lake water, S2 = nearby water circulated by an electric power plant since 1952. 17 IV '86 = 17th April, 1986, Depth = depth in meter, numbers in circles = individual density per liter (as to diameter of circles see Carlin, 1943), + = individual density less than 0.1.

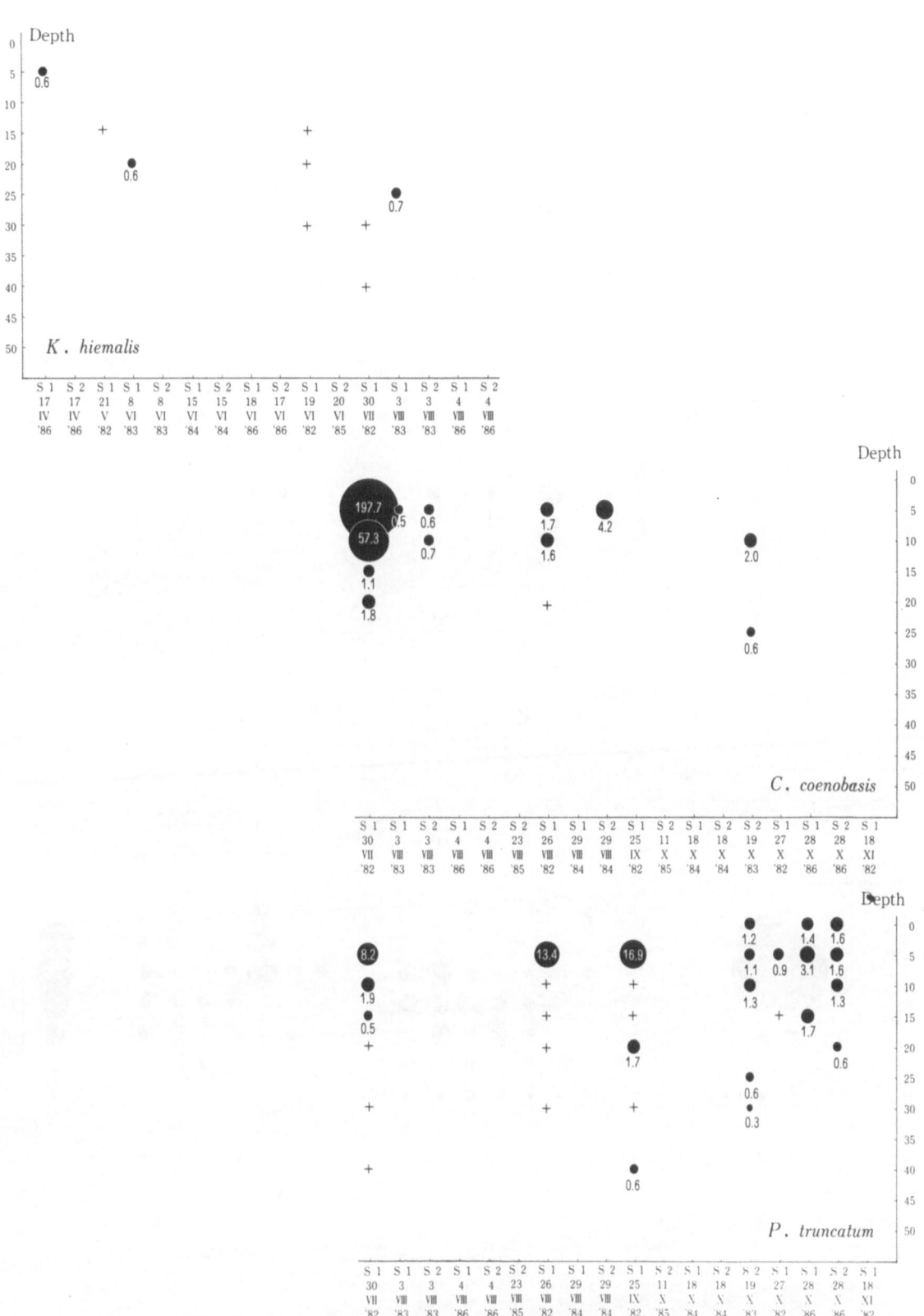

Fig. 2. Seasonal and vertical occurrences of 3 rotifers in lake, Numasawa during 1982-86.

250

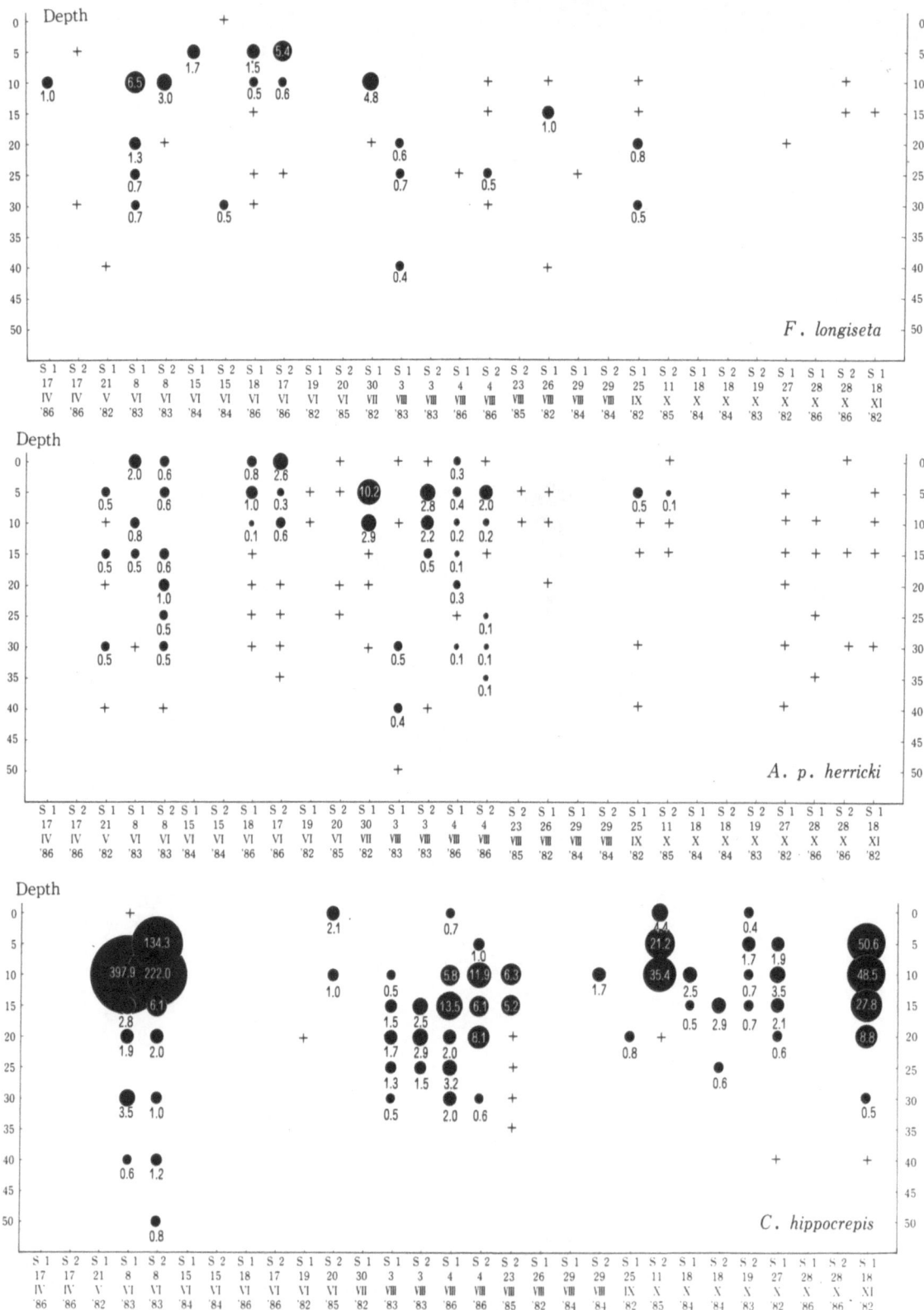

Fig. 3. Seasonal and vertical occurrences of 3 rotifers in lake, Numasawa during 1982-86.

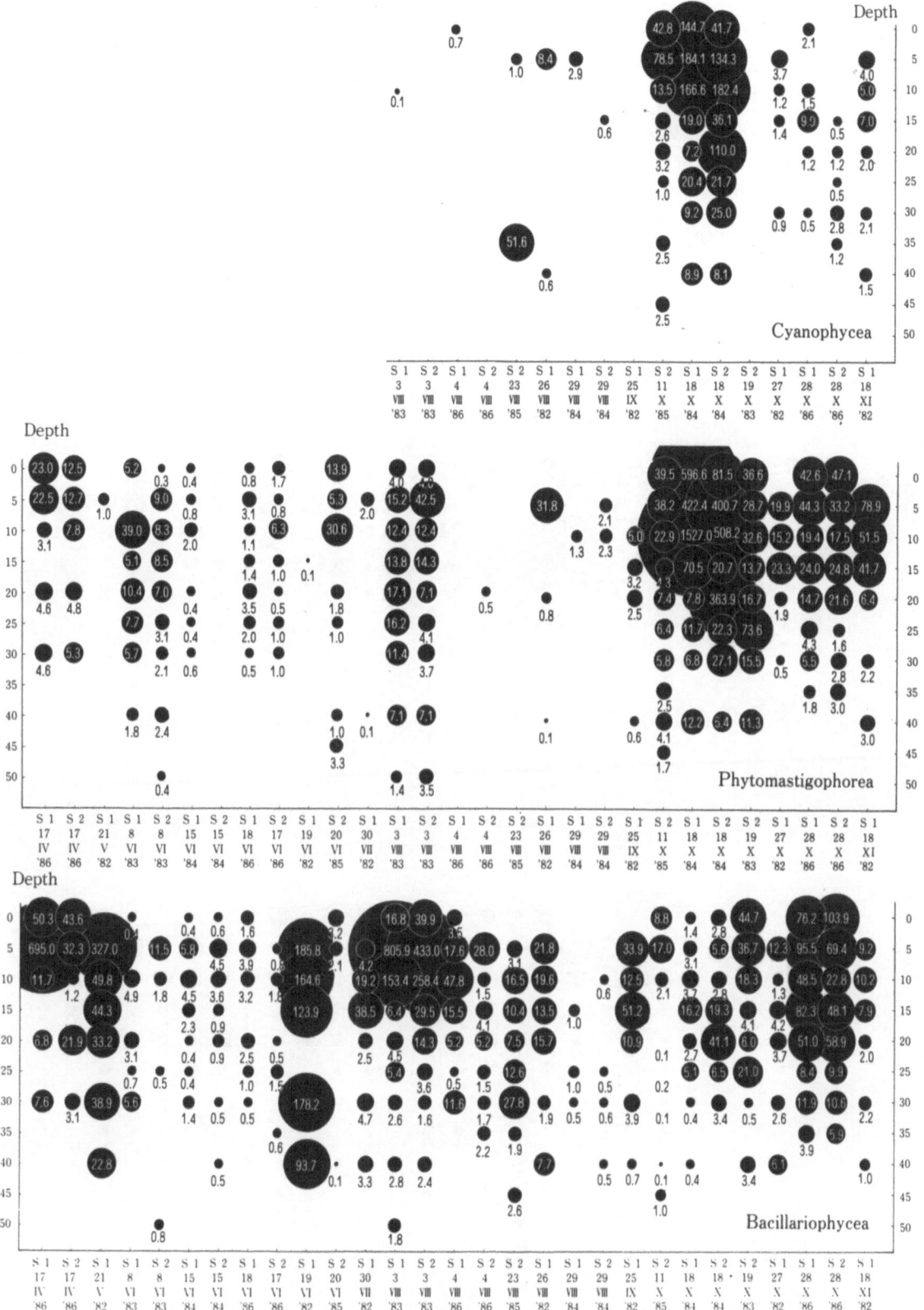

Fig. 4. Seasonal and vertical occurrences of phytoplankton in lake, Numasawa during 1982-86.

252

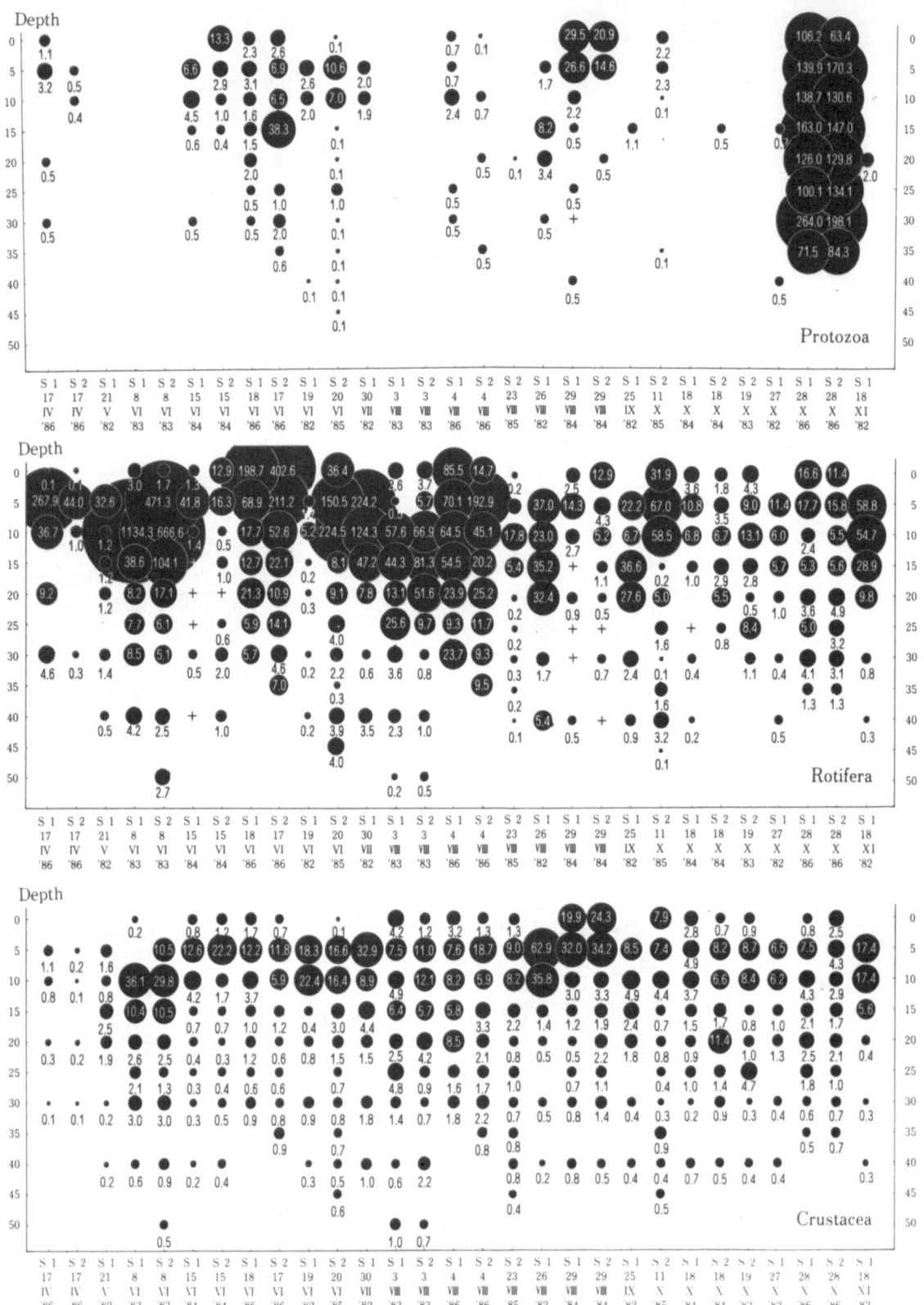

Fig. 5. Seasonal and vertical occurrences of zooplankton in lake, Numasawa during 1982-86.

Table 1. List of taxa investigated.

Rotifera
Polyarthra trigla dolichoptera, P.t. vulgaris, Synchaeta sp. 1, *S.*
sp. 2, *Ploesoma hudsoni, P. truncatum, Ascomorpha ecaudis,
Chromogaster ovalis, Kellicottia longispina, Notholca labis,
Anuraeopsis* sp., *Keratella cochlearis, K. hiemalis, Brachionus
calyciflorus, B. urceolaris, Lepadella patella, Lecane inermis,
L. nodosa, L.* sp., *Monostyla quadridentata, Proalides* sp.,
Diurella sp., *Filinia longiseta, Asplanchna priodonta herricki,
Conochiloides coenobasis, Conochilus hippocrepis, Collotheca*
sp., Bdelloidea, etc.
Cyanophycea
Gomphosphaerium sp., etc.
Mychophycea
Bacillariophycea
Asterionella spp., *Cyclotella* spp., *Tabellaria* spp., etc.
Chlorophycea
Oocystis spp., *Chroococcus* spp., *Dictyosphaerium* spp,.
Pandorina spp., *Eudorina* spp., *Uroglenopsis* spp., *Closterium*
spp., etc.
Phytomastigophorea
Peridineae, *Dinobryon* spp., *Mallomonas* spp., *Ceratium* sp.,
etc.
Protozoa
Zoomastigophorea, Actinopodea, *Centropyxis* spp., *Difflugia*
spp., *Cyphoderia* sp., *Euglypha* spp., Holotricha, *Epistylis* spp.,
Zoothamnium spp., other Spirotricha, etc.
Trematoda (larva), Nematoda, *Chaetonotus* spp.,
Crustacea
Ceriodaphnia sp., *Daphnia* spp., *Alona* spp., *Bosmina* spp.,
Bosminopsis sp., *Holopedium* sp., *Polyphemus* sp.,
Diaphanosoma sp., Copepoda, Nauplii, shrimps
Insecta
Chironomus spp., Collembola
Arachnida
Acarina, Araneae.

high abundance in Numasawa-numa. Nevertheless, *P. hudsoni* increases year after year. It is regarded as a summer form (Carlin, 1943; Pejler, 1957) or oligo-mesotrophic (Mäemets, 1983), although Sladeček (1973) regarded it as oligo-saprobic. 5) *Kellicottia longispina* is considered to be eurythermal and perennial. 6) Varga (1954) mentioned that *K. longispina* is a glacial relict in Lake Balaton and adapted to rising water temperatures. However, Larsson (1971) classified it as a perennial summer form even though it has not been detected from tropical waters (Sudzuki, 1989) nor ponds in temperate regions (Sudzuki, 1964). 7) *K. longispina* abundance is negatively correlated with *P. t. vulgaris*, a fact which was observed by Laxhuber (1987). 8) *Keratella hiemalis* is treated as a winter or cold tolerant stenothermal species from the hypolimnion (Carlin, 1943; Pejler, 1961). In contrast, it rarely is encountered in Numasawa-numa. 9) *Conochiloides coenobasis* is considered to be a warm water hypolimnion inhabitant. 10) *Ploesoma truncatum* is considered to be a summer inhabitant in the hypolimnion. 11) *P. hudsoni* appears as a successor of *P. truncatum*, since *truncatum* was common until 1983 but a shift from *truncatum* to *hudsoni* happened in 1984-85. In 1986 *hudsoni* became dominant. 12) *Filinia longiseta* is considered to be the eurythermal-perennial of the hypolimnion. Nearly the same results were obtained (Pejler, 1957); George & Fernando, 1969; Larsson, 1971 & Herzig, 1987). The species studied by Carlin (1943), Pejler (1961), Berner-Fankhauser (1983) and Laxhuber (1987) appeared seldom or never in summer. 13) *Asplanchna p. herricki* also is eurythermal and perennial. The same kind of result was reported (Berner-Fankhauser, 1983) in contrast to Carlin's findings (1943). 14) *Conochilus hippocrepis* is considered to be eurythermal and perennial, but concentrates in the hypolimnion. 15) A correlation between the occurrence of Rotifera and other plankton groups is not always simple. For example, *P. t. vulgaris* occurrence is correlated with Phytomastigophorea but varies inversely with Bacillariophycea; *F. longiseta* abundance is negatively correlated with Bacillariophycea; *A. p. her-*

epilimnion as already depicted by Carlin (1943), Sudzuki (1955), George & Fernando (1969), Larsson (1971), Berner-Fankhauser (1983), Herzig (1987) and Laxhuber (1987). 2) *Keratella cochlearis* was not perennial in Numasawa-numa contrary to Carlin (1943), Sudzuki (1955, 1962), Pejler (1961), George & Fernando (1969), Larsson (1971), Berner-Fankhauser (1983), Herzig (1987) and Laxhuber (1987). 3) When *K. cochlearis* was common in 1986, it suggested certain pollution (beta-oligosaprobity: Sladeček, 1973 or meso-eutrophic: Mäemets, 1983, but a conflicting case in Neusiedlersee in 1979: Herzig, 1987) 4) *Ploesoma hudsoni* has not yet reached

Table 2. Seasonal and vertical occurrences of uncommon rotifers. Density = Number of individuals per liter, Date = day/month/year, Depth = depth in meter, S1 = site 1 (natural lake water), S2 = site 2 (circulated water by an electric power plant).

Taxa	Density	Date	Depth	Site
Synchaeta sp. 1	0.1–27.7	IV–X/84–86	0–45	S1S2
S. sp. 2	3.0–11.6	20/IV/85	5–10	S2
Ascomorpha ecaudis	0.1– 0.7	4/VIII/86	10–35	S2
Chromogaster ovalis	0.5– 4.2	28/X/86	0–30	S1
Notholca labis	0.6	17/IV/86	5	S1
Anuraeopsis sp.	0.5	4/VIII/86	15	S1
Collotheca sp.	0.5– 5.6	VIII–X/83–85	0–30	S1S2
Brachionus calyciflorus	0.5– 1.7	17–18/VI/86	0–30	S1S2
B. *urceolaris*	2.6	17/VI/86	0	S2
Lepadella patella	0.7	28/X/86	0	S1
Lecane nodosa	0.4– 0.6	VI–VIII/83	0–15	S1S2
L. *inermis*	0.6– 5.3	VI–X/85–86	0–30	S2
L. sp. 1	1.0– 5.3	20/VI/85	0–30	S2
Monostyla quadridentata	0.1– 1.0	VI,VIII/85	0–40	S2
Proalides sp.	1.0– 1.1	17–18/VI/86	20	S1S2
Diurella sp.	0.5– 0.6	28/X/86	20–30	S1S2
Bdelloidea	0.2– 4.3	VI–X/86	0–35	S1S2

ricki is inversely related to Phytomastigophorea and so on.

From Tab. 2 we may make some observations: 1) The following species have appeared mainly in S1 or natural lake water: *Chromogaster ovalis, Notholca labis, Lepadella patella* and *Anuraeopsis* sp. 2) The following mainly in S2 or the water circulated by an electric power plant: *Brachionus urceolaris, Ascomorpha ecaudis, Monostyla quadridentata, Synchaeta* sp.2, *Lecane* sp.1. 3) The following have a tendency to increase their individual densities at both sites (S1,S2): *Brachionus calyciflorus, Synchaeta* sp.1, *Proalides* sp., *Diurella* sp. and Bdelloidea. 4) The following have a tendency to decrease at both sites: *Lecane nodosa, Collotheca* sp. 5) From the fact above we may consider that the occurrence of Rotifera may be affected by circulation of an electric power plant.

References

Berner-Fankhauser, H., 1983. Abundance, dynamics and succession of planktonic rotifers in Lake Biel, Switzerland. Hydrobiologia 104: 349–352.

Carlin, B., 1943. Die Planktonrotatorien des Motalaström. Medd. Lunds Univ. Limnol. Inst. 5: 1–256.

George, M. G. & C. H. Fernando, 1969. Seasonal distribution and vertical migration of planktonic rotifers in two lakes. Verh. int. Ver. Limnol. 17: 817–829.

Herzig, A., 1987. The analysis of planktonic rotifer populations. Hydrobiologia 147: 163–180.

Larsson, P., 1971. Vertical distribution of planktonic rotifers in a meromictic lake. Norw. J. Zool. 19: 47–75.

Laxhuber, R., 1987. Abundance and distribution of pelagic rotifers in a cold, deep oligotrophic lake (Königssee). Hydrobiologia 147: 189–196.

Mäemets, A., 1983. Rotifers as indicators of lake types in Estonia. Hydrobiologia 104: 357–361.

Pejler, B., 1957. Taxonomical and ecological studies on planktonic Rotatoria from northern Swedish Lapland. K. svensk Vetensk Akad. Handl. 4: 3–52.

Pejler, B., 1961. The zooplankton of Osbysjön, Djursholm. I.Oikos 12: 225–248.

Sladeček, V., 1973. System of Water quality from the biological point of view. Ergeb. Limnol. 7: 218 p.

Sudzuki, M., 1955. Life histories of some Japanese rotifers *I. Polyarthra trigla*. Sci. Rep. Kyoiku Daigaku 118: 41–64.

Sudzuki, M., 1962. Class Rotatoria. In: Uchida Ed. Systematic Zoology 2: 9–74 (Nakayama-shoten, Tokyo).

Sudzuki, M., 1964. New systematical approach to the Japanese planktonic Rotatoria. Hydrobiologia 23: 1–124.

Sudzuki, M., 1983. Occurrence of Rotifera in the field under natural and conditions. Hydrobiologia 104: 341–347.

Sudzuki, M., 1989. Rotifera from the Oriental region and their characteristics. Special Issue for Centennial Anniversary of Nihon Daigaku 3: 301–343.

Varga, L., 1954. A relict of the ice age in the water fauna of lake Balaton. Annal. Biol. Tihany 22: 227–234.

Hydrobiologia **186/187**: 255–278, 1989.
C. Ricci, T. W. Snell and C. E. King (eds), Rotifer Symposium V.
© 1989 *Kluwer Academic Publishers.*

The skeletal muscles of rotifers and their innervation

Pierre Clément[1] & Jacqueline Amsellem[2]
[1]*Laboratoire Ethologie, Equipe Neuro-Ethologie, IASBSE,* [2]*Laboratoire Histologie Expérimentale
UA-CNRS 244, ICBMC, Université Lyon I, bâtiment 403, 69622 Villeurbanne, France*

Key words: muscles, innervation, behavior, cytology, ultrastructure, rotifers, smooth muscles, striated muscles, motor units, phylogeny

Abstract

The skeletal muscles of rotifers are monocellular or occasionally bicellular. They display great diversity of cytological features correlated to their functional differentiation. The cross-striated fibers of some retractors are fast contracting and relaxing, with A-band lengths of 0.7 μm to 1.6 μm, abundant sarcoplasmic reticulum and dyads. Other retractors and the circular muscles are tonic fibers (A band > 3 μm), stronger (large volume of myoplasm) or with greater endurance (superior volume of mitochondria/myoplasm). All of these retractor muscles are coupled by gap junctions and are innervated at two symmetrical points; they constitute two motor units implicated in withdrawal behaviour.

The muscles inserted on the ciliary roots of the cingulum control swimming. They are multi-innervated and each of them constitute one motor unit. They have characteristics of very fast fibers; the shortest A-band length is 0.5 μm in *Asplanchna*.

All the skeletal muscles of bdelloids are smooth or obliquely striated as are some skeletal muscles of monogononts. These muscles are well suited for maximum shortening and are either phasic or tonic fibers.

All rotifer skeletal muscles originate from ectoderm and contain thin and thick myofilaments whose diameters are identical to those of actin and myosin filaments in vertebrate fast muscles or in insect flight muscles. There are no paramyosinic features in the thick myofilaments. The insertion, innervation, coupling by gap-junctions and other cytological differentiations of rotifer skeletal muscles are reviewed and their phylogeny discussed.

Introduction

Muscles of rotifers can be grouped into two classes. The visceral muscles lie along the digestive tract and surround other internal organs. The second class are the skeletal muscles which control body form, movement, and spatial displacement. This paper reviews only skeletal muscles; visceral muscles will be treated in another review.

The first rotiferologists described skeletal circular muscles as anucleate (see review by Remane,

1929–1932; Hyman, 1951; de Beauchamp, 1965). Electron microscopy showed that every rotifer muscular fiber has a nucleus (Clément, 1977) and that each skeletal muscle is uni- or bicellular.

Studying skeletal muscles and their innervation is essential to analyze the behavior of rotifers. Their small size (generally from 0.1 to 1 mm) makes an electrophysiological approach difficult and this has not yet been attempted. Nevertheless, this technique would be of great interest because each skeletal muscle is made of only one

256

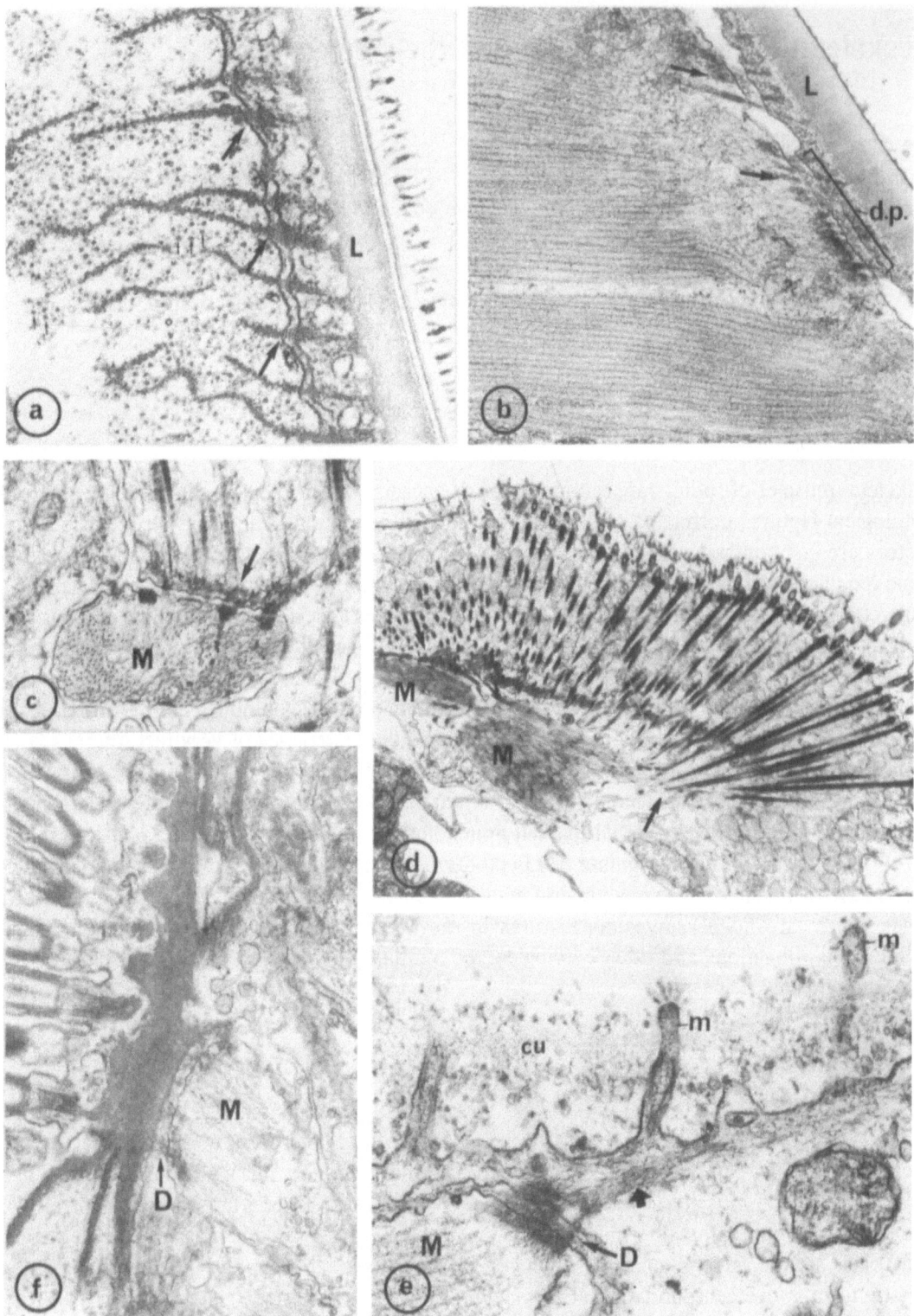

Plate I: Insertions of the skeletal muscles.

Fig. a. Philodina roseola. The oblique arrows show some of the musculo-epithelial desmosomes. The ribbon-like material is inserted on the desmosomes and goes through the transversally sectioned myofilaments (small arrows). At right, a dense material

or two fibers. The fiber lengths are as much as $2/3$ of the body lengths and this review will show that each fiber is differentiated and can be observed in living animals (light microscopy) through their transparent integument.

From the beginning of the century, the transparence of the integument has permitted localization of the musculature in monogononts (Remane, 1929–1932) and bdelloids (Brakenhoff, 1937). However, no functional analysis was done because the cyto-histological observations with light microscopy were not precise enough.

Electron microscopy increased our knowledge of skeletal muscles, with two kinds of observations:

1) Precise cytology from serial ultrathin sections permitted quantification of cytological parameters which have been correlated with muscular characteristics (Josephson, 1975; Hoyle, 1983; Nicaise & Amsellem, 1983). These include speed and strength of contraction, endurance, maximum shortening and speed of relaxation.

2) Precise localization of innervations or gap junctions permits definition of motor units. There are very few motor units in rotifers and their identification is essential for the analysis of behavior (Clément et al., 1983; Clément, 1987).

Ultrastructural observations of rotifer muscles were first available from analysis of other organs, like the integument (Koehler, 1965, 1966; Clément, 1969, 1977, 1981, 1985, 1987; Schramm, 1979). An exhaustive study of muscles

in the monogonont *Trichocerca rattus* was made from seriated ultrathin sections of contracted or relaxed animals (Amsellem & Clément, 1977, 1981, 1988). This review gives information about this work which is largely unpublished and on unpublished results obtained in our laboratory on several other species of rotifers: the bdelloid *Philodina roseola* and six monogononts: *Brachionus urceolaris sericus* (female and male) *B. calyciflorus* (collaboration with A. Cornillac), *B. plicatilis* (collabortion with A. Luciani) *Asplanchna brightwelli* (collaboration with A. Cornillac), *B. plicatilis* (collaboration with E. Wurdak), *Rhinoglena frontalis, Notommata copeus.*

Phylogenetic relationships are also proposed, based on ultrastructural analysis of rotifer skeletal muscles. Clément (1981) has used several other ultrastructural criteria to show that rotifers are very primitive metazoa. Finally, we present new evidence showing the extraordinary diversity of specializations of rotifer skeletal muscles.

The epithelial origin of the skeletal muscles of rotifers

Vertebrate cross-striated muscle fibers come from mesoderm whereas most smooth fibers and cardiac muscle cells come from the mesenchyme. A few smooth muscular cells, like those around the acini of the salivary glands, come from ectoderm and are thus called 'myo-epithelial' cells. In sponges myocytes which are the first expression

---

joins each desmosome to the skeletal lamina densa (L) which is intracytoplasmic in the syncitial integument of the trunk. The cuticle is external. × 60 000.

Fig. b. Trichocerca rattus. Muscle dense bodies are juxtaposed to desmosomial plates (d.p.) where thin myoflimaments are also inserted. Thick myofilaments are on the left. In the syncitial integument, in front of each dense body, dense material joins the desmosome to the skeletal intracellular lamina (L) of the trunk. × 32 000.

Fig. c. Brachionus plicatilis. Cingulum ciliary roots are inserted on a muscle cell (M) (transversally sectioned): at the level of the desmosome, the cross-striated ciliary roots insert in a submembranar fibrillar material (arrow). × 21 000.

Fig. d. Brachionus calyciflorus. The cingulum cell contains oblique and axial sections of ciliary roots associated with numerous mitochondria. The muscle (M) inserts on a basal fibrillar lamina (arrow at left) and directly on the basis of the ciliary roots (arrow at right). × 5000.

Fig. e. Brachionus calyciflorus. Apex of the corona: a desmosome (D) joins the muscle (M) to the microfibrillar subepithelial layer (thick arrow). Microvilli (m) cross the cuticle (cu). × 39 000.

Fig. f. Trichocerca rattus. The desmosome (D) joins the muscle to the dense material of a buccal cell. Buccal cilia are on the left. × 36 000.

258

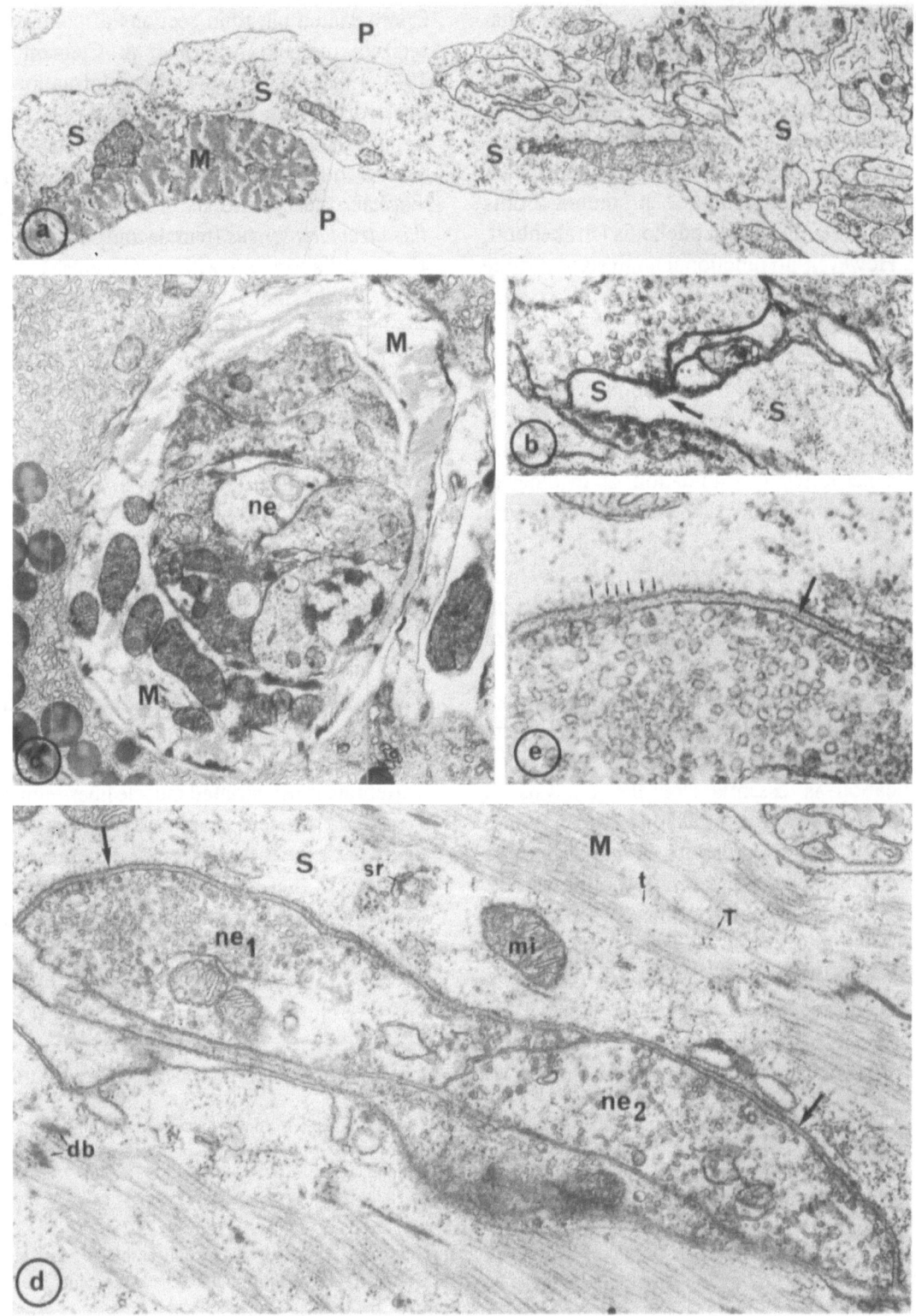

Plate II: Innervation

Fig. a. Brachionus plicatilis. Cross section of a CRM, showing the myoplasm (M) and the sarcoplasmic extension (S) which penetrates into the brain (at right). P = pseudocoel. × 5800.

of a muscular differentiation are mesenchymatous cells (Hoyle, 1983; Pavans de Ceccatty, 1974). In Cnidaria and Ctenaria, myoepithelial cells are juxtaposed to mesoglean muscular cells (e.g. the giant smooth muscular fiber in the ctenophore *Beroe*: Hernandez-Nicaise & Amsellem, 1980).

In rotifers, there is neither mesenchyme nor typical conjunctive tissue (Clément, 1980). All the skeletal muscle cells come from ectoderm. Contrary to nematodes, rotifer muscle cells only have a muscular function, but there are some exceptions to this rule. They can have an epithelial function when adjacent to neurons. For instance, around the rotifer brain there are bordering epithelial cells with associated muscle cells (rotifers have no glial cells, Clément 1977, 1980). More spectacular is a cylindrical muscle in which a part of a nerve is lodged (Plate II–c). This socket cell is a classical specialization of epithelial cells around a sensory nerve as in the nematode *Caenorhabditis elegans* (Ward *et al.*, 1975; Ware *et al.*, 1975).

Rotifer muscle fibers have another characteristic resembling the muscle cells from the body wall of the nematode *Ascaris* (Rosenbluth, 1965) where each cell sends an extension to the peripheral longitudinal nerve cord for innervation. Paired rotifer central retractor muscles each send a sarcoplasmic arm inside the brain where the two symmetric arms contact by a large gap junction and are innervated by cerebral neurons (Plate II, Clément, 1987; Amsellem & Clément, 1988). Similar muscle extensions with neuromuscular synapses have been observed in Platyhelminths (Chien & Koopowitz, 1972) in the polychaete *Syllis*, the onychophore *Peripatus*, a starfish and in the prochordate *Amphioxus* (Hoyle, 1983).

In rotifers, even in the absence of a real sarcoplasmic extension, the muscular synapses occur often on the peripheral sarcoplasm near the nucleus when it is peripheral (Clément, 1987). Nevertheless, numerous neuro-muscular synapses occur near the myoplasm of the skeletal fibers even in bdelloïds (Plate II – d).

The insertions of the skeletal muscles of rotifers

Insertions on the peripheral skeleton

Each ending of circular muscles is inserted on the skeletal syncytial integument of the trunk and sometimes at the level of muscle Z material. Longitudinal muscles constitute two groups: one where muscle fibers join the trunk to the articulations of the foot and toes (Clément, 1987) and another were muscle fibers join the anterior rotatory apparatus to the trunk (Plate I – c, d, e).

Muscle insertions on the skeletal integument always implicate the thin filaments which terminate in the dense peripheral material of a desmosomial structure. From the epithelial part of the desmosome, a fibrillar dense material goes through the epithelium to join the internal sheet of the intracytoplasmic internal skeletal lamina (Koehler, 1965; Clément, 1969, 1985). In the primitive insertions of bdelloids, punctual desmosomes are regularly spaced (Plate I – a; Koehler, 1966; Schramm, 1978; Clément, 1981, 1985) and the Z dense material (but is it Z dense material or others structural proteins?) corresponding with the desmosomes penetrates deeply into the muscle cell (Plate I – a). In monogononts, punctual desmosomes are grouped in desmosomial plates (Plate I – b). Such juxtapositions have

---

Fig. b. Trichocerca rattus. Innervation (arrow) at the end of the sarcoplasmic extension of a CRM, inside the brain. × 66 400.

Fig. c. Philodina roseola. A cross section of a cylindrical muscle (M) which is a socket cell for a nerve (ne = a transversal section of neurite). Part of the gastric gland at left. × 13 300.

Fig. d. Philodina roseola. Multi-innervation (arrows) of a skeletal muscle (M = myoplasm, s = sarcoplasm) by two nerve endings (ne1 and ne2). In the muscle fiber sections surrounding the nerves, note thin (t) and thick (T) myofilaments, dense bodies (d.b.), sarcoplasmic reticulum (sr) and some mitochondria (mi). × 33 200.

Fig. e. Philodina roseola. Detail of Fig. d., showing the neuromuscular synapse. Note the numerous synaptic vesicles, the characteristic dense line intercellular (thick arrow) and the post-synaptic submembranar material (little arrows). × 66 400.

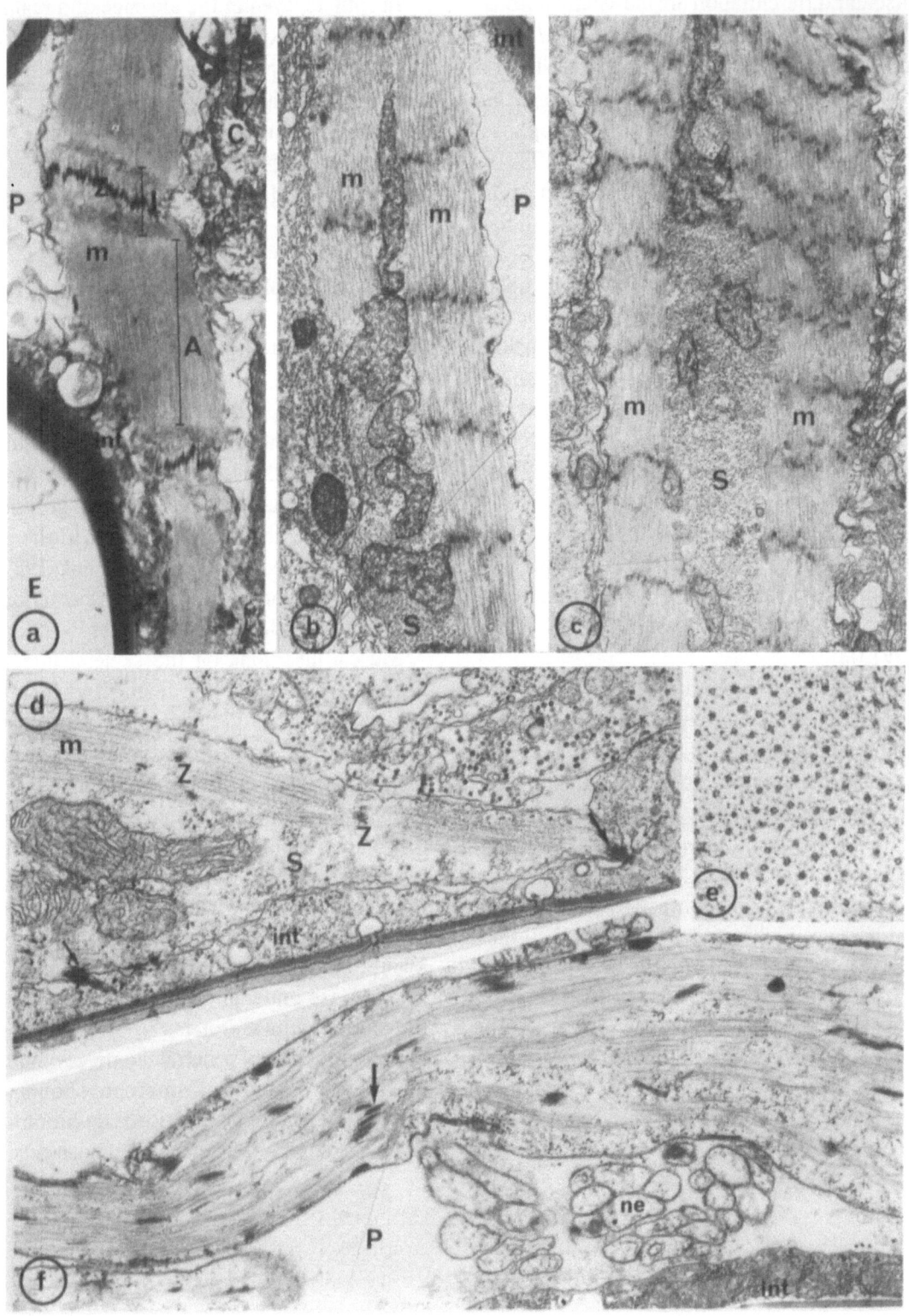

Plate III: Fig. a, b, c. Axial sections of the retractors of *Trichocerca rattus.*

    Fig. a. Relaxed LRM (lateral retractor muscle): A band (A) and I band (I) and lined Z bodies in the myoplasm (m). × 12000.

    Fig. b. Contracted VRM (ventral retractor muscle). The sarcoplasm (s) with mitochondria and glycogen particles, is juxta-

been described in *Sagitta* (Duvert, 1971). The insertions on the supple integument (anterior integument surrounding the rotatory apparatus, or constituting the pedal articulations) are of the same type but with less spectacular desmosomes.

Insertions on the ciliary of the rotatory apparatus

Muscle insertions on the ciliary cells of the cingulum and of the pseudotrochus are unique. Ciliary roots or a fibrillar material which is in close contact with them ends in the epithelial part of the musculo-epithelial desmosomes (Plate I – d, c and Clément, 1987; Plate II, Clément & Fouillet, 1971; Clément, 1977, 1985). With this structure, the muscle can control the ciliary beat.

The retractors establish classic single desmosomes with the anterior cells or desmosomes in relation with the microfilamentous network which lies at the base of the microvilli (Plate I – e) or desmosomes related with the uniformly dense material which is at the base of the buccal cilia (Plate I – f). These musculo-epithelial desmosomes have a mechanical role in the retraction of the rotatory apparatus by these retractor muscles.

Musculo-muscular junctions

Rotifer skeletal muscles are generally unicellular but some are bicellular as the central retractors (CRM) of *Trichocerca*, *Brachionus* and other species; the anterior cell is short and ramified.

The two cells are closely joined by two types of junctions: mechanical ones, which are desmosomes and hemidesmosomes (Plate IV – a) and gap-junctions which are electrical couplings of the two muscular cells (Plate IV – a). In other cases, the retractor is inserted on other muscles such as the ventral retractor muscle (VRM) on the peripheral muscles of the mastax.

In conclusion, each skeletal muscle insertion plays a precise functional role. Some mechanical insertions are involved in animal deformations like coronal withdrawal behavior, movements of the foot or of spines and paddles. More functionally complex are the insertions on the cingulum and pseudotrochus ciliary roots involved in the control of ciliary beat.

Cross-striated muscles implied in the withdrawal behavior of *T. rattus*

The most extensive study of skeletal muscles in rotifers was made on the retractor muscles involved in the withdrawal of the ciliary corona in *T. rattus* (Amsellem & Clément, 1977, 1988). Transversal serial ultrathin sections of one animal were used to quantify cytological features of each muscle cell and longitudinal, oblique or transversal sections on several other animals. The different muscles implicated in withdrawal behavior, which is the most common in rotifers (Clément *et al.*, 1983; Clément, 1987) are the following.

1) Three pairs of retractors: central (CRM),

---

posed to a lamina of myoplasm (m) which reply on itself at the upper part of the picture. Note the supercontraction at the level of Z line. int: rigid integument of the trunk. × 12 000.

Fig. c. Contracted CRM (central retractor muscle). Axial sarcoplasm (S) in a cylindrical myoplasm (m). Supercontraction is more pronounced at the upper part of the picture. × 12 000.

Thick myofilaments are shorter in b and c than in a.

Fig. d. Axial section of the base of a CRM of *Brachionus urceolaris sericus* male; the myoplasm (m) is cylindrical around the sarcoplasm (s) and attached to the integument by two desmosomes (arrows). × 20 000.

Fig. e. Transverse section of a CRM of *T. rattus*. Thick and thin myofilaments. Note the hollow center of the thick ones. × 100 000.

Fig. f. Axial section of a tonic smooth muscle of *Philodina roseola*. Thick myofilaments are very long. Some dense bodies can be juxtaposed (arrow) but they are generally scattered in the myoplasm. × 19 500.

C: ciliated cell; E: external medium; P: pseudocoel; ne: neurites (ventral nerve); int: integument; m: myoplasm; S: sarcoplasm; Z: dense bodies.

262

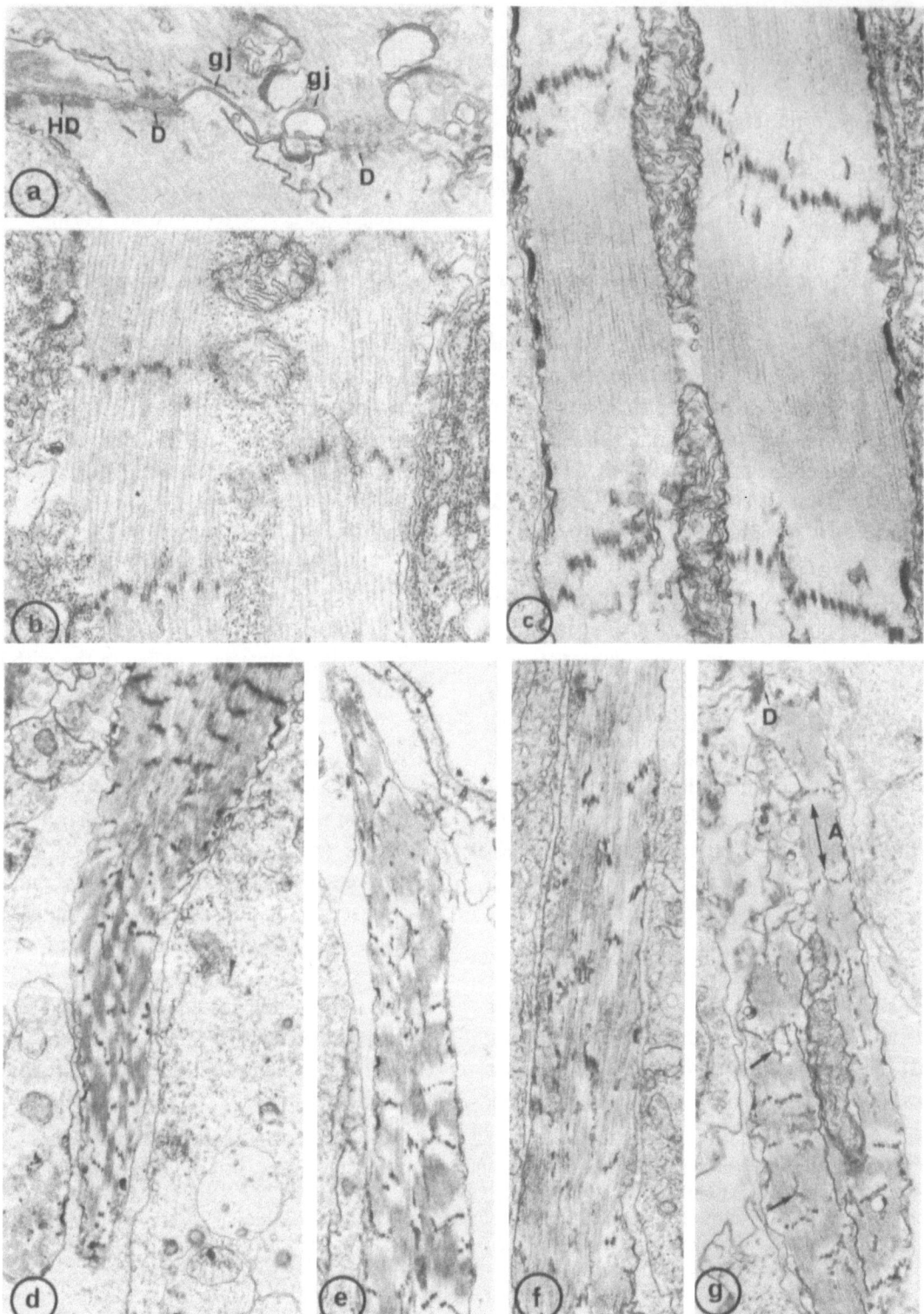

Plate IV. Phasic cross-striated retractor muscles.

Fig. a, b, c. Trichocerca rattus. a) Junctional complex between the two muscle cells of CRM, with gap-junctions (gj), desmosomes (D) and hemidesmosomes (HD). × 29 000. b) axial section of the supercontracted CRM. × 21 000. c) axial section of the relaxed CRM. × 26 000. Sarcoplasm is axial in a and b.

Fig. d, e, f. Brachionus calyciflorus. d-e): CRM (central retractor muscle): axial sections. f): DRM (dorsal retractor muscle): it is more obliquely striated than the cross-striated CRM of d and e. × 8500.

Fig. g. Asplanchna brightwelli. Note the shortness of the A bands (A) and the abundance of sarcoplasmic reticulum (arrows). This muscle is inserted to a cingulum cell by a desmosome (D) and controls swimming. × 12 400.

ventral (VRM) and lateral (LRM). The CRM are bicellular; VRM and LRM are monocellular but the VRM are partly inserted on small peripheral muscles of the mastax or on anterior buccal muscles.

2) Two anterior dorsal and ventral circular muscles constitute a spincter involved in the protection of the retracted rotatory apparatus (Fig. 1).

These four paired muscles are cross-striated but reveal great differences in their cytological organization (Fig. 2).

It is generally believed that the activity of a muscle cell is determined by the organization of its contractile apparatus, the development of its

sarcoplasmic reticulum and the density of mitochondria. The different concepts about the structure-function relation in cross-striated muscles has been reviewed by Josephson (1975).

The *speed of shortening* of a muscle cell is inversely proportional to the length of its A-band, directly proportional to the number of sarcomeres in series and to the density of dyads (sites for excitation-contraction couplings). The density of sarcoplasmic reticulum surrounding the myofilaments is also related to the speed of contraction and relaxation. Regarding these parameters, the CRM and VRM are considered fast muscle cells. The circular and LRM would be of a slower type (Table 1).

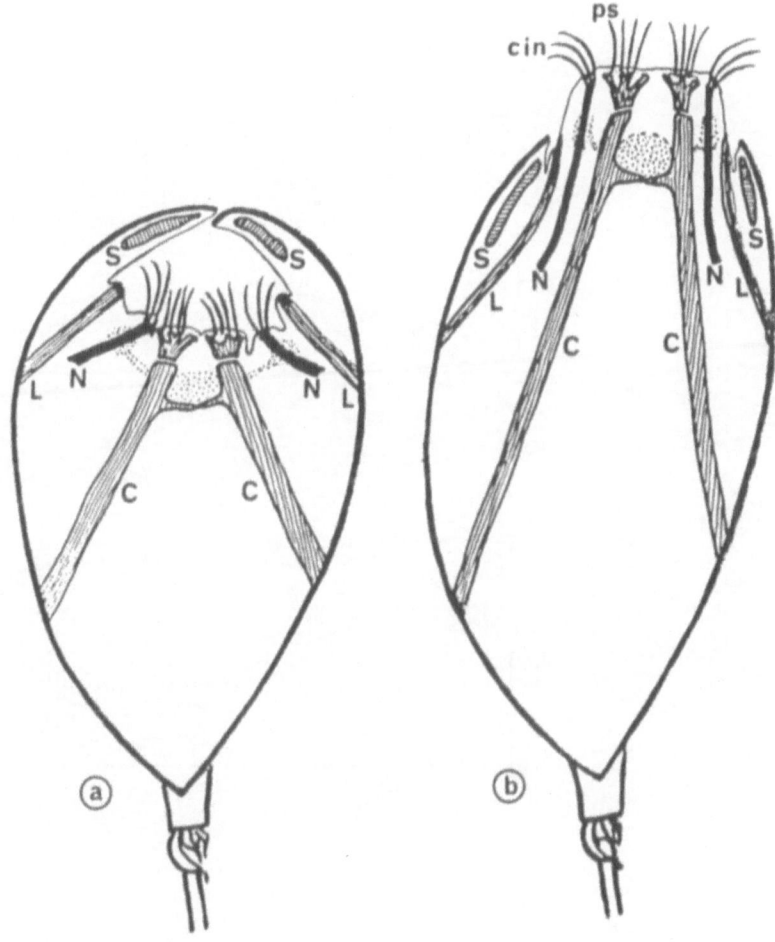

Fig. 1. Diagram of the muscles involved in withdrawal of the corona in *Trichocerca rattus*. Dorsal views, the brain is stippled. a: contracted animal; b: relaxed animal. Location of the central retractors and some other muscles is indicated. C: central retractors (bicellular) S: sphincter circular muscles; L: lateral retractors; cin: cingulum; ps: pseudotrochus; N: swimming muscles.

264

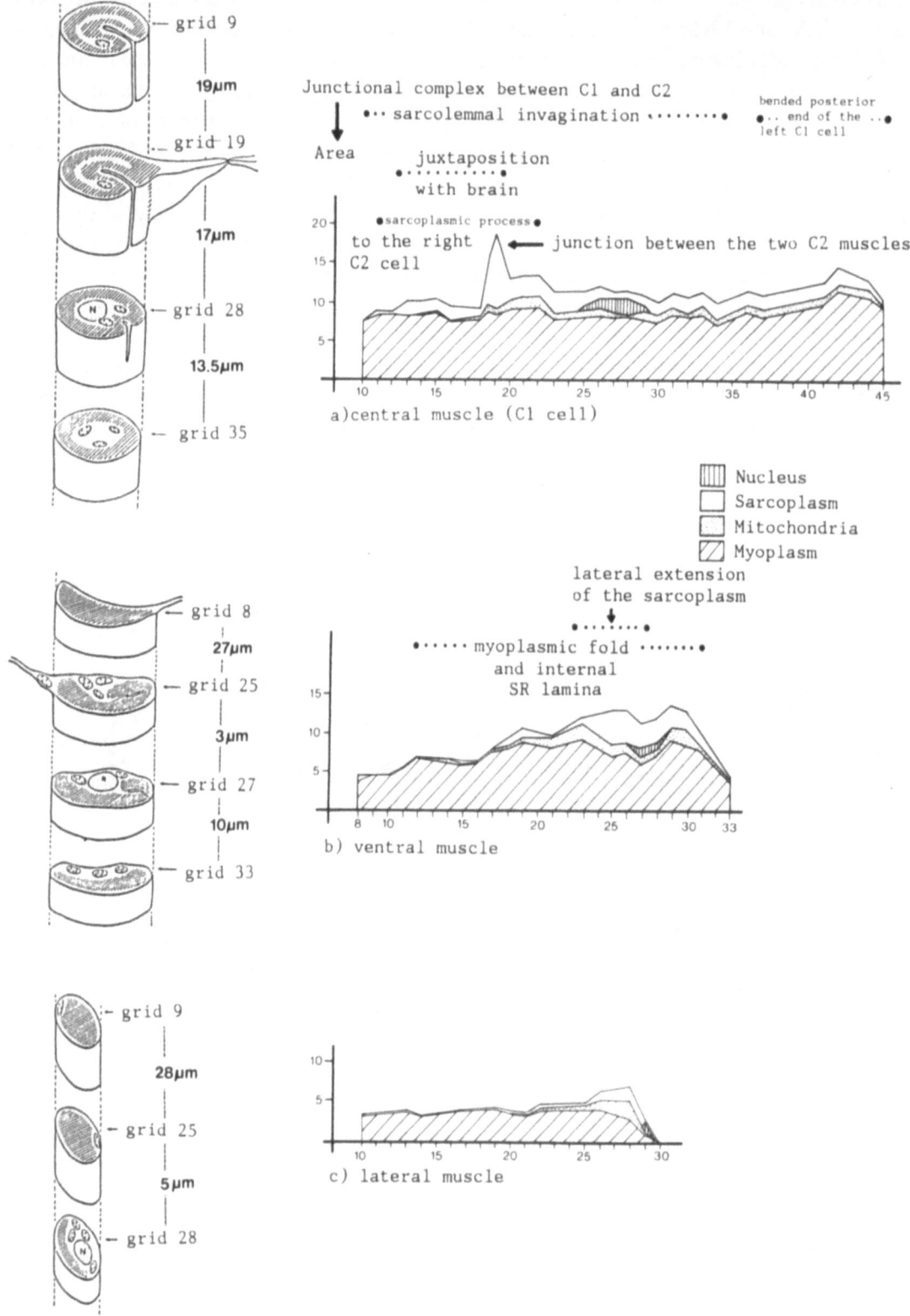

Fig. 2. Diagrams of the progressive characteristic levels of the central muscle (a), ventral muscle (b) and lateral muscle (c) of *T. rattus* representatives of the morphological heterogeneity of muscle fibers. Quantitative cytology for the myoplasm, mitochondria and sarcoplasm surface areas from serial transverse sections in the left CRM C1 cell, VRM cell and LRM cell. Beyond the curves, details about the morphology of the fiber are indicated between the precise levels where they appear and terminate.

Table 1. Morphometric study. Characteristic cytological parameters of central (CRM) ventral (VRM), lateral (LRM) retractors and anterior circular (Cir) muscles of *T. rattus*.

	CRM	VRM	LRM	Cir.
Speed of shortening	+ + +	+ +	+	+
Maximum force	+	+	+	+ + +
Endurance	+	+	+ +	+
$\dfrac{\text{Volume of mitochondria}}{\text{volume of myoplasm}}$ (%)	6	8.5	10.7	3.1
Number of sacromers	30	12–14	6–7	4
Length of the fiber (μm)	81	44	37.5	20
Mean surface of myoplasm (μm^2) (transverse section)	9	6.5	3.3	30
A-band length (thick myofilaments) (μm)	1.5	1.7	3	3.3
Thick myofilament diameter (nm)	15–20	15–20	15–20	15–20
Thin filaments diameter (nm)	7	7	7	7
Ratio f/F = thin (f), thick (F)	4 : 1	4 : 1	4 : 1	3 : 1
Z dense body length (nm)	90	90	140–200	90–120
Number of dyads/μm^2	2.7	1.5	1.7	–

The *maximum force* of a muscle cell varies with the number of bridges made between the two types of myofilaments during sliding. Force would be proportional to the surface of the myoplasm in a cross section fiber, the density of thick myofilaments in this section, the ratio of thin/thick myofilaments and the length of these latter myofilaments. Regarding these parameters, the three retractor muscles do not differ significantly. But the circular muscle is clearly stronger than the others which is due to the large volume of its myoplasm (Table 1). The function of each muscle is consistent with these characteristics.

Endurance is measured by the volume of mitochondria/volume of myoplasm. It is very low in the for types of muscles (3 to 11%), compared to the mastax muscles (where this relationship is around 50%); these retractors obviously do not have great endurance. This result is consistent with the fact that the withdrawal response is generally short. When contraction persists, the LRM seem to be primarily involved (Amsellem & Clément, 1977).

The *maximum shortening* of a muscle cell is the difference in length between its resting and maximum contraction states. This value is maximum for the C1 cell of the CRM, which has the longest retractor fibers (80 μm for the left). Less maximum shortening was found in VRM and LRM (Table 1). In these three cases, the maximum shortening is around half of the length of the resting fiber, due to the capacity of supercontraction where thick filaments cross over the Z bodies lines and go into the two adjacent sarcomeres (Plate IV – b). The network appearance (like a perforated disc) of Z material (Fig. 29, Amsellem & Clément, 1988) in rotifers is similar to that found in supercontracting muscles of insects (Hardie & Hawes, 1982).

The maximum shortening is less important in the two circular anterior muscles which are shorter and have a small number of sarcomeres. They are thick laminar muscles and the number of sarcomeres varies from 2 to 3 depending of the position of the myofibrils in the lamina.

In *conclusion* – muscles implicated in the withdrawal behavior can be classified in two categories: (1) the phasic muscles: CRM and VRM, with a high speed of shortening and relaxation, the CRM being the fastest with the maximum capacity of shortening, and (2) the tonic muscles: LRM and anterior circular muscles, with slower and more sustained contractions, the circular muscles being the strongest and the LRM proba-

266

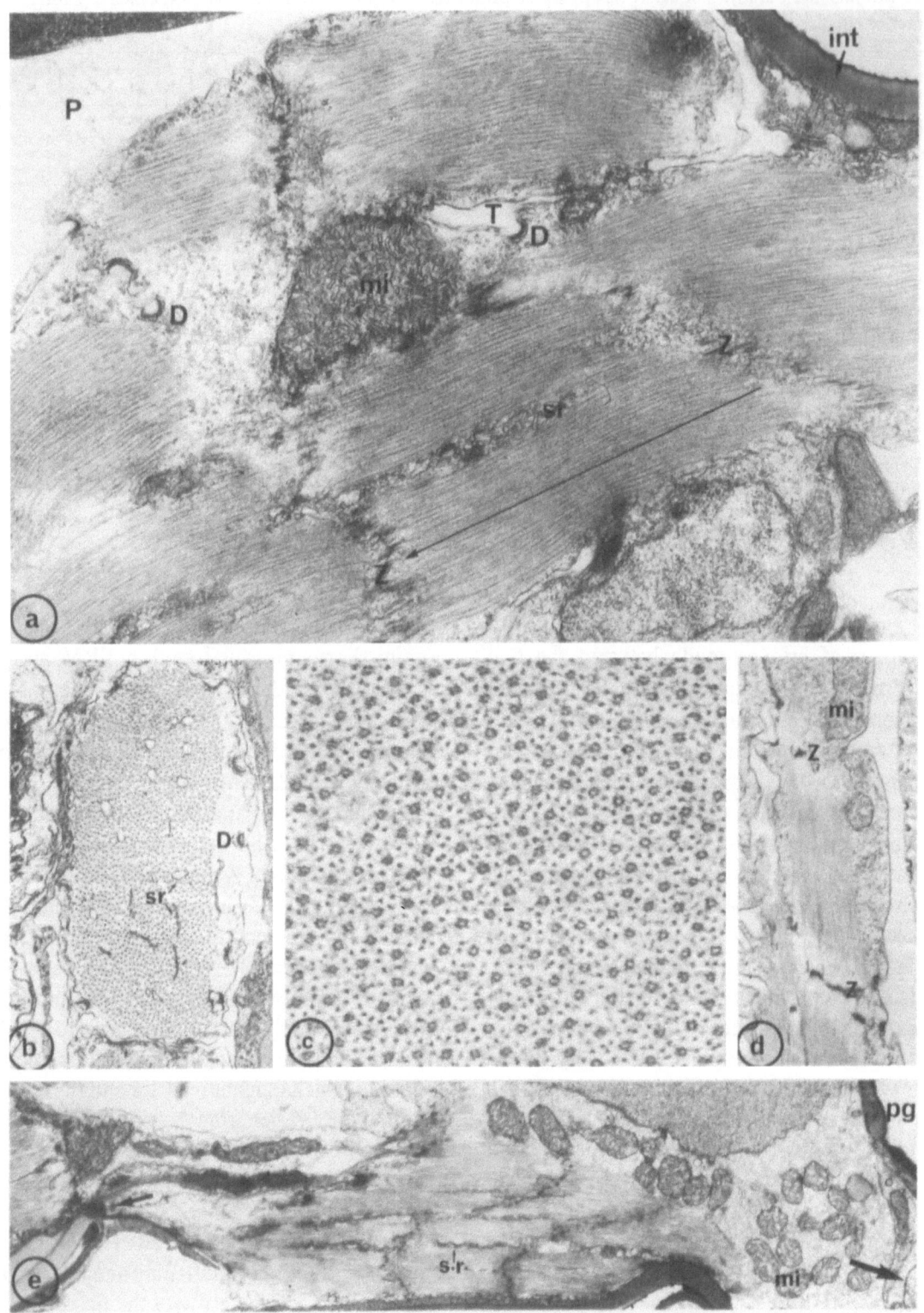

Plate V. Tonic skeletal muscles (a,c,d) and pedal muscles (b, e)

Fig. a. Axial section of the circular anterior dorsal muscle in *T. rattus*. Myoplasm is very abundant and organized in juxtaposed myofibrils where the sarcomeres are well defined; thick myofilaments (A-band) are 3 μm long (arrow) and myofibrils are

bly having greater endurance. Among the parameters which differentiate phasic and tonic muscles, the most useful are the thick myofilament length and the Z dense bodies length. The presence of invaginated sarcolemmal tubules making contacts with the sarcoplasmic reticulum is related to a large volume of myoplasm: such invaginations of the sarcolemma are numerous in some parts of the circular muscle (Plate V – a) and absent in LRM (Plate III – a).

Innervation of the muscles involved in the withdrawal behavior

The innervation of the muscles involved in this behavior was studied in *T. rattus* (Clément & Amsellem, 1987; Clément, 1987). Results are summarized in Fig. 3 and 4.

The two CRM are innervated into the brain by way of the sarcoplasmic arms sent in the third part of their longest CRM cells (Plate II – a, b). The two cells of each CRM are coupled (Plate IV – a) and the two CRMs are coupled by a large gap junction (Amsellem & Clément, 1988). So, any stimulation, lateralized or not, provokes the simultaneous contraction of this motor-unit (Fig. 3).

The other muscles (VRM, LRM, anterior sphincter) seem not to be directly innervated (we did not find obvious synapses). They are nevertheless coupled by gap junctions and constitute one motor unit. This motor unit is innervated by the anterior ganglia by way of a pair of little lateral

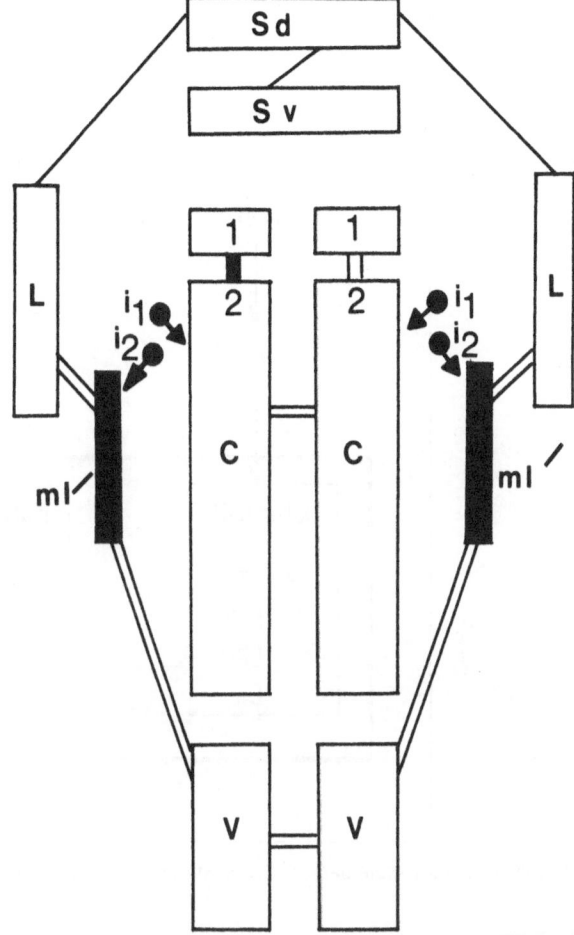

Fig. 3. Diagram of the gap junctions present between the muscles. A double line between 2 cells indicates that the junctions have been observed; a single line indicates that such junctions are very likely to be present (the membranes are in close contact) but have not been actually observed because the angle of sectioning was unfavorable. Sd: sphincter circular dorsal; Sv: sphincter circular ventral; L: lateral retractor LRM; C: central retractor muscles CRM; V: ventral retractor muscles VRM; ml: secondary lateral muscles; i1, i2: stimuli.

separated by sarcoplasm which contains mitochondria (mi), sarcoplasmic reticulum (s.r.); dyads (D) are visible between the SR and tubular invaginations (T) of the sarcolemma int: integument; P: pseudocoele; Z: alignment of dense Z bodies. × 21 000.

Fig. b. Transversal section of a pedal muscle of *T. rattus*. Note the density of longitudinal tubules of sarcoplasmic reticulum (s.r.) and the peripheral dyads (D) × 20 000.

Fig. c. Transversal section of a circular muscle in *T. rattus*. Note the regular arrangement of myofilaments and the hollow structure of the thick myofilaments. × 172 000.

Fig. d. Axial section of a tonic circular muscle of *B. calyciflorus*. Sarcomeres are well defined with a A-band of 3.5 μm. × 7300.

Fig. e. Axial section in a pedal muscle of *T. rattus* with its insertion on the integument (little arrow) and its innervation (large arrow); the pedal ganglia innervates the other pedal muscles and the muscular lamina surrounding the pedal gland (p.g.). The muscle is striated and the A-band length is around 3 μm. The densities of SR (s.r.) and mitochondria (mi) are higher than in tonic skeletal muscles of the same species. × 8900.

268

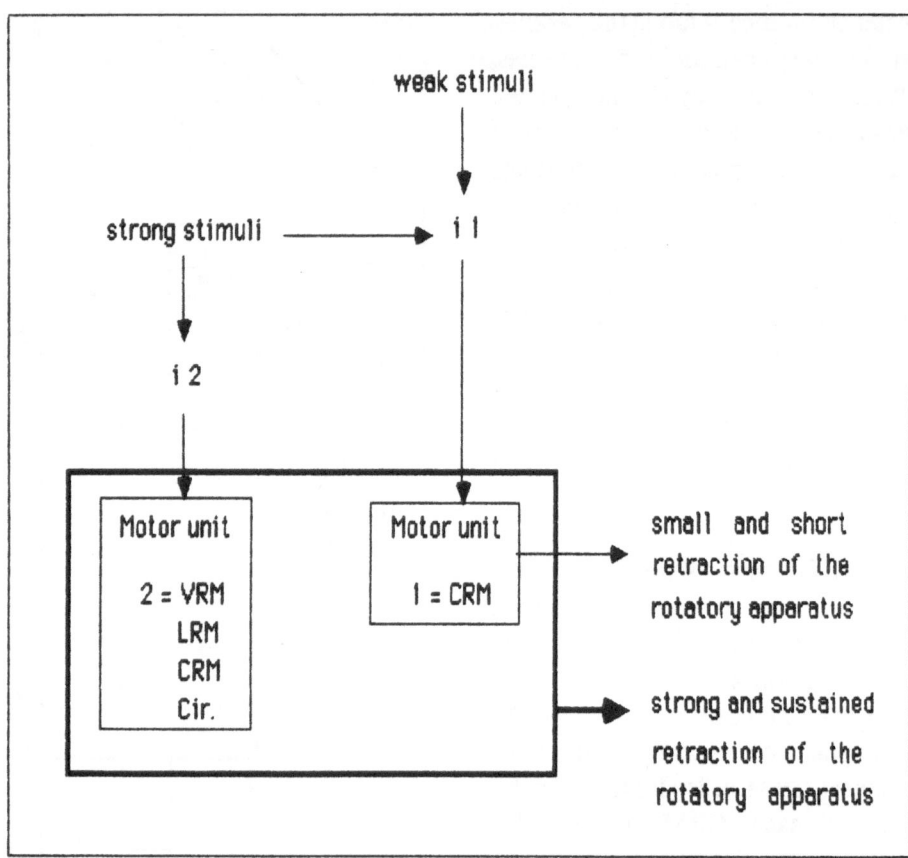

Fig. 4. Schematic summary of the presumed involvement of successive motorunits depending of the intensity of the stimuli.

muscles, each one being innervated and coupled to VRM and LRM.

In conclusion, the probable mechanism of the withdrawal behavior can now be more precisely explained (Fig. 4). Following stimulation, the innervation i1 of the CRM is activated; the contractile response is brief and fast, followed by fast relaxation. If the stimulus is stronger, innervation i2 is also activated: coronal retraction thus depends on the characteristics of all the muscles involved in this response. The phasic CRM retract the corona in the deepest part of the trunk; the phasic VRM retract the mastax and the buccal field, but not so deep; the tonic LRM and the anterior circular muscles then maintain this contraction until their own relaxation. The corona can then emerge because the CRM and VRM are already relaxed, there is no antagonist muscle working for this emergence, which seems to be a passive consequence of relaxation of all retractor muscles and of internal pressure.

The cross-striated muscles which control the ciliary beat for swimming

Swimming is a constant behavior of pelagic rotifers and a frequent behavior of other rotifers. Since the work of Viaud and others (review in Clément *et al.*, 1983), it is known that different stimuli (light, vibrations, chemical substances) influence swimming characteristics like linear speed, angular speed and direction. These precise motor responses to stimuli are supported by sensory organs and the nervous system (Clément *et al.*, 1983; Clément, 1987), but the ciliary cells which are responsible for swimming (the cingulum cells) are not innervated. Muscle cells are inserted on the cingulum ciliary roots (Clément, 1987 – 'N' on Fig. 1) (see above and plates I – c, d; VI – a, c). Each of these muscle cells is monocellular and constitutes an independent motorunit which is multi-innervated. Two independent motor-units, one at right and one at left, control swimming behavior so that the right part of the cingulum can be slowed down apart from the left part.

Cytologically, these muscles (Plate VI) are similar to small phasic muscles. In *T. rattus*, a cingulum muscle cell has the structural characteristics of a CRM cell with more sarcoplasmic reticulum and shorter A-bands (1.5 μm) (Plate VI – d).

The cingulum muscles are more specialized in some rotifers. Male *Brachionus* have shorter A-bands (1.3 μm) and the sarcoplasmic reticulum is well developed (Plate VI – c). In *Asplanchna brightwelli*, the swimming path is not helicoidal and Clément (1987) described the double orientation of the cingulum ciliary roots which looks like an inverted V at the base of each cilia. Each extremity of the V is anchored in a fibrillar material on which are inserted two opposite cingulum muscle cells. The structure of these cells is exceptional in that the length of the A-band is the shortest known in animals (0.5 μm; 1 μm for the sarcomere length in resting state) (Plate VI – a, b). In invertebrates, the shortest A-band lengths (<2 μm) always correspond to fast fibers (Hoyle, 1983). Values like those of the cingulum A band of *Asplanchna* occasionally have been observed. The thick myofilament length of the subumbrellar muscles of Scyphomedusa (Cnidaria) is 0.58 μm long (Chapman *et al.*, 1962; Mackie & Singla, 1975); that of the fast cross-striated muscles in the tail of cercaria (Platyhelminths) is about 0.5 μm (Rees, 1975) and in the chromatophore muscle cells of *Loligo* (Cephalopod), this value is 0.5 to 1 μm (Weber, 1970). Generally, in the fast fibers of invertebrates, A-band length averages 2.1 to 4.8 μm, but 1.3 μm was found in the mantle of *Sepia*. In vertebrates, this parameter varies from 1.7 to 1.9 μm for the same muscle type (Hoyle, 1983).

The other cross-striated skeletal muscles of rotifers

An axial section of a cross-striated muscle of *Asplanchna* is seen on Plate IV – g; A-band length is 0.7 μm. This muscle is inserted on the fibrillar material of the anterior ciliary cells. We do not actually know if this fibrillar material is connected with the ciliary roots of the cingulum. If connected, this muscle would be an another type of cingulum muscle, closely related to those illustrated Plate VI. If not, this muscle would be a retractor, faster than those found in other rotifers.

In female *Brachionus calyciflorus* or *plicatilis*, some retractors are cross-striated, but more irregularly than in *T. rattus* (Plate IV – d, e) and the Z dense-bodies are shorter. Plate IV – d illustrates axial sections of *B. calyciflorus* CRM where different states of contraction are seen in the same fiber. The supercontracted part occurred near the middle of the sarcoplasmic arm of the fiber which goes into the brain to provide innervation. In supercontracted fibers, the superposition of the A bands of two contiguous sarcomeres is the origin of a pseudostriation which is easily observed (Plate IV – d, upper part).

Intermediate situations exist between transversal and oblique striation in retractor muscles. The dorsal retractor of the female of *Brachionus* (Plate IV – f) is representative of such a situation. Some small dense body alignments also exist in the more primitive fibers of *Philodina roseola* where all skeletal muscles are smooth or obliquely-striated (Plate VII).

We need observations on the most developed cross-striated skeletal muscles in jumping rotifers (*Polyarthra, Hexarthra, Triarthra, Filinia*). This type of locomotion is very different than ciliary locomotion. Gilbert (1987) studied the escape response of *Polyarthra* with high speed cinematography. Using this precise analysis, it would be interesting to know the characteristics of the muscles involved in these movements.

The only muscles supporting articulate movements in rotifers that we observed in electron microscopy are those controling movement of the foot. In *Philodina roseola* they are smooth fibers (Clément, 1987) like other smooth skeletal muscles described below. In *T. rattus*, the tail is very developed and its movements are controled by specialized original muscles (Plate V – b, e). Their sarcomeres are well defined, delimited

270

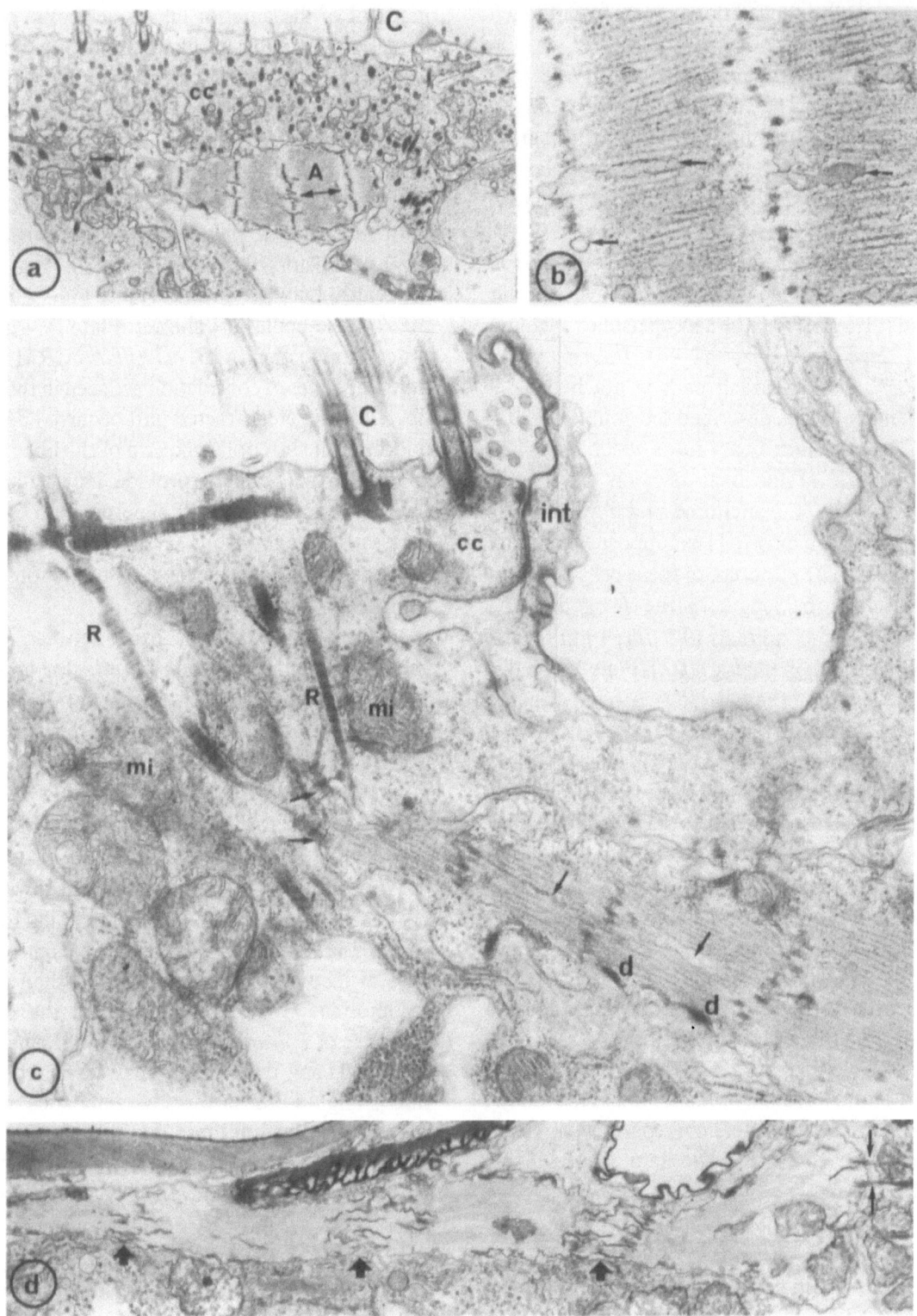

Plate VI: Muscles of cingulum

Fig. a, b. Asplanchna brightwelli. a) Axial section of a muscle ending, inserted (arrow) on a cingulum cell (cc) ciliary roots associated with mitochondria. C = cilia of cingulum. The muscle is cross-striated with a very short A band (A): 0.5 μm × 14 000.

either by cross or oblique striation (Plate V – e). Their myoplasm is laminar with numerous large longitudinal tubules of sarcoplasmic reticulum inside the myoplasm. Sarcoplasm mainly extends from the anterior ending of the muscle cell where it contains the nucleus, glycogen particles and some mitochondria (Plate V – e). Innervation occurs by the pedal ganglion which also innervates other muscles involved in the movement of the foot and the muscular lamina surrounding the pedal gland (Plate V – e). Thus, all these pedal muscles constitute an unique motor-unit.

Obliquely striated fibers

Obliquely striated muscle cells are largely found in Annelida, Nematoda (Rosenbluth, 1972), Cephalopoda (Amsellem & Nicaise, 1980), and have also been observed in the slow fibers of cercaria (Plathyhelminths: review in Hoyle, 1983). All of these fibers contain paramyosin in their thick myofilaments. In rotifers, we never observed the structural characteristics of the paramyosin in the obliquely striated fibers. In these muscle cells, the dense bodies are obliquely lined up in relation to the direction of the myofilaments (fiber axis). When a cross-striated fiber contracts, the maximum shortening is limited by the sarcomere length, even if the thick myofilaments get through the Z material (supercontraction: Plate IV – b, d). The obliquely striated fibers have a greater capacity of length changes than cross striated ones; in addition to the classical sliding mechanism, Rosenbluth (1967, 1972) and Knapp & Mill (1971) (Fig. 5) suggested that the sliding would be increased by a 'shearing' mechanism (sliding of thick filaments on themselves) during con-

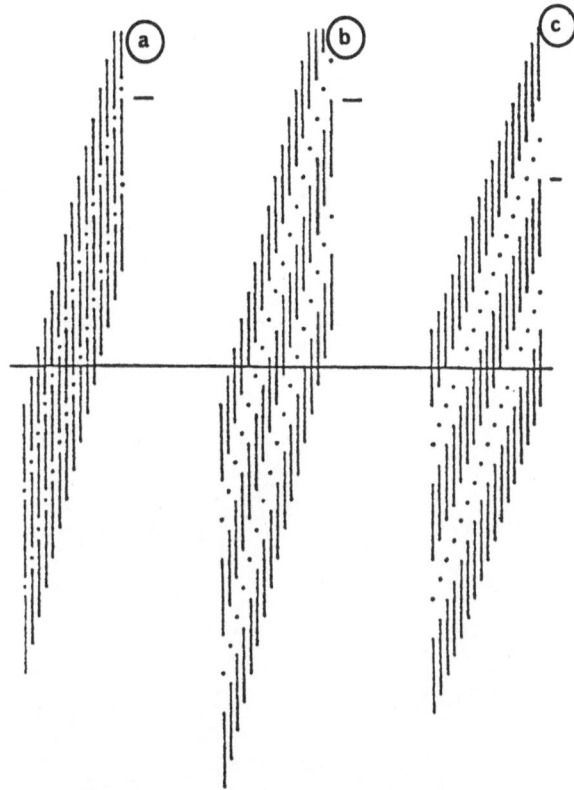

Fig. 5. The two types of shortening occurring in oblique striated muscle cells (sliding: -a (contracted) and -b (relaxed) without altered degree of obliquity and shearing: -c (contracted) with altered degree of obliquity; Rosenbluth, 1972) which permits the muscle to adjust to a wide range of lengths passively (from Knapp & Mill, 1971).

traction; but the recent studies of Lanzavecchia (1977), Lanzavecchia & Arcidiacono (1981) and Lanzavecchia *et al.* (1985) rather confirm the change of partner hypothesis between thick and thin myofilaments (during elongation) introduced by Miller (1975) to explain the length variation range in these fibers. It is interesting to note that such a length variation range is mainly found in invertebrates with soft bodies. In a previous paper about obliquely striated muscle in molluscs

---

b) Detail of the same muscle: sarcoplasmic reticulum (arrows). × 42000.

Fig. c. Male of *Brachionus urceolaris sericus.* The muscle is inserted (horizontal arrows) at the base of the striated ciliary roots (R) which are associated to mitochondria (mi). The ciliary cell of the cingulum (cc) bears the cilia (c) responsible for the swimming and is juxtaposed to the syncitial integument (int). The muscle is cross-striated, contracted, with short A-bands (1,3 μm), intramyoplasmic sarcoplasmic reticulum (oblique arrows) and peripheral dyads (d). x26000.

Fig. d. Axial section of the relaxed three anterior sarcomeres of the cingulum muscle in *T. rattus.* Insertion on the basis of the ciliary roots (small arrows). A-bands are 1.7 μm and the sarcoplasmic reticulum is abundant, particularly in the I band (large arrows). × 16800.

272

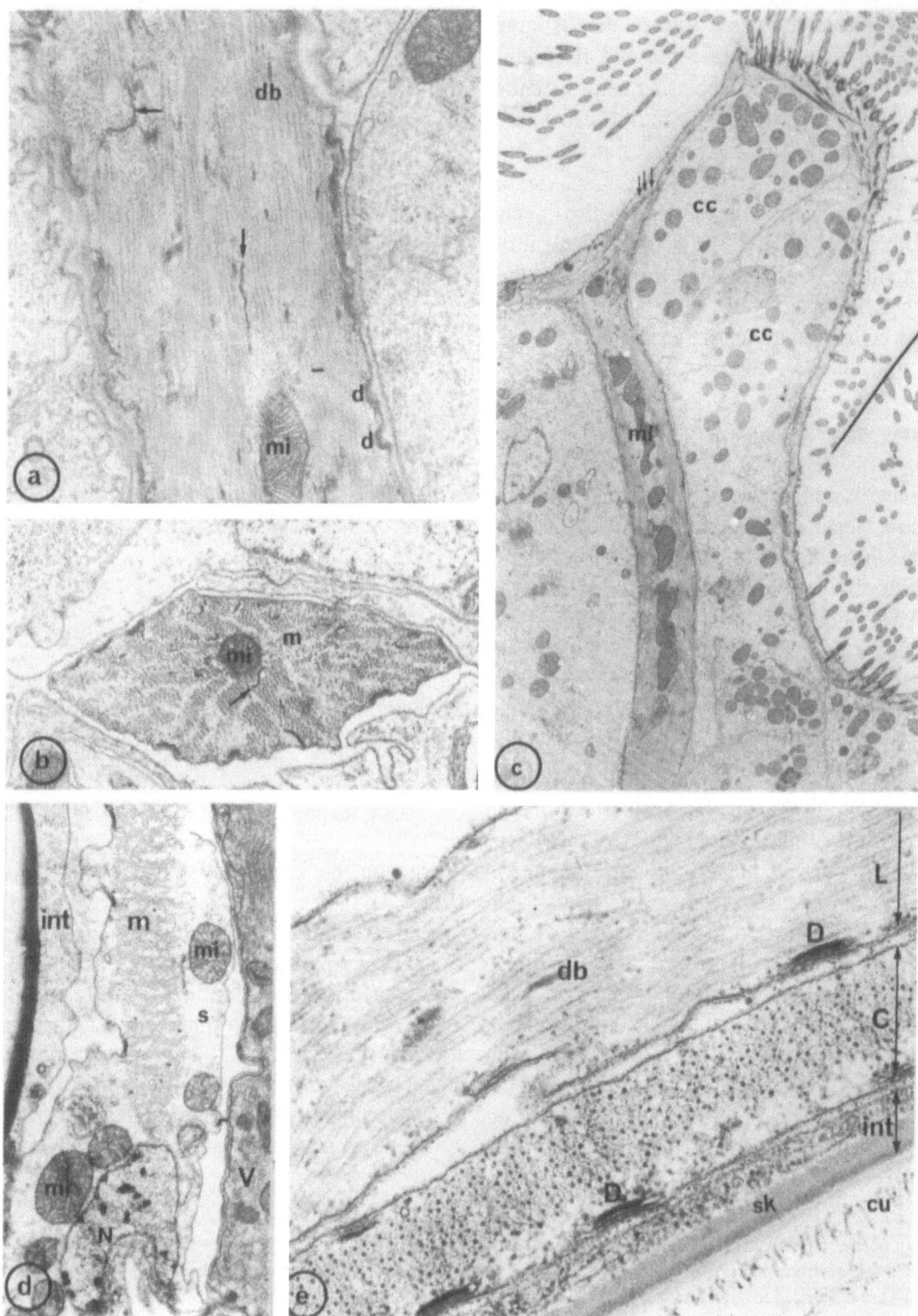

Plate VII: Skeletal muscles of a bdelloid *Philodina roseola*

Fig. a, b, c. The central retractor muscle (CRM). a) Axial section showing myofilaments, dense bodies (d.b.), sarcoplasmic reticulum (arrows), mitochondria (mi) and peripheral dyads (d). × 23 000. b) Transversal section; the organisation of the

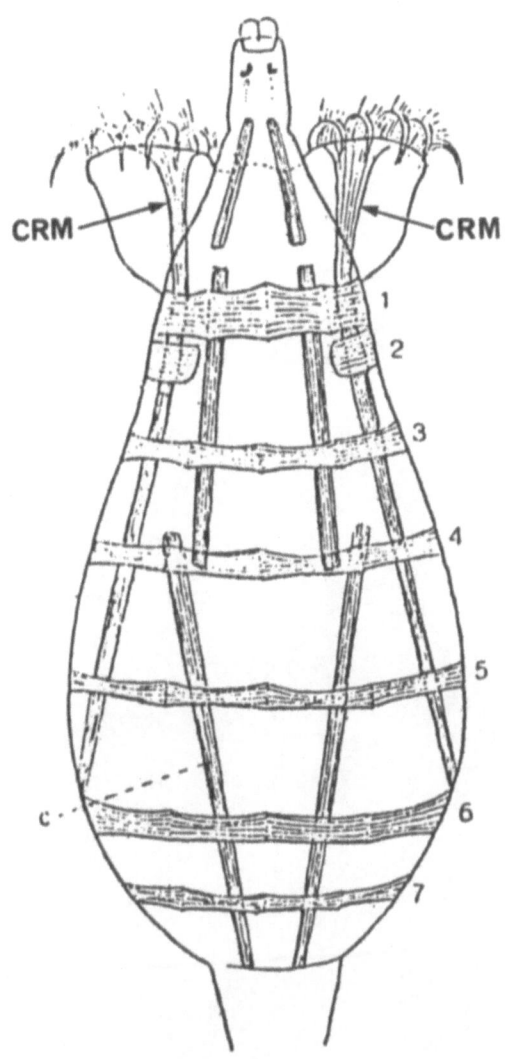

Fig. 6. *Rotaria crurata*. Skeletal muscles (modified from Brakenhoff, 1937). CRM central retractor muscles; 1 to 7: circular muscles; c: pedal muscle.

(Amsellem & Nicaise, 1980), we noted that 'it is puzzling that the most highly evolved molluscs' *Sepia* and *Octopus*, (cephalopods) 'but not the primitive ones, developed to that extent the use of muscle cells which are mostly found in worms'. Hence, this perhaps would not be a problem of phylogeny and Lanzavecchia & Arcidiacono (1981) considered this striation as an evolved response to the problem of locomotion with hydraulic systems.

An oblique striation occurs in the longitudinal retractors of *Philodina*, where the movements of shortening and elongation of the body seem to overlap a variation range greater than that only performed by the sliding mechanism in muscles. The central retractor of *Philodina* inserts near the cingulum (Fig. 6, Plate VII – b). Its contraction provokes the withdrawal of the rotatory apparatus deeply into the trunk of the animal. The dense bodies (average length: 120 nm) are either obliquely or transversely lined up (Plate VII – a, c) and can be the sign of diversely contracted myofibrils. A transverse section (Plate VII – b) shows the typical arrangement of the myoplasm in irregular obliquely-striated fibers. The sarcoplasm lying between the arrays of myofilaments contains glycogen like particles, mitochondria and sarcoplasmic reticulum; peripheral dyads are abundant. On some favorable longitudinal sections, the thick myofilament length is around 1.3 μm. Thus, this muscle seems to have the structural characteristics of a fast muscle.

Oblique striation occurs also in some retractors

contractile apparatus is visible. m: myofibrils; arrow: sarcoplasmic reticulum. × 21 600. c) Axial section of the anterior part of this CRM, showing its insertion (arrows) on the thin apical integument juxtaposed to the cingulum cells (cc). The myoplasm is cylindrical around the axial sarcoplasm which contains the mitochondria (mi) x5000..

Fig. d. Transversal section of a skeletal longitudinal muscle, located in the pseudocoel between the integument (int) and the vitellarium (V); the myoplasmic lamina is organized in parallel myofibrils (m): this disposition is of intermediate organization between the obliquely striated CRM (Fig. b) and the smooth circular muscles (Fig. e). Note the localization of the sarcoplasm (s) along the myoplasmic lamina (m) and in the sarcoplasmic extensions which contain mitochondria (mi) and nucleus (N). × 12.000.

Fig. e. Transversal section of a circular muscle (C) and axial section of a longitudinal muscle (L). The circular muscle is a lamina, just under the integument (int) -skeletal intracytoplasmic lamina (sk) and a thin extracellular cuticle (cu). The myoplasm of the circular lamina is not arranged in myofibrils. The cross-section aspect of thin and thick myofilaments is similar to that found in monogononts. The dyads (D) are peripheral. The longitudinal muscle organization is similar to that seen in Fig. d). Dense bodies (d.b.) are scattered in the myoplasm.

274

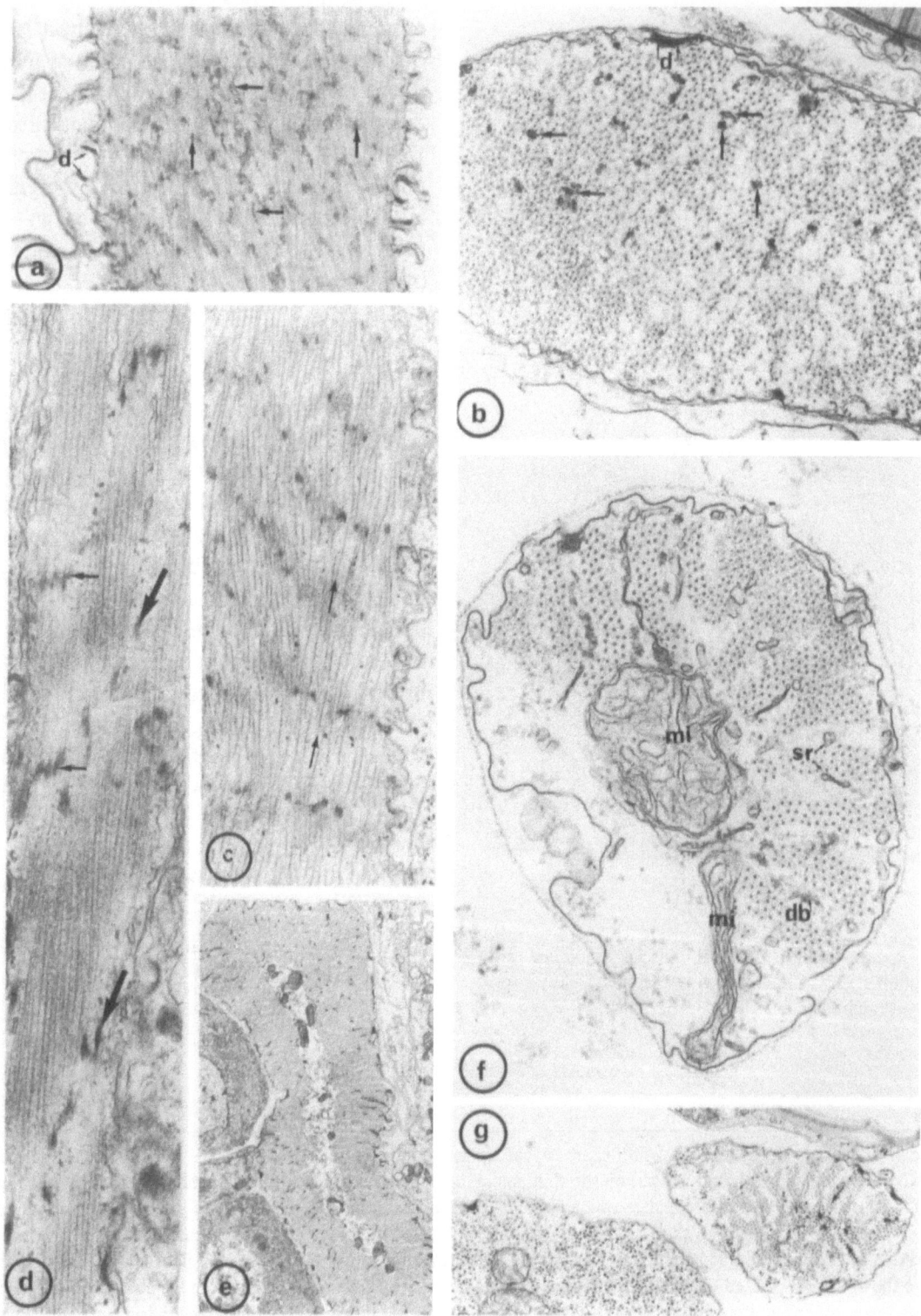

Plate VIII. Smooth and obliquely striated skeletal muscles of Monogononta

Fig. a. Axial section in one of the two ventro-buccal circular muscle in *Asplanchna brightwelli.* These two muscles are smooth: note the irregular distribution of the dense Z bodies (vertical arrows) and SR (horizontal arrows) in this contracted muscle. The dyads (d) are peripheral. × 15000.

of monogononts, for example in male and female *Brachionus*, *Rhinoglena* (Plate VIII – g) and *Asplanchna* (Plate VIII – f). In *Notommata copeus*, large retractors look like smooth fibers in longitudinal sections but aspects of the contractile apparatus in transverse section (Plate VIII – e) suggest that these fibers could be irregular obliquely-striated.

Smooth muscles

In rotifers, we called 'smooth muscles' the muscle cells with irregularly scattered dense bodies either in transversal or longitudinal section (Plate VIII – -a, b). These cells possess two types of myofilaments with actin-like filaments attaching to the dense bodies. Sarcoplasmic reticulum is present, mainly peripheral, forming dyads with the sarcolemma. Intermediate organizations exist between the typical obliquely striated type and the typical smooth type depending of the state of contraction. For instance, there is often a beginning of alignment of the dense bodies in supercontracted smooth fibers where the position of thick myofilaments makes dark bands at the Z level, producing a pseudostriation (Plate VIII – c; IV – d).

Rotifer smooth fibers have features nearer to those of molluscan smooth fibers than those of coelenterates and are of different categories. In *Philodina roseola*, the circular muscles possess a laminar myoplasm without intramyoplasmic sarcoplasmic reticulum. This localization is peripheral with some dyadic contacts with the sarcolemma (Plate VII – e). The sarcoplasm is scarce and limited in the perinuclear zone with the mitochondria (Plate VII – d). The dense bodies are often ribbon-like, attached to the sarcolemma and penetrating deeply in the fiber (Plate I – a). The longitudinal retractors of *Philodina roseola* are more or less cylindrical fibers with a more structurated contractile apparatus. Small rare mitochondria are distributed along the myofilaments and, in some fibers, dense bodies are relatively of small size (70 to 150 nm length) (Plate VII). They can however reach more than 200 nm in others. In this case, the thick myofilaments are more than 3 μm long, but the micrographs did not allow more precise measurements (Plate II – f). These differences are probably expressions of functional specializations, but we need more data about behavior and structure.

Smooth fibers are observed in the long and big central retractors of *N. copeus* (Plate VIII – e) as well as the short symmetrical anterior ventral circular muscle in *Asplanchna* (Plate VIII – a). In some loricate monogononts as *T. rattus*, the smooth skeletal fibers are lacking: the only smooth muscles are visceral (ex: the cloacal sphincter).

Fig. b. Transversal section of a skeletal smooth muscle of *Brachionus calyciflorus*. Dense bodies (vertical arrows) and sections of sarcoplasmic reticulum (horizontal arrows) are scattered in the myoplasm, where the myofilaments are not arranged in myofibrils. × 28 000.

Fig. c. Axial section of a contracted longitudinal muscle of the male of *B. urceolaris sericus*. Note the alignment of the dense bodies and a zone of characteristic supercontraction where thick myofilaments go deeply into the neighboring sarcomeres (arrows). × 30 000.

Fig. d. Axial section of a relaxed longitudinal muscle of the male of *B. urceolaris sericus*. Note the insertion of the thin myofilaments on dense bodies and the transversal (small arrows) or oblique (large arrows) orientation of dense Z bodies. × 37 000.

Fig. e. Transversal section of a CRM (central retractor muscle) of *Notommata copeus*. Note the cylindrical disposition of the myoplasm around the central sarcoplasm; the myoplasm is nearly fragmented in radial myofibrils. This CRM is large because *N. copeus* is a big rotifer (length = 1 mm). × 4500.

Fig. f. Transversal section of a retractor muscle of *Asplanchna brightwelli*. The myoplasm is fragmented in radial distinct myofibrils separated by sarcoplasm with sarcoplasmic reticulum (s.r.): this is a typical arrangement of oblique striated muscle. mi: mitochondria; d.b.: dense Z bodies. × 43 000.

Fig. g. Transversal section of a retractor muscle in the rostrum of *Rhinoglena frontalis*. The myoplasm displays also the typical arrangement of an oblique striated muscle. × 14 000.

Conclusions

Skeletal muscles in rotifers are very diverse in form, type and function. They can be classified by different cytological criteria as: smooth, obliquely or cross-striated; phasic and tonic; strong or not; endurant or not.

The precise localization of their insertions are important data to understand their functions which can be grouped in three behaviors: withdrawal of the corona, control of ciliary beat in the corona; articulated movements (foot, tail, paddles).

Their coupling by gap-junctions and their innervations allow definition of motor-units responsible of these behaviors. Each cingulum muscle is a motor-unit which controls precisely swimming behavior. The different kinds of phasic and tonic retractors and the anterior circular muscles constitute two motor-units. One is involved in the quick withdrawal behavior and the other with the complete and prolonged retraction of the corona. All of the pedal muscles in *T. rattus* are a single motor-unit which controls the movement of the foot and secretions of the pedal gland.

Each muscle is either monocellular or bicellular and functionally adapted to a precise participation in the behavior. Some specializations are extreme as in the cingulum muscle controling swimming in *Asplanchna* (0.5 μm A bands).

The bdelloid *Philodina roseola* does not possess typical cross-striated skeletal muscles. All are smooth or obliquely striated muscles without obvious phasic or tonic characteristics. Smooth and obliquely striated skeletal muscles are also present in monogononts, principally in those where the skeletal integument is soft (the 'illoricates'). They are present in *N. copeus*, but absent in *T. rattus* and more abundant in *Rhinoglena* than in *Brachionus*.

The most primitive muscles seem to be those with a homogeneous laminar myoplasm. This lamina could evolve by folding on itself to become a cylinder and by fragmentation of adjacent myofibrils.

Rotifer skeletal muscles originate from ectoderm. They all possess thick and thin myofila-

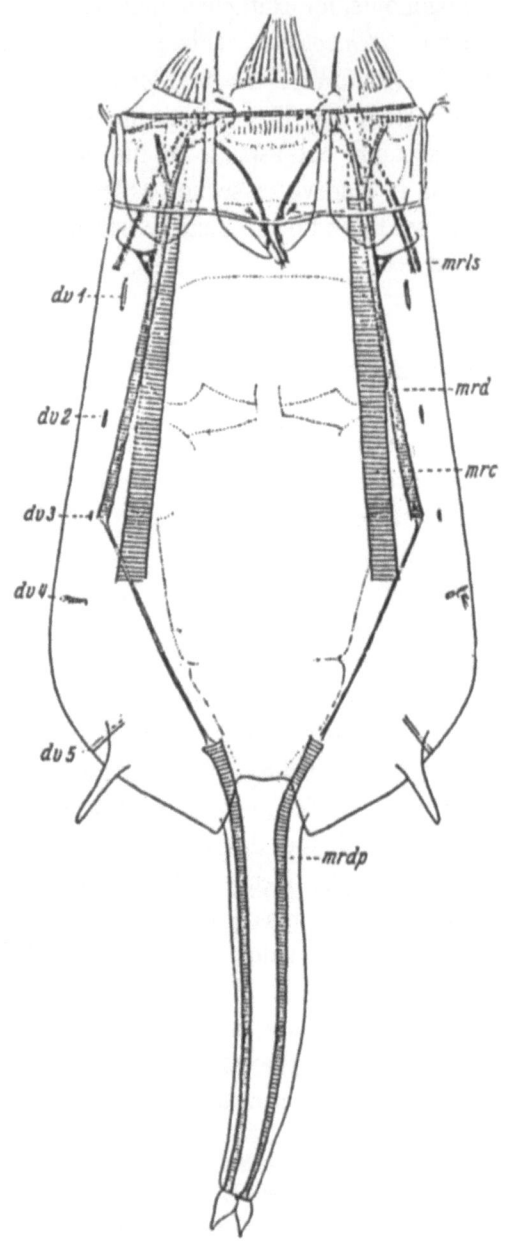

Fig. 7. Brachionus. Skeletal muscles (modified from Stossberg, 1932) dv1 to dv5 = circular muscles; mrls: cingulum muscles; mrd: dorsal retractor muscles; mrc: central retractor muscles; mrdp: pedal retractor muscle.

ments, structurally similar to the myosin and actin of the fast fibers of arthropods and vertebrates. Thick myofilaments never show the structural aspect of paramyosin. Dense Z bodies are present even in smooth muscles. Some smooth muscles of Cnidaria (Perkins *et al.*, 1971; Wood, 1961) and of Ctenaria (Hernandez-Nicaise & Amsellem, 1980) have thick and thin myofilaments, but no dense body.

Platyhelminth muscles (Rees, 1975; Hoyle, 1983) show some features and a diversity which can be compared with those of rotifers. Nevertheless, they contain paramyosin as in nematodes, annelids and other groups of worms.

These observations are consistent with other ultrastructural data, from which Clément (1981) suggested that rotifers share a recent common ancestor with platyhelminths. The overall features of rotifer musculature illustrate that they constitute a relative homogeneous group. With regard to muscle characteristics, bdelloids seem to be more primitive than monogononts. Yet even in bdelloids, the observed diversity of specializations is related to precise functions and behaviors.

Acknowledgements

We acknowledge the researchers of the Neuroethology team who worked in electron microscopy: E. Wurdak on *Asplanchna brightwelli*; A. Cornillac on *Brachionus calyciflorus*; Anne Luciani on *Brachionus plicatilis*. All the electron microscopy was done at the C.M.E.A.B.G., Université C. Bernard, Lyon I. We also acknowledge Terry Snell who read this manuscript to correct its English.

References

Amsellem, J. & P. Clément, 1977. Correlations between ultrastructural features and contraction rates in rotiferan muscle. Cell Tissue Res., 181: 81–90.

Amsellem, J. & P. Clément, 1988. The muscle of a monogonont rotifer *Trichocerca rattus*. II. The central retractor muscles. Tissue and cell, 20(1): 89–108.

Amsellem, J. & G. Nicaise, 1980. Ultrastructural study of muscles cells and their connections in the digestive tract of *Sepia officinalis*. J. submicrosc. cytol., 12(2): 219–231.

Beauchamp, P. de, 1965. Classe des Rotifères. In: Traité de Zoologie, Anatomie, Systématique, Biologie. P. P. Grassé, IV, 3: 1225–1379.

Brakenhoff, H., 1937. Zur morphologie der Bdelloidea. Zool. Jahrb. Abt. Anat., 63.

Chapman, R. A., C. F. A. Pantin & E. A. Robson, 1962. Muscle in Coelenterates. Rev. Canad. Biol., 21: 267–278.

Chien, P. & H. Koopowitz, 1972. The ultrastructure of neuromuscular systems in *Notoplana acticola*, a free-living polyclad flatworm. Z. Zellforsh., 133: 277–288.

Clément, P., 1969. Premières observations sur l'ultrastructure comparée des téguments de rotifères. Vie et Milieu A, 20: 461–482.

Clément, P. & X. Fouillet, 1970. Les jonctions cellulaires d'un organisme (rotifère). 7ème Cong. Intern. Micr. Electr., Grenoble, 1: 7–8.

Clément, P., 1977. Ultrastructural research on rotifers. Arch. Hydrobiol. Beih. Ergebn. Limnol., 8: 270.

Clément, P., 1980. Phylogenetic relationships of rotifers and derived from photoreceptor morphology and other ultrastructural analysis. Hydrobiologia, 73: 93–117.

Clément, P., 1985. The relationships of rotifers, as deduced from their ultrastructure and behavior. In: The origin and relationships of lower Metazoa. (eds Conway Morris *et al.*) 224–247. Clarendon Press, Oxford.

Clément, P., 1987. Movements in rotifers: correlations of ultrastructure and behavior. Hydrobiologia, 147: 339–359.

Clément, P. & J. Amsellem, 1986. Ultrastructures et comportements: neuro-éthologie des rotifères. In: Neuro-éthologie (ed. R. Campan). Comportements. 39–57. CNRS. Paris.

Clément, P., E. Wurdak & J. Amsellem, 1983. Behavior and ultrastructure of sensory organs in rotifers. Hydrobiologia, 104: 89–130.

Duvert, M., 1971. Ultrastructure de la jonction myo-épidermique dans les muscles du tronc de *Sagitta cetosa* (Chaetognates). C.R. Acad. Sci., Paris, Ser. D, 272: 2575–2577.

Hardie, J. & C. Hawes, 1982. The three dimensional structure of the Z-disc in Insect supercontracting muscle. Tissue and Cell, 14(2): 219–231.

Hernandez-Nicaise, M. L. & J. Amsellem, 1980. Ultrastructure of the giant smooth muscle fiber of the Ctenophore *Beroe ovata*. J. Ultrastruct. Res., 72: 151–168.

Hyman, L. H., 1951. Class Rotifera. In: The Invertebrates. Mc Graw-Hill Book Company, Inc, New-york, Toronto, London, 3: 59–151.

Hoyle, G., 1983. Muscles and their neural control. (Wiley, J. and Sons eds.), New-york, Chichester, Brisbane, Toronto, Singapour.

Josephson, R. K., 1975. Extensive and intensive factors determining the performance of striated muscles. J. exp. Zool., 194: 135–154.

Knapp, M. F. & P. J. Mill, 1971. The contractile mechanism in obliquely striated body wall muscle of the earthworm *Lumbricus terrestris*. J. Cell Sci., 8: 413–425.

Koehler, J. K., 1965. A fine study of the rotifer integument. J. exp. Zool., 162: 231–243.

Koehler, J. K., 1966. Some comparative fine structure relationships of the rotifer integument. J. exp. Zool., 162: 231–243.

Lanzavecchia, G., 1977. Morphological modulations in heli-

278

cal muscles (Aschelminthes and Annelida). Int. Revue Cytol. 51: 133–86.

Lanzavecchia, G. & G. Arcidiacono, 1981. Contraction mechanism of helical muscles: experimental and theoretical analysis. J. submicroscopic cytol., 13(2): 253–266.

Lanzavecchia, G., M. De Eguileor & R. Valvassori, 1985. Superelongation in helical muscles of leeches. J. Musc. Res. Cell Motil., 6: 569–584.

Mackie, G. O. & C. L. Singla, 1975. Neurobiology of stomatoca. I. Action systems. J. Neurobiol. 6: 339–356.

Nicaise, G. & J. Amsellem, 1983. Cytology of muscle and neuromuscular junction. In: The Mollusca., Vol. 4. Physiology, Part I. 1–31.

Pavans de Ceccatty, M., 1974. The origin of the integrative systems: a change in view derived from research on Coelenterates and Sponges. Perspectives in Biol. and Medic. 17(3): 379–390.

Ress, G., 1975. The arrangement and ultrastructure of the musculature, nerves and epidermis in the tail of the cercaria of Cryptocotyle lingua (Creplin) from *Littorina Littorea* (L.). Proc. R. Soc. Lond., Ser. B, 190: 165–186.

Remane, A., 1929–32. Rotatoria. In: H. G. Bronn's klassen und ordnungen des TierReichs, IV 1 (Rotatorien Gastrotrichen und Kinorhynchen), 2 (Verms), (Aschelminth), 5: 336–370.

Rosenbluth, J., 1965. Ultrastructure of somatic muscle cells in *Ascaris lumbricoides*. J. Cell Biol. 26: 579–591.

Rosenbluth, J., 1967. Obliquely striated muscle. III. Contraction mechanism of *Ascaris* body muscle. J. Cell Biol., 34: 15–33.

Rosenbluth, J., 1972. Obliquely striated muscle. In: The structure and function of muscle. 2nd edition. Vol. I. Part 1 (Bourne G. H. Ed.) Academic Press, N.Y., London.

Schramm, U., 1978. Studies on the ultrastructure of the integument of the rotifer *Habrotrocha rosa* Donne (Aschelminth). Cell Tissue Res. 189; 167–172.

Ward, S. N., J. G. Thomson, White & S. Brenner, 1975. Electron microscopical reconstruction of the anterior sensory anatomy of the nematode *Caenorhabditis elegans*. J. Comp. Neurol., 160: 317–338.

Ware, R. W., B. Clark, K. Crossland & R. L. Russel, 1975. The nerve ring of the nematode *Caenorabditis elegans*: sensory input and motor output. J. Comp. Neurol. 162: 71–110.

Weber, W., 1970. Zur ultrastruktur des chromatophoren muskel zellen von *Loligo vulgaris*. Z. Zellforsch. Mikrosk. Anat., 108: 446–450.

Wood, R. L., 1961. In: The biology of Hydra and some other coelenterates (eds. Loomis W. F. & Lenhoff H. M.) Univ. of Miami Press: 51–64.

Hydrobiologia **186/187**: 279–284, 1989.
C. Ricci, T. W. Snell and C. E. King (eds), Rotifer Symposium V.
© 1989 *Kluwer Academic Publishers.*

Classical taxonomy and modern methodology

W. Koste[1] & R.J. Shiel[2]
[1] *Ludwig-Brill-Strasse 5, Quakenbrück, D-4570, F.R.G.*; [2] *Murray-Darling Freshwater Research Centre, P.O. Box 921 Albury, N.S.W. 2640, Australia*

Key words: Rotifera, species definition, taxonomy, biogeography, nomenclature, ecotypes, polymorphism

Abstract

Classical rotifer taxonomy and recent approaches are reviewed. α-taxonomy (morphology) remains the most widely-used technique, and most prone to subjective errors. Computer technology, and genetic or biochemical methods are not widely available in areas of most need – those with developing limnology programs. Strict adherence to the International Code of Zoological Nomenclature, uniform treatment of ecotypic variation and polymorphism, communication with and between systematists, and establishment of a centralized data base are among the points discussed. A standardized nomenclature is proposed.

Introduction

The credibility of ecological, genetic, biochemical and other studies relies heavily on the correct identification of the organisms involved. The need for a review of the taxonomic bases of the Rotifera has become increasingly urgent as a new generation of rotifer workers, particularly in areas with developing limnology programs, find that their taxa are not always in accord with the available (largely European or North American) keys. Established workers, on the other hand, are finding distinct, separable morphotypes within populations previously considered to be single, albeit cyclomorphic, species.

In the first instance, an indeterminate number of good species undoubtedly has been synonymised with the recognized taxon which they most closely resemble, either to conform to an authorative taxonomic reference work, or by a reviewer faced with inadequate descriptions and figures, or no type material. Several of these were encountered in our review of the Australian Brachionidae (Koste & Shiel, 1987), e.g. *Brachionus dichotomus*

and *B. lyratus*, both Australian endemics, in the absence of evidence to the contrary, had been synonymised with *B. falcatus* and *B. angularis* respectively (Ahlstrom, 1940; Voigt, 1957; Koste, 1978). To resolve such synonymy on a global basis, in the manner of Harring's (1913) 'Synopsis' will be a daunting task.

In the second instance, a distinct taxon may be masked by the range of morphological variation (seasonal di- or polymorphism) in a co-occurring congener. Several morphs of the polytypic Formenkreis *Keratella cochlearis* discussed by Ruttner-Kolisko (1974) are now considered valid species, e.g. *K. tecta* (cf. Ridder, 1973; Hofmann, 1983; Koste & Poltz, 1984). A similar situation is evident in the Canadian *Keratella cochlearis* populations described by Chengalath & Koste (1987), and *Polyarthra* populations from Mali (Koste & Tobias, 1989).

Many species have one or more subspecies, varieties, forms, or morphotypes. In only a few cases is this morphological variation attributable to environmental variables, e.g. the responses of brachionids and *Asplanchna* to predatory

Asplanchna (cf. Gilbert & Stemberger, 1984). Comparable responses to predators or other environmental variables are poorly known for most genera.

Resolution of erroneously synonymised or 'hidden' taxa as noted above will add appreciably to the rotifer record, and undoubtedly add strength to the increasing evidence for non-cosmopolitanism (cf. Dumont, 1983). Conversely, a simultaneous intensive effort is required to determine the identity of species described from single individuals, aberrant forms, ecotypes, and *inter alia*, preservation artefacts (see Shiel & Koste, 1985), which have been named without the slightest information on individual species' variability. Naming of the slightest variation from the 'norm' (indeterminate!) has only promoted confusion.

In the following discussion we reiterate points made 25 years ago, when Ruttner-Kolisko (1963) summarised the inter-relationships of the Rotifera, and suggested a set of taxonomic guidelines. Had these been consistently followed much of the present systematic uncertainty within the group would have been obviated, as would the lax approach to rotifer systematics in some investigations, noted by Dumont (1980). Attention also is drawn to points raised in Ruttner-Kolisko (1974), Pejler (1977) and taxonomic/biogeographic papers in previous Symposium proceedings (King, 1977; Dumont & Green, 1980; Pejler *et al.*, 1983 & May *et al.*, 1987).

The species concept

The predominance of parthenogenetic reproduction in rotifers renders the biological concept of species unacceptable (i.e. *groups of interbreeding natural populations reproductively isolated from other such groups*) (see Ruttner-Kolisko, 1974, and for bdelloid rotifers, Ricci, 1987). The evolutionary species of Simpson (1961) remains valid for unisexual (bdelloid) and bisexual (Monogonont) rotifers:

'An evolutionary species is a lineage (an ancestral-descendant sequence of populations)

evolving separately from others and with its own unitary evolutionary rôle and tendencies'.

In this context, the criterion of the species is the genetic isolation of populations.

This definition encompasses the great diversity demonstrated within apparently contiguous populations (e.g. King & Zhao, 1987) as well as phenotypic plasticity evident in some taxa (cf. King, 1977; Pejler, 1980). It does, however, raise the problem of the status of geographically isolated populations, such as the *Brachionus plicatilis* of Soda Lake, Nevada described by King & Zhao, and those of western Victoria described by Walker (1981). These populations clearly have evolved in isolation from each other, to the extent where a recognizable morph, presently identified as the ssp. *colongulaciensis* Koste & Shiel occurs in Victoria. If the evolutionary species concept is followed to the letter, these two populations should be accorded specific status. Comparative genome analysis (e.g. King, this volume) is necessary to determine the extent of interpopulation genetic variation, then (given that there is already considerable intrapopulation or clonal variation), the acceptable degree of genetic dissimilarity must be defined before specific status is accorded. The intraspecific morphological variability in *B. plicatilis* discussed by Sudzuki (1987) may reflect such dissimilarity. Resolution, on a global scale, appears to be a long-term prospect!

Because neither the expertise nor the facilities for such genetic analyses, or those involving biochemical specificity (e.g. Snell, this volume) are readily available to the 'average' rotifer taxonomist, there is a continuing dependence on comparative morphology. Before considering some of the basés of α-taxonomy, it is appropriate to mention current nomenclatural considerations.

The Code

The International Code of Zoological Nomenclature was established 'to promote stability and universality in the scientific names of animals and to ensure that the name of each taxon is unique and distinct' (Ride *et al.*, 1985: 3). To ensure uniformity, the conventions of the Code should be observed, including conformity to Latin form, and the rules of Latin grammar (Jeffrey, 1977). Rotifer taxonomy has not conformed consistently to the requirements of the Code, particularly in adequacy of description and/or illustration, nomination and location of type material, and concurrent description of nearest related taxa with designation of features differentiating a new taxon. Even basic errors of gender have passed referees and editors through several generations of major revisions, (e.g. *Brachionus bidentatus* is intermittently cited as *B. bidentata*, depending on the revision available).

It is beyond the scope of this paper to detail correct nomenclatural orthography, however 'rotiferologists' should be aware that incorrect terminations, generic and/or specific names, figures, authors or citations are to be found in the rotifer literature. If these are not to be perpetuated, some degree of careful research is necessary when transcribing from original descriptions or previous works. Correct procedures are detailed in Mayr (1973) and Jeffrey (1977), and if in doubt, the Code should be consulted (Ride *et al.*, 1985).

Nomenclature

The proliferation of rotifer names includes a large number of trinomina, and occasionally, especially in the bdelloids (cf. Ricci 1987), quadrinomina. Although quadrinomina are admissable under the Botanical Code, they are not under the Zoological Code, and only one category below species (subspecies) is recommended under the present Code (Article 45e). In the past, trinomens have been subspecies or infrasubspecific descriptors, i.e. forma or varietas (var.). The Code recommends that they should take subspecific rank (with exceptions, which are detailed in the Code) (Ride *et al. loc. cit.*).

This is the nomenclature used by Kutikova (1970), however difficulties immediately become evident when considering ecotypic variation such as shown by *Brachionus calyciflorus* Pallas. This polymorphic species has variants which have been named: *anuraeiformis* (Brehm), *amphiceros*

(Ehrenberg), *dorcas* (Gosse), *spinosa* (Beauchamp), *monstruosa* (Ridder), *inter alia*, with intergrades between some of these. Two or more of these morphs often may occur together, i.e., spine development shows intrapopulation variation, and ssp. status clearly is inappropriate. Subspecies, as defined by Mayr (1973), are geographically defined aggregates of local populations which differ taxonomically from other such subdivisions of the species.

Such morphs were termed the '*calyciflorus* group' by Ruttner-Kolisko (1972, 1974), while Koste (1978), in the context of an 'Artgruppe', used subspecies, varietas and forma, which encompassed morphotypic variations in sympatric populations. For such variable, closely-related sympatric (sometimes allopatric) monophyletic groups, the term superspecies was applied by Mayr (1931). This is more precise than the category 'Artgruppe' (species group) in Koste (1978). Other superspecies exhibiting wide variation within populations include *Anuraeopsis fissa*, *Asplanchna sieboldi*, *Brachionus angularis*, *B. bidentatus*, *B. quadridentatus* s.l., *Keratella cochlearis*, *K. slacki*, *Lepadella acuminata*, *Notholca squamula*, to name a few!

To communicate information on morphotypes of a superspecies which may occur in the same population, a more detailed nomenclature, such as that used by algologists, is desirable, e.g.:

B. calyciflorus Pallas, 1766 (f. anuraeiformis after Brehm, 1909)

uses parentheses to designate an infrasubspecific taxon, which has no systematic status *per se*, but designates which morphotype is involved. 'Form' in this context is ecotypic, seasonal or sexual variation, and conforms to the recommendations of the Code. In other than superspecies, more stringent applications of the rules of binomial nomenclature are required than has been the case in the past. New names should be avoided until the most complete possible information on intra- and interpopulation variation for rotifer taxa is available. We believe a more detailed nomenclature is necessary for both ecological studies and better communication between rotifer workers.

α-taxonomy

The description of new species of Rotifera, and their organization into genera and higher categories often has been based on morphological features which are at best variable, and at worst, useless as taxonomic identifiers. In loricate taxa, e.g. *Keratella* and *Brachionus*, lorica facettation appears to be species-specific in some taxa, e.g. *K. cochlearis* but variable in others, e.g. *B. angularis*. Number and placement of anterior lorica spines appears to be characteristic for each species, but there is considerable variation in their size, also of the ventral lorica margin. Similarly, caudal spines may be highly variable or absent (cf. Fig. 1, *K. slacki* from Tasmania), and of no taxonomic value.

Illoricate taxa present even more difficulties. Size and shape of most species is variable, with overlap between congeners. Comparative morphology of appendages such as bristles (*Filinia*) and fins or paddles (*Polyarthra*), and location of lateral antennae (*Testudinella*), have been used diagnostically with some success, although there may be inter- and intrapopulation variation. Morphologically similar congeneric species may cooccur and not be identified. A more detailed account of the difficulties in each of the monogonont genera is included in Ruttner-Kolisko (1974) and Koste (1978), and for the bdelloids, in Donner (1965).

Trophi structure has been used successfully to separate congeners; Guiset (1977) separated several *Polyarthra* species on the basis of fulcri length. *Synchaeta* species were differentiated by Stemberger (1979). *Filinia hofmanni* and *F. australiensis* were distinguished from congeners by Koste (1980), and more recently, trophi structure separated species of *Polyarthra* in west African waters (Koste & Tobias, 1989). Trophi structure has been inadequately studied for many taxa, yet appears to be species-specific, and therefore a valuable taxonomic discriminator.

The small size of most trophi has been a deterrent to intensive study. The trophi in most taxa are < 100 μm long, in some species, e.g. *Encentrum*, < 20 μm. These tiny structures are at the limit of light microscopy, and can be resolved only with

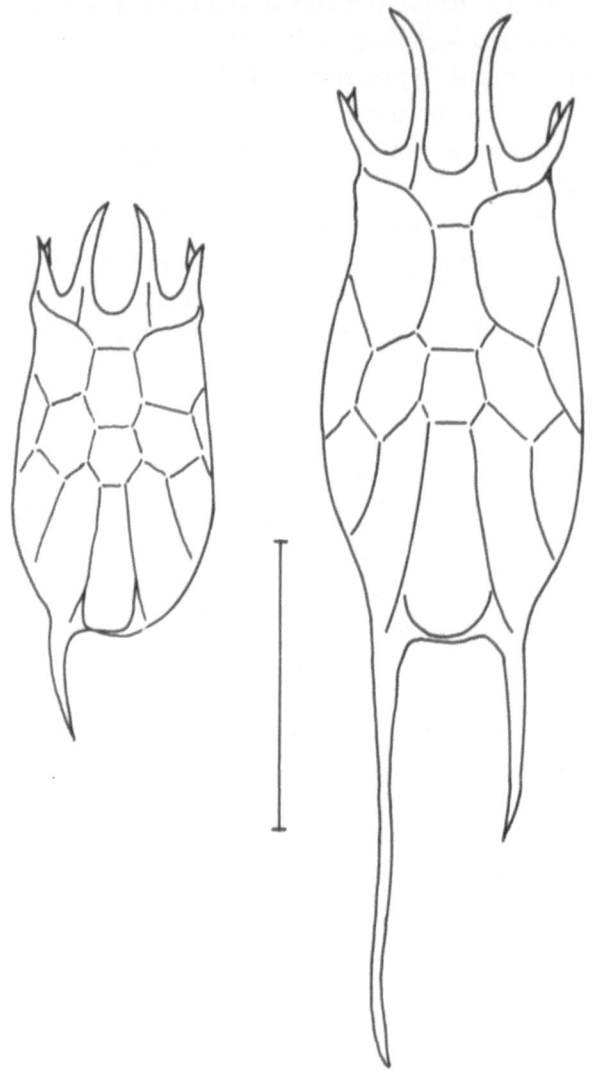

Fig. 1. Ecotypic variations in lorica morphometry, *Keratella slacki* (Berzins) from Tasmania, spring 1987 [see Shiel *et al.* this volume]. Scale bar 100 μm.

difficulty under oil immersion. Scanning electron microscopy (SEM) provides the necessary resolution, however until recently had been used successfully only with the larger trophi of *Asplanchna* (e.g. Salt *et al.*, 1978). *Asplanchna* trophi may reach 340 μm (Koste & Shiel, 1980), are flat, simple in structure, and can be treated for SEM examination reasonably easily. To treat the smaller trophi of other genera requires extreme care in handling (see Markevich & Koreneva, 1981). Ideally, detailed trophi structure determination for each species is necessary to clarify the present systematic confusion. The recent work of

Markevich (1985 and this volume) provides the resolution of the problem of 'lumping' morphologically similar but nonetheless distinct species (Dumont, 1983: 21), and enables more precise diagnosis of higher categories.

Modern methodology

Technological advances in optics, computing, analytical techniques, *inter alia*, have been in evidence in each successive symposium volume. We see these as valuable adjuncts to morphological taxonomy which will continue to enhance rather than replace traditional methodology. Improved methods have played a significant role in the renaissance of systematics in this generation (cf. Mayr, 1973), permitting the finer resolution or information processing capacity previously lacking. SEM methods, for example, will undoubtedly clarify the systematic problems; the recent work of Kutikova & Markevitch (this volume) represents a major impetus to a revision of current rotifer taxonomy. SEM is not widely available however, and its use is in practice restricted to research institutions. Similarly, image-analysis systems enable rapid morphometric computations, with image and data storage on disk facilitating exchange of information between workers, but again are not widely available.

More accessible for image/data storage and exchange are increasingly inexpensive video systems. The resolution obtainable by video recording is approaching that of a microscope-mounted 35 mm camera; we have used a high resolution National F-10 system in ecological studies on rotifer/bacteria and rotifer/algae interactions (e.g. *Tetrasiphon hydrocora* grazing on *Staurastrum*) in Murray-Darling waters, but see a wider use as a potential means of communication between rotifer workers. For a taxonomist located far from source material (which generally arrives preserved and less than life-like), a recording of the organisms before preservation would remove the difficulties encountered with preservation artefacts, provide behavioural, ecological information and so on which previously has been unavailable. Although video is more widely available, com-

patability problems between various systems have yet to be resolved.

Computer analytical techniques have not yet become widely available in rotifer taxonomy, although their validity has been established for field population studies. Available analytical programs permit rapid handling of amounts of data which previously have been prohibitive; a number of programs are applicable to species morphometric analysis. An example of these methods applied to rotifer community data is found in Nogrady (1988). Details of methodology are given in Romesburg (1984).

Examples of biochemical and molecular techniques which represent future directions from which taxonomy must benefit are given by Snell (this volume) and King (this volume).

It is clear from the range of papers presented at this symposium that the expertise exists to resolve the taxonomic uncertainties of the Rotifera, but it is widely dispersed. It cannot be expected that those with the expertise and facilities to run gels, extract DNA, operate SEM, TEM and computer analytical systems can do so on a global basis. Comparative morphology must, therefore, remain the principal tool of taxonomic workers while the specialized techniques become more widely disseminated. We are hopeful that the global cooperation which results from the rotifer meetings, will include more communication between widely-dispersed taxonomists. To this end we would like to see a centralised data base established which ultimately could include morphological and ecological information for all rotifer taxa. Such a data base would extend beyond the lifetime of individuals contributing to it; housing in a research institution would be desirable, with provision for continued contributions. We consider the benefits of centralized records to far outweigh the costs of setting up a workable system.

References

Ahlstrom, E. H., 1940. A revision of the Rotatorian genera Brachionus and Platyias with descriptions of one new species and two new varieties. Bull. am. nat. Hist. Mus. 77: 148–184.

284

Chengalath, R. & W. Koste, 1987. Rotifera from North-western Canada. Hydrobiologia 147: 49–56.

Donner, J., 1965. Ordnung Bdelloidea. Bestimmungsb. Bodenf. Eur. 6: 1–297.

Dumont, H. J., 1980. Workshop on taxonomy and biogeography. Hydrobiologia 73: 205–206.

Dumont, H. J., 1983. Biogeography of rotifers. Hydrobiologia 104: 19–30.

Dumont, H. J. & J. Green, 1980. Rotatoria. Proceedings of the 2nd International Rotifer Symposium. Hydrobiologia 73: 1–263.

Gilbert, J. J. & R. S. Stemberger, 1984. Asplanchna-induced polymorphism in the rotifer Keratella slacki. Limnol. Oceanogr. 29: 1309–1316.

Guiset, A., 1977. Some data on variation in three planktonic genera. Arch. Hydrobiol. Beih. 8: 237–239.

Harring, H. K., 1913. Synopsis of the Rotatoria. Bull. am. Nat. Hist. Mus. 81: 1–226.

Hofmann, W., 1983. On temporal variation in the rotifer Keratella cochlearis (Gosse): the question of 'Lauterborncycles'. Hydrobiologia 101: 247–254.

Jeffrey, C., 1977. Biological Nomenclature. 2nd edit. E. Arnold, London: 1–70.

King, C. E., 1977. Proceedings of the First International Rotifer Symposium. Arch. Hydrobiol. Beih. 8: 1–315.

King, C. E., 1989. Molecular genetics of rotifers: preliminary restriction mapping of the mitochondrial genome of Brachionus plicatilis. Hydrobiologia, this volume.

King, C. E. & Y. Zhao, 1987. Coexistence of rotifer (Brachionus plicatilis) clones in Soda Lake, Nevada. Hydrobiologia 147: 57–64.

Koste, W., 1978. Die Rädertiere Mitteleuropas (Ü.-Ordn. Monogononta). Begr. von M. Voigt. 2 Vols. Borntraeger, Stuttgart.

Koste, W. & W. Poltz, 1984. Über die Rädertiere (Rotatoria, Phylum Aschelminthes) des Dümmers, NW-Deutschland. Osnabrücker Naturwiss. Mitt. 11: 91–125.

Koste, W. & R. J. Shiel, 1980. Preliminary remarks on the characteristics of the rotifer fauna of Australia (Notogaea). Hydrobiologia 73: 221–227.

Koste, W. & R. J. Shiel, 1987. Rotifera from Australian inland waters. II. Epiphanidae and Brachionidae (Rotifera, Monogononta). Invert. Taxon. 1: 949–1021.

Koste, W. & W. Tobias, 1989. Zür Rädertierfauna der Lélingué Talsperre in Mali, Westafrika (Aschelminthes: Rotatoria). Senckenbergiana biologica 69 (4/6) (in press).

Kutikova, L. A., 1970. Rotifer fauna of the U.S.S.R. Subclass Eurotatoria. Fauna CCCP 104: 1–744.

Kutikova, L. A. & G. I. Markevich. 1989. The basic principles of constructing a system of rotifers. Hydrobiologia, this volume.

Markevich, G. I., 1985. The main trends of idioadaptive evolution of rotifers: Jaws. IN Proceedings of the 2nd All-Union symposium on rotifers, Leningrad.: 17–37.

Markevich, G. I. & E. A. Koreneva, 1981. On the technique of preparing the mastax of rotifers for electron microscopy. Zool. Zh. 60: 1562–1564.

May, L., R. Wallace & A. Herzig, 1987. Rotifer Symposium IV. Proceedings of the IVth International Rotifer Symposium. Hydrobiologia 147: 1–381.

Mayr, E., 1931. Birds collected during the Whitney South Sea expedition. XII. Notes on Halcyon chloris and some of its subspecies. Am. Mus. Nov. 469: 1–10.

Mayr, E., 1967. Artbegriff und Evolution. Paul Parey, Hamburg.: 1–617.

Mayr, E., 1973. Priciples of Systematic Zoology. McGraw-Hill, New Delhi.: 1–428.

Nogrady, T., 1988. The littoral rotifer plankton of the Bay of Quinte (Lake Ontario) and its horizontal distribution as indicators of trophy. Arch. Hydrobiol. Suppl. 79: 145–165.

Pejler, B., 1977. On the global distribution of the family Brachionidae (Rotatoria). Arch. Hydrobiol. Suppl. 53: 255–306.

Pejler, B., P. Starkweather & T. Nogrady, 1983. Biology of Rotifers. Proceedings of the 3rd International Rotifer symposium. Hydrobiologia 104: 1–396.

Ricci, C., 1987. Ecology of bdelloids: how to be successful. Hydrobiologia 147: 117–127.

Ridder, M. De, 1973. Atlas provisoire de Rotiferès de Belgique. Fac. Sci. Agron., Gembloux.

Ride, W. D., C. W. Sabrosky, G. Bernardi & R. V. Melville (eds), 1985. International Code of Zoological Nomenclature. #3rd edit. Univ. of Calif. Press, Berkeley.: 1–321.

Romesburg, H. C., 1984. Cluster analysis for researchers. L.L.P., California.: 1–334.

Ruttner-Kolisko, A., 1963. The inter-relationships of the Rotatoria. In L. Hyman (ed.) The Lower Metazoa. Univ. Calif. Press.: 263–272.

Ruttner-Kolisko, A., 1972. Rotatoria. Die Binnengewässer 26: 99–234.

Ruttner-Kolisko, A., 1974. Plankton Rotifers. Biology and taxonomy. Die Binnengewässer Suppl.: 1–146.

Salt, G. W., G. F. Sabbadini & M. L. Commins, 1978. Trophi morphology relative to food habits in six species of rotifers (Asplanchnidae). Trans. am. Microsc. Soc. 97: 469–485.

Shiel, R. J. & W. Koste, 1985. New species and new records of Rotifera (Aschelminthes) from Australia. Trans. r. Soc. S. Aust. 109: 1–15.

Simpson, G. G., 1961. Principles of Animal Taxonomy. Columbia Univ. Press., N.Y.: 1–247.

Snell, T., 1989. Systematics, reproductive isolation and species boundaries in Monogonont rotifers. Hydrobiologia, this volume.

Stemberger, R. S., 1979. A guide to the rotifers of the Laurentian Great Lakes. E.P.A., Ohio.: 1–186.

Sudzuki, M., 1987. Intraspecific variability of Brachionus plicatilis. Hydrobiologia 147: 45–47.

Voigt, M., 1957. Rotatoria. Die Rädertiere Mitteleuropas. Borntraeger, Berlin: 1–508.

Walker, K. F., 1981. A synopsis of ecological information on the saline lake rotifer Brachionus plicatilis Müller, 1786. Hydrobiologia 81: 159–167.

Hydrobiologia **186/187**: 285–289, 1989.
C. Ricci, T. W. Snell and C. E. King (eds), Rotifer Symposium V.
© *1989 Kluwer Academic Publishers.*

Mastax morphology under SEM and its usefulness in reconstructing rotifer phylogeny & systematics

G.I. Markevich[1] & L.A. Kutikova
[1] *The Institute for the Biology Inland Waters, USSR Academy of Sciences, Borok, 152742, USSR; The Zoological Institute, USSR Academy of Sciences, Leningrad, 199034, USSR*

Abstract

Scanning electron micrographs have been prepared of mastaxes of more than 120 species of rotifers. The basic directions of the evolutionary transformation of mastax sclerites and their significance for reconstructing rotifer phylogeny are illustrated.

Introduction

Dissatisfaction with the current taxonomy of rotifers is growing since it is constructed as phenetic classification based solely upon the typological species concept (Dumont, 1980; Kutikova, 1983). So far, it has been difficult to apply the techniques of phylogenetic systematics to rotifers. This is despite the advantages of this approach for describing the evolutionary history of this group.

Analysis of the prevailing classification systems for rotifers suggests that, in spite of some differences, they have much in common. For example, the separation of rotifer classes by gonad number is used by both de Beauchamp (1907) and Remane (1929-1933). The main groups of rotifers on the order level are actually identical in both classification systems. The main disagreements among these taxonomies result not from the classification of rotifers, but from different interpretations of their evolution due to different assumptions about ancestral groups. Previous attempts to reconstruct the phylogeny of rotifers on the basis of comparative, anatomical and morphological data have lead to contradictory conclusions even by such eminent zoologists as de Beauchamp and Remane.

What are the characters that are best for classifying rotifers and revealing their phylogeny? Kutikova (1970, 1983) suggested that these should be characteristics of the structure and functioning of the corona and mastax (locomotory-trophic approach). Other rotiferologists, however, do not expect to find answers about rotifer evolution by using conventional morphological analysis of females. Rather, they pin their hopes on male morphology (Sudzuki, 1964, 1977) or on tissue ultrastructural organization (Clement, 1980). Previous efforts to reconstruct rotifer phylogeny with morphological methods have not succeeded because of the insufficient resolving power of optical microscopy. In principle, morphological analysis is sufficient to describe phylogenetic relationships within this phylum. Our work uses the greater resolution of scanning electron microscopy to examine the mastaxes of several rotifer species and employs this information to describe a phylogeny.

Results

Methods for preparing inner sclerite parts of the mastax for scanning electron microscopy (SEM)

286

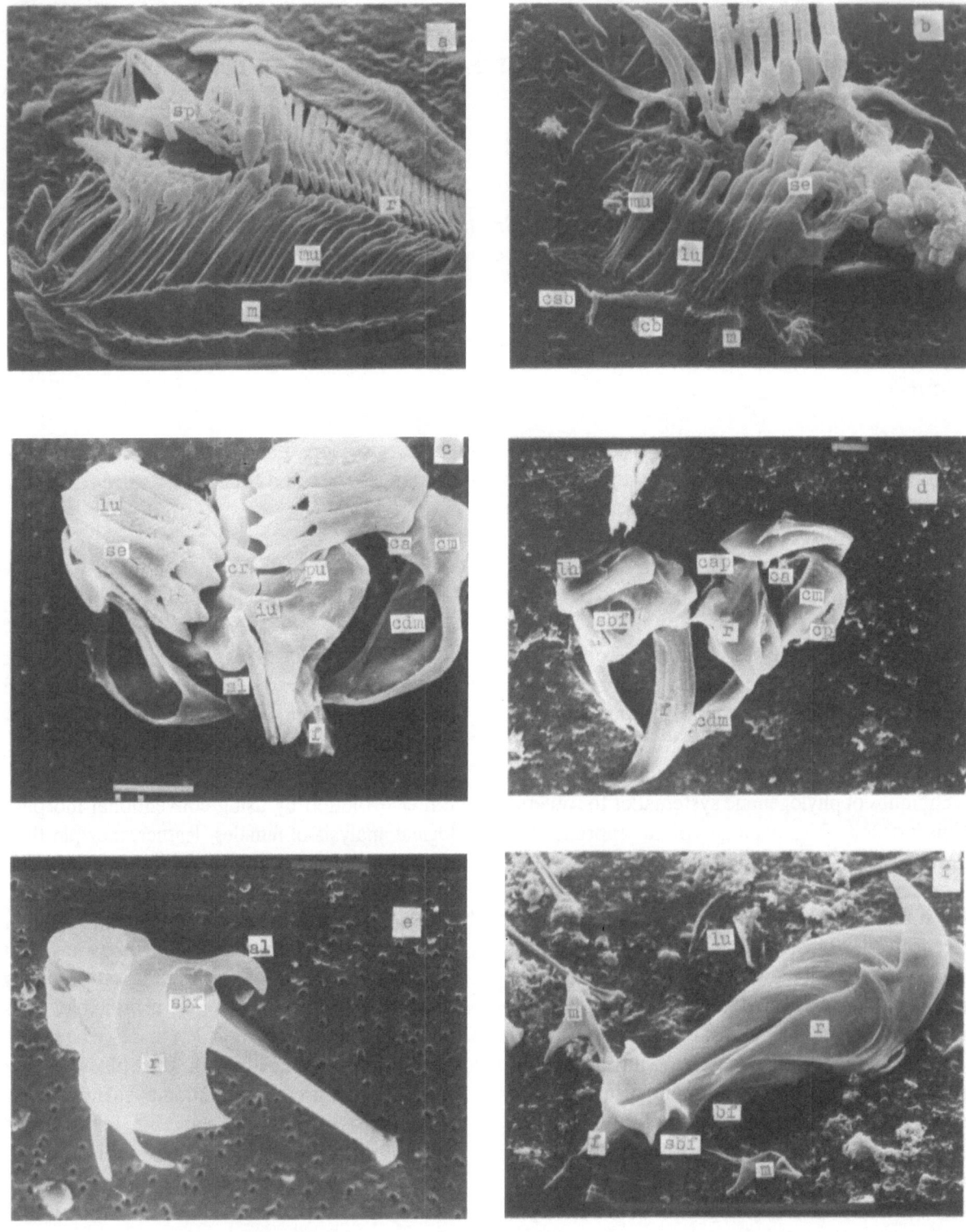

Fig. 1. Sclerite systems of mastaxes of some rotifers and details of their structure. a – *Rotaria tardigrada*; b,h,i – *Conochiloides natans*; c – *Brachionus calyciflorus*; d – *Notommata allantois*; e – *Trichocerca sulcata*; f – *Asplanchna herricki*; g – *Floscularia decora*; j – *Epiphanes brachionus*; d – *Euchlanis dilatata*; l – *Dicranophorus forcipatus*. -a-d,f,l – sclerite systems; e,g-i – rami; j – unci; k

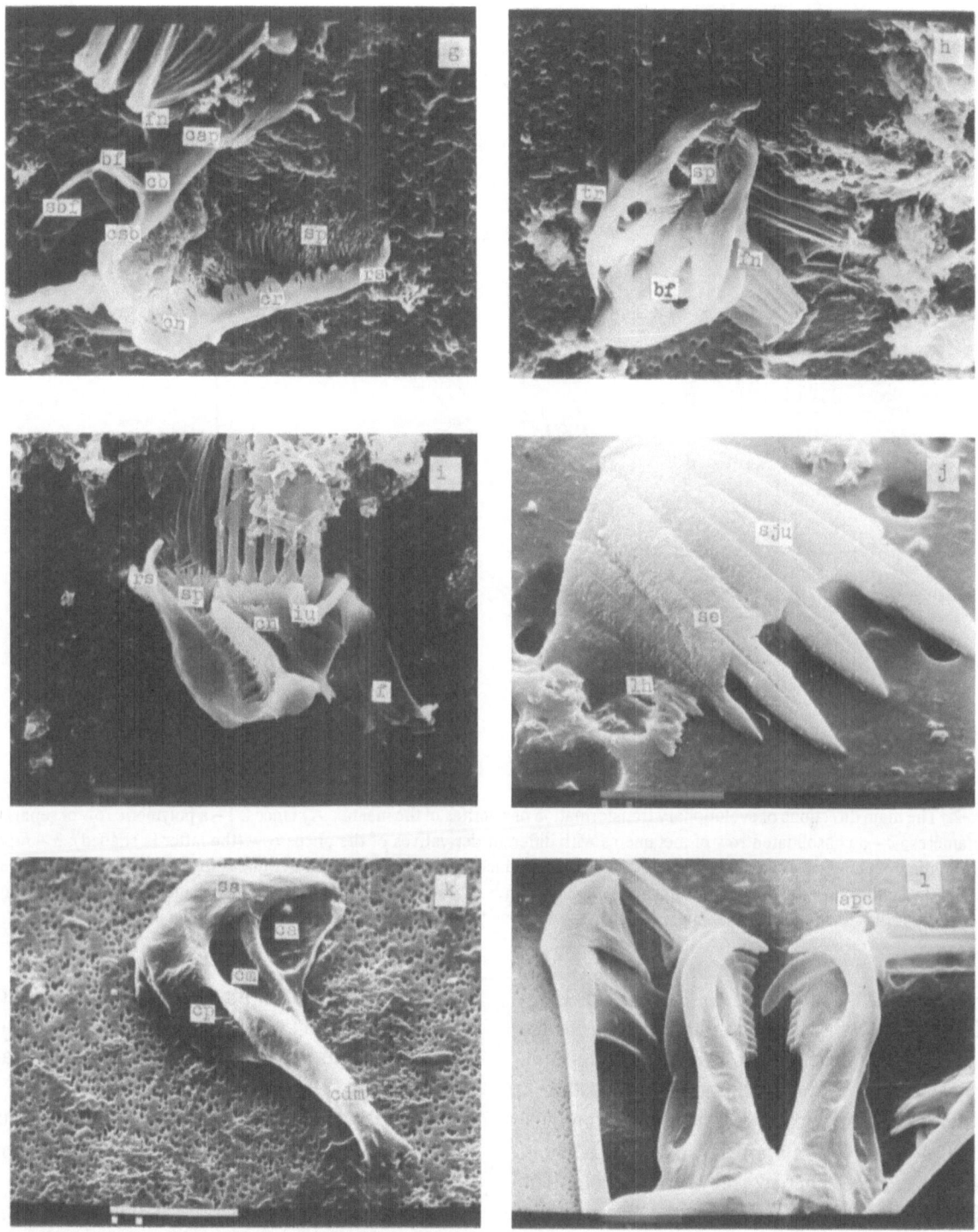

– manubrium. – al – alula; apc – apophysis cardalis; bf – basifenestra; ca – camera anterior; cap – camera apicalis; cb – camera basalis; cdm – cauda manubrialis; cm – camera medialis; cn – constrictor; cp – camera posterior; cr – crista; csb – camera subbasalis; fn – fenestrula; fu – fulcrum; iu – impressio uncinalis; lh – lamina hilaris; lu – lamina uncinalis; m – manubrium; mu – metamerae unci; pu – praeuncus; r – ramus; rs – rostellum; sa – sutura anterior; sbf – subbasifenestra; se – sutura externa; sl – satelliti; sp – scleropili; sju – sutura jugalis; tr – trabecula. Scale bars, marked by one quadrant – 1 mkm; by two – 10 mkm; by three – 100 mkm.

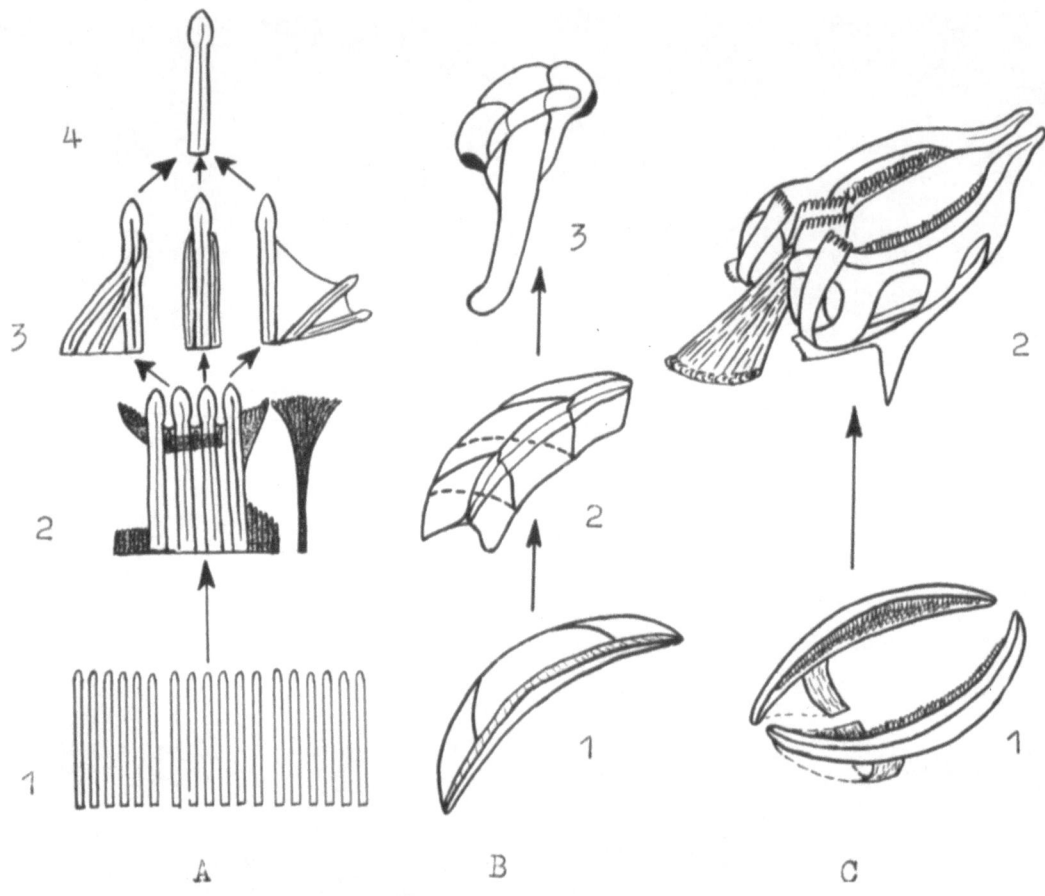

Fig. 2. The main directions of evolutionary transformation of sclerites of the mastax. A) Uncus: 1 – a polymeric row of separated metameres; 2 – a consolidated row of metameres with different derivatives of the uncus row (the latter is shaded); 3 – modes of formation of uniapical uncus – contraction, confluence, abduction; 4 – mono-elementary uncus. B) Manubrium: 1 – navicular form of the lamina; 2 – growth of lamina and formation of chambers; 3 – formation of the cauda by the medial chamber. C) Rami: 1 – are separated lamellar; 2 – growth of ramus trabecules and formation of scleropilar fulcrum.

are available (Markevich & Koreneva, 1981; Markevich, 1984). These allowed Markevich (1987a, b) to successfully apply morpho-functional analysis to the evolution of the mastax. A rotifer phylogeny based on these data favors a close phylogenetic relationship between bdelloid rotifers and cercomer worms (Monogenea). The SEM method was used to study the sclerite complexes of the mastax in 120 species of freshwater rotifers representing practically all morpho-functional types (Markevich, 1983, 1985a & b, 1987b). The detailed study of sclerites of the mastax has revealed the following structures which were previously unknown: sutura: exterior, jugalis, anterior; rostellum, crista, scleropilli,

trabecula, fenestrula, basifenestra, subbasifenestra, constrictor; camera: apicalis, posterior, medialis, anterior; and lamina: hilaris, uncinalis (Markevich, 1989, Fig. 7).

Comparative analysis of mastax fine structure suggests that the reduction in mastax elements common in Eurotatoria are derived from an ancestral polymeric sclerite system, present only in the Bdelloida (Markevich, 1983, 1985a&b, 1987b). These transformations of the mastax can be accomplished by changes in single elements rather than a fundamental change in mastax organization.

The following basic trends of morphological transformations of jaws of the mastax have been revealed: oligomerization of rows of uncus metamers, 'ramatization' of the sclerite system, i.e. a progressive transfer of functional load from unci to rami, intersclerite integration of elements in the mastax, and development of close spatial and functional relations between the mastax and corona (intra- and intermodule integration) (Fig. 2). The latter is accompanied by a shift of the mastax closer to the mouth, turning on its axis in the pharynx and other complex allometric transformations. Combinations of intramastax characters provides significant information about trends in sclerite system transformation and is useful in rotifer systematics in general.

Conclusions

On the basis of our analysis, the phylogeny of rotifers has been revised. We suggest that the Bdelloidea be placed in a new subclass called the Archeorotatoria. We also suggest that the Ploima be divided into four new orders: Transversiramida, Saltiramida, Saeptiramida, Antrorsiramida.

Therefore, the first results of a comparative SEM morphological study of the mastax support the usefulness of morphological analysis and Remane's criteria (1956). The morphological approach to rotifer phylogeny based on SEM of mastax fine structure of the complex sclerite muscular apparatus appears to be sufficient for reconstruction of a rotifer phylogeny. A successful application of SEM methods to morphological analysis in a study of rotifer polygeny suggests that further analysis of their locomotory and trophic systems (corona, mastax) is likely to yield interesting insights into rotifer evolution.

References

Beauchamp, P. de, 1907. Morphologie et variations de l'appareil rotateur dans la serie des Rotiferes. Arch. Zool. exptl. et gen. 4,6: 1–29.

Beauchamp, P. de, 1965. Classe des Rotiferes. Traite de Zoologie, Anatomie, Systematique, Biologie IV, 3: 1225–1379.

Clement, P., 1980. Phylogenetic relationships of rotifers, as derived from photoreceptor morphology and other ultrastructural analyses. Hydrobiologia 73: 93–117.

Kutikova, L. A., 1970. Rotifers of the USSR. Leningrad. 744 pp. (In Russian).

Kutikova, L. A., 1983. Pecularities of diagnostics of taxa in rotifers. In: The Rotifers. Proceedings of the Second All-Union Symposium on Rotifers. Leningrad, pp. 4–17. (In Russian).

Markevich, G. I., 1984. Ultrastructure of chitionoid structures of the coracidia Triaenophorus nodulogus (Pallas, 1781). DAN 278, 4: 1022–1024. (In Russian).

Markevich, G. I., 1985a. The main trends of idioadaptive evolution of rotifers. Jaws. In: Proceedings of the Second All-Union Symposium on Rotifers. Leningrad 17–37. (In Russian).

Markevich, G. I., 1985b. Ultrathin morphology of mastax of rotifers. I. Bdelloida. Bulletin of the Institute of the Biology Inland Waters 68: 31–35. (In Russian).

Markevich, G. I., 1987a. The main trends of idioadaptive evolution in rotifers, Locomotion and locomotory behavior. In: Fauna and biology of freshwater organisms. Leningrad. Proceedings of the Institute of the Biology Inland Waters: 54(57): 175–190. (In Russian).

Markevich, G. I., 1987b. Functional morphology of the jaw apparatus in rotifers. Synopsis of a thesis for the degree of candidate of biological sciences. Leningrad. 23 pp. (In Russian).

Markevich, G. I., 1989. (in press). Morphology and the principle organization of sclerite system of the mastax in rotifers. In: Biologia and functional morphology of freshwater animal. Leningrad. Proceedings of the Institute of the Biology Inland Waters (In Russian).

Markevich, G. I. & E. A. Koreneva, 1981. On the technique of preparing of the mastax of rotifers for electron microscopy. Zool. zhurn. 60(10): 1562–1564. (In Russian).

Remane, A., 1929-33. Rotatorien. Dr. H. G. Bronn's Klassen and Ordnungen d.Tierreichs, Leipzig IV,II,1,1-4: 576 S.

Remane, A., 1956. Die Grundlagen des naturlichen Systems der vergleichenden Anatomie une der Phylogenetik. Leipzig. 364 S.

Sudzuki, M., 1964. New Systematical Approach to the Japanese Planktonic Rotatoria. Hydrobiol. Acta Hydrobiol., Hydrogr. et Protist. 23: 1–124.

Sudzuki, M., 1977. Classification based on the male. Arch. Hydrobiol. Beih. Ergebn. Limnol. 8: 221.

Hydrobiologia **186/187**: 291–298, 1989.
C. Ricci, T. W. Snell and C. E. King (eds), Rotifer Symposium V.
© 1989 *Kluwer Academic Publishers.*

Problems in taxonomy of rotifers, exemplified by the *Filinia longiseta – terminalis* complex

Agnes Ruttner-Kolisko
Biological Station of the Austrian Academy of Science

Key words: rotifers, taxonomy, variation, genetics, parthenogenesis, resting eggs, ecology

Abstract

The validity of taxonomical categories in parthenogenic groups is discussed. Special problems of rotifer taxonomy are caused by: facultative or obligatory parthenogenic reproduction, high morphological and genetic variability and paucity of morphological characteristics.

Examples of these problems are provided by the *Filinia terminalis-longiseta* group. The different ecological properties of various populations belonging to this group are emphasized. Suggestions concerning the creation of new taxa are made; in particular, the importance of using ecological data is stressed.

Taxonomy is an artificial grid, imposed by scientists on the continuous flow of adaptations, variations and mutations, which is characteristic of nature and results in evolution. There is only one point where this artificial grid has a biological meaning; that is the concept of *species*, which is defined biologically as 'a group of organisms, which breed to produce viable offspring' (Mayr, 1942). All other categories, above or below the species level, are arbitrary conventions. From an evolutionary perspective (Simpson, 1951), a species is 'a lineage, an ancestral-descendent sequence of populations, evolving separately from others'. This latter definition applies better to rotifers or other facultative or obligatory asexual organisms, where breeding, in the sense of exchange of genes, occurs rarely or not at all.

So, if taxonomy is artificial and 'does not correspond with anything in nature' (Simpson, 1967, 3.ed. pg. 17), why do we accept it as part – and even an important part – of biological sciences? The answer is obvious: without taxonomical names and categories and the man-made hierar-chy of the animal and plant kingdoms, we would be hopelessly lost in a bewildering diversity whose direction of flow we understand poorly or not at all. But it must be kept strictly in mind that our taxonomical grid – whether based on morphology, physiology, genetics or ecology, and whether coarse or fine – is not a property of nature as such. It is, in fact, only our necessary means to converse with each other about the facts, that we have found in nature.

I don't want to dwell any longer on these ideas; they are not new or original. Instead I shall try to apply these ideas to the special conditions of rotifers and use, as an example, the *terminalis-longiseta* complex of the genus *Filinia*.

The problems of rotifer taxonomy have various inter-related sources, namely:

1. The facultative (heterogonic) or obligatory parthenogenetic reproduction;
2. The short generation time and high mutation rate;
3. The paucity of morphological characteristics;
4. The high variability.

1. Parthenogenetic reproduction implies transfer of the parent's genotype to offspring without change. New mutations, if not lethal, may be immediately multiplied within the population. Both facts are partly responsible for the high degree of variability observed in rotifers. Obligatory parthenogenetic groups are potentially even more variable than heterogonic ones, since they lack the sexual recombination of genes during mixis, which unifies the genetic pool of a heterogonic population. Such groups are: the Bdelloidea, some planktonic populations which have never been found to undergo a mictic period, and populations which produce pseudosexual resting eggs, a phenomenon that probably occurs more often than we know.

Very likely, *Filinia terminalis* produces pseudosexual resting eggs in Esthwaite Water, a eutrophic lake of 15 m depth in the English Lake District. A peculiar situation concerning the vertical distribution of this population was found in 1979. At spring homothermy, in April, *F. termi-*nalis was scarce in the whole water column, but, when thermal stratification and oxygen depletion started to build up, the population became denser in the bottom layer. On May 15th we found an enormous increase just above the lake bottom (14 m) that could not be explained by reproduction at the low temperature of 9 °C (Fig. 1). Vertical migration was out of the question because there was no epilimnic population. The only explanation was an explosive hatching of resting eggs, but what triggered the hatching is not known.

Such an explosive hatching implies a massive production of resting eggs during the life cycle of the population. There are, in fact, reports of very high resting egg production; among others, Hofmann (1979) mentions 36-45%, Loose (1987) reports 50-100% females carrying resting eggs (referred to as 'mictic rate') in the Plußsee. Pejler (1961) indicates several mictic events on his graph of *F. terminalis* in Ösbysjön, but in the text he speaks of resting eggs only. Neither mictic females

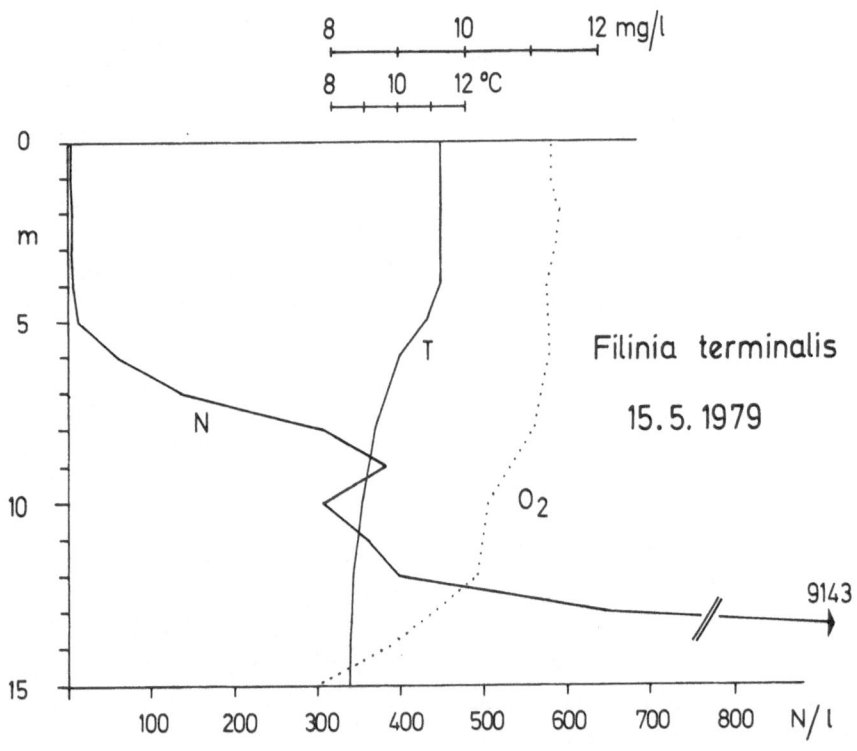

Fig. 1. Vertical distribution of *Filinia terminalis* in Esthwaite Water, English Lake District, on 15.5.1979. *N* = number of individuals; T = Temperature; O$_2$ = Oxygen content.

carrying the small eggs from which males hatch, nor males as such are mentioned by these authors. Moreover, a resting egg ratio of over 50% makes it very unlikely, and of 100% virtually excludes a 'normal' mictic phase. Both the unusual high percentage of resting eggs and the lack of males corroborate my suspicion that the resting eggs of *F. terminalis* are not fertilized, but pseudosexual. If this hypothesis is correct, the expression 'mictic females' and 'mictic rate' are inadequate. Moreover, the likelihood of extensive variation – genetical as well as morphological – would, theoretically, increase.

2. The short generation time of rotifers, which allows for approximately 100 generation in one year under favorable conditions (for instance in the tropics), multiplies the spreading of any genetic change in the population. King (1980) made a rough estimate of mutations per locus/year; he gives a figure of approximately 4000 for *F. terminalis*, which, in my opinion, is too high since *F. terminalis*, living at 4-6 °C has a generation time about four to five times longer than assumed in that paper.

3. The paucity of useful morphological characteristics poses – together with variability – the most serious difficulties for rotifer taxonomy. Rotifers are small and their morphological features, below the generic level, are hard to detect, at least with the means that ecologists are forced to apply in their work. Encouraged by modern microscopical techniques, particularly scanning electron microscopy, taxonomists have recently tended to use more and more minute morphological features to create new taxa in order to cope with an increasing diversity of biological data. In practice, such inconspicuous features are ascertained on very few specimens (often only one) and the results are applied to the whole population.

An example is again available from the *F. terminalis-longiseta* complex. Koste (1980) described a new taxon, *F. hofmanni*, from the Plußsee, N. Germany, on the grounds of a symmetrical tooth formula (15/15), setae measurements, and the

posture of the caudal seta in preserved condition. The tooth formula is similar to *F. terminalis* (15/16) and the teeth are not easy to count; seta measurements vary widely. The posture of the caudal seta, whether it is straight or abduced, depends on the insertion of the retractor muscles on the body wall, behind or before the seta (Fig. 2). It is also influenced by the intensity of contraction during preservation; if a specimen is narcotized, the caudal seta is not abduced. Therefore the new taxon is difficult to distinguish. Even

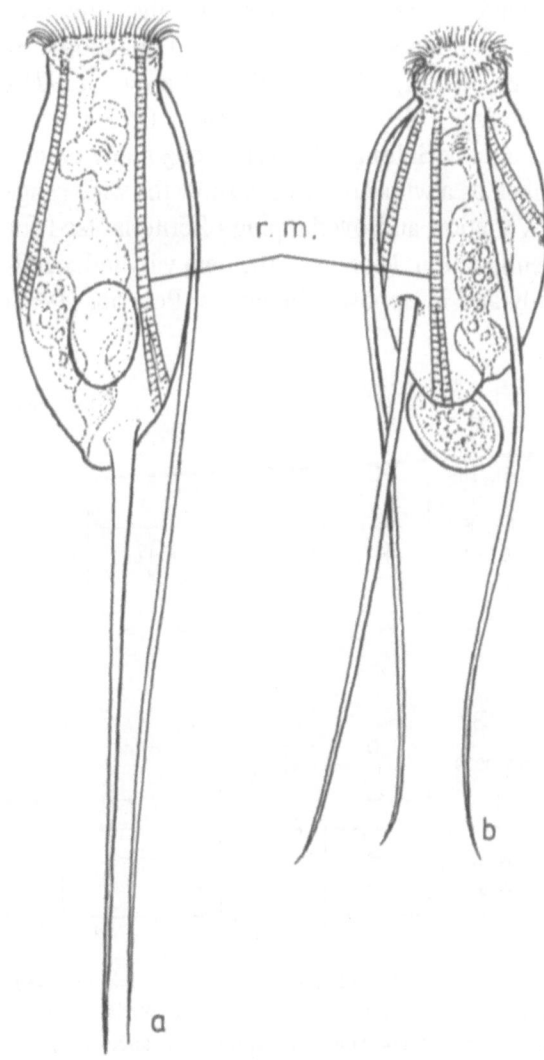

Fig. 2. Insertion of the retractor muscles on the body wall. a = *Filinia terminalis* (Plußsee, 11.5.1970); b = *Filinia terminalis* = *hofmanni* (Lunzer Obersee, 20.8.1979). r.m. = retractor muscle.

294

Loose (1987), working also on Plußsee material, could not distinguish *F. hofmanni* from *F. terminalis* neither 'living nor dead' (pg. 21). In fact, most investigators rely on the dubious, but easily visible, criterion of seta posture.

4. The taxonomical difficulties arising from variability are convincingly exemplified in the genus *Filinia*. To differentiate *F. longiseta* and *F. terminalis*, absolute measurements or ratios of body and appendices have until recently been used, although the variability of these features has been documented for at least 30 years. Ratios of lateral to caudal setae and the insertion of the latter have been computed in graphs and tabulations by Pejler (1957a), Parise (1961), Larrson (1971), Hofmann (1974), Stemberger (1977), and Schaber & Schrimpf (1984) among many others. Taken as a whole these data show the wide range of variation and overlapping of criteria used for identification. Moreover, the data vary independently and in opposed directions. Pourriot (1964)

even found that the setae of *Filinia* increased in length by about 50% within a few days in cultures containing the well-known 'Asplanchna substance'. Slonimski (1962) has shown experimentally that the insertion of the caudal seta is shifted ventrally with increasing temperature.

To add one more set of data to the host already existing, Fig. 3 shows the variation of *F. terminalis*[x] in 4 lakes in the English Lake District. The result of these measurements indicates no clear trend, except for a decrease of size from spring to summer and from eutrophic to oligotrophic lakes. In summary, the efforts of many scientists over a long time have shown that measures of body or appendices are poor means to distinguish taxa or to describe new ones within the genus *Filinia*. A number of measurements sufficient for statistical analysis is, nonetheless, helpful; mean setae ratios may differ, and regression lines on plots show different slopes for various populations or taxa (Schaber & Schrimpf, 1984).

In contrast to the picture that emerges from

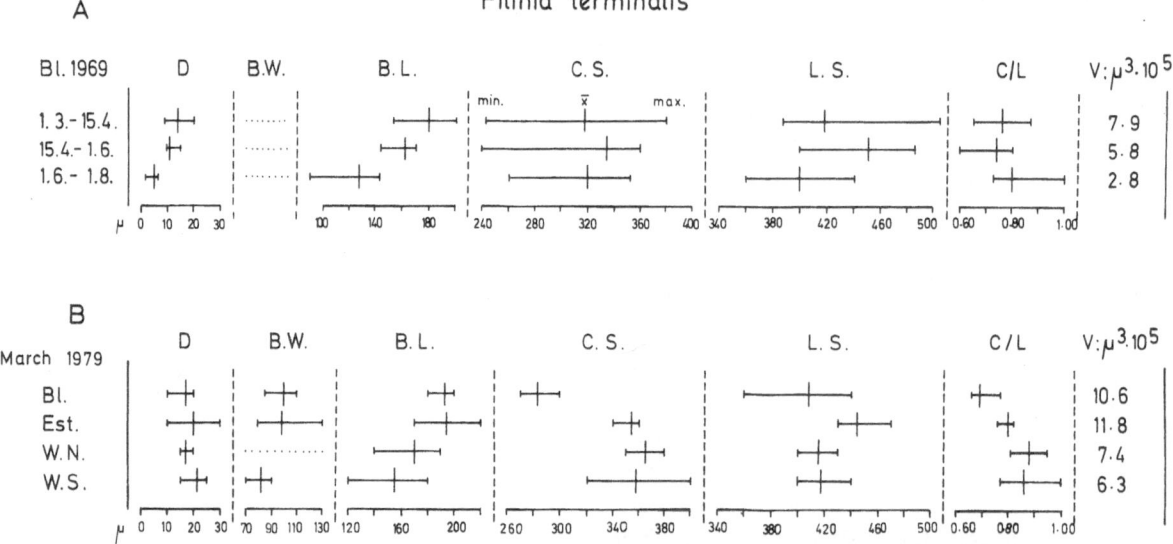

Fig. 3. Measurements of *Filinia terminalis* in 4 lakes of the English Lake District. A. Variation in time in one lake; B. Variation in different lakes at the same time (March 1979). Bl = Blelham Tarn; Est. = Esthwaite Water; W.N. = Windermere North Basin; W.S. = Windermere South Basin. *D* = Distance of caudal seta to end of body; B.W. = Body width; B.L. = Body length; C.S. = Caudal seta; L.S. = Mean of lateral setae; C/L = Ratio of caudal to lateral setae; V = Volume calculated from mean body length; Each graph represents at least 10 measurements.

---

[x] Determination according to Pourriot (1965) and Pejler (1957a) and confirmed by Hofmann (letter of 11.7.79).

morphological variation, from the ecological standpoint, at least in palaearctic lakes, two clear-cut groups occur within the genus *Filinia*. One group lives in summer at temperatures above approximately 18 °C in the epilimnion of lakes, ponds and slowly flowing rivers. The other group lives at temperatures below approximately 12 °C in the hypolimnion of eutrophic or meromictic lakes. The latter group can tolerate, but is *not restricted to* very low or nearly zero oxygen levels. The epilimnic, thermophilous group, with all its variations, is comprised under the name *F. longiseta* (sensu lato). The predominantly hypolimnic, cold stenothermous group with all its variations goes under the name of *F. terminalis* (sensu lato).

Within the cold stenothermous group the situation is rather complicated. Various populations occupy definite ecological niches according to the lake type, as we have shown in the 4 lakes of the English Lake District (Ruttner-Kolisko, 1980). The ecological behaviour of these populations at their maximal density is computed in Table 1. In the two eutrophic lakes the population maximum occurs in the nearly deoxygenated layer just above lake bottom. In the two much deeper meso- to oligotrophic lakes, the maximal abundance occurs in the metalimnion at oxygen saturation and is about 2-3 orders of magnitude lower. Only the temperature required is identical in all 4 lakes. Obviously, food availability regulates the location of the maximum of these bacteria-feeding rotifers. This assumption corresponds fairly well with the quantitative distribution of bacteria known from the respective lakes (Jones, 1972).

In meromictic lakes the *Filinia* populations show a different distribution, as depicted in Fig. 4; similar situations are described by Pejler (1957a) and Schaber & Schimpf (1984). In these cases the population is perennial and stays in a water layer of approximately 4 °C and between 1 mg l and zero O_2-content. This layer is just above the plate of chemotrophical bacteria. In the Lunzer Obersee (Fig. 4A) this layer moves up to 3-4 m depth under ice and down to 11-12 m in the short summer period. In Blankvatn (Fig. 4B) low

Table 1. Ecological data of *Filinia terminalis* populations at maximal density.

	Blelham Tarn (14 m) (eutrophic)	Esthwaite Water (15 m) (eutrophic)	Windermire South-Basin (40 m) (mesotrophic)	Windermire North-Basin (60 m) (oligotrophic)
Abundance at max. density (Indiv./l)	1800	1500	100	3
Time of max. density	May-June	May-June	June	June-Juley
Depth of max. density (m)	9–12 (hypolimn.)	12–15 (hypolimn.)	10–15 (metalimn.)	10–15 (metalimn.)
Temperature at max. density (°C)	<10	<10	10	10–12
Oxygen content (%-saturation)	10–40	10–40	>90	>90
Mean indiv. body size ($\mu^3 \cdot 10^5 = V$)	10.6	11.8	6.3	7.4
Occurrence of the population during the year	spring/ summer	spring/ summer	perennial	perennial

296

A LUNZER OBERSEE

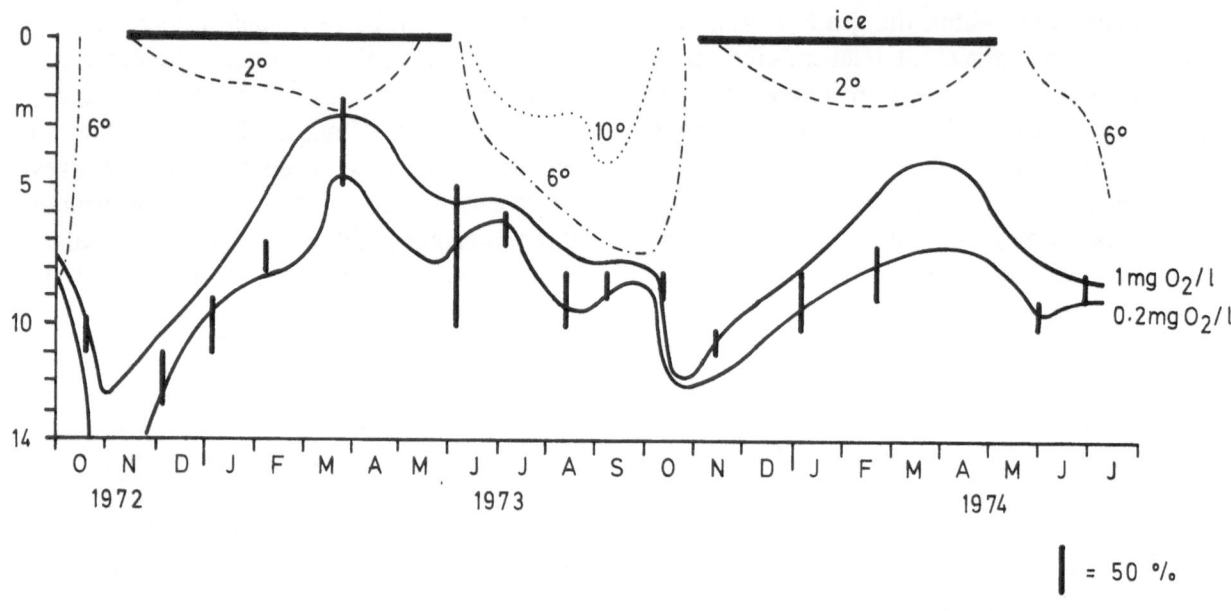

RUTTNER - KOLISKO 1974

FILINIA TERMINALIS = F. HOFMANNI

B BLANKVATN

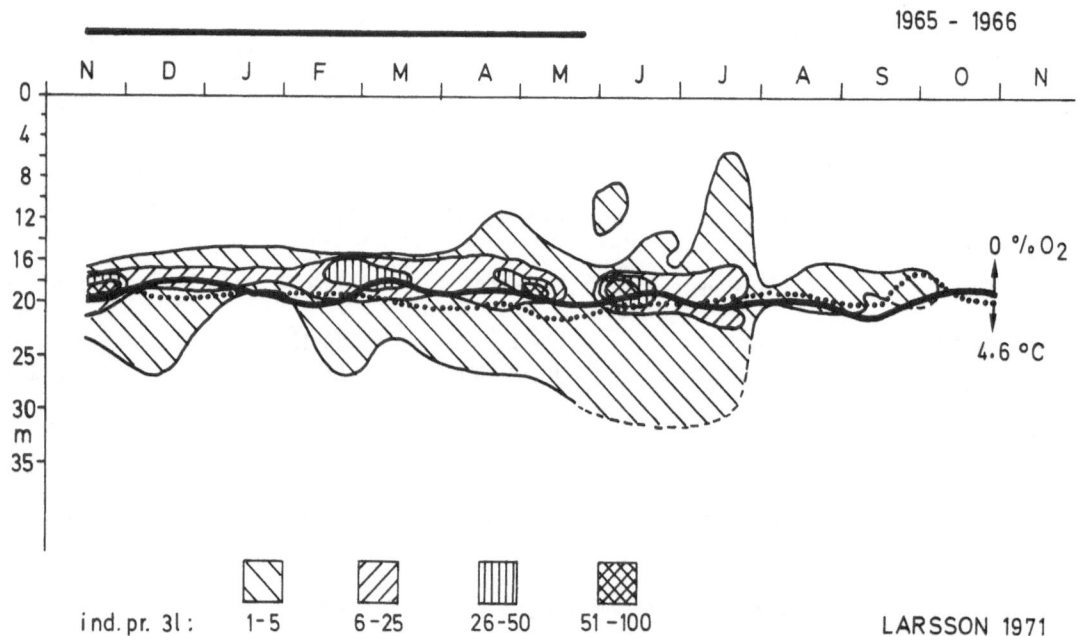

LARSSON 1971

Fig. 4. The vertical distribution of *Filinia terminalis = hofmanni* in two meromictic lakes. A. Lunzer Obersee, Austria (after Ruttner-Kolisko 1974); black bars indicate the location of 50% of the population. B. Blankvatn, Norway (after Larsson 1971); the black line indicates zero oxygen, the dotted line 4.6° temperature, the striped areas show the number of individuals per 3/l.

O_2-content and low temperature are found continuously between 18-20 m depth. Under such ecological conditions *Filinia* populations[x] can occur all year, while they are eliminated because of too high temperature and zero O_2-level in Esthwaite Water (Ruttner-Kolisko, 1980) and other eutrophic lakes, where the anoxic layer reaches temperatures above 12-16 °C in summer.

Hofmann investigated the Plußsee in N. Germany over many years (1972-1978). This eutrophic lake completely overturns only at irregular intervals and stays meromictic in between. Hofmann found both types of *Filinia* distribution in this lake and also a part of the population showing a deduced caudal seta.[xx] Reluctantly Hofmann called it *F.c.f. longiseta*, but fortunately we have now the name *F. hofmanni* for these populations, thanks to Koste (1980). Although the taxon is not discernible on morphological criteria only, the name is now, rightly, used for any population of meromictic distribution.

The taxonomical difficulties and errors in the *F. terminalis/longiseta* group have been described by Pejler (1957b) from their beginning until the time of his own investigations. At present, the situation is such that it is sometimes impossible to discern what someone has seen when he uses a particular name. The more papers that are written, the more confused the situation becomes. There already exist approximately 50 publications on the *F. terminalis/longiseta* problem; they all go around in more or less the same circles. This is an incredible waste of time and effort, not resulting in any new knowledge of natural facts. The state of affairs is not much better in the genus *Hexarthra* and will be the same – I am quite certain – in the genus *Kellikottia* as soon as rotiferologists start to look closer into the ecological data already available on what we call today *Kellikottia longispina*.

Coming back to my view of taxonomy as an artificial grid imposed on the continuous flow of nature, we must still discuss the mesh size of the grid and the question of who imposes the grid. Geneticists use the finest grid; their mesh size is defined as a succession of genetical units which can change in time and some consider a population, in the sense of ecologists, not as a unit, but as an assemblage of units with no more affinity to each other than 'griffins, unicorns and mermaids' (King, 1980). The grid of ecologists, like me, is much coarser; they call the mesh size a *population* – at least in the case of predominantly parthenogenetic groups. A population consists of a considerable number of similar individuals of the same, clearly discernible unit, living together in the same ecological niche. The niche must be defined by as many environmental parameters as are available. In my opinion, this discernible unit is, in the case of rotifers, what is called 'genus'. The mesh size of taxonomists is arbitrary and depends entirely on the investigator. Because of this arbitrariness I have purposely used only the general term 'taxon' in this paper, to avoid any position in the taxonomical hierarchy.

With my criticisms of creating taxa, I can rely on a very renowned authority in hydrobiology, Wesenberg-Lund, who wrote as early as the year 1900 about non-scientific species making, in this own German words 'diese recht unwissenschaftliche Artmacherei'. Although I do not want to go that far, I would like to make a few suggestions, which – I hope – may be considered when new rotifer taxa are created.

Suggestions:

1. Preserved material from an ecologically unknown biotop should not be considered sufficient to describe a new taxon.
2. The description of a new taxon should be based on easily discernible features. Minute characteristics, which cannot be determined during ecological work, should be avoided or used only as supplemental evidence.

---

[x] Named *F. terminalis* by both authors according to the nomenclature prevailing at that time.

[xx] The same phenomenon is described by Schaber (1980), who found up to 25% of the so called *F.c.f. longiseta* in some of his Tyrolian lakes.

298

3. In order to ascertain possible variation of the criteria used to characterize the new taxon, a fair number of individuals (50-100) should be available.

4. The description of a new taxon using *only* variable morphological features should not be considered; if the variation overlaps with already existing taxa, the criterion is useless.

5. Ecological data are to be incorporated in the description of the new taxon and should be given a special emphasis.

6. A new taxon should only be created if discrimination from already known taxa is needed for some scientific reason – never just for the sake of a new name.

References

Hofmann, W., 1972. Zur Produktionsbiologie des Zooplanktons im Plußsee. Verh. int. Ver. Limnol. 18: 410–418.

Hofmann, W., 1974. Zur Taxonomie und Verbreitung von Filinia-Arten..... Faun. ökol. Mitt. 4: 437–444.

Hofmann, W., 1979. Untersuchungen zur ökologischen Nische zweier nahe verwandter Rotatorien Arten des Plußsees. Habilitationsschrift, Kiel.

Hofmann, W., 1982. On the coexistence of two pelagic Filinia species in lake Plußsee. Arch. Hydrobiol. 95: 125–137.

Hofmann, W., 1987. Population dynamics of hypolimnetic rotifers in the Plußsee. N. Germany. Hydrobiologia 147: 197–201.

Jones, J. G., 1972. Studies of Freshwater bacteria..... J. Ecol. 60: 59–75.

King, C. E., 1980. The genetic structure of zooplankton populations. Evolution and Ecology of Zooplankton Communities. Univ. Press of New England.

Koste, W., 1980. Über zwei Plankton-Rädertiertaxa, Filinia australiensis, n.sp. und Filinia hofmanni, n.sp. Arch. Hydrobiol. 90: 230–256.

Larsson, P., 1971. Vertical distribution of planktonic rotifers in a meromictic lake. Norw. J. Zool. 19: 47–75.

Loose, C., 1987. Populationsökologische Untersuchungen an hypolimnischen Rädertieren des Plußsee. Diplomarbeit, Kiel.

Mayr, E., 1942. Systematics and the origin of species. Columbia Univ. Press, N. York.

Parise, A., 1961. Sur les genres Keratella, Synchaeta, Polyarthra et Filinia d'un lac italien. Hydrobiologia 18: 121–135.

Pejler, B., 1957a. On variation and evolution in planktonic rotatoria. Zool. Bidr. Uppsala Univ. 32: 1–66.

Pejler, B., 1957b. Taxonomical and ecological studies on planktonic rotatoria from northern Swedish Lapland. Kungl. Svenska Vetens. Akad. Handl. 6: 1–67.

Pejler, B., 1961. The zooplankton of Ösbysjön I. Seasonal and vertical distribution. Oikos 12: 225–248.

Pourriot, R., 1964. Etude experimentale des variations morphologiques chez certaines espèces de rotifères. Bull. Soc. Zool. France 89: 555–561.

Pourriot, R., 1965. Notes taxonomiques sur quelques rotifères planktoniques. Hydrobiologia 16: 579–604.

Ruttner-Kolisko, A., 1974. The vertical distribution of plankton rotifers in a small alpine lake. Verh. int. Ver. Limnol. 19: 1286–1294.

Ruttner-Kolisko, A., 1980. The abundance and distribution of Filinia terminalis in various types of lakes. Hydrobiologia 73: 169–175.

Schaber, P., 1980. Die Rädertiergattung Filinia in Tiroler Badeseen. Jb. Abt. Limnol. Innsbruck 6: 159–171.

Schaber, P. & Schrimpf, A., 1984. On morphology and ecology of the Filinia-terminalis-longiseta group in Bavarian and Tyrolian lakes. Arch. Hydrobiol. 101: 147–257.

Simpson, G. G., 1951. The species concept. Evolution 5: 285–298.

Simpson, G. G., 1967. Principles of Animal Taxonomy. 3. ed. Columbia Univ. Press, N. York.

Slonimski, P., 1926. Sur la variation saisonière chez Filinia longiseta. C. R. Soc. Biol. 94: 543–545.

Stemberger, R. S., 1976. Systematics of the Rotatorian genera Synchaeta and Filinia. Proposal to National Science Foundation. Manuscript.

Wesenberg-Lund, C., 1900. Von dem Abhängigkeitsverhältnis zwischen dem Bau der Planktonorganismen und dem spez. Gewicht des Wassers. Biol. Zentr. Blatt 20: 606.

Hydrobiologia **186/187**: 299–310, 1989.
C. Ricci, T. W. Snell and C. E. King (eds), Rotifer Symposium V.
© 1989 *Kluwer Academic Publishers.*

Systematics, reproductive isolation and species boundaries in monogonont rotifers

Terry W. Snell
Division of Science, University of Tampa, Tampa, FL 33606, USA

Key words: mating, reproductive isolation, Rotifera, sexuality, systematics

Abstract

The typological concept of rotifer species and the morphological basis of rotifer systematics is reviewed and alternatives proposed. Occasional sexuality in the cyclical parthenogenetic life cycle of monogononts permits application of the biological species concept to this group. Data from cross-mating experiments with *Asplanchna, Brachionus* and *Epiphanes* illustrate the usefulness of reproductive isolation as a criterion for species boundaries. Populations from different geographic regions are often interfertile indicating that rotifer species are genetically integrated over wide areas. The main types of isolating mechanisms operating in monogononts are reviewed. The role of behavioral reproductive isolation in maintaining species boundaries is examined. The use of a mate recognition bioassay which estimates the probability of copulation and quantifies the degree of isolation is described. Recent work of the mechanism of mate recognition is reviewed. It is concluded that the biological species concept is applicable to rotifers and that a more experimental approach to determining species boundaries is both feasible and desirable.

Introduction

Although the concept of species is essential in systematics and evolution, it lacks a universal definition. The most widely accepted species concept was developed by Dobzhansky and Mayr beginning in the 1930's (see Dobzhansky, 1970 and Mayr, 1970 for reviews). The hallmark of the biological species concept is that species are distinguished by their reproductive isolation from other species. Because of its prominent role, reproductive isolation has been the subject of intense study (Littlejohn, 1981). Despite its widespread acceptance for several years, the biological species concept recently has been vigorously attacked by a number of authors. Hauser (1987) reviews the objections to the biological species concept and examines 15 recently published alter-

native views. He concludes that the biological species concept provides an objective means for separating species and represents the causal factor which produces and maintains discrete evolutionary units. Hauser refutes the major criticisms of the biological species concept which are based on its lack of universality, impracticality and inapplicability to species separated in time. It is concluded that the biological species concept does not need to be changed or rejected on the basis of the prevailing criticisms.

Perhaps the strongest stimulus for re-examining the species concept has come from Paterson (1982, 1985). He argues that the prevailing biological species concept is actually an 'isolation concept'. Paterson reasons cogently that speciation occurs as a side-effect of adaptations that promote fertilization and is therefore an evolutionary effect rather than a cause. Selection acts to enhance fertilization systems in populations, a consequence of which is speciation. Defined in this way, species are the most inclusive population of biparental organisms which share a common fertilization system. Paterson's 'recognition concept' of species more clearly separates evolutionary causes from effects and does not rely on relational characteristics to define species. The recognition concept of species is gaining wide attention, but it is not without critics (Templeton, 1987).

Against this background of controversy, the classification of rotifers proceeds, as it does in all organisms, without a unanimous notion of species. Yet the question of how to best organize rotifer diversity into meaningful taxonomic and evolutionary units persists. There is a basic impression that some populations are similar due to genetic integration, probably resulting from gene flow, stabilizing selection, developmental homeostasis or some combination of these factors. Other populations are decidedly different, probably due to evolutionary divergence in isolation. But how can these ideas be forged into a systematics of rotifers consistent with modern evolutionary theory?

As presently conceived, rotifer species are groupings based on morphological similarity of easily measured characters. These are for the most part visually distinguished at the light or electron microscope level. Data are available for only a few species on biochemical traits or on reproductive isolation from cross-mating experiments. Discontinuities among rotifer species are identified by the judgement of systematists, so rotifer taxonomy is based on the classical typological concept of species. Species boundaries occur where phenotypic variation between groups is too large to fit comfortably into one group. In theory, rotifer taxonomy is based on evolutionary systematics, rather than numerical phenetics or cladistics (Mayr, 1981). But in practice it is based almost entirely on morphology. Rotifer systematics is therefore more phenetic than evolutionary or cladistic, but has not utilized the modern numerical taxonomic methods of this

approach. A first attempt to apply cladistic methodology to rotifers appears in this volume (Wallace & Colburn, 1989).

For the most part, rotifer genera are morphologically well defined as taxa, but establishing species boundaries has been a very difficult problem for rotifer systematists (Ruttner-Kolisko, 1963). The use of morphology to define rotifer species is confounded by extensive morphological variation, like cyclomorphosis, that is typical of the group (Hutchinson, 1967). In spite of these difficulties with morphology, rotifer systematics has not employed approaches that have been helpful in establishing species boundaries in other animals. Reproductive isolation, perhaps the most natural marker of a species boundary, has seldom been utilized in rotifers. While theoretically attractive, there is some question whether the criterion of reproductive isolation is applicable to rotifers. The problem is rooted in the question asked by Pejler (1977a): is parthenogenesis essentially obligatory in monogonont rotifers? Obligatory parthenogenesis is well known in the class Bdelloidea where males have never been observed and the cytological evidence supports the idea that bdelloids are agamospecies. If most monogononts are also obligatorily parthenogenetic, reproductively definable boundaries between species do not exist and rotifer systematics can only be based on subjective decisions. Whether sexual reproduction occurs occasionally, rarely, or never in monogononts therefore will determine if reproductive isolation is useful as a criterion for rotifer species boundaries. Holman (1987) has compared the recognizability of species in bdelloids and monogononts.

My objective in this paper is to explore the feasibility of establishing a more experimental definition of rotifer species. The use of cross-mating experiments and mate recognition bioassays as reproductive isolation tests is demonstrated. It is argued that these tools make it possible to define rotifer species boundaries empirically. The small data base from cross-mating experiments between species and among strains within species is reviewed. The most common types of reproductive isolating mechanisms operating in rotifers is characterized and the role of behavioral isolation in defining and maintaining species boundaries discussed. The use of the probability of copulation for quantifying the degree of isolation is described as is recent work on mechanisms of mate recognition.

Do monogononts reproduce sexually?

Males have not been observed in many monogonont species, but are they really absent or are sampling programs so inadequate as to make male detection improbable? Long term studies by Carlin (1943) provide some insight into the intermittent nature of male production by *Polyarthra* and *Notholca* in the Motala River in Sweden. Male production did not occur in every population in every year. In *P. vulgaris*, for example, male production did not occur in 1935 and occurred during only one week in 1936, despite the near continuous presence of this species. Similar patterns were observed for *P. major* and *N. caudata*. Of the 23 monogonont species closely monitored by Carlin, 52% produced males at some time during the five years of sampling. Although many species eventually produced males, these were only detectable with a long-term sampling program with frequent observations.

The difficulties of confirming sexual reproduction in natural rotifer populations were further elucidated by King & Snell (1980) who quantitatively sampled *Asplanchna girodi* in Golf Course Pond (Florida) daily or bidaily from April through July of 1977. In that study, hundreds of liters were sampled each time, providing one of the most finely detailed views of population dynamics for any natural rotifer population. Males were present (> 0.001 per liter) for 15 days in April, 35 days in May-June and 12 days in July. Male density exceeded 1 per liter for only 13 days in April and 5 days in June. If the sampling program had not been so intensive, it is unlikely that males would have been detected in this population. The presence of males only suggests sexual reproduction. Resting egg (cyst) bearing females also were

detected to verify the completion of sexual reproduction. Cyst bearing female density exceeded 1 per liter for 8 days in April and 5 days in June. These observations clearly illustrate the ephemeral nature of sexual reproduction in natural rotifer populations and the rigorous sampling program required for detection. It is therefore not surprising that males and cyst-bearing females are not often observed in natural rotifer populations.

Although sexual reproduction has not been observed in many natural rotifer populations, sex is common in lab populations, especially when rotifers are first isolated from the field. Males and cyst production have been observed for many species of *Asplanchna* and *Brachionus*. More species will have to be cultured, however, before it is known whether these genera typify monogononts. At present, it seems safe to conclude that at least many *Asplanchna* and *Brachionus* species are not obligate parthenogens and that sexual reproduction is an important feature of their life cycle.

Defining species boundaries with mating experiments

The classical approach to defining species boundaries as prescribed by the biological species concept is the mating experiment. Several mating experiments using monogonont rotifers have been completed, verifying the feasibility of this approach with species amenable to laboratory culture.

Early attempts to cross different strains of a single rotifer species suggested that strains from distant geographical locations could be mated. Shull (1911) crossed an *Epiphanes senta* strain from New York with one from Maryland and later (Shull, 1915) crossed a strain from England with one from Lincoln, Nebraska. Also working in the same area of Nebraska, Hertel (1942) successfully crossed four *E. senta* strains collected from surrounding ponds.

Gilbert (1963) attempted to mate three brachionid species, all collected from a small, slightly alkaline pond near New Haven, Connecticut. *Brachionus calyciflorus* and *B. angularis* were planktonic whereas *B. quadridentatus* was sometimes planktonic, but usually attached to submerged objects. All three species commonly co-occur in alkaline ponds in temperate regions throughout the world. Gilbert found that the sexual periods of *B. calyciflorus* and *B. angularis* overlapped, but all attempts to mate these species failed. Interspecific matings with *B. quadridentatus* also failed indicating that the reproductive isolation of these species is complete.

Mating experiments between geographically separated *Asplanchna brightwelli* populations were carried out by Birky (1967). Populations collected from Bloomington, Indiana were mated with populations from Paris, Tennessee and produced viable F_1, F_2 and backcross progeny. These populations clearly share a common gene pool, but intra- and interpopulational crosses were not equally fertile. Intrapopulational crosses (self fertilization) were 77% successful, whereas crosses between Indiana and Tennessee populations had only a 48% success rate, suggesting partial reproductive isolation. Birky also attempted to cross *A. brightwelli* from Tennessee with *A. girodi* from the same pond. No interspecific matings were ever observed although both species self fertilized with >60% success. *Asplanchna brightwelli* and *A. girodi* were therefore completely reproductively isolated despite the fact that both species occupied the same region of the pond and both reproduced sexually at the same time. No hybrids were observed verifying the effectiveness of the isolating mechanisms.

Mating barriers between rotifer species are not always complete. Ruttner-Kolisko (1969) successfully crossed *B. uroeolaris* with *B. quadridentatus*. Males of both species copulated with conspecific as well as heterospecific females, although the probability of copulation was not quantified so male preferences may have gone undetected. However, while pre-mating barriers were absent, post-mating barriers were observed. Crosses between *B. quadridentatus* females and *B. uroeolaris* males produced cysts but none hatched. The reciprocal cross, in contrast, was fertile but the small number of mictic females utilized precluded quantification of post-mating barriers. Differences in the outcome of reciprocal crosses illustrates the asymmetry common in post-mating reproductive isolation (Giddings & Templeton, 1983). The species boundary between *B. uroeolaris* and *B. quadridentatus* therefore is not clearly defined. Even though these species appear to be morphologically distinct, further analysis of their reproductive isolation is warranted.

Similar experiments have been conducted with *A. brightwelli*, *A. intermedia* and *A. sieboldi* (Gilbert et al., 1979). Crosses between *A. brightwelli* females and *A. intermedia* or *A. sieboldi* males yielded a 39% fertilization rate in 74 attempts indicating little if any pre-mating isolation among these species. However, since none of the cysts hatched, zygote inviability was the likely cause of the complete post-mating isolation. The reciprocal cross with *A. intermedia* or *A. sieboldi* females and *A. brightwelli* males yielded a fertilization rate of 47% in 47 crosses, but again none of the cysts hatched. Crosses between *A. intermedia* and *A. sieboldi* indicated no pre-mating isolation, but too few crosses were completed to conclude anything about post-mating barriers. The general conclusions from this study were that *A. brightwelli* is reproductively isolated from *A. intermedia* and *A. sieboldi* by real, quantifiable post-mating barriers. Additional crosses are necessary to determine if *A. intermedia* and *A. sieboldi* are likewise completely reproductively isolated. Only a few strains were used in this study so more populations need to be examined to assess the range of intraspecific variation for reproductive isolation. Results of interspecific crosses among rotifer species are summarized in Table 1.

Two unpublished works provide additional information on reproductive isolation among geographically separated *Asplanchna* strains. Birky (unpublished, reported in Birky & Gilbert, 1971) attempted to cross his Tennessee and Indiana strains of *A. brightwelli* with a German strain which was morphologically identical and physiologically similar. No cysts were produced from any of these crosses. Snell (unpublished, reported in King, 1977) attempted to cross

Table 1. A summary of interspecific crosses attempted with monogonont rotifers. *Brachionus* species: caly – *calyciflorus*. ang – *angularis*. quad – *quadridentatus*. plic – *plicatilis*, rubens – *rubens*, urcea – *uroeolaris*. *Asplanchna* species: bright – *brightwelli*, girodi – *girodi*, inter – *intermedia*, sieboldi – *sieboldi*.

Cross	Pre-mating barrier	Post-mating barrier	Reference
♂ X ♀		*Brachionus*	
caly X ang	no mating	–	Gilbert 1963
caly X quad	no mating	–	Gilbert 1963
caly X *Synchaeta*	no mating	–	Gilbert 1963
caly X *Euchlanis*	no mating	–	Gilbert 1963
ang X caly	no mating	–	Gilbert 1963
plic X *Synchaeta*	no mating	–	Snell & Hawkinson 1983
plic X rubens	no mating	–	Snell unpublished
urceo X quad	none*	no cyst hatching	Ruttner-Kolisko 1969
quad X urceo	none*	none?	Ruttner-Kolisko 1969
		Asplanchna	
bright X girodi	?	no cysts produced	Birky 1967
bright X inter	none*	no cysts hatching	Gilbert *et al.* 1979
bright X sieboldi	none*	no cysts hatching	Gilbert *et al.* 1979
sieboldi X inter	none*	?	Gilbert *et al.* 1979

* Probability of copulation not quantified, subtle differences in male mating preferences could exist.

A. brightwelli from Florida with Birky's strains from Tennessee and Indiana and the German strain. None of these crosses produced cysts, yet all strains were successfully self fertilized.

Reproductive isolation among geographically separated *Brachionus plicatilis* strains was investigated by Ruttner-Kolisko (1983, 1985). A strain cultured in Ruttner-Kolisko's lab in Austria for 20 years was crossed with strains from Scotland and Colorado. Austria, Scotland and Colorado males did not attempt to mate with Austria females. Austria females apparently have lost the ability to elicit male mating responses after many years of parthenogenetic reproduction in the laboratory. Austria males, in contrast, were able to fertilize Scotland and Colorado females and produce viable cysts. Crosses between Scotland males and Colorado females produced viable cysts, but the reciprocal cross with Scotland females and Colorado males was infertile. Self fertilization in Scotland and Colorado strains was highly successful. Gene flow therefore is possible among all three strains, but all are at least partially reproductively isolated from one another. These experiments illustrate that reproductive isolation is often asymmetrical between strains and reproductive incompatibilities can be due to either males or females. Results of intraspecific crosses among strains of a single rotifer species are summarized in Table 2.

These studies permit several conclusions about reproductive isolation in rotifers. (1) It is possible to identify absolute species boundaries, consistent with the biological species concept, among closely related rotifer species. Cross-mating experiments have accomplished this for *B. calyciflorus*, *B. angularis* and *B. quadridentatus*, *A. brightwelli*, *A. sieboldi* and *A. intermedia*. These taxa are good biological species which are not exchanging genes and are pursuing independent evolutionary courses. (2) Absence of hybrids suggests that reproductive isolation is usually complete even in species with overlapping sexual periods (e.g. *B. calyciflorus* and *B. angularis*, *A. brightwelli* and *A. girodi*). (3) Populations within a species often are genetically integrated over a wide range and retain the ability to exchange genes. This is demonstrated with *A. brightwelli* from Indiana and Tennessee and *B. plicatilis* from Austria, Scotland and Colorado. (4) Postmating barriers effectively separate closely related species as in *A. brightwelli*, *A. sieboldi*

Table 2. A summary of intraspecific crosses attampted with monogonont rotifers. *Asplanchna* abbreviations: Ind – Indiana, Tenn – Tennessee, Fla – Florida, 4 temporal pops – all from Lake Thonotosassa, Florida. *Brachionus* abbreviations: 3 temporal pops – all from McKay Bay, Florida; geographic pops – from different countries worldwide.

Cross	Pre-mating barrier	Post-mating barrier	Reference
Epiphanes senta			
New York X Maryland	none*	none	Shull 1911
England X Nebraska	none*	none	Shull 1911
4 Nebraska pops	none*	none	Hertl 1942
Asplanchna brightwelli			
Ind X Tenn	none*	partial	Birky 1967
Ind X Germany	complete isolation, barrier unknown		Birky unpublished
Tenn X Germany	complete isolation, barrier unknown		Birky unpublished
Fla X Ind	complete isolation, barrier unknown		Snell unpublished
Fla X Tenn	complete isolation, barrier unknown		Snell unpublished
Fla X Germany	complete isolation, barrier unknown		Snell unpublished
4 temporal pops	none*	partial	King 1977
Brachionus plicatilis			
Austria X Scotland	none*	partial	Ruttner-Kolisko 1983
Austria X Colorado	none*	partial	Ruttner-Kolisko 1983
Scotland X Colorado	none*	partial	Ruttner-Kolisko 1983
3 temporal pops	none	–	Snell & Hawkinson 1983
8 geographic pops	partial	–	Snell & Hawkinson 1983

* probability of copulation not quantified, subtle differences in male mating preferences could exist.

and *A. intermedia*. (5) Sibling species may be common in monogont rotifers as suggested by the reproductive isolation of Tennessee and Indiana *A. brightwelli* from morphologically identical strains from Florida and Germany.

Reproductive isolating mechanisms

Mating experiments illustrate that many monogonont rotifer species can be defined as reproductively isolated evolutionary units. An important consideration is how reproductive isolation among species arises and how it is maintained. Several examples of reproductive isolating mechanisms in rotifers will provide insight into this question.

Reproductive isolating mechanisms are broadly classified as pre-mating or post-mating (Mayr, 1963). Pre-mating barriers include seasonal, habitat, behavioral and mechanical mechanisms; post-mating barriers include gametic incompatability, hybrid inviability and hybrid

sterility. Examples of seasonal reproductive isolation in rotifers are found in Carlin (1943) who quantitatively sampled the Motala River for five consecutive years. *Polyarthra vulgaris* and *P. remata* produced males from September through November in each of the five years sampled. *Polyarthra major* consistently produced males July through October and *P. dolichoptera* produced males April through June. Even though females of these species sometimes co-occurred in the plankton, the sexual periods of *P. major* and *P. dolichoptera* did not overlap one another or those of *P. vulgaris* and *P. remata*. Non-overlapping sexual periods enforce reproductive isolation by preventing males and mictic females from meeting.

In some types of environments seasonal reproductive isolation can breakdown. Pejler (1956) investigated the reproductive isolation between *P. vulgaris* and *P. dolichoptera* in Lake Luossajarvi in Sweden. *Polyarthra vulgaris* was eurythermal, preferred oxygen rich surface waters and had a sexual period stretching from summer to early winter. *Polyarthra dolichoptera*, in contrast, preferred deep, cold water that was oxygen poor and reproduced sexually in early spring. This seasonal and ecological isolation was very effective in maintaining species boundaries, as no hybrids were ever detected in the lake. These species also co-occurred in tarns, small ponds 1-3 meters deep that are abundant in the region. Here, *P. vulgaris* dominated, but intermediate forms were common. Pejler argued that the environment in the shallow tarns broke down the reproductive isolation between these two species leading to hybridization. This example illustrates the importance of the environment in maintaining species separation.

Examples of habitat isolation can be found in various *Brachionus* species. Ruttner-Kolisko (1974) described the ecological characteristics of several brachionid species in the *uroeolaris* group. *Brachionus uroeolaris* is found primarily in freshwater where it is mainly benthic, often attaching to submerged vegetation; *B. plicatilis* is found in brackish waters and soda lakes where it is usually planktonic; and *B. rubens* is also found in freshwater, often epizooic on *Daphnia*. The habitat differences of these three species promotes their reproductive isolation. Additional examples of habitat isolation among six species of *Brachionus* are provided by Miracle *et al.* (1987). They analyzed the distribution of these species using discriminant analysis of 17 physical and chemical parameters. Separation was achieved on the basis of sulphate, temperature, chloride, alkalinity, conductivity and O_2 concentration, suggesting considerable habitat specialization by these brachionids. Studies like this and that of Bogdan & Gilbert (1987) suggest that rotifers finely partition the environment providing ample opportunity for habitat isolation.

Behavioral reproductive isolating mechanisms are among the most important in animals (Dobzhansky, 1970). Mate recognition in monogononts and its role in the development of reproductive isolation is an area of active research. This work will be described in detail in the next section.

Examples of mechanical isolation have been reported for rotifers, but there is some doubt about their significance. Ruttner-Kolisko (1983) described a series of mating experiments with *B. plicatilis* where some interstrain crosses failed. She attributed this failure to differences in lorica thickness of newborn females and the males ability to penetrate the lorica with hyperdermic insertion of the penis. If this accurately represents copulation, it would be an example of mechanical reproductive isolation based on lorica thickness. There is some doubt, however, whether males penetrate the lorica since most copulations occur in the coronal region which is not covered by the lorica (Gilbert, 1963; Snell & Hoff, 1987).

Examples of post-mating reproductive isolation in rotifers are not as abundant as premating barriers. There is some evidence of hybrid inviability in crosses among *A. brightwelli*, *A. sieboldi* and *A. intermedia* (Gilbert *et al.*, 1979). King (1977) reported partial post-mating isolation among *A. brightwelli* populations collected at different times from Lake Thonotosassa, Florida. Four discrete populations were collected in 1974: April, June, September and November. The hatchability of cysts produced from intrapopulational crosses of was much higher than that of interpopulational crosses.

Mate recognition and species boundaries

Mate recognition is a crucial behavior with wide ranging evolutionary consequences. Mate recognition systems restrict gene exchange between species and promote divergence among populations. Divergence in mate recognition systems can be an important component of speciation (Paterson, 1982; Thornhill & Alcock, 1983; Giddings & Templeton, 1983; Nevo & Capranica, 1985; Ryan & Wilczynski, 1988). Incorrect recognition leads to gamete wastage which is a serious error considering the small number of gametes produced by male rotifers (Snell & Hoff, 1987). Inviable or reduced fertility hybrids lower reproductive output and overall fitness. As a result, strong selection for precise mate recognition systems is expected.

Mate recognition in rotifers is based on coronal contact chemoreception by the male of a chemical on the body surface of females (Gilbert, 1963). Mating begins when a male makes head-on contact with a conspecific female. Chemosensory receptors in the male's coronal region (Clement et al., 1983) sense a glycoprotein present on females (Snell et al., 1988). If a male senses the correct glycoprotein, he begins copulatory behavior by circling the female, skimming over her body while maintaining coronal contact. After several seconds of circling, the male moves toward the female's corona where he then inserts his penis into her pseudocoelom. The entire copulatory sequence takes an average of 1.2 minutes (Snell & Hoff, 1987). Males are smaller, faster swimming than females and actively discriminate conspecific females. Females swim about randomly, taking no apparent role in mate recognition or copulation.

Mate recognition behavior of male rotifers can be a useful tool for establishing species boundaries. Males discriminate conspecifics using a precise biochemical mechanism that has evolved over thousands of generations. This recognition system can be used as a simple bioassay to probe the dimensions of a species' gene pool and its boundaries with adjacent species. Mate recognition occurs at the instant of male-female contact. A conspecific contact usually results in circling of the female by the male. Circling behavior is easily detected as it differs markedly from normal male swimming behavior. A heterospecific contact results in males changing direction and swimming away. Observation of 50-100 encounters provides enough data to calculate a probability of copulation (P_c), the liklihood that males will initiate mating behavior upon contacting a female. If no mating attempts are observed after a suitable number of encounters ($P_c = 0$), reproductive isolation is assumed to be complete between the populations tested. If P_c is greater than 0, the species are potentially able to exchange genes, but further tests are necessary to determine if mating is consumated or if post-mating barriers exist. Calculation of P_c allows detection of partial reproductive isolation by quantitatively comparing homogamic and heterogamic pairings (Snell & Hawkinson, 1983). The absence of mating behavior is therefore sufficient evidence for reproductive isolation, but if mating occurs additional experiments are necessary to establish species boundaries.

This approach to determining species boundaries is appealing because it relies on rotifer males rather than human systematists to discriminate species. Mate recognition is a natural bioassay for species boundaries and can be carried out quickly with high reliability. The protocol is simple and requires minimal equipment. The application of this bioassay to laboratory populations should not be difficult because male production is common in many cultured species. The mating bioassay will be most useful in sorting out sibling species where morphological criteria are not discriminating. Sibling species are a common problem in cladocera (Hebert, 1987) which are sexually similar to rotifers. Another application will be in cases where cyclomorphosis confuses phenotypic variation. Although, the mate recognition bioassay is not a panacea, it is a useful tool that, when properly applied, can uncover biologically meaningful discontinuities in rotifer diversity.

Using mate recognition to discriminate rotifer species is appealing for other reasons. Mate recognition is a central element of the recognition concept of species (Paterson, 1985) and is gaining considerable attention. Specific mate-recognition systems (SMRS) (Paterson 1978, 1980) are involved in signaling between mating partners and are therefore a subset of the fertilization system. In Paterson's view, species boundaries are set by the limits of shared fertilization systems, a primary component of which is the SMRS. Species are genetically cohesive units that are resistant to change (homeostatic) because of stabilizing selection of the SMRS. Speciation occurs as an incidental effect when the SMRS breaks down in allopatry. In the case of rotifers, the limits of the SMRS can be defined experimentally by the mate recognition bioassay. The recognition concept of species makes several predictions (Lambert & Paterson, 1983), many of which are testable with a rotifer model.

Mate recognition bioassays have been used to examine species boundaries in several brachionid rotifers. Gilbert (1963) mated *Brachionus calyciflorus* and *B. angularis* males with *B. calyciflorus*, *B. angularis*, *B. quadridentatus*, *Synchaeta* sp. and *Euchlanis* sp. females. Males of these two species attempted to mate only with conspecific females demonstrating the strict species specificity of the mating reaction. *Brachionus plicatilis* males are also strictly species-specific in their mating activity when exposed to *Synchaeta bicornis* females (Snell & Hawkinson, 1983) or *B. rubens* females (Snell, unpublished). Snell and Hawkinson also examined mating reactions among temporally isolated *B. plicatilis* populations from the same bay as well as spatially and geographically separated populations. No differences in P_c were detected among the temporally separated populations. However, spatially and geographically separated populations showed mating preferences. Then strains from around the world were characterized for probabilities of copulation. P_c values ranged from 78% to 6.2% in all possible pairings. In no case was $P_c = 0$, indicating that all strains have retained their mating compatability despite wide geographical separation. This suggests that a globally distributed species like *B. plicatilis* (Pejler, 1977b) can maintain an integrated gene pool. Post-mating barriers need to be investigated, but an intriguing question is raised. How is mate recognition maintained in geographically isolated populations separated for thousands of generations? Theoretical models of the evolution of reproductive isolation suggest that this is sufficient time for the establishment of reproductive barriers (Nei *et al.*, 1983). Several possible causes for the maintenance of species integrity in the absence of gene flow have been postulated (e.g. Hutchinson, 1968; Van Valen, 1982).

Strict species specificity of the mating reaction may not be characteristic of all rotifers. In *Asplanchna*, Gilbert *et al.* (1979) report that male *A. brightwelli*, *A. intermedia* and *A. sieboldi* attempt to mate with females of all three species. Only when post-mating barriers were examined, along with six morphological traits, could these species be discriminated. *Asplanchna* males also often attempt to copulate with conspecific males, a phenomenon that is not observed in brachionids. This further suggests that the mating reaction in *Asplanchna* may not be as specific as in brachionids.

Molecular aspects of mate recognition

The female's signal and the male's reception are the basis of mate recognition on rotifers. More knowledge of the chemical nature of the signal and how it varies among rotifer species will provide insight into how reproductive isolation develops.

Gilbert (1963) showed that mate recognition in rotifers is based on contact chemoreception and provided initial biochemical observations of the mate recognition factor on female *B. calyciflorus*. Using biochemical techniques considered crude by today's standards, Gilbert concluded that the mate recognition factor is a small, amphoteric molecule less than 4000 daltons with aromatic characteristics. More recently, Snell *et al.* (1988) examined the biochemical basis of mate recognition in *B. plicatilis*. They showed that females heated to 100 °C for 5 minutes lost their ability to elicit a male mating reaction. In contrast, freeze-killed females retained their ability to elicit male mating activity at a level indistinguishable from live females. Snell *et al.* (1988) also showed that the mate recognition factor was destroyed by the proteases proteinase K, pronase E and chymotrypsin and the glycohydrolase β-amylase. These observations clearly demonstrated that the mate recognition factor is a heat labile glycoprotein and that the carbohydrate portion of the molecule is necessary for mate recognition.

The mate recognition glycoprotein (MRG) of *B. plicatilis* females has been further characterized by Snell & Nacionales (1989a). They showed that male mating activity is blocked by exposing females to the lectin concanavalin A (con A) and a lectin isolated from the lentil *Lens culinaris*. These lectins selectively bind to glycoproteins containing terminal -D-mannosyl, -D-glucosyl or

-N-acetylglucosamine residues. Their binding to the MRG suggests the presence of these residues and their blocking of mating activity reinforces the conclusion that the carbohydrate portion of the MRG is important in mate recognition. When males are exposed to these lectins, mate recognition also is blocked, suggesting that lectins bind to the male receptor of the MRG, causing disruption of mate recognition.

Using fluoresence microscopy, localization of the MRG on females and its receptor on males has been accomplished (Snell & Nacionales, 1989b). Con A conjugated to the fluorochrome FITC (fluoroisothiocyanate) produced fluoresence at sites where Con A bound to the MRG. On females the MRG was highly localized in the buccal field and corona, dorsal and lateral antennae, pedal sphincter and tip of the foot. The MRG receptor on males is also highly localized in the coronal region, dorsal and lateral antennae, tip of the foot, tip of the penis and the accessory gland. The sites of highest MRG concentration on females correspond to the sites where the most copulations occur (Snell & Hoff 1987). It appears that the MRG guides males to a site on the female's body surface that is favorable for penetration.

The goal of research on the biochemical features of mate recognition in rotifers is to better understand the molecular basis of reproductive isolation. Since mate recognition is the cornerstone of reproductive isolation, it is important to know how barriers to cross-mating develop. What kind of molecular changes cause reproductive isolation? Are the mutations responsible simple or complex? This could indicate how difficult it is for behavioral reproductive isolation to evolve. Isolation and characterization of the MRG will provide opportunities to develop biochemically based, rapid screening assays of reproductive isolation. Information like this will make it easier to sort out discontinuities in rotifer diversity and provide better tools for reconstructing rotifer evolution.

Future research

In my view, there are several areas that would benefit from a focused research effort. These include the following:

1) There is more information to be extracted from cross-mating experiments. Although these are tedious, they yield a large amount of information. A better understanding of intraspecific variation in reproductive barriers is needed. Do intraspecific barriers to gene flow exist and how do they compare to interspecific barriers? Enough strains within a single species should be tested to characterize these barriers. We need to clearly establish species boundaries for at least one model rotifer species using the criterion of reproductive isolation. Brachionids are good candidates for this.

2) The mate recognition bioassay should be used as a rapid screening technique to examine species boundaries in as many rotifer species as possible. Closely related species in the families Brachionidae (genera like *Epiphanes*, *Brachionus*, *Euchlanis*) or Synchaetidae (*Synchaeta*, *Polyarthra*) are good candidates because several species can be cultured, males are common and many species have been described. Several predictions of the recognition concept of species could be tested.

3) More work is needed on mate recognition in rotifers. Are glycoproteins the basis of recognition phenomena in all species? What molecular changes in the carbohydrate moieties lead to reproductive isolation? Is it possible to construct a phylogeny of monogononts using molecular differences in mate recognition glycoproteins?

Acknowledgments

I thank Mike Childress and Lisa Nacionales for expert technical assistance. Bob Wallace made several suggestions that improved the manuscript. This material is based upon work supported by the National Science Foundation under Grant

OCE-8600305 and by the National Institutes of Health under grant ES-04749.

References

Birky, C. W., Jr. & J. J. Gilbert, 1971. Parthenogenesis in rotifers: The control of sexual and asexual reproduction. Amer. Zool. 11: 245–266.

Birky, C. W., Jr., 1967. Studies on the physiology and genetics of the rotifer *Asplanchna*. III. Results of outcrossing, selfing and selection. J. exp. Zool. 165: 104–116.

Bogdan, K. G. & J. J. Gilbert, 1987. Quantitative comparison of food niches in some freshwater zooplankton. Oecologia 72: 331–340.

Carlin, B., 1943. Die Planktonrotatorien des Motalastrom. Medd. Lunds Univ. Limnol. Institut., 258 pp.

Clement, P., E. Wurdak & J. Amsellem, 1983. Behavior and ultrastructure of sensory organs in rotifers. Hydrobiologia 104: 89–130.

Dobzhansky, T., 1970. Genetics of the Evolutionary Process. Columbia University Press, New York, 505 pp.

Giddings, L. V. & A. R. Templeton, 1983. Behavioral phylogenies and the direction of evolution. Science 220: 372–378.

Gilbert, J. J., 1963. Contact chemoreceptors, mating behavior and reproductive isolation in the rotifer genus *Brachionus*. J. Exp. Biol. 40: 625–641.

Gilbert, J. J., C. W. Birky, Jr. & E. S. Wurdak, 1979. Taxonomic relationships of *Asplanchna brightwelli*, *A. intermedia* and *A. sieboldi*. Arch. Hydrobiol. 87: 224–242.

Hauser, C. L., 1987. The debate about the biological species concept – a review. Z. zool. Syst. Evolut.-forsch. 25: 241–257.

Hebert, P. D. N., 1987. Genotypic characteristics of the cladocera. Hydrobiologia 145: 183–193.

Hertel, E. W., 1942. Studies on vigor in the rotifer *Hydatina senta*. Physiol. Zool. 15: 304–324.

Holman, E. W., 1987. Recognizability of sexual and asexual species of rotifers. Systematic Zoology 36: 381–386.

Hutchinson, G. E., 1967. A Treatise on Limnology. Volume 2. Introduction to Lake Biology and Limnoplankton. John Wiley & Sons, New York, 1115 pp.

Hutchinson, G. E., 1981. When are species necessary? In: Population Biology and Evolution, R. C. Lewontin (ed.), Syracuse University Press, Syracuse, NY: 177–186.

King, C. E. & T. W. Snell, 1980. Density-dependent sexual reproduction in natural populations of the rotifer *Asplanchna girodi*. Hydrobiologia 73: 149–152.

King, C. E., 1977. Genetics of reproduction, variation, and adaptation in rotifers. Arch. Hydrobiol. Beih. 8: 187–201.

Lambert, D. M. & H. E. H. Paterson, 1983. On bridging the gap between race and species: The isolation concept and an alternative. Proc. linn. Soc. N.S.W. 107: 501–514.

Littlejohn, M., 1981. Reproductive isolation: a critical review. In: Evolution and Speciation, W. R. Atchley & D. S. Woodruff (eds.), Cambridge University Press, Cambridge: 298–334.

Mayr, E., 1963. Animal Species and Evolution. Belknap Press, Harvard University, Cambridge, MA.

Mayr, E., 1970. Populations, Species and Evolution. Belknap Press, Harvard University, Cambridge, MA.

Mayr, E. 1981. Biological classification: Toward a synthesis of opposing methodologies. Science 214: 510–516.

Miracle, M., M. Serra, E. Vicente & C. Blanco, 1987. Distribution of *Brachionus* species in Spanish mediterranean wetlands. Hydrobiologia 147: 75–81.

Nei, M., T. Maruyama & C. Wu, 1983. Models of evolution of reproductive isolation. Genetics 103: 557–579.

Nevo, E. & R. R. Capranica, 1985. Evolutionary origin of ethological isolation in cricket frogs, *Acris*. Evol. Biol. 19: 147–214.

Paterson, H. E. H., 1978. More evidence against speciation by reinforcement. S. African J. Science 74: 369–371.

Paterson, H. E. H., 1980. A comment on 'mate recognition systems'. Evolution 34: 330–331.

Paterson, H. E. H., 1982. Perspectives on speciation by reinforcement. S. African J. Science 78: 53–57.

Paterson, H. E. H., 1985. The recognition concept of species. In: Species and Speciation, E. S. Vrba (ed.), Transvaal Museum Monograph No. 4, Transvaal Museum, Pretoria, Rep. S. Africa: 21–29.

Pejler, B., 1956. Introgression in planktonic Rotatoria with some points of view on its causes and conceiveable results. Evolution 10: 246–261.

Pejler, B., 1977a. General problems on rotifer taxonomy and global distribution. Arch. Hydrobiol. Beih. 8: 212–220.

Pejler, B., 1977b. On the global distribution of the family Brachionidae (Rotatoria). Arch. Hydrobiol. Beih. Supplement 2: 255–306.

Ruttner-Kolisko, A., 1963. The interrelationships of the Rotatoria. In: The Lower Metazoa, E. C. Dougherty, (ed.), Univ. Calif. Press, Berkeley, Calif: 263–272.

Ruttner-Kolisko, A., 1969. Kreuzungexperimente zwischen *Brachionus urceolaris* and *Brachionus quadridentatus*, ein Beitrag zur Fortpflanzungbiologie der heterogonen Rotatoria. Arch. Hydrobiologie 65: 397–412.

Ruttner-Kolisko, A., 1974. Plankton Rotifers. Binnengewasser 26 suppl.: 1–146.

Ruttner-Kolisko, A., 1983. The significance of mating processes for the genetics and for the formation of resting eggs in monogonont rotifers. Hydrobiologia 104: 181–190.

Ruttner-Kolisko, A., 1985. Results of individual cross-mating experiments in three distinct strains of *Brachionus plicatilis* (Rotatoria). Verh. int. Ver. Limnol. 22: 2979–2982.

Ryan, M. J. & W. Wilczynski, 1988. Coevolution of sender and receiver: Effect on local mate preference in cricket frogs. Science 240: 1786–1788.

Shull, A. F., 1911. Studies in the life cycle of *Hydatina senta*. II. The role of temperature, of chemical composition of the medium, and of internal factors on the ratio of parthenogenetic to sexual forms. J. exp. Zool. 10: 117–166.

Shull, A. F., 1915. Inheritance in *Hydatina senta*. IV. Characters of females and their parthenogenetic eggs. J. exp. Zool. 18: 145–186.

Snell, T. W. & C. A. Hawkinson, 1983. Behavioral reproductive isolation among populations of the rotifer *Brachionus plicatilis*. Evolution 37: 1294–1305.

Snell, T. W. & F. H. Hoff, 1987. Fertilization and male fertility in the rotifer *Brachionus plicatilis*. Hydrobiologia 147: 329–334.

Snell, T. W., M. J. Childress & B. C. Winkler, 1988. Characteristics of the mate recognition factor in the rotifer *Brachionus plicatilis*. Comp. Biochem. Phys. 89A: 481–485.

Snell, T. W. & M. A. Nacionales, 1989a. Sex pheromone communication in *Brachious plicatilis* (Rotifera). Submitted.

Snell, T. W. & M. A. Nacionales, 1989b. Localization of the mate recognition glycoprotein on the rotifer *Brachionus plicatilis*. Submitted.

Templeton, A. R., 1987. Species and speciation. Evolution 41: 233–235.

Thornhill, R. & J. Alcock, 1983. The Evolution of Insect Mating Systems. Harvard University Press, Cambridge, MA.

Van Valen, L., 1982. Integration of species: Stasis and biogeography. Evol. Theory 6: 99–112.

Wallace, R. L. & R. A. Colburn, 1989. Phylogenetic relationships within the phylum Rotifera: orders and genus *Notholca*. Hydrobiologia 186/187: 311–318.

Hydrobiologia **186/187**: 311–318, 1989.
C. Ricci, T. W. Snell and C. E. King (eds), Rotifer Symposium V.
© 1989 *Kluwer Academic Publishers.*

Phylogenetic relationships within phylum Rotifera: orders and genus *Notholca*

Robert Lee Wallace & Rebecca Arlene Colburn
Biology Department, Ripon College, Ripon, WI, 54971-0248, U.S.A.

Key words: cladistics, computers, evolution, evolutionary trees, *Notholca*, orders, phylogeny, Rotifera, synapomorphies

Abstract

We investigated evolutionary relationships among orders in phylum Rotifera and among species in genus *Notholca* (Rotifera) by computing parsimonious cladograms. All of the most-parsimonious cladograms generated for the ordinal level confirm the view that class Monogononta, superclass Eurotatoria, and phylum Rotifera are monophyletic. Species within the genus *Notholca* were separated into six groups (clades), but some species have been defined based on highly variable characters not reliably studied using cladistics. Therefore, phenetic studies are warranted, especially for species possessing caudal processes.

Introduction

Cladograms (phylogenetic trees or networks) generated with the aid of computer programs are powerful tools for investigation of plausible evolutionary relationships among organisms. However, to our knowledge these techniques have never been applied to taxa within phylum Rotifera below the level of class although they could help elucidate difficult taxonomies: especially because cladistical analyses focus on the evolution of shared derived characters (synapomorphies) (Ridley, 1986).

In general, distinctions among taxa within the Rotifera are well-defined, leaving little ambiguity. However, as Ruttner-Kolisko (1963) indicated, variation below the level of genus can be so great that clear designation of some species has become extremely difficult. This problem is due, in part, to phenotypic variation (e.g., dietary- and predatory-induced polymorphisms, differences in resting egg hatchlings from subsequent generations, and dwarfing).

Variation in body form and size has led to confusion in the definition of species in some genera. For example, the taxonomy of *Notholca* has been described as hopelessly confused and chaotic, and the morphological variation as extraordinarily great (e.g., Björklund, 1972; Koste & Shiel, 1987; Pejler, 1977; Ruttner-Kolisko, 1974:77). To help rectify this problem we initiated a cladistic analysis of the genus *Notholca* by developing parsimonious cladograms with the aid of several computer programs. *Notholca* was selected among other genera that suffer from difficult taxonomies for two reasons. (1) *Notholca* was the subject of a detailed study in which 53 morphological characters plus seven ecological and biogeographical features were analyzed (Kutikova, 1980). (2) About 40 putative species and subspecies have been described for this genus, thereby providing sufficient information for a cladistical study without the necessity of manipulating an intractable data set. We began our study by analyzing the phylum Rotifera at the ordinal level as a simple test of our procedures.

312

Methods

At the ordinal level information about eight characters (ovaries, vitellarium, males, resting eggs, desiccation ability, type of trophi, spermatophores, and prostate glands) was obtained from four sources (Edmondson, 1959; Gilbert, 1988; Koste, 1978; Ruttner-Kolisko, 1974). For 37 *Notholca* species (including subspecies and variations) data on nine characters (four types of spines, spurs, keels, ornamentation of the surface, type of caudal process, and habitat) were obtained from 18 reports (Amrén, 1964; Björklund, 1972; Carlin, 1943; Chengalath, 1978; Chengalath & Koste, 1987; Daday, 1913; Dartnall & Holloway, 1985; Focke, 1961; Koste, 1974; Koste, 1978; Koste & Shiel, 1987; Kutikova, 1980, 1986; Lair & Koste, 1984; de Paggi, 1982; Pejler, 1977; Ruttner-Kolisko, 1972; Stemberger, 1976).

Using these data, cladograms were calculated by employing computer programs available in PHYLIP© (Phylogeny Inference Package, v3.0) which use numerical procedures that are applied to morphological data (character states), ordering them into the most parsimonious cladogram(s). Cladograms were then examined more closely (corroboration) for treelength and homoplasy (Wiley, 1981:12) using MacClade© (v2.1). For a review of PHYLIP© and MacClade© and information concerning their availability consult Fink (1986). During corroboration we amended our assumptions concerning possible evolution among character states by using a combination of the following transformation types: unordered (u), any change in character state is possible (Fitch parsimony); reversible (r), forward and reverse changes may occur in binary characters (Wagner parsimony); Dollo (d), increases in state may occur only once; irreversible (i), no losses are possible (Camin-Sokal parsimony) (Maddison & Maddison, 1987). All characters were weighted equally throughout both analyses. Copies of the data sets used in both analyses are available from the senior author.

Results and discussion

Phylogeny of rotiferan orders

Our phylogenetic analysis of rotifer orders produced several equally parsimonious cladograms. We selected one to display here to illustrate three points (Fig. 1). (1) Three characters created some difficulty in our study (type of trophi, presence or absence of vitellaria, and presence or absence of prostate glands) because we were unsure how to define their ancestral character states and/or transformation types. Therefore, we decided to use an unordered transformation type for the trophi and report alternative positions on the cladogram where evolutionary events may have occurred for the other two characters; neither compromise alters our basic conclusions. (2) Treelength was independent of the transformation type used for all of the most parsimonious cladograms: 11 steps were required using either Wagner parsimony for all characters (except trophi, unordered) or a combination of transformation types (Mix-1) based on assumptions of how each character might evolve. (See the legend of Fig. 1 for a list of these assumptions.) (3) No characters demonstrated homoplasy using either Wagner or Mix-1 parsimony.

All available information was not used in constructing this cladogram (e.g., certain ultrastructural information, physical development of males within class Monogononta, and presence or absence of sessile forms). However, inclusion of such data does not significantly improve discrimination among the five orders. Nevertheless, differences in trophi, coronae, and feeding types in orders Collothecaceae and Flosculariaceae suggests that the sessile condition has evolved twice, at least once in each order. In some forms, subsequent evolution has led to planktonic existence once again (e.g., *Collotheca mutabilis*, *Ptygura libera*).

Our analysis of the phylogeny of rotifer orders is very similar to one proposed by Epp & Lewis (1979), except that their phylogeny emphasizes different taxonomic levels within the Monogononta. In contrast, Kutikova (1983) proposed

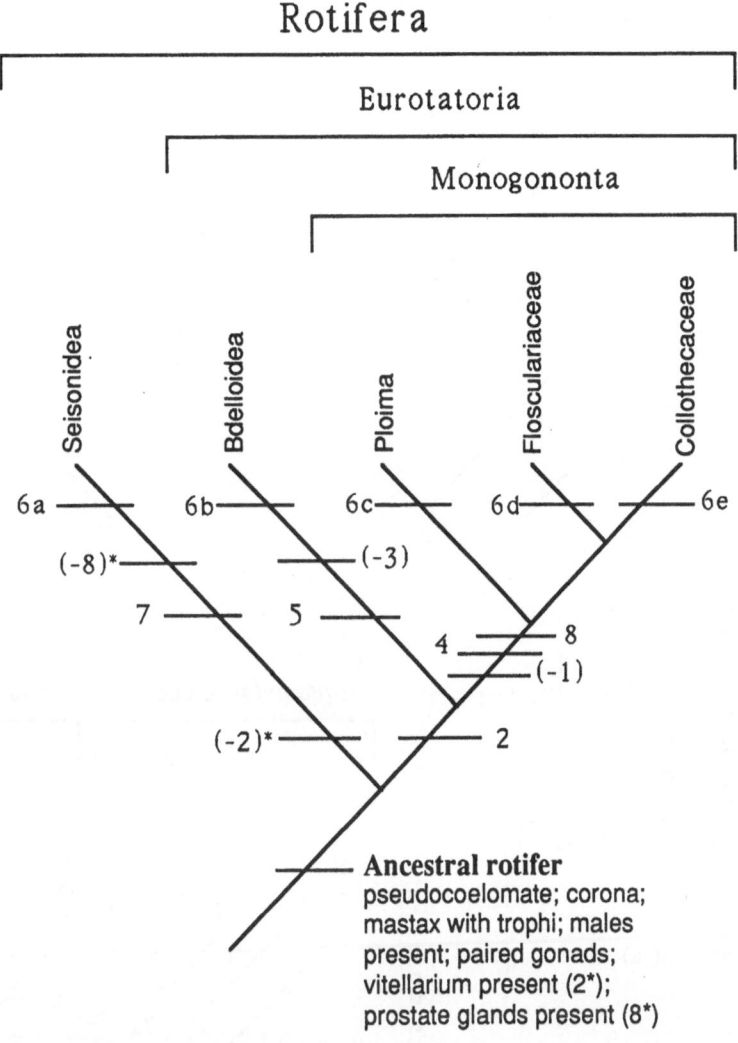

Fig. 1. Cladogram of phylum Rotifera at the ordinal level. Numbers refer to changes in character state as gains or losses (–). Numbers with asterisks (*) indicate alternative positioning of evolutionary events when the status of the ancestral character state has not been ascertained. Characters and transformation types used (Mix-1) are as follows: 1 = paired gonads (i); 2 = vitellarium (r); 3 = male (i); 4 = resting egg innovation (d); 5 = desiccation (r); 6 = trophi (u): a – aberrant, b – ramate, c – others, d – malleoramate, e – uncinate; 7 = spermatophores (d); 8 = prostate glands (r). See text for an explanation of the transformation types. (NB: Four steps are required for the evolution of the five types of trophi described here, assuming one is ancestral.)

a very different phylogeny based on use of the corona in locomotion and feeding, and direction of ciliary metachronal movement. Kutikova's phylogeny may be represented using the standard convention for tree representation as follows:

(Seisonidea, (Ploima, ((Bdelloidea, Flosculariaceae), Collothecaceae))); (eq. 1).

Our view of phylogeny at this taxonomic level (Fig. 1) differs significantly from Kutikova's view. Whereas Kutikova's tree (eq. 1) suggests that the Bdelloidea and Flosculariaceae are sister groups, ours places bdelloids more distant, thus:

(Seisonidea, (Bdelloidea, (Ploima, (Flosculariaceae, Collothecaceae)))); (eq. 2).

We suggest that our cladogram (Fig. 1, eq. 2) is more plausible for several reasons. It is based on a larger number of unequivocal anatomical features and is more parsimonious: 11 (eq. 2) verses 14 steps (eq. 1), using these eight characters and the transformation type Mix-1. Even using Wagner parsimony (except trophi, unordered), eq. 1 requires 13 steps. Further, eq. 1 entails homoplasous evolution of both ovarian number and resting egg innovation and predicts

314

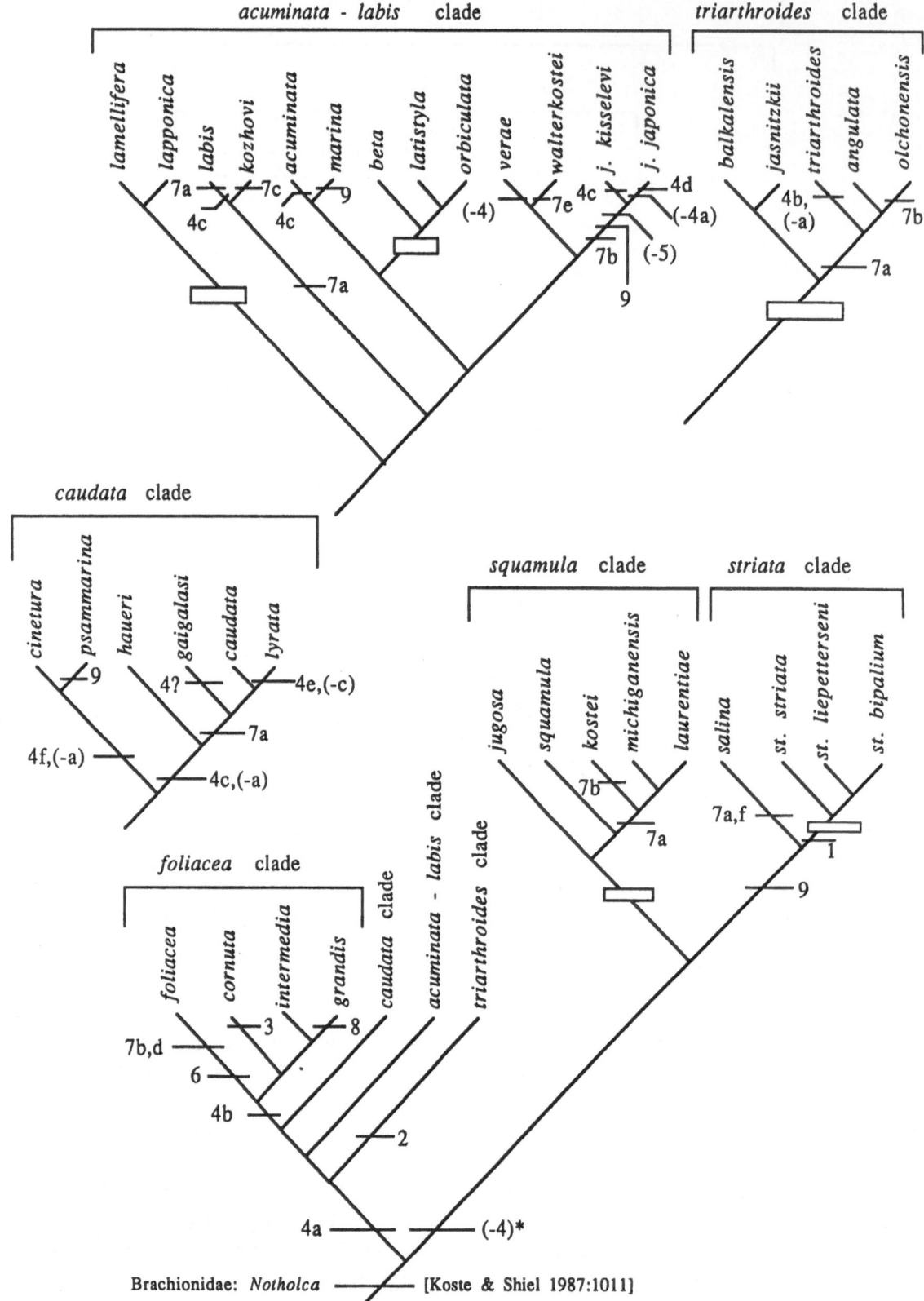

Brachionidae: *Notholca* [Koste & Shiel 1987:1011]

polyphyly for class Monogononta. Based on the number of evolutionary steps required and the lack of homoplasy, our analysis provides strong evidence for monogonont monophyly.

The cladogram shown in figure 1 is only one of several most parsimonious cladograms possible: minor variation in placement of monogonont orders within that clade is possible without increasing the number of evolutionary steps. Further differentiation among cladograms will not be possible without additional information, especially of the type provided by ultrastructural research (e.g., Clément, 1980, 1985). Presence of a unique syncytial epidermis comprised of a intracytoplasmic lamina (Clément, 1985; Lorenzen, 1985) suggests a relationship between the Rotifera and the Acanthocephala, perhaps with the acanthocephalan line diverging before the ancestral rotifer. However, we disagree with Lorenzen's view that the lemnisci and proboscis of acanthocephalans are synapomorphic with seemingly similar structures in bdelloids, and that acanthocephalans and bdelloids are sister groups.

Phylogeny of Notholca

Using a user-defined transformational type (Mix-2), our analysis produced several equally parsimonious cladograms for the 37 putative species of *Notholca* examined here (Fig. 2). This cladogram separates the taxa into six groups of species or clades, named for prominent member(s) contained in each: four clades possessing caudal processes (*foliacea, caudata, acuminata – labis, triarthroides*), and two lacking caudal processes (*squamula, striata*). We selected this

cladogram over other equally parsimonious ones because it separated species into clades of 4-13 taxa, a size convient for further comparisons. Suprisingly, we found that the first four clades can be combined into a single larger one or separated further without increasing treelength or homoplasy. These clades are defined based on a few important characters, including presence or absence and type of lateral spines (movable or rigid) and presence or absence and type of caudal processes. Other characters that were useful in separating these clades included number of anterior spines and habitat (fresh- or salt water). Characters concerning size and shape of anterior spines were much too variable to be of value in this study and therefore were not included.

Three synapomorphies were congruent and easily traceable: movable and rigid, lateral spines (*striata* and *triarthroides* clades, respectively), and reduced number of anterior spines (i.e., *japonica*). However, homoplasy was found in the characteristic of inhabiting salt water, indicating multiple invasions of the marine habitat (Fig. 2).

Two characters clearly require greater study before they may be resolved. Both type of caudal process and ornamentation on the surface of the lorica routinely are provided in descriptions of species, but neither have been defined precisely. Once these characters have been examined more critically, cladistic studies may eventually prove useful in separating species possessing them. However, in those species possessing a caudal process misintrepetation of exuberent morphotypes for solid species, especially in the *acuminata – labis* clade, should not be overlooked as a possible explanation for some forms.

Differences in the phenetic characters of size

Fig. 2. Cladogram of the genus *Notholca*. Numbers refer to changes in character state as gains or losses (–). Numbers with asterics (*) indicate alternative positioning of evolutionary events when the status of the ancestral character state has not been ascertained. Characters and transformation types used (Mix-2) are as follows: 1 = moveable lateral spines (d); 2 = rigid lateral spines (d); 3 = central spur (d); 4 = caudal processes (u): a – variable spatulate, b – thin, c – broad, d – forked, e – crescent, f – moveable,? – unknown; 5 = six anterior spines (i); 6 = keeled (d); 7 = lorica surface (u): a – stripes, b – dotty, c – granular, d – ridges, e – cup-like, f – smooth; 8 = ventral plate appendage (d); 9 = marine (r). See text for an explanation of the transformation types. Polymorphic forms for character states within characters 4 and 7 are present in some taxa (e.g., *kostei* possesses both characters 7a and b). Taxa above open rectangles are separated based on characters best studied using phenetic analysis.

and shape of the body or spines account for the separation of 19 taxa, especially species within the *acuminata – labis*, *squamula*, and *striata* clades (Fig. 2). Because these characteristics do not lend themselves to cladistic analysis, a detailed phenetic analysis is required before these species may be assigned a permanent position. However, size and shape may vary according to nutritional status and season (Björklund, 1972), therefore, the examinations of a few samples from natural populations may lead to improper separation of species. Any phenetic study must examine specimens collected throughout the year and should include populations from laboratory cultures grown under a wide variety of conditions.

Several authors have separated *Notholca* into groups comprising various species (Björklund, 1972; Koste, 1978; Kutikova, 1980; Ruttner-Kolisko, 1974), but each has emphasized slightly different criteria, the most frequently used being type of caudal process and lateral spines, and size and shape of the lorica. Our analysis is not directly comparable to any of these investigations, as none of them attempted a phylogenetic study, but a general comparison shows that we differ significantly from these other works only in the placement of six species. Koste (1978) and Ruttner-Kolisko (1974) place *psammarina* in the *striata* clade and Kutikova (1980) places *verae* in the *squamula* clade. Both relationships are possible without increasing the number of evolutionary steps required or homoplasy. However, we arbitrarily elected to place *psammarina* in the *caudata* clade to emphasize the presence of a caudal process in *psammarina*, and we placed *verae* in the *acuminata – labis* clade to stress the putative relationship between *verae* and *walterkostei* (Dartnall & Hollowday, 1985). Koste (1978), Kutikova (1980), and Ruttner-Kolisko (1974) place *lapponica* in the *squamula* clade, but this requires an extra evolutionary step. The salt water form *salina* is routinely considered to be related to *squamula* (e.g., Björklund, 1972), but surprisingly an extra evolutionary step is required for it to be placed there. Placement of *salina* as a sister species within the *striata* clade, therefore, is more parsimonious and less homoplasous. If this senario is correct, then invasion of the marine environment took place in these forms before the evolution of moveable, lateral spines by members of the *striata* clade. Placement of *cornuta* in the *triarthroides* clade and *angulata* in the *acuminata – labis* clade (Kutikova, 1980) requires two or more evolutionary steps and a concomitant increase in homoplasy.

We do not suggest that all species shown in Figure 2 are 'good' species, rather our analysis suggests several places where additional study might lead to combining several taxa or separating them. For example, *acuminata* and *marina* differ only in habitat and development of the caudal process (see Björklund, 1972). Thus, the fundamental question remains; to what extent do species within the various subgroups represent true species, incipient speciation within a broadly distributed population, or simply phenotypic responses to different environments (i.e., ecotypes)? Further differentiation of the *Notholca* will not be possible without additional information, especially concerning the caudal processes, surface ornamentation, and size and shape of the lorica. Data concerning interspecific mating will be exceedingly important, if sexuality can be induced in laboratory cultures [e.g., Snell, in this volume]. Finally, little information is available on the trophi of *Notholca* species. This inadequacy also must be rectified before significant progress may be made in clearing up confusion within this genus.

Conclusions

The computer programs used in this study can greatly expedite phylogenetic analyses, so that relationships among rotifers may be examined using cladistic analysis even with a recondite data set containing more than 100 elements. However, species with great morphological variability are not well resolved with cladistics. Further, such efforts are only as good as the data available. Therefore, we recommend that taxonomists pay close attention to the suggestions of Ruttner-Kolisko (1963) concerning rotifer taxonomy and we restate them here in a slightly revised form. (1) Formal descriptions of new species should not be based on fragmentary information of '... a few individuals...' or the '... naming of morphological pecularities', or even morphological analysis of specimens from a single habitat. Such practices only serve to obfuscate, not elucidate taxonomic relationships. (2) Detailed statistical analyses of morphological characteristics ought to be provided along with taxonomic descriptions whenever appropriate (i.e., phenetic analyses [e.g., Björklund, 1972; Stemberger, 1976]). (3) If at all possible, several typical specimens from each season should be cultured under a wide variety of conditions to define the developmental plasticity of the species in question. (4) Where possible physiological, ecological, and genetic parameters must be investigated along with morphological ones. Finally, although we may be accused of adopting a reductionist position, we suggest that studies of form and function (i.e., functional morphology) be undertaken with the goal of better understanding the structural significance of the characters used in species designation.

Acknowledgements

We wish to thank Dr. Joe Felsenstein (Department of Genetics, University of Washington) who supplied us with a copy of PHYLIP© and Drs. Wayne and David Maddison (Museum of Comparative Zoology, Harvard University) who supplied us with a copy of MacClade©, version 2.1. We also thank Drs. Terry Snell, George Wittler, and William Brooks who read and improved this manuscript. Help from Oussama El-Hilali, Larry Fahnoe, Mike Post, Tom Oyster (Computer Services, Ripon College), especially their loan of a Macintosh Plus™, was greatfully appreciated. This work was supported, in part, by a grant from Ripon College funds for Faculty Development.

References

Amrén, H., 1964. Ecological and taxonomical studies on zooplankton from Spitsbergen. Zool. Bidrag. Uppsala Bd 36: 209–276.

Björklund, B. G., 1972. Taxonomic and ecological studies of species of Notholca (Rotatoria) found in sea- and brackish water, with description of a new species. Sarsia 51: 25–66.

Carlin, B., 1943. Die Planktonrotatorien des Motalaström. Zur Taxonomie und Ökologie der Planktonrotatorien. Medd. Lunds Univ. Limnol. Instn. 5: 1–256.

Chengalath, R., 1978. A new species of the genus Notholca Gosse, 1886, (Brachionidae: Rotifera), from Great Slave Lake, N.W.T. Can. J. Zool. 56: 363–364.

Chengalath, R. & W. Koste, 1987. Rotifera from Northwestern Canada. Hydrobiologia 147: 49–56.

Clément, P., 1980. Phylogenetic relationships of rotifers, as derived from photoreceptor morphology and other ultrastructural analyses. Hydrobiologia 73: 93–117.

Clément, P., 1985. The relationships of rotifers. In S. Conway Morris, J. D. George & H. M. Platt (eds.), The Origins and Relationships of Lower Invertebrates. System. Assoc. Volume 28. Clarendon Press, Oxford, U.K.: 224–247.

Daday, E., 1913. Beiträge zur Kenntnis der Mikrofauna des Kossogol-Beckens in der Nordwestlichen Mongolei. Mathmatische und naturwissenschaftliche Berichte aus Ungarn 26: 274–360.

Dartnall, H. J. G. & E. D. Hollowday, 1985. Antarctic rotifers. Brit. Antarctic Sur. Sci. Reports 100: 1–46.

Edmondson, W. T., 1959. Rotifera. In W. T. Edmondson (ed.), Fresh-Water Biology, 2nd ed., John Wiley & Sons, Inc., N.Y.: 420–494.

Epp, R. W. & W. M. Lewis, Jr., 1979. Sexual dimorphism in Brachionus plicatilis (Rotifera): evolutionary and adaptive significance. Evolution 33: 919–928.

Fink, W. L., 1986. Microcomputers and phylogenetic analysis. Science 234: 1135–1139.

Focke, E., 1961. Die Rotatoriengattung Notholca und ihr Verhalten im Salzwasser. Kieler Meeresforschungen 17: 109–205.

Gilbert, J. J., 1988. Rotifera. In K. G. & R. G. Adiyodi (eds.), Reproductive Biology of Invertebrates, Volume III. Accessory Sex Glands. Oxford & IBH Publ., Co., New Delhi.: 73–80.

Koste, W., 1974. Das Rädertier-Porträt. Die Rädertiergattung Notholca. Mikrokosmos 63: 48–52.

Koste, W., 1978. Rotatoria. Die Rädertier mitteleuropas. 2 vols, Gebrüder Borntraeger, Berlin-Stuttgart, 673 pp, 234 plates.

Koste, W. & R. J. Shiel, 1987. Rotifera from Australian Inland Waters. II. Epiphanidae and Brachionidae (Rotifera: Monogononta). Invertebr. Taxon. 7: 949–1021.

Kutikova, L. A., 1980. On the evolutionary pathways of speciation in the genus Notholca. Hydrobiologia 73: 215–220.

Kutikova, L. A., 1983. Parallelism in the evolution of rotifers. Hydrobiologia 104: 3–7.

318

Kutikova, L. A., 1986. Taxonomic review of the rotifer fauna of the Lake Baikal. USSR Acad. Sci. Proc. Zool. Instit. 152: 89–105. (in Russian, with English summary)

Lair, N. & W. Koste, 1984. The rotifer fauna and population dynamics of Lake Studer 2 (Kergulelen Archipelago) with description of *Filinia terminalis kergueleniensis* n.ssp. and a new record of *Keratella sancta* Russel 1944. Hydrobiologia 108: 57–64.

Lorenzen, S., 1985. Phylogenetic aspects of pseudocoelomate evolution. In S. Conway Morris, J. D. George & H. M. Platt (eds.), The Origins and Relationships of Lower Invertebrates. System. Assoc. Volume 28. Clarendon Press, Oxford, U.K.: 210–223.

Maddison, W. P. & D. R. Maddison, 1987. MacClade, version 2.1. An interactive, graphical program for analyzing phylogenies and studying character evolution. Harvard Univ., Cambridge, MA. (distributed by the authors).

de Paggi, S. B. J., 1982. *Notholca walterkostei* sp. nov. y otros Rotiferos dulceacuicolas de la Peninsula Potter, Isla 25 de Mayo (Shetland del sur, Antartida). Rev. Asoc. Cienc. Nat. Litoral 13: 81–95.

Pejler, B., 1977. On the global distribution of the family Brachionidae (Rotatoria). Archiv für Hydrobiologie (Suppl.) 53: 255–306.

Ridley, M., 1986. Evolution and Classification. Longman, New York, 201 pp.

Ruttner-Kolisko, A., 1963. The interrelationships of the Rotatoria. E. C. Dougherty (ed.), The Lower Metazoa, Comparative Biology and Phylogeny. Univ., California Press, Berkeley, CA.: 263–272.

Ruttner-Kolisko, A., 1974. Planktonic rotifers biology and taxonomy. Binnengewässer (Suppl.) 26: 1–146.

Snell, T. W., 1989. Systematics, reproductive isolation and species boundaries in monogonont rotifers. Hydrobiologia (in this volume).

Stemberger, R. S., 1976. *Notholca laurentiae* and *N. michiganensis*, new rotifers from the Laurentian Great Lakes region. J. Fish. Res. Bd Can. 33: 2814–2818.

Wiley, E. O., Phylogenetics: the Theory and Practice of Phylogenetic Systematics. John Wiley and Sons, NY, 439 pp.

Hydrobiologia **186/187**: 319–324, 1989.
C. Ricci, T. W. Snell and C. E. King (eds), Rotifer Symposium V.
© *1989 Kluwer Academic Publishers.*

A re-appraisal of two members of the genus *Notholca* from the Andes, with a note on the fine structure of the lorica of *N. foliacea* (Ehrenberg)

Eric D. Hollowday[1] & Charles G. Hussey[2]
[1] *45, Manor Road, Aylesbury, Buckinghamshire HP20 1JB, England;* [2] *British Museum (Natural History), Cromwell Rd., London SW7 5BD, England*

Key words: Rotifera, *Notholca*, fine structure, Andes, South America, taxonomy

Abstract

Rotifers described from the Andes by Murray (1913) and De Beauchamp (1939) as *Notholca foliacea* (Ehrenberg) are reviewed and re-assessed as *Notholca walterkostei* De Paggi, 1982. They are compared with *N. foliacea* and details of the lorica of this species seen with the scanning electron microscope are presented.

Introduction

A rotifer collected in 1912 from Lake Titicaca, Peru/Bolivia, was described by Murray (1913) as *Notholca foliacea* (Ehrenberg, 1838). In 1937 the Percy Sladen Trust Expedition to Lake Titicaca brought back a *Notholca* from Lagunilla Saracocha, a lake 80 km east of Lake Titicaca at an altitude of 4130 m. This single specimen, virtually an empty lorica, was briefly described, without a figure, by De Beauchamp (1939), also as *N. foliacea*. He noted its similarity to Murray's material but expressed uncertainty regarding its true identity.

De Beauchamp's specimen has been located by the senior author in the collections of the British Museum (Natural History). With access to this material and the more recent literature on *Notholca*, we are in a position to redetermine these specimens.

Materials and methods

In addition to the published literature, the following material was examined:

(1) A permanent slide preparation labelled '*Notholca foliacea* (Ehr.) sensu Murray. Exped. Titicaca 225/4. Lag. Saracocha 4.IX.37.' in the collections of the British Museum (Natural History), London. Registration No. 1940.4.1.6.

(2) Living and preserved specimens of *N. foliacea* from Aylesbury, Buckinghamshire.

The preserved specimens were stored in 2% formalin and mounted unstained. Light microscopy included use of differential interference contrast (Nomarski). Specimens of *N. foliacea* examined by scanning electron microscopy were dehydrated in acetone, critical point dried, coated with gold/palladium and viewed in a Hitachi S800 Field Emission Scanning Microscope operated at an accelerating voltage of 8kv.

Systematic section

Both Murray and De Beauchamp appear to have identified their specimens with *N. foliacea* because of the presence of numerous small circular depressions or fossettes on the dorsal plate. In considering the taxonomic status of the Titicaca and Saracocha specimens, it is necessary to consider this, and other, characteristics of *N. foliacea*.

Taxonomic position of N. foliacea

We wish to endorse the current view that this species belongs in the genus *Notholca*. Gillard (1948) erected a new genus, *Argonotholca*, for this species, chiefly because of the presence of a ventral keel and a median dorsal ridge. The former structure is not however, a permanent feature. It appears when the flexible posterior portion of the ventral plate expands to allow for the displacement of the internal organs when the head is retracted within the lorica. When swimming and feeding, the ventral plate appears little different, if at all, to that of other species we have observed alive.

In species such as *N. squamula* (Müller, 1786), *N. labis* Gosse, 1887, *N. bipalium* (Müller, 1786) and, to a slightly lesser extent, *N. acuminata* (Ehrenberg, 1832) such displacement is accommodated by the bilateral expansion of the lorica. It would seem that *N. foliacea*, together with other species which exhibit a postero-ventral keel, (i.e. *N. walterkostei* De Paggi, 1982; *N. verae* Kutikova, 1958), lack this bilateral flexibility. Unless preserved specimens have died with the head fully extended (as does happen on very rare occasions), such material will invariably exhibit this so-called 'keel'. We have not seen living material of *N. walterkostei* or of *N. verae* but believe our remarks on this phenomenon also apply to these species.

Aspects of the structure of N. foliacea

Most members of the genus are strongly convex in cross-section. The areas on each side of the dorsal midline slope down to the lateral margins, but in *N. foliacea* this is accentuated by the occurrence of a three-ridged median dorsal crest, the outer pair of which arise on the dorsal surface of the antero-dorsal median spines, converging slightly below these spines and then proceeding along a fairly parallel course until a little over half way along the dorsal plate, almost level with the apertures of the lateral antennae. Depending on the degree of compression or crumpling of the lorica, these outer ridges may then unite to form a single ridge extending as far as the base of the postero-dorsal spine, or they may remain separate as far as the base of this spine (Fig. 1a). From the margin of the antero-dorsal median sulcus, two more ridges arise (Fig. 1c), uniting about 10 μm below the sulcal margin, and running between the two outer ridges as far as the point at which these widen, and then unite just below the median transverse line.

Examination with the scanning electron microscope shows that the three ridges forming the median crest consist of closely-packed smoothly-rounded tubercles, varying in size from approximately 0.4-0.8 μm (Fig. 1c). Extensive areas on each side of the median dorsal crest are free of longitudinal ridges, but have long been known to be decorated with randomly scattered pimples. SEM examination shows these also to be smoothly-rounded tubercles of similar size to those forming the median dorsal crest. They also occur, more sparingly, on the ventral plate.

Examination of these areas between the median and lateral ridges of the dorsal plate with the light microscope at high magnification, using brightfield illumination, reveals what appears to be a finely reticulate pattern. SEM examination shows this pattern to consist of very minute dimples or fossettes varying in diameter from approximately 0.8-1.75 μm (Fig. 1d). In addition, the whole surface of the lorica, both dorsal and ventral, is densely covered with extremely small tubercles no larger than 0.1 μm (Fig. 1d).

Towards the lateral margins on each side of the median dorsal crest are at least four dorso-lateral longitudinal ridges, two and sometimes three of which can usually be seen from the dorsal aspect

321

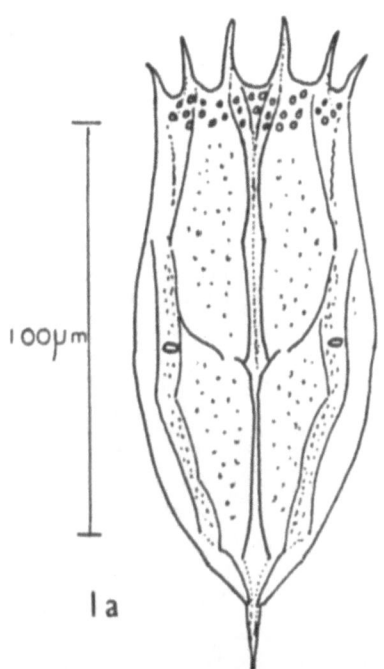

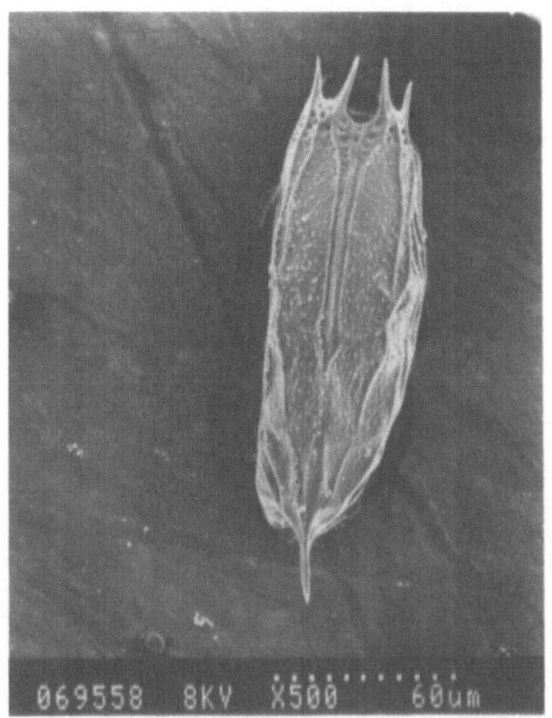

(b)

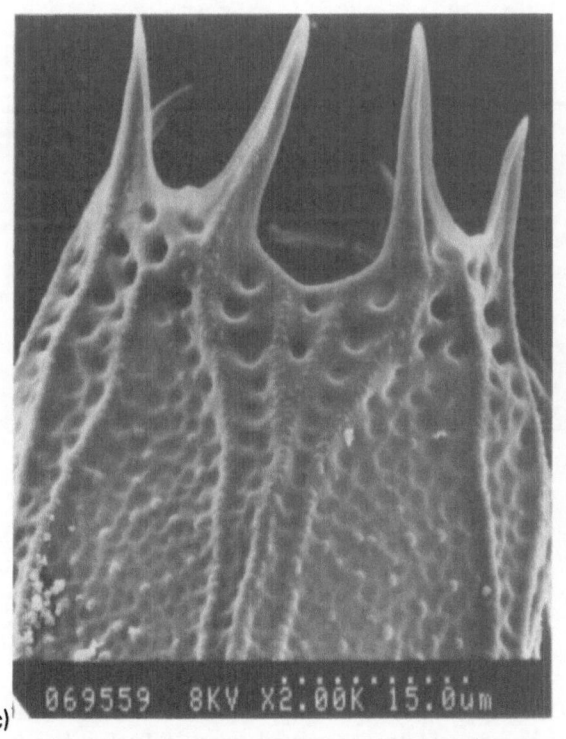

(c)

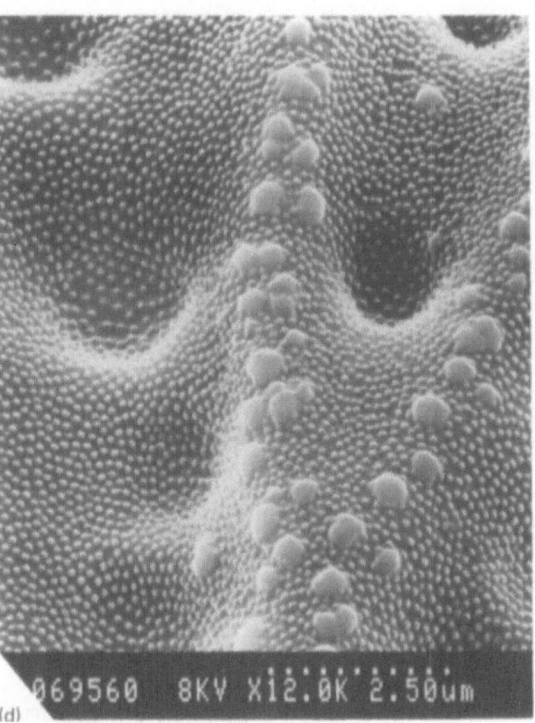

(d)

Fig. 1. Notholca foliacea (Ehrenberg). (a) Dorsal aspect (drawn from Aylesbury specimen); (b) Entire animal, dorsal aspect, S.E.M. × 500; (c) Antero-dorsal area of lorica S.E.M. × 2000; (d) Small fossettes and tubercles. S.E.M. × 12000.

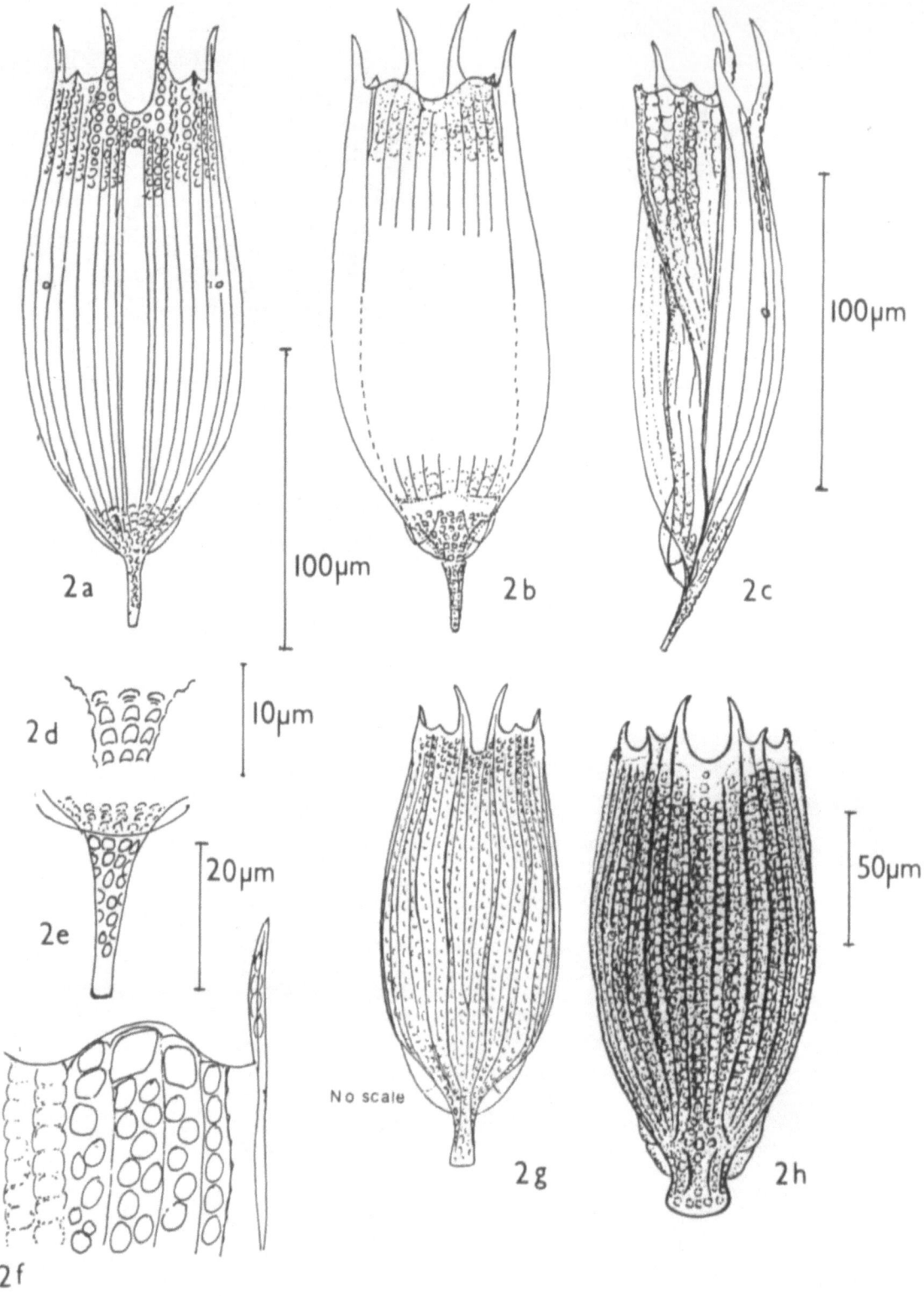

Fig. 2. Notholca walterkostei De Paggi, from L. Saracocha. (a) Dorsal aspect; (b) Ventral aspect; (c) Left ventro-lateral aspect. (d) Postero-dorsal extension, detail of base, dorsal aspect; (e) Postero-dorsal extension, ventral aspect. (f) Shallow fossettes on anterior of ventral plate. (g) *Notholca walterkostei* from Lake Titicaca. (after Murray 1913); (h) *N. walterkostei* from South Shetland. (Unpublished drawing by Walter Koste). Note: Figs. 2a-f drawn from British Museum (Nat. Hist.). slide Figs. 2e and f share common scale.

(Fig. 1a), and these also consist of unbroken rows of rounded tubercles, like those forming the median dorsal crest. The ventral plate has at least eight longitudinal ridges, although these might more accurately be described as pleats, the edges of which are furnished with tubercles, more widely spaced than those forming the dorsal and dorso-lateral ridges. It lacks the extensive dimpling so characteristic of the dorsal plate.

Although it has not been possible to make a comparative SEM study of the Andean specimens, it would appear, on the basis of light microscopy alone, that neither can be considered to belong to *N. foliacea*.

Titicaca specimens

From Murray's figure (Fig. 2g), one can see that the antero-dorsal intermediate spines are almost obliterate and although he shows the two central longitudinal ridges uniting in the posterior half of the dorsal plate, there is no central ridge comparable with the crest of *N. foliacea*. Instead, the whole of the dorsal plate is ridged, with no relatively smooth areas between the median longitudinal line and the lateral margins. The posterior prolongation of the dorsal plate is widened and flattened at its extremity and is spatula-like rather than a spindle-shaped spine as in *N. foliacea*. Most strikingly, the grooves between the longitudinal ridges are pitted with rows of large circular fossettes extending the whole length of the dorsal plate and along its posterior extension.

All these features agree with *N. walterkostei* De Paggi, 1982. It is unfortunate that Murray provides no indication of scale. There are slight differences in the shape of the postero-dorsal extension and in the length of the antero-dorsal median spines, but we have no hesitation in assigning Murray's material to this species.

Saracocha specimen

This specimen differs from *N. foliacea* in much the same details as the Titicaca material and should

also be assigned to *N. walterkostei*. There are, however, some peculiarities that may make it a distinct form or even subspecies, although this cannot be determined on the basis of a single specimen. The arrangement of depressions, or fossettes, is unique. They extend backwards along the dorsal surface, in decreasing size, from the anterior dorsal margin for nearly a quarter of the length of the dorsal plate and re-occur on the posterior tenth of the dorsal plate, continuing, in reduced size, on all surfaces of the posterior extension and some way along the ventral surface of the dorsal plate. They also occur on the antero-dorsal median spines. These spines are long (34 μm) and are separated by a deep sulcus. The intermedian spines are obliterate and the dorso-lateral spines are approximately 17 μm long. There is no median dorsal crest but at least sixteen longitudinal ridges, similar to *N. walterkostei*. The posterior extension is a bluntly terminated narrow tube, 23 μm long. Total length is 200 μm and maximum width is 75 μm.

Discussion

Koste & De Paggi (1982), in their checklist of Monogononta from Neotropis, list De Beauchamp's record from L. Saracocha as *N. foliacea* without comment and do not include any records from Murray (1913). *N. walterkostei* was described by De Paggi (1982) from the South Shetland Islands. It is also known from the South Orkney Islands (Dartnall, 1983 – as *Notholca* sp.), South Georgia (Dartnall & Hollowday, 1985) and as far south as Alexander Island off the Grahamland Peninsular (Hollowday, unpublished). A subspecies, *N. walterkostei reducta* Dartnall & Hollowday, 1985, occupies a restricted distribution on the South Orkney Islands.

Our redetermination of the Andean specimens therefore considerably extends the known range of *N. walterkostei*. Assuming a Laurasian origin for the genus (Dumont, 1983), and a gradual southerly migration into South America after that continent had attained its present position, it is not difficult to envisage a further southward movement of *N. walterkostei*. Although the Andean specimens do exhibit some variation with regards to *N. walterkostei sensu stricto*, it is not unreasonable to suppose that they may have had a common origin. Even so, parallelism cannot be entirely ruled out (Kutikova, 1983). There is a striking resemblance between the postero-dorsal extension of *Notholca lamellifera jashnovi* Kutikova, 1986 from Lake Baikal and that of *N. walterkostei*.

Another Antarctic species, *N. verae* Kutikova, 1958, with its apparently bilaterally rigid dorsal plate and postero-ventral flexibility seems to have affinities with the group, especially the sub-species *reducta*, wherein the postero-dorsal extension is almost obliterate and the antero-dorsal spines exceedingly short. Kutikova (1980) tends to associate *N. verae* with the *N. squamula* species group but a recent analysis by Wallace & Colburn (this volume) suggests a relationship between *N. verae* and *N. walterkostei*. Furthermore, their work separates *N. foliacea* and *N. walterkostei* as occurring in different clades.

Acknowledgements

We wish to thank Dr. Walter Koste for permission to use his unpublished drawing of *Notholca walterkostei* De Paggi (Fig. 2h).

References

Dartnall, H. J. G., 1983. Rotifers of the Antarctic and subantarctic. Hydrobiologia 104: 57–60.

Dartnall, H. J. G. & E. D. Hollowday, 1985. Antarctic rotifers. Scient. Rep. Br. Antarct. Surv. 100: 1–46.

De Beauchamp, P., 1939. The Percy Sladen Trust Expedition to Lake Titicaca in 1937. V. Rotifères et turbellariés. Trans. Linn. Soc. Lond. 3rd ser., 1(1): 51–79.

De Paggi, S. J., 1982. *Notholca walterkostei* sp. nov. y otros rotiferos dulceacuicolas de la Pensinsula Potter, isla 25 de Mayo (Shetland des Sur, Antartida). Rev. Asoc. Cienc. nat. litoral 13: 81–95.

De Paggi, S. J. & W. Koste, 1984. Checklist of the rotifers recorded from Antarctic and Subantarctic areas. Senckenbergiana biol. 65: 169–178.

Dumont, H. J., 1983. Biogeography of rotifers. Hydrobiologia 104: 19–30.

Gillard, A. A. M., 1948. De Brachionidae (Rotatoria) van België met Beschouwingen over de Taxonomie van de Familie. Natuurwet. Tijdschr. 30: 159–218.

Koste, W. & S. J. De Paggi, 1982. Rotifera of the Super Order Monogononta recorded from Neotropis. Gewass. Abwass. 68/69: 71–102.

Kutikova, L. A., 1980. On the evolutionary pathways of speciation in the genus *Notholca*. Hydrobiologia 73: 215–220.

Kutikova, L. A., 1983. Parallelism in the evolution of rotifers. Hydrobiologia 104: 3–7.

Murray, J., 1913. Notes on the natural history of Bolivia and Peru. Scottish Oceanographical Laboratory, Edinburgh, 45 pp.

Wallace, R. L. & R. A. Colburn, 1989. Phylogenetic relationships within the phylum Rotifera: orders and genus *Notholca*. Hydrobiologia 186/187: 311–318.

Hydrobiologia **186/187**: 325–330, 1989.
C. Ricci, T. W. Snell and C. E. King (eds), Rotifer Symposium V.
© 1989 *Kluwer Academic Publishers.*

Protein patterns in rotifers: the timing of aging

Maria José Carmona, Manuel Serra & Maria Rosa Miracle
Departamento de Ecología, Facultad de Biología, Universidad de Valencia, 46100-Burjasot, Valencia, Spain

Key words: aging, rotifers, protein electroforetic patterns, individual growth

Abstract

Single rotifer individuals have been characterized biochemically to obtain a fingerprint of their physiological state using a modified ultrasensitive silver-stain procedure to detect total proteins in polyacrylamide gels. Patterns are completely uniform for young isogenic individuals raised in the same culture, but they start to change when these individuals reach a certain age. This age is close to the mean lifespan and to both the cessation of body growth and reproduction. Variability is greatest among individuals of the same chronological age, thus the rate of aging is different even among individuals having identical genotypes and experiencing the same environment.

Introduction

Aging is the age-dependent acquisition of alterations in both morphology and physiology. It is a complex problem which can be more easily investigated by studying the lower metazoa. Rotifers are ideal for this kind of study because they are simple eutelic organisms with a few cells, a short lifespan, and parthenogenetic reproduction. High numbers of isogenic individuals can be obtained for experimental purposes. For many years rotiferologists have observed aging effects, like the maternal age effect on offpring senescence reported by Jennings & Lynch (1928) and Lansing (1947, 1948). These studies have stimulated a large number of investigations, not only on rotifers (reviewed by King, 1969, 1980 and 1983), but also on other organisms (reviewed by Lints, 1978). Most rotifer aging studies have been based on laboratory population dynamics and almost no attempt has been made to characterize biochemi-

cally the different ages, except for a few data on the age-variation of certain enzyme activities (Fanestil & Barrows, 1965; Meadow & Barrows, 1971). The aim of this paper is to obtain a characterization of aging through the study of protein patterns.

Methods

A clone CU of *Brachionus plicatilis* (Serra & Miracle, 1983 and 1985) was cultivated since 1981 in our laboratory at 25 °C, 12 ppt salinity, under continuous illumination (PAR: approx. 35 μE/m^2 s), and fed with *Tetraselmis* sp. Experiments were initiated by collecting 500 ovigerous parthenogenetic females from log phase cultures. These females produced 430 newborn females which were isolated in individual glass wells and transfered daily into fresh medium with 5.3 × 10^5 *Tetraselmis* cells ml^{-1}. Twenty F$_1$ females, all

326

parthenogenetic, were collected at random 2, 4, 6, 8, and 9 days after birth. A total of 100 individuals were collected, and the proteins of each were separated electrophoretically. This procedure was repeated twice over one year.

For protein extraction, single individuals of *B. plicatilis* were rinsed with diluted sea water (12 ppt salinity), separated from their eggs, and isolated in 5 μl of this water on a hydrophobic surface. Rotifers were then treated with 20 μl of a protein dissolving solution containing SDS and mercaptoethanol (Semancik, 1976). The samples, once checked for complete dissolution, were collected and boiled for 5 minutes. Molecular weight protein markers were treated in the same way. Proteins were separated by vertical SDS-discontinuous polyacrylamide gel electrophoresis on a 1 mm gel. Gels made according to established procedures (Conejero & Semancik, 1977) had the following acrylamide concentrations: 4% (pH:

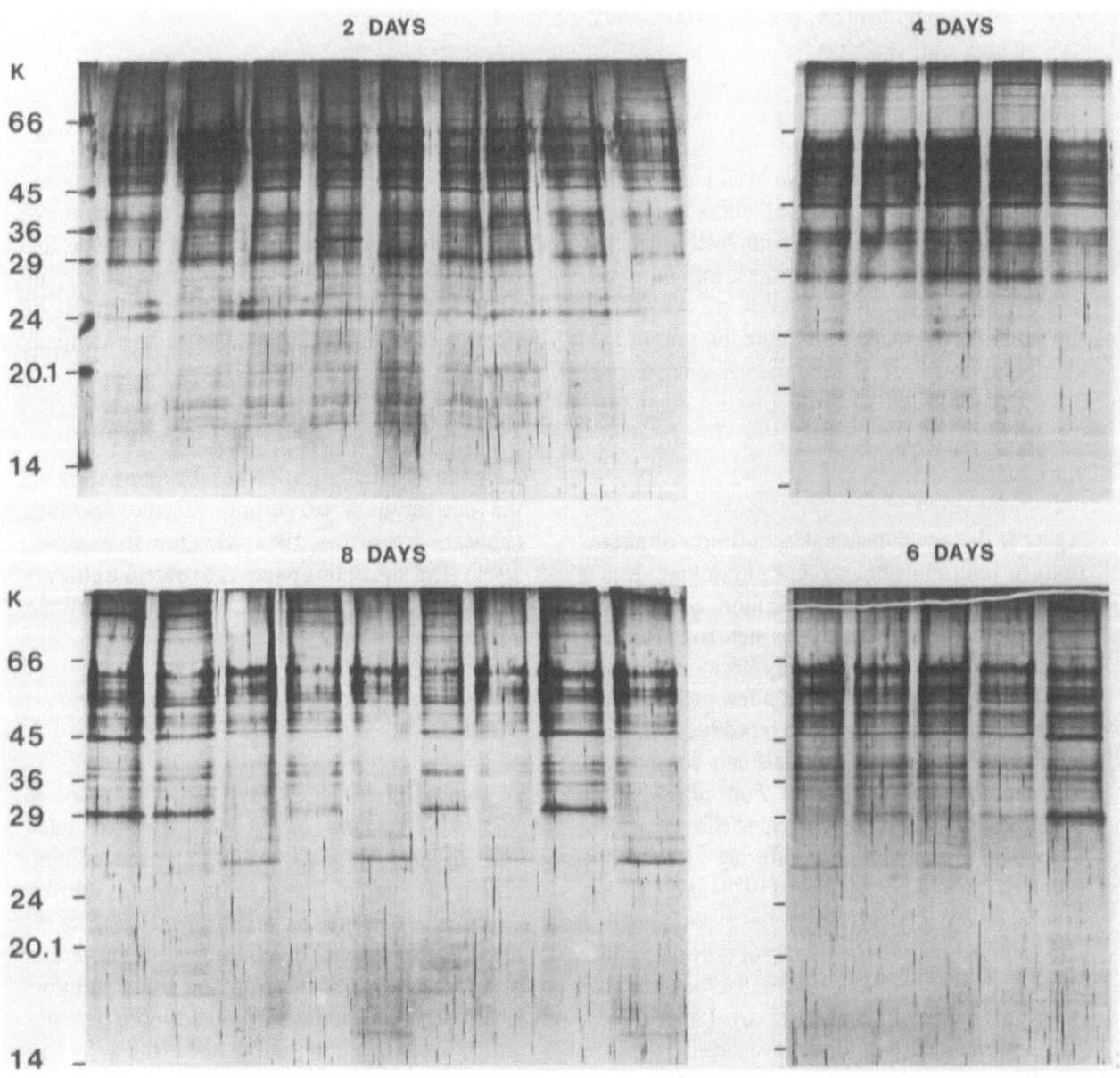

Fig. 1. Protein patterns of *Brachionus plicatilis* (clone CU at 25 °C at 12% salinity). Molecular weight protein markers are shown at the left.

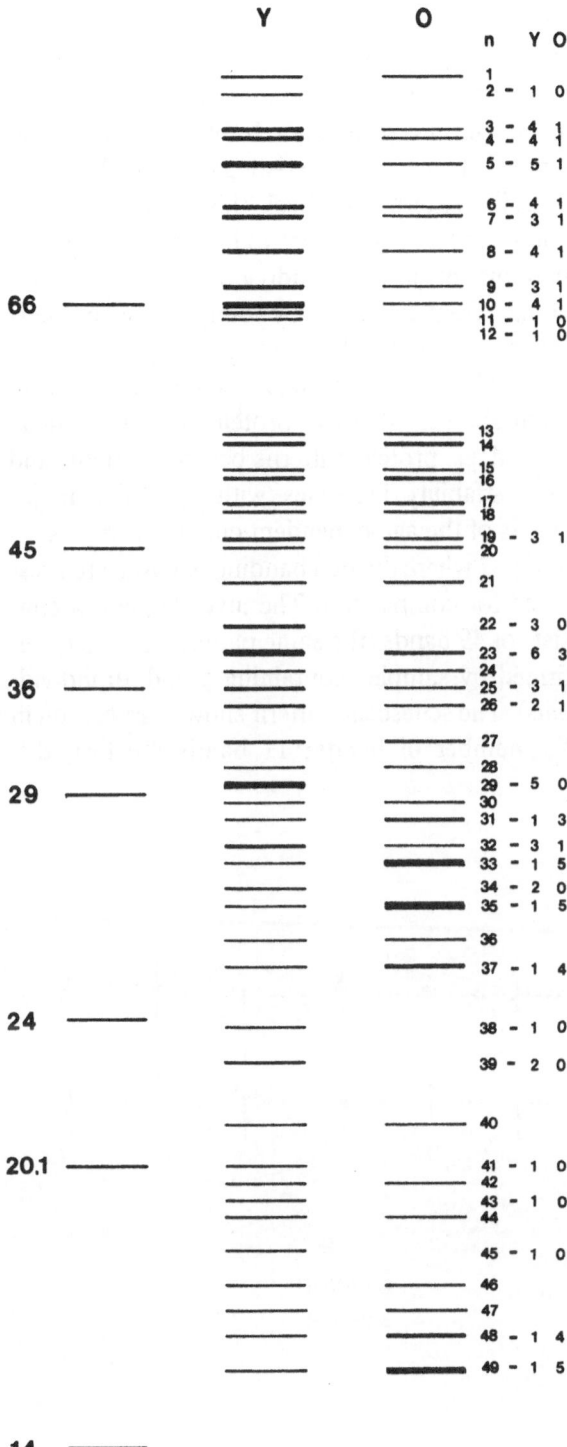

8), 6.5% (pH: 8.8) and 14% (pH: 8.8) in the stacking, spacer and resolution gels, respectively. Since an ultrasensitive method is required to detect the proteins of a single rotifer, an improved silver stain technique was applied, after pre-staining with Coomassie Brilliant Blue. We have modified the ultrasensitive silver stain procedure described by Oakley *et al.* (1980) to obtain both lower levels of background and maximum staining sensitivity. Proteins were fixed overnight with a solution of trichloroacetic acid (12.5%, w/v) and isopropyl alcohol (25%, v/v). It was pre-stained in 5% Coomassie Brilliant Blue over-night and then washed with a solution of 10% acetic acid and 10% isopropyl alcohol until the background was colourless. The gel was then stained using the modified silver stain procedure. The gel was soaked in a solution of 40% methanol and 10% acetic acid for 3 hours and then washed twice for 15 minutes in a solution of 10% methanol and 5% acetic acid. Once fixed with 10% glutaraldehyde for 30 minutes (washed after-wards with water overnight and 3 additional times of 1 hour each), the gel was stained with a freshly prepared ammoniacal silver solution (0.4%, w/v $AgNO_3$; 0.35%, v/v NH_4OH; 0.072%, w/v NaOH) for one hour. Afterwards, three washes in water for 2 minutes were performed. Then, a citric acid-formaldehyde developer (Oakley *et al.*, 1980) was applied at 40 °C until the bands appeared. (The agitation of the gel is mandatory especially in staining and developing.) Development was stopped by discarding developer and adding acetic acid (1%, w/v), for 5 minutes. The gel was then washed with water and hot-vacuum dried.

The basic modifications to obtain both lower levels of background and maximum staining sen-sitivity are: (1) a longer application of the ammoniacal silver-stain solution (1 hour, instead of 15 minutes) but at half the $AgNO_3$ concen-tration, (2) more washes after application of the

Fig. 2. Outline of the electrophoretograms corresponding to young (Y) and old (O) females. Position and molecular weight (Kd) of protein markers is shown at the left. The pattern of a young individual consists of 49 bands. The bands are numbered according to their mobility, at the right hand side. For each of the bands that showed differences, a number indicating their absence (O) or intensity, when present, is registered for young and old patterns. Intensity is a subjective number, ranging from 1 to 6 for increasing intensity, esti-mated for each band independently (this estimation is derived from many electrophoretograms from this and other works on protein variability different clones and females).

328

above mentioned solution, and (3) a much higher developing temperature (40 °C, instead of room temperature).

Moreover, in order to evaluate the individual somatic growth, 100 ovigerous parthenogenetic females were collected and their offspring were cultured individually. They were transferred daily into 0.5 ml of fresh culture medium with 5.3×10^5 *Tetraselmis* sp. cells ml^{-1}. At 4, 6, 8, 12 hours of life and daily from then on, 10 females, randomly collected, were fixed with lugol's and measured with a Zeiss Invertoscop at 40 × magnification. The number of measured females was slightly less for the last three days of the experiment due to the survival structure of the population.

Results

The protein patterns of single individuals of clone CU in the two replicated experiments show homo-

geneity among young individuals (Fig. 1). However as age increases, the patterns change and show great variability among individuals of the same chronological age. Individuals of 2 and 4 days of age characteristically have the same juvenile pattern. To check this homogeneity we conducted an experiment to compare the protein patterns of juvenile individuals (aged 2 days) coming from eggs layed by young and old mothers (aged 2 and 6 days). No maternal age effect was detected in these juvenile individuals, as all of them showed identical protein patterns. But at age 6 days, protein patterns become variable and the variability increases with age. The major trends of the age-dependent changes can be seen in Fig. 2 where the two banding patterns are presented for comparison. The juvenile pattern consists of 49 bands (the same pattern has been confirmed by samples containing 5 and 10 individuals). The senescent pattern shows a reduction in the number of bands; 11 bands are lost, dis-

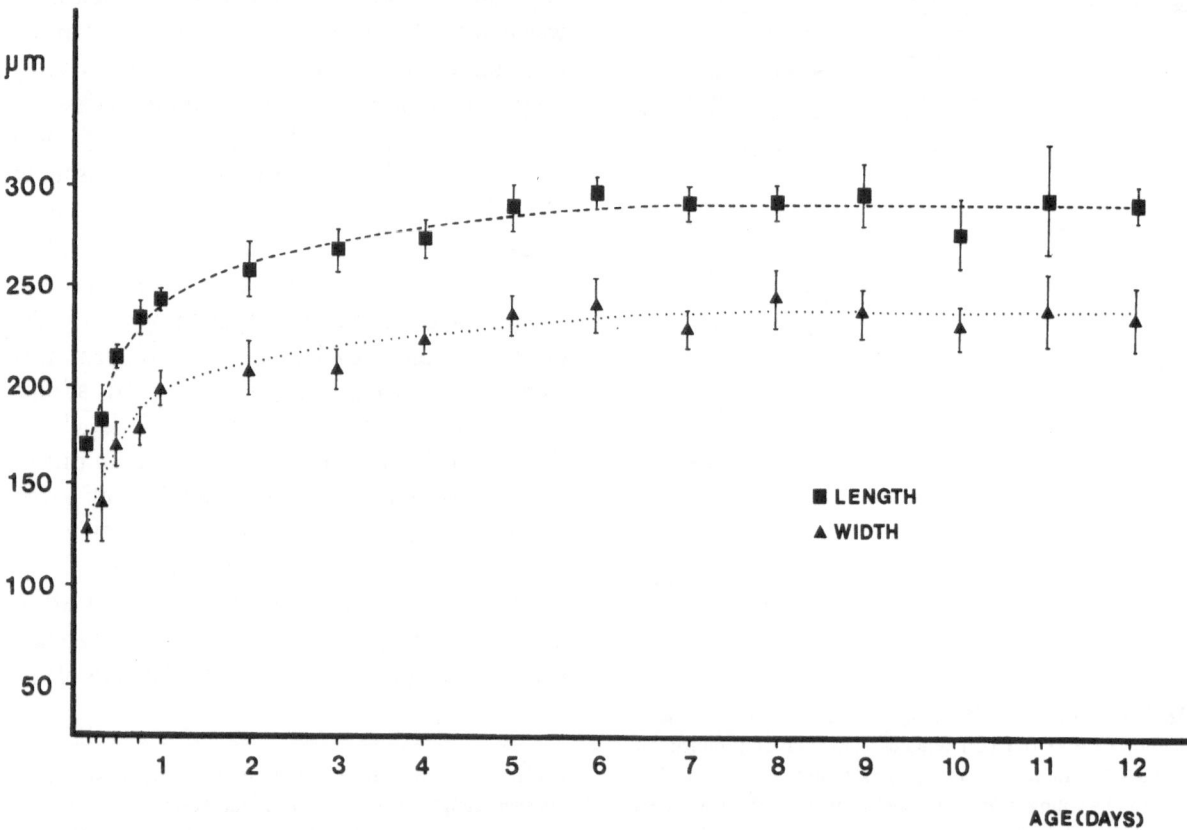

Fig. 3. Growth of length and width during the life of *Brachionus plicatilis* at 25 °C. The vertical bands represent 95% confidence limits calculated from Student's t distribution.

tributed among the whole range of relative molecular mass (Mr). These bands represent more than the 22% of the total number of polypeptide subunits. Moreover, 13 bands show a remarkable decrease in intensity, corresponding to polypeptides of high Mr, usually bands over Mr 66 KD and a few others over Mr 24 KD. On the other hand, 6 bands of low Mr, almost imperceptible in the juvenile pattern, are intense in senescent individuals. These are four 24-29 KD bands and the two fastest proteins. The most distinctive features of the senescence patterns are the two intense protein bands at 26 and 27 KD, respectively, and the reduced intensity of the highest Mr protein components.

Additionally, some direct observations were carried out on the old females during the experiments. Aged females had degenerate digestive organs and gonads and very low or nonexistent egg production. Also evident was decreased movement of cilia and increased pigmentation.

Study of the individual somatic growth (Fig. 3) showed that growth, with increasing age, follows approximately the Bertalanffy function (Bertalanffy, 1948), as found in other rotifer species (Lebedeva & Gerasimova, 1985). The individuals of clone CU have at birth already 58% of their maximum length and 54% of their maximum width. They reach adult size – 90% of maximum length, 85% of maximum width – between the first and second day of life, which is when they begin reproduction. Maximum size is more or less reached by the sixth day of life, and the maximum reproductive age is 8 days, while maximum longevity is 12 days.

Results from population dynamic experiments (Serra, 1987) showed that clone CU has a mean lifespan of 6.86 days and a mean age of last reproduction of 6.5 days old.

Discussion

The directly observable features associated with senescence that we have found have been described by others authors (Lansing, 1947). Slower swimming with aging has been quantified by Luciani et al. (1983) and the fertility decrease with the senescence is a well-known (e.g. King, 1967; Aronovich & Spektorova, 1974; Snell & King, 1977; Ricci, 1983). The reduction of high Mr proteins in senescent individuals may be related to the reduction of gonads in senescence and degenerative processes of other kinds. On the other hand, the intensification of some low Mr bands could be associated with post-translational changes in proteins, such as spontaneous racemization or deamination in certain amino acid residues, leading to increased proteolytic degradation, (McKerrow, 1979; Finch, 1987).

Aging patterns were observed from age 6 days in our experiment, the same age when growth ceases. This agrees with Lansing's conclusion (Lansing, 1948) that aging is a consequence of growth cessation in rotifers, which is also true for other organisms (Lints, 1978). The effects of senescence, leading to death, became increasingly apparent after growth cessation. Thus, the mean lifespan is almost coincident with the average age of last reproduction. Nevertheless, there is great variability in individual lifespans and the maximum lifespan is almost double the mean. The cessation of reproduction is much less variable. In accordance with this, our data illustrate the lack of relationship of physiological age and chronological age, even among isogenic individuals under the same environmental conditions. Individuals of the same chronological age show different protein patterns. In spite of this individual variability, life-history traits respond to environmental factors in predictable ways. Thus, if these eutelic organisms are grown under lower temperature, the lifespan is longer (Miracle & Serra, in this volume) and body size is larger (King & Miracle, 1980; Levedeva & Gerasimova, 1985; Serra & Miracle, 1987), but the relation between mean and maximum lifespan, reproductive and growth periods, is maintained.

From these results, it may be inferred that rotifer aging is a consequence of development, growth and/or differentiation, which implies that the number of genes potentially capable of expressing themselves becomes less and less. This could be associated with the reduction in the num-

330

ber of bands. However, considerable variability in the timing of developmental aging events can occur, which is due not only to genetic or general environmental conditions, but also to other causes. These causes could be: (1) random factors involved in the reduction of gene expression in non-dividing cells, (2) other factors such as maternal age or infective agents. Both causes could be due to small differences between the individuals' environment and its effects on gene expression or to random allocation of non-nuclear genetic material among isogenic individuals. Aging could amplify these differences dramatically.

Acknowledgements

We are most grafetul to Dr. C. Dawson who corrected the English text. M. J. Carmona was supported by a grant from 'Ministerio de Educación y Ciencia'.

References

Aronovich, T. M. & L. V. Spektorova, 1974. Survival and fecundity of *Brachionus calyciflorus* in water of different salinities. Hydrobiol. J. 10: 71–74.

Bertalanffy, L., 1948. Das organische Wachstum und seine Gesetzmssigkeiten. Experientia 4: 255–269.

Conejero, V. & J. S. Semancik, 1977. Analysis of the proteins in crude extracts by polyacrylamide slab gel electrophoresis. Phytopathology 66: 1424–1426.

Fanestil, D. D. & C. H. Barrows, 1965. Aging in the rotifer. J. Gerontol. 20: 462–469.

Finch, C. E., 1987. The ordely decay of order in the regulation of aging processes. In F. E. Yates (ed.), Self-organizing systems. The emergence of order. Plenum Press, New York: 213–236.

Jennings, H. S. & R. S. Lynch, 1928. Age, mortality, fertility and individual diversities in the rotifer *Proales sordida* Gosse. I. Effect of the age of the parent on characteristics of the offspring. J. exp. Zool. 50: 345–407.

King, C. E., 1967. Food, age, and the dynamics of a laboratory population of rotifers. Ecology 48: 111–128.

King, C. E., 1969. Experimental studies of ageing in rotifers. Exp. Gerontol. 4: 63–79.

King, C. E., 1983. A re-examination of the Lansing Effect. Hydrobiologia 104: 141–146.

King, C. E. & M. R. Miracle, 1980. A perspective on aging in rotifers. Hydrobiologia 73: 13–19.

Lansing, A. I., 1947. A transmissible, cumulative and reversible factor in aging. J. Gerontol. 2: 228–239.

Lansing, A. I., 1948. Evidence for aging as a consequence of growth cessation. Proc. nat. Acad. Sci. U.S.A. 34: 304–310.

Lebedeva, L. I. & T. N. Gerasimova, 1985. Peculiarities of *Philodina roseola* (Ehrbg.) (Rotatoria Bdelloida). Growth and reproduction under various temperature conditions. Int. Revue. ges. Hydrobiol. 70: 509–525.

Lints, F. A., 1978. Genetics and aging, Interdisciplinary Topics in Gerontology-Karger, S., Basel 129 pp.

Luciani, A., J. Chasse & P. Clement, 1983. Aging in *Brachionus plicatilis*: The evolution of swimming as a function of age a two different calcium concentrations. Hydrobiologia 104: 141–146.

McKerrow, J. H., 1979. Non-enzymatic, post-translational anmino acid modifications in ageing. A brief review. Mech. Ageing Dev. 10: 371–377.

Meadow, N. D. & C. H. Barrows, 1971. Studies on aging in a bdelloid rotifer I. The effect of various culture systems on longevity and fecundity. J. exp. Zool. 176: 303–313.

Oakley, B. R., D. R. Kirsch & N. R. Morris, 1980. A simplified ultrasensitive silver stain for detecting proteins in polyacrylamide gels. Anal. Biochem. 105: 361–363.

Ricci, C., 1983. Life histories of some species of Rotifera Bdelloidea. Hydrobiologia 104: 175–180.

Semancik, J. S., 1976. Structure and replication of plants viroids. In: Animal Virology, ICN-UCLA, Symposia on Molecular and cellular Biology. Baltimore, D., A. S. Huang & C. F. Fox (eds.), pp. 529–545 Acad. Press, New York: 523–545.

Serra, M., 1987. Variación morfométrica, isoenzimática y demográfica en poblaciones de *Brachionus plicatilis*: diferenciación genética y plasticidad fenotípica. Ph. D. Thesis, Universidad de Valencia.

Serra, M. & M. R. Miracle, 1983. Biometric analysis of *Brachionus plicatilis* ecotypes from Spanish lagoons. Hydrobiologia 104: 279–291.

Serra, M. & M. R. Miracle, 1985. Enzyme polymorphism in *Brachionus plicatilis* populations from several Spanish lagoons. Verh. int. Ver. Limnol. 22: 2991–2996.

Serra, M. & M. R. Miracle, 1987. Biometric variation in three strains of *Brachionus plicatilis*. Hydrobiologia 147: 83–89.

Snell, T. W. & C. E. King, 1977. Lifespan and fecundity patterns in rotifers: The cost of reproduction. Evolution 31: 882–890.

Hydrobiologia **186/187**: 331–337, 1989.
C. Ricci, T. W. Snell and C. E. King (eds), Rotifer Symposium V.
© *1989 Kluwer Academic Publishers.*

Brachionus plicatilis tolerance to low oxygen concentrations

Angeles Esparcia, María R. Miracle & Manuel Serra
Dep. de Ecología, Fac. C. Biológicas, Universidad de Valencia, 46100-Burjasot (Valencia), Spain

Key words: rotifers, oxygen, extreme conditions, population dynamics

Abstract

Tolerance to low oxygen concentrations is expected in *Brachionus plicatilis*, a rotifer adapted to live in saline warm waters. The population dynamics of a clone of this species, isolated from an endorreic saline lake, was studied under controlled laboratory conditions. Although their growth and metabolism is extremely reduced, *B. plicatilis* populations are able to maintain relatively high-density populations (a mean of 35 ind ml^{-1}) in oxygen concentrations below 1 mg l^{-1}, for more than one month. Major features of population growth related to oxygen are discussed.

Introduction

Oxygen concentration may play an important role in distribution of the rotifers both in different lakes and in the vertical profile of stratified lakes, where oxygen is frequently correlated with temperature. Rotifer communities adapted to both low oxygen and low temperature (Pejler, 1957; Edmondson, 1965; Larsson, 1971; Miracle, 1976; Hoffman, 1977) are found in deep waters. Oxygen could also be a limiting factor in brackish waters of warm latitudes, as well as in deep waters of meromictic lakes near the oxycline (Pourriot, 1965; Ruttner-Kolisko, 1980; Miracle & Vicente, 1983). The studied species, *Brachionus plicatilis*, is adapted to high salinity and temperature, conditions that determine low oxygen solubility coefficients. Moreover it is cultivated in mass cultures for applied purposes, where its extremely high densities may cause oxygen depletion.

There are very few experimental studies on rotifer tolerance to low oxygen concentrations (e.g., Ruttner-Kolisko, 1980; Schlüter, 1980). The aim of the present work is to evaluate the tolerance of *Brachionus plicatilis* to limited oxygen levels and to characterize the main features of such tolerance.

Material and methods

Clone CU of *Brachionus plicatilis* (Serra & Miracle, 1983) was cultivated in diluted sea water with a salinity of 12 g l^{-1}, at 25 °C, and under light. The rotifers were fed with *Tetraselmis* sp. thermally treated at 65–70 °C, for 90 min, in order to inactivate them. The efficiency of both thermic treatment and feeding with the obtained food has been studied by Esparcia & Serra (1988). All cultures were grown in flasks with three mouths. Two of these inputs were for N$_2$ gas inflow into the culture and into the gas chamber over the culture; the third opening was for access to the culture (see Fig. 1). A magnetic stirrer was used to stir the flasks.

Experiment 1 was initiated by placing 500

332

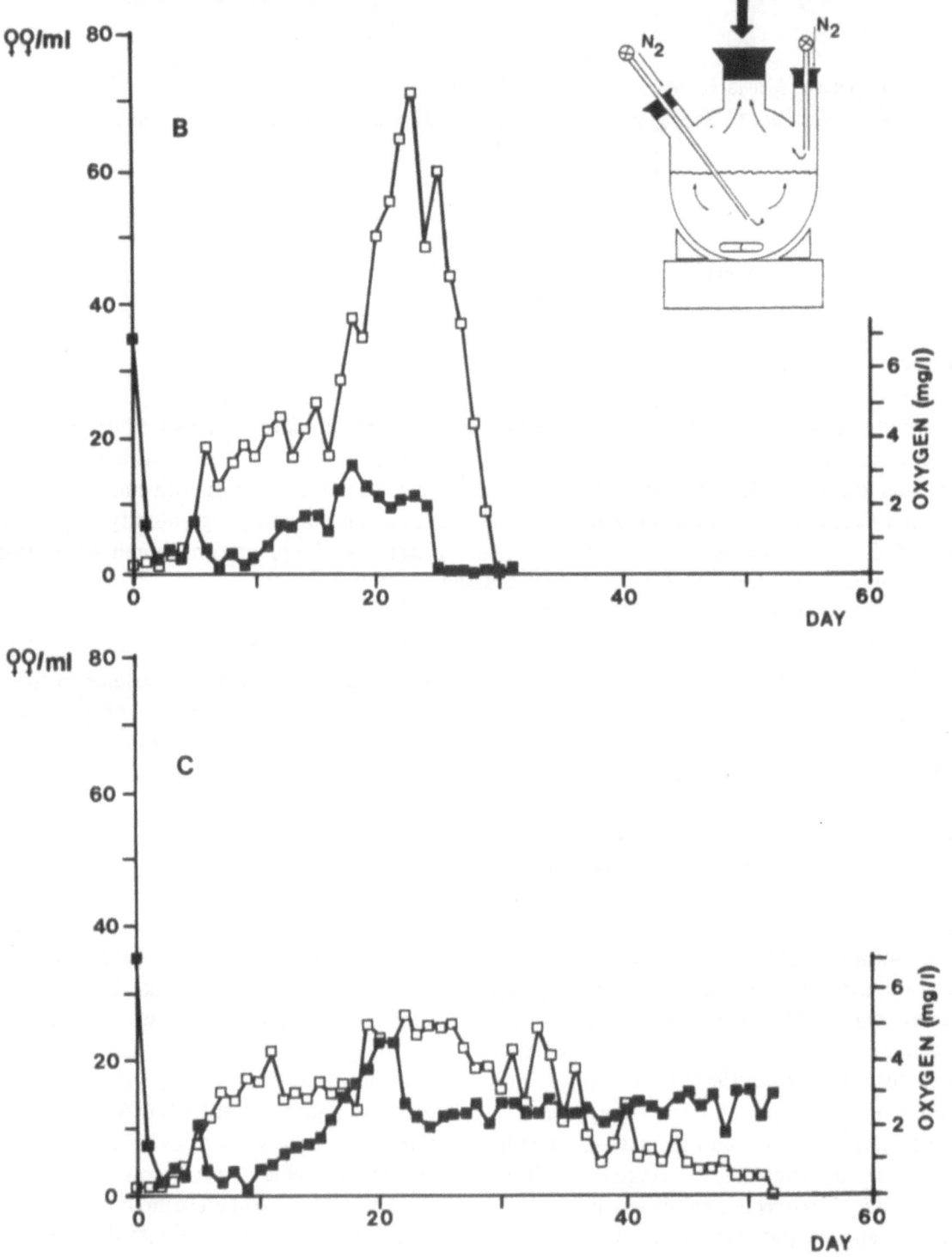

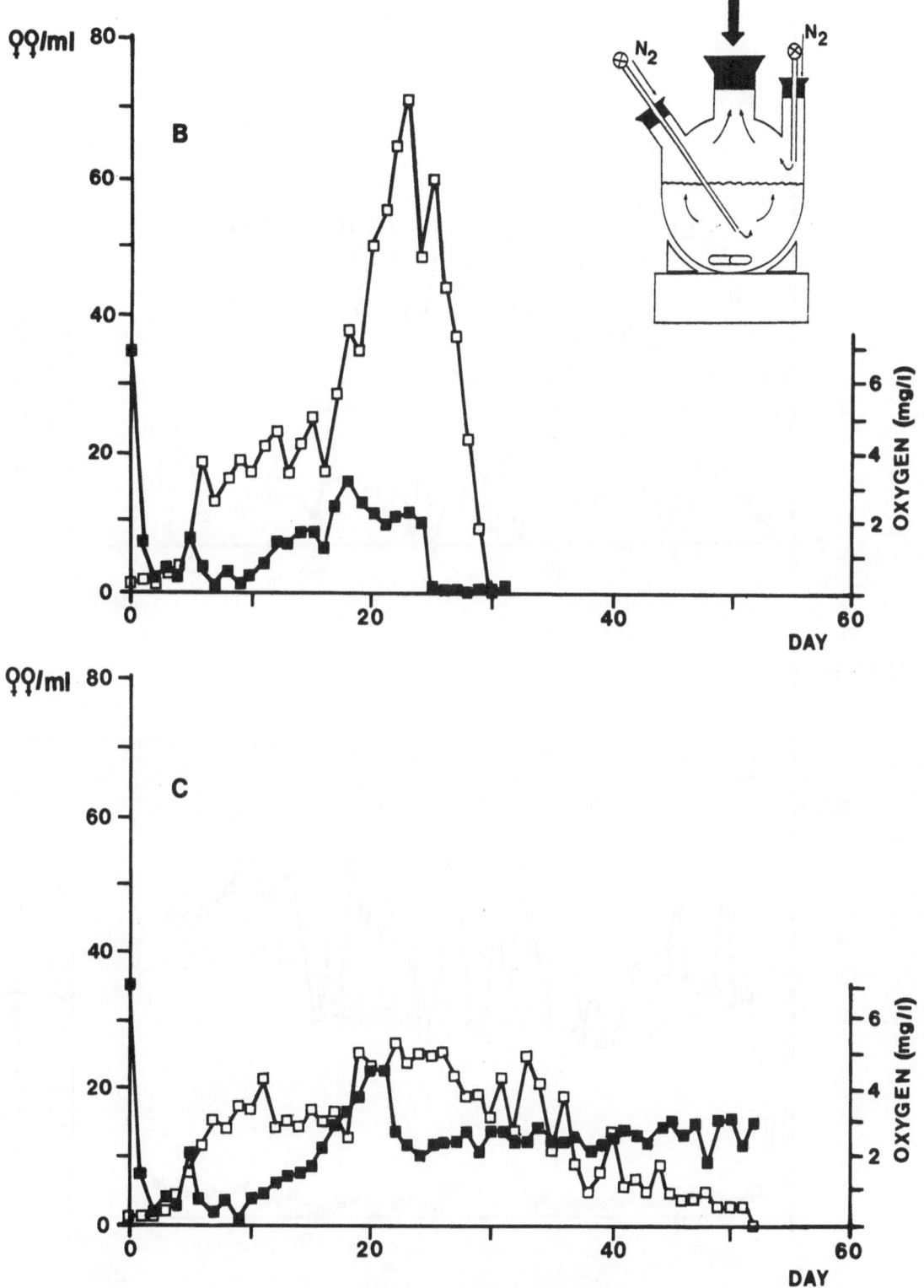

Fig. 1. Daily variation in population density of *Brachionus plicatilis* (open symbols) and oxygen concentration (closed symbols). A, B and C are populations grown in hermetically closed flasks; the control is a population not grown in a hermetically closed flask. The four populations were followed until their extinction except for population A. The control population died at the 81st day. A drawing of the culture flask is shown in the upper right corner.

334

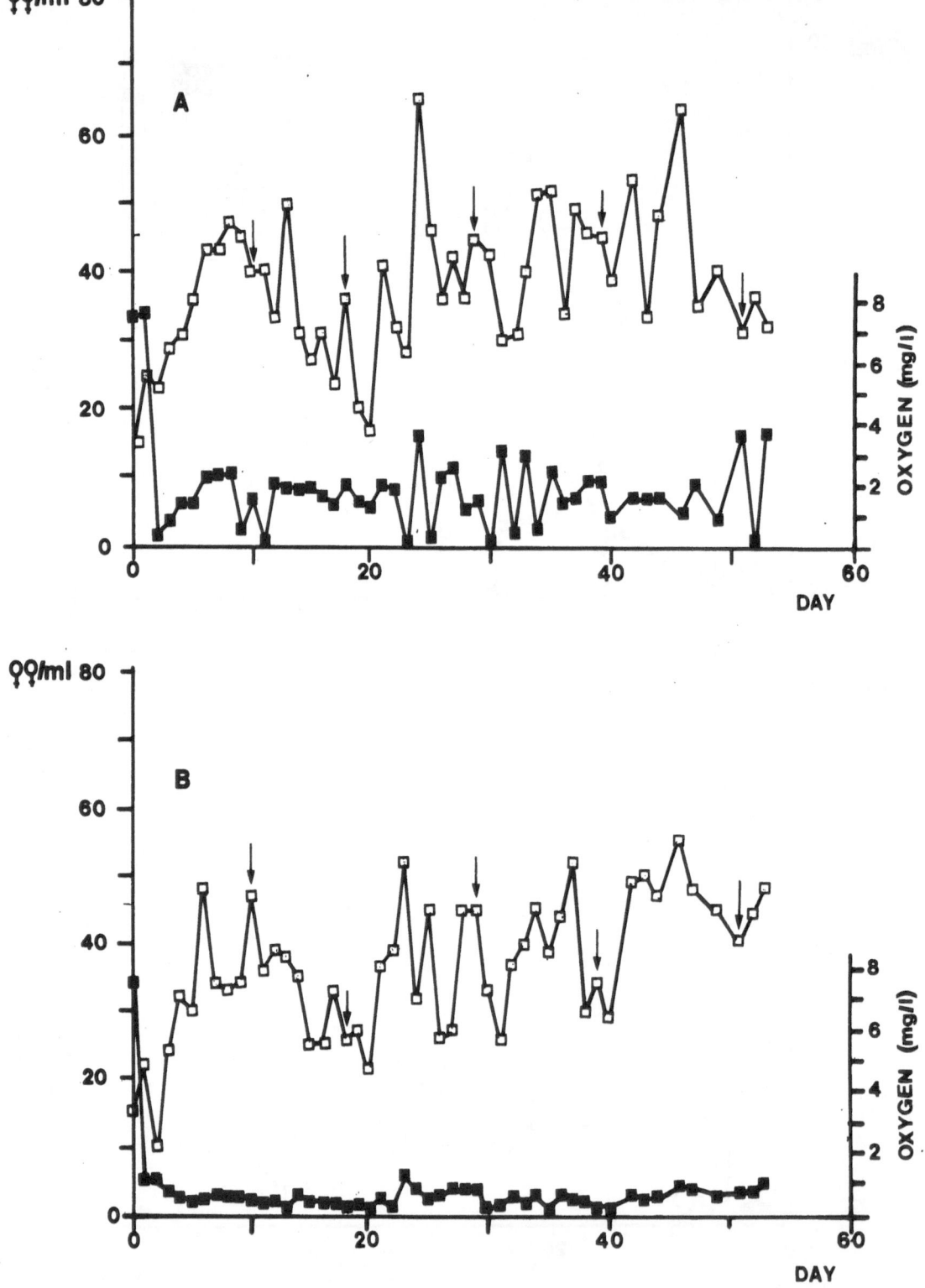

Fig. 2. Daily variations in population density of *Brachionus plicatilis* (open symbols) and oxygen concentration (closed symbols). The oxygen concentrations were daily adjusted to $1-2$ mg l^{-1} (A) and to $0-1$ mg l^{-1} (B). The arrows indicate the times of food supply; dilutions of rotifer density were $1:1.03$ (10th day), $1:1.02$ (18th day), $1:1.03$ (29th day), $1:1.13$ (39th day) and $1:1.12$ (51st day) respectively.

females, from a population previously adapted to the above indicated conditions, into each of four flasks. These flasks contained 400 ml of culture medium with 0.8×10^6 cells ml^{-1} treated algae, (food concentration was determined with a ZM Coulter Counter). Three flasks were hermetically closed and the oxygen was reduced by the respiration of the rotifers themselves, the fourth flask was maintained as a control and stoppered with a cotton plug.

Both population density and oxygen concentration were monitored daily. Rotifers from one or two samples of 1 ml were counted using a dissecting microscope. Oxygen was measured with an electrode (Orion 97-08-99), placed directly into the medium. During culture handling, oxygenation was prevented with N_2 flow into the air chamber, except in the control. Food was never added and the experiment ended when the population died, (except for population A which could not be followed to completion).

In experiment 2, two cultures were initiated each with 6000 females (15 females ml^{-1}) and food in excess. The oxygen level was adjusted to approximately 1–2 mg l^{-1} in one culture and to 0–1 mg l^{-1} in the other by bubbling N_2 through

the culture medium. The cultures were fed periodically (arrows in Fig. 2), by adding medium with 0.8×10^6 cells ml^{-1} of *Tetraselmis* sp. The experiment ended on the 53rd day.

Instantaneous growth rates (r) were estimated using the regression equation $\ln N_t = (\ln N_o) + rt$, where N_o and N_t are the number of individuals at the initiation and the end, respectively, of the exponential growth phase at the beginning of the experiment. The integration of the number of rotifers with respect to time, was calculated as $\sum_i (N_i + N_{i+1})/2$, where N_i and N_{i+1} are the daily measurements of population density.

Results

Experiment 1 showed that *Brachionus plicatilis* is able to tolerate oxygen concentrations lower than 1 mg l^{-1}, at least for 13 days (Fig. 1). The three replicates behaved differently. The B population reached a high density with a maximum similar to that of the control, but after this maximum, the oxygen level dropped quickly due to rotifer respiration and the population died when the oxygen level reached zero. However, the A and C popu-

Table 1. Some population parameters and oxygen concentrations for low aeration cultures. The values in parentheses indicate the oxygen concentrations (mg l^{-1}) at the time interval when the corresponding parameter was evaluated. r is calculated for the exponential growth phase at the begining of the experiment. $N \times time$ is the integrated number of rotifers through time.

	r, days^{-1}	Maximum density, Ind $\cdot$ ml^{-1}	Average $N \times time$ per day, Ind $\cdot$ ml^{-1}	$N \times time$ till population extinction, Ind $\cdot$ d $\cdot$ ml^{-1}
Experiment 1 (low initial density)				
CONTROL[a]	0.773 (4.77)	78 (8.90)	22.7 (8.70)	1836.7
A	0.701 (0.95)	29 (0.30)	12.3 (0.91)	-
B	0.528 (0.90)	71 (2.30)	26.3 (1.20)	787.9
C	0.564 (0.93)	27 (2.70)	13.4 (2.18)	698.8
Experiment 2 (high initial density)				
A	0.147 (1.20)	65 (3.57)	38.4 (1.95)	-
B	0.336 (0.78)	55 (1.00)	37.1 (0.79)	-

(a). Without N_2 flow.

lations did not reach high densities, while the oxygen concentration was maintained at a lower level it was not exhausted and the populations survived longer.

Oxygen concentration appeared to have an effect on both instantaneous rates of population increase and maximum densities (Table 1); both decrease when oxygen levels go down. However, the r values were measured in the initial culture phase and a time lag from the oxygenated cultures could be present. For a closed culture, without addition of food, the integration of rotifers with respect to time, till the extinction of the population (Table 1), is an estimation of the efficiency of exploitation of the medium's resources (this is similar to the carrying capacity concept). This parameter has been calculated and it shows a clear decrease with the oxygen concentration decrease (Table 1).

Experiment 2, with addition of food, confirmed that rotifer populations are able to live at oxygen concentrations lower than $1 \, mg \, l^{-1}$ for long periods of time (Fig. 2). For the oxygen concentrations assayed, the calculated population parameters also showed a clear inverse relationship with oxygen concentration.

Discussion

Brachionus plicatilis inhabits waters where low oxygen concentrations might be expected because of its strong tolerance of high salinities and temperatures (Ito, 1960; Ruttner-Kolisko, 1974; Walker, 1981). In fact, abundant populations of this species have been found at oxygen concentrations lower than $0.37 \, ml \, l^{-1}$ in saline and brackish waters (Serra, 1987), and in saline meromictic lakes near the oxycline (Swift & Hammer, 1979; Miracle & Vicente, 1983).

Salinity and temperature have different effects on oxygen requirements. High temperature increases the metabolic rate and the oxygen consumption in rotifers (e.g., Pourriot & Deluzarches, 1970; Galkovskaya, 1987). Thus, as temperature increases, an opposition between oxygen availability and oxygen requirements might be expected.

In contrast, high salinities could decrease (Epp & Winston 1978) *Brachionus plicatilis* oxygen consumption as has been observed in other osmoconformers. However no difference was observed by Ruttner-Kolisko (1972) between 17 and 36 g l^{-1} salinity.

Our results give experimental proof of *Brachionus plicatilis* tolerance to low oxygen concentrations for long time periods, independent of other enviromental factors. However, the cultures living at low oxygen levels grow at very low rates. A decrease in metabolic rate seems to be the main feature induced by low oxygen availability, as has been observed in other invertebrates (e.g., Zwaan & Wijsman, 1976; Gäde, 1984; Hochachka, 1986). Because oxygen depletion slows down growth and thus increases the survival time of the population, it could function as a regulating mechanism. In one of the studied cases (Experiment 1, population B) in which the density of *Brachionus plicatilis* reached high values, the population died off immediately after the period of uncontrolled growth.

Finally, a poorer use of resources has been found at lower oxygen concentration, as could be derived from the integrater number of rotifer through time, supported at the different conditions.

In conclusion, *Brachionus plicatilis* could be maintained at reduced metabolic rates in microaerophilic conditions. However individual replacement occurred at low rates due to poor reproduction and extremely long generation times. This species also seems to be capable of activating fermentative metabolic pathways (Esparcia, unpublished data), thus reducing the oxygen requirements but also reducing the energetic efficiency.

Acknowledgments

The authors wish to thank Dr. C. A. Dawson for language revision. A. Esparcia is the recipient of a grant from the Ministerio de Educatión y Ciencia.

References

Edmondson, W. T., 1965. Reproductive rate of planktonic rotifers as related to food and temperature in nature. Ecol. Monogr. 35: 61–111.

Epp, R. W. & P. W. Winston, 1978. The effect of salinity and pH on the activity and oxygen consumption of *Brachionus plicatilis*. Comp. Biochem. Physiol. 59A: 9–12.

Esparcia, A. & M. Serra, 1988. Efecto del alimento tratado térmicamente en el crecimiento poblacional de *Brachionus plicatilis* Müller 1786 (Rotifera: Brachonidae). Inv. Pesq. 52: 345–353.

Gäde, G., 1984. Effects of oxygen deprivation during anoxia and muscular work on the energy metabolism of the crayfish *Orconetes limosus*. Comp. Biochem. Physiol. 77A: 495–502.

Galkovskaya, G. A., 1987. Plancktonic rotifers and temperature. Hydrobiologia 147: 307–317.

Hochachka, P. W., 1986. Defense strategies against hypoxia and hypothermia. Science 231: 234–241.

Hofmann, W., 1977. The influence of abiotic environmental factors on population dynamics in planktonic rotifers. Arch. Hydrobiol. Beih. 8: 77–83.

Ito, T., 1960. On the culture of mixohaline rotifer *Brachionus plicatilis* O. F. Müller in the sea water. Rep. Fac. Fish. prefect. Univ. Mie. 3: 708–740.

Larsson, P., 1971. Vertical distribution of planktonic rotifers in a meromictic lake, Blankvatn near Oslo, Norway. Norwerg. J. Zool. 19: 47–75.

Miracle, M. R., 1976. Distribución en el espacio y en el tiempo de las especies del zooplancton del lago de Banyoles. Instituto Nacional para la Conservación de la Naturaleza. (Ministerio de Agricultura). Monografías 5, Madrid 270 pp.

Miracle, R. & E. Vicente, 1983. Vertical distribution and rotifer concentrations in the chemocline of meromictic lakes. Hydrobiologia 104: 259–267.

Pejler, B., 1975. Taxonomical and ecological studies on planktonic Rotatoria from Northern Swedish Lappland. Kungl. Svensk. Vetensk. Handl. 6(5): 1–68.

Pourriot, R., 1965. Recherches sur l'écologie des rotifères. Supplément 21 a 'Vie et milieu'. Masson, Paris. 224 pp.

Pourriot, R. & M. Deluzarches, 1970. Sur la consommation d'oxigene par les rotiferes. Ann. Limnol. 6: 229–248.

Ruttner-Kolisko, A., 1972. Der Einfluss von Temperatur und Salzgehalt des Mediums auf Stoffweschsel-und Vermehrungs-intensität von *Brachionus plicatilis* (Rotatoria). Dt. Zool. Ges. 65: 89–95.

Ruttner-Kolisko, A., 1974. Plankton rotifers. Biology and taxonomy. Biennengewässer 26, 146 pp.

Ruttner-Kolisko, A., 1980. The abundance and distribution of *Filinia terminalis* in various types of lakes as related to temperature, oxygen, and food. Hydrobiologia 73: 169–175.

Schlüter, M., 1980. Mass culture experiments with *Brachionus rubens*. Hydrobiologia 73: 45–50.

Serra, M. & M. R. Miracle, 1983. Biometric analysis of *Brachionus plicatilis* ecotypes from spanish lagoons. Hydrobiologia 104: 279–291.

Serra, M., 1987. Variabilidad morfológica, isoenzimática y demográfica en poblaciones de Brachionus plicatilis. Diferenciación genética y plasticidad fenotípica. Tesis doctoral. Universidad de Valencia.

Swift, M. C. & U. T. Hammer, 1979. Zooplankton population dynamics and *Diaptomus* production in Waldsea lake, a saline 'meromictic' lake in Saskatchewan. J. Fish. Res. Bd Can. 36: 1431–1438.

Walker, K. F., 1981. A synopsis of ecological information on the saline lake rotifer *Brachionus plicatilis* O. F. Müller 1786. Hydrobiologia 81: 159–167.

Zwaan, A. De & T. C. M. Wijsman, 1976. Anaerobic metabolism in bivalvia (Mollusca). Characteristics of anaerobic metabolism. Comp. Biochem. Physiol. 54B: 313–324.

Hydrobiologia **186/187**: 339–346, 1989.
C. Ricci, T. W. Snell and C. E. King (eds), Rotifer Symposium V.
© 1989 *Kluwer Academic Publishers.*

Analysis of protein, carbohydrate and lipid in rotifers

Cástor Guisande & Laura Serrano
Departamento de Ecología, Facultad de Bilogía, Universidad de Sevilla, 41080 Sevilla, Spain

Key words: rotifers, protein, carbohydrate, lipid

Abstract

Protein, carbohydrate and lipid amounts were determined for several rotifer species collected directly from the field. *Brachionus calyciflorus* was the most abundant species; therefore making possible more data for it. An increase in protein content of this species occurred when its concentration in food (μg protein/ml) also increased. *Keratella tropica* showed a similar pattern, but *Asplanchna brightwelli* did not.

Carbohydrate proved to be the main form of storage in rotifers. In *Brachionus calyciflorus* females bearing no egg, 8% of the total biomass was carbohydrate; in females bearing one egg, 15% carbohydrate was found. Lipid does not appear to be used for storage since no increase in the amount of lipid was detected in females bearing eggs or embryos. This suggests that lipid has a structural function. Finally, a relationship between rotifer body volume and protein content at a given food concentration was obtained. The cladoceran *Daphnia magna* follows the same pattern.

Introduction

Rotifers can be viewed as the most significant community within the freshwater zooplankton in regard to the shifts that take place in species dominance. Temperature plays an important role in the separation of species into ecological niches and the time of year may determine their presence or absence. Feeding plays a significant part in this separation as species differ in their ability to ingest different food sizes (Bogdan & Gilbert, 1987). Energy requirements also vary among species (Guisande & Toja, 1988). The importance of rotifer feeding has produced a wealth of published material, such as clearance rate experiments (Starkweather & Gilbert, 1977; Starkweather, 1980; Bogdan & Gilbert, 1982; Walz, 1983) and feeding selectivity studies (Dumont, 1977; Pourriot, 1977; Gilbert & Bogdan, 1984). More recently differences between food threshold levels have been examined (Stemberger & Gilbert, 1985). These factors may help to explain rotifers species distribution and community structure in nature.

The aim of this research was to determine the relationship between existing food concentration and nutritive state of rotifer populations. A total protein analysis was performed on both rotifers and their natural medium. Carbohydrate and lipid contents were also quantified in order to estimate rotifer investment in stored reserves and the relative importance of each component.

Material and methods

Animals were collected with a 35 μm closing net from several ponds throughout the year. Samples

Table 1. Minimum and maximum values of protein content (protein range) and each mean range value for different rotifer species, measured at several food concentrations.

Species	Food concentration		Protein range	Mean range value	N
Body volume (μm^3) Body lenght (μm)	(μg protein ml^{-1}) <10 μm	<0.2 μm	(μg ind^{-1})	(μg ind^{-1})	
Asplanchna brightwelli	27.25	26.94	0.40–0.82	0.61	10
	22.45	19.94	0.28–1.40	0.84	12
$\bar{V} = 1.725 * 10^7$	15.95	-	0.26–2.55	1.40	10
	12.86	9.66	0.85–1.25	1.05	8
$\bar{L} = 491.41$	9.09	-	0.45–1.47	0.96	13
	8.78	8.06	0.11–0.36	0.24	21
Hexarthra fennica	7.25	-	0.09–0.24	0.16	14
$\bar{V} = 7.63 * 10^5$					
$\bar{L} = 156$					
Brachionus urceolaris	25.57	-	0.16–0.51	0.33	25
$\bar{V} = 6.99 * 10^5$					
$\bar{L} = 171.1$					
Brachionus calyciflorus f. *anuraeiformis*	27.25	26.94	0.38–0.45	0.41	7
	22.23	11.5	0.21–0.60	0.40	17
$\bar{V} = 4.68 * 10^5$	18.06	-	0.33–0.55	0.44	18
	15.95	-	0.13–0.54	0.34	26
$\bar{L} = 150.35$	12.90	-	0.04–0.63	0.33	32
	12.86	9.66	0.17–0.58	0.37	20
	8.78	8.06	0.06	0.06	4
	7.28	7.85	0.04–0.18	0.10	15
Keratella tropica	27.25	26.94	0.26–0.44	0.35	10
	22.45	19.94	0.29–0.49	0.39	14
$\bar{V} = 2.57 * 10^5$	22.23	11.5	0.43–0.46	0.44	8
	18.06	5.3	0.20	0.20	3
$\bar{L} = 103.31$	15.95	-	0.10–0.27	0.18	18
	14.56	10.94	0.12–0.41	0.26	14
Filinia terminalis	22.23	11.5	0.08–0.17	0.12	15
	18.06	5.3	0.02–0.18	0.10	18
$\bar{V} = 1.72 * 10^5$	15.95	-	0.05–0.21	0.13	8
$\bar{L} = 102.43$					
Trichocerca sp.	27.25	26.94	0.17–0.18	0.17	9
	22.45	19.94	0.15–0.35	0.25	8
$\bar{V} = 9.47 * 10^4$					
$\bar{L} = 105.86$					

were also collected from Doñana National Park (D.N.P.). After examination under a binocular microscope, the animals were transferred to distilled water so that errors from protein, carbohydrate and lipid in the natural water could be reduced. Under a Nikon inverted microscope, the animals were taken from the distilled water with a micropipette in order to get the same volume of 10 μl for each individual in protein and carbohydrate analysis and 5 μl in lipid analysis. Measurements of rotifer body lengths were taken from living, field-collected individuals. The sensitivity of the analysis allowed the use of few individuals for each sample, making the use of laboratory cultures unnecessary; the samples were taken directly from the field.

Experimental animals were grouped into different categories depending on their reproductive state: C0 (females bearing no egg), C1 (females bearing one egg), C2 (females bearing two eggs), etc. Neither males nor females with mictic eggs were collected. *Asplanchna brightwelli* fell into only three categories: C0 (females with no embryo); C1 (females with embryo) and C2 (females with resting egg).

In the protein analysis 2–5 individuals were

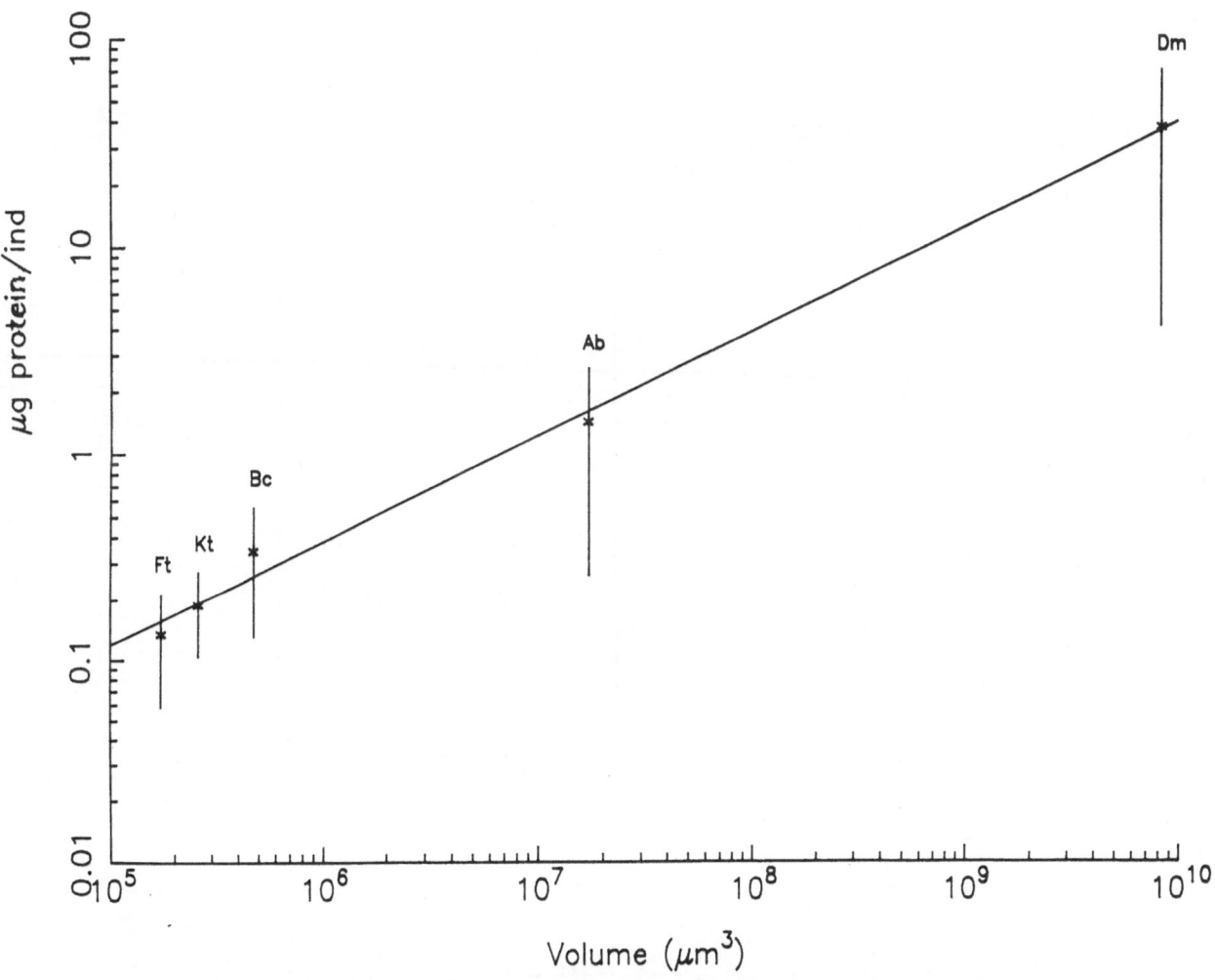

Fig. 1. The relationship of volume (*V*) and the mean value of the protein range (*P*) for various zooplankton species at a given food concentration. Vertical lines designate the protein range between the minimum and maximum protein content values for each species. Solid line designates the predicted log-log least squares linear regression, which fits the equation: $P = 3.6*10^{-4} *V^{0.5036}$. Ab = *Asplanchna brightwelli*; Bc = *Brachionus calyciflorus*; Dm = *Daphnia magna*; Ft = *Filinia terminalis*; Kt = *Keratella tropica*.

used; 2–10 were used for carbohydrate and 15–20 for lipid analysis. Lowry's *et al.* method (1951), modified by Markwell *et al.* (1976) was used for the protein analysis, reaching a final volume of 1.075 ml. BSA up to 18 μg was used for the linear standard. For the carbohydrate experiments Dubois *et al.* method (1956) was followed, reaching a final volume of 2.410 ml; glucose was used for the linear standard. Böeringher's standard was used for lipid quantifications, using a final volume of 2.5 ml and Böeringher's solution for the linear standard.

In the case of protein analysis, SDS was employed to degrade the rotifers. This detergent did not affect the mastax and this portion of

undegraded protein was disregarded determining total protein content.

Protein and carbohydrate analysis of the medium were also performed in order to estimate the amount of food available for rotifer populations. The water was passed through a 10 μm net and a 0.2 μm filter in some cases to determine the portion of dissolved protein and carbohydrate in the medium.

It was found to be more convenient to calculate the rotifer body volume using Ruttner-Kolisko's method (1977), since body volume as measurement of biomass is more reliable than body size.

Temperature was also measured in every pond. Values varied between 15 and 23 °C.

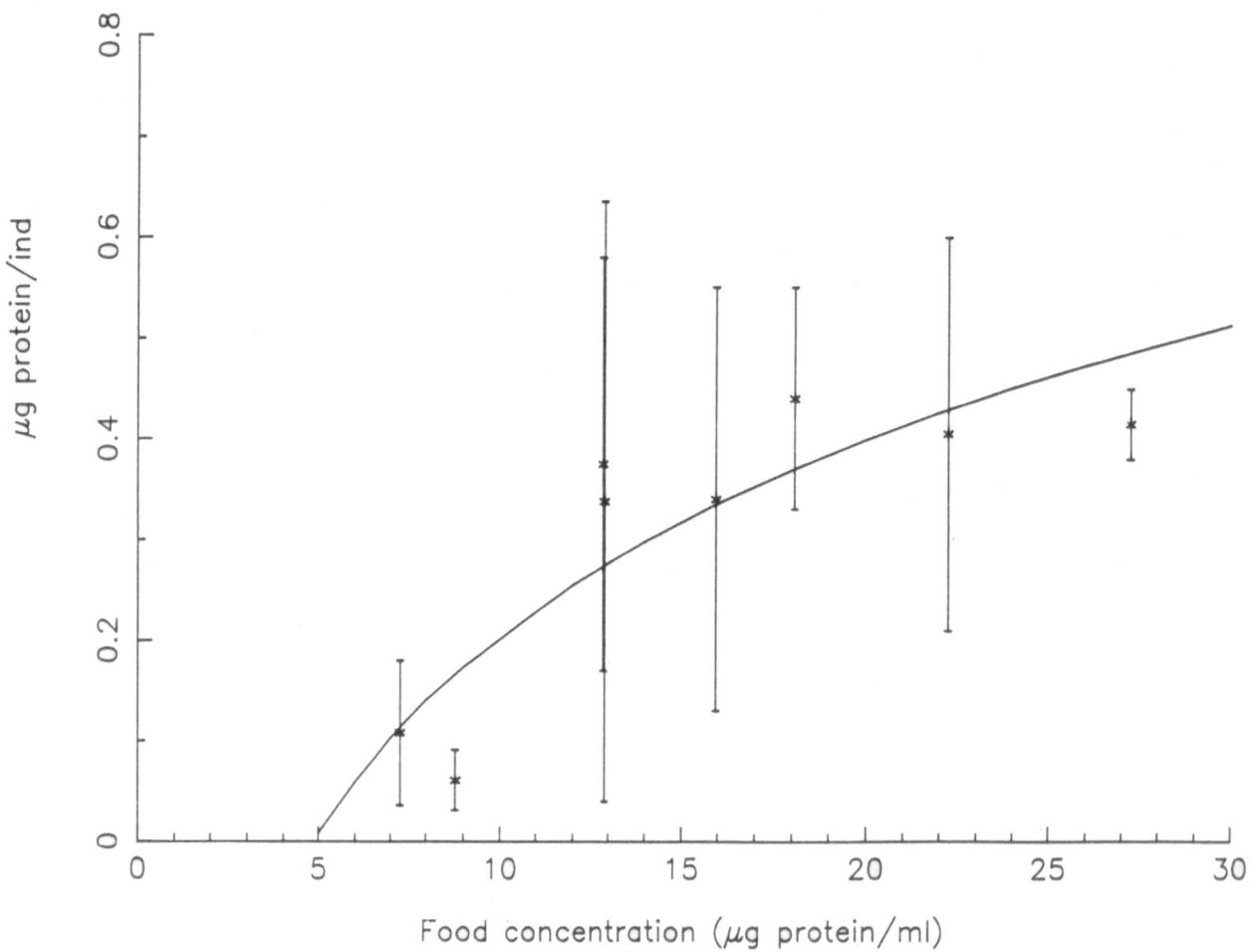

Fig. 2. The relationship of food concentration (*C*) and the mean value of the protein range (*P*) for *Brachionus calyciflorus*. Vertical lines designate the protein content range between the minimum and maximum protein content values at each food concentration. Curve fits the equation: $P = -0.446 + 0.282 * Ln\ (C)$.

Results

Table 1 lists the protein values obtained for each species. Some other species were collected in the samples, such as *B. angularis*, *B. budapestinensis*, *Polyarthra major* and *Synchaeta pectinata*, but the populations of these species were not sufficiently large for the experiment.

At a food concentration of 15.95 μg protein ml^{-1} reached in one of the ponds (Table 1), four species appeared in sufficient numbers for the analysis. A relationship between body volume (V) and mean protein value (P) was found (Fig. 1). But even more significant was that the protein amount predicted at this food concentration for

Daphnia magna (Guisande, unpublished) matched the equation estimated for rotifers: $P = 3.6*10^{-4} *V^{0.5036}$ ($r^2 = 0.99$; $p < 0.001$). When body size, excluding spines, was considered, instead of body volume, the equation was: $P = 4.33*10^{-5}*L^{1.758}$ ($r^2 = 0.99$; $p < 0.001$), which is very similar to the one given by Stemberger & Gilbert (1987) between body mass (M) and body length (L) for several rotifer species at optimum food level: $M = 3.1*10^{-5}*L^{1.574}$.

B. calyciflorus was the most widely-represented species in the ponds tested. Figure 2 shows the variation of protein content in *B. calyciflorus* for several food concentrations. Individual mean protein content (P) as a function of protein con-

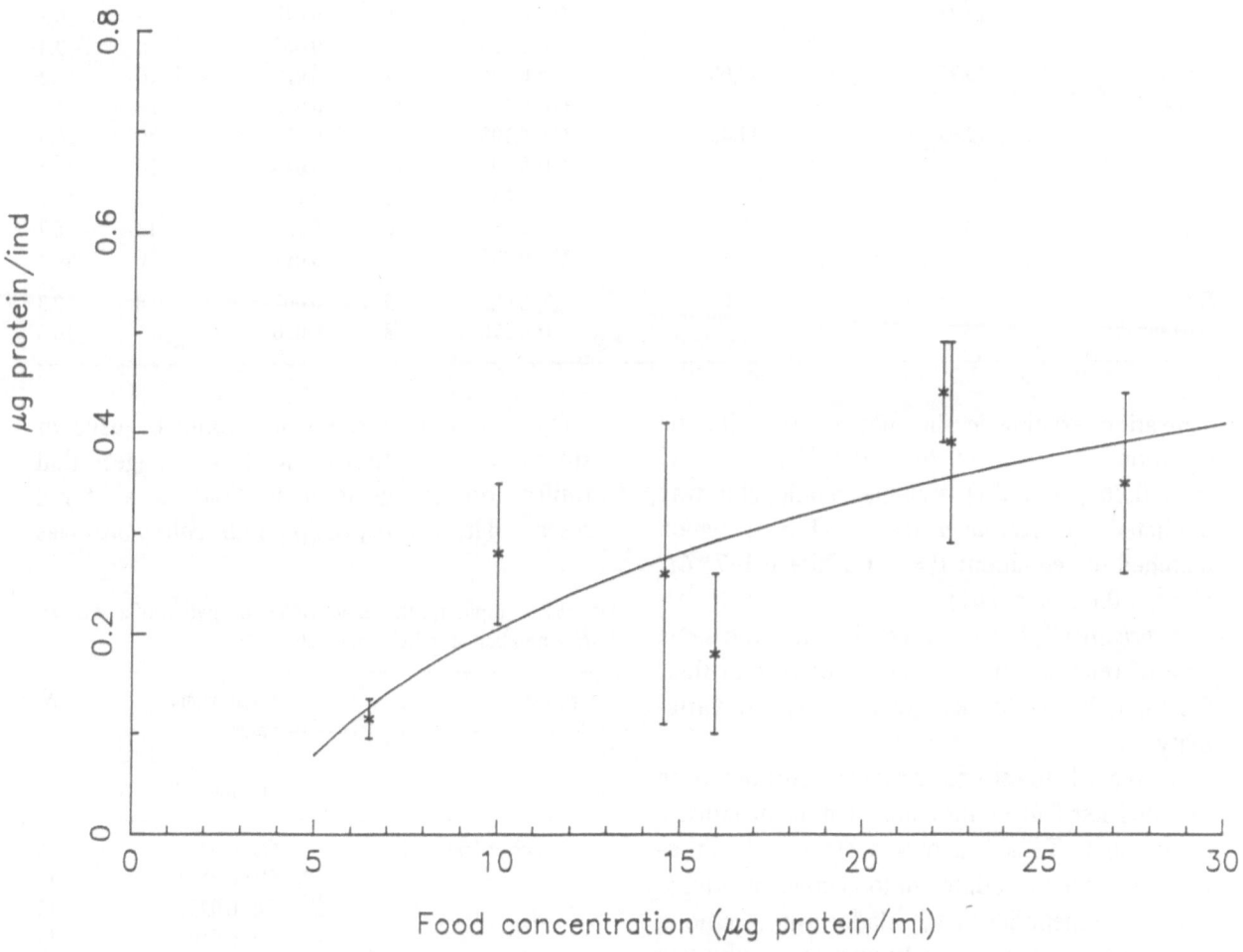

Fig. 3. The relationship of food concentration (C) and the mean value of the protein range (P) for *Keratella tropica*. Vertical lines designate the protein content range between the minimum and maximum protein content values at each food concentration. Curve fits the equation: $P = -0.225 + 0.187*Ln (C)$.

Table 2. Relation between protein and carbohydrate (ch) content at several food concentrations for different rotifers. This food concentration mentioned here, is the one measured in the water filtered through 10 μm. C0: rotifers with no egg; C1: rotifers with one egg; C2: rotifers with two eggs; C3: rotifers with three eggs. *N* = number of individuals tested.

Species	Food concentration		Mean protein content	*N*	Mean CH content	*N*	CH
	(μg protein ml^{-1})	(μg ch ml^{-1})					(%)
			(μg ind^{-1})		(μg ind^{-1})		
Asplanchna	15.95	13.00	C0: 1.173	8	0.059	5	4.7
brightwelli			C1: 0.768	2	0.116	2	13.1
	12.86	7.33	C0: 0.850	4	0.056	5	6.2
			C1: 1.190	4	0.167	9	12.3
Hexarthra	7.25	8.46	C0: 0.166	10	0.029	22	15.1
fennica							
Brachionus	26.38	23.28	C0: -	-	0.027	17	-
calyciflorus			C1: -	-	0.052	16	-
f. anuraeiformis			C2: -	-	0.123	3	-
	22.23	10.01	C0: 0.260	8	0.018	7	6.5
			C3: 0.600	4	0.062	8	9.4
	15.95	13.00	C0: 0.275	12	0.028	16	9.2
			C1: 0.380	14	0.066	24	14.8
	12.90	11.46	C0: 0.108	20	0.039	30	26.5
			C1: 0.191	11	0.045	24	19.1
			C2: 0.631	1	0.113	2	15.1
	12.86	7.33	C0: 0.230	8	0.022	13	8.7
			C1: 0.405	8	0.047	6	10.4
Filinia	22.23	10.01	C0: 0.086	9	0.002	8	2.3
terminalis			C1: 0.170	8	0.020	6	10.5

centration existing in the medium (*C*) fits the equation: $P = -0.446 + 0.282*LN$ C ($r^2 = 0.76$, $p < 0.001$). A similar relationship was predicted for *Keratella tropica* (Fig. 3) which matched the equation: $P = -0.225 + 0.187*LN$ C ($r^2 = 0.66$, $p < 0.01$).

A. brightwelli, however, did not show the same type of relationship. This could be due to their feeding habits or to their greater body-size variability.

In regard to stored reserves, carbohydrate revealed itself to be the most significant storage component. Table 2 shows protein and carbohydrate contents at different food concentrations.

Only copepodes and cladocerans, among zooplankton species, are known to accumulate lipid (Goulden & Henry, 1984). Wurdak *et al.* (1978) found lipid droplets in *B. calyciflorus* and *Asplanchna sieboldi* eggs and suggested the possi-

bility that these components could be used as stored reserves. In this work we suggest that rotifers do not accumulate lipid as a stored reserve. The quantity of lipid in *B. calyciflorus* was

Table 3. Lipid mean range value for two rotifer species. *N* = number of individuals tested.

Species	Mean range value	*N*
	(μg ind^{-1})	
Asplanchna brightwelli	C0: 0.025	15
	C1: 0.019	15
	C0: 0.023	15
	C2: 0.010	15
Brachionus calyciflorus	C0: 0.009	25
f. anuraeiformis	C1: 0.003	17
	C2: 0.008	15

the same regardless of the presence or absence of eggs; this was also the case for *A. brightwelli* (Table 3). Lipid represents 1.06% of the total biomass of *B. calyciflorus* and approximately 1.5% of *A. brightwelli*.

Discussion

The results of protein analysis for *B. calyciflorus* (Fig. 2) suggest that upon increasing food concentration, the individual protein content increases as well, though this increase becomes less pronounced when the food level is increased in large amounts. *Keratella tropica* shows a similar pattern (Fig. 3). Although there is not enough available data for the rest of species, a similar relationship could be expected.

These results indicate that the food concentration required to reach the optimum protein content is lower for smaller species. This is seen by observing the behaviour of *Filinia terminalis* (Table 1). Its protein range does change when food concentration varied from 15.95 to 22.23 μg ml^{-1}. However, *K. tropica* varied from 0.293 to 0.355 and *B. calyciflorus* from 0.335 to 0.428 μg protein ind^{-1} at those food concentrations. Therefore, larger species required more food to reach their optimum nutritive state and their protein content changed over a greater food range than smaller species. This was also found by Stemberger & Gilbert (1985; 1987), who show that the food level required to increase population was lower for smaller species.

Carbohydrate is the main storing substance in rotifers. As can be seen in Table 2, the average *B. calyciflorus* carbohydrate content for females bearing no egg is 0.027 μg ind^{-1} (SE = 0.003); for females bearing one egg is 0.052 μg ind^{-1} (SE = 0.0004) and 0.118 μg ind^{-1} (SE = 0.025) for females bearing two eggs. The average *Asplanchna brightwelli* carbohydrate content is 0.057 μg ind^{-1} (SE = 0.001) for females with no embryo and 0.141 μg ind^{-1} (SE = 0.025) for females with embryo. Also *A. brightwelli* showed a greater amount of carbohydrate than *B. calyciflorus*, the relationship of protein/carbo-

hydrate content is similar in both species (Table 2). There is not enough information to assess the relative expenditure of stored reserves for either species.

Carbohydrate analysis of *B. calyciflorus* was only performed at food concentrations above 12 μg protein ml^{-1}. At low food levels, individuals were scarce and few animals had eggs. No data are available below this food concentration, thus whether or not a further decrease in stored reserves exists at lower food levels has not been determined. It would be extremely interesting to know if there is a decrease in the stored reserves invested in reproduction at food limiting conditions or if, on the contrary, the amount of carbohydrate allocated stays constant and a reduction in the number of eggs then occurs.

In regard to the relationship obtained between body volume and the amount of protein existing in the environment (Fig. 1), it would be interesting to determine if other filtering species, of different sizes, also fit this equation. Larger species show a greater change under varying food concentrations than smaller species. The existence of this kind of relationship in filtering species could indicate that feeding strategy variability for different species does not reflect protein content variation. A variability of the amount stored reserves invested in reproduction and of the consumption of these compounds by embryos may also occur (Goulden & Henry, 1987).

To summarize, our results suggest that protein content increases with increasing food level. On reaching the optimum food level, further protein increase is less pronounced. When food concentration increases greatly, a decrease in the protein content can be expected, since above a certain food concentration the animals spend more energy and time rejecting excess food than they obtain through ingestion. Carbohydrate is the main stored reserve in the rotifers tested, as no significant lipid accumulation was detected.

In future research, it would be interesting to estimate the minimum food concentration required for each species, as well as determining whether stored reserves decrease or disappear at low food levels. Studies on the determination of the amount of protein and carbohydrate appear to be a valuable approach to the study of rotifer feeding processes.

Acknowledgement

This research was supported by CAICYT Project no. 0940/87.

References

Bogdan, K. G. & J. J. Gilbert, 1982. Seasonal patterns of feeding by natural populations of *Keratella*, *Polyarthra*, and *Bosmina*: Clearance rates selectivities and contributions to community grazing. Limnol. Oceanogr. 27: 918–934.

Bogdan, K. G. & J. J. Gilbert, 1987. Quantitative comparison of food niches in some freshwater zooplankton. Oecologia. (Berlin), 72: 331–340.

Dubois, M., K. A. Gilles, J. K. Hamilton & F. Smith, 1965. Colorimetric method for determination of sugars and related substances. Anal. Chem. 28: 350–356.

Dumont, H. J., 1977. Biotic factors in the population dynamics of rotifers. Arch. Hydrobiol. Beih. Ergeb. Limnol. 8: 98–122.

Gilbert, J. J. & K. G. Bogdan, 1984. Rotifer grazing: in situ studies in selectivity and rates. In D. G. Meyers & J. R. Strickler, editors. Trophic interactions within aquatic ecosystems. AAAS Selected Symposium series, 85: 97–133.

Goulden, C. E. & L. Henry, 1984. In D. G. Meyers & J. R. Strickler, editors. Lipid energy reserves and their role in cladocera. AAAS Selected symposia series 85: 167–185.

Goulden, C. E., L. Henry & D. Berrigan, 1987. Egg size, postembryonic yolk, and survival ability. Oecologia, 72: 28–31.

Guisande, C. & J. Toja, 1988. The dynamics of various species of the genus *Brachionus* (Rotatoria) in the Guadalquivir River. Arch. Hydrobiol. 112: 579–595.

Lowry, O. H., N. J. Rosenbraugh, A. L. Farr & R. J. Randall, 1951. Protein measurements with the Folin phenol reagent. J. Biol. Chem. 193: 256–275.

Markwell, M. A. K., S. M. Haas, L. L. Bieber & N. E. Tolbert, 1978. A modification of the Lowry procedure to simplify protein determination in membrane and lipoprotein samples. Analyt. Biochem. 87: 206–210.

Pourriot, R. 1977. Food and feeding habits of Rotifera. Arch. Hydrobiol. Beih. Ergeb. limnol. 8: 98–122.

Ruttner-Kolisko, A., 1977. Suggestions for biomass calculation of plankton rotifers. Arch. Hydrobiol. Beih. 8: 71–76.

Starkweather, P. L. & J. J. Gilbert, 1977. Feeding in the rotifer *Brachionus calyciflorus* II. Effect of food density on feeding rates using *Euglena gracilis* and *Rhodotorula glutinis*. Oecologia (Berlin) 28: 133–139.

Starkweather, P. L., 1980. Aspects of the feeding behaviour and trophic ecology of suspension feeding rotifers. Hydrobiologia 73: 63–72.

Stemberger, R. S. & J. J. Gilbert, 1985. Body size, food concentration, and population growth in planktonic rotifers. Ecology 66(4): 1151–1159.

Stemberger, R. S. & J. J. Gilbert, 1987. Rotifer threshold food concentration and the size-efficiency hypothesis. Ecology 68(1): 181–187.

Walz, N., 1983. Continuous culture of the pelagic rotifers *Keratella cochlearis* and *Brachionus angularis*. Arch. Hydrobiol. 98: 70–92.

Wurdak, E. S., J. J. Gilbert & R. Jagels, 1978. Fine structure of resting eggs of the rotifers *Brachionus calyciflorus* and *Asplanchna sieboldi*. Trans. Amer. Micros. Soc. 97(1): 49–72.

Hydrobiologia **186/187**: 347–354, 1989.
C. Ricci, T. W. Snell and C. E. King (eds), Rotifer Symposium V.
© 1989 *Kluwer Academic Publishers.*

A laboratory study of phosphorus and nitrogen excretion of *Euchlanis dilatata lucksiana*

R. D. Gulati[1], J. Ejsmont-Karabin[2], J. Rooth[1] & K. Siewertsen[1]
[1] *Limnological Institute, Vijverhof Laboratory, Rijksstraatweg 6, 3631 AC Nieuwersluis, The Netherlands;*
[2] *Polish Academy of Sciences, Institute of Ecology, Hydrobiological Station, ul. Lésna 13, 11-730 Mikolajki, Poland*

Key words: rotifers, *Euchlanis*, phosphorus, nitrogen, excretion, body size

Abstract

Phosphorus (PO_4-P) and nitrogen (NH_4-N) excretion rates of *Euchlanis dilatata lucksiana*, a rotifer, isolated from Lake Loosdrecht (The Netherlands) and cultured in the lake water at 18–19 °C, were measured in the laboratory.

In a series of experiments, the effects of experiment duration on the P and N excretion rates were examined. The rates measured in the first half-hour were about 2 times higher for P and 2–4 times for N than the rates in the subsequent three successive hours which were quite comparable.

Eight experiments were carried out in triplicate, 4 each for P and N excretion measurements, using animals of two size ranges: 60–125 μm and > 125 μm. The specific excretion rates varied from 0.06 to 0.18 μg P.mg^{-1} DW.h^{-1} and 0.21 to 0.76 μg N.mg^{-1} DW.h^{-1}. Generally an inverse relationship was observed between the specific excretion rates and the mean individual weight. The excretion rates of *Euchlanis* measured by us are lower than those reported for several other rotifer species, most of which are much smaller than *Euchlanis*.

Extrapolating the excretion rates of *Euchlanis* to the other rotifer species in Lake Loosdrecht, and accounting for their density, size and temperature, rotifer excretion appears to be a significant, potential nutrient (N,P) source for phytoplankton growth in the lake. The excretion rates for the rotifers appear to be about two thirds of the total zooplankton excretion, even though the computed rotifer mean biomass is about one-third of the total zooplankton biomass.

Introduction

In a recent study on phosphorus and nitrogen excretion rates of the zooplankton community from the Loosdrecht lakes, it was demonstrated that the P excreted by zooplankton is substantial in comparison with P-demand of phytoplankton (den Oude & Gulati, 1988). This study was, however, restricted to the large-sized (> 150 μm) zooplankton, mainly the crustaceans. Studies on the composition, densities and biomass of zooplankton of the Loosdrecht lakes (Gulati, in press) and calculations of P and N regeneration rates of zooplankton based on Ejsmont-Karabin (1983) show that the rotifers in summer, when they are abundant, may contribute from 45 to 95% of the P regeneration by total zooplankton. These release rates, if included in the P-flow scheme of Lake Loosdrecht (van Liere *et al.*, in press), will largely meet the primary production requirements of phytoplankton.

The mean annual densities of rotifers in the Loosdrecht lakes are between 3 and 8 times higher than the crustacean densities. Because the mean individual weight of the rotifers (0.05 μg C ind^{-1}) in the lakes is an order of magnitude lower than that of the crustaceans (0.44 μg C ind^{-1}) (Gulati, in press), the mean specific excretion rates of rotifers will be about 2.5 times higher than the rates of crustaceans, based on the studies of Ejsmont-Karabin (1984).

Considering the aforementioned arguments, the role of P and N regeneration by rotifers in the Loosdrecht lakes needs to be investigated, and its quantitative importance in the lake ecosystem assessed. This paper deals with N and P excretion rates of *Euchlanis dilatata lucksiana* and is the first attempt to examine rotifer excretion in these lakes. The choice of *Euchlanis* sp. for measuring excretion rates, like in the feeding studies of this euchlanid (Gulati *et al.*, 1987), is based on the ease of culturing and handling this species, mainly because of its relatively large size. The animals used for the excretion studies were obtained from the same cultures as for the studies on feeding and assimilation rate (Gulati *et al.*, 1987). Besides the excretion rates experiments, the effect of experimental duration on the excretion rates was measured. The results obtained are discussed in relation to: 1) the N and P contents of the body and food of *Euchlanis*; and 2) their significance in the nutrient regeneration by the zooplankton community as a whole.

Material and methods

Isolation, culture and biomass of Euchlanis

Euchlanis was isolated from Lake Loosdrecht and identified as *E. dilatata lucksiana* (Gulati *et al.*, 1987). It was cultured in the laboratory at 18–19 °C in lake water filtered with a 33 μm mesh sieve; Therefore, the *in situ* food (< 33 μm), dominated by filamentous cyanobacteria, was the exclusive food source in the cultures. The animals were removed from the cultures using 60 μm and 125 μm sieves and categorized as small and large, respectively. On termination of each experiment, the lengths of the animals used in the replicates were measured and the mean dry weights computed from the length-weight regression formula given by Dumont *et al.* (1975). For deriving the carbon biomass, the computed dry weight was multiplied by a factor of 0.427 (Gulati *et al.*, 1987).

Excretion experiments

The excretion rates of *Euchlanis* were measured in the dark at 18–19 °C, the culture temperature. About 200 animals were transferred to a Winkler bottle containing 50 ml fresh, filtered (GFF Whatman filter) lake water. The experiments were carried out in triplicate; the controls contained only the filtered lake water.

In the 'time-course experiments' the effect of incubation time on the N and P excretion rates was studied in separate experiments using two size classes of *Euchlanis*. The excretion rates were measured after 30 min of incubation in the filtered lake water and in the subsequent 1st, 2nd and 3rd h, i.e. after each hour removing the animals and re-transferring them into the freshly filtered lake water in the Winkler bottles.

In the second set of experiments the excretion rates were measured. The animals were incubated for 30 min in the glass-fibre filtered lake water to allow them to clear their guts of the unassimilated food, and thereafter transferred to fresh filtered lake water again and excretion rates measured after 2 h of incubation. Four such experiments were carried out each for N as well as P excretion measurements separately, using two size categories. At the end of each experiment the animals were separated by filtering, killed and fixed in 4% formalin for counting and length measurements.

N and P analyses

The water from which the animals had been removed was used for N and P analyses.

Ammonium nitrogen (NH_4-N) was determined by the bis-pyrazolone method (Procházková, 1964) in which the coloured complex is extracted in chloroform and its extinction measured at 450 nm. Phosphorus (PO_4-P: molybdenum-blue reactive phosphate) was measured according to Stephens (1963), using isobutanol as the extracting solvent for the coloured complex, and the extinction values at 690 nm as measure of the P concentration. Both these methods have been used successfully in recent other excretion studies of zooplankton (e.g. Blažka *et al.*, 1982; Ejsmont-Karabin, 1984).

For calculating the nutrient recycling rates, body nitrogen and phosphorus contents of both *Euchlanis* and its food in the cultures were measured. Micro-Dumas method was used for the analysis of N using an automated C,H,N analyzer (Perkin Elmer, No. 240); for P, a known amount of the dry matter was combusted to ash, hydrolyzed and the concentration measured according to Murphy & Riley (1962).

Results

Time course experiments

The observed excretion rates of both N and P, for large as well as small *Euchlanis*, are markedly higher in the first 30 min than in the subsequent 3 h (Fig. 1; Table 1). The excretion rates in the first half-hour can be apparently higher if the egestion products contribute significantly to the excretion losses. The egestion, however, appears to contribute negligibly to the excretion because

Table 1. A comparison of the specific NH_4-N and PO_4-P excretion rates (μg N or P·mg^{-1} DW·h^{-1}) of *Euchlanis dilatata lucksiana* in the first 30 min of incubation (A) with the means of the three successive one-hour incubations (B); coefficients of variation (%) are given in the parentheses.

Animal size (μg DW·ind^{-1})	A	B	A/B
	N excretion rates		
1.07	0.61	0.26	2.35
(12.1)	(14.2)	(9.7)	
0.51	3.31	0.77	4.30
(33.7)	(24.7)	(5.3)	
	P excretion rates		
1.01	0.17	0.078	2.05
(17.7)	(34.3)	(22.3)	
0.79	0.72	0.18	4.00
(25.7)	(20.8)	(16.4)	

assimilation efficiencies of *Euchlanis* are rather high, invariably 100% (Gulati *et al.*, 1987). Small animals excreted N and P 5 times and 4 times faster, respectively, than the large animals. Since there was no trend in the fluctuations in the 3 successive one-hour measurements, it was implied that the excretion rates stabilized after transfer to the filtered lake water within the first 30 min.

Excretion rates

The mean rates of N excretion in the first three of the four experiments were quite comparable, with daily recycling via excretion of 5–6% body N (Table 2). The specific excretion rates for the smallest animals (0.514 μg DW.ind^{-1}: Exp. 4)

Table 2. Specific NH_4-N excretion rates (μg N·mg^{-1} DW·h^{-1}) of *Euchlanis dilatata lucksiana* of different sizes; coefficients of variations (%) of the replicates are given in parentheses; also daily recycling of the body N is given.

Exp. No.	Size of the animal (μg DW·ind^{-1})	Excretion rate	Recycling (d^{-1})
1	1.09 (8.6)	0.24 (5.1)	0.06
2	1.06 (12.1)	0.25 (0.6)	0.06
3	0.62 (26.7)	0.21 (15.7)	0.05
4	0.51 (33.7)	0.76 (5.3)	0.18

350

were, however, about 3 times the rates of largest animals. Interestingly, these N excretion rates are almost identical to the rates measured in the time-course experiments excluding the first half-hour (Fig. 1 and Table 1).

The P recycling rates of about 14% d^{-1} for the animals weighing about 1 μg DW.ind^{-1} were on the average 2.4 times faster than of N recycling (Tables 2 and 3: Exp. 1 and 2). Such a comparison of the N and P recycling is, however, not

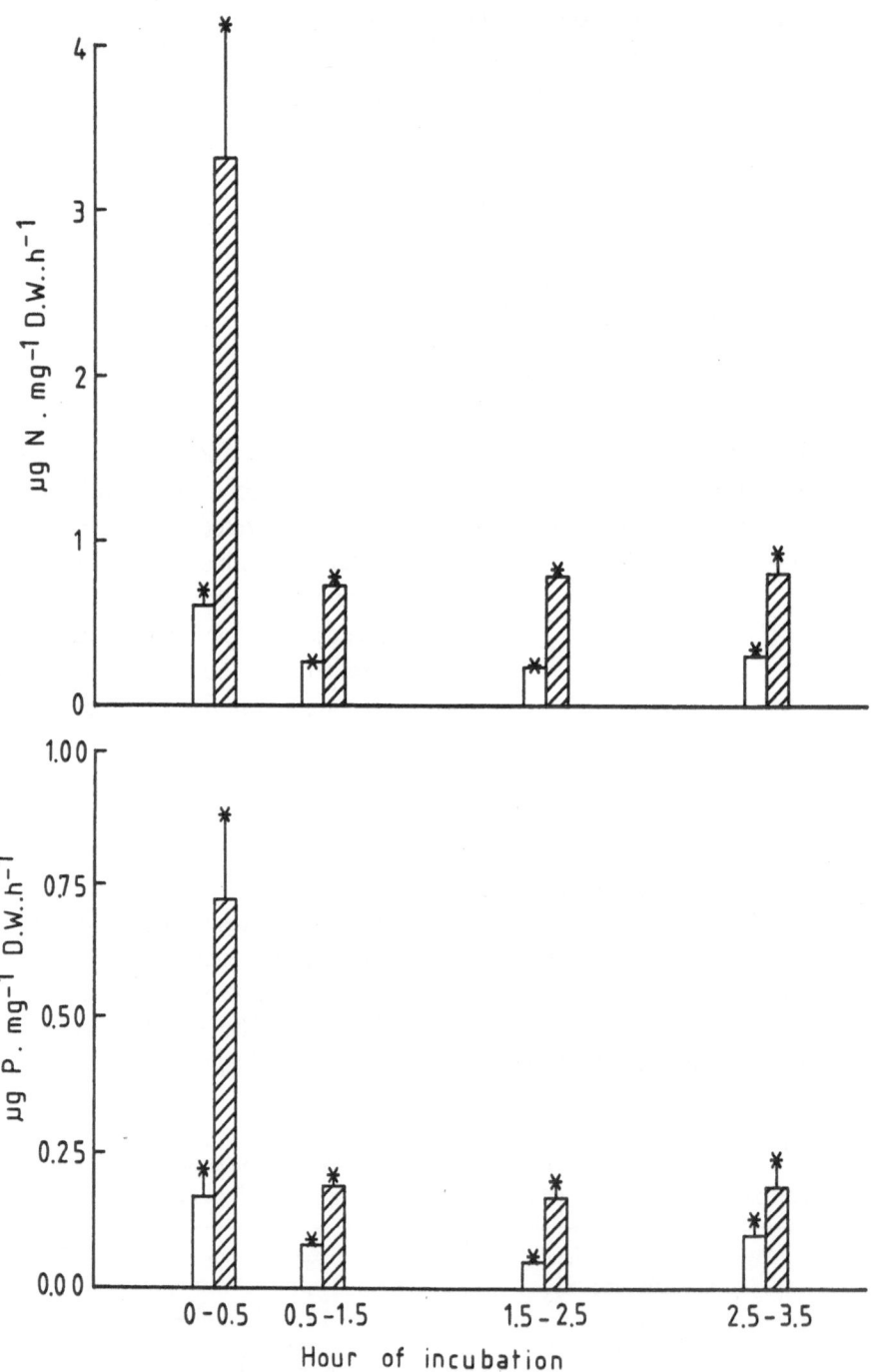

Fig. 1. Specific NH$_4$-N and PO$_4$-P excretion rates of *Euchlanis dilatata lucksiana* in the course of 3.5 h; incubation time of the first experiment was 30 min and of the subsequent three experiments 1 h each. The two bars for each experiment represent: blank bar, large animals; and hatched bars, small animals; the asterisk represents the + S.D. For more details see Table 1.

Table 3. Specific PO_4-P excretion rates ($\mu g \ P \cdot mg^{-1} \ DW \cdot h^{-1}$) of *Euchlanis dilatata lucksiana* of different sizes; coefficients of variation (%) of the replicates are given in parentheses; also daily recycling of the body P is given.

Exp. No.	Size of the animal ($\mu g \ DW \cdot ind^{-1}$)	Excretion rate	Recycling (d^{-1})
1	1.01 (17.7)	0.068 (6.8)	0.14
2	0.95 (7.3)	0.063 (24.8)	0.14
3	0.81 (25.7)	0.182 (13.8)	0.39
4	0.65 (30.4)	0.170 (17.3)	0.36

possible for the smaller animals: first, because of the greater weight differences between the animals used for N and P in the Exp. 3 and 4 (Tables 2 and 3); and secondly, also due to the high rates of P excretion for the smaller animals, both in Exp. 3 and 4 (Table 3) but of N excretion only in Exp. 4 (Table 2).

The coefficients of variation in the replicates of N and P excretion rate measurements, with one exception (Exp. 2, Table 3), were less than the variations in the size replicates of *Euchlanis*. The replicates of N excretion rates varied less than the replicates of P excretion rates in 3 out of the 4 experiments.

Discussion

This study restricts the excretion measurements to PO_4-P and NH_4-N. The measured rates of P excretion will include, besides the orthophosphates, the phosphates formed by hydrolysis of dissolved organic phosphorus compounds. These latter comprise about 20% of the total P excretion (Peters & Lean, 1973; Ejsmont-Karabin, 1984). For N, ammonia is the predominant break-down product of aquatic invertebrates including the zooplankton, which are ammoniotelic.

Factors affecting the excretion rates

Several factors may affect the rate of zooplankton excretion as well as comparability of the different studies. Important among these are: a) the water

temperature; b) animal size; c) feeding history; d) concentration or extent of crowding; and e) length of the incubation period.

The temperature which has generally a significant and direct effect on rotifer excretion, with a Q_{10} of about 2.6 (Ejsmont-Karabin, 1984), was kept at 18.5 $\pm$ 0.5 °C both in the culture medium and during the incubations for excretion measurements. The excretion rates tended to increase with the decreasing weight of *Euchlanis*. The significance of the differences, however, cannot be tested because of the limited number of size classes studied. This narrow size range within a clone of a rotifer species in culture may be because of the short life span and constancy of cell number in rotifers after the embryonic development is completed (Ruttner-Kolisko, 1972). Moreover, our computed dry weights of *Euchlanis* based on the L-W regressions of Dumont (1975) are about twice the weights based on the volume-to-weight conversion (Ruttner-Kolisko, 1977), assuming dry weight to be 10% of the wet weight (see e.g. Bottrell *et al.*, 1976). Consequently, our specific excretion rates calculated using the regression-based dry weights are half as high as the rates based on the volume. Such discrepancies will affect the comparability of literature data adversely, like the errors relating to methodological and behavioural differences.

As regards feeding history and starving of the animals during the experiments, the stability of excretion rates after the first 30 min shows that starving most likely did not alter the physiology such that the rates were reduced (see also Ikeda, 1971; Mayzaud, 1976). Also, release of N and P through bacterial activity and increased mineralization, which may mask the decrease in excretion rates (Le Borgne, 1979), does not appear to have played a role because we used filtered lake water and corrected our results for changes in the controls.

In our experiments we used concentrations (*ca.* 4000 ind.l^{-1}) of *Euchlanis* that are quite similar to the mean density (3770 ind.l^{-1}) of total rotifers encountered in the Loosdrecht lakes (Gulati, in press). Moreover, they are much lower than the densities (40000 ind.l^{-1}) in our *Euchlanis* stock cultures (Ejsmont-Karabin *et al.*, 1989). Therefore, crowding-related factors do not appear to be relevant in our study.

The exposure-time related changes in the specific excretion rates of *Euchlanis* differed in their magnitude for the small and large animals. Both, the N and P excretion rates are higher during the first 30 min than in the subsequent 3 h, roughly 4 times for the small animals and 2 times for the large animals (Table 1). In the first 30 min the small animals excrete both N and P 4–5 times more rapidly than the large animals. The size-dependent difference in the animal behaviour is too great to be explained on the basis of metabolism alone.

For the stabilization of the excretion rates after the first 30 min there are two possible explanations. First, the stress effects (Le Borgne, 1979) on transferring the animals from the cultures to fresh, filtered lake water without food may stimulate the animal to filter maximally in a medium devoid of food. However, this increased activity is short-lasting, tending to stabilize even before the 3 successive hourly transfers into the fresh food-less medium each time. Second, the animals apparently get quickly adapted to the short-term starvation, with no more decrease in metabolic activity than that observed in the first hour after the first 30 min in the time course experiments. In short, like in several other studies on zooplankton excretion, but especially those on the rotifers (e.g., Ejsmont-Karabin, 1983, 1984; Ejsmont-Karabin *et al.*, 1983), no symptoms of starvation, or other adverse effects, were observed, which may lead to decreased excretion in the initial few hours of incubation. Lastly, the rates in the 0.5–3.5 h of

incubation were not only comparable with one another (Fig. 1), but also with those measured subsequently on the basis of 2-h incubations (Tables 2 and 3).

Ejsmont-Karabin (1984) has clearly demonstrated that the mean specific excretion rates of rotifers increase markedly with decreasing weight, e.g. P excretion rates increase from 0.19 to 35.8 μg P.mg^{-1} DW.h^{-1} corresponding to a decrease in individual weights from 0.776 to 0.013 μg DW. The range of variations is even much larger if all the extreme values are included. P excretion rates of *Euchlanis* measured by us (Table 3) are very comparable with the minimal rates (range: 0.06–0.42 μg P.mg^{-1} DW.h^{-1}; and mean: 0.19 mg P.mg^{-1} DW.h^{-1}) reported for a rotifer, namely *Asplanchna priodonta* (Ejsmont-Karabin 1984: Table 1). *Asplanchna* has a range of weights, 0.5–1.07 μg DW.ind^{-1} (Ejsmont-Karabin, 1984), that are quite similar to the weights of our *Euchlanis* (0.5–1.0 μg DW.ind^{-1}). Interestingly, these *Euchlanis* weights converted to carbon do not differ much from the mean weight (0.44 μg C.ind^{-1}) of crustacean community of the Loosdrecht lakes, in which small cladoceran species of *Bosmina* and *Chydorus* predominate (Gulati, in press). Also P excretion rates of crustaceans in these lakes of 0.17 μg P.mg^{-1} DW.h^{-1} (Den Oude & Gulati, in press: Table 1) are almost identical with the rates of *Euchlanis* measured by us (Exp. 3 and 4: Table 3).

Elemental ratios

The N : P ratios of 3.4–3.6 for the large animals and 4.2–4.5 for the small animals, based on ammonia and phosphate excretion, were constant and comparable with the ratios derived from the generalized regression formulae for the rotifers of Ejsmont-Karabin (1984), namely 3.6 for the large and 4.0 for the small animals. The higher N : P ratios for the smaller animals may be because of the relatively higher protein catabolism of these animals; moreover, the smaller animals (0.51 μg DW.ind^{-1}) in the N-experiment were much smaller than the animals (0.79 μg DW.ind^{-1}) in the P experiments (Table 1).

On comparing the elemental ratios, namely the C:N:P ratio of the *Euchlanis* food, *Euchlanis* body and the excretion products summarized below, some interesting discrepancies emerge:

Ratio	C	:N	:P
Food	100	: 15.4	: 0.9
Euchlanis	100	: 23.7	: 2.6
Excretion products	100	: 16.0	: 4.0

Euchlanis contains about 3 times more P and 1.5 times more N than its food. Also, per unit carbon, the excretion products contain about 50% less N, but 50% more P than N and P in the animal body. Evidently, the animal is able to maintain a lower C:N ratio in the body than its food; however, P losses as indicated by the both lower C:P and N:P ratios of the excretory products – than of the animal and its food – are difficult to explain.

Role of rotifer excretion

The *Euchlanis* excretion rates are not quite representative of the rotifer community of the Loosdrecht lakes. The rotifers in these lakes are dominated by *Keratella cochlearis* and *Anuraeopsis fissa*, with individual weights (0.01–0.02 μg DW.ind^{-1}) that are an order of magnitude lower than that of *Euchlanis*. Their excretion rates (Ejsmont-Karabin, 1984: Table 1) are, therefore, probably two orders of magnitude greater than the rates of *Euchlanis* measured under similar temperatures (± 20 °C), using the same methods. Our computed individual mean weight of rotifers in the Loosdrecht lakes of *ca.* 0.05 μg DW.ind^{-1} (Gulati, in press) is about one-tenth the individual mean weight of crustaceans. One would, therefore, expect the rotifer specific excretion rates to be much higher and their contribution to the P and N excretion of total zooplankton in the lakes much greater than based on the crustacean:rotifer mean biomass ratio of *ca.* 2:1 calculated from the lake data of 1982–'87. Thus, the rotifers that contribute about 35% to the total zooplankton biomass will contribute roughly twice as much

(*ca.* 65%) to the mean zooplankton excretion rates, with a range of values between 45 and 90% in summer (Gulati, in press; Van Liere *et al.*, in press).

The excretion rates of *Euchlanis* found by us, though in the range of literature values that vary widely, are low. This, like for the crustacean excretion rates (Den Oude & Gulati, 1988), may be attributed to high C:P ratios (± 125) of sestonic lake food of *Euchlanis*, because of P-limitation of cyanobacteria that dominate the seston. Evidence of the direct effects of P-content of zooplankton food on excretory turnover rate of P has been presented in some recent studies (e.g. Lehman & Naumoski, 1985).

Finally, our study on *Euchlanis* and some comparisons of the literature data on rotifer excretion have demonstrated: 1) the specific excretion rates of rotifers – even within a species – vary enormously because of the inverse relationship between the rates and individual weights; 2) the range of excretion rates is very wide if one compares the different rotifer species, which differ in weight about 60 fold; and 3) in shallow and eutrophic lakes like Loosdrecht Lake, both the much smaller size and relatively higher densities of rotifers than of the crustaceans may contribute significantly to the importance of rotifers in nutrient regeneration, especially in summer.

Acknowledgements

We thank Drs. B. Z. Salomé, Dr. J. Vijverberg & Miss A. L. Wilms for critical reading and Miss Cecilia Kroon for typing of the manuscript. The second author (J. Ejsmont-Karabin) is grateful to the Limnological Institute (Nieuwersluis) for funding her travel to and short stay in The Netherlands in 1984 when the *Euchlanis* investigation was carried out.

References

Blažka, P., Z. Brandl & L. Procházková, 1982. Oxygen consumption and phosphate excretion in pond zooplankton. Limnol. Oceanogr. 27: 294–303.

354

Bottrell, H. H., A. Duncan, Z. M. Gliwicz, E. Grygierek, A. Herzig, A. Hillbricht Ilkowska, H. Kurasawa, P. Larrson & T. Weglenska, 1976. A review of some problems in zooplankton production studies. Norw. J. Zool. 24: 419–456.

Den Oude, P. & R. D. Gulati, 1988. Phosphorus and nitrogen excretion rates of zooplankton from the eutrophic Loosdrecht lakes, with notes on other P sources for phytoplankton requirements. Hydrobiologia 169: 379–390.

Dumont, H. J., I. van de Velde & S. Dumont, 1975. The dry weight estimate of biomass in a selection of Cladocera, Copepoda and Rotifera from the plankton, perifyton and benthos of continental waters. Oecologia 19: 75–97.

Ejsmont-Karabin, J., 1983. Ammonia nitrogen and inorganic phosphorus excretion by the planktonic rotifers. Hydrobiologia 104: 231–236.

Ejsmont-Karabin, J., 1984. Phosphorus and nitrogen excretion by zooplankton (rotifers and crustaceans) in relation to individual bodyweights of the animals, ambient temperature and presence or absence of food. Ekol. pol. 32: 3–42.

Ejsmont-Karabin, J., R. D. Gulati & J. Rooth, 1989. Is food availability the main factor controlling Euchlanis dilatata lucksiana Hauer abundance in a shallow, hypertrophic lake? Hydrobiologia, 186/187: 29–34.

Ejsmont-Karabin, J., L. Bownik-Dylinska & W. A. Godlewska-Lipowa, 1983. Biotic structure and processes in the lake system of R. Jorka Watershed (Masurian Lakeland, Poland). VII. Phosphorus and nitrogen regeneration by zooplankton as the mechanism of the nutrient supplying for bacterio- and phytoplankton. Ekol. pol. 31: 719–746.

Gulati, R. D., in press. Zooplankton structure in Loosdrecht lakes in relation to the trophic status and the recent restoration measures. Hydrobiologia.

Gulati, R. D., J. Rooth & J. Ejsmont-Karabin, 1987. A laboratory study of feeding and assimilation in Euchlanis dilatata lucksiana. Hydrobiologia 147: 289–296.

Ikeda, T., 1971. Changes in respiration rate and in composition of organic matter in Calanus cristatus (Crustacea, Copepoda) under starvation. Bull. Fac. Fish. Hokkaido Univ. 21: 280–298.

Lehman, J. T. & T. Naumoski, 1985. Content and turnover of phosphorus in Daphnia pulex: Effect of food quality. Hydrobiologia 128: 119–125.

Le Borgne, R. P., 1979. Influence of duration of incubation on zooplankton respiration and excretion results. J. exp. mar. Biol. Ecol. 37: 127–137.

Mayzaud, P., 1976. Respiration and nitrogen excretion of zooplankton. IV. The influence of starvation on the metabolism and biochemical composition of some species. Mar. Biol. 37: 47–58.

Murphy, J. & J. P. Riley, 1962. A modified single solution method for the determination of phosphate in natural waters. Anal. Chim. Acta 27: 31–36.

Peters, R. & D. Lean, 1973. The characterization of soluble phosphorus released by limnetic zooplankton. Limnol. Oceanogr. 18: 270–279.

Procházková, L., 1964. Spectrophotometric determinations of ammonia as rubazoic acid with bispyrazolone reagent. Anal. Chem. 365: 865–871.

Ruttner-Kolisko, A., 1972. Rotatoria. In: Das Zooplankton der Binnengewässer Teil 1. Schweizerbart, Stuttgart, 1972. (Series) Die Binnengewässer 26 Teil 1. P 99–234.

Ruttner-Kolisko, A., 1977. Suggestions for biomass calculation of plankton rotifers. Arch. Hydrobiol. Beih. 8: 71–76.

Stephens, K., 1963. Determinations of low phosphate concentrations in lake and marine waters. Limnol. Oceanogr. 8: 361–362.

Van Liere, L., R. D. Gulati, F. G. Wortelboer & E. H. R. R. Lammens, in press. Phosphorus dynamics following restoration measures in Loosdrecht lakes. Hydrobiologia.

Hydrobiologia **186/187**: 355–361, 1989.
C. Ricci, T. W. Snell and C. E. King (eds), Rotifer Symposium V.
© 1989 *Kluwer Academic Publishers.*

A new method to estimate individual dry weights of rotifers*

Hans-Rainer Pauli
Limnologisches Institut, Universität Konstanz, FRG

Key words: rotifers, dry weight, biomass, productivity

Abstract

Body and egg volumes of 12 rotifer species of Lake Constance were calculated from more than 18 000 individually measured animals using geometric approximations. Body (V_A) and egg (V_E) volumes of various rotifer species have an allometric relationship described by: $V_E = 1272 \cdot V_A^{0.3379}$, while the relation V_E/V_A for each species was constant ranging from 0.016 to 0.87. Dry weight of adult rotifers was assumed to be the sum of the egg dry weight plus a corrected weight increment from egg to adult. It was estimated with the formula $W_A = W_E \cdot (1 + a \cdot F_A/F_E)$ where (a) was varied between 0 and 0.1. Egg dry weight content was assumed to be 35% of fresh weight. The estimates of dry weight to wet weight ratios ranged from 0.57% (*Asplanchna priodonta*) to 29.4% (*Kellicottia longispina*). Dry weight estimates obtained with this method are comparable to those obtained with direct measurements of dry weight or carbon content.

Introduction

A reliable estimate of productivity of natural rotifer communities is still difficult. While final calculation of production is relatively simple to perform (Downing & Rigler, 1984), the collection of the appropriate data needed for these calculations poses several problems. The estimate of rotifer production requires the following data: (1) population census in short temporal and dense spatial intervals (including numbers of individuals as well as number of eggs, sorted by species categories), (2) conversion of numbers of individuals and eggs to appropriate biomass units, (3) determination of rates for somatic growth and growth of the population.

When estimating biomass or production of natural rotifer populations from field data the most disputable step is the conversion of body volume to dry weight or other biomass units. Only very few direct measurements of dry weight or carbon content exist. The common practice to derive individual body mass from values published in the literature is unsatisfactory, because size variation within a given rotifer species can be significant (Dumont *et al.*, 1975; Latja & Salonen, 1978; Salonen & Latja, 1988). Similarly, transformation of body volumes, which are commonly used as an estimate of fresh weight, with a general conversion factor of 10 percent dry weight of body volume, does not reflect the differences among rotifer species. Large bodied genera like

---

*This investigation was supported by the Deutsche Forschungsgemeinschaft within the Special Collaboration Program (SFB) 'Cycling of Matter in Lake Constance'.

Asplanchna contain more than 90 percent water (Salonen & Latja, 1988) while small genera like *Keratella* contain less (Walz, 1987).

This study proposes a method for conversion of volumes to biomass incorporating some simple but probably not unrealistic assumptions.

Material and methods

The material for this study was collected during two years (1984, 1985) from Lake Constance-Überlingersee (West Germany). The sampling intervals ranged from two days during spring to 14 days during winter. The upper 20 meters of water were sampled completely with 10 samples using a 2 meter long sampling device. Three additional samples were taken at 30, 40, and 50 meters depth. Samples were filtered through steel gauze of 50 μm mesh size and preserved in 4% sugar formalin (Haney & Hall, 1973). Counting of individuals and eggs was carried out following the Utermöhl method (1958). Measurements of morphometric parameters (body length, body width, length of spines and appendices, egg length, egg width) were performed using an inverted microscope combined with a semi-automatic image analysis system. Body and egg volumes were calculated from these individual size measurements using geometric approximations as proposed by Ruttner-Kolisko (1978). These volumes were then used as estimates of fresh weight, assuming a density of 1.

The following abbreviations and definitions are used in this paper: V (μm^3): volume, F: fresh weight, W (μg): dry weight, a: increment factor, $_E$: egg, $_A$: adult, $_I$: increment, p: proportion dry weight of fresh weight (W/F), $F = V$, $W = F \cdot p$.

For the conversion of carbon content to dry weight, the carbon content of organic material was assumed to be 50% of dry weight, according to Latja & Salonen (1978).

Results

An allometric relationship ($y = a \cdot x^b$) between egg and body volume of 12 rotifer species was

calculated from more than 18 000 measured animals:

$$V_E = 1272 \cdot V_A^{0.3379} \quad r^2 = .73. \qquad (1)$$

The equivalent graphic regression line is shown in Fig. 1. The portion of explained variance is 73%, standard error of the mean is 0.7% for the intercept and 1.3% for the slope, respectively. The square symbols represent mean values for different rotifer species from Lake Constance. Additional values for other species are included in this figure, recalculated from literature data on length and width of animals and of eggs (e.g. Ruttner-Kolisko, 1972; Koste, 1978). Even these very rough estimates fit well into the confidence limits of the calculated regression line. Only one value, reported for *Keratella quadrata* from Walz (1987), falls outside the confidence limits for predicted values. Equation (1) was used to estimate the fresh weight of eggs of those species where no eggs could be found or assigned to without doubt.

In Fig. 2, the relationship between body and egg volumes for species of different size is compared with four relationships calculated for four groups of species of similar dimensions. The

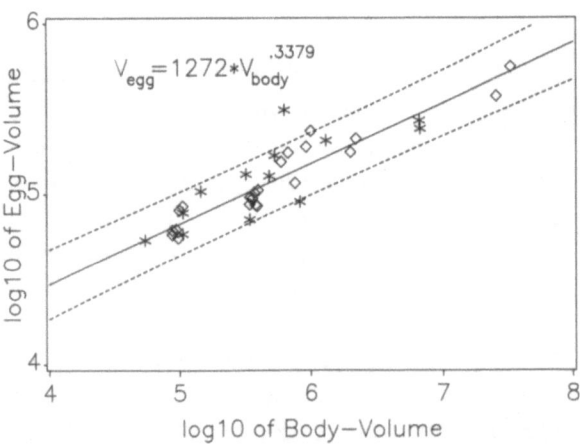

Fig. 1. Volumes of rotifer eggs versus body volumes: Regression line of the base-10 logarithms of egg and body volumes (μm^3) of 12 rotifer species from Lake Constance-Überlingersee (squares). Stars represent values recalculated from literature data. The dashed lines represent 95% confidence limits for the predicted single values.

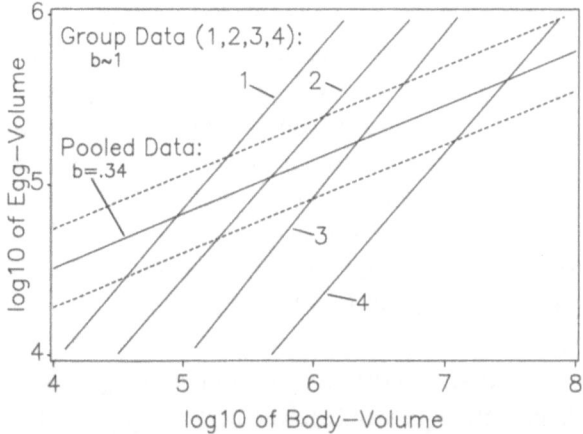

Fig. 2. Volumes of rotifer eggs versus body volumes: Regression line of the base-10 logarithms of egg and body volume (μm^3) of 12 rotifer species from Lake Constance-Überlingersee and separately calculated regression lines for four groups of rotifer species with similar body size. For composition of the groups see text.

groups are composed of the species: (1) *Keratella cochlearis, Kellicottia longispina, Pompholyx sulcata,* (2) *Polyarthra vulgaris, P. dolychoptera, P. major, Brachionus angularis, Filinia terminalis, Keratella hiemalis, K. quadrata,* (3) *Synchaeta pectinata,* (4) *Asplanchna priodonta.* Body size relationships for each group were calculated separately and their slopes were tested for significant differences. The slopes of the regression lines for the four groups ranged from 1.065 to 0.905. None was significantly different from any other at the 95% significance level. These slopes with numerical values close to one indicate a linear proportionality of egg to body volumes among rotifers of similar size or within species.

The following formulas and assumptions are proposed for the transformation of fresh weight to dry weight. Dry weight of an adult rotifer (W_A) was assumed to be the sum of egg dry weight (W_E) and a weight increment (W_I) from egg to adult:

$$W_A = W_E + W_I. \tag{2}$$

Under the assumption that dry weight to wet weight ratios of adult animals are smaller than those of their eggs ($W_A/F_A < W_E/F_E$), and therefore also the dry weight to wet weight ratio of the

weight increment (W_I), a correction factor (c) was introduced to obtain:

$$W_I/F_I = W_E/F_E \cdot c. \tag{3}$$

Replacing (W_I) with ($W_A - W_E$) and F_I with ($F_A - F_E$) we get:

$$W_A = W_E + W_E \cdot c \cdot F_A/F_E - W_E \cdot c \tag{4}$$

Under the simplified assumption that the amount of biomass accumulated from egg to adult is related to the final body size, F_I in equation (3) can be replaced by F_A, with a modified correction factor (a):

$$W_I/F_A = W_E/F_E \cdot a. \tag{5}$$

The final dry weight of the adult animal (W_A) can be calculated from:

$$W_A = W_E \cdot (1 + a \cdot F_A/F_E), \tag{6}$$

and the relative proportion of dry weight of the adult (p_A) as

$$p_A = p_E \cdot (a + F_E/F_A). \tag{7}$$

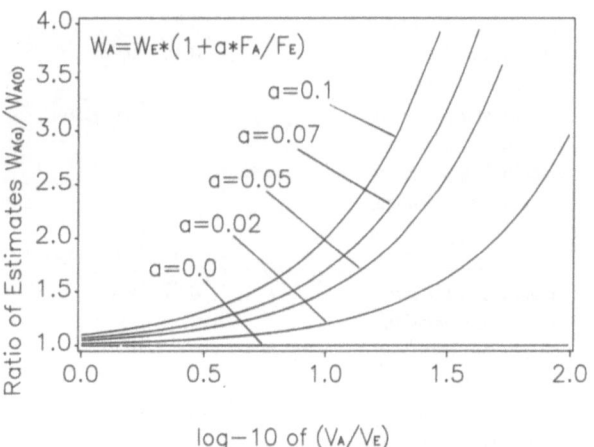

Fig. 3. Ratios $W_{A(a)}/W_{A(o)}$ of dry weight (W_A) of adult rotifers estimated with $W_A = W_E \cdot (1 + \cdot F_A/F_E)$ among species with different V_A/V_E ratios. $W_{A(a)}$ was calculated for different values of (a) (0.02, 0.05, 0.07, 0.1), $W_{A(o)}$ was calculated for $a = 0$.

Table 1. Mean body volumes, mean egg volumes and mean egg to body volume ratios for 12 rotifer species from Lake Constance-Überlingersee (W-Germany).

Species	Body volume	Egg volume	V_E/V_A
	μm^3	μm^3	
Asplanchna priodonta	27270000	424600	0.016
Brachionus angularis	852000	192500	0.23
Filinia terminalis	443000	100700	0.23
Kellicottia longispina	99700	86600	0.87
Keratella cochlearis	87500	61000	0.70
Keratella hiemalis	350200	92600	0.27
Keratella quadrata	733700	183500	0.25
Polyarthra dolychoptera	407800	103500	0.25
Polyarthra major	826000	140300	0.17
Polyarthra vulgaris	443200	105800	0.24
Pompholyx sulcata	85600	60200	0.70
Synchaeta pectinata	2338000	188000	0.08

Because the dry weight to fresh weight ratio (W_A/F_A) of an adult rotifer is expected to be smaller than that of its eggs (W_E/F_E), the value of the weight increment factor (a) has to be smaller than $1 - F_E/F_A$. A value for (a) equal to $1 - F_E/F_A$ would result in an identical value for the proportion of dry weight in both eggs and adults. To obtain a reasonable value for (a), it was varied arbitrarily between 0 and 0.1, which means a volume dependent weight increase between 0 and 10 percent from egg to adult. The values obtained with equation (4) differ from those estimated with equation (6) by 0 to 10% of the egg dry weight. Figure 3 shows the effect of different values of (a) on dry weight estimates (W_A) calculated with equation (6). The curves represent the ratio of W_A for $a = 0.02$, $a = 0.05$, $a = 0.07$, $a = 0.1$ to W_A for $a = 0$ among species with different V_A/V_E ratios. The effect on the estimates when varying (a) is greater in species with large values for the ratio V_A/V_E. Table 1 summarizes egg and body volumes and ratios V_E/V_A, which cover the range from 0.016 (*Asplanchna priodonta*) to 0.87 (*Kellicottia longispina*). Table 2 contains dry weight estimates obtained under the assumption that egg dry weight content of rotifers is comparable to that of *Daphnia* in which it is 35%

Table 2. Measurements and estimates of rotifer dry weights. All values are transformed to dry weight. Original units were dry weight (DW), body volume (V), and carbon content (C).

Species or Genus	% Dry weight in fresh weight	Dry weight $\mu g/Ind$	Original unit	Reference
Asplanchna brightwelli	3.90	0.28 −1.50	DW	1
Asplanchna herricki	0.34− 0.96	2.10 −3.20	C	2
Asplanchna priodonta		10.00	V	3
Asplanchna priodonta		1.70	V	4
Asplanchna priodonta		0.60	DW	5
Asplanchna priodonta		0.48 −0.52	DW	1
Asplanchna priodonta		0.44	DW	6
Asplanchna priodonta		0.12 −0.50	C	7
Asplanchna priodonta	0.42− 1.9	0.30 −1.30	C	2
Asplanchna priodonta	0.57− 4.1	0.04 −3.80	V	11
Brachionus angularis		0.40 −0.54	DW	1
Brachionus angularis		0.05	DW	6
Brachionus angularis	15.1	0.11	C, V	8
Brachionus angularis	8.3 −11.8	0.08 −0.12	V	11
Brachionus calyciflorus		0.11 −0.47	DW	1
Brachionus calyciflorus		0.20	DW	6
Brachionus calyciflorus	2.2 − 5.7	0.06 −0.37	V	11
Conochilus unicornis		0.084	C	7
Conochilus unicornis	10.4 −13.9	0.03 −0.06	V	11
Kellicottia longispina		0.03 −0.06	C	7
Kellicottia longispina	43.2	0.04	C, V	8

Table 2. Continued.

Species or Genus	% Dry weight in fresh weight	Dry weight µg/Ind	Original unit	Reference
Kellicottia longispina	25.9 –29.4	0.02 –0.04	V	11
Keratella cochlearis		0.038	DW	9
Keratella cochlearis		0.11	DW	1
Keratella cochlearis		0.07	DW	6
Keratella cochlearis		0.04 –0.05	C	7
Keratella cochlearis	43.2	0.04 –0.05	C, V	8
Keratella cochlearis	24.2 –27.7	0.02 –0.04	V	11
Keratella quadrata	11.9	0.075	DW, V	10
Keratella quadrata		0.32 –0.35	DW	1
Keratella quadrata		0.07	DW	6
Keratella quadrata	7.7 –11.2	0.03 –0.11	V	11
Polyarthra vulgaris		0.043	DW	6
Polyarthra vulgaris		0.066	DW	7
Polyarthra vulgaris	8.7 –12.2	0.02 –0.06	V	11
Polyarthra spp.		0.74	DW	1
Polyarthra dolychoptera	9.4 –12.9	0.03 –0.06	V	11
Pompholyx sulcata	22.8 –26.4	0.02 –0.03	V	11
Synchaeta spp.		0.26 –0.27	DW	1
Synchaeta spp.	9.8 –13.3	0.03 –0.06	DW	11
Synchaeta oblonga		0.02	DW	6
Synchaeta stylata		0.20	DW	6
Synchaeta stylata	2.7 – 6.2	0.06 –0.15	V	11
Synchaeta pectinata	3.2 – 6.7	0.03 –0.23	V	11
Ascomorpha ecaudis	12.4 –15.9	0.03 –0.04	V	11
Ascomorpha ovalis	16.7 –20.2	0.02 –0.03	V	11
Filinia terminalis	9.2 –12.7	0.03 –0.06	V	11
Gastropus stylifer	9.0 –12.5	0.02 –0.04	V	11
Keratella hiemalis	9.5 –13.0	0.02 –0.06	V	11
Notholca caudata	4.1 – 7.6	0.05 –0.08	V	11
Polyarthra major	5.7 – 9.2	0.04 –0.11	V	11
Trichocerca birostris	12.2 –15.7	0.03 –0.04	V	11
Trichocerca capucina	5.2 – 8.7	0.05 –0.08	V	11
Trichocerca porcellus	10.2 –13.7	0.03 –0.05	V	11
Trichocerca pusilla	24.2 –27.7	0.02	V	11

Reference: (1) Dumont *et al.* (1975), (2) Salonen & Latja (1988), (3) Naulapää (1966), (4) Aasa (1970), (5) Narita & Mori (1975), (6) Bottrell *et al.* (1976), (7) Latja & Salonen (1978), (8) Walz (1987), (9) Schindler & Novén (1971), (10) Doohan & Rainbow (1971), (11) present study

(Bikar, 1986). Included in this table (2) are literature values of direct measurements. The lower values of the given ranges are obtained by setting (*a*) to 0, the upper bound reflects a value for (*a*) of 0.1. The resulting estimates of adult dry weight content of fresh weight ranges from 0.57% (*Asplanchna priodonta*) to 29.4% (*Kellicottia longispina*) among the species included in this study.

Discussion

The common use of allometric equations in biology in general has good theoretical grounds (Schmidt-Nielsen, 1984; Peters, 1983). Even if not completely understood in a particular case, empirical use of such equations can be very helpful.

In rotifers, egg volume obviously represents a

constant fraction of the body volume in a given species, where larger animals carry larger eggs. For rotifer species of increasing size, the relation of egg volume to body volume decreases. The exponent of 0.34 in equation (1) is relatively small when compared to those for other groups of invertebrates. Peters (1983) suggested that the reproductive investment of invertebrates rises with body mass to the power of 0.75. The considerable difference between these two values is possibly due to the fact that rotifer species differ more in their dry biomass content than species of other taxa.

Constant egg to body volume ratios of particular rotifer species have been reported by Walz (1987). These values found by Walz are comparable to those of the present study although his values are smaller for *Kellicottia longispina* (0.75) and *Keratella cochlearis* (0.56) and higher for *Keratella quadrata* (0.50) and *Polyarthra* (0.27) species. Because these values are based on a relatively small number of observations, they are probably biased by seasonal and individual variations. They fall, however, into the limits of the observed range of egg to body volume relationships of these species.

A comparison of dry weight estimates obtained by the method proposed here with the few existing direct measurements of dry weight or carbon content is difficult for several reasons. The necessity of expressing the results in equal units is not yet realized by most authors. Some results are expressed as dry weight or carbon weight for single individuals, others as dry weight recalculated from body volumes with the general assumption of 10% dry weight content, and in few cases as percentage of body volume. Body volume, even when calculated from a large number of measurements of linear size with suitable geometric models, is only a rough estimate of fresh weight. Nevertheless, it is the best available one. Body size measurements are easy to perform and provide additional information on seasonal changes and individual variation among species. Data on dry weight or carbon content of individuals should be applicable for the conversion of abundance data to biomass. Therefore, data on absolute dry weight or carbon content of single specimens are of little use without additional information on size or fresh weight of the analyzed individuals. Body size of defined species can vary in a range of one order of magnitude. Therefore, dry weight or carbon content should be expressed in relation to body size.

The assumption of 10% dry weight of body volume for all rotifer species can no longer be accepted. Large, soft-bodied and transparent genera like *Asplanchna* and *Synchaeta* have dry weight content lower than 10%, while the values are higher in smaller genera with thick loricas or long spines like *Keratella* or *Kellicottia*. For several species of *Asplanchna* dry weight to fresh weight ratios between 3.9% (Dumont *et al.*, 1975) and 0.34% (Salonen & Latja, 1988) have been measured, while the proposed method for the estimation of dry weight results in values between 0.57% and 4.1%, depending on the value for the factor (*a*). For *Keratella cochlearis* a carbon content of 21.4% was measured by Walz (1987). This value suggests a dry weight content of 40% under the general assumption that 50% of dry weight consists of carbon. The values of 24.2–27.7% estimated in the present study agree better to that value than to the common 10% concept. Latja & Salonen (1978) reported for singular individuals of *Keratella cochlearis* carbon weights of 0.024 µg and for singular individuals of *Kellicottia longispina* carbon weights between 0.015 and 0.03 µg. When recalculating these values, assuming a mean body size similar to those found by other authors, carbon content amounts to almost 50% for *Keratella cochlearis* and it ranges from 30% to 60% for *Kellicottia longispina*. Although these values seem to be rather high, they support the assumption of a higher dry weight content than 10% for these species. The estimated weights for particular individuals in this study, ranging from 0.02 to 0.04 µg for both *Keratella cochlearis* and *Kellicottia longispina*, agree well. The value of 11.9% dry weight content of *Keratella quadrata* reported by Doohan & Rainbow (1971) was calculated from the given mean length of 142 µm using Ruttner-Kolisko's (1977) formula and the given individual dry weight of 0.075 µg. The corresponding values of this study cover the range between 7.7% and 11.2%, the dry weights of single individuals calculated from this value are similar to the 0.075 µg from Doohan & Rainbow (1971) and to the 0.07 µg given in Bottrell et al. (1976). The only values which do not match well are those reported by Dumont et al. (1975), which are considerably higher than most other values in the literature. For example, their value for *Keratella quadrata* (0.35 µg) exceeds by seven times those of other authors (0.07 µg). For *Brachionus angularis*, their values (0.4–0.54 µg) are about 10 times higher than the value of 0.05 µg cited in Bottrell et al. (1976).

The method proposed in this study does not, of course, estimate 'true' values of dry weights in rotifers, but the estimates are quite comparable to those obtained by direct measurements and can be calculated for any species, with a defined, consistent method.

References

Aasa, R., 1979. Plankton i Lilla Ullevifjärden. Nat. Swedish Envir. Prot. Bd, Limnol. Surv. Uppsala, Rep. 33: 1–62.

Bikar, K., 1986. Dauer der Eientwicklung von Planktoncrustaceen des Überlingersees (Bodensee) in Abhängigkeit von der Saisonalen Variation der Umgebungstemperatur, der Adultgröße und des Eigewichtes. Diplomarbeit, Universität Freiburg (W-Germany).

Bottrell, H. H., A. Duncan, Z. M. Gliwicz, E. Grygierek, A. Herzig, A. Hillbricht-Ilkowska, H. Kurasawa, P. Larsson & T. Weglenska, 1976. A review of some problems in zooplankton production studies. Norw. J. Zool. 24: 419–456.

Doohan, M. & V. Rainbow, 1971. Determination of dry weights of small Aschelminthes (<.1 µg). Oecologia (Berl.) 6: 380–383.

Downing, J. A. & F. H. Rigler, 1984. A manual on Methods for the Assessment of Secondary Productivity in Fresh Waters, 2., Blackwell Scientific Publications, Oxford, 501 pp.

Dumont, H. J., I. Van de Velde & S. Dumont, 1975. The dry weight estimate of biomass in a selection of Cladocera, Copepoda, and Rotifera from the plankton, periphyton, and benthos of continental waters. Oecologia (Berl.) 19: 75–97.

Haney, J. F. & D. J. Hall, 1973. Sugar-coated Daphnia: A preservation technique for Cladocera. Limnol. Oceanogr. 18: 322–333.

Koste, W., 1978. Rotatoria. Die Rädertiere Mitteleuropas. 1. Textband. Gebrüder Bornträger, Berlin, Stuttgart, 673 pp.

Koste, W., 1978. Rotatoria. Die Rädertiere Mitteleuropas. 2. Tafelband. Gebrüder Bornträger, Berlin, Stuttgart, 234 Tafeln.

Latja, R. & K. Salonen, 1978. Carbon analysis for the determination of individual biomasses of planktonic animals. Verh. int. Ver. Limnol. 20: 2556–2560.

Naulapää, A., 1966. Eräiden Suomessa esiintyvien planktereiden tilavuuksien keskiarvoja [Mean volumes for some plankters found in Finland]. Vesiensuojelutoimiston tiedonantoja 21: 1–26.

Peters, R. H., 1983. The ecological implications of body size. Cambridge University Press, Cambridge, England.

Ruttner-Kolisko, A., 1972. Rotatoria. In: Elster, H.-J. & W. Ohle (eds.), Die Binnengewässer. Bd. 26. Das Zooplankton der Binnengewässer. 1. Teil. Schweizer-bart'sche Verlagsbuchhandlung, Stuttgart: 99–234.

Ruttner-Kolisko, A., 1977. Suggestions for biomass calculation of planktonic rotifers. Arch. Hydrobiol. Beih./Ergebn. Limnol. 8: 71–76.

Salonen, K. & R. Latja, 1988. Variation in the carbon content of two Asplanchna species. Hydrobiologia 162: 79–87.

Schmidt-Nielsen, K., 1984. Scaling: Why is animal size important? Cambridge University Press.

Utermöhl, H., 1958. Zur Vervollkommnung der quantitativen Phytoplankton-Methodik. Mitt. int. Ver. Limnol. 9: 1–38.

Walz, N., 1987. Stoffumsatz und Kinetik von Regulationsprozessen bei Zooplankton-Populationen. Habilitationsschrift, Universität München, W-Germany, 225 pp.

Hydrobiologia **186/187**: 363–369, 1989.
C. Ricci, T. W. Snell and C. E. King (eds), Rotifer Symposium V.
© 1989 *Kluwer Academic Publishers.*

Temperature aspects of ecological bioenergetics in *Brachionus angularis* (Rotatoria)

Norbert Walz, Tanja Gschloessl, Ulli Hartmann
Zoologisches Institut, Seidlstr. 25, D-8000 München 2, FRG

Key words: rotifers, temperature, food concentration, bioenergetic parameters

Abstract

The influence of temperature and food quality was studied on the following energy balance parameters of *B. angularis*: ingestion, production, growth and mortality. The ingestion rate rises to an optimum at 15 and 20 °C and decreases at 25 °C. The other rates increase continuously over the 5–25 °C range. The Q_{10}-values of production rate are higher than those of ingestion rate. Temperature also modifies the relationship between food concentration and bioenergetic rates. They react according to a Monod function (production at all temperatures, growth at 10 °C) or decrease at high concentrations (growth at 15° and 20 °C).

Introduction

Bioenergetic metabolism is ecologically important for all species in their environment. Two factors have especially important influences on energy balance: food conditions and temperature. From studies on energy balance we can draw conclusions on the food and temperature demands a species makes to its environment.

The energy balance can be formulated as follows (Walz, 1987b):

ingestion – feces = *production* + respiration
Production = *growth* + mortality

(parameters in italics were measured).

Studies of the food demands of *Brachionus angularis*, a rotifer common in eutrophic lakes were made by Walz (1987b) and Walz & Gschloessl (1988). Temperature has such a great influence on this eurythermic species, that it even modifies the dependence of some bioenergetic parameters on food density.

Methods

Brachionus angularis was cultured with the unicellular green alga *Stichococcus bacillaris*. The culture conditions were further described by Walz (1983) and Walz (1987). Rotifers were acclimated to experimental temperatures for at least three weeks before the tests. The ingestion experiments were conducted by [14]-C-labelled *Stichococcus* according to Walz & Gschloessl (1988). The rotifers in the growth-experiments were transferred daily into 20 ml glass vessels with fresh suspensions of algae of defined concentration. To avoid sedimentation the vessels were fixed on a rotation wheel (1 rpm) and exposed to the desired temperature ('rotating cultures'). The number of rotifers and their eggs were controlled daily. The specific production rate was measured as birth rate according to Paloheimo (1974) and Romanovsky & Polishchuck (1982). For more details see Hartmann (1987).

Results

Bioenergetic rates

At the experimental food concentrations, the ingestion rate of a *B. angularis* population increases with temperature (Fig. 1a). The 5 μg C ml^{-1} ingestion rate reaches a maximum at 15 °C, double the value at 2.5 μg C ml^{-1}, and decreases further on. At 2.5 μg C ml^{-1}, the maximum is achieved at 20 °C. The 20 °C rates are on a plateau at all food concentrations between 2 and 5 μg C ml^{-1} (Walz & Gschloessl, 1988). Therefore, the 2.5 and 5 μg C ml^{-1} values are not significantly different. If temperature is increased to 25 °C, the 5 μg C ml^{-1} ingestion rate remains unchanged. The decrease of the 2.5 μg C ml^{-1} ingestion rate at 25 °C is not significant, but the 25 °C ingestion rates at both food concentrations differ significantly ($P = 0.05$). At temperatures <20 °C no plateau was found.

In contrast to these curves of food ingestion, the output bioenergetic rates all increase with temperature (Fig. 1, b–d). Specific production achieves a maximum value of about 0.7 d^{-1}, growth rate about 0.5 d^{-1} and mortality rate about 0.3 d^{-1}. Food concentration also has a distinct influence on these rates. In general, rates rose with increasing food concentration (see

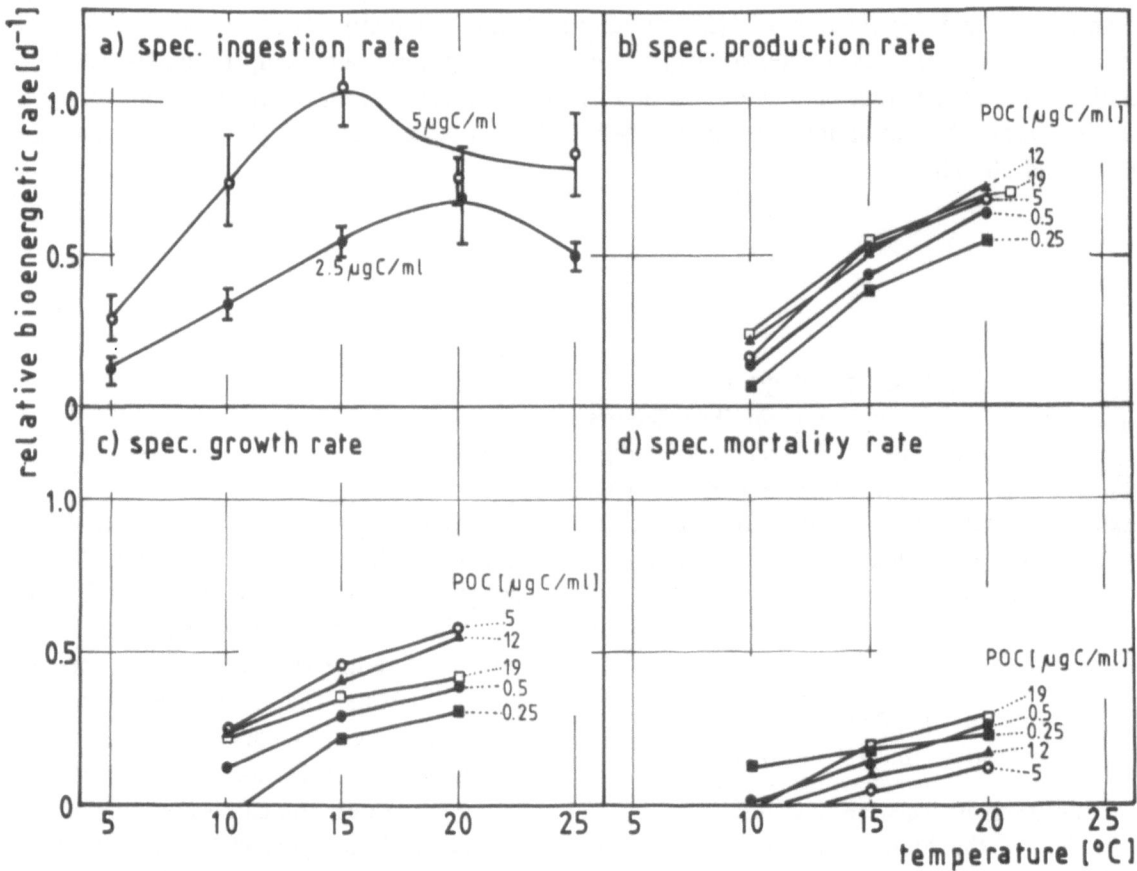

Fig. 1. Temperature relationship of individual bioenergetic parameters at different food concentrations (POC = particulate organic carbon = μg C ml^{-1}). $\pm 95\%$ confidence intervals. Solid squares: 0.25 μg C ml^{-1}, solid circles: 0.5 μg C ml^{-1}, open triangles: 2.5 μg C ml^{-1}, open circles: 5.0 μg C ml^{-1}, solid triangles: 12.0 μg C ml^{-1}, open squares: 19.0 μg C ml^{-1}.

Table 1. The food utilization efficiencies of *B. angularis.* (PE = production/ingestion, Y = growth/ingestion.

Food concentration		Temperature (°C)		
		10	15	20
$2.5 \mu g$ C ml^{-1}	PE	0.49	-	0.42*
	Y	0.48	-	0.48*
$5.0 \mu g$ C ml^{-1}	PE	0.22	0.49	0.28*
	Y	0.30	0.44	0.18*

* From chemostat experiments (Walz 1987b).

section: kinetics), but this effect was not found for mortality rate.

Food utilization efficiency

The efficiencies (production efficiency, PE; yield = growth/ingestion, Y) at $5 \mu g$ C ml^{-1} increase with temperature, climbing from 10 to 15 °C where a turnover of little less than 0.5 is observed (Table 1). With a temperature increase to 20 °C the efficiencies decrease again. Food utilization in the lower food concentration ($2.5 \mu g$ C ml^{-1}) is higher than at $5 \mu g$ C ml^{-1} (see also Walz, 1987b) and is a little less than 0.5 at 10 and 20 °C. At 15 °C, no experiments were made at this food concentration.

Q_{10}-values

The Q_{10} values in the temperature range of 10 to 20 °C were different for each individual bioenergetic parameters. Each parameter increased at a different rate with increasing temperature (Table 2). Production rate shows the greatest increase with a mean Q_{10} of 4.0. The temperature dependencies of some coupled life cycle parameters, when treated as reciprocals to time, are similar. Life span Q_{10} is 3.3, embryonic development Q_{10} is 4.0, juvenile development Q_{10} is 4.9, and development of egg laying interval Q_{10} is 3.1 (according to Walz, 1987a).

Population growth rate and mortality rate have lower Q_{10}s of 2.7 and 3.0 respectively. Respiration rate Q_{10}s, calculated from Galkovskaya (1987) are also 2.7. The ingestion rate Q_{10} is still lower = 2.3. Although it is possible that the assimilation quotient increases with temperature, the energy available for utilization is lower at higher temperatures. The Q_{10} also is dependent on the food supply and, in general, it is higher at lower food densities.

Temperature modification of food – bioenergetic rate kinetics

With rising food concentration (= POC = particulate organic carbon) the specic production rate (b) increases at all studied temperatures according to a Monod saturation function ($b = b_{max} \cdot POC/K_s + POC$) (Fig. 2a). In the lower concentration range the rate accelerates steeply and reaches a plateau where additional food does not produce further rate increases. Plateau values increase with temperature and the increase is higher between 10 and 15 °C than between 15 and 20 °C, corresponding to the slightly flattened curve in Fig. 1c.

Monod function, regression and correlation

Table 2. The Q_{10}-values of bioenergetic parameters between 10° and 20 °C. SE = Standard error.

Parameter	Food concentration (μg C ml^{-1})							mean Q_{10} ± SE
	0.5	2.5	5.0	8.5	12.0	15.5	19.0	
Ingestion rate	-	2.6	2.0	-	-	-	-	2.3 ± 0.3
Production rate	4.7	4.4	4.2	3.8	3.5	4.1	3.4	4.0 ± 0.2
Growth rate	3.1	4.1	2.5	2.7	2.4	1.9	2.0	2.7 ± 0.3
Mortality rate	3.6	-	4.8	-	2.6	-	2.2	3.0 ± 0.6

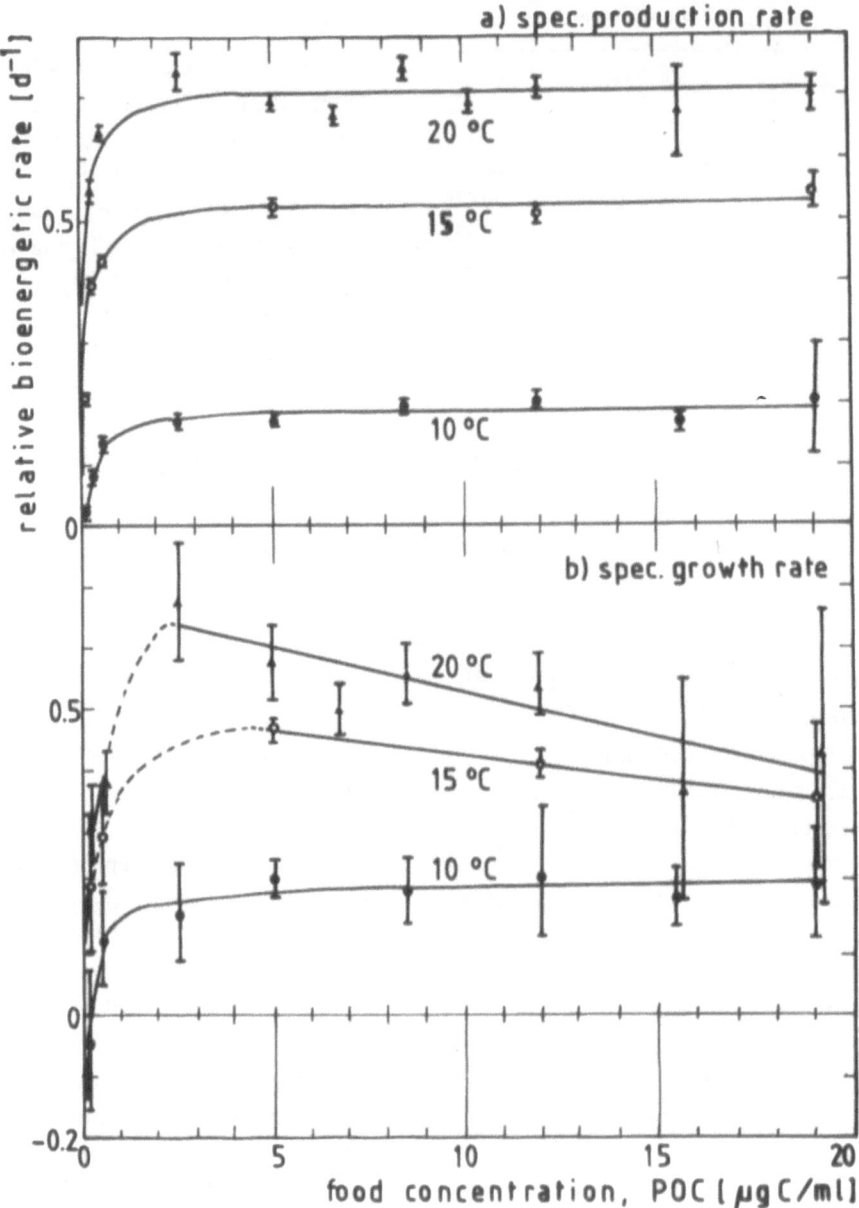

Fig. 2. Modification of the kinetics of individual bioenergetic rates in dependence on temperature and food concentration. The dashed lines are fitted by eye. $\pm 95\%$ confidence intervals.

coefficients are shown in Table 3. They were fitted with a non-linear regression according to Bliss & James (1966). With rising temperature the b_{max} values (plateau levels) increase and the K_s values decrease.

The specic growth rates (Fig. 2b) take contrasting courses. Only the 10 °C curve is in accordance with a Monod model and shows a saturation function. At higher temperatures the growth rate accelerates at the lower concentrations up to a maximum and decreases linearly thereafter. The linear regression coefficients (Table 3) at 15° and 20 °C are significantly different from zero, but they do not differ from each other ($P = 0.05$).

The mortality rate decreases with increasing food concentration in the lower concentration range and increases further at 15° and 20 °C.

Table 3. Coefficients of the Kinetics of production and growth rate. * coefficients of Monod function, ** coefficients of linear regression. POC, food concentration (μg C ml^{-1}), bioenergetic rate (d^{-1}), R = correlation coefficient.

Temperature	Spec. production rate	Spec. growth rate
'rotating cultures' 10 °C	$b^* = \dfrac{0.196 \cdot POC}{0.313 + POC}$ $R = 0.94, n = 29$	$r^* = \dfrac{0.239 \cdot POC}{1.344 + POC}$ $R = 0.89, n = 29$
15 °C	$b^* = \dfrac{0.529 \cdot POC}{0.097 + POC}$ $R = 0.99, n = 17$	POC > 2.5 μg C/ml: $r^{**} = 0.504 - 0.008 \cdot POC$ $R = -0.99, n = 8$
20 °C	$b^* = \dfrac{0.712 \cdot POC}{0.073 + POC}$ $R = 0.73, n = 49$	POC > 2.5 μg C/ml: $r^{**} = 0.676 - 0.015 \cdot POC$ $R = -0.73, n = 37$
Chemostat culture 20 °C (WALZ 1987b)	$b^* = \dfrac{0.548 \cdot POC}{0.775 + POC}$ $R = 0.84, n = 22$	$r^* = \dfrac{0.579 \cdot POC}{4.045 + POC}$ $R = 0.94, n = 25$

This increase is in contrast to chemostat experiments (Walz, 1987b), where the mortality rates only decline asymptotically with food concentration from high values at low concentrations.

Discussion

The individual bioenergetic rates of *B. angularis* show characteristic differences in their temperature course. They either increase continuously with temperature or reach a maximum and then decrease at higher temperatures. For *B. angularis* ingestion rates, a maximum was reached at 15° or 20 °C. In contrast, all other rates increase continuously, but this is not always the case for other species. In *B. calyciflorus*, ingestion rate increases continuously with temperature (Galkovskaya, 1987). Respiration rate seems to be the only rate that consistently increases continuously with temperature in all species studies (Galkovskaya, 1987; Lampert, 1984). As with *B. angularis*, *Keratella cochlearis* production rate also continuously rises with temperature (Edmondson, 1965; Walz, 1983). In contrast, egg-ratio in both

species, which is the basis for calculation of production rate (Paloheimo, 1974), has a maximum at intermediate temperatures and decreases thereafter. This is also the case for *B. calyciflorus* (Halbach, 1970). The decreasing egg-ratio of *B. angularis* and *K. cochlearis* at higher temperatures is probably more than compensated by the increased rate of embryonic development. In contrast to the above species, the egg-ratio of *B. rubens* declines with climbing temperature (Vuckovic, 1981).

According to the portion of production going into mortality or growth (see equation in the Introduction), growth rate increases continuously, if mortality rate rises constantly (Fig. 1) (*B. calyciflorus*, Halbach, 1970; *B. dimidiatus*, Pourriot & Rougier, 1975). Alternatively, in cases mortality rate increases progressively, growth rate declines after reaching a maximum (*K. cochlearis*, Walz, 1983; *B. plicatilis*, Hirayama & Kusano, 1972). *B. rubens* follows a more complicated pattern. At high food levels growth rate decreases with temperature and increases at low food conditions (Vuckovic, 1981).

In *B. angularis* we showed that the energy

balance grows more unfavorable at higher temperatures (higher Q_{10} values for production than ingestion). At 25 °C, our population approached its thermal tolerance limits. At this point, it was difficult to keep the culture going and no growth experiments were carried out. *B. calyciflorus*, an even more pronounced warm-eurythermal rotifer, was possible to be cultured up to 40 °C (Galkovskaya, 1987). In contrast, the net reproductive rate (R_o) of the cold-stenothermal *Notholca caudata* falls dramatically at temperatures above 10 °C (Laxhuber & Hartmann, 1988). Also *N. squamula* could be maintained in culture only under 10 °C (May, 1987).

Temperature effects on bioenergetic rates are modified by food concentration. In rotifers, this relationship is often represented as a saturation function according to the Monod model (e.g. the production rates in *B. angularis* at all temperatures and growth rate at 10 °C). The egg-ratio of *B. rubens* increases either according to a saturation function (Vuckovic, 1981) or continuously (Rothhaupt, 1985). In contrast, production rate is often found to decrease at higher food concentrations (e.g. *B. rubens*, Pilarska, 1977; and *B. plicatilis*, Hirayama *et al.*, 1974). The same is also true for growth rates (*Euchlanis dilatata*, King, 1967 and *B. calyciflorus*, Halbach & Halbach-Keup, 1974).

Usually this decrease is explained by algae toxins, but this cannot explain our observations in *B. angularis*. First, a toxic effect also would have affected production rate. Second, growth rate at 10 °C follows a Monod function. Only at 15° and 20 °C did a decline of growth rate at higher food concentrations occur.

This decrease of growth rate was not found in chemostat experiments at 20 °C with the same food algae (Walz, 1987b) (Table 3). Only in long-term experiments with residence times longer than the daily transfer-times of the 'rotating cultures' could such a toxic effect be detected.

As rotifer growth rates decreased at higher temperatures, increasing respiration of algae (dark experiments) could have led to a critical O_2 supply for the rotifers. Young rotifers seemed to have suffered more from this, as the egg laying of the adults was not affected. However, the O_2 content in 'rotating cultures' was not measured. Though we tried to prevent O_2 depletion by leaving air bubbles in the vessels, this may not have been sufficient at high temperatures and high algae densities. Although in chemostats good aeration is provided.

Alternatively, the algae in 'rotating cultures' could have clumped together in higher concentrations and become unavailable as food for the young rotifers. Currently, no decision can be made between these hypotheses. Differences between chemostat and 'rotating culture' production coefficients at 20 °C (Table 3) was discussed by Walz (1987b).

References

Bliss, C. I. & A. T. James, 1966. Fitting the rectangular hyperbola. Biometrics 22: 573–602.

Edmondson, W. T., 1965. Reproductive rates of planktonic rotifers as related to food and temperature in nature. Ecol. Monogr. 35: 61–112.

Galkovskaya, G. A., 1987. Planktonic rotifers and temperature. Hydrobiologia 147: 307–317.

Gschloessl, T., 1985. Die Nahrungsaufnahme von *Brachionus angularis* (Rotatoria) in Abhängigkeit von der Futterart, der Futterkonzentration, der Temperatur und der Fütterungsdauer. Diplomarbeit Faculty of Biology, Univ. Munich, 127 pp.

Halbach, U., 1970. Einfluß der Temperatur auf die Populationsdynamik des planktischen Rädertierchens *Brachionus calyciflorus* Pallas. Oecologia 4: 176–207.

Halbach, U. & G. Halbach-Keup, 1974. Quantitative Beziehungen zwischen Phytoplankton und der Populationsdynamik des Rotators *Brachionus calyciflorus* Pallas. Befunde aus Laboratoriumsexperimenten und Freilanduntersuchungen. Arch. Hydrobiol. 73: 273–309.

Hartmann, U., 1987. Die Populationsdynamik der pelagischen Rotatorien *Brachionus angularis* und *Notholca caudata* in Abhängigkeit von der Futterkonzentration und der Temperatur. Diplomarbeit Faculty of Biology, Univ. Munich, 91 pp.

Hirayama, K. & T. Kusano, 1972. Fundamental studies on physiology of rotifer for its mass culture. II. Influence of water temperature on population growth of rotifers. Bull. Jap. Soc. Sci. Fish. 38: 1357–1363.

Hirayama, K., K. Watanabe & T. Kusano, 1973. Fundamental studies on physiology of rotifer for its mass culture. III. Influence of phytoplankton density on population growth. Bull. Jap. Soc. Sci. Fish. 39: 1123–1127.

King, C. E., 1967. Food, age, and the dynamics of a laboratory population of rotifers. Ecology 48: 111–128.

Lampert, W., 1984. The measurement of respiration. In: Downing, J. A. & F. H. Rigler (eds), A manual on methods for the assessment of secondary production in fresh waters. 2nd ed. Blackwell Scientif. Publ., Oxford. IBP-Handbook 17: 413–468.

Laxhuber, R. & U. Hartmann, 1988. The influence of temperature on the life-cycle of the cold-stenothermal rotifer *Notholca caudata*. Verh. int. Ver. Limnol. 23: 2016–2018.

May, L., 1987. Culturing freshwater planktonic rotifers on *Rhodomonas minuta* var. nannoplanktica Skuja and *Stichococcus bacillaris* Nägeli. J. Plankton Res. 9: 1217–1223.

Paloheimo, J. E., 1974. Calculation of instantenous birth rates. Limnol. Oceanogr. 19: 692–694.

Pilarska, J., 1977. Ecophysiological studies on *Brachionus rubens* Ehrbg. (Rotatoria). II. Production and respiration.- Pol. Arch. Hydrobiol. 24: 329–341.

Pourriot, R. & C. Rougier, 1975. Dynamique d'une population experimentale de *Brachionus dimidatus* (Bryce) (Rotifere) en fonction de la nourriture et de la temperature. Ann. Limnol. 11: 125–143.

Romanovsky, Yu. E. & L. Polishchuck, 1982. A theoretical approach to calculation of secondary production at the population level. Int. Revue. ges. Hydrobiol. 67: 341–359.

Rothhaupt, K. O., 1985. A model approach to the population dynamics of the rotifer *Brachionus rubens* in two-stage chemostat culture. Oecologia 65: 252–259.

Vuckovic, M., 1981. Einfluß der Temperatur und der Nahrungsquantität auf die Populationsdynamik des planktischen Rädertiers *Brachionus rubens* Ehrenberg. Diplomarbeit Dep. Biology, Univ. Frankfurt/M., 75 pp.

Walz, N., 1983. Individual culture and experimental population dynamics of *Keratella cochlearis* (Rotatoria). Hydrobiologia 107: 35–43.

Walz, N., 1987a. Comparative population dynamics of *Brachionus angularis* and Keratella cochlearis. Hydrobiologia 147: 209–213.

Walz, N., 1987b. Stoffumsatz und Kinetik von Regulationsprozessen bei Zooplankton-Populationen. Analysen und Modelle in Rotatorien-Chemostaten und im Plankon eines Sees. Habilitationsschrift Faculty of Biology, Univ. Munich, 225 pp.

Walz, N. & T. Gschloessl, 1988. Functional response of ingestion and filtration rate of the rotifer *Brachionus angularis* to the food concentration. Verh. int. Ver. Limnol. 23: 1993–2000.

Hydrobiologia **186/187**: 371–374, 1989.
C. Ricci, T. W. Snell and C. E. King (eds), Rotifer Symposium V.
© 1989 *Kluwer Academic Publishers.*

Biomass and elemental composition (C.H.N.) of the rotifer *Brachionus plicatilis* cultured as larval food

M. Yúfera & E. Pascual
Instituto de Ciencias Marinas de Andalucia (C.S.I.C.), Apdo. oficial, 11510 Puerto Real. Cádiz. Spain

Abstract

Changes in dry weight, carbon, nitrogen, hydrogen and ash content of the rotifer *Brachionus plicatilis* were investigated during population growth in culture. Also caloric equivalent was estimated from carbon content. Body dry weight and weight of C, N, and H increased with increasing the egg/female ratio of the population, while ash content remained constant. As percentage of dry weight, C content increased with fecundity, H and N remained constant and ash content decreased. The caloric equivalent ranged from 1278 μcal to 4406 μcal for rotifers from 314 μg to 846 μg of dry weight.

Introduction

Rotifer *Brachionus plicatilis* is widely used as first food for marine fish larvae in aquaculture. The strains commonly cultured range from 90–140 μm to 200–300 μm in length and differ considerably in female body mass. However, variations of biomass may also be found in a single strain as a consequence of the culture system and the demographic characteristics of the population. Such variations may affect the nutritional quality of rotifers as prey due to its caloric content. Knowledge of this energy value is needed in feeding and metabolic studies of larval fish and crustacea. This paper studies changes in biomass and C, N, H content through population growth of *B. plicatilis* in culture.

Materials and methods

Brachionus plicatilis strain S-1 (Yúfera, 1982) was cultured at 25 °C. Two batch cultures (A and B) and two semicontinuous cultures (C and D) with a partial harvest of 10–20% daily, were carried out to determine the variations in body mass and composition. Cultures A, B and C were fed on the Eustigmatophyceae *Nannochloropsis gaditana*, and the culture D was maintained with bakers yeast *Saccharomyces cerevisiae*.

Dry weight (DW) was determined by drying samples of about 400–800 individuals (females with eggs) at 85 °C to constant weight. Elemental analyses (C.N.H.) were carried out with an Elemental Analyzer Model 1106 (Carlo Erba Science) using cyclohexane 2.4-dinitrophenylhydrazone as standard; samples weighed about 1 mg. Ash content was determined by incinerating dried samples at 550 °C during 24 h. Three replicates per sample were used to determine ash and C.N.H. content. Energetic content was estimated from carbon content using the factor of 10.9 cal per mg C (Salonen *et al.*, 1976).

Results

Batch cultures A and B had a characteristic peak of fecundity on day 2 and 3 due to synchronous deposition of the first egg (Fig. 1). The semicon-

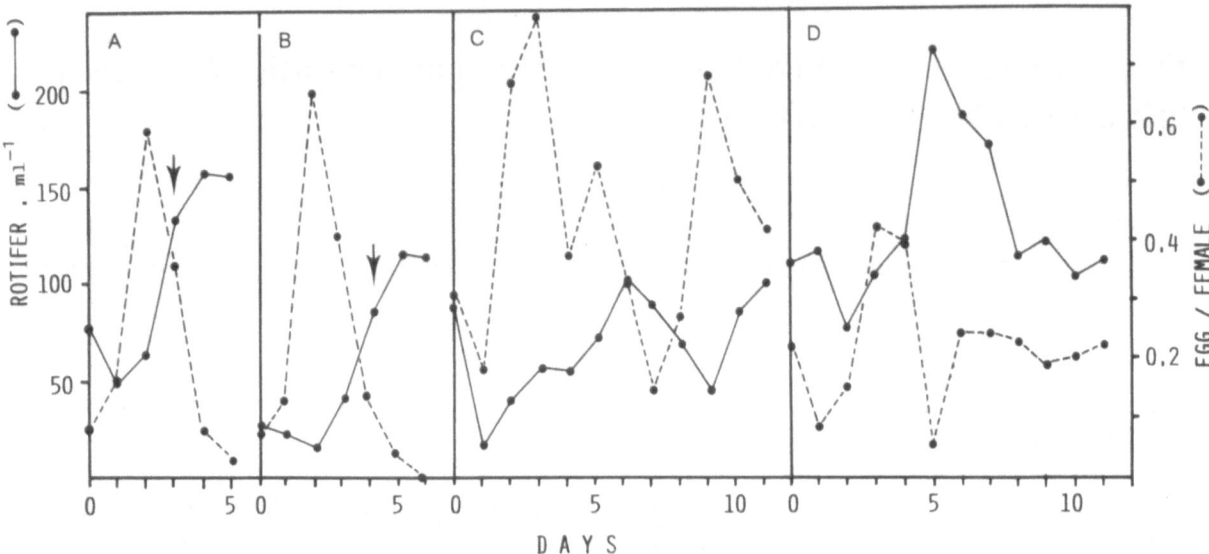

Fig. 1. Population growth curves. Cultures A, B, and C with *Nannochloropsis gaditana* and culture D with *Saccharomyces cerevisiae* as food. Arrows indicate the exhaustion of food.

tinuous cultures (C and D) had more erratic variation in fecundity. Changes in individual dry weight were related to fecundity. Body dry weight were positively correlated to eggs/female (E/F) ratio (Fig. 2). In contrast, ash content varied little, ranging from 60–120 ng (average 86 ng $\pm$ SD 19) and was not correlated with E/F ratio ($r = 0.189$). Data of rotifers fed microalgae and those fed on

yeast have been combined because no significant differences were found between them.

The H and N contents were practically constant through population growth (Fig. 3). H accounted for about $6.3 \pm 0.3\%$ and N $9.2 \pm 0.65\%$ of the body dry weight, indicating a mean protein level ($N \times 6.25$) of about 57.5%. The C percentage increased with the E/F ratio

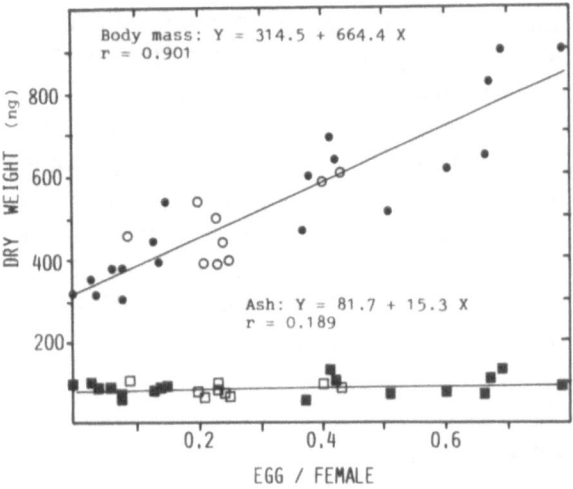

Fig. 2. Body dry weight (circles) and ash weight (squares) in relation to egg/female ratio. Open symbols = rotifer fed on algae; dark symbols = rotifer fed on yeast.

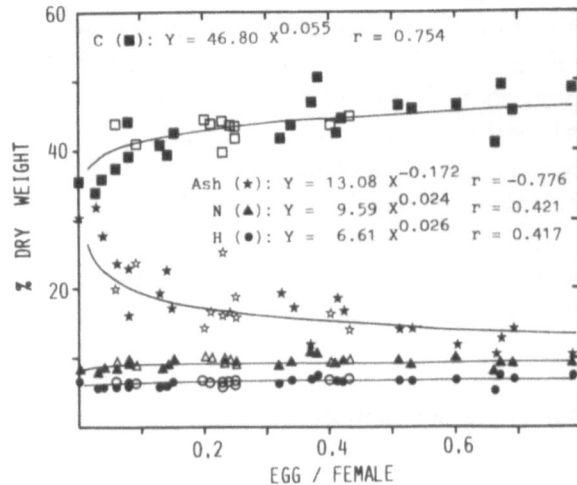

Fig. 3. Regressions of percent C, N, H and ash on egg/female ratio. Open symbols = rotifer fed on algae; dark symbols = rotifer fed on yeast.

Table 1. Caloric data of rotifer biomass as a function of egg/female ratio. Values estimated from regression equations.

E/F ratio	0.0	0.1	0.2	0.4	0.6	0.8
cal/rotifer	1278	1669	2060	2842	3624	4406
cal/mg DW	4.06	4.39	4.61	4.89	5.08	5.21
cal/mg AFDW	5.49	5.61	5.69	5.77	5.83	5.87

according to a power curve. This increase occurs until E/F reaches 0.2 egg per female. Percentage of ash increases markedly when the fecundity is very low and it may reach up to 32% of body weight.

The C, N and H contents in weight per individual follow a similar pattern that the body dry weight and show high correlations with E/F ratio (Fig. 4). The value of these parameters are three or four times higher in the samples showing highest fecundity than in the populations without eggs.

The calorimetric data estimated from carbon content has values ranging from 1278 to 4406 μcal per individual and between 4.06 and 5.21 cal per mg of dry matter (Table 1).

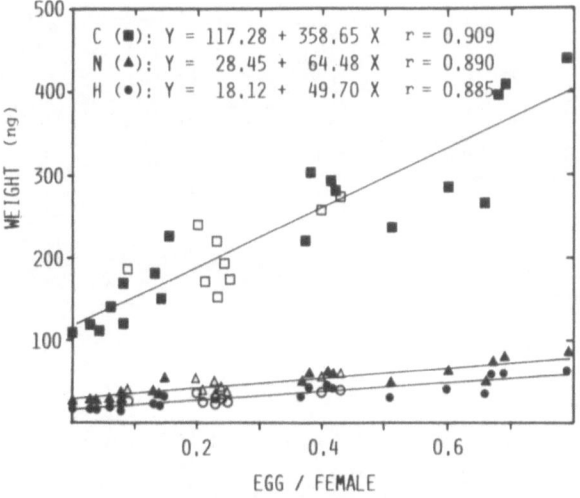

Fig. 4. Regressions of C, H, and N content per rotifer (dry weight · individual^{-1}) on egg/female ratio. Open symbols = rotifer fed on algae; dark symbols = rotifer fed on yeast.

Discussion

Rotifer dry weight obtained in this study is higher than that obtained for the same strain fed on the Chlorophyceae *Nannochloris oculata* (Yúfera *et al.*, 1983). This agrees with the bigger size of females and eggs observed when fed on *N. gaditana* at the same culture conditions (Yúfera, 1987).

Relationships between biomass and E/F ratio has been reported by Scott & Baynes (1978) and Yúfera *et al.* (1983). The present work reveals that C content, and therefore its caloric equivalent, are also related to E/F. This supports the usefulness of E/F as a good index to assess the nutritional quality and growth potential of population (Yúfera & Pascual, 1984; Snell *et al.*, 1987).

Theilacker & McMaster (1971) determined a caloric content of 5.34 cal · mg^{-1} ash free DW for *B. plicatilis* (females without eggs). Doohan (1973) measured 790–1290 μcal per rotifer (DW = 158–250 ng · ind^{-1}, assuming 50% of C content). From C data reported by Schlosser & Anger (1982), a caloric content of 1264 μcal per rotifer (DW = 280 ng · ind^{-1}) and 4.5 cal · mg^{-1} DW can be estimated. More recently Theilacker & Kimball (1984) determined 440–2070 μcal (4.4 cal · mg^{-1} DW and 4.8 cal · mg^{-1} ash free DW, assuming 8.7% of ash content) for non-ovigerous females ranging from 100 to 470 ng · ind^{-1} DW. In general, values reported in this study (4.06–5.21 cal · mg^{-1} DW) are similar to those presented for these authors and agree with 5.4–5.7 cal · mg^{-1} AFDW estimated for aquatic invertebrates (Wacasey & Atkinson, 1987). The differences among caloric values presented by different authors are attributable to variations in ash content.

It is interesting to note that caloric equivalent

of the organic matter (cal·mg^{-1} AFDW) increases with increasing fecundity, which is associated with the yolk content in the population. In routine semicontinuous mass culture the E/F ratio usually ranges between 0.2–0.4. Caloric content per rotifer in these conditions ranged from 1871 to 2574 μcal (4.61–4.89 cal·mg^{-1} DW). When food was exhausted fecundity fell quickly and the caloric content decreased 1275 μcal per individual and 4.06 cal·mg^{-1} DW.

Acknowledgements

We thank Olimpio Montero and José Mª Espigares for technical assistance and Manuel Arjonilla for its collaboration in the CHN analyses.

References

Salonen, K., J. Sarvala, I. Hakala & M. L. Viljanen, 1976. The relation of energy and organic carbon in aquatic invertebrates. Limnol. Oceanogr. 21: 424–730.

Schlosser, H. J. & K. Anger, 1982. The significance of some methodological effects on filtration and ingestion rates of the rotifer *Brachionus plicatilis*. Helgoländer wiss. Meeresunters. 35: 215–225.

Scott, A. P. & S. M. Baynes, 1978. Effect of algal diet and temperature on the biochemical composition of the rotifer, *Brachionus plicatilis*. Aquaculture 14: 247–260.

Snell, T. W., M. J. Childress, E. M. Boyer & F. H. Hoff, 1987. Assessing the status of rotifer mass cultures. J. World Aqua. Soc. 18: 270–277.

Theilacker, G. H. & A. S. Kimball, 1984. Comparative quality of rotifers and copepods as food for larval fishes. CalCOFI Rep. 25: 80–86.

Theilacker, G. H. & M. F. McMaster, 1971. Mass culture of the rotifer *Brachionus plicatilis* and its evaluation as a food for larval anchovies. Mar. Biol. 10: 183–188.

Wacasey, J. W. & E. G. Atkinson, 1987. Energy values of marine benthic invertebrates from the Canadian Artic. Mar. Ecol. Prog. Ser. 379: 243–250.

Yúfera, M., 1982. Morphometric characterization of a small-sized strain of *Brachionus plicatilis* in culture. Aquaculture 27: 56–61.

Yúfera, M., 1987. Effect of algal diet and temperature on the embryonic development time of the rotifer *Brachionus plicatilis* in culture. Hydrobiologia 147: 319–322.

Yúfera, M. & E. Pascual, 1984. Influencia de la dieta sobre la puesta del rotifero *Brachionus plicatilis* en cultivo. Inv. Pesq. 48: 549–556.

Yúfera, M. & L. M. Lubián & E. Pascual, 1983. Efecto de cuatro algas marinas sobre el crecimiento poblacional de dos cepas de *Brachionus plicatilis* (Rotifera: Brachionidae) en cultivo. Inv. Pesq. 47: 325–337.

Hydrobiologia **186/187**: 375–380, 1989.
C. Ricci, T. W. Snell and C. E. King (eds), Rotifer Symposium V.
© *1989 Kluwer Academic Publishers.*

Molecular genetics of rotifers: preliminary restriction mapping of the mitochondrial genome of *Brachionus plicatilis*

Charles E. King
Dept. of Zoology, Oregon State University, Corvallis, OR 97331, U.S.A.

Key words: mitochondrial DNA, genetic variation, restriction mapping, rotifers, *Brachionus plicatilis*

Abstract

Methods are presented to extract and purify mitochondrial DNA from the rotifer *Brachionus plicatilis*. The mtDNA obtained is of sufficient purity for digestion with restriction endonucleases. *Eco*R I restriction patterns are presented for 4 geographically separated clones. A restriction map based on digestion with 5 different restriction enzymes is included for one of these clones. Finally, use of mtDNA analysis for studies on the population structure and biogeography of rotifers is discussed.

Introduction

Over the past decade a change of revolutionary proportions has taken place in evolutionary biology. The methodology of molecular genetics has profoundly impacted the study of such diverse areas as biogeography, interspecific gene flow, systematics, phylogenetic relationships and the genetic structure of populations. Problems in these areas that have long been sequestered in the realm of speculation are now yielding to objective analysis (Avise *et al.*, 1987; Hillis, 1987; Moritz *et al.*, 1987). Much of this progress has been made possible by molecular analysis of mitochondrial DNA (mtDNA).

The mtDNA of metazoans is composed of double-stranded, circular molecules that vary from about 15.5 to 20 kbp (kilobase pairs) in size. Mitochondria are maternally inherited and their DNA lacks introns and has few or no spacers between genes. The mtDNA base sequences of humans, cattle, mice and 2 species of *Drosophila* are now known and in each case the gene pro-

ducts are restricted to 2 rRNAs (12S and 16S), 22 tRNAs and 13 polypeptides that serve as subunits for the enzymes of the inner mitochondrial membrane (Brown, 1985; Garesse, 1988). Rates of mtDNA evolution in vertebrates appear to be substantially higher that those of single-copy nuclear DNA, but this relationship may not hold for some invertebrates (Caccone *et al.*, 1988).

The present paper is the first to present information on the mtDNA of rotifers. In fact the entire literature on the molecular biology of rotifers consists of only two papers; Kumazaki *et al.* (1982) sequenced the 5S rRNA of *Brachionus plicatilis* and Ishikaea (1977) studied the thermal stability of the rRNA of the same species. An important qualification is required for the present paper. Successful extraction of mtDNA and its digestion with restriction enzymes has only recently occurred and additional replication and confirmatory tests are necessary before the detailed quantitative results of this research can be accepted with complete confidence. Never-the-less, the power of this approach and its potential applicability to many of the problems being discussed at this symposium make publication of the techniques and the preliminary results appropriate.

Preparation of mtDNA

Mitochondrial DNA was extracted and purified using a modification of the methods employed by MacNeil & Strobeck (1987). Because the success and yield of the extraction is dependent upon closely following the protocol, procedures will be presented in detail. Except as noted, all steps were carried out at 4 °C using autoclaved labware.

Brachionus plicatilis clones were cultured for mtDNA analysis in 20 l glass carboys in a 25 °C constant temperature room. *Saccharomyces cerevisiae* was used as food and autoclaved 2/3 Pacific Ocean sea water as the medium. When ready for DNA extraction, individuals were removed from their medium by filtration through a 60 μm nitex screen and resuspended overnight in 2 l sea water without food to permit the rotifers to clear previously ingested food. The following morning the rotifers were again filtered from the medium, a wet weight was taken, and the animals were washed off the filter and into a 15 ml Dounce tissue grinder with approximately 10 ml of ice cold pH 8.5 H (homogenization) medium (0.07 M sucrose, 0.21 M manitol, 1.9 mM HEPES, 0.5% bovine serum albumen). Approximately 2–3 g wet weight of rotifers were used for each extraction.

Animals were homogenized on ice with five up and down strokes of the grinder. A small aliquot of the homogenate was then checked under a dissecting scope to ensure adequacy of homogenization. The homogenate was transferred to a 15 ml corex tube and centrifuged in a Sorvall SS-34 rotor at 3000 rpm (1085 × g) for 4 min to pellet debris and nuclei. Mitochondria remain suspended in the H medium during this centrifugation because of their density. The supernatant was transferred to a 50 ml Oak Ridge tube and held on ice while the pellet was resuspended in 10 ml H medium and recentrifuged as above. The supernatant from this second centrifugation was combined with the first and the pellet resuspended in 10 ml H medium and centrifuged a third time. The combined supernatants from the three centrifugations were then centrifuged at 3000 rpm for 4 min and the supernatant was transferred to a clean Oak Ridge tube. The pellet was discarded. This procedure was repeated 3 to 5 times until no visible pellet was formed.

The clean supernatant was then centrifuged at 7500 rpm (6780 × g) for 15 min to pellet the mitochondria. The mitochondrial pellet should be a uniform pale yellow color – a brown ring in the pellet indicates that additional slow spins should be made to further clean the preparation before proceeding. The supernatant from this centrifugation was discarded and the mitochondrial pellet was resuspended in 10 ml H medium using a pasteur pipette. The mitochondrial suspension was then centrifuged for 10 min at 13000 rpm (20200 × g) and the supernatant was again discarded.

The pellet from this second high speed spin was resuspended in 1.4 ml H medium and divided between two 1.5 ml microfuge tubes. To digest any contaminating nuclear DNA in the mitochondrial suspension, $MgCl_2$ (1 M stock) and DNase I (10 mg ml^{-1} stock) were added to final concentrations of 0.01 M and 40 μg ml^{-1} and incubated on ice for 30–60 min. At this point the mitochondria should still be intact thereby protecting the mtDNA from degradation by the DNase. Following the incubation the tubes were microfuged at 4 °C for 10 min and the pellets were resuspended in 1 ml H medium and again pelleted as above.

The washed pellets were gently resuspended with a pasteur pipette in 950 μl of pH 7.5 L (lysis) buffer (0.1 M EDTA, 0.1 M Trizma-HCl, 0.2 M NaCl) and placed on ice for 10 min. Proteinase K from a 10 mg ml^{-1} stock was then added to a concentration of 50 μg ml^{-1} and incubated at 37 °C for 1 hr. SDS from a 30% stock was added to a final concentration of 1% and the incubation continued for another 30 min. The microfuge tubes were then placed on ice for 15 min and microfuged for 5 min. The resulting pellet was discarded. Contaminating proteins and SDS were removed from the supernatants by three extractions with 0.1 M, pH 8.0 Trizma-HCl buffer-saturated phenol. An additional extraction was performed with a 1:1 mixture of phenol: chloroform (containing 4% isoamyl alcohol here and subsequently) and another extraction was performed with the chloroform. Traces of phenol were then removed with two ether extractions.

After residual ether was removed using a gentle stream of air, the DNA was precipitated overnight with 1/10 volume of 3 M NaOAc at pH 8.0 and two volumes of 100% ethyl alcohol at $-20\ °C$. The DNA was then pelleted by microfuging for 15 min and the pellets were washed twice in 70% EtOH and desiccated under vacuum. 100 μl of TE buffer (10 mM trizma, 1 mM EDTA, pH 8.0) were added to each of the dried pellets and the tubes were placed at 4 °C until the DNA dissolved.

RNase T1 (from a stock of 1000 units ml^{-1}) and RNase A (from a stock of 10 mg ml^{-1}) were added to the samples at concentrations of 40 units and 100 μg per ml and incubated for 1 hr at 50 °C. Samples were then extracted once each with phenol, phenol-chloroform, chloroform and then twice with ether. A second overnight mtDNA precipitation was carried out with 1/2 volume of 7.5 M NH$_4$OAc and 2 volumes of 100% EtOH at $-20\ °C$. After pelleting, washing twice with 70% EtOH, and desiccation, the mtDNA in each microfuge tube was redissolved in 30 μl of water.

To check for purity and yield of mtDNA, a 2 μl aliquot was electrophoresed on a 0.75% agarose minigel. *Hind* III digested bacteriophage lambda DNA was used for a size marker. Ideally, nuclear DNA should be absent and the rotifer mtDNA should be seen as two bands, i.e., one for the relaxed and one for the supercoiled mtDNA. Restriction digests were performed according to the recommendations of the enzyme supplier with the following important exception. Each digest was carried out by diluting the desired volume of mtDNA solution to 20 μl with water, adding 2.5 μl of 10× buffer, and then adding 1 μl of 0.1 M spermidine. This solution was placed on ice for 10 min and then 1 μl of restriction enzyme was added. Incubation for the digestion took place at

the recommended temperature for at least 3 hr (and usually overnight). Three μl of stop mixture were then added and the electrophoresis was carried out at 30–40 V with a 0.7–0.9% agarose gel.

This protocol differs from most DNA isolation procedures not employing density-gradient centrifugation in two regards. The first of these is the DNase treatment. In the absence of this step it is not unusual to see substantial quantities of nuclear DNA contaminating the preparation. This contamination is either from nuclei that ruptured early in the procedure or from nuclei that were not removed by the initial 3000 rpm centrifugations. While this nuclear DNA may not interfere with the analysis, its removal by DNase treatment was desirable since specific mitochondrial probes have not been used in this study.

The second and more critical difference is the use of spermidine prior to addition of the restriction enzymes. A potent inhibitor of restriction enzyme activity is liberated from the rotifers during the isolation procedure. This inhibitor coprecipitates with the mtDNA and produces a complete or partial inhibition of the digestion. An alternative method to remove the inhibitor – spermine precipitation – was also successful, but was judged to be more time consuming than the spermidine treatment.

Results and discussion

Preliminary analyses have been carried out on four clones of *Brachionus plicatilis*. Clone Q16 was collected in October, 1984 from Dong Feng Saltern on the Shandong Peninsula near Qingdao, People's Republic of China (PRC). Additional information on this site is available in King *et al.* (1988a, b). Clone G8 was collected at Shenzhen Saltern near Guangzhou (Canton) in the Southern PRC. Clone AS is from Scotland (but was not mentioned by May, 1983) and was given by J.M. Scott to A. Ruttner-Kolisko who supplied it to me. This clone was designated S in Ruttner-Kolisko (1983). Clone BS3 was collected

378

by the author in December, 1987 from Big Soda Lake in Nevada, U.S.A. (King & Zhao, 1987).

Figure 1 presents restriction patterns of the four clones obtained after digestion with the enzyme *Eco* R I. Variation among clones is of two types. First, there are obvious restriction fragment length polymorphisms (RFLPs) within clones that can be used to distinguish between clones. Second, the estimated sizes of the four mitochondrial genomes vary from 17.5 kb (Q16) to 19.9 kb (AS). As discussed by Brown (1985), most size variation in mtDNA is attributable to changes in the displacement loop (D-loop) region that is presumed to act as the origin of H strand replication in vertebrates. The same region also appears to have more base sequence variation. Thus the variability among clones seen in Fig. 1 may be nonrandomly distributed over the entire mitochondrial genome. In particular each of the four clones has a fragment of approximately 5.4 kb in size (the apparent variation is less than 300 bp). Mitochondrial probes could be used to determine whether or not these bands are the same. Unfortunately, however, a mitochondrial genome library has not been constructed for *B. plicatilis* or any other rotifer.

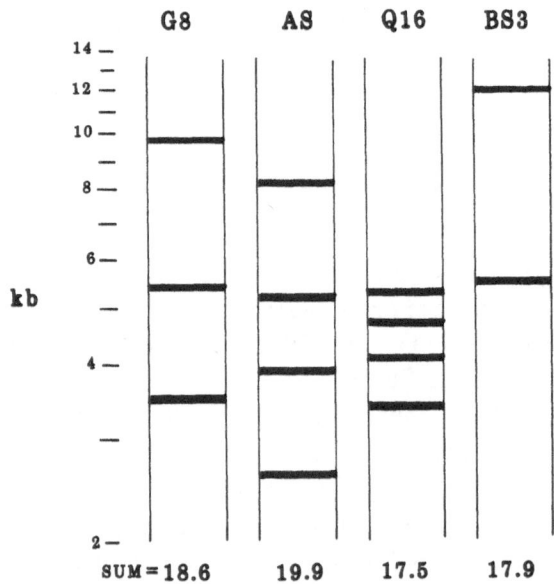

Fig. 1. Eco R I restriction patterns for the mtDNA of four strains of *Brachionus plicatilis*. The vertical scale is fragment size in kilobases (kb) and the total mitochondrial genome sizes are given at the bottom of the figure.

While the RFLP patterns in Fig. 1 are obviously quite different, for such comparisons to be useful intrapopulation variability must also be measured. This step will require identification of more map sites, i.e., genetic markers, so that the clonal structure of a given population can be specified with greater confidence. Two methods are available for increasing the number of map sites per clone. *Eco*R I and the other restriction enzymes used in this study have hexanucleotide targets – that is, a specific sequence of six base pairs is required for cleavage. Under the assumption of random ordering of four bases, hexanucleotide targets are expected to occur at intervals of $4^6 = 4096$ bases. A mitochondrial genome of 18.6 kb is therefore expected to have 4.5 target sites. An obvious way to increase the number of genetic markers is to use restriction enzymes that recognize smaller targets. Enzymes that recognize pentanucleotide targets are expected to produce 18.2 cleavages in the same molecule, those that recognize tetranucleotide targets are expected to produce 72.7 cleavages.

An alternative method to obtain more genetic markers per clone is to increase the number of restriction enzymes employed. This latter technique also has the advantage of making it possible to physically map the mitochondrial genome.

Of the four clones discussed in this paper, only G8 has been studied with more than a single mtDNA preparation. To date, the mitochondrial genome of G8 has been treated with five restriction enzymes in both single and double digestions. The resultant data have been used to construct the preliminary restriction map presented in Fig. 2. Twenty-two sites have been identified for the five enzymes: four for *Bam*H I, six for *Bgl*II, three for *Eco*R I, four for *Hin*d III, and five for *Xba* I.

The map in Fig. 2 has sufficient detail to permit reasonable determinations of clonal identity. The strategy to be employed in future work is to determine the amount of map variability among clones from the same lake. Findings based on allozyme differences suggest relatively little variation within populations of *Asplanchna girodi* (King, 1977, 1980), but substantially more variation within the

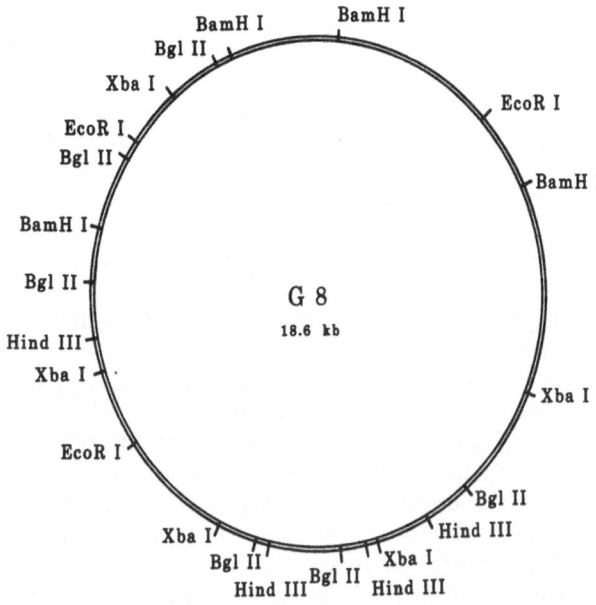

Fig. 2. Mitochondrial DNA restriction map of *Brachionus plicatilis* clone G8 based on digestions with five restriction endonucleases.

compare the mtDNA patterns of nearby and distant (intercontinental) populations to determine patterns of colonization. That is, homogeneity of populations within an area and heterogeneity between areas would support a vicariance model based on little effective gene flow. The counter finding – as much variation within an area as between areas – would provide powerful evidence for the role of gene flow in tying together disjunct populations of a species.

The role of dispersal in determination of species distributions constitutes but one example of an important problem in rotifer biology that might yield to mtDNA analysis. Many others occur in fields as diverse as hybridization and ecotypic vs genetic determinants of morphological variation (Pejler, 1980; Serra-Galindo, 1987). It thus seems likely that the revolution referred to in the first sentence of this paper will likewise profoundly impact the study of rotifer biology.

B. plicatilis of Soda Lake (King & Zhao, 1987; Zhao & King *in press*). If the mtDNA restriction maps of this latter species show a comparable amount of variation, it is likely that the Soda Lake rotifers had multiple independent origins. On the other hand, homogeneity of the mtDNA would suggest a small number of founders. Thus the analysis of mtDNA variation is potentially a powerful tool to unravel details of population structure and dispersal.

At our 1982 Uppsala symposium, Dumont (1983) presented an elegant treatment of rotifer biogeography. The core of this paper was a consideration of the relative significance of dispersal vs. vicariance in determining the distributions of rotifers. That rotifers have a remarkable capacity for passive dispersal via their resting eggs is beyond question. At an earlier rotifer symposium, Lair (1980) reported collecting *Brachionus havanaensis* (a Caribbean species) and *Lecane papuana* (restricted to tropical and subtropical areas of the eastern hemisphere) immediately downstream from the heat exchangers of nuclear power plants in central France. If such dispersal is both frequent and successful it should be possible to

References

Avise, J. C., J. Arnold, R. M. Ball, E. Bermingham, T. Lamb, J. E. Neigel, C. A. Reeb & N. C. Saunders, 1987. Intraspecific phylogeography: The mitochondrial bridge between population genetics and systematics. Ann. Rev. Ecol. Syst. 18: 489–522.

Brown, W. M., 1985. The mitochondrial genome of animals. In: MacIntyre, R. J. (Ed.). Molecular Evolutionary Genetics. Plenum Press, New York: 95–130.

Caccone, A., G. D. Amato & J. R. Powell, 1988. Rates and patterns of scnDNA and mtDNA divergence within the *Drosophila melanogaster* subgroup. Genetics 118: 671–683.

Dumont, H. J., 1983. Biogeography of rotifers. Hydrobiologia 104: 19–30.

Garesse, R., 1988. *Drosophila melanogaster* mitochondrial DNA: Gene organization and evolutionary considerations. Genetics 118: 649–663.

Hillis, D. M., 1987. Molecular versus morphological approaches to systematics. Ann. Rev. Ecol. Syst. 18: 23–42.

Ishikaea, H., 1977. Comparative studies on the thermal stability of animal ribosomal RNA's. IV. Thermal stability and molecular integrity of ribosomal RNA's from several protostomes (rotifers, round-worms, liver-flukes, and brine shrimps). Comp. Biochem. Physiol. 56: 229–234.

King, C. E., 1977. Genetics of reproduction, variation, and adaptation in rotifers. Arch. Hydrobiol. Ergebn. Limnol. 8: 187–201.

King, C. E., 1980. The genetic structure of zooplankton populations. In: Kerfoot, W. C. (Ed.). Evolution and Ecology of Zooplankton Communities. New England University Press: 315–329.

King, C. E. & Y. Zhao, 1987. Coexistence of rotifer (*Brachionus plicatilis*) clones in Soda Lake, Nevada. Hydrobiologia 147: 57–64.

King, C. E., Y. Zhao, X. Liu & M. R. Li, 1988a. Polyacrylamide gel electrophoresis of isozyme variation in *Artemia* from Chinese salterns. J. Shandong College of Oceanography 18: 64–69.

King, C. E., Y. Zhao, X. Liu & M. R. Li, 1988b. Genetic variation in parthenogenetic *Artemia* from the Shandong Peninsula, P.R.C. Oceanology and Limnology 6: 179–185.

Kumazaki, T., H. Hori, S. Osawa, N. Ishii & K. Suzuki, 1982. The nucleotide sequences of 5S rRNA's from a rotifer, *Brachionus plicatilis*, and two nematodes, *Rhabditis tokai* and *Caenorhabditis elegans*. Nucleic Acids Res. 10: 7001–7004.

Lair, N., 1980. The rotifer fauna of the river Loire (France), at the level of the nuclear power plants. Hydrobiologia 73: 153–160.

MacNeil, D. & C. Strobeck, 1987. Evolutionary relationships among colonies of Columbian ground squirrels as shown by mitochondrial DNA. Evolution 41: 873–881.

May, L., 1983. Rotifer occurrence in relation to water temperature in Loch Leven, Scotland. Hydrobiologia 104: 311–315.

Moritz, C., T. E. Dowling & W. M. Brown, 1987. Evolution of animal mitochondrial DNA: relevance for population biology and systematics. Ann. Rev. Ecol. Syst. 18: 269–292.

Pejler, B., 1980. Variation in the genus *Keratella*. Hydrobiologia 73: 207–213.

Ruttner-Kolisko, A., 1983. The significance of mating processes for the genetics and for the formation of resting eggs in monogonont rotifers. Hydrobiologia 104: 181–190.

Serra-Galindo, M., 1987. Variación morfométrica, isoenzimática y demográfica en poblaciones de *Brachionus plicatilis*: Diferenciación genética y plasticidad fenotípica. Ph.D. Thesis, Universitat de Valencia.

Zhao, Y. & C. E. King. Ecological genetics of the rotifer *Brachionus plicatilis* in Soda Lake, Nevada, U.S.A. Hydrobiologia *in press*.

Hydrobiologia **186/187**: 381–386, 1989.
C. Ricci, T. W. Snell and C. E. King (eds), Rotifer Symposium V.
© 1989 Kluwer Academic Publishers.

Size variation in *Brachionus plicatilis* resting eggs

Laura Serrano, Manuel Serra & Maria R. Miracle
Dep. Ecologia. Universidad de Valencia, 46100-Burjasot (Valencia), Spain

Key words: biometric analysis, resting eggs, rotifers, salinity, temperature, genetic variation

Abstract

The effect of temperature and salinity on resting egg size of two *Brachionus plicatilis* (Rotifera) clones was investigated. Clones were selected according to their different behaviour in laying resting eggs: one clone ejects them, whereas they remain inside the females body in the other clone. The difference in resting eggs size between the two clones is noticeable, although the difference is not as great as that between female body size. An important temperature-salinity interaction on resting egg size has been observed. The general inverse relationship between size and temperature is only true at lower temperatures. At high temperatures size varies around the mean although could be greater than at intermediate temperatures. This is more evident at the intermediate salinity tested which is considered to be the closest to the optimum in our experiments. This pattern of variation suggests that mean size is bigger than expected, in relation to temperature and salinity, when these factors have values close to the extremes of their range, normally found in nature, and to which adaptative mechanisms can evolve. Size is bigger at the salinity – temperature low – low and high – high combinations which are the most commonly found in the temperate environments.

Introduction

In a previous paper (Serra & Miracle, 1986) several clones of *Brachionus plicatilis*, a cosmopolitan species from a fluctuating environment, were used to clarify the degree of morphological plasticity that this genome expresses. We examined plasticity by studying the effect of temperature and salinity on body size.

We have extended that work by analyzing the direct response of the same clones to the same sets of experimental conditions, but this time studying the size of the resting eggs produced. Our intention is to evaluate the extent of genetic control and of environmental variation on the size of the sexual cyst, which has dormancy and dispersal capabilities. Moreover, the evaluation of these effects may be useful in exploring the recent history of natural rotifer populations by monitoring resting eggs in the sediment.

Materials and methods

Biometrical experiments on resting eggs were carried out on two clones of *Brachionus plicatilis*, CU, collected from a continental saline lake, and SPO, collected from a coastal marsh, which have been maintained in laboratory culture since 1981. The clones have been characterized by electrophoretic analysis as well as morphometrically (Serra & Miracle, 1983, 1985 and 1986).

382

The two clones of *B. plicatilis* were cultured in diluted sea water enriched with f/2 medium of Guillard & Ryther (1962) in flasks of 300 ml, using *Tetraselmis* sp. as food at concentrations greater than 10^6 cells ml^{-1}. Rotifers were cultured at different salinities (9, 12 and 24‰) and temperatures (20, 25 and 30 °C).

For each salinity, temperature and clone, two replicate flasks (blocks) were prepared; more than three months elapsed between the replicate experiments. The cultures were kept under the experimental conditions for three or more weeks until resting eggs appeared. Each week, half of the culture volume was renewed with new medium containing more than 10^6 cells ml^{-1}. Then, 25 resting eggs were collected from each replicate flask at random.

Resting egg size was determined by measuring length and width, and for clone SPO the width of the lorica containing the eggs was also measured (Fig. 1). All measurements were made with a Wild M40 microscope at 40 ×. Egg volume was estimated representing the egg as a composite figure of two spheres connected by a cylinder of the same diameter. This shape is more similar to that of a resting egg than a revolution ellipsoid. Volume was computed as $V = (\pi/12)(3WL - W)$ where L is the length and W the width of the egg.

To determine the main effects and interactions of the different experimental conditions and clones on resting egg size, a multivariate analysis of variance (MANOVA) was performed and their global significance was tested with Wilkis' criterion (Cooley & Lohnes, 1971; Kres, 1983). The significance of each effect on individual measurements was determined with a univariate analysis of variance (ANOVA) as described by Cooley & Lohnes (1971).

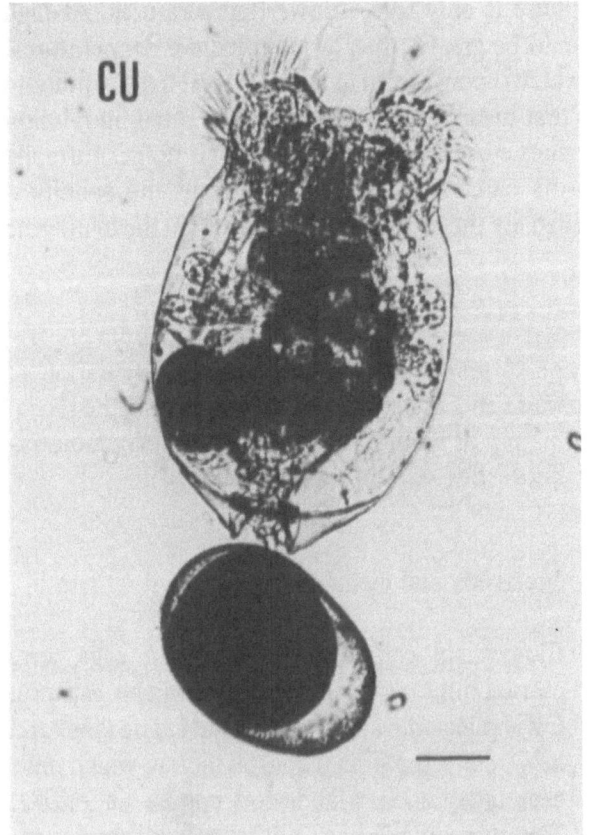 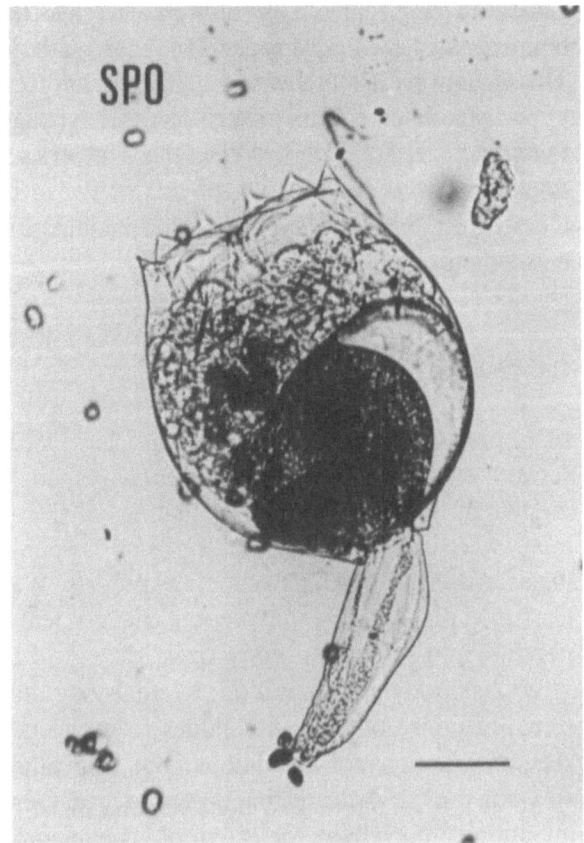

Fig. 1. Photomicrograph of clone CU and SPO. Bar is 40 μm.

Results

Two clones were selected to study the effect of environmental conditions on resting egg size, in clone CU which lays resting eggs outside the female and SPO which deposits resting eggs inside the animal where they remain until its death, with release occuring when the lorica decomposes (Fig. 1). Resting eggs of both clones are viable and their hatching levels (40-70%) were similar to that obtained by Minkoff *et al.* (1983).

Figures 2, 3 and 4 show means and standard

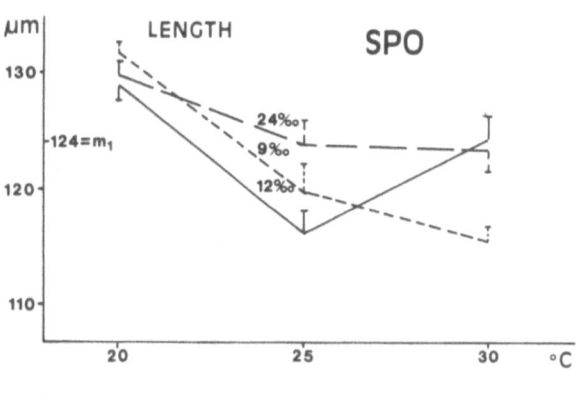

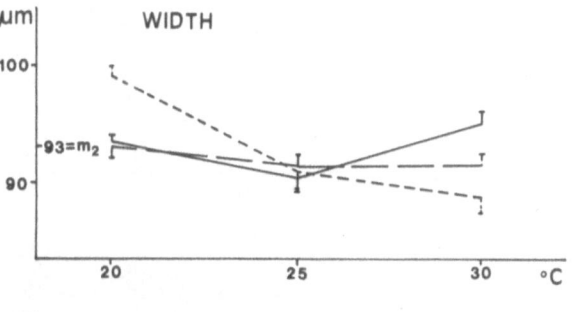

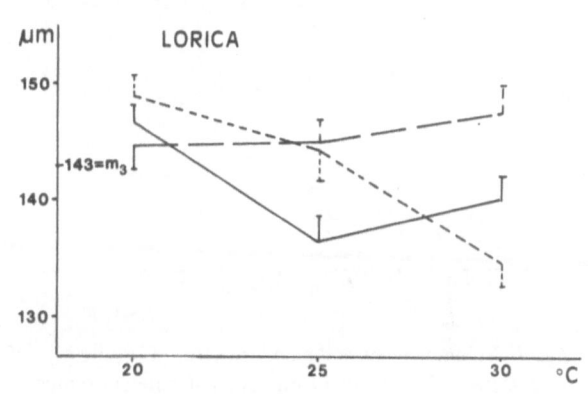

Fig. 3. Length and width means and standard error of resting eggs of clone SPO and width of the lorica containing a resting egg (ordinates) as a function of temperature (20, 25 and 30 °C, abscissas) and salinity (9‰, 12‰ and 24‰). The global means for each measurement are also indicated (m^1, m^2, m^3).

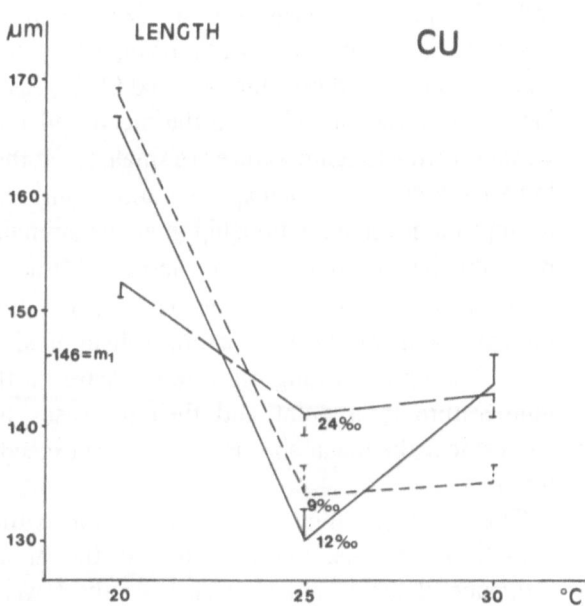

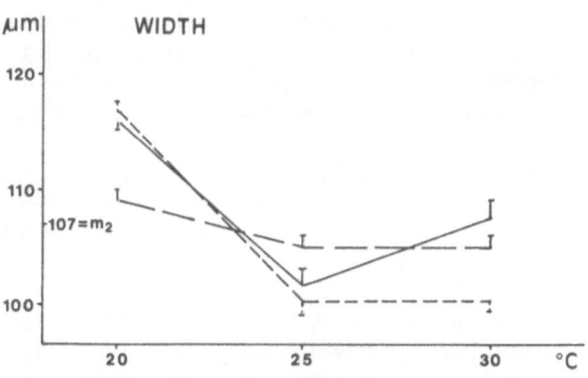

Fig. 2. Length and width means and standard errors of resting eggs of clone CU (in ordinates) at different conditions of temperature (20, 25 and 30 °C, abscissas) and salinity (9‰, 12‰ and 24‰). The global means for each measurement are also indicated (m^1, m^2).

errors of resting egg length, width and volume at the different temperatures and salinity conditions.

The results of the multivariate analysis of variance (Table 1) show significant differences between clones and the significant effect of temperature and blocks in all measurements. The interactions clone-temperature and temperature-salinity are also significant in terms of length. So the genotype largely determines the response to

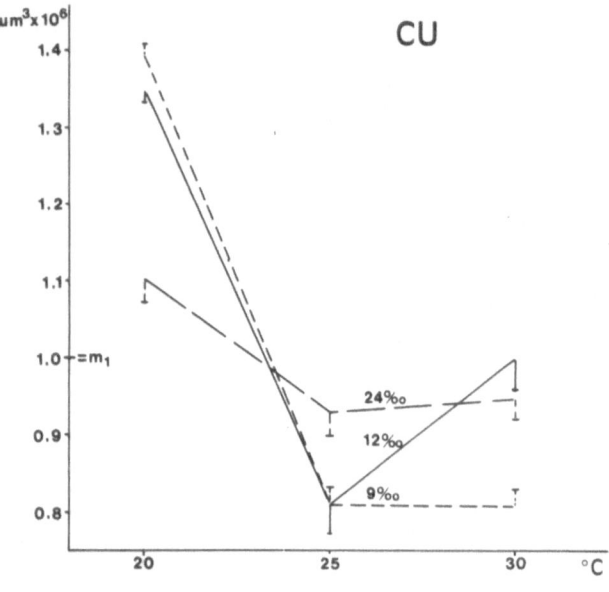

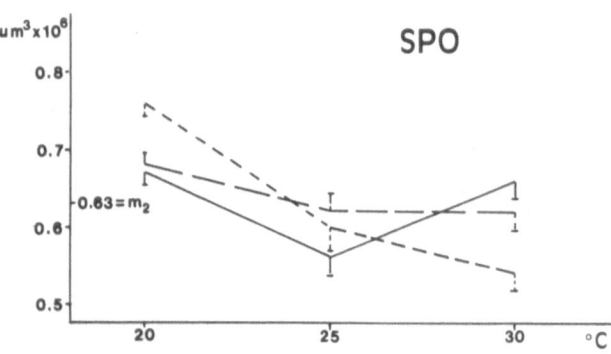

Fig. 4. Volume means and standard error of resting eggs of both clones CU and SPO (ordinates) at different temperatures (20, 25 and 30 °C, abscissas) and salinity (9‰, 12‰ and 24‰). The global means are also indicated (m^1, m^2).

temperature, and salinity influences this response greatly.

Differences in resting eggs size between the two clones are noticeable, the average length and width of the CU resting eggs being 1.18 and 1.16 times bigger respectively than the average length and width of the SPO resting egg. However, these differences are much smaller than the differences between amictic female body sizes. CU females are approximately 1.4 times bigger than SPO females (the relationship between the average lengths of the two clones, at different conditions is 1.41, and between their widths it is 1.38 (Serra & Miracle, 1986).

The effect of temperature on the width of the mictic females of clone SPO was different from the effect on parthenogenetic females of this clone. Parthenogenetic females showed a decrease in width at higher temperatures at all salinities and an increase at higher salinities (Serra & Miracle, 1986). In contrast, the response to temperature-salinity conditions of mictic females with resting eggs had a complex pattern similar to that of resting eggs (Fig. 1). The differences in the width of these females were much more attenuated than those of amictic females.

The general effects of temperature on resting egg size for both clones can be summarized in the following points: (1) variation in size is mainly due to variation in length, width being more constant, (2) at the highest salinity tested (24‰) resting egg size remains close to the mean and the variations due to temperature are small, (3) at the lowest salinity (9‰) the expected and confirmed asymptotic inverse relationship in active animals between size and temperature (Serra & Miracle, 1986) is also true for resting eggs, (4) at the medium salinity (12‰) temperature effect on size is very complex, resting egg size decreases with temperature to a point and then increases to values near the mean size, in most cases exceeding it.

The effect of salinity is not significant in the statistical analysis, perhaps due to the great influence of temperature on size. At the lowest

Table 1. MANOVA and ANOVA results of three factorial fixed effects (clone, temperature and salinity) and a random nested effect (blocks) on two resting eggs measurements ($N = 25$,* is $p < 0.05$ and ** is $p < 0.01$).

Source	Wilks' λ	df		F ratio	
		n_1	n_2	Length	Width
Clone (C)	0.0547**	1	18	298.3**	75.7**
Temperature (T)	0.0848**	2	18	80.0**	8.5**
Salinity (S)	0.7900	2	18	0.9	0.3
$C \times T$	0.4183*	2	18	10.4**	1.3
$C \times S$	0.8120	2	18	0.8	0.2
$T \times S$	0.2862**	4	18	8.8**	1.9
$C \times S \times T$	0.7243	4	18	1.4	0.3
Blocks	0.06830**	18	864	3.4**	15.6**

Table 2. Linear correlations between egg length and width at different temperatures and salinities.

Clone	Salinity	Temperature		
		20 °C	25 °C	30 °C
CU				
	9‰	0.361	0.856	0.699
	12‰	0.181	0.676	0.796
	24‰	0.625	0.557	0.367
SPO				
	9‰	0.152	0.788	0.489
	12‰	0.171	0.529	0.460
	24‰	0.301	0.551	0.661

temperature, resting eggs are bigger at the lowest salinity levels and smaller at the highest in both clones. However, at the highest temperature they are smaller at the lowest salinity level and they are bigger at the intermediate optimum salinity 12‰.

Since the ratios length/width of the two clones are rather similar, we investigated the degree of correlation between these measurements under different conditions (Table 2). The two measurements are similar, but the correlation coefficients are not very high, approximately 0.5. In a few cases correlation is very low. These cases are the same for both clones and correspond to rotifers reared at the lowest assayed temperature, where length greatly exceedes the mean, but width closely approximates the mean.

The comparison of these results with those derived from the effect of temperature and salinity on the body size of active parthenogenetic females (Table 3) shows that the effect of these factors on resting egg length is similar to the effect on body length. However, in active females the effect on

Table 3. Ratios of the maximum and minimum mean values at the different temperatures and salinities (L: length; W: width).

	CU		SPO		
	Body size	Resting egg	Body size	Resting egg	Lorica
L	1.28	1.29	1.15	1.14	–
W	1.36	1.16	1.22	1.11	1.11

width is greater than on length, while in resting eggs width shows little variation.

Discussion

The two clones differ much more in amictic female size than in resting eggs size. However resting eggs size in each clone is only slightly less variable than body size of parthenogenetic females. This variation in resting eggs size has been observed in nature and on laboratory cultures whenever it has been measured. For instance, polymorphism has been observed in rotifer resting eggs (Gilbert, 1974), and a significant size variation of *B. plicatilis* resting eggs in relation to their vertical location in sediments has been recorded (Snell *et al.*, 1983).

The two clones selected for this study originated from contrasting environments: CU comes from a shallow highly fluctuating, endorreic saline, inland lagoon, which shows an euryoic behaviour. Clone SPO comes from a less fluctuating karstic coastal lagoon (Serra, 1987). Resting eggs of SPO from a more stable environment are bigger in relation to the size of the maternal females than those produced by clone CU. Moreover, the variation in resting eggs size of the euryoic clone CU with salinity and temperature variation is higher. When body size was analyzed (Serra & Miracle, 1986), clone CU again showed an important increase after a decrease with rising temperature when salinity was high. Clone SPO showed the expected inverse size-temperature relationship at all tested salinities.

The tendency to follow the inverse size-temperature relationship has been confirmed in a great variety of organisms, including amictic eggs of *Brachionus calyciflorus* (Pourriot, 1973). High salinity reduces this response and the volume of resting eggs closely approximates the clone mean. At intermediate salinities, high as well as low temperatures determine a bigger size. Intermediate salinity (12‰) is common in spanish mediterranean wetlands (Miracle *et al.*, 1986) and is the salinity we used in maintaining all clones. According to the literature, half-strength sea water is

386

optimal for *B. plicatilis* (Ruttner-Kolisko, 1974; Walker, 1981), as well as for resting egg hatching (Minkoff *et al.*, 1983), who recorded a optimal salinity of 16‰, among the 9, 16, 24, 32 and 40‰ tested. On the other hand, population dynamics of the same clones (Serra, 1987) showed that SPO and CU are better adapted to the low range of salinities studied (9‰ and 12‰).

At the latitudes when the clones were collected, high temperatures correspond to summer periods of reduced food or desiccation. Furthermore, hatching levels, according to Minkoff *et al.* (1983), show an inverse relationship with temperature, which has been attributed to lower dissolved oxygen concentration at higher temperatures. For all these reasons, larger resting egg size at high temperatures could be adaptive. High temperatures correspond, in temperate aquatic ecosystems, to high salinities, the tolerance to extreme values of the latter parameter being highly affected by the other.

From the global observation of the results, two conclusions seem logical with respect to resting eggs: (1) width is less variable because egg is oriented longitudinally at its passage through oviducts which are restrictal narrow openings; (2) mean size is bigger in extreme environments: low salinity – low temperature, high salinity – high temperature. On the other hand, mean size is smaller in favorable conditions where hatching is not delayed by diapause. At intermediate salinities rotifers seem to be able to regulate egg size against direct environmental effects such as those due to temperature.

References

Cooley, V. W. & P. R. Lohnes, 1971. Multivariate data analysis. John Wiley & Sons, N.Y., 364 p.

Gilbert, J. J., 1974. Dormancy in rotifers. Trans. Amer. Micros. Soc., 93 (4): 490–513.

Guillard, R. L. & J. M. Ryther, 1962. Studies of marine planktonic diatoms. Ca. J. Microbiol. 8: 229–239.

Kres, H., 1983. Statistical tables for multivariate analysis. Springer-Verlag. New York. 504 p.

Minkoff, G., E. Lubzens & D. Kahan, 1983. Environmental factors affecting hatching of rotifer *Brachionus plicatilis* resting eggs. Hydrobiologia 104: 61–69.

Miracle, M. R., M. Serra, E. Vicente & C. Blanco, 1987. Distribution of Brachionus species in the Spanish Mediterranean Wetlands. Hydrobiologia 147: 75–81.

Pourriot, R., 1973. Rapports entre la température, la taille des adultes, la longueur des œufs et le taux de développement embryonneire chez *Brachionus calyciflorus* Pallas (Rofitère). Ann. Hydrobiol. 4: 103–115.

Ruttner-Kolisko, A., 1974. Plankton Rotifers. Biology and taxonomy. Binnengewasser. 26: 146 p.

Serra, M., 1987. Variabilidad morfológica, isoenzimática y demográfica en poblaciones de *Brachionus plicatilis*. Diferenciación genética y plasticidad fenotípica. Tesis doctoral. Universidad de Valencia.

Serra, M. & M. R. Miracle, 1983. Biometric analysis of *Brachionus plicatilis* ecotypes from spanish lagoons. Hydrobiologia 104: 279–291.

Serra, M. & M. R. Miracle, 1985. Enzyme polymorphism in *Brachionus plicatilis* populations from several spanish lagoons. Verh. int. Ver. Limnol. 22: 2991–2996.

Serra, M. & M. R. Miracle, 1987. Biometric variation of *Brachionus plicatilis* strains as a direct response to abiotic parameters. Hydrobiologia 147: 83–89.

Snell, T. W., B. E. Burke & S. D. Messur, 1983. Size and distribution of resting eggs in a natural population of the rotifer *Brachionus plicatilis*. Gulf Research Reports. 7 (3): 285–287.

Walker, K. F., 1981. A synopsis of ecological information on the saline lake rotifer *Brachionus plicatilis*, Muller, 1786. Hydrobiologia 81: 150–167.

Hydrobiologia **186/187**: 387–400, 1989.
C. Ricci, T. W. Snell and C. E. King (eds), Rotifer Symposium V.
© 1989 *Kluwer Academic Publishers.*

Rotifers as food in aquaculture

E. Lubzens[1], A. Tandler[2] & G. Minkoff[3]

[1] *National Institute of Oceanography, Israel Oceanographic & Limnological Research, Tel-Shikmona, P.O.B. 8030, Haifa 31080, Israel*; [2] *National Center for Mariculture, Israel Oceanographic & Limnological Research, P.O.B. 1212, Eilat, Israel*; [3] *Tinamenor S.A., Pesues, Cantabria, Spain*

Key words: rotifers, marine larvae, nutrition

Abstract

The rotifer *Brachionus plicatilis* (O.F. Muller) can be mass cultivated in large quantities and is an important live feed in aquaculture. This rotifer is commonly offered to larvae during the first 7-30 days of exogenous feeding. Variation in prey density affects larval fish feeding rates, rations, activity, evacuation time, growth rates and growth efficiencies. *B. plicatilis* can be supplied at the food concentrations required for meeting larval metabolic demands and yielding high survival rates. Live food may enhance the digestive processes of larval predators. A large range of genetically distinct *B. plicatilis* strains with a wide range of body size permit larval rearing of many fish species. Larvae are first fed on a small strain of rotifers, and as larvae increase in size, a larger strain of rotifers is introduced. Rotifers are regarded as living food capsules for transferring nutrients to fish larvae. These nutrients include highly unsaturated fatty acids (mainly 20:5 n-3 and 22:6 n-3) essential for survival of marine fish larvae. In addition, rotifers treated with antibiotics may promote higher survival rates. The possibility of preserving live rotifers at low temperatures or through their resting eggs has been investigated.

Introduction

The history of larval rearing of marine fish can clearly be divided into two distinct phases. Lack of adequate first feeds caused limited success in larval rearing of marine fish up to the sixties (Dannevig, 1897; Blaxter, 1968; Lasker *et al.*, 1970). This was before rotifers, primarily the species *Brachionus plicatilis*, were established as first larval feed. Most of the effort in this first phase was limited to small scale experiments of rearing larvae of different species for identification purposes or for fish stock assessment in fisheries (Houde & Palko, 1970). In these experiments, small prey organisms for the early larvae

were ciliates, dinoflagellates, trochophores and wild plankton (Okamoto, 1969; Harada, 1970). The second phase started when rotifers were found to be suitable as first feed for marine fish larvae in Japan in the late sixties and early seventies. Rotifers became ubiquitous in all mass rearing trials after their successful use in the mass rearing of the red seabream (*Pagrus major*) in Japan (Fujita, 1973, 1979). The introduction of rotifers marked the first regular successes in the mass larval rearing of several marine species of economic value such as grey mullet (*Mugil cephalus*) (Nash *et al.*, 1974), sole (*Solea solea*) (Howell, 1973; Girin, 1974; Fuchs, 1978, 1982; Dendrinos & Thorpe, 1987), gilthead seabream

388

(*Sparus aurata*) (Person-Le Ruyet & Verillaud, 1980, 1981; Tandler & Helps, 1985), sea bass (*Dicentrarchus labrax*) (Barnabe, 1974; Girin, 1975), turbot (*Scophthalmus maximus*) (Kuhlmann *et al.*, 1981; Olsen & Minck, 1983; Witt *et al.*, 1984), and flounder (*Paralichthys olivaceus*) (Fukusho *et al.*, 1985), milkfish (*Chanos chanos*) (Liao *et al.*, 1979; Juario *et al.*, 1984) and others (Morales, 1983). These successes resulted from the development of rotifer culturing methods (Ito, 1960; Theilacker & McMaster, 1971; and reviews by Hirata, 1980; Fukusho, 1983; Hirano, 1987; Lubzens, 1987). Today, most marine fish are raised on basically the same methods, whereby *B. plicatilis* is provided as the first food during the first days of exogeneous feeding. The length of the period depends on the fish species.

Marine fish larvae are usually small at hatching (Theilacker & Dorsey, 1980; Kissil, 1984/85), and except for a few species, their size ranges between 2-7 mm. Rotifers offered to them must meet their nutritional requirements for optimization of growth and survival. These include: (1) the size; (2) the distribution and concentration of rotifers in the larval tanks; (3) the total amount available; (4) digestability and absorption; and (5) nutritional quality.

Size of rotifers

The size of the prey eaten by the fish larvae is a function of the larval mouth width. Within a fish species, mouth width is related to length, but it varies greatly between species (Arthur, 1976; Beyer, 1980; Hunter, 1980; Hunter & Kimbrell, 1980). Although mouth width limits the maximum prey size, in nature the mean diameter of the prey

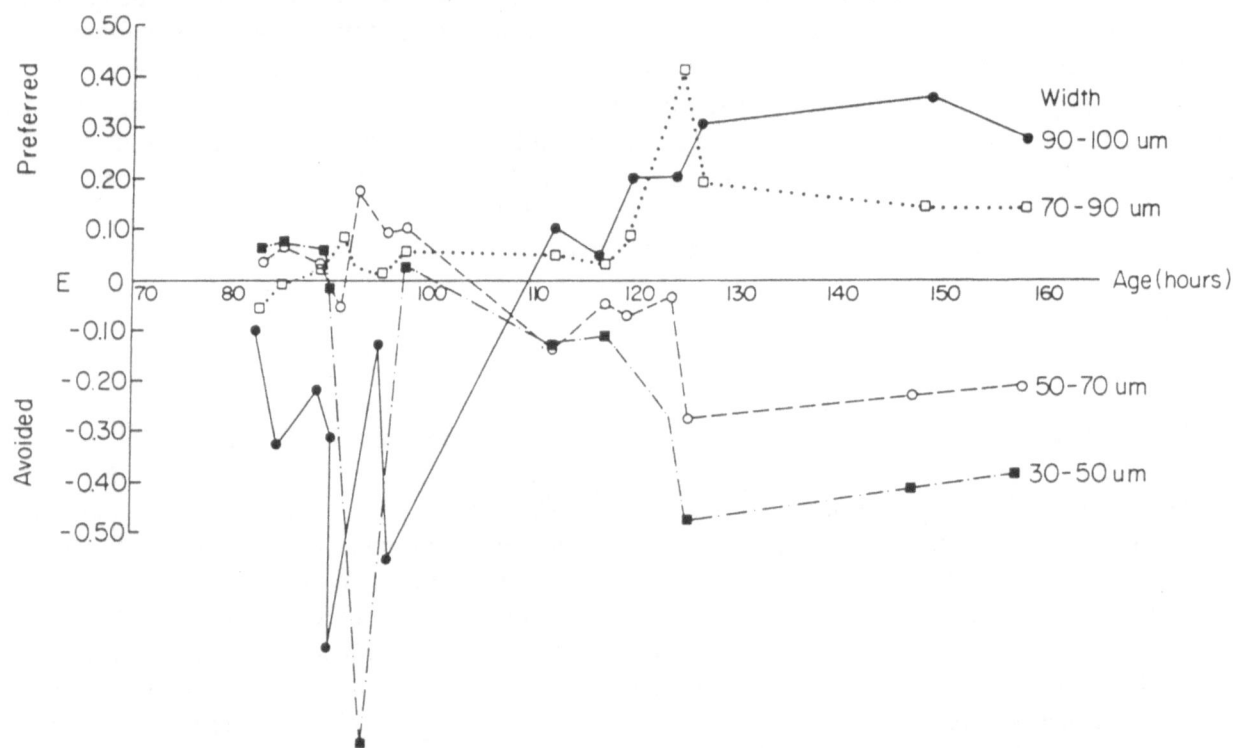

Fig. 1. Changes in the electivity indices (*E*) for larval sea bream (*Sparus aurata*) feeding on rotifers of various sizes as a function of larval age (redrawn from Helps [1982]). Electivity index (*E*) was calculated, based on Ivlev [1961], in the following way:

$$E = (r - p)/(r + p)$$

where *r* and *p* are the relative occurrence of a given rotifer size in populations exposed and unexposed to larval predation pressure, respectively.

consumed by fish larvae was only 38% of their mouth width (Hunter, 1980). There is a tendency for larvae in the sea to feed on progressively larger prey as they grow. In the laboratory it was found that the size of the prey fed to larvae must increase for larvae to grow at maximum rates (Lasker *et al.*, 1970; Hunter, 1980; Hunter & Kimbrell, 1980). Beyer & Laurence (1981) concluded that as larvae reach certain sizes, the energetic cost of each attack on prey exceeds the gain from ingesting smaller food particles. This size depends upon the larva's metabolic requirements, which are imposed both genetically and environmentally. It means that the larvae develop selection related to particle size both in natural populations and laboratory-reared species (Stepiens, 1976; Hunter, 1980).

The effect of age of gilthead seabream larvae (*Sparus aurata*) on their preference for rotifer strains of different sizes was examined in the laboratory (Helps, 1982). In this study, larvae showed a clear age effect on their feeding preference of different size rotifers (Fig. 1). Young larvae, up to 85 h after hatching, preferred small rotifers (30–70 μm in width) and avoided feeding on large rotifers (90–100 μm in width), while in older larvae (160 h) the pattern was reversed. The effect of rotifer size in long-term experiments was further tested, and it was found that the presence of small rotifers for the first 12 days of feeding of gilthead seabream larvae was associated with an improved growth rate (Tandler, pers. com.). This was especially pronounced in the size structure of the population of 32-day-old larvae. In these larvae, the percent of the population with a wet weight greater than 12 mg was 51.5 and 36, respectively, when small and large rotifers were offered as first food. In these studies, two genetically different

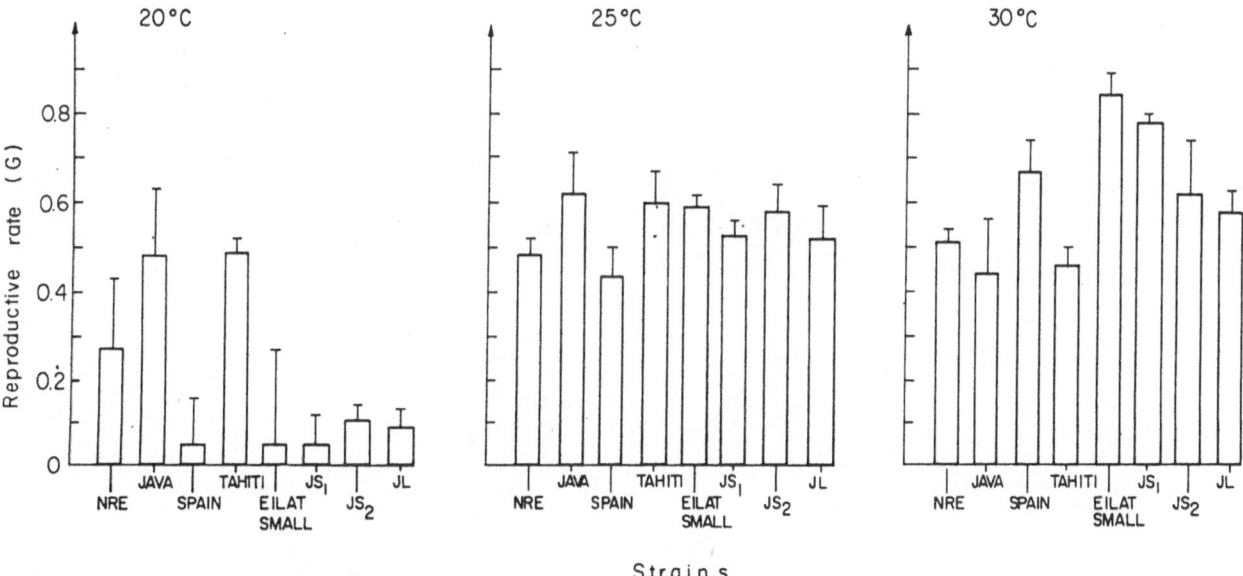

Fig. 2. The effect of temperature on the amictic reproductive rates of eight rotifer strains or clones cultured on *Nannochloris* sp. Reproductive rate was calculated from the equation:

$$G = 1/T \ln N_t/N_o,$$

where G_0 = experimental growth coefficient,
N_o = initial estimated population size,
N_t = final estimated population size,
T = time in days.

The mean values of five replicates and the standard deviations are presented by vertical bars.
Strains Java and Tahiti were obtained from INRA/IFREMER, France; strain Spain from Dr. E. Yufera; clones N.R.E. and Eilat Small from Israel; and clones JS_1, JS_2 and JL were hatched from resting eggs collected by Dr. M. Okauchi in Japan (from Lubzens *et al.*, in preparation).

size strains of rotifers were used. There is interest in these different size strains because they allow different larval stages to select prey of the most effective size (Fukusho, 1983; Korunuma & Fukusho, 1987).

Two methods of obtaining different sizes of rotifers have been considered; (1) sieving rotifer cultures through small size meshes which will allow the separation of small size rotifers from the larger ones; (2) culture of rotifer strains whose size is genetically determined. The first method does not guarantee full control over rotifer size. By sieving a rotifer culture, the small neonates will be separated from the adults, but within a short period they will eventually increase in size, while staying in the larval tanks. The culture of geneti-

cally different size rotifers offers a better solution (see Lubzens, 1987). However, this requires aquaculturists to maintain stocks of at least two rotifer strains in the hatchery.

Obtaining optimal mass cultures of different rotifer size strains may mean culturing them at different temperatures and offering them different types of food. For example, the small size strain in Japan (S-type) occurs in the summer when high temperatures prevail, while the large size rotifer strain (L-type) predominates in the winter (Fukusho & Iwamoto, 1980; Fukusho, 1983). Laboratory studies under controlled conditions of food and temperature show that the reproductive rate of rotifer strains is not always related to size (Fig. 2). Some of the small size strains (Java,

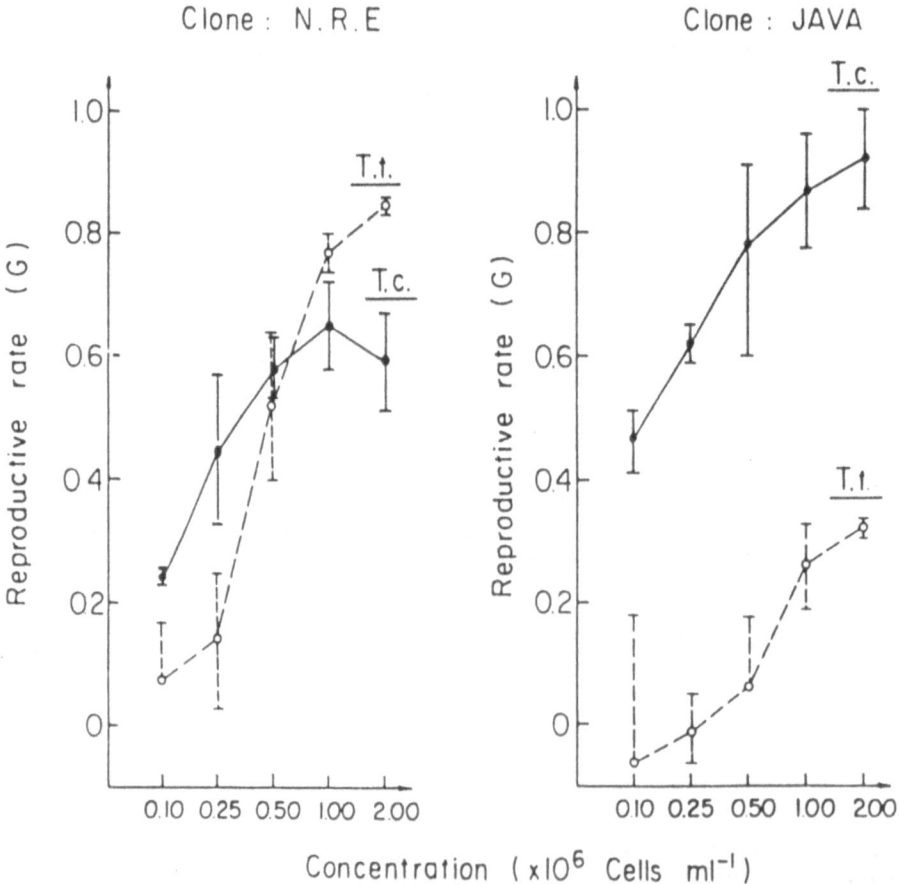

Fig. 3. The effect of two algae differing in size (*Tetraselmis tetrathele* and *T. chuii*) on the reproductive rates of large (clone N.R.E.) and small (strain Java) rotifers cultured at 25 °C. The average dimensions (in μm) of the algae were 15.3 × 8.5 × 10.5 and 12.0 × 6.5 × 6.0 for *T. tetrathele* and *T. chuii*, respectively. The average length and width (in μm) of the rotifers were 196.0 × 139.2 and 173.5 × 115.6 for clone N.R.E. and the Java strain, respectively (from Lubzens *et al.*, in preparation). The average reproductive rates and standard deviations were calculated as described in Fig. 2.

Tahiti) reproduce equally well at 20°, 25° or 30 °C while others (Spain, Eilat small, JS1 and JS2) clearly prefer high temperatures. Similarly, the large size strains (N.R.E. and JL) higher reproductive rates were found at 25 °C or 30° than those at 20 °C. Furthermore, a positive correlation was found between body size and maximum size of particles ingested (Hino & Hirano, 1980). It is likely that large size algae (e.g. *Tetraselmis tetrathele*), which are currently popular in feeding rotifers, will not easily be ingested by small rotifers. Small rotifers cultured on such algae will probably have low reproductive rates (Fig. 3). Consequently, choosing the appropriate rotifer strains to be mass cultured will depend on the environmental conditions prevailing at the hatchery and the size of larval predators. Moreover, in offering different rotifer size strains, their dry weight must be taken into account in evaluating the number of rotifers required by the predatory larvae (see below).

The distribution and concentration of rotifers in larval tanks

The distribution of the rotifers in the water column of larval tanks depends mainly on the salinity and to a lesser extent on water quality. Exposing rotifers to sudden changes in salinity will result in their adhesion to the bottom or sides of containers (Gatesoupe & Luquet, 1981; Lubzens, 1987), making them unavailable to the larvae. Other factors affecting the distribution of rotifers are oxygen and ammonia levels which affect the swimming speed (Epp & Winston, 1978; Snell *et al.*, 1987) and the type of food offered to rotifers.

The nutrition of fish larvae depends primarily on the probability of encounter between the food and the larva as well as its suitability in terms of size and nutrient composition. The probability of encounter between the larva and the rotifer depends on concentration. In turn, the perception of food organisms affects larval swimming behavior (Hunter, 1980). Several studies showed a direct correlation between food concentration and larval survival (Theilacker & Dorsey, 1980).

The optimal food concentration depends on the life stage and temperature. First feeding larvae have relatively slow swimming speeds (Fukuhara, 1983) and low capture success at onset of feeding (2-10%) (Hunter, 1980), so they may require a high rotifer concentration. This concentration could be reduced up to the stage where the larvae may demand higher quantities in order to meet energetic requirements. Up to a rotifer concentration of 10 ml^{-1}, a direct relationship between survival, growth and food concentration was shown in gilthead seabream (Tandler & Sherman, 1981; Peguin, 1984), and in bream (*Archosargus rhomboidalis*) (Dowd & Houde, 1980). Beyond this concentration, a sharp drop in both growth and survival of seabream larvae was observed (Tandler & Sherman, 1981; Peguin, 1984). These results, however, were not related to an increase in the metabolite concentration (NH_3-NH_{4+}) in the water body. In a study in which rotifers, labelled with ^{14}C, were offered to seabream larvae, at concentrations of 10 and 40 ml^{-1}, it was shown that feeding rate of these larvae was directly correlated to rotifer concentration (Tandler & Mason, 1984). Therefore, it can be proposed that the negative effect of elevated rotifer concentrations on growth were associated with reduced efficiency of the digestive process at high rotifer concentrations. Similar observations were made by Boehlert & Yoklavich (1984), who found that the assimilation efficiency of herring larvae (*Clupea harengus*) was negatively and logarithmically correlated to rotifer concentrations between 1 and 10 ml^{-1}.

Almost no information is available on the effect of larval density in the culture tanks on rotifer concentration. Okamoto (1969) showed that survival of larvae is clearly related to larval stocking density, but it is not known whether this is related also to the shortage in food which may occur at high larval densities.

Marine fish larvae depend primarily on vision to find their food, as a result of the pure cone retina found in the majority of marine larvae (Blaxter & Staines, 1970; Blaxter, 1975). Photoperiod, light intensity and color of background of rearing containers are therefore of paramount importance for their hunting success.

The effect of photoperiod in combination with rotifer concentration is intimately related to the probability of encounter between the larva and its food (Dowd & Houde, 1980; Peguin, 1984). In both studies it was found that at low rotifer concentrations, there was a direct relationship between the duration of the photoperiod and growth. At high rotifer concentrations of 25 ml^{-1} in gilthead seabream (Peguin, 1984) and 100 to 500 l^{-1} in seabream (Dowd & Houde, 1980), there was a mid-range photoperiod which supported best growth, beyond which a further increase was associated with no improvement in the former and with reduced growth rates in the latter species. A similar response of growth to a mid-range photoperiod of 18 h was reported by Barahona-Fernandes (1979) for sea bass (*Dicentrarchus labrax*). On the other hand, Fuchs (1978) obtained best larval growth in sole (*Solea solea*) under continuous illumination. Survival of both gilthead seabream (Peguin, 1984) and seabream (Dowd & Houde, 1980) peaked at mid-range photoperiods of 15 and 13 h, respectively. In seabream, the effect of photoperiod was independent of rotifer concentration, while in gilthead seabream, a direct correlation between photoperiod duration and survival was observed only at low food concentrations of 1.0 ml^{-1}.

The number and concentration of rotifers supplied to fish cultures is expected to vary with temperature. Temperature will affect several processes in the growth and survival of larvae, among them metabolic rate and swimming activity (Theilacker & Dorsey, 1980). This will result in higher predation and energy (food) requirements.

Availability of rotifers

The large quantities of rotifers required for raising marine fish larvae (40000–173000 per surviving larva) (Okauchi et al., 1980 Kafuku & Ikenoue, 1983) are of constant concern to the aquaculturist. The development of high density mass culture techniques (Hirata, 1980; Lubzens, 1987; Yamasaki et al., 1988) allows for constant production of rotifers, but does not insure against

sudden shortages in supply as a result of unforeseen collapses. Preventive measures may be taken, to some extent, by routine daily monitoring of swimming speed, proportion of eggs (Snell et al., 1987), and physical parameters (pH, oxygen level, etc.). Preserved rotifers could be helpful in the management of supply and demand in the hatchery. In this respect, the possible use of artificially produced rotifer resting eggs has been evaluated in recent years (Lubzens, 1981, 1989). It was found to be too expensive for routine use, in the way that *Artemia* cysts are used today in aquaculture. In extreme cases, these eggs could be used to initiate small-scale new cultures. Preservation of live rotifers at low temperatures has been examined recently (Berghahn et al., 1989; Lubzens, 1989; Lubzens et al., 1989). This method would permit the preservation of different strains or clones, including those that do not produce resting eggs. In those cases where *B. plicatilis* is abundant, it could also be used for culturing of freshwater fish larvae, e.g. ayu (*Plecoglossus altivelis*) or cyprinids (*Cyprinus carpio*) (Kanazawa et al., 1982; Lubzens et al., 1987).

Ingestion, digestion and conversion rates of rotifers by fish larvae

Ingestion

As already mentioned above, fish larvae tend to increase both the quantity and the size range of particles upon which they feed throughout their ontogeny. The increase in the number of food items ingested daily by fish larvae is an exponential function of larval age (Stepiens, 1976; Buckley & Dillmann, 1982) or length (Okauchi et al., 1980; Barahona-Fernandes & Conan, 1981). In absolute terms, larval daily rations increase as a linear function of their weight (Laurence, 1977; Houde & Schekter, 1983; Klumpp & von-Westernhagen, 1986; Minkoff, 1987). The numbers of rotifers consumed daily (Rn) as a function of larval length (L) has been determined for porgy (*Acanthopagrus schlegeli*) as $Rn = 2.43 L^{3.05}$ (Okauchi et al., 1980) and red sea bream (*Pagrus major*) as $Rn = 0.39\ L^{3.67}$ (Fujita, 1979). Likewise, the dry weight of rotifers consumed daily, Rw (daily ration) as a function of larval dry weight (Wt, in mg) has been evaluated for the blenny (*Blennius pavo*) as

$$Rw = 314.5\ Wt + 26.3$$

(Klumpp & von-Westernhagen, 1986), herring (*Clupea harengus*) as

$$Rw = 0.13\ Wt + 49.67$$

and turbot (*Scophthalmus maximus*) as

$$Rw = 0.43\ Wt + 38.45$$

(Minkoff, 1987). Concomitant to the increase in feeding rates, the growing larvae tend to require larger food particles. This factor has led to the use of different rotifer strain sizes in rearing of fish larvae (see above).

Conversion

Rotifers were found to contain acid proteinase, alkaline proteinase and two kinds of alkaline proteases (Hara *et al.*, 1979a, b, 1984). All the current data show that larvae at first feeding seem to possess the necessary complement of enzymes for digesting their prey (e.g. Govoni *et al.*, 1986).

The assimilation of ingested rotifers, or any other food organism, by fish larvae has been shown to be very rapid. ^{14}C originating from labeled rotifers has been found to be respired by fish larvae 3 h following their ingestion (Govoni *et al.*, 1982).

The gross growth efficiency, which is the proportion of the ingested food invested in growth (Ki) (Brett & Groves, 1979), has been evaluated for larvae feeding on rotifers; for herring at 20–60%, turbot 20–40% (Minkoff, 1987) and blenny 50–56% (Klumpp & von-Westernhagen, 1986). The values obtained for the first two species are similar to values reported for other fish species feeding on other types of prey (Govoni *et al.*, 1986; Klumpp & von-Westernhagen, 1986).

One factor which influences the assimilation efficiency in the early larval stages is that of food density. As already discussed before, larvae, being 'number maximizers', tend to consume prey in relation to availability rather than according to satiation. Therefore, at high prey densities they will tend to consume food at a rate which decreases its residence time in the gut with a concomitant reduction in the assimilated efficiency (Boehlert & Yoklavich, 1984).

Nutrition

The current knowledge pertaining to marine fish larval nutrition has been derived through the analysis of the performance of the larvae while feeding on different rotifers 'enriched' with various nutrients and through testing of formulated inert diets. It was found that rotifers cultivated on baker's yeast alone could not support larval growth or survival. The failure of these rotifers was attributed to their lack of the long chain highly unsaturated fatty acids (HUFA) on the n-3 series, mainly 18:3 n-3, 20:5 n-3 and 22:6 n-3 (Owen *et al.*, 1975; Yone & Fujii, 1975; Cowey *et al.*, 1976; Fujii & Yone, 1976; Gatesoupe *et al.*, 1977; Watanabe *et al.*, 1983; Dendrinos & Thorpe, 1987). These fatty acids are found in some of the algae of marine origin. This has led to an extensive search for species of algae which are most suitable for enrichment of rotifers. From a practical aspect, using yeast to raise rotifers instead of algae reduces greatly the production costs in the hatchery. Furthermore, replacing the algal enrichment step with inert rotifer diets will assist in curtailing the production price of nutritionally adequate rotifers.

Rotifer biochemical composition in relation to larval requirements

The nutritional value of rotifers to larvae depends on their dry weight, caloric content and biochemical composition. The dry weight of a single rotifer, excluding eggs, as measured by Doohan (1973), Tandler & Mason (1984), Theilacker &

Kimball (1984), Yufera & Pascual (1984) and Minkoff (1987), ranges between 120-360 ng/ind, depending on the nutritional state and body size. Scott & Baynes (1979) estimated that during periods of high growth rates and ample food supplies, the rotifer's weight could increase up to 620 ng/ind. Furthermore, the latter authors have also demonstrated that at 24-26 °C, in the absence of food, the rotifer loses daily between 18-26% of its body weight. The influence of the dietary condition on the rotifer's dry weight has also been noted by Minkoff (1987), who found that rotifers raised on baker's yeast alone increased their dry weight by 12-23% following prolonged (48 h) enrichments with algae. Conversely, rotifers lost 30% of their body weight following a 24 h starvation period at 20 °C. The rapid loss of organic material from rotifers which are deprived of food is generally perceived as one of the main factors causing poor growth and high mortalities in fish larval cultures. This is probably the reason for the use of the 'green water' by most larval culturists, which maintains the rotifers in a healthy nutritious state.

The other prominent factor influencing the rotifer's weight is that of the animal's size, which, as pointed out previously, is mainly strain-dependent. Theilacker & Kimball (1984) have described the relationship between the width (X, in μm) and dry weight (Y, in ng) of *B. plicatilis* to be $Y = 1.4 \times 10^{-5} X^2$.

Direct caloric measurements show that the ash-free caloric value of a rotifer ranges from 4.8 cal/mg (Theilacker & Kimball, 1984) to 6.46 cal/mg (Minkoff, 1987). Again, the value per individual will depend on both the size and nutritious state of the organism. Moreover, it was reported (Kentouri & Divanach, 1982; Minkoff, 1987), and also found in *Sparus aurata* cultures (Tandler, pers. com.), that rotifer eggs and the rotifer lorica are not digested by the fish larvae. Therefore, it is essential to correct for egg/female ratios in the estimates made of the nutritional requirements of larvae (Minkoff, 1987).

Most researchers have noted that the species of algae or yeast had only minor effects on the biochemical composition of rotifers fed on them,

and the values reported fall within the range of the experimental error. Both Scott & Baynes (1978) & Minkoff (1987) have found protein in the range of 50-58% of the rotifer's dry weight. Watanabe *et al.* (1983) have set protein levels at 65%, while Dendrinos & Thorpe (1987) found these levels to be 34-36%. Contrary to this, Ben-Amotz *et al.* (1987) have reported a much wider range of crude protein from 28-51%. Lipids usually range from 9-16% of the dry weight (Scott & Baynes, 1978; Minkoff, 1987), and up to 23.5% and 28% (Watanabe *et al.*, 1983; Dendrinos & Thorpe, 1987) (Table 1). One observation seems to be consistent throughout the literature: rotifers fed only on baker's yeast are poorer in their total lipid content than their counterparts that had either been cultivated only on algae or had received some algae in their diet. This low level of lipids can be elevated by prolonged feeding on algae (Minkoff, 1987) or by feeding rotifers with lipid-enriched yeasts (Dendrinos & Thorpe, 1987).

Inasmuch as marine fish require a diet containing 40-60% protein (Castell *et al.*, 1986) and 13-16% lipid, there is no doubt that rotifers fulfill these requirements. However, for the rapidly developing larval stages, it is probable that the balance of dietary amino acids, fatty acids, vitamins and minerals determines the success of organogenesis rather than the gross composition of the diet.

The amino acid profiles of rotifers enriched on a variety of marine algae or yeast have been determined by a number of researchers (Watanabe *et al.*, 1983; Dendrinos & Thorpe, 1987; Minkoff, 1987; Rezeq & James, 1987). These analyses show no significant differences in rotifer amino acid profiles and therefore fail to explain the advantage of certain algae used for rotifer enrichment. Furthermore, the amino acid profiles of rotifers do not seem to deviate significantly from those of other zooplanktonic organisms either cultured or of wild origin, which are recognized as being beneficial for growth of larvae (Castell *et al.*, 1986).

Undoubtedly, the most significant dietary factors to influence the growth and survival of marine fish larvae are the highly unsaturated fatty acids

Table 1. Proximate dry weight values of *B. plicatilis* cultured on different diets

Diet	Protein %	Lipid %	Carbo-hydrate %	Ash %	Authors
Isochrysis galbana	36.3	21.1	–	–	Dendrinos & Thorpe, 1987
Nannochloris oculata	34.5	21.4	–	–	Dendrinos & Thorpe, 1987
Saccharomysis cerevisiae	41.6	14	–	–	Dendrinos & Thorpe, 1987
Saccharomysis cerevisiae L4	33.17	24	–	–	Dendrinos & Thorpe, 1987
Saccharomysis cerevisiae L5	42.3	28.5	–	–	Dendrinos & Thorpe, 1987
Chaetoceros gracilis	32	20.1	44.9	–	Ben-Amotz *et al.*, 1987
Chlorella stigmatophora	28.1	7.4	10	–	Ben-Amotz *et al.*, 1987
Isochrysis galbana	29.1	11.1	7	–	Ben-Amotz *et al.*, 1987
Nannochloropsis salina	28	16.4	24.1	–	Ben-Amotz *et al.*, 1987
Phaeodactylum tricornutum	54.4	8.9	13.5	–	Ben-Amotz *et al.*, 1987
Yeast	55.4	4.5	28	–	Ben-Amotz *et al.*, 1987
Yeast	65–67	15–21	–	10	Watanabe *et al.*, 1983
Chlorella minutissima	58–72	21–31	–	4–9	Watanabe *et al.*, 1983
Chlorella + yeast	63–67	20–22	–	4–6	Watanabe *et al.*, 1983
Nannochloropsis oculata	50–53	9–13	–	–	Minkoff, 1987
Isochrysis galbana	50–54	10–12	–	–	Minkoff, 1987
Yeast	53–57	6–9	–	–	Minkoff, 1987

(HUFA) of the n-3 series. These fatty acids most likely form a part of the larvae's cellular membranes and therefore are crucial in determining the rates of enzymatic processes taking place at these sites. It is well established that in rotifers, the fatty acid profile is chiefly determined by diet (Scott & Middleton, 1979; Watanabe *et al.*, 1983) and that only minor alterations in the fatty acid composition can be achieved through *de novo* synthesis by the rotifer itself (Lubzens *et al.*, 1985). These lipids are digested by the rotifer and incorporated into its cellular phospholipids (Lubzens *et al.*, 1985). Rotifers which are raised on baker's yeast alone are inadequate as food for larvae due to their lack of these fatty acids. Therefore, rotifer enrichment procedures were developed to insure the ready supply of nutrients required by the developing larva. The enrichment regimes are based on the immersion of rotifers for 8-24 hours in a medium rich in essential nutrients, especially n-3 HUFA. The enrichment medium may contain one of the following constituents: (1) different algal species (Kitajima *et al.*, 1979), for example: *Isochrysis galbana* (Howell, 1979; Tandler & Helps, 1985); *Nannochloropsis* sp. and *Tetraselmis* sp. (Fukusho *et al.*, 1984); (2) a spe-

cial yeast enriched with (n-3)HUFA (Kitajima *et al.*, 1980). (3) emulsions which are based on (n-3)HUFA rich oil from marine fish or cuttlefish extracts (Watanabe *et al.*, 1983). (4) a dry diet for rotifers which both supports their culture and improves their quality as food for sea bass larvae (Gatesoupe & Luquet, 1981; Gatesoupe & Robin, 1982). (5) microcapsules (Walford & Lam, 1987).

The dietary HUFA requirements for most marine fish larvae have not yet been determined in a way that will differentiate between a need for either 20:5 n-3 or 22:6 n-3 or both. Red sea bream larvae have very similar growth rates when fed on rotifers containing the C:20 HUFA incorporated from *Chlorella*, as on rotifers containing both the C:20 and C:22 HUFA incorporated from yeasts cultured in the presence of marine lipids (Kitajima *et al.*, 1980). The ayu, *Plecoglossus altivelis*, which spends most of its early life history in fresh water, has a dietary requirement for 18:3 n-3 (Watanabe *et al.*, 1983) which is very similar to what is known for salmonids (Kanazawa, 1985). However, the ayu can benefit from the input of both the C:20 and C:22 FA into its diet (Oka *et al.*, 1980). A recent attempt to differentiate between the requirements for C:20

and C:22 FA in a range of larvae suggests that only turbot have an essential fatty acid (EFA) requirement for both C:20 and C:22 HUFA, while herring and plaice thrive equally well when only the C:20 HUFA or both the C:20 and the C:22 are present in the diet (Minkoff, 1987).

Can rotifers be replaced by inert diets?

At present the aquaculturist views the rotifer mainly as a 'living food capsule' through which nutrients essential to the fish larva can be transferred (Watanabe et al., 1983). Special rotifer feeding techniques were developed for their 'enrichment' with various essential nutrients. Lately, larvae are being offered rotifers which have been enriched with antibiotics. Gatesoupe (1982), for example, found that larval turbot given rotifers which were previously immersed in a 48 mg l^{-1} Tribrissen for 0.5 hr grew faster and survived better than the untreated control.

The general dependence on live feeds, mainly rotifers, is a constraint as far as dietary manipulation of feed body composition is concerned. Therefore, efforts were recently made by various laboratories to develop inert microdiets in which the composition is easy to alter. At present, however, microdiets cannot completely replace rotifers in first feeding marine fish larvae. In most of the reports, microdiets are offered to marine fish larvae as a partial replacement only 8-10 days after hatching (Kanazawa et al., 1982; Kanazawa, 1985). Kissil (1984/85) correlated the larval size of first feeding larvae of 10 species and found that the minimum length of fish reported to feed on formulated diets is 7 mm. This suggests that the minimum age/size recommended for successful use of inert diets is associated with the stage of development of the digestive system. A similar conclusion was reached by Dabrowski (1984), who proposed that the advantage of live feeds results from the presence of digestive enzymes, and the activation of zymogens in the developing larval digestive tract.

Culture of crustacean larvae important in aquaculture with B. plicatilis

Several reports show that rotifers may fill the gap between the period phytoplankton and Artemia are offered to developing shrimp larvae (Mock et al., 1980). Rotifers are consumed by zoea stages 2-3 of Penaeus japonicus (Hirata et al., 1985), P. kerathurus (Yufera et al., 1984), and P. semisulcatus (Watanabe, 1980; Samocha et al., 1988). P. indicus has been shown to consume rotifers as early as Z 1 (Emmerson, 1984). It has been recently challenged by Samocha et al. (1988) that since rotifers are less adequate and more expensive than Artemia (Tandler, 1984/85), they are not essential in culturing commercially important crustaceans, but if available may replace Artemia for short periods (Yamasaki & Hirata, 1982).

Summary and conclusions

Rotifers of the species Brachionus plicatilis were found to be primarily an essential live food source for larvae of several species of marine fish and to a lesser extent for larvae of shrimp or crabs. Rotifers possess several characteristics that make them attractive as live food in mariculture: (1) They are relatively small, ranging in size from 0.06–1.00 mm, depending on zoogeographical strain and stage of development; (2) They are slow swimmers, maintaining their position in the water column; (3) They can be cultured at high densities; (4) They reproduce rapidly, making them available in large quantities within a relatively short period; (5) They can be enriched with fatty acids or antibiotics which are required for growth and survival of larvae.

The development of standardized mass culture techniques insures, to some extent, an adequate supply of the large numbers of rotifers required to raise commercially important fish species. The number of rotifers required depends on the size of the rotifer strain and the duration it is supplied in the fish larval tanks. The temperature prevailing in the culture site, the salinity of the water, and the food offered to rotifers will determine to a large extent the rotifer strain chosen for cultivation. Choosing the appropriate strain is important for obtaining high reproductive rates, thus reducing the rotifer 'standing stock' in the hatchery.

In order to ensure adequate amounts of essential lipids, rotifers must be enriched with either an appropriate alga, emulsified fresh oils or inert particles such as microcapsules. The nutritional quality of the rotifers, which depends on their state of satiation, is maintained in the presence of algae in the larval tanks. The duration of providing live food to larval tanks is constantly reduced with the concomitant development of inert foods. The possible future development of preserved live rotifers may improve the management of supply and demand of rotifers in the hatchery and may open the possibility for using *B. plicatilis* as food for raising problematic freshwater fish species.

Acknowledgements

We would like to thank the U.S. Agricultural Research and Development Fund (BARD) and the National Council for Research and Development for providing grants (I-186 and AQ-16, respectively) supporting our research.

References

Arthur, D. K., 1976. Food and feeding of larvae of three fishes occurring in the California current, *Sardinops sagax*, *Engraulis mordax* and *Trachurus symmetricus*. Fish. Bull. USA. 74: 517–530.

Barahona-Fernandes, M. H., 1979. Some effects of light intensity and photoperiod on the sea bass larvae, *Dicentrarchus labrax* (L.), reared at the Centre Oceanographique de Bretagne. Aquaculture 17:311–322.

Barahona-Fernandes, M. H. & G. Conan, 1981. Daily food intake of reared larvae of the European sea bass (*Dicentrarchus labrax* L.); statistical analysis and modelling. Rapp. P.-v. Reun. Cons. Int. Explor. Mer 178:41–44.

Barnabe, G., 1974. Mass rearing of the bass *Dicentrarchus labrax* L. In J. H. S. Blaxter (ed), The Early Life History of Fish. Springer-Verlag, Berlin: 749–753.

Ben-Amotz, A., R. Fishler & A. Schneller, 1987. Chemical composition of dietary species of marine unicellular algae and rotifers with emphasis on fatty acids. Mar. Biol. 95: 31–36.

Berghahn, R., S. Euteneuer & E. Lubzens, 1989. High density storage of rotifers (*Brachionus plicatilis*) in cooled and undercooled water. Spec. Publ. Europ. Aquacult. Soc. (in press).

Beyer, J. E., 1980. Feeding success of clupeoid fish larvae and stochastic thinking. Dana 1: 65–91.

Beyer, J. E. & G. C. Laurence, 1981. Aspects of stochasticity in modelling growth and survival of clupeoid fish larvae. Rapp. P.-v. Reun. Cons. Int. Explor. Mer 178: 17–23.

Blaxter, J. H. S., 1968. Rearing herring larvae to metamorphosis and beyond. J. mar. biol. Ass. UK. 48: 17–28.

Blaxter, J. H. S., 1975. The eyes of larval fish. In: M. A. Ali (ed), Vision in Fishes. Plenum Press, New York: 427–443.

Blaxter, J. H. S. & M. Staines, 1970. Pure-cone retinae and retinomotor responses in larval teleosts. J. mar. biol. Ass. UK. 50: 449–460.

Boehlert, G. W. & M. M. Yoklavich, 1984. Carbon assimilation as a function of ingestion rate in larval Pacific herring, *Clupea harengus pallasi* Valenciennes. J. exp. mar. Biol. Ecol. 79: 251–262.

Brett, J. R. & T. D. D. Groves, 1979. Physiological energetics. In: W. S. Hoar, D. J. Randall & J. R. Brett (eds), Fish Physiology, Vol. 8. Academic Press, New York: 279–352.

Buckley, L. J. & D. W. Dillmann, 1982. Nitrogen utilization of larval summer flounder, *Parallichthys dentatus* (Linnaeus). J. exp. mar. Biol. Ecol. 59: 243–256.

Castell, J. D., D. E. Conklin, J. S. Craigie, S. P. Lall & K. Norman-Boudreau, 1986. Aquaculture nutrition. In M. Bilio, H. Rosenthal & C. J. Sindermann (eds), Realism in Aquaculture; Achievements, Constraints, Perspectives. European Aquaculture Society, Bredene, Belgium: 251–308.

Cowey, C. B., J. M. Owen, J. W. Adron & C. Middleton, 1976. Studies on the nutrition of marine flatfish. The effect of dietary fatty acids on the growth and fatty acid composition of turbot (*Scophthalmus maximus*). Br. J. Nutr. 36: 479–486.

Dabrowski, K., 1984. The feeding of fish larvae: present 'state of the art' and perspectives. Reprod. Nutr. Develop. 24: 807–833.

Dannevig, H., 1897. On the rearing of larval and post larval stages of plaice and other flatfishes. Rep. Fish. Bd Scot. 1896: 175–193.

Dendrinos, P. & J. P. Thorpe, 1987. Experiments on the artificial regulation of the amino acid and fatty acid contents of food organisms to meet the assessed nutritional requirements of larval, post-larval and juvenile Dover sole (*Solea solea* L.). Aquaculture 61: 121–154.

Doohan, M., 1973. An energy budget for adult *Brachionus plicatilis* Muller (Rotatoria). Oecologia (Berl.) 13: 351–362.

Dowd, C. E. & E. D. Houde, 1980. Combined effects of prey concentration and photoperiod on survival and growth of larval sea bream, *Archosargus rhomboidalis* (Sparidae). Mar. Ecol. Prog. Ser. 3: 181–185.

Emmerson, W. D. 1984. Predation and energetics of *Penaeus indicus* (Decapoda: Penaeidae) larvae feeding on

398

Brachionus plicatilis and *Artemia* nauplii. Aquaculture, 38: 201–209.

Epp, R. W. & P. W. Winston, 1978. The effects of salinity and pH on the oxygen consumption and activity of *Brachionus plicatilis* (Rotatoria). Comp. Biochem. Physiol. 59: 9–12.

Fuchs, J., 1978. Influence de la photoperiode sur la croissance et la survie de la larve et du juvenile de sole (*Solea solea*) en elevage. Aquaculture 15: 63–74.

Fuchs, J., 1982. Production de juveniles de sole (*Solea solea*) en conditions intensives. 1. Le premier mois d'elevage. Aquaculture 26: 321–337.

Fujii, M. & Y. Yone, 1976. Studies on nutrition of red sea bream. XII. Effect of dietary linolenic acid and ω 3 polyunsaturated fatty acids on growth and feed efficiency. Bull. Jap. Soc. Sci. Fish. 42: 583–588.

Fujita, S., 1973. Importance of zooplankton mass culture in producing marine fish seed for fish farming. Bull. Plankton Soc. Jpn. 20: 49–53.

Fujita, S., 1979. Culture of red sea bream, *Pagrus major*, and its food. Cultivation of fish fry and its live food. Spec. Publ. Europ. Maricult. Soc. 4: 183–197.

Fukuhara, O., 1983. Effect of prey density on the swimming behaviour of larval black porgy, *Acanthopagrus schlegeli* (Bleeker). Bull. Nansei Reg. Fish. Res. Lab. (15): 97–101.

Fukusho, K., 1983. Present status and problems in culture of the rotifer *Brachionus plicatilis* for fry production of marine fishes in Japan. Symp. Int. Aquacult., Coquimbo, Chile, Sept. 1983: 361–374.

Fukusho, K. & H. Iwamoto, 1980. Cyclomorphosis in size of the cultured rotifer *Brachionus plicatilis*. Bull. Natl. Res. Inst. Aquacult. 1: 29–37. (in Japanese, English summary).

Fukusho, K., M. Okauchi, S. Nuraini, A. Tsujigado & T. Watanabe, 1984. Food value of rotifer *Brachionus plicatilis*, cultured with *Tetraselmis tetrathele* for larvae of red sea bream *Pagrus major*. Bull. Jpn. Soc. Sci. Fish. 50: 1439–1444.

Fukusho, K., M. Okauchi, H. Tanaka & S. I. Wahyuni, 1985. Food value of rotifer *Brachionus plicatilis*, cultured with *Tetraselmis tetrathele* for larvae of a flounder *Paralichthys olivaceus*. Bull. Natl. Res. Inst. Aquacult. 7: 29–36.

Gatesoupe, F. J., 1982. Nutrition and antibacterial treatments of live food organisms: the influence on survival, growth rate and weaning success of turbot (*Scophthalmus maximus*). Ann. Zootech. 31: 353–368.

Gatesoupe, F. J. & P. Luquet, 1981. Practical diet for mass culture of the rotifer *Brachionus plicatilis*: application to larval rearing of sea bass *Dicentrarchus labrax*. Aquaculture 22: 149–163.

Gatesoupe, F. J. & J. H. Robin, 1982. The dietary value for sea bass larvae (*Dicentrarchus labrax*) of the rotifer *Brachionus plicatilis* fed with or without a laboratory-cultured alga. Aquaculture 27: 121–127.

Gatesoupe, F. J., C. Leger, R. Metailler, P. Luquet & M. Malaval, 1977. Alimentation lipidique du turbot (*Scophthalmus maximus* L.). I. Influence de la longueur de chaine de acides gras de la serie ω 3. Ann. Hydrobiol. 8: 89–97.

Girin, M., 1974. Regime alimentaire et pourcentage de survie chez la larve de sole (*Solea solea* L.). Actes de Colloques, C.N.E.X.0. (1): 175–185.

Girin, M., 1975. La ration alimentaire dans l'elevage larvaire du bar, *Dicentrarchus labrax* (L.). 10th Europ. Symp. Mar. Biol., Ostend, Belgium, Sept. 17-23, 1: 171–188.

Govoni, J. J., D. S. Peters & J. V. Merriner, 1982. Carbon assimilation during larval development of the marine teleost *Leiostomus xanthurus* Lacepede. J. exp. mar. Biol. Ecol. 64: 287–299.

Govoni, J. J., G. W. Boehlert & Y. Watanabe, 1986. The physiology of digestion in fish larvae. Envir. Biol. Fishes 16: 59–77.

Hara, K., T. Ishihara, H. Arano & M. Yasuda, 1979a. Studies on protease of the rotifer, *Brachionus plicatilis*. I. Some properties of proteinase in the crude extracts. Bull. Fac. Fish. Nagasaki Univ. 46: 31–35.

Hara, K., T. Ishihara, H. Arano & M. Yasuda, 1979b. Studies on protease of the rotifer, *Brachionus plicatilis*. II. Hydrolytic properties on some synthetic substrates. Bull. Fac. Fish. Nagasaki Univ. 46: 37–42.

Hara, K., H. Arano & T. Ishihara. 1984. Some enzymatic properties of alkaline proteases of the rotifer *Brachionus plicatilis*. Bull. Jpn. Soc. Sci. Fish. 50: 1611–1616.

Harada, T., 1970. The present status of marine fish cultivation research in Japan. Helgolander wiss. Meeresunters. 20: 594–601.

Helps, S., 1982. An examination of prey size selection and its subsequent effect on survival and growth of larval gilthead seabream (*Sparus aurata*). M. Sc. thesis, Plymouth Polytechnic, UK., 50 pp.

Hino, A. & R. Hirano, 1980. Relationship between body size of the rotifer *Brachionus plicatilis* and the maximum size of particles ingested. Bull. Jpn. Soc. Sci. Fish. 46: 1217–1222.

Hirano, K., 1987. Studies on the culture of the rotifer (*Brachionus plicatilis* O. F. Muller). Bull. Fac. Agric. Miyazaki Univ. 34: 57–122.

Hirata, H., 1980. Culture methods of the marine rotifer *Brachionus plicatilis*. Min. Rev. Data File Res. 1: 27–46.

Hirata, H., M. Anastasios & S. Yamasaki, 1985. Evaluation of the use of *Brachionus plicatilis* and *Artemia* nauplii for rearing prawn *Penaeus japonicus* larvae on a laboratory scale. Mem. Fac. Fish. Kagoshima Univ. 34: 27–36.

Houde, E. D. & B. J. Palko, 1970. Laboratory rearing of clupeid fish *Harengula pensacolae* from fertilized eggs. Mar. Biol. 5: 354–358.

Houde, E. D. & R. C. Schekter, 1983. Oxygen uptake and comparative energetics among eggs and larvae of three subtropical marine fishes. Mar. Biol. 72: 283–293.

Howell, B. R., 1973. Marine fish culture in Britain. VIII. A marine rotifer, *Brachionus plicatilis* Muller, and the larvae of the mussel, *Mytilus edulis* L., as foods for larval flatfish. J. Cons. int. Explor. mer 35: 1–6.

Howell, B. R., 1979. Experiments on the rearing of larval turbot, *Scophthalmus maximus* L. Aquaculture 18: 215–225.

Hunter, J. R., 1980. The feeding behaviour and ecology of

marine fish larvae. In J. E. Bardach, J. J. Magnuson, R. C. May & J. M. Reinhart (eds), Fish Behaviour and Its Use in the Capture and Culture of Fishes. ICLARM Conf. Proc. 5, Manila, Philippines: 287–330.

Hunter, J. R. & C. M. Kimbrell, 1980. Early life history of Pacific mackerel *Scomber japonicus*. Fish. Bull. 78: 89–102.

Ito, T., 1960. On the culture of the mixohaline rotifer *Brachionus plicatilis* O. F. Muller, in sea water. Rep. Fac. Fish. Prefect. Univ. Mie 3: 708–740.

Ivlev, V. S., 1961. Experimental ecology of the feeding of fishes. Yale University Press, New Haven, CT, 302 pp.

Juario, J. V., M. N. Duray, V. M. Duray, J. F. Nacario & J. M. E. Almendras, 1984. Induced breeding and larval rearing experiments with milkfish *Chanos chanos* (Forskal) in the Philippines. Aquaculture 36: 61–70.

Kafuku, T. & H. Ikenoue, 1983. Modern methods of aquaculture in Japan. Developments in Aquaculture and Fisheries Science, 11. Kodansha Ltd., Tokyo, and Elsevier, Amsterdam, 216 pp.

Kanazawa, A., 1985. Essential fatty acid and lipid requirement of fish. In C. B. Cowey, A. M. Mackie & J. G. Bell (eds), Nutrition and Feeding in Fish. Academic Press, London: 281–294.

Kanazawa, A., S. I. Teshima, S. Inamori, S. Sumida & T. Iwashita, 1982. Rearing of larval red sea bream and ayu with artificial diets. Mem. Fac. Fish. Kagoshima Univ. 31: 185–192.

Kentouri, M. & P. Divanach, 1982. Comportement et regime alimentaire des larves de marbre *Lithognathus mormyrus* (Poisson, Teleost., Sparidae) elevees dans des conditions de choix trophique polyspecifique et pluridimensionnel. C.R. Acad. Sci. Paris, Ser. III, 294: 859–861.

Kissil, G. Wm., 1984/85. Overview: rearing larval stages of marine fish on artificial diets. Israel J. Zool. 33: 154–160.

Kitajima, C., S. Fujita, F. Ohwa, Y. Yone & T. Watanabe, 1979. Improvement of dietary value for red sea bream larvae of rotifers *Brachionus plicatilis* cultured with baker's yeast *Saccharomyces cerevisiae*. Bull. Jpn. Soc. Sci. Fish. 45: 469–471.

Kitajima, C., T. Arakawa, F. Oowa, S. Fujita, O. Imada, T. Watanabe & Y. Yone, 1980. Dietary value for red sea bream larvae of rotifer *Brachionus plicatilis* cultured with a new type yeast. Bull. Jpn. Soc. Sci. Fish. 46: 43–46.

Klumpp, D. W. & H. von-Westernhagen, 1986. Nitrogen balance in marine fish larvae; influence of developmental stage and prey density. Mar. Biol. 93: 189–199.

Korunuma, K. & K. Fukusho, 1987. Rearing of Marine Fish Larvae in Japan. IDRC, Ottawa, Canada: 109 pp.

Kuhlmann, D., G. Quantz & U. Witt, 1981. Rearing of turbot larvae (*Scophthalmus maximus* L.) on cultured food organisms and postmetamorphosis growth on natural and artificial food. Aquaculture 23: 183–196.

Lasker, R., H. M. Feder, G. H. Theilacker & R. C. May, 1970. Feeding, growth, and survival of *Engraulis mordax* larvae reared in the laboratory. Mar. Biol. 5: 345–353.

Laurence, G. C., 1977. A bioenergetic model for the analysis

of feeding and survival potential of winter flounder *Pseudopleuronectes americanus* larvae during the period from hatching to metamorphosis. Fish. Bull. U.S. 75: 529–546.

Liao, I. C., J. V. Juario, S. Kumagai, H. Nakajima, M. Natividad & P. Buri, 1979. On the induced spawning and larval rearing of the milkfish *Chanos chanos* (Forskal). Aquaculture 18: 75–93.

Lubzens, E., 1981. Rotifer resting eggs and their application to marine aquaculture. Spec. Publ. Europ. Maricult. Soc. 6: 163–179.

Lubzens, E., 1987. Raising rotifers for use in aquaculture. Hydrobiologia 147: 245–255.

Lubzens, E., 1989. Possible use of rotifer resting eggs and preserved live rotifers (*Brachionus plicatilis*) in aquaculture and mariculture. In N. De Paw, E. Jaspers & H. Ackeford (eds), Aquaculture – A Biotechnology in Progress. European Aquaculture Society (in press).

Lubzens, E., A. Marko & A. Tietz, 1985. *De novo* synthesis of fatty acids in the rotifer *Brachionus plicatilis*. Aquaculture 47: 27–37.

Lubzens, E., S. Rothbard, A. Blumenthal, G. Kolodny, B. Perry, B. Olund, Y. Wax & H. Farbstein, 1987. Possible use of *Brachionus plicatilis* (O. F. Muller) as food for freshwater cyprinid larvae. Aquaculture 60: 143–155.

Lubzens, E., B. Perry, S. Euteneuer & R. Berghahn, 1989. Preservation of rotifers for use in aquaculture. Spec. Publ. Europ. Maricult. Soc. (in press)

Minkoff, G., 1987. The effect of secondarily enriched rotifers on growth and survival of marine fish larvae. Ph. D. thesis, University of Stirling, UK.

Mock, C. R., D. B. Revera & C. T. Fontaine, 1980. The larval culture of *Penaeus stylirostris* using modifications of the Galveston Laboratory technique. Proc. World Maricult. Soc. 11: 102–117.

Morales, J. C. 1983. Acuicultura, Marina Animal. Ediciones Mundi-Prensa, Madrid, 670 pp.

Nash, C., C. M. Kuo & S. C. McConnel, 1974. Operational procedures for rearing larvae of the grey mullet (*Mugil cephalus* L.). Aquaculture 3: 15–24.

Oka, A., N. Suzuki & T. Watanabe, 1980. Effect of fatty acids in rotifers on growth and fatty acid composition of larval ayu *Plecoglossus altivelis*. Bull. Jpn. Soc. Sci. Fish. 46: 1413–1418.

Okamoto, R., 1969. Rearing of red sea bream larvae. Bull. Jpn. Soc. Sci. Fish. 35: 563–566.

Okauchi, M., T. Oshiro, S. Kitamura, A. Tsujigado & F. Fukusho, 1980. Number of rotifer *Brachionus plicatilis* consumed daily by a larva and juvenile of porgy, *Acanthopagrus schlegeli*. Bull. Natl. Res. Inst. Aquacult. 1: 39–45.

Olsen, J. O. & F. Minck, 1983. A technical solution to the mass culturing of larval turbot. Aquacult. Eng. 2: 1–12.

Owen, J. M., J. W. Adron, C. Middleton & C. B. Cowey, 1975. Elongation and desaturation of dietary fatty acids in turbot *Scophthalmus maximus* L. and rainbow trout *Salmo gairdneri* Rich. Lipids 10: 528–531.

Peguin, C. L., 1984. The effect of photoperiod and prey den-

400

sity on the growth and survival of larval gilthead seabream, *Sparus aurata* L. (Perciformes, Teleostei). M. Sc. thesis, Hebrew University, Jerusalem: 93 pp.

Person-Le Ruyet, J. & P. Verillaud, 1980. Techniques d'elevage intensif de la Daurade doree *Sparus aurata* (L.) de la naissance a l'age de deux mois. Aquaculture 20: 351–370.

Person-Le Ruyet, J. & P. Verillaud, 1981. Techniques d'elevage intensif de la Daurade doree (*Sparus aurata*) de la naissance a l'age de 2 mois. Rapp. P.-v. Reun. Cons. Int. Explor. Mer 178: 527–529.

Rezeq, T. A. & C. M. James, 1987. Production and nutritional quality of the rotifer *Brachionus plicatilis* fed *Chlorella* sp. at different cell densities. Hydrobiologia 147: 257–261.

Samocha, T. M., N. Uziel & C. L. Browdy, 1988. Evaluation of animal protein sources for the culture of penaeid shrimp larvae. Aquaculture (in press).

Scott, A. P. a S. M. Baynes, 1978. Effect of algal diet and temperature on the biochemical composition of the rotifer, *Brachionus plicatilis*. Aquaculture 14: 247–260.

Scott, A. P. & S. M. Baynes, 1979. The effect of unicellular algae on survival and growth of turbot larvae (*Scophthalmus maximus* L.). In J. E. Halver & K. Tiews (eds), Finfish Nutrition and Fishfeed Technology. Proc. World Symp., Hamburg, 20–23 June 1978, Vol. I: 423–433.

Scott, A. P. & C. Middleton, 1979. Unicellular algae as food for turbot (*Scophthalmus maximus*) larvae – the importance of dietary long chain polyunsaturated fatty acids. Aquaculture 18: 227–240.

Snell, T. W., M. J. Childress, F. M. Boyer & F. H. Hoff, 1987. Assessing the status of rotifer mass cultures. J. World Aquacult. Soc. 18: 270–277.

Stepiens, W. P. Jr., 1976. Feeding of laboratory reared larvae of the sea bream *Archosargus rhomboidalis* (Sparidae). Mar. Biol. 38: 1–16.

Tandler, A. 1984/85. Overview: food for the larval stages of marine fish; live or inert. Israel J. Zool. 33: 161–166.

Tandler, A. & S. Helps, 1985. The effects of photoperiod and water exchange rate on growth and survival of gilthead sea bream (*Sparus aurata*, Linnaeus; Sparidae) from hatching to metamorphosis in mass rearing system. Aquaculture 48: 71–82.

Tandler, A. & C. Mason, 1984. The use of ^{14}C labelled rotifers (*Brachionus plicatilis*) in the larvae of gilthead seabream (*Sparus aurata*): Measurements of the effect of rotifer concentration, the lighting regime and seabream larval age on their rate of rotifer ingestion. Europ. Maricult. Soc. 8: 241–259.

Tandler, A. & R. Sherman, 1981. Food organism concentration, environmental temperature and survival of the gilthead bream (*Sparus aurata*) larvae. Spec. Publ. Europ. Maricult. Soc. 6: 237–248.

Theilacker, G. & K. Dorsey, 1980. Larval fish diversity; a summary of laboratory and field research. IOC Workshop Rep. 28: 105–142.

Theilacker, G. H. & A. S. Kimball, 1984. Comparative quality of rotifers and copepods as foods for larval fishes. CalCOFI Rep. 25: 80–86.

Theilacker, G. H. & M. F. McMaster, 1971. Mass culture of the rotifer *Brachionus plicatilis* and its evaluation as food for larval anchovies. Mar. Biol. 10: 183–188.

Walford, J. & T. Lam, 1987. Effect of feeding with microcapsules on the content of essential fatty acids in live foods for the larvae of marine fishes. Aquaculture 61: 219–229.

Watanabe, T., 1980. Studies on the improvement of feeding techniques for rearing the larvae of *Panaeus semisulcatus*. Kuwait Inst. Scient. Res., Spec. Publ. KISR/PP 1012/FRM-RT-R-8001. 24 pp.

Watanabe, T., C. Kitajima & S. Fujita, 1983. Nutritional values of live organisms used in Japan for mass propagation of fish: a review. Aquaculture 34: 115–143.

Witt, U., G. Quantz & D. Kuhlmann, 1984. Survival and growth of turbot larvae *Scophthalmus maximus* L. reared on different food organisms with special regard to long-chain polyunsaturated fatty acids. Aquacult. Eng. 177–190.

Yamasaki, S. & H. Hirata, 1982. Rearing of the prawn, *Penaeus japonicus*, fed on frozen and living rotifers. Min. Rev. Data File Fish. Res. 2: 87–89.

Yamasaki, S., S. H. Cheuh, K. J. Ang, H. Hirata, A. Z. Abidin & A. Z. Alias, 1988. Manual of hatchery management based on bio-physico-chemical control. Min. Rev. Data File Fish. Res. 5: 1–102.

Yone, Y. & M. Fujii, 1975. Studies on nutrition of red sea bream. XI. Effects of $\omega 3$ fatty acid supplement in acorn oil diet on growth rate and feed efficiency. Bull. Jpn. Soc. Sci. Fish. 41: 73–77.

Yufera, M. & E. Pascual, 1984. La produccion de organismos zooplanctonicos para la alimentacion larvaria en acuicultura marina. Inf. Tecn. Inst. Inv. Pesq. 119: 27 pp.

Yufera, M., A. Rodriguez & L. M. Lubian, 1984. Zooplankton ingestion and feeding behaviour of *Penaeus kerathurus* larvae reared in the laboratory. Aquaculture 42: 217–224.

Hydrobiologia **186/187**: 401–408, 1989.
C. Ricci, T. W. Snell and C. E. King (eds), Rotifer Symposium V.
© 1989 *Kluwer Academic Publishers.*

Nitrogen flow through a *Brachionus*/*Chlorella* mass culture system

Warren D. Nagata
Present address: Aquaculture Department, Malaspina College, 900 Fifth Street, Nanaimo, B.C., Canada, V9R 5S5

Key words: mariculture, rotifers, batch culture, nitrogen, recycling

Abstract

A model of nitrogen flow is presented through the *Brachionus plicatilis/Chlorella saccharophila* mass (batch) culture system, from the initial input of inorganic nitrogen to the algal culture medium to the final production of rotifers, dissolved nitrogen and particulate nitrogen.

A nitrogen budget was first formulated for *B. plicatilis* relating ingestion, excretion, egestion, somatic growth and reproductive growth. Measurements were made on rotifers from 20° and 10° cultures.

The calculated model of nitrogen flow through the rotifer/algal batch culture system estimates the percentages of the original input nitrogen which will be incorporated into algal nitrogen, rotifer nitrogen and the particulate and dissolved nitrogen pools. It is suggested that the dissolved nitrogen pool could be recycled directly for use in subsequent algal culture.

Introduction

Since its initial application as a live food organism for the larval sweet smelt, the euryhaline rotifer *Brachionus plicatilis* Müller has become a widely-used food organism in the larval culture of over 60 marine finfish and 18 crustacean species throughout the world (Fujita, 1983; Lubzens, 1987). Although mass culture techniques for this rotifer have been greatly improved over the last 25 years (Hirata, 1980; Lubzens, 1987), unexplained decline or inhibited growth in the cultured populations is still periodically encountered using the standard batch culture mode. Self-pollution of the culture environment through accumulation of waste products has been deemed a major contributing factor to this culture 'crash' phenomenon (Hirata, 1980; Hirayama, 1987). A more thorough understanding of the nutrient dynamics of this culture system may lead to improvement of its present modes of operation and the future

development of more efficient modes for this species.

The object of this study was to examine the dynamics of nitrogen cycling within the *B. plicatilis* mass culture system in the batch culture mode of operation. Initially, nitrogen budgets were formulated for this species. These budgets were then expanded into general models of nitrogen flow within the culture system.

Materials and methods

Cultures of *Brachionus plicatilis* and its food organism *Chlorella saccharophila* var. *saccharophila* Kruger were obtained from the same sources and cultured as previously described (Nagata, 1985a).

The following formula modified from Conover (1978) was used to formulate the nitrogen budgets for *B. plicatilis*.

$$R = G + U + E, \tag{1}$$

where R = ration, food consumed per unit time; G = growth or production, the change in weight per unit time; U = excretion, material released as urine or through the body surface per unit time; and E = egestion, the amount of feces produced per unit time. The growth term (G) was also subdivided into somatic growth (G_B) and reproductive growth (G_R).

Assimilation efficiency (A'), the percentage of the ingested food digested and absorbed into the body was calculated as

$$A' = (G + M)/R. \tag{2}$$

Since *B. plicatilis* does not produce coherent fecal pellets collection of feces was not feasible. Egestion was estimated by difference ($E = R - G - U$).

Carbon and nitrogen contents of dried algal and rotifer samples were analyzed by a Hitachi Model 026 CHN Analyzer, coupled to a Takeda Riken TR-2217 integrator and were calibrated against both caffeine and acetanilide standards.

Both rotifer ingestion and excretion were measured concurrently. Algae were centrifuged then resuspended in ASP low nitrogen medium (Hirata & Nagata, 1982) and allowed to acclimate in the dark. Rotifers were filtered from their culture medium, rinsed several times with ASP medium and resuspended into the algal/ASP suspension. They were maintained at a density of about 100 ind ml^{-1}, and were never concentrated to densities greater than their stock cultures. The combined rotifer/algal culture was then acclimated for an additional 2 h period, transferred to experimental flasks and tightly stoppered. The flasks were attached to a plankton rotor within an incubator and rotated over their longitudinal axes through a radius of 10 cm at 5 rpm for an incubation period of 3 h at 10 or 20 ± 0.2 °C. Two control flasks correcting for nutrient uptake and cell increase by the algae, and changes in the nutrient levels due to bacterial and physical processes were also prepared. Prior to and following the incubation period samples were taken for algal cell counts and dissolved nutrient analyses.

Algal cell counts were carried out by the method of Nagata (1985b), while clearance and ingestion rates were calculated by the equation of Schlosser & Anger (1982). Nutrient analyses for ammonium, urea and total dissolved nitrogen were carried out by the methods of Solorzano (1969), McCarthy (1970), and Solorzano & Sharp (1980) and by automated modifications of these methods adapted to run on the Technicon Autoanalyzer IIc system. Intracellular ammonium pools of the algae were also measured by the C-2 method of Thoreson *et al.* (1982) and later used to correct the rotifer excretion data.

Results and discussion

A mean dry weight of 5.8 ± 1.2 pg cell^{-1} was obtained for *C. saccharophila* ($n = 38$) from cultures ranging in density from 1.5 to 17.5×10^6 cells ml^{-1}. Mean carbon and nitrogen composition of this alga were found to be 54.58 ± 8.38 and $5.34 \pm 1.64\%$, respectively.

Dry weight data for eggs, neonates and adult rotifers at 10° and 20 °C, are presented in Table 1. The neonate mean dry weights at the two temperatures were 0.32 ± 0.06 μg ind^{-1} (20 °C) and 0.38 ± 0.02 μg ind^{-1} (10 °C). Mean dry weights of egg-bearing and non-egg-bearing females at the two temperatures were not significantly different, so pooled means of 0.82 (1 egg) and 0.55 (no egg) μg ind^{-1} were used in subsequent calculations.

Mean carbon and nitrogen contents of *B. plicatilis* were: $6.56 \pm 0.61\%$ N and $39.57 \pm 2.24\%$ C at 20 °C, and $6.23 \pm 0.21\%$ N and $42.68 \pm 1.65\%$ C at 10 °C. Carbon and nitrogen contents of eggs and neonates at 20 °C were determined to be 5.60% N and 37.09% C, and 2.69% N and 25.47% C, respectively.

From the mean calculated dry weights (Table 1), nitrogen and carbon content of individual eggless adult rotifers were: 28.54 ng N ind^{-1} and 172.13 ng C ind^{-1} at 20 °C, and 28.97 ng N ind^{-1} and 198.46 ng C ind^{-1} at 10 °C (Table 2). Those of neonates and amictic eggs were 8.61 ng N and 81.50 ng C ind^{-1}, and 14.56 ng N and

Table 1. Mean dry weight of *B. plicatilis* at 20° and 10° and the results of variance ratio testing between data from the two temperatures. Each sample was comprised of 50 individuals. Egg dry weight was determined by difference.

Stage	Temp.	n	Dry weight (μg)	Variance ratio test results
Neonate	20	8	0.32 ± 0.06	$F_o = 6.86 > F0.05(2)7,6 = 5.70$
Neonate	10	7	0.38 ± 0.02	$0.02 < P < 0.05$
Adult	20	20	0.55 ± 0.08	$F_o = 2.29 > F0.05(2)19,13 = 2.96$
(no egg)	10	14	0.55 ± 0.05	$0.01 < P < 0.02$
Adult	20	15	0.81 ± 0.05	$F_o = 1.71 > F0.05(2)14,7 = 4.60$
(1 egg)	10	8	0.83 ± 0.04	$0.20 < P < 0.50$
Egg	20	–	0.26	
	10	–	0.28	

96.43 ng C egg^{-1}, respectively at 20 °C. At 10 °C these values were slightly higher. Droop (1976) obtained a similar carbon content of this species of 180 ng C ind^{-1}, while Schlosser & Anger (1982) found values of 9.3% N or 26 ng N ind^{-1} and 41.5% C or 116 ng C ind^{-1}.

In terms of elemental composition the growth from neonate to adult at 20 °C results in a gain of 27.47 ng N and 136.14 ng C ind^{-1} (Table 2). At 10 °C, similar values were 24.05 ng N and 137.95 ng C ind^{-1}. As for total reproduction (22.5 eggs ind^{-1} lifetime^{-1}, from lifetable in: Nagata, 1985a) 327.60 ng N and 2169.68 ng C are partitioned into eggs at 20 °C, while 276.50 ng N and 1809.94 ng C at 10 °C (16.1 eggs ind^{-1} lifetime^{-1}). Thus, about 12 times as much nitrogen and 16 times the carbon are diverted into reproduction as opposed to somatic growth over the lifetime of an amictic female of this species at 20 °C. The same values for 10 °C females were $10.5 \times N$ and $12 \times C$.

A summary of the clearance and ingestion rates of this rotifer is presented in Table 3. Clearance rates ranged from 0.90 to 2.44 μl ind^{-1} h^{-1} at 20 °C and 0.99 to 1.99 μl ind^{-1} h^{-1} at 10 °C. Ingestion rates, on the other hand ranged from 45 to 345 cells ind^{-1} min^{-1} at 20 °C, and 35 to 77 cells ind^{-1} min^{-1} at 10 °C. Clearance rates appeared to be little affected by algal concentration and temperature within the tested ranges, while ingestion rates seem to be greatly affected by these parameters. Mean clearance rates obtained in the present study were within the range of 1 to 10 μl ind^{-1} h^{-1} obtained for other rotifers at 20-25 °C (Starkweather, 1980) and 0.38 to 10.00 μl ind^{-1} h^{-1} for *B. plicatilis* at 20-28 °C (Hirayama, 1983).

As a conservative mean estimate of daily consumption, an adult rotifer weighing 0.44 μg dry weight at 20 °C, or 0.46 μg at 10 °C, could ingest 174 960 cells or 1.01 μg dry weight of *C. saccharophila* per day at 20 °C; and 98 640 cells or 0.57 μg

Table 2. Elemental composition (C,N) of *B. plicatilis* from egg to adult at 20 °C and 10 °C.

Temp. (°C)	Stage	Dry weight (ng ind^{-1})	N (ng ind^{-1})	C (ng ind^{-1})
20	egg	260	14.56	96.43
20	neonate	320	8.61	81.50
20	adult	550	36.08	217.64
10	*egg	280	15.68	103.85
10	*neonate	380	10.22	96.79
10	adult	550	34.27	234.74
	$\bar{X}$ 20 °C =	435	28.54	172.13
	$\bar{X}$ 10 °C =	465	28.97	198.46

* Assuming same carbon and nitrogen percentage composition as 20 °C.

Table 3. Summary of clearance and ingestion rates of *B. plicatilis* at 20 °C and 10 °C and various algal densities.

Temp. (°C)	n	Algal density (10^4 cells ind^{-1})	Clearance rate (μl ind^{-1} h^{-1})	Ingestion rate (cells ind^{-1} h^{-1})
20	6	16.72	1.46 $\pm$ 0.31	290 $\pm$ 49
20	8	4.21	1.78 $\pm$ 0.53	91 $\pm$ 29
10	5	2.50	1.60 $\pm$ 0.37	52 $\pm$ 16
Total $\overline{X}$ =			1.63 $\pm$ 0.47	143 $\pm$ 108

dry weight per day at 10 °C. In terms of carbon and nitrogen: 54 ng N and 550 ng C at 20 °C, and 30 ng N and 310 ng C at 10 °C. Thus the amount of food required per rotifer at 10 °C is about 56% that required at 20 °C.

A summary of the derived excretion rates of *B. plicatilis* is also presented (Table 4). Nitrogenous excretion in this species was comprised mainly of ammonium and urea-N as previously observed by Hirata & Nagata (1982). The proportions of urea excreted ranged from 27.27 to 77.78%, with a mean of 48.59% of the total persulfate nitrogen (TPN). Proportions of ammonium ranged from 16.33 to 72.53% with a mean of 43.34%. Despite the great variation in proportions of these two nutrients, they always comprised together the majority of the TPN excretion (90.24 $\pm$ 9.45%).

Mean TPN excretion rates at 20 °C of 3.94 $\pm$ 0.76 and 0.67 $\pm$ 0.15 ng N ind^{-1} h^{-1} were obtained for the high and low algal trials, respectively, indicating a dependence of excretion rates of *B. plicatilis* on algal density. The excretion rates of *B. plicatilis* at 10 °C of NH$_4$-N, urea-N, and TPN at 10 °C were considerably lower than those at 20 °C under similar food regimes.

The only comparable rotifer excretion rates to those of this study are seen in the studies on *B. calyciflorus* of Galkovskaya & Ejsmont-Karabin (1981) and Ejsmont-Karabin (1983), in which only the rates of ammonium excretion were measured. Excretion rates of 0.26 $\times$ 10^{-3} μg N ind^{-1} h^{-1} or 16.91 μg N mg dry wt^{-1} day^{-1} at 20 °C were obtained in the former, while a rate of 24 μg N mg dry wt^{-1} day^{-1} was obtained in the latter at 20 °C.

Nitrogen budgets were calculated for three experimental groups of *B. plicatilis* which varied in acclimation temperature and food concentration (Table 5). The values for excretion labeled (a) are the initial uncorrected values, while those labeled (b) and (c) are those corrected for 5 and 10% leakage (Dagg, 1976) from the intracellular ammonium pools of the algae, respectively.

From Table 5, at 20 °C and low algal density (1.20 μg N ml^{-1}) about 83% of the ingested algal nitrogen was assimilated. Of this, 66% was channeled into growth, while 34% was excreted. At the higher algal density (4.35 μg N ml^{-1}) 85% of the ingested algal nitrogen was assimilated, of which 28% was used for growth and 72% excreted. At 10 °C, assimilation was decreased to 61%, of which 52% was channeled into growth and 49% was excreted.

Table 4. Summary of nitrogenous excretion in *B. plicatilis* at 20 °C and 10 °C.

Temp. °C	Algal density (μg dry wt ml^{-1})	Rotifer dry wt (ng ind^{-1})	Rotifer N	Excretion (μg N mg dry wt^{-1} day^{-1})		
				NH$_4$-N	Urea-N	Total-N
20	81.4	0.44	28.54	82.64	68.48	151.19
20	22.5	0.44	28.54	15.66	13.89	31.42
10	15.5	0.47	28.97	5.24	11.97	19.41

Table 5. Daily nitrogen budget of B. plicatilis at 20 °C and 10 °C. Figures in brackets indicate percentage of assimilated algal ration. G_B, G_R, U, E, and R expressed as: ng N ind^{-1} day^{-1}. (a)-uncorrected excretion values. (b)-excretion values corrected for 5% spillage. (c)-excretion values corrected for 10% spillage. Refer to text for further details of correction method.

Temp. (°C)	Algal density (μg N ml^{-1})	Somatic growth (G_B)	Reproductive growth (G_R)	Excretion (U)	Egestion (E)	Ingestion (R)	Assimilation (%)
20	1.20	2.39(5.1)	28.49(60.7)	(a)16.08(34.2)	9.12	56.08	83.74
20	1.20	2.39(5.1)	28.49(61.3)	(b)15.62(33.6)	9.58	56.08	82.92
20	1.20	2.39(5.2)	28.49(61.9)	(c)15.15(32.9)	10.05	56.08	82.08
20	4.35	2.39(1.9)	28.49(22.7)	(a)94.56(75.4)	3.18	128.62	97.53
20	4.35	2.39(2.2)	28.49(25.9)	(b)79.08(71.6)	18.66	128.62	85.49
20	4.35	2.39(2.5)	28.49(30.0)	(c)63.60(67.0)	34.14	128.62	73.46
10	0.83	0.91(4.4)	9.53(45.9)	(a)10.32(49.7)	12.42	33.18	62.57
10	0.83	0.91(4.5)	9.53(47.0)	(b) 9.86(48.6)	12.88	33.18	61.18
10	0.83	0.91(4.6)	9.53(47.9)	(c) 9.44(47.5)	13.30	33.18	59.92

The assimilation efficiencies from this study are comparable to values of 64.4 and 74.1% for this species (Droop & Scott, 1978), and of 78% for *B. calyciflorus* (Galkovskaya, 1971).

In Table 6 the above terms are expressed as gross (K_1) and net (K_2) growth efficiencies. These efficiencies were calculated as: $K_1 = G_B + G_R/R$ and $K_2 = G_B + G_R/G_B + G_R + U$. The K_1 value of 55.06% for this rotifer is high, but comparable to those of other zooplankton, (i.e. *Artemia* sp., K_1 = 53% (Gibor, 1952); *Rhinocalanus nasutus* K_1 = 55% (Mullin & Brooks, 1970).

The K_2 value of 66.41% is comparable to those of other brachionid rotifers, i.e. *B. calyciflorus*, K_2 = 69% (Galkovskaya, 1963); and K_2 = 74% (Galkovskaya & Ejsmont-Karabin, 1981).

From the preceding information, the nitrogen flow through the culture system as a whole can be estimated. A model of nitrogen flow through the *B. plicatilis/C. saccharophila* culture system operating in the batch culture mode at 20 °C is illustrated in Fig. 1. Transfer efficiencies were assumed to be at maximum levels, such that: assimilation of medium N to *Chlorella* N = 100%; rotifer ingestion efficiency = 90%; rotifer assimilation efficiency = 83.0%, and rotifer net growth efficiency = 66.4%. The original input of nitrogen contained in the Yashima 1-A medium of 5.46 g per m^3 will under ideal growth conditions yield an equivalent amount of

Chlorella cellular nitrogen. This in turn will yield 2.70 g of rotifer nitrogen. A rotifer mortality rate of 1.62% day^{-1} would result in 0.24 g N being lost to the particulate pool. Thus the net output of rotifer biomass will contain 45.05% of the input nitrogen, while 29.85 and 25.09% of the remaining nitrogen will enter the particulate and dissolved nitrogen pools, respectively.

A similar model was constructed for this culture system at 10 °C (Fig. 2). The major differences between the models at the two temperatures was seen in the lower final yield of rotifer nitrogen (1.15 g N), which contained only 21.06% of the original input nitrogen, and increase in the particulate nitrogen pool (2.85 g N) at 10 °C. The latter comprised 52.2% of the original input nitrogen. Of this particulate nitrogen pool, rotifer fecal material was the major contributor (1.91 g) while 0.39 g N was contained in dead rotifer nitrogen. The dissolved nitrogen pool, however, was very similar to that at 20 °C comprising 26.74% of the original input nitrogen. Thus it can be seen that the overall efficiency of the culture system at 10 °C can be expected to be less than half that at 20 °C.

Although the nitrogen contained in the particulate N pool cannot be readily reutilized, the amount of nitrogen contained in the dissolved nitrogen pool is almost entirely in remineralized forms which can be utilized directly as N sources by *Chlorella* sp. Nevertheless, this amount of

Table 6. Gross (K_1) and net (K_2) growth efficiencies of *B. plicatilis* at 20 °C and 10 °C.

Temp. (°C)	Algal density (μg dry wt ml^{-1})	Assimilation (%)	K_1 ($G_B + G_R/R$)	K_2 ($G_B + G_R/G_B + G_R + U$)
20	22.5	83.74	55.06	65.76
20	22.5	82.92	55.06	66.41
20	22.5	82.08	55.06	67.09
20	81.4	97.53	24.01	24.62
20	81.4	85.49	24.01	28.08
20	81.4	73.46	24.01	29.60
10	15.5	62.57	31.46	50.29
10	15.5	61.18	31.46	51.43
10	15.5	59.92	31.46	52.52

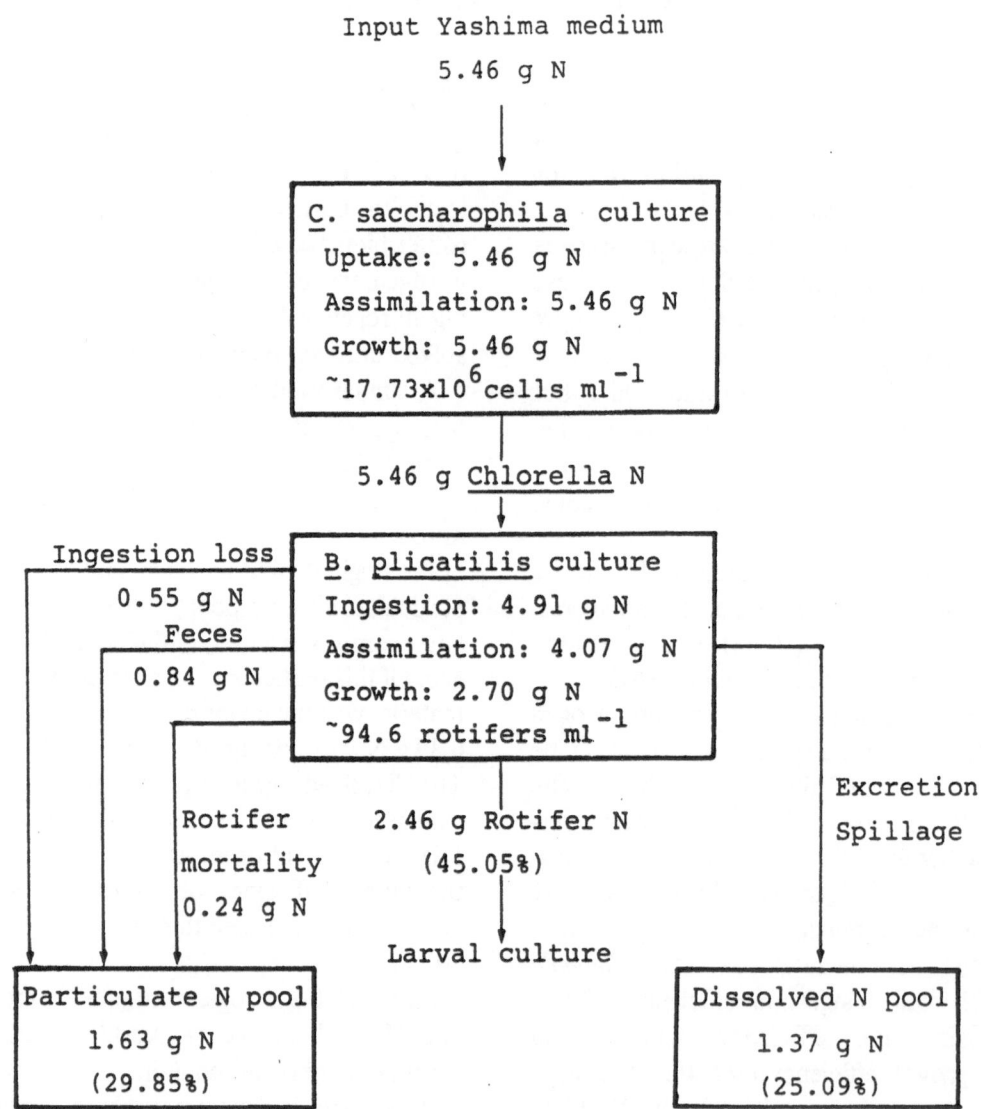

Fig. 1. Model of nitrogen flow through the *B. plicatilis/C. saccharophila* mass culture system at 20 °C. Figures in brackets indicate the percentage of the input nitrogen. All nitrogen values are expressed as N per m³.

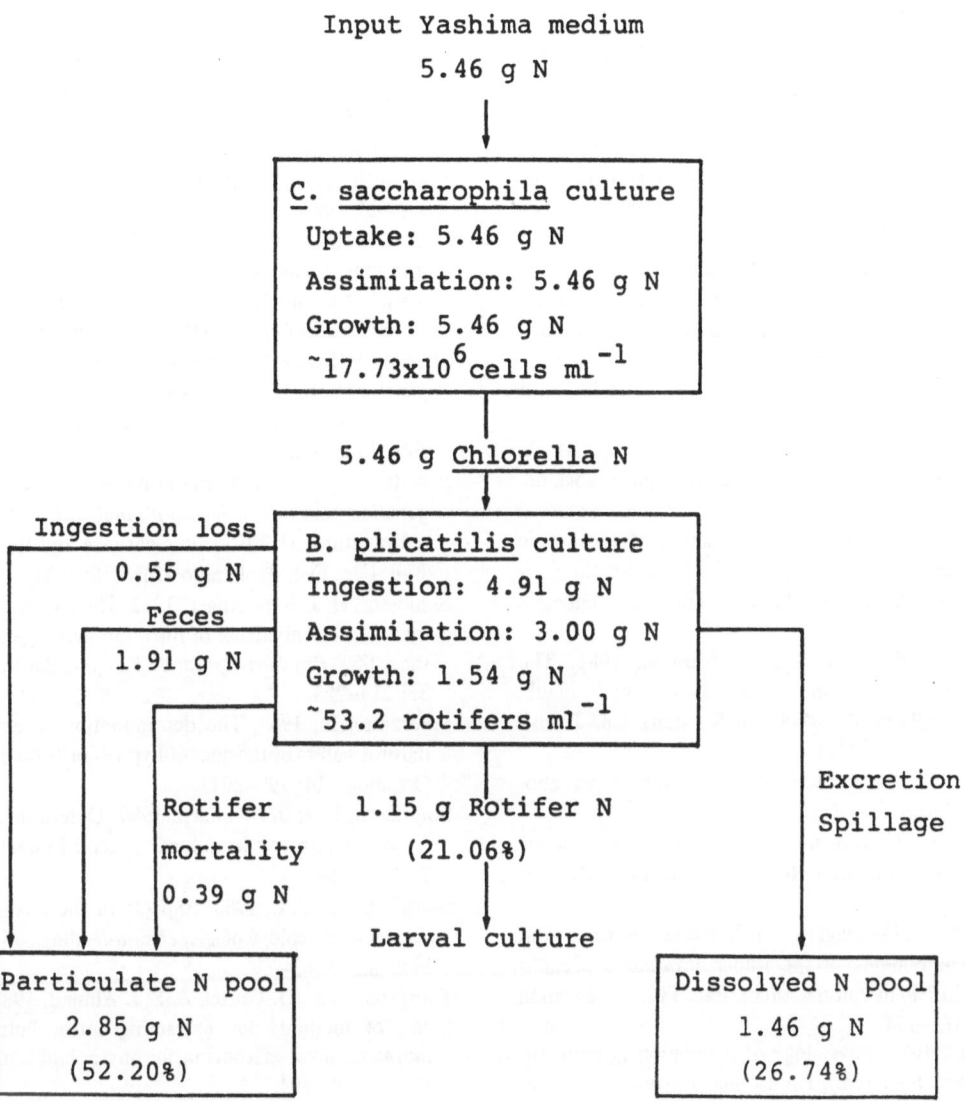

Input Yashima medium
5.46 g N

C. saccharophila culture
Uptake: 5.46 g N
Assimilation: 5.46 g N
Growth: 5.46 g N
~17.73x10^{6}cells ml^{-1}

5.46 g Chlorella N

Ingestion loss
0.55 g N
Feces
1.91 g N

B. plicatilis culture
Ingestion: 4.91 g N
Assimilation: 3.00 g N
Growth: 1.54 g N
~53.2 rotifers ml^{-1}

Excretion
Spillage

Rotifer
mortality
0.39 g N

1.15 g Rotifer N
(21.06%)

Larval culture

Particulate N pool
2.85 g N
(52.20%)

Dissolved N pool
1.46 g N
(26.74%)

Fig. 2. Model of nitrogen flow through the *B. plicatilis/C. saccharophila* mass culture system at 10 °C. Figures in brackets indicate the percentage of the input nitrogen. All nitrogen values are expressed as N per m^3.

nitrogen can only account for about 25% of the nitrogen requirement for suitable algal biomass production.

Although these simplified models overlook several potential fluxes of nitrogen, they may facilitate the development of more efficient mass culture systems for *B. plicatilis*.

Acknowledgements

The author wishes to express his deepest gratitude to Prof. T. Minoda of the Plankton Laboratory, Hokkaido University for his encouragement and guidance throughout this study. The author is also indebted to the Ministry of Education of the Japanese Government for financial support.

References

Conover, R. J., 1978. Transformation of organic matter., In: O. Kinne (ed.), Marine Ecology Vol. IV Dynamics, Wiley, London: 221–499.
Dagg, M. J., 1976. Complete carbon and nitrogen budgets for the carnivorous amphipod, *Calliopius laeviusculus* (Kroyer). Int. Revue ges. Hydrobiol. 61: 297–357.

408

Droop, M. R., 1976. The chemostat in mariculture., In: C. Persoone & E. Jaspers (eds.), Proc. 10th Europ. Symp. Mar. Biol. 1: 71–93.

Droop, M. R. & J. M. Scott, 1978. Steady-state energetics of a planktonic herbivore. J. mar. biol. Ass. U.K. 58: 749–772.

Ejsmont-Karabin, J., 1983. Ammonia nitrogen and inorganic phosphorus excretion by the planktonic rotifers. Hydrobiologia 104: 231–236.

Fujita, S., 1983. The rotifer as living food for seedling production., In: Jpn. Soc. Sci. Fish. (ed.) The Rotifer Brachionus plicatilis, Biology and Mass Culture. Koseisha-Koseikaku, Tokyo, Vol. 44: 9–21, (In Japanese).

Galkovskaya, G. A., 1963. On the utilization of food for growth and conditions for the maximum production of the rotifer Brachionus calyciflorus PALLAS. Zool. zh. Mosk. 42: 506–512, (In Russian), Fish. Res. Bd Can. Transl. no. 997.

Galkovskaya, G. A., 1971. The production of planktonic Rotatoria. In: G. G. Winberg (ed.), Methods for the Estimation of Production of Aquatic Animals, Academic Press, London: 123–128.

Galkovskaya, G. A. & J. Ejsmont-Karabin, 1981. The metabolic O:N ratio in rotifers. Dok. Akad. Nauk. belorussk. SSR 25: 472–474, (In Russian), Ca. Transl. Fish. aquat. Sci. no. 4811.

Gibor, A., 1957. Conversion of phytoplankton to zooplankton. Nature, Lond. 179: 1304.

Hirata, H., 1980. Culture methods of the marine rotifer, Brachionus plicatilis. Min. Rev. Data File Fish. Res., 1: 27–46.

Hirata, H. & W. D. Nagata, 1982. Excretion rates and excreted components of the rotifer Brachionus plicatilis O. F. MÜLLER in culture. Mem. Fac. Fish. Kagoshima Univ. 31: 161–174.

Hirayama, K., 1983. Physiology of population growth. In: Jpn. Soc. Sci. Fish. (ed.), The Rotifer Brachionus plicatilis, Biology and Mass Culture. Koseikosha-Koseikaku, Tokyo, Vol. 44: 52–68 (in Japanese).

Hirayama, K., 1987. A consideration of why mass culture of the rotifer Brachionus plicatilis with baker's yeast is unstable. Hydrobiologia 147: 269–270.

Lubzens, E., 1987. Raising rotifers for use in aquaculture. Hydrobiologia 147: 245–255.

MacCarthy, J. J., 1970. A urease method for urea in seawater. Limnol. Oceanogr. 15: 309–313.

Mullin, M. M. & E. R. Brooks, 1970. Growth and metabolism of two planktonic, marine copepods as influenced by temperature and type of food. In: J. H. Steele (ed.) Marine Food Chains. Oliver & Boyd, Edinburgh: 74–95.

Nagata, W. D., 1985a. Long-term acclimation of a parthenogenetic strain of Brachionus plicatilis to subnormal temperatures. I. Influence on size, growth and reproduction. Bull. Mar. Sci. 37: 716–725.

Nagata, W. D., 1985b. Long-term acclimation of a parthenogenetic strain of Brachionus plicatilis Müller to subnormal temperatures II. Effect on clearance and ingestion rates. Bull. Fac. Fish. Hokkaido Univ. 36: 1–11.

Schlosser, H. J. & K. Anger, 1982. The significance of some methodological effects of filtration and ingestion rates of the rotifer, Brachionus plicatilis. Helgolander Meeresunters. 35: 215–225.

Solorzano, L., 1969. The determination of ammonium in natural waters by the phenol-hypochlorite method. Limnol. Oceanogr. 14: 799–801.

Solorzano, L. & J. H. Sharp, 1980. Determination of total dissolved nitrogen in natural waters. Limnol. Oceanogr. 25: 751–754.

Starkweather, P. L. 1980. Aspects of the feeding behavior and trophic ecology of suspension-feeding rotifers. Hydrobiologia 73: 63–72.

Thoreson, S. S., Q. Dortch & S. I. Ahmed, 1982. Comparison of methods for extracting intracellular pools of inorganic nitrogen from marine phytoplankton. J. Plankton Res. 4: 695–704.

Hydrobiologia **186/187**: 409–413, 1989.
C. Ricci, T. W. Snell and C. E. King (eds), Rotifer Symposium V.
© 1989 *Kluwer Academic Publishers.*

Kinetics of n-3 fatty acids in *Brachionus plicatilis* and changes in the food supply

Yngvar Olsen, Jose Rodriguez Rainuzzo[1], Olav Vadstein & Arne Jensen[1]
SINTEF, Division of Applied Chemistry, Aquaculture Group, N-7034 Trondheim, Norway; [1]*Laboratory of Biotechnology, Norwegian Institute of Technology, University of Trondheim, N-7034 Trondheim, Norway*

Key words: *Brachionus plicatilis*, live feed, enrichment, n-3 fatty acids

Abstract

Moderately starved rotifers exhibited a two-phased increase in n-3 fatty acids when they were fed a diet rich in these fatty acids. The first 20–30 min of enrichment, the increase in n-3 fatty acids was primarily due to increased gut content. The subsequent slow increase was due to an incorporation of n-3 fatty acids into rotifers tissues. Saturation was achieved before 24 h of exposure and the saturation level was independent of the initial content of n-3 fatty acids in the rotifers.

Starvation and limited feeding of the enriched rotifers for additional 4–8 h at 10–20 °C did not affect the accumulated fatty acids significantly. This was found for rotifers with high and low initial content of n-3 fatty acids. The n-3 fatty acids were assimilated with high efficiency from the feed and were not metabolized faster than other groups of fatty acids.

Enriched rotifers retained their nutritional value for a sufficient period after enrichment to serve well as live feed for marine fish larvae.

Introduction

The use of the rotifer *Brachionus plicatilis* as live feed for marine fish larvae has been successful (Ito, 1960; Lubzens, 1987). This is primarily because it can be efficiently cultured and enriched with essential fatty acids (Watanabe *et al.*, 1983) to satisfy the high demand of most marine larvae for polyunsaturated, long-chain n-3 fatty acids (Owen *et al.*, 1972; Cowey *et al.*, 1976).

When enriched rotifers are transferred to larval rearing tanks with particle-free seawater, the rotifers experience a sudden decrease in food concentration. If the enriched fatty acids were primarily stored in the gut, a rapid decrease in n-3 fatty acids would result because of reduced feeding rate and defecation. In addition, metabo-lism of the long-chain fatty acids could dominate over assimilation if the ingestion rate of the fatty acids was reduced. The goal of this work was to study the kinetics of the n-3 fatty acids in *B. plicatilis* after shifts in the ingestion rate of these fatty acids.

Materials and methods

The rotifer *Brachionus plicatilis* was grown in 250 l conical, fiberglass vessels in 20‰ seawater at 20 °C. Each 200 l culture was aerated and fed 20–50 g bakers yeast (wet weight) plus 2–5 g capelin oil (10% by weight) once per day. Oil and yeast were mixed with 0.5 l freshwater in a blender for 10–15 s. The cultures were used in the experiments when they had reached a density of 200–250 ind ml^{-1}.

Two feeding experiments were undertaken, each with two phases. At time zero, moderately starved rotifers were fed a diet rich in n-3 fatty acids, obtained from 25 μm micronized herring roe powder. After 4–8 h the animals were removed and starved or fed a low concentration (5 mg l^{-1}) of the roe powder, respectively.

In each feeding experiment two replicate 100 l cultures of rotifers at density of 200 ind ml^{-1} were given 0.1 g l^{-1} of blended roe powder under standard growth conditions. Five liter samples for lipid analysis were removed from one culture just before and just after addition of the feed and thereafter 5-7 times through the following 4-8 h. The second culture was used for post-feeding analysis. One liter was washed and transferred to a 20 l plastic vessel. Five such vessels contained 19 liter 5 μm filtered, 20‰ seawater and five vessels contained 19 liter 5 μm filtered seawater plus 5 mg roe powder l^{-1}. The final concentration of rotifers through post-feeding was approximately 10 ind ml^{-1} and the added food was sufficient to keep rotifer guts filled for several hours. Samples for lipid analysis were collected at different times through the following 4–8 h by harvesting two entire vessels. In addition, the culture was sampled at start of the post-feeding phase.

The two feeding experiments differed in various ways summarized in Table 1. In Exp. 1, the capelin oil was excluded from the diet through 5 d before start of the experiment. Moreover, the post-feeding phase of Exp. 1 was initiated 6 h after the addition of the roe powder diet, whereas it was initiated after 24 h in Exp. 2. The rotifers were cooled from 20 to 10 °C between 8 and 24 h in Exp. 2. This cooling procedure is routinely applied for preparation of enriched live feed in our laboratory.

Samples of rotifers (0.1-2 g dry weight) were collected on 70 μm screens, washed with 20‰ seawater, frozen on dry ice, transferred to test tubes, freeze dried, and stored at −80 °C in the dark under a nitrogen atmosphere.

Total lipids were extracted from 25 mg freeze dried samples using a modified Bligh & Dyer

Table 1. Summary of the experimental conditions of the feeding experiments in 20‰ seawater

	Exp. 1	Exp. 2
Cultivation feed	Baker's yeast	Baker's yeast + capelin oil
Feeding phase		
Density. ind ml^{-1}	200	200
Temp,. °C	20	20*
Duration. h	4	8
Post-feeding phase		
Density. ind ml^{-1}	10	10
Time from start of feedingphase, h	6	24
Temp,. °C	20	10
Duration. h	4	8

* Cooled from 20 to 10 °C between 8 and 24 h after start of feeding.

(1959) method. The fatty acids were methylated with 12% BF$_3$ in methanol (Metcalfe *et al.*, 1966) after alkaline hydrolysis (0.5 N NaOH), and the methyl esters were determined by capillary gas chromatography (Carlo Erba HRGC 5160 series equipped with a SP-2330 glass capillary column, splitless injection and flame ionization detector) fitted with a Shimadzu-Chromatopac C-R3A computing integrator. Fatty acid identifications were based on standard references (Nu-Chek Prep., Elysium Minn.). Total lipids and total methylated fatty acids were determined gravimetrically in separate samples.

Results

The content of total lipid and the lipid composition of the rotifers are shown in Table 2. The concentration of n-3 fatty acids was higher in animals fed bakers yeast + capelin oil (yeast-oil rotifers) than in those which were given bakers' yeast through the five days preceding the experiment (yeast rotifers). The latter animals still contained a significant amount of 20 : 5 n-3 from earlier feeding on yeast + oil. The content of total lipids was the same in both groups. The fatty acid fraction of total lipids was not dependent on

Table 2. Content of n-3 fatty acids and polyunsaturated fatty acids (PUFA)* in rotifers cultured on yeast (yeast rotifers) or yeast + oil (yeast-oil rotifers) and in the rotifer feeds.

Fatty acids, mg (g dry wt)$^{-1}$	Rotifers		Feeds		
	Yeast	Yeast + oil	Yeast	Capelin oil	Roe powder
20 : 5 n-3	2.2	3.3	–	54	12
22 : 6 n-3	0.63	2.3	–	42	35
Sum n-3	4.3	8.4	0.2	153	52
Sum PUFA	7.0	15	2.0	164	54
Total lipid, mg (g dry wt)$^{-1}$	100	94	21	–	150

* Sum of fatty acids with more than one double bond.

rotifer feed in the present experiments and averaged $0.57 \pm 0.02\%$ (mean $\pm$ SE, n = 20) of total lipids.

Lipid composition of the feeds was rather different. Baker's yeast was practically free from n-3 fatty acids, whereas both capelin oil and herring roe powder were very rich. The concentration of 22 : 6 n-3 in the roe powder is notable. The content of lipids was also much higher in the roe powder than in baker's yeast.

The fatty acid composition of both yeast-oil rotifers and yeast rotifers changed rapidly when the animals were fed the roe powder. The concentration of 22 : 6 n-3 and the sum of n-3 fatty acids are shown in Fig. 1A. The rate of increase in the n-3 fatty acids was high through the first half hour, whereafter it leveled off and continued increasing at low rate for at least 4 h. The concentration of 22 : 6 n-3 and the total n-3 fatty acids remained uniformly higher in the yeast-oil rotifers than in the yeast rotifers, but became equal within start of the post-feeding phase of the experiments (Table 1 and Fig. 1B).

The concentrations of 22 : 6 n-3 and n-3 fatty acids in the rotifers through post-feeding are shown as a function of time in Fig. 1B. The contents remained constant in both yeast-oil rotifers and yeast rotifers. The addition of roe powder at a concentration sufficient to maintain maximum ingestion rate and filled guts (Korstad *et al.*, 1989) did not have any significant effect upon their fatty acid content through post-feeding. The decrease

in total lipid in the rotifers as also insignificant ($P < 0.01$) through post-feeding, which is not surprising because this phase was short and temperature was low.

The average concentration of some n-3 fatty acids through the post-feeding period are shown in Table 3. The concentration of n-3 fatty acids were slightly, but significantly ($P < 0.05$), higher in the yeast-oil rotifers than in the yeast rotifers, whereas the content of polyunsaturated fatty acids remained lower. These differences are regarded as accidental because some variations occurred in fatty acid content from one experiment to another.

Discussion

The main conclusion of the present study is that the increased content of highly unsaturated n-3 fatty acids obtained by feeding rotifers for short

Table 3. Average content ($\pm$ SE) of n-3 fatty acids and PUFA in the rotifers through the post-feeding phase. Data for rotifers with and without food added are pooled (n = 11).

Fatty acids, mg (g dry wt)$^{-1}$	Yeast rotifers	Yeast-oil rotifers
20 : 5 n-3	4.80 ± 0.08	5.33 ± 0.08
22 : 6 n-3	6.60 ± 0.14	7.04 ± 0.08
Sum n-3	13.3 ± 0.26	14.1 ± 0.18
PUFA	18.3 ± 0.37	15.5 ± 0.29

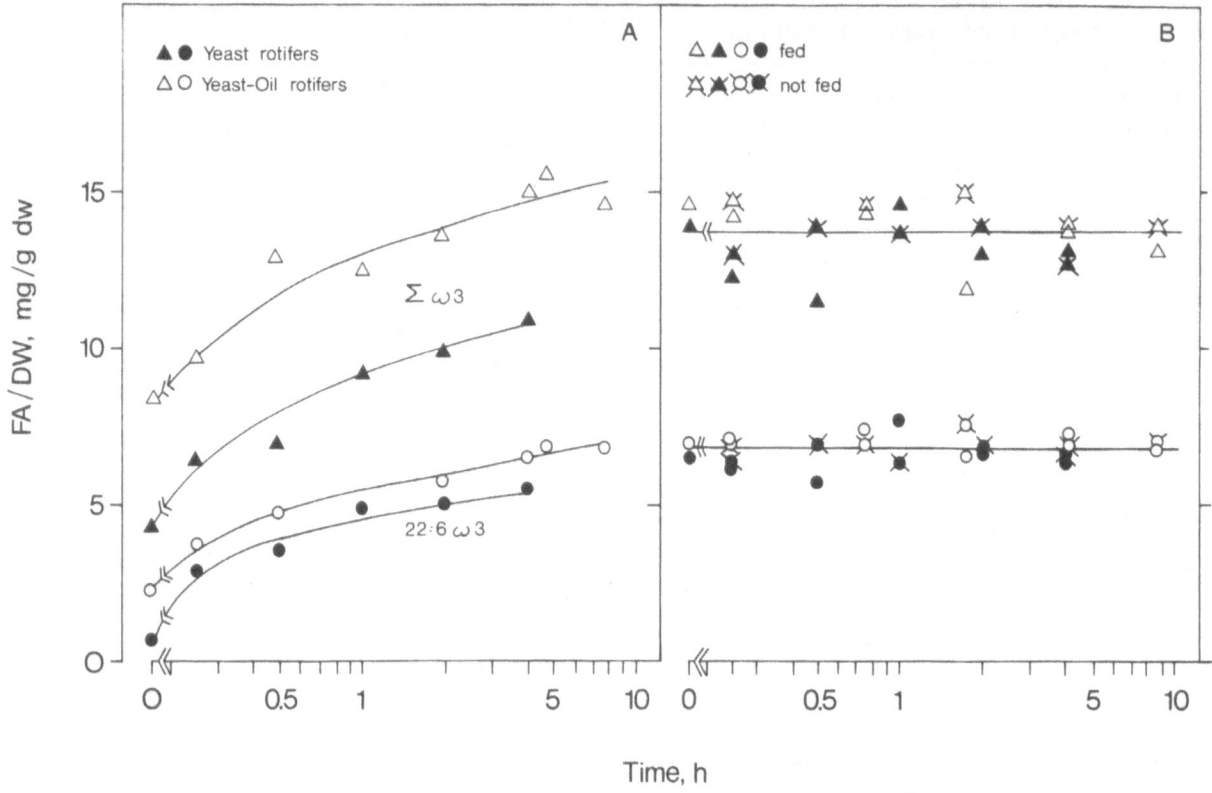

Fig. 1. The kinetics of some n-3 (w3)fatty acids of yeast fed and yeast + oil fed rotifers. A: Feeding phase. Variation in 22:3 n-3 and the sum of n-3 fatty acids upon addition of a diet rich in n-3 fatty acids at time zero. B: Post-feeding phase. Variation in 22:6 n-3 and the sum of n-3 fatty acids after transferring the rotifers to low food concentration. Crossed values represent starved groups and the remaining values represent groups fed 5 mg roe powder l^{-1}.

time (6–24 h) with a feed rich in n-3 fatty acids (enrichment) is stable for hours after transferring the rotifers to conditions of very low food concentration. This was found in rotifers which exhibited different initial contents of n-3 fatty acids in their tissues and at different temperatures through post-feeding (10–20 °C). This conclusion suggests that enriched *B. plicatilis* will maintain their improved nutritional value for several hours after application as live food for marine fish larvae

The kinetics of fatty acids through the feeding period indicate a two-step process of fatty acid accumulation. The rotifers filled their empty guts in about 20–30 min and this resulted in a rapid initial increase in the principal n-3 fatty acids of the roe powder (cf. Table 2). The slower increase which followed probably represented an accumulation of the actual n-3 fatty acids in the body tissues of the rotifers.

The accumulation rate of 22:6 n-3 and other n-3 fatty acids in the rotifers tissue must necessarily level off at some level dependent on the composition of the fed and the metabolic processes in the rotifers. In the present experiments, an apparent saturation level was obtained within 24 h. Other feeding experiments have indicated that only minor changes take place in the fatty acid distribution of rotifers after 24 h of feeding on a given diet. These findings are also in agreement with the fact that the same saturation level of n-3 fatty acids was obtained for rotifers with different initial content of n-3 fatty acids in the present experiments (cf. Table 2). However, further additions of roe powder could theoretically have resulted in significantly higher content of n-3 fatty acids after 24 h, because the rotifers probably reduced the food level below saturation over that time in Exp. 2.

413

The stability of the fatty acids in the rotifers in the post-feeding period was surprising. The guts of the rotifers were filled when they were transferred to post-feeding in both experiments, even in Exp. 2 where the concentration of food was low after 24 h. The groups of rotifers which were fed low concentrations of roe powder through post-feeding were expected to maintain their guts filled and to have a constant content on n-3 fatty acids (cf. Fig. 1B). The fact that the starved groups also kept a constant and high level of n-3 fatty acids implies that the assimilation efficiency of the present fatty acids were high and that insignificant amounts were released through defecation. It also follows that the assimilated n-3 fatty acids were not metabolized at higher rates than other fatty acids in the post-feeding phase.

Acknowledgements

This study forms part of a research project on live feed for marine fish larvae (HB4925.18213) financed by The Royal Norwegian Council for Scientific and Industrial Research. We thank dr. Grethe Rosenlund for access to the herring roe powder and Marte Morkved for technical assistance.

References

Bligh, E. G. & W. J. Dyer, 1959. A rapid method of total lipid extraction and purification. Can. J. Biochem. Physiol. 37: 911–917.

Cowey, C. B., J. M. Owen, J. W. Adron & C. Middelton, 1976. Studies on the nutrition of marine flatfish. The effect of different diatary fatty acids on the growth and fatty acid composition of turbot (*Scophthalmus maximus*). Br. J. Nutr. 36: 479–486.

Ito, T., 1960. On the culture of the mixohaline rotifer *Brachionus plicatilis* O. F. Muller in the sea water. Rep. Fac. Fish Pref. Univ. Mie. 3: 708–740.

Korstad, J. E., O. Vadstein & Y. Olsen, 1988. Feeding kinetics of *Brachionus plicatilis* fed *Isochrysis galbana*. This volume.

Lubzens, E., 1987. Raising rotifers for use in aquaculture. Hydrobiologia. 147: 245–255.

Metcalfe, L. D., A. A. Schimtz & J. R. Pelka, 1966. Rapid preparation of fatty acid esters from lipids for gas chromatography. Analyt. Chem. 38: 514–515.

Owen, J. M., J. W. Adron, J. R. Sargent & C B. Cowey, 1972. Studies on the nutrition of marine flatfish. The effect of different diatary fatty acids on the tissue fatty acids of the plaice *Pleuronectes platessa*. Mar. Biol. 13: 160–166.

Watanabe, T., C. Kitajima & S. Fujita, 1983. Nutritional values of live food organisms used in Japan for mass propagation of fish: a review. Aquaculture. 34: 115–143.

Hydrobiologia **186/187**: 415–421, 1989.
C. Ricci, T. W. Snell and C. E. King (eds), Rotifer Symposium V.
© 1989 *Kluwer Academic Publishers.*

Effect of incubation and preservation on resting egg hatching and mixis in the derived clones of the rotifer *Brachionus plicatilis*

Atsushi Hagiwara[1], Akinori Hino[2]
[1]*Graduate School of Marine Science & Engineering, Nagasaki University, Bunkyo 1–14, Nagasaki 852, Japan*; [2]*Department of Fisheries, Faculty of Agriculture, The University of Tokyo, Yayoi, Bunkyo, Tokyo 113, Japan*

Key words: Rotifera, resting eggs, light, temperature, hatching, mixis

Abstract

The marine rotifer *Brachionus plicatilis typicus* (Clone 8105A, Univ. of Tokyo) was cultured in 500 ml beakers to form resting eggs. *Tetraselmis tetrathele* was used as a culture food. Just after formation, resting eggs were exposed to various temperature (5-25 °C) and light regimes (24L : OD and OL : 24D). When eggs were exposed to light just after formation, the eggs hatched sporadically over a month. No hatching was observed for six months when eggs were preserved under dark conditions regardless of the temperature. These eggs hatched simultaneously after being exposed to light and eggs preserved at 5 °C showed twice as high hatching rate (40%) as that of eggs preserved at 15-25 °C (24%). Clones from resting eggs that were kept under different temperature and light regimes were reared individually to the third generation. Incubation at 25 °C with lighting produced the highest (5.4% and 5.2%) rate of mictic females during their 2nd and 3rd generations, respectively. The lowest rates (0 and 1.5%) were found when the eggs were kept at 5 °C in total darkness for six months. A lower rate of amictic female production was found in clones with higher rates of mixis.

Introduction

Sexual reproduction in monogonont rotifer is affected by external and internal factors (reviewed by Gilbert, 1974, 1977; Pourriot & Snell, 1983; Hino, 1983). Of these, internal factors have been less investigated. *Brachionus plicatilis* stocks collected from different sites show variable characteristics in the sexual reproduction (Snell & Hoff, 1985; Hagiwara *et al.*, 1988). These stocks sometimes show distinctive genetic differences among each other (Snell & Winkler, 1984; Fu *et al.*, 1989). The differences of mictic reproduction among clones in a stock may be smaller than that among stocks, but these differences are still

noticeable (Gilbert, 1977; Hino & Hirano, 1977). Thus, genetic differences among stocks are the primary determinants of sexual reproduction in *B. plicatilis*. The age of the parental female is also a significant internal factor that affects the production of mictic daughters (Pourriot & Rougier, 1976; Clement *et al.*, 1976: Rougier & Pourriot, 1977). Lately, Hino & Hirano (1985, 1988) reported that external factors such as temperature and salinity during the resting egg formation affect the internal factors such as the mictic reproduction in derivative clones, although the physiological mechanism of this action is unknown.

Gilbert (1974) and Pourriot & Snell (1983) reviewed the hatching of rotifer resting eggs. For

B. plicatilis, it is known that incubation temperature and salinity significantly affect the hatching rate and incubation time (Ito, 1960; Minkoff *et al.*, 1983; Hagiwara *et al.*, 1985). Culture conditions during the resting egg formation affect the optimum salinity for hatching (Ito, 1960; Gilbert, 1980).

In this report, we examined the effect of external conditions during the resting stage and incubation period of resting eggs on the hatching pattern (sporadic or simultaneous), hatching rate and sexual and asexual reproductive characteristics in the hatched clones.

Materials and methods

Brachionus plicatilis (Clone 8105A, Univ. of Tokyo) was mass cultured in 500 ml beakers to produce resting eggs. The temperature was $25 \pm 1\,°C$, salinity was 14.5 ppt obtained by diluting natural sea water with distilled water and the food was *Tetraselmis tetrathele*. Resting eggs were removed from the culture every day. At the late-exponential growth stage when mixis rate reached to the highest level, the resting eggs deposited in the last 24 hours were collected for the hatching experiment. Just after the collection, samples of 100 eggs were exposed to three different conditions of temperature and lighting: 1) 25 °C and 24L:OD since formation (condition EA), 2) 25 °C and OL:24D for the first 15 days after the formation and 25 °C and 24L:OD from day 16 (condition EB), 3) 5 °C and OL:24D for the first month and 25 °C and 24L:OD thereafter (EC) and 4) 5 °C and OL:24D for the first six months and 25 °C and 24L:OD thereafter (ED). Light intensity was 2000 lux with white fluorescence bulbs. The hatching of resting eggs was observed directly and number of hatched eggs was monitored every day until no more hatching occurred.

The hatched clones were classified into four groups according to the conditions of resting egg incubation and their hatching patterns. These were: 1) incubation in condition EA and hatching on day 11-14 (RA), 2) incubation in condition EA and hatching on day 26-48 (RB), 3) incubation in

condition EB (RC) and 4) incubation in condition EC (RD).

Five to ten clones were randomly selected from four groups and reared individually (1 individual*$0.2\,ml^{-1}$) from the first to the third generation according to the method of Hino & Hirano (1977). Culture salinity was 14.5 ppt and the medium was changed daily. *Tetraselmis tetrathele* and *Chlamydomonas* sp. grown in Miquel medium (Allen & Nelson, 1910) was suspended in fresh medium at a density of 5×10^5 cells*ml^{-1} and fed to rotifers. The female type (mictic or amictic) was determined and the life-span and fecundity monitored for all the females of the 1st and 2nd generation. As an index of amictic female viability, the number of offspring per day of reproductive life was calculated (Snell & King, 1977).

A Chi square test was used to evaluate the effects of incubation and preservation conditions on the hatching rate of resting eggs and the rate of mixis in the derivative clones. Analysis of variance was calculated to compare reproductive characteristics among experimental clones. Regression analysis was applied to determine if the rate of mixis was related to the viability of a clone.

Results

Cumulative percentages of resting egg hatching in three different conditions are presented in Fig. 1. When resting eggs were exposed to light just after their formation at 25 °C (EA), they started hatching on Day 11. These eggs continued to hatch sporadically for 40 days from the first hatch. The resting eggs kept in the dark for the first 15 days (EB) hatched simultaneously 2-3 days after being exposed to light. The final hatching rates of resting eggs incubated in conditions EA or EB, however, remained 26 and 22%, respectively. On the other hand, when resting eggs were kept at 5 °C for 1 and 6 months (condition EC or ED), hatching rates were 38 and 42%, respectively. The hatching rates of the resting eggs incubated in conditions EC or ED were significantly higher than those incubated in EA or EB (Chi-square

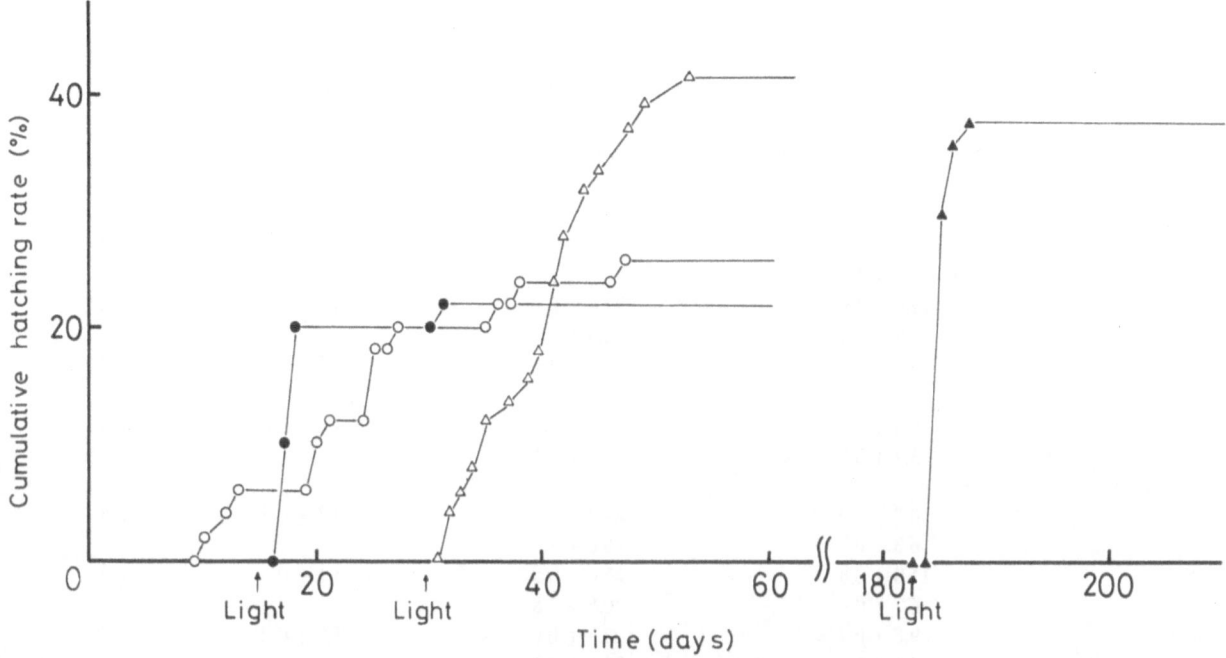

Fig. 1. Effect of conditions of the preservation and incubation on the hatching of rotifer resting eggs. O: Condition EA. ●: EB.
△: EC. ▲: ED. See text for explanation of conditions.

test, $P < 0.05$). The resting eggs kept at 5 °C for one month hatched sporadically, whereas those kept at 5 °C for six months hatched simultaneously.

The reproductive results of the 1st and 2nd generations of clones in the four different groups are presented in Table 1. There were no signifi-

cant differences in the lifespan, reproductive period, fecundity and number of offspring among clones from the four different incubation and hatching conditions.

The rate of mictic female production in the 2nd generation of the stem mothers is shown in Fig. 2. The rats were: RA- 5.03%, n = 199; RB- 5.75%,

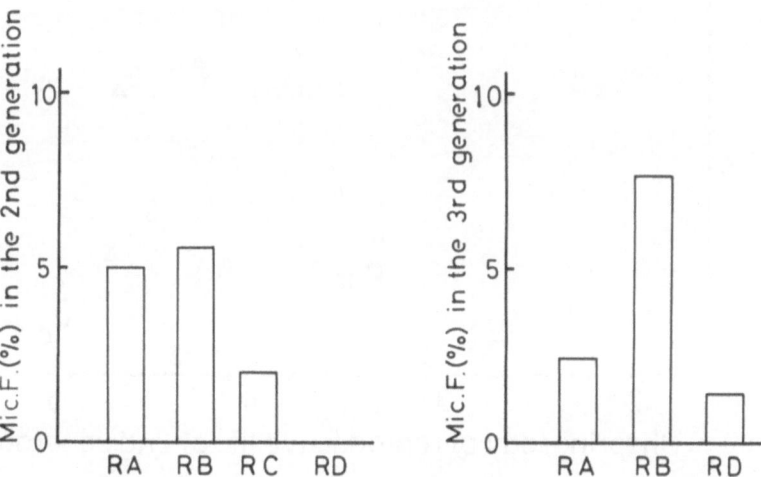

Fig. 2. Percent mictic females produced in the 2nd and 3rd generations hatched from resting eggs preserved and incubated in different conditions. See text for explanation of RA, RB, RC and RD.

Table 1. Reproductive characteristics of the 1st and 2nd generation of *B. plicatilis* clones hatched from resting eggs which are exposed to the different conditions of preservation and incubation.

Condition	Lifespan*	Fecundity**	Viability index***	Mic. F. production (%)
	Reproductive days	No. of offspring		
1st generation				
RA	11.3 ± 4.0	23.2 ± 3.0	4.1 ± 1.0	5.0
	5.8 ± 1.1	22.7 ± 3.3		
RB	12.4 ± 3.9	23.4 ± 0.9	3.7 ± 0.2	5.8
	6.2 ± 0.4	22.8 ± 0.4		
RC	12.8 ± 4.1	21.2 ± 3.6	3.3 ± 0.9	2.0
	6.2 ± 0.8	20.2 ± 4.0		
RD	10.3 ± 3.2	25.7 ± 3.1	4.6 ± 0.6	0.0
	5.7 ± 0.6	25.7 ± 3.1		
2nd generation				
RA	10.5 ± 1.0	27.9 ± 0.6	4.3 ± 0.5	2.6
	6.3 ± 0.7	26.6 ± 1.2		
RB	11.0 ± 0.8	26.6 ± 0.5	3.1 ± 0.5	8.0
	6.7 ± 0.9	20.5 ± 1.8		
RC	9.8 ± 0.4	26.3 ± 0.6	3.9 ± 0.2	–
	6.2 ± 0.5	22.6 ± 1.7		
RD	9.0 ± 0.8	28.3 ± 0.4	5.0 ± 0.0	0.7
	5.4 ± 0.1	27.1 ± 0.7		

* Days; ** No. of offspring plus No. of unhatched egg; *** Offspring per day of reproductive life. Each value shows average ± standard error. See text for the explanation of RA, RB, RC and RD.

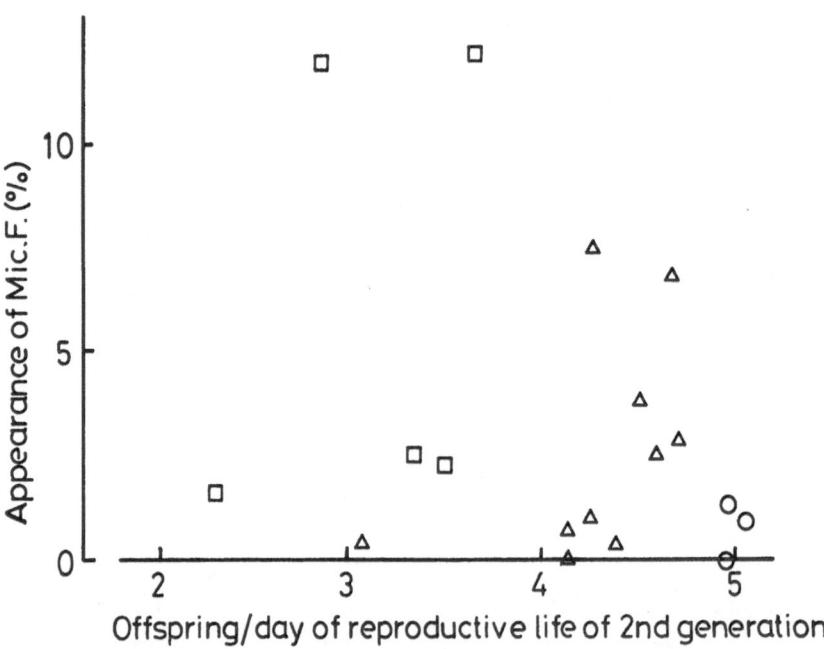

Fig. 3. Relation between the viability index of amictic females in the 2nd generation and the percent mictic female produced by clones hatched from resting eggs preserved and incubated in different conditions. ○: Hatched after preservation at 5 °C for 1 month. △: Hatched after preservation at 25 °C for 15 days. □: Hatched without preservation.

n = 226; RC- 2.04%, n = 98; RD- 0%, n = 136. Differences in the rate of mictic female production was significant among all groups except RA and RB (Chi-square, $P < 0.05$ between RC and RD, $P < 0.01$ among other groups). In the offspring of the 2nd generation, the rate was 2.57% ($n = 4483$) and 7.96% ($n = 4348$) in RA and RB, respectively. These are significantly different with $P < 0.01$. The mixis rate of the RD group was 1.45% ($n = 2746$). The rate was highest in the group RB, followed by RA and RD in both 2nd and 3rd generations. In groups RB and RD, the rates in the 3rd generation were significantly higher than in the 2nd generation. In group RA, however, the rate was higher in the 2nd generation.

The relation between the conditions of preservation and incubation and the index of clonal viability in the 2nd generation is presented in Fig. 3. Regression analysis did not reveal a significant relation between the viability index and mixis rate. But highest rate of the mictic female production was observed in the RD group (1.6-12.2%, 5 clones), followed by RA (0-7.46%, 10 clones) and RB (0-1.2%, 3 clones).

When clones were fed *Chlamydomonas*, there was a more noticeable trend (Fig. 4). The results of regression analysis on RD (3 clones) and RB group (5 clones) indicate a highly significant negative relationship ($P < 0.01$, $y = -8.6 \times + 29.9$, $r = -0.93$).

Discussion

Environmental conditions during preservation and incubation significantly affected the hatching rates and hatching patterns of *B. plicatilis* resting eggs. Resting eggs exposed to lower temperature showed higher hatching rates (EC, ED). The highest hatching rate of the eggs used for this experiment was 40%. We have conducted preliminary experiments with resting eggs from a different clone (clone 8401, University of Tokyo) and obtained 90% hatching rate without exposing them to low temperature. This illustrates the variability among clones in storage requirements and hatching rates.

Two hatching patterns, sporadic and simultaneous, were observed depending on the conditions of preservation and incubation. The control mechanism of hatching pattern seems to be species specific in monogonont rotifers (Pourriot & Snell 1983). When resting eggs of *B. plicatilis* were incubated just after their formation (EA), or when the preservation period was comparatively shorter (EC), resting eggs hatched sporadically. This suggests that each resting egg has a different diapause period. Accordingly, it seems to be contradictory that one group (EA; no preservation) showed sporadic hatching for over 50 days while the other group (EB; preserved in darkness for 15 days at 25 °C) finished hatching in almost 2 days. One possible explanation is that each resting egg has a different level of sensitivity to the environ-

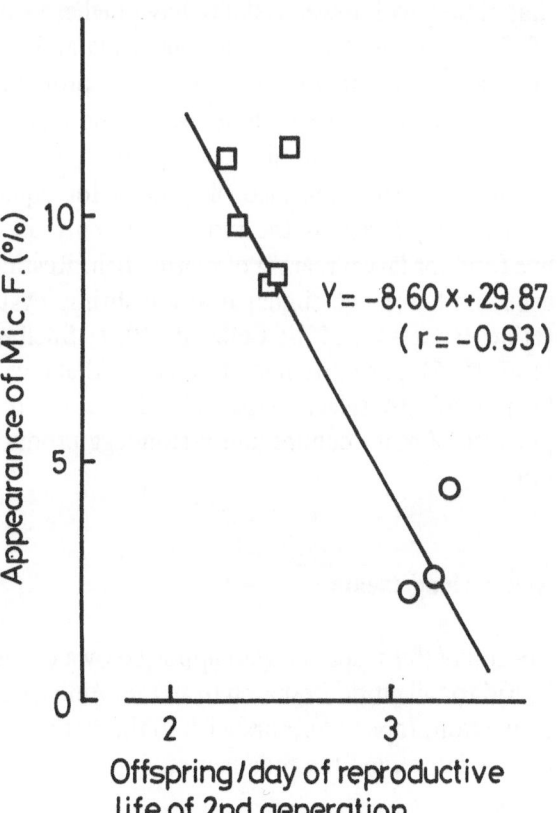

$$Y = -8.60 \times + 29.87$$
$$(r = -0.93)$$

Fig. 4. Relation between the viability index of amictic females in the 2nd generation and the percent mictic female produced by clones hatched from resting eggs preserved and incubated in different conditions. ○: Hatched after preservation at 5 °C for 1 month. □: Hatched without preservation.

ment. Changes of environmental conditions, such as from darkness to light, or lower to higher temperature, seems to stimulate hatching of resting eggs. Resting eggs hatched simultaneously when preserved for 15 days at 25 °C (EB) and for 6 months at 5 °C (ED). This indicates that the diapause period of resting eggs increases with a decrease of temperature. We incubated resting eggs just after the formation at various temperatures (5-30 °C) and salinity levels (7.2-28.9 ppt), and observed that they started to hatch earlier when they were incubated at higher temperature or lower salinity (Hagiwara & Hino, in preparation). This suggests that the completion of diapause by rotifer resting eggs depends on the environmental conditions. Development of embryos in resting eggs probably stops at some point if resting eggs are kept in darkness. Light triggers the resumption of egg development, followed by synchronous hatching. Study of rotifer embryological development will be of interest to examine these hypotheses.

No mictic females appeared in stem females hatched from resting eggs. The rate of mictic female appearance in the 2nd and 3rd generation varies depending on conditions during the period from the egg formation to hatching. The percent of mixis in the 3rd generation (groups RB and RD) was higher than that in the 2nd generation and this is consistent with the result of Hino & Hirano (1977). Current results showed that the higher mixis rate was observed in clones hatched from resting eggs which were incubated just after their formation (RA, RB). Moreover, resting eggs requiring longer incubation time hatched into clones with the highest mixis rate (RB). When eggs were moderately preserved at 25 °C and darkness, they hatched into clones with moderate mixis rate. Resting eggs experiencing extreme preservation conditions had the lowest mixis rates upon hatching (RD).

The viability index was inversely related to mixis rates. The viability of a clone tended to be high when its mixis rate is low. This suggests that rotifer clones experiencing inhibitory conditions for hatching will show higher population growth rates and less mixis. If resting eggs were not exposed to inhibitory conditions, they hatched into clones with higher mixis rates and lower population growth rates. *Brachionus plicatilis* populations in nature may be controlled by such mechanisms. When resting eggs in the sediments hatch after a winter season of diapause, the hatching clones will probably show fast parthenogenetic growth. In contrast, if resting eggs hatch without experiencing inhibition, the clones will show high mixis rates and produce a large number of resting eggs. These resting eggs can be the seeds for initiating population next season.

It has been suggested that sexual reproduction of *B. plicatilis* occurs when parthenogenetic reproduction is vigorous in the moderate environmental conditions (Lubzens *et al.*, 1985; Snell, 1986; Hagiwara *et al.*, 1988). The highest mixis rate in *B. plicatilis* populations is observed when they are in exponential growth. Current results suggest that clones with lower viability have higher rates of mixis. These clones are less vigorous and the population growth of these clones is probably slower. But even these clones show the highest mixis during exponential growth phase.

Current results are also of interest for aquaculture. *B. plicatilis* is the most commonly used live food for larval rearing of marine fish. Resting eggs can be seeded similar to brine shrimp cysts (Hino & Hirano, 1976; Lubzens, 1981; Snell & Hoff, 1985). The current study suggests that it will be possible to select favourable clones for the purpose of mass culture and resting egg production.

Acknowledgements

Portion of this research was supported by a Grant in Aid to Scientific Research from the Ministry of Education, Japan. Authors wish to thank reviewers for improving the manuscript.

References

Allen, E. & E. Nelson, 1910. On the artificial culture of marine plankton organisms. J. mar. biol. Ass. UK. 8: 421–474.

Clement, P., R. Pourriot & C. Rougier, 1976. Les facteurs exogenes et endogenes qui controlent l'apparition des males chez les Rotiferes. Bull. Soc. zool. Fr. 101: 86–95.

Fu, Y., Y. Natsukari & K. Hirayama, 1989. A preliminary study on genetics of two types of the rotifer *Brachionus plicatilis*. NOAA Technical Report NMFS, in press.

Gilbert, J. J., 1974. Dormancy in Rotifers. Trans. am. microsc. Soc. 93: 490–513.

Gilbert, J. J., 1977. Mictic female production in monogonont rotifers. Arch. Hydrobiol. Beih. 8: 142–155.

Gilbert, J. J., 1980. Some effects of diet on the biology of the Rotifer *Asplanchna* and *Brachionus*. In: Nutrition in the lower Metazoa. Pergamon Press, Oxford: 57–71.

Hagiwara, A., A. Hino & R. Hirano, 1985. Combined effects of environmental conditions on the hatching of fertilized eggs of the rotifer *Brachionus plicatilis* collected from an outdoor pond. Nippon Suisan Gakkaishi. 51: 755–758.

Hagiwara, A., A. Hino & R. Hirano, 1988. Comparison of resting egg formation among five Japanese stocks of the rotifer *Brachionus plicatilis*. Nippon Suisan Gakkaishi. 54: 577–580.

Hino, A. & R. Hirano, 1976. Ecological studies on the mechanism of bisexual reproduction in the rotifer *Brachionus plicatilis*-I. General aspects of bisexual reproduction. Nippon Suisan Gakkaishi. 43: 1147–1155.

Hino, A. & R. Hirano, 1977. Ecological studies on the mechanism of bisexual reproduction in the rotifer *Brachionus plicatilis*-II. Effects of cumulative parthenogenetic generation. Nippon Suisan Gakkaishi. 43: 1147–1155.

Hino, A., 1983. Life cycle – especially on the bisexual reproduction inducing factors. In: The rotifer *Brachionus plicatilis* – Biology and mass culture, pp 22–34. Ed. by Nippon Suisan Gakkai. Tokyo: Koseisha Koseikaku. (In Japanese).

Hino, A. & R. Hirano, 1985. Relationship between the temperature given at the time of fertilized egg formation and bisexual reproduction pattern in the deriving strain of the rotifer *Brachionus plicatilis*. Nippon Suisan Gakkaishi. 51: 511–514.

Hino, A. & R. Hirano, 1988. Relationship between water chlorinity and bisexual reproduction rate in the rotifer *Brachionus plicatilis*. Nippon Suisan Gakkaishi. 54: 1329–1332.

Ito, T., 1960. On the culture of the mixohaline rotifer *Brachionus plicatilis* O. F. Müller in the sea water. Rep. Fac. Fish. Prefect. Univ. Mie 3: 708–740.

Lubzens, E., 1981. Rotifer resting eggs and their application to marine aquaculture. European Maric. Soc., Spec. Publ. 6: 163–179.

Lubzens, E., G. Minkoff & S. Marom, 1985. Salinity dependence of sexual and asexual reproduction in the rotifer *Brachionus plicatilis*. Mar. Biol. 85: 123–126.

Minkoff, G., E. Lubzens & D. Kahan, 1983. Environmental factors affecting hatching of rotifer (*Brachionus plicatilis*) resting eggs. Hydrobiologia, 104: 61–69.

Pourriot, R. & R. Rougier, 1976. Influence de l'age des parents sur la production de femelles mictiques chez *Brachionus calyciflorus* (Pallas) et *B. rubens* Ehr. (Rotiferes). C.R. Acad. Sc. Paris 283: 1497–1500.

Pourriot, R. & T. Snell, 1983. Resting eggs in rotifers. Hydrobiologia, 104: 213–224.

Rougier, C. & R. Pourriot, 1977. Aging and control of the reproduction in *Brachionus calyciflorus* (Pallas) (Rotatoria). Exp. Gerontol. 12: 137–151.

Snell, T., 1986. Effect of temperature, salinity and food level on sexual and asexual reproduction in *Brachionus plicatilis* (Rotifers). Mar. Biol., 92: 157–162.

Snell, T. & F. Hoff, 1985. The effect of environmental factors on resting egg production in the rotifer *Brachionus plicatilis*. J. World Maricul. Soc. 16: 484–497.

Snell, T. & C. King, 1977. Lifespan and fecundity patterns in rotifers: the cost of reproduction. Evolution 31: 882–890.

Snell, T. & B. Winkler, 1984. Isozyme analysis of rotifer proteins. Biochem. System. Ecol. 12: 199–202.

Hydrobiologia **186/187**: 423–430, 1989.
C. Ricci, T. W. Snell and C. E. King (eds), Rotifer Symposium V.
© 1989 Kluwer Academic Publishers.

Intensive rotifer cultures using chemostats *

Charles M. James & T. Abu Rezeq
Kuwait Institute for Scientific Research, Mariculture and Fisheries Department, P.O. Box: 1638, 22017 Salmiya, Kuwait

Key words: *Brachionus plicatilis*, production, *Chlorella*, *Nannochloropsis*, fatty acids

Abstract

Continuous production of the rotifer *Brachionus plicatilis rotundiformis* (S-type) in an intensive chemostat culture system has been investigated. The production dynamics of rotifers in relation to different flow rates and feed regimes show that the growth rate and production depends on the type of algal feed and flow rate utilized in the culture system. It was possible to achieve a mean production of up to 318.84×10^6 rotifers $m^{-3} d^{-1}$ at a flow rate of $6 l h^{-1}$ in 100 l chemostats and up to 261.21×10^6 rotifers $m^{-3} d^{-1}$ at a flow rate of $40 l h^{-1}$ while using 1 m^3 capacity rotifer chemostats as production units. The $\omega 3$ fatty acid composition of rotifers while using *Chlorella* and *Nannochloropsis* in the culture system has been described. The results of this investigation show that the rotifer productivity in the continuous culture system is considerably higher than in any of the conventional culture systems described to date for aquacultural purposes.

Introduction

One of the problems in aquaculture is the adequate production of nutritionally enriched live microorganisms such as rotifers at the lowest possible cost for feeding the larval stages of fin fishes and crustaceans. The success of a marine fish hatchery, especially during early larval rearing stages, exclusively depends on the high quality live microorganisms such as microalgae (eg. *Chlorella*) and rotifers (Scott & Middleton, 1979; Hirata, 1980; Gatesoupe & Luquet, 1981; Watanabe *et al.*, 1983; James *et al.*, 1983, 1987; James & Rezeq, 1989). In recent years, many attempts have been made to improve rotifer culture conditions and the nutritional quality of the rotifers produced for aquaculture.

The usual technique for producing microalgae such as *Chlorella* and rotifers consists of a multistep backup batch culture system involving several tanks that occupy a great deal of hatchery space and is labour intensive (Lubzens, 1987). Synchronized continuous culture systems have been given importance in recent years for the culture of planktonic rotifers, and many investigators have discussed the production dynamics of rotifers under small scale experimental conditions (Droop, 1976; Droop & Scott, 1982; Boraas, 1983; Trotta, 1980; Walz, 1983; Rothhaupt, 1985). Taub (1980) stated that continuous cultures would not only provide constant yield, but also control the nutritional quality of the cultured organisms. However, most of the work done to date on continuous culture systems has been

*This research was financed by the Kuwait Foundation for the Advancement of Sciences (KFAS), Kuwait, under a contract research project code 86-04-02.

424

largely restricted to small-scale experiments to study the population dynamics of various species of microalgae and rotifers. The present investigation was aimed at developing an automated, intensive continuous culture system using the chemostat principle for producing rotifers for aquaculture. The study examines the production dynamics and fatty acid composition of rotifers under different feed regimes and culture conditions.

Materials and methods

The investigations were carried out using the S-type rotifer *Brachionus plicatilis rotundiformis* (size 95-160 μm) in a two-stage intensive continuous culture system (Fig. 1). The stage-1 continuous algal culture (using filtered and diluted sea water enriched with nutrients at 30‰ salinity) has been described by James *et al.* (1988). The stage-2 rotifer culture system consisted of three 100 l ca-

pacity and two 1 m^3 capacity chemostats provided each with 50 l and 500 l capacity mixing reactors. The desired food level in the rotifer chemostats were synchronized by metering pumps from the mixing reactors. The washout from the rotifer chemostats was collected using 100 l and 1 m^3 capacity rotifer concentration tanks to facilitate daily harvest. The rotifer chemostats were kept in a temperature controlled room provided with aeration and temperature controllers to maintain the temperature at 25 °C.

The experiments were carried out in two parts. The first part of the experimental program investigated the effect of different flow rates and cell densities of two species of microalgae (*Chlorella* strain MFD-1 and *Nannochloropsis* strain MFD-2) isolated from local sea water for producing S-type rotifers using 100 l capacity chemostats. The algal flow rates (from the mixing reactor) at desired algal cell densities (10 and 20×10^6 cells ml^{-1}) were maintained at 2, 4, 6, 8 and 10 l h^{-1} to determine the optimum food

Stage 1: Algal Chemostat   Stage 2: Rotifer Chemostat

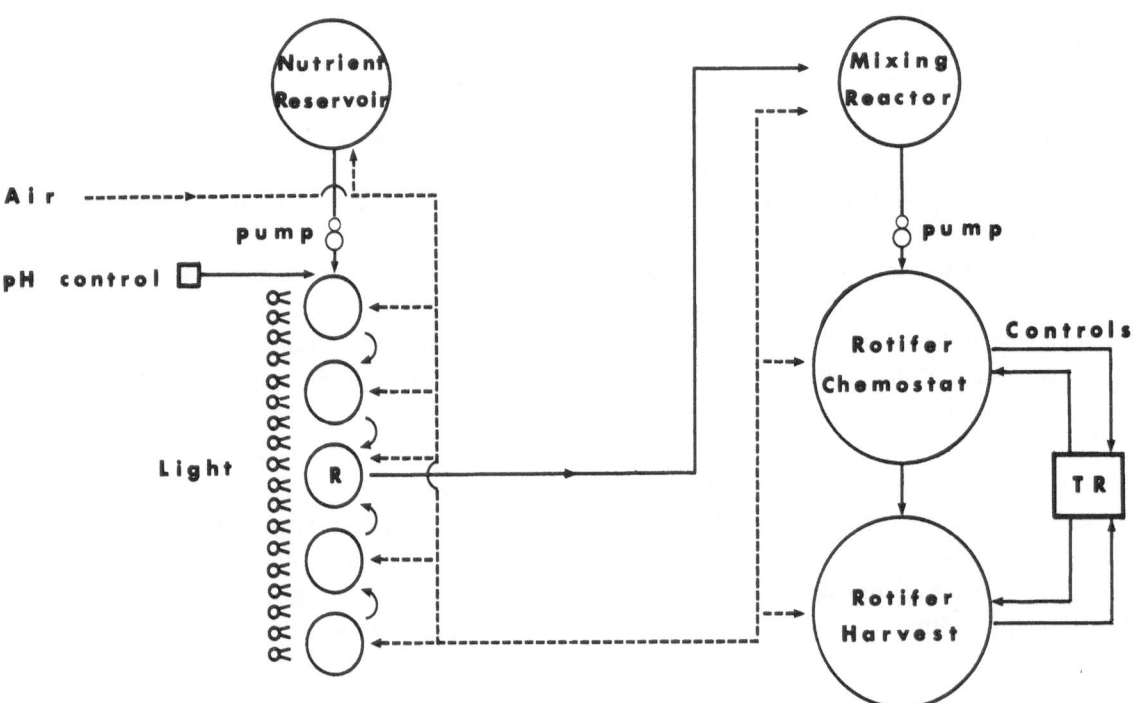

Fig. 1. Schematic diagram of the two stage rotifer chemostat system. R-Algal reservoir; TR-Temperature monitor and regulator.

level requirements. Baker's yeast was used four times a day, as a feed supplement which varied from 38 to 144 mg 10^{-6} rotifers d^{-1}. The dilution rates were synchronized with the population growth rate using a three channelled metering pump and a timer. During the observation period, *Chlorella* strain MFD-1 and *Nannochloropsis* strain MFD-2 from the algal chemostats were diluted with 30‰ sea water in the mixing reactor to make up the desired cell densities of 10 and 20×10^6 cells ml^{-1} for different treatments. Amictic rotifers acclimated in each algal species were used to seed the cultures to obtain an initial density of about 50 rotifers ml^{-1}. The population counts, pH and DO were monitored every 24 h. The production dynamics of rotifers were calculated according to the equation discussed by James *et al.* (1986).

Based on the results obtained in 100 l capacity rotifer chemostats, the second part of the experimental program investigated the use of *Chlorella* strain MFD-1 and *Nannochloropsis* strain MFD-2 at 10 and 20×10^6 cells ml^{-1} using 1 m^3 capacity rotifer chemostats. The algal flow rate was maintained at 40 l h^{-1}. The dilution rates were adjusted once in 24 h according to the growth rate of rotifers in the culture system. Dry weights of rotifers were calculated from freeze-dried and oven dried samples and averaged 0.22 μg per rotifer. Feeding efficiency equaled the total biomass of food provided d^{-1} divided by biomass (same units) of rotifers produced d^{-1}. The fatty acid composition in rotifers were analyzed as described by James *et al.* (1987, 1988).

Results

Production dynamics of rotifers in 100 l chemostats

Table 1 shows the production dynamics of rotifers in relation to different flow rates and algal feeds in 100 l capacity chemostats. While using different flow rates from 2 to 10 l h^{-1}, a flow rate of 6 l h^{-1} appears to be most conducive for producing rotifers in the culture system. Also while using *Chlorella* and *Nannochloropsis* at 10×10^6

cells ml^{-1}, the *Nannochloropsis* fed rotifers give significantly higher ($P < 0.001$) rotifer production at a flow rate of 6 l h^{-1} compared to rotifers fed with *Chlorella* at 10×10^6 cells ml^{-1}. Furthermore, the rotifer productivity significantly increased ($P < 0.01$) by feeding *Nannochloropsis* at 20×10^6 cells ml^{-1} as compared to rotifers fed 10×10^6 cells ml^{-1}. The results of these experiments show that rotifer productivity increase with increasing cell density and flow rates of up to 6 l h^{-1} used in the culture system. The dilution rate varied up to 0.6 d^{-1} (60% of culture volume per day). Rotifer production declined significantly ($P < 0.001$) at flow rates above 6 l h^{-1} (*Nannochloropsis* at 10 and 20×10^6 cells ml^{-1}). Rotifer population growth rate and doubling time were similarly reduced. Rotifer production at flow rates of 4 and 6 l h^{-1}, was not significantly different ($P > 0.05$).

Production dynamics of rotifers in 1 m^3 chemostats

Table 2 shows the production dynamics of rotifers in 1 m^3 chemostats using *Chlorella* and *Nannochloropsis* at 10 and 20×10^6 cells ml^{-1}. The flow rate was constant at 40 l h^{-1}. The results are similar to those of the 100 l rotifer chemostats. Rotifer production significantly increased ($P < 0.001$) using *Chlorella* at 20×10^6 cells ml^{-1} as compared to 10×10^6 cells ml^{-1}. Furthermore, rotifer productivity significantly increased ($P < 0.001$) using *Nannochloropsis* at 10×10^6 cells ml^{-1} as compared to *Chlorella* at 10×10^6 cells ml^{-1}. However, this difference disappeared at 20×10^6 cells ml^{-1}.

Fatty acid composition of rotifers

Studies carried out using different cell densities and flow rates of *Nannochloropsis* in 100 l rotifer chemostats show that the fatty acid content in rotifers depends on algal flow rate and cell density. Rotifer ω3 PUFA content significantly increases ($P < 0.01$) using *Nannochloropsis* at 20×10^6 cells ml^{-1} as compared to 10×10^6 cells

Table 1. Production dynamics of rotifers in relation to different flow rates and algal feeds in 100 l capacity chemostats (B.Y. = Bakers' yeast, n = 14)

Algal strain ($\times 10^6$ cells ml^{-1})	Flow rate (1 h^{-1})	Feed amount (g d^{-1} dry wt.) Algae Mean ± S.D.	B.Y. Mean ± S.D.	Growth rate (ml d^{-1}) Mean ± S.D.	Doubling time (d) Mean ± S.D.	Egg ratio (E) Mean ± S.D.	Birth rate (ml d^{-1}) Mean ± S.D.	Production ($\times 10^6$ m^{-3} d^{-1}) Mean ± S.D.	Feed efficiency Mean ± S.D.
MFD-1, at 10×10^6	2	3.7 ± 0.9	14.5 ± 5.8	1.0 ± 0.4	0.8 ± 0.3	0.4 ± 0.2	0.7 ± 0.3	87.4 ± 44.4	12.2 ± 4.0
	4	3.7 ± 0.9	15.6 ± 5.7	1.3 ± 0.3	0.6 ± 0.2	0.5 ± 0.1	0.8 ± 0.1	108.8 ± 41.3	9.4 ± 2.6
	6	3.6 ± 0.9	16.2 ± 5.3	1.3 ± 0.3	0.6 ± 0.2	0.4 ± 0.1	0.7 ± 0.1	114.5 ± 47.2	9.1 ± 2.6
	8	3.3 ± 0.9	18.9 ± 5.6	1.3 ± 0.6	0.9 ± 0.9	0.4 ± 0.1	0.6 ± 0.2	94.0 ± 35.4	13.8 ± 6.7
	10	3.2 ± 0.9	18.6 ± 5.4	1.3 ± 0.4	0.7 ± 0.5	0.4 ± 0.1	0.7 ± 0.1	75.3 ± 28.6	17.0 ± 9.1
MFD-2, at 10×10^6	2	3.4 ± 1.6	20.1 ± 7.5	1.0 ± 0.5	0.4 ± 1.2	0.4 ± 0.1	0.7 ± 0.2	148.3 ± 61.9	9.3 ± 5.0
	4	4.2 ± 1.5	22.6 ± 7.0	1.6 ± 0.4	0.5 ± 0.4	0.4 ± 0.1	0.6 ± 0.3	231.0 ± 20.9	5.7 ± 0.5
	6	4.2 ± 1.5	21.6 ± 6.5	1.8 ± 0.4	0.5 ± 0.4	0.4 ± 0.1	0.7 ± 0.2	239.0 ± 36.5	5.4 ± 0.6
	8	4.2 ± 1.7	11.5 ± 5.2	0.3 ± 0.6	−1.9 ± 8.1	0.3 ± 0.1	0.5 ± 0.2	189.6 ± 118.6	6.0 ± 4.4
	10	3.7 ± 1.5	12.5 ± 5.0	0.6 ± 0.3	2.5 ± 3.7	0.3 ± 0.1	0.5 ± 0.2	140.5 ± 59.8	6.4 ± 2.4
MFD-2, at 20×10^6	2	6.5 ± 2.0	13.9 ± 6.0	1.1 ± 0.9	0.8 ± 5.5	0.4 ± 0.1	0.7 ± 0.1	224.5 ± 74.2	4.7 ± 0.9
	4	9.0 ± 3.0	11.2 ± 4.3	0.6 ± 1.0	1.3 ± 0.4	0.3 ± 0.1	0.6 ± 1.0	259.7 ± 67.8	3.1 ± 1.0
	6	8.2 ± 2.7	12.3 ± 4.8	0.8 ± 0.2	1.0 ± 0.4	0.4 ± 0.2	0.6 ± 0.3	318.8 ± 80.4	3.2 ± 0.7
	8	8.1 ± 2.8	11.0 ± 4.8	0.6 ± 0.1	1.2 ± 0.3	0.4 ± 0.2	0.6 ± 0.3	225.6 ± 67.3	4.4 ± 1.4
	10	8.1 ± 2.8	10.7 ± 4.4	0.6 ± 0.2	0.6 ± 1.8	0.4 ± 0.2	0.7 ± 0.3	182.4 ± 35.2	5.0 ± 1.0

Table 2. Production dynamics of rotifers in relation to different algal feeds in 1 m³ capacity chemostats (B.Y. = Bakers' yeast, n = 14)

Algal strain (×10⁶ cells ml⁻¹)	Flow rate (l h⁻¹)	Feed amount (g d⁻¹ dry wt.)		Growth rate (ml d⁻¹)	Doubling time (d)	Egg ratio (E)	Birth rate (ml d⁻¹)	Production (×10⁶ m⁻³ d⁻¹)	Feed efficiency
		Algae Mean ± S.D.	B.Y. Mean ± S.D.	Mean ± S.D.	Mean ± S.D.	Mean ± S.D.	Mean ± S.D.	Mean ± S.D.	Mean ± S.D.
MFD-1, at 10 × 10⁶	40	31.9 ± 10.0	140.3 ± 68.4	1.1 ± 0.3	0.7 ± 0.2	0.4 ± 0.1	0.7 ± 0.1	128.4 ± 38.6	7.7 ± 2.1
MFD-1, at 20 × 10⁶	40	67.3 ± 18.1	206.3 ± 77.7	1.4 ± 0.2	0.5 ± 0.1	0.4 ± 0.0	0.7 ± 0.1	246.9 ± 77.3	6.1 ± 1.6
MFD-2, at 10 × 10⁶	40	43.5 ± 14.9	177.5 ± 54.5	0.6 ± 0.2	1.2 ± 0.3	0.4 ± 0.1	0.7 ± 0.1	255.9 ± 59.5	4.5 ± 1.0
MFD-2, at 20 × 10⁶	40	85.5 ± 27.3	154.3 ± 75.5	1.0 ± 0.1	0.9 ± 0.4	0.4 ± 0.1	0.7 ± 0.1	261.2 ± 23.5	4.9 ± 1.2

ml^{-1}. A highly significant increase ($P < 0.001$) in EFA $20:5\ \omega 3$ was also observed on this feeding regime and at a flow rate of $6\,l\,h^{-1}$.

Rotifers produced in $1\ m^3$ rotifer chemostats show that the fatty acid composition depends on the cell density and the type of microalgae used in the culture system (Fig. 2). A cell density of 20×10^6 cells ml^{-1} of *Chlorella* yielded significantly higher ($P < 0.01$) rotifer $\omega 3$ PUFA content than a density of 10×10^6 cells ml^{-1}. However, the total rotifer $\omega 3$ PUFA content on *Nannochloropsis* at 10×10^6 cells ml^{-1} was significantly higher ($P < 0.01$) than *Chlorella* at 20×10^6 cells ml^{-1}. Total $\omega 3$ PUFA content of rotifers fed *Nannochloropsis* at 20×10^6 cells ml^{-1} ($18.9 \pm 3.62\%$) was significantly higher ($P < 0.01$) than a feeding level of 10×10^6 cells ml^{-1}.

Discussion

The rotifer production achieved during this investigation ($318 \times 10^6\ m^{-3}\ d^{-1}$ in $100\,l$ units and $261 \times 10^6\ m^{-3}\ d^{-1}$ in $1\ m^3$ tanks) is considerably higher than any of the conventional rotifer production systems reported to date for aquacultural purposes. Lubzens (1987), while reviewing the conventional rotifer production systems, stated that low yields were mostly associated with large culture tanks (10-20 m^3). The highest rotifer production achieved was in $60\,l$ containers and only 122 rotifers $ml^{-1}\ d^{-1}$ were recorded (Gatesoupe & Luquet, 1981). This is considerably lower than our mean rotifer production of 318 rotifers ml^{-1} d^{-1}. Furthermore, no significant difference ($P < 0.05$) in rotifer production was observed using $100\,l$ capacity and $1000\,l\,(1\ m^3)$ chemostats.

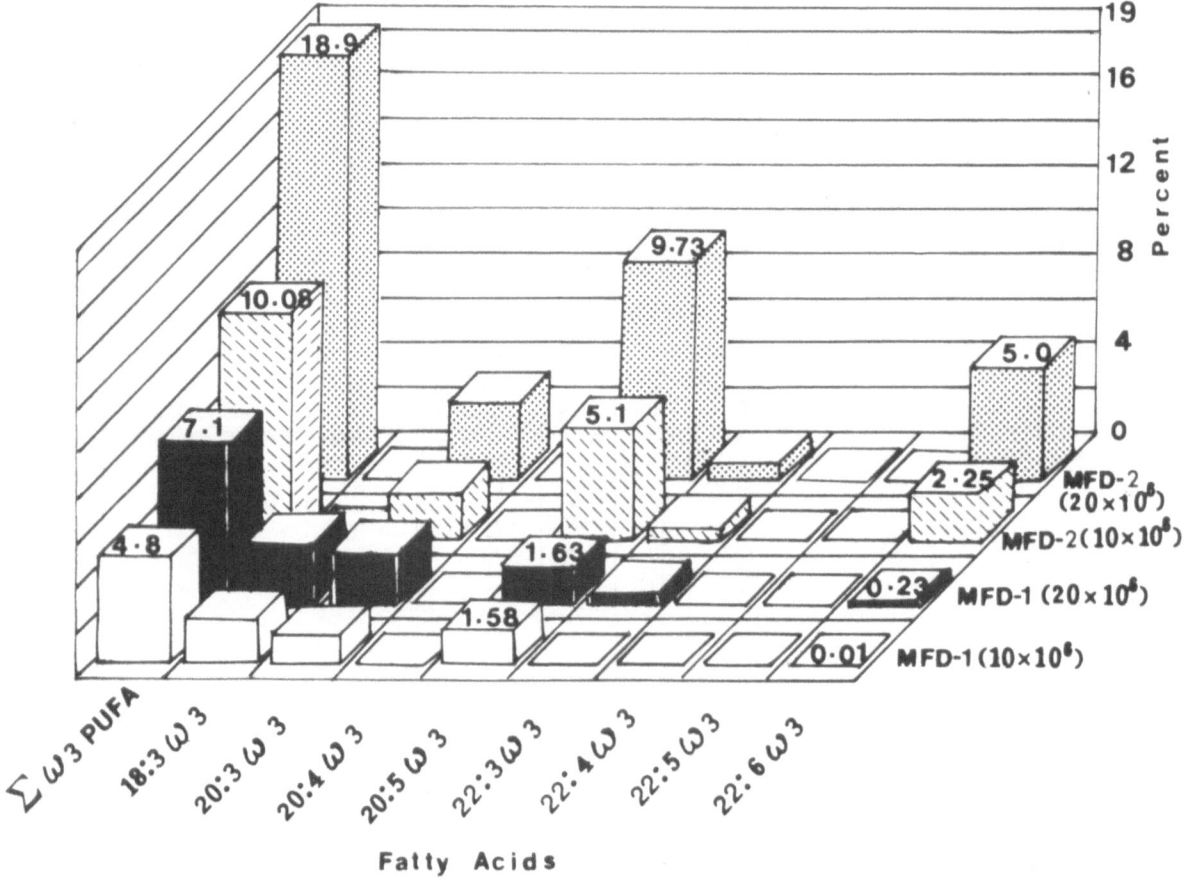

Fig. 2. Fatty acid composition in rotifers in relation to different cell densities of *Chlorella* strain MFD-1 and *Nannochloropsis* strain MFD-2.

This illustrates that $1 m^3$ capacity rotifer chemostats could be adapted as suitable units for the large scale production of rotifers in aquaculture.

Boraas (1983) cultured *B. calyciflorus* in a two-stage chemostat system and obtained a dry weight biomass of $0.0202 g l^{-1} d^{-1}$ at a dilution rate of $0.83 d^{-1}$ and $0.0252 g l^{-1} d^{-1}$ at a dilution rate of $1.08 d^{-1}$. In our investigation, a dilution rate of $0.6 d^{-1}$ yielded a dry-weight rotifer biomass of $0.0699 g l^{-1} d^{-1}$ ($318 \times 10^6 m^{-3} d^{-1}$) in $100 l$ chemostats and up to $0.0574 g l^{-1} d^{-1}$ ($261 \times 10^6 m^{-3} d^{-1}$) in $1 m^3$ capacity chemostats. This is higher rotifer production than any previously described culture systems (Hirata, 1979; Groeneweg & Schluter, 1981; James *et al.*, 1987).

It is generally known that different algal foods can yield substantially different reproductive rates (Hirayama *et al.* 1979; Yufera *et al.* 1983; Snell *et al.* 1983; James & Rezeq, 1988). Furthermore, rotifer growth and productivity also depends on the cell densities of *Chlorella* utilized in the culture system (Hirayama *et al.*, 1973; Halbach & Halbach-Keup, 1974; Pilarska, 1977a, b; Endo & Mochizuki, 1979; Yamasaki & Hirata, 1985; Rezeq & James, 1987) as does nutritional quality (James *et al.*, 1986; 1987; 1988). The results of our investigation are in accordance with the above observations showing increased rotifer productivity with higher cell densities of microalgae. Also the studies show that increased rotifer productivity could be achieved while using *Nannochloropsis* strain MFD-2 as compared to *Chlorella* strain MFD-1. This is in accordance with the observations of James & Rezeq (1988) who used a different species of *Chlorella*.

The presence of $\omega 3$ PUFA in rotifers, which are indispensable for rearing marine fish larvae (Watanabe *et al.*, 1983), shows that the level of these essential fatty acids in rotifers depends on the cell density and species of microalgae used in the culture system. Also, the flow rate of algal feed in the culture system has an impact on the availability of EFA in rotifers. The slight reduction in the $\omega 3$ PUFA content observed in the $1 m^3$ chemostat may be due to the flow rate ($40 l h^{-1}$)

as compared to the $100 l$ capacity chemostats where maximum rotifer $\omega 3$ PUFA content was observed at a flow rate of $6 l h^{-1}$ ($= 60 l h^{-1}$ in $1 m^3$ tank).

James *et al.* (1989) while investigating the growth and $\omega 3$ fatty acid composition of marine algal species under different temperature regimes concluded that fatty acid composition, especially the EPA (eicosapentaenoic acid, $20:5 \omega 3$) profile, is species dependent and varies according to different temperatures prevailing in the culture system. In our investigation the temperature was controlled at $25 °C$ for all the treatments. Hence, variations in EFA content observed in rotifers while using the two species of microalgae appears to be due to the type of algal feed utilized (James & Rezeq, 1988, James *et al.*, 1989) in the chemostat culture system. Moreover, adequate quantities of EFA are present in rotifers fed with *Nannochloropsis* at 20×10^6 cells ml^{-1}, so that further nutritional enrichment is not required for feeding marine fish larvae (Watanabe *et al.*, 1983).

Acknowledgement

This research was financially supported by Kuwait Foundation for the Advancement of Sciences, Kuwait, under Project Code 86-04-02. Our sincere thanks are due to Dr. Ziad Shehadeh, Manager, Aquaculture, Kuwait Institute for Scientific Research who provided invaluable insight and guidance through all phases of this study.

References

Boraas, M. E., 1983. Population dynamics of food limited rotifers in two-stage chemostat culture. Limnol. Oceanogr. 28: 546–563.

Droop, M. R., 1976. The chemostat in mariculture. In: A. Persoone & E. Jaspers (ed.). Proceedings of the 10th European Symposium on Marine Biology. Belgium University Press, Ostend, pp. 71–93.

Droop, M. R. & J. M. Scott, 1982. A steady state approach to some micro-plankton problems. Annales de l'Institut Oceanographique 58: 47–54.

Endo, E. & T. Mochizuki, 1979. Relation between specific

430

growth rate of rotifer, *Brachionus plicatilis*, and concentration of food, marine species of *Chlorella*. J. Fermentation Technol., Osaka 57: 372–374.

Gatesoupe, F. J. & P. Luquet, 1981. Practical diet for mass culture of the rotifer *Brachionus plicatilis*: application to larval rearing of sea bass, *Dicentrarchus labrax*. Aquaculture 22: 149–163.

Groeneweg, J. & M. Schluter, 1981. Mass production of freshwater rotifers on liquid wastes. II. Mass production of Brachionus rubens Ehrenberg 1838 in the effluent of high-rate algal ponds used for the treatment of piggery waste. Aquaculture 25: 25–33.

Halbach, V. & G. Halbach-Keup, 1974. Quantitative Beziehungen zwischen Phytoplankton und der Populationsdynamik des Rotators *Brachionus calyciflorus* Pallas. Befunde aus Laboratoriumsexperiment und Freilanduntersuchungen. Arch. Hydrobiol., 73: 273–309.

Hirata, H., 1979. Rotifer culture in Japan. Spec. Publ. Europ. Maricult. Soc. 4: 361–388.

Hirata, H., 1980. Culture methods for the marine rotifer *Brachionus plicatilis*. Min. Rev. Data File Fish Res. 1: 27–46.

Hirayama, K., K. Watanabe & T. Kusano, 1973. Fundamental studies on physiology of rotifer for its mass culture III. Influence of phytoplankton density on population growth. Bull. Jpn. Soc. Sci. Fish. 39: 1123–1127.

Hirayama, K., K. Takagi & H. Kimura, 1979. Nutritional effect of eight species of marine phytoplankton on population growth of the rotifer *Brachionus plicatilis*. Bull. Jpn. Soc. Sci. Fish. 45: 11–16.

James, C. M., M. Bou-Abbas, A. M. Al-Khars, S. Al-Hinty & A. E. Salman, 1983. Production of the rotifer *Brachionus plicatilis* for aquaculture in Kuwait. Hydrobiologia 104: 77–84.

James, C. M., T. Abu-Rezeq, P. A. Dias & A. E. Salman, 1986. Production dynamics and nutritional quality of the rotifer *Brachionus plicatilis* under different feed regimes. Technical Report KISR2183: 1–30.

James, C. M., P. Dias & A. E. Salman, 1987. The use of marine yeast (*Candida* sp.) and bakers' yeast (*Saccharomyces cerevisiae*) in combination with *Chlorella* sp. for mass culture of the rotifer *Brachionus pliatilis*. Hydrobiologia 147: 263–268.

James, C. M. & T. Abu-Rezeq, 1988. Effect of different cell densities of *Chlorella capsulata* and a marine *Chlorella* sp. for feeding the rotifer *Brachionus plicatilis*. Aquaculture 69: 43–56.

James, C. M. & T. A. Rezeq, 1989. Production and nutritional quality of two small-sized strains of the rotifer *Brachionus plicatilis*. Journal of the World Aquaculture Society 20: 1–15.

James, C. M., A. M. Al-Khars & P. Chorbani, 1988. pH dependent growth of *Chlorella* in a continuous culture system. Journal of the World Aquaculture Society 19: 27–35.

James, C. M., S. Al-Hinty & A. E. Salman, 1989. Growth and ω3 fatty acid and amino acid composition of marine microalgae under different temperature regimes. Aquaculture 77: 337–351.

Lubzens, E., 1987. Raising rotifers for use in aquaculture. Hydrobiologia 147: 245–255.

Pilarska, J., 1977a. Ecophysiological studies on *Brachionus rubens* Ehrbg (Rotatoria). I. Food selectivity and feeding rate. Pol. Arch. Hydrobiol. 24: 319–328.

Pilarska, J., 1977b. Ecophysiological studies on *Brachionus rubens* Ehrbg (Rotatoria). II. Production and respiration. Pol. Arch. Hydrobiol. 24: 329–341.

Rezeq, T. A. & C. M. James, 1987. Production and nutritional quality of the rotifer *Brachionus pliatilis* fed marine *Chlorella* sp. at different cell densities. Hydrobiologia 147: 257–261.

Rothhaupt, K. O., 1985. A model approach to the population dynamics of the rotifer *Brachionus rubens* in two-stage chemostat culture. Oecologia 65: 252–259.

Scott, A. P. & C. Middleton, 1979. Unicellular algae as food for turbot (*Scophthalmus maximus* L.) larvae – the importance of dietary long – chain polyunsaturated fatty acids. Aquaculture 18: 227–240.

Snell, T. W., C. J. Bieberich & R. Fuerst, 1983. The effects of green and blue-green algal diets on the reproductive rate of the rotifer *Brachionus plicatilis*. Aquaculture 3: 21–30.

Taub, F. B., 1980. Use of continuous culture technique to control nutritional quality. In: G. Shelef & C. J. Soeder (ed.) Algae Biomass. Amsterdam: Elsevier/North-Holland Biomedical Press, pp. 708–721.

Trotta, P., 1980. A simple and inexpensive system for continuous monoxenic culture of *Brachionus pliatilis* Muller as a basis for mass production. In G. Shelef & C. J. Soedor (ed.) Algae Biomass. Amsterdam: Elsevier/North-Holland Biomedical Press, 307–313.

Walz, N. 1983. Continuous culture of the pelagic rotifers *Keratella cochlearis* and *Brachionus angularis*. Arch. Hydrobiol. 98: 70–92.

Watanabe, T., C. Kitajima & S. Fujita, 1983. Nutritional values of live organisms used in Japan for mass propagation of fish. A review. Aquaculture 34: 115–143.

Yamasaki, S. & H. Hirata, 1985. Efficient rates of food supply for cultivation of rotifer *Brachionus plicatilis* fed on marine *Chlorella* sp. Suisan Zoshoku 32: 225–229.

Yufera, M., L. M. Lubian & E. Pascula, 1983. Effecto de cuatro algas marinas sobre el crecimiento poblacional de dos cepas de *Brachionus plicatilis* (Rotifera: Brachionidae) en cultivo. Invest. Pesq. 47: 325–337.

Hydrobiologia **186/187**: 431, 1989.
C. Ricci, T.W. Snell and C.E. King (eds), Rotifer Symposium V.

Author Index

Hydrobiologia **186/187**: 433, 1989.
C. Ricci, T.W. Snell and C.E. King (eds), Rotifer Symposium V.

Subject Index